INTERSTELLAR MOLECULES

INTERNATIONAL ASTRONOMICAL UNION

UNION ASTRONOMIQUE INTERNATIONALE

SYMPOSIUM No. 87

HELD AT MONT TREMBLANT, QUÉBEC, CANADA
AUGUST 6 – 10, 1979

INTERSTELLAR MOLECULES

EDITED BY

B. H. ANDREW

Herzberg Institute of Astrophysics,
National Research Council of Canada,
Ottawa, Canada

D. REIDEL PUBLISHING COMPANY

DORDRECHT : HOLLAND / BOSTON : U.S.A. / LONDON : ENGLAND

Library of Congress Cataloging in Publication Data
Main entry under title:

Interstellar molecules.

(Symposium – International Astronomical Union ; no. 87)
Sponsored by Commission 14, 34, and 40 of IAU
Includes indexes.
    1.   Interstellar matter–Congresses.   I.   Andrew, Bryan H.
II.   International Astronomical Union.Commission 14.   III.   International
Astronomical Union. Commission 34.   IV.   International Astronomical Union.
Commission 40.   V.   Series: International Astronomical Union. Symposium ;
no. 87.
QB790.I58   523.1'12   80–20663
ISBN-13: 978-94-009-9099-9   e-ISBN-13: 978-94-009-9097-5
DOI: 10.1007/978-94-009-9097-5

Published on behalf of
the International Astronomical Union
by
D. Reidel Publishing Company, P.O. Box 17, 3300 AA Dordrecht, Holland

Sold and distributed in the U.S.A. and Canada
by Kluwer Boston Inc.,
190 Old Derby Street, Hingham, MA 02043, U.S.A.

In all other countries, sold and distributed
by Kluwer Academic Publishers Group,
P.O. Box 322, 3300 AH Dordrecht, Holland

D. Reidel Publishing Company is a member of the Kluwer Group.

TABLE OF CONTENTS

PREFACE

In the course of editing this volume I discovered that I am temper-
amentally unsuited to the task, an impediment that in no small way
accounts for the fact that its publication is later than I would have
hoped or anticipated. I am, at heart, a pedant, particularly with
respect to the use and abuse of the English language. Many of my
scientific colleagues are a good deal less punctilious; they take the
reasonable view that if the message is clear, the manner of expression
is unimportant. Experience has taught me that what is clear to the
author is often quite unclear to the reader unless the author takes the
trouble to express himself with precision. I have also found that most
scientists are unwilling to devote to the presentation of their results
the care that they lavish on obtaining them. But I found myself slipping
beyong this often justifiable complaint into a state of inexcusable
self-righteousness.

It began innocently enough. Conscious of the need for speed, I
read the first two or three texts with murmurs of approval, and scribbled
'no changes needed' on the title page of the copy. Then I began to
notice one or two grammatical mistakes, plural verbs with singular nouns,
mixtures of tenses, and so on, which had to be corrected; not serious, I
told myself, a matter only of changing a word here, retyping a line
there. Quickly I sank deeper. Sentences were reversed to provide
clearer or less clumsy expression. Repetitive chunks were cut out.
Unsatisfying idiom was frowned upon and altered. Colloquialisms were
ruthlessly expunged. Soon red ink flowed like blood upon the pages.
The massacre gathered momentum. I spent hours in the library immersed
in Fowler. Whole mornings were devoted to the difference between 'that'
and 'which'; sadly, it remained as before, clear in Mr. Fowler's mind
but not in mine. I wallowed in the distinctions between 'shall' and
'will' and between 'should' and 'would.' I became embroiled in unresolved arguments over the merits of fused participles. I was like a
pyromaniac let loose in a napalm factory, overwhelmed by the possibilities
for destruction. Every comma was subjected to vicious scrutiny; some
were removed, then replaced, then removed again in endless agonising.
Colons and semi-colons were thrust upon unwilling prose. Texts were
retyped in dozens. I was reviled by secretaries. All this without a
word to the innocent victims, the authors, who were going blissfully
about their daily business secure in the knowledge that publication was
proceeding smoothly. Time, I told myself,did not permit; God did not
create the world in six days by engaging in consultations. Mea culpa.

So, I offer my apologies to those of you who do not recognise what

B. H. Andrew (ed.), Interstellar Molecules, xvii–xx.

you have written. To the others go my regrets for the delay. My only
excuse is that I did what I did for the common good, by keeping in mind
the lonely reader in Ulan Bator who does not speak the NRAO jargon. I
made every effort not to alter the scientific sense of what was said,
and in fact avoided altering passages that no amount of rereading
could help me understand.

    The order of presentation of articles in this book is not the order
in which papers were presented at the Symposium itself. The reasons
are twofold: first, it was agreed that the volume should include the
poster papers that were such a prominent and gratifying part of the
Symposium; second, there were a number of late papers presented, not
all of which fitted the context of the session to which time and con-
venience constrained them. For those who are interested, the programme
for the Symposium is included elsewhere in this volume.

    At first I tried to arrange the contents of this book into clearly
defined sections. I quickly gave up. As is common in astronomy, my
material defied attempts to divide and simplify it. Instead it presented itself as a prism spectrum, wherein the individual colours were
obvious but the boundaries between them were not. I have avoided
arbitrary division by arranging the articles so that each, where
possible, follows from the preceding; there are obvious groupings
within this continuum that are readily identifiable with various areas
of research.

    The question period that followed each talk was recorded in two
forms: on tape, by means of a fearsome-looking parabolic microphone,
for which we are grateful to the Canadian Broadcasting Corporation who
loaned it to us on the understanding that it be returned in time for
the next Ottawa Rough Rider football game; and in written form, each
questioner being asked to paraphrase his question after the event, and
each speaker to render once more his response. The comparison between
written and recorded forms was often interesting. We also accepted
questions in writing that shortness of time precluded in verbal form.
On the other hand some questions have been omitted because they are
answered in the published version of the talk or because they relate
to diagrams that are not included in the published version. Finally
I rearranged many of the questions in what seemed to me a more logical
order than that in which they occurred.

    In the Preface to the Proceedings of Symposium No. 90 my colleagues
Ian Halliday and Bruce McIntosh write that they did not abdicate their
responsibility to summarise the Symposium, but instead transferred it
to a worthy recipient, Peter Millman. I, braver than they, have no
such qualms. I abdicate my responsibility, resoundingly, fearlessly,
and without heavy heart; partly because I was too busy locally organising while at the Symposium to hear much of what was going on, and too
preoccupied to understand much of what I did hear; partly because the
mind, while editing, tends to absorb words without digesting their
meaning; but mostly because I have no insights to offer. I have the

feeling that at Symposium No. 87 the tide of scientific progress flowed inexorably forward, making obvious headway without engulfing any familiar landmarks.

The Symposium itself had its roots in the decision in 1976 to hold the XVII General Assembly of the IAU in Montreal. There was a general feeling at the Herzberg Institute of Astrophysics that we ought to 'do something' appropriate to the occasion. A symposium was the ready answer, and a symposium on interstellar molecules was clearly apt, since just two years previously the Herzberg Institute had been established with interstellar spectroscopy very much in mind, combining as it did the diverse astronomical and laboratory spectroscopic interests of the National Research Council. The permission of the IAU was sought and granted.

Symposium No. 87 on Interstellar Molecules lasted five days, had two hundred and twenty participants, plus wives, children, and travelling companions, nine sessions, twenty-one invited talks, forty-eight presented papers, about seventy poster papers, a round table discussion, and several papers submitted to the Proceedings that would have been posted had the authors only been able to attend the Symposium. Luckily for everyone's sanity the organisers chose to hold the event in relaxing surroundings at a resort hotel in the scenic Laurentian Mountains. Several members of the Local Organising Committee found it an agreeable part of their duties to investigate in detail the competing attractions of the various resort sites throughout Québec and Ontario. It was their good fortune to choose Mont Tremblant Lodge, and thanks are due to its management and staff, who far exceeded the level of hospitality and efficiency that might reasonably have been expected. Legion are the astronomers who pushed themselves from the dinner table glassy-eyed but smiling beatifically.

Obviously, in an undertaking of this size there are many to thank, all of whose contributions were invaluable. The members of the Local and Scientific Organising Committees were, of course, a sine qua non. Their names are listed elsewhere in these preliminary pages. They all contributed eagerly, but I hope the other members of the Scientific Committee will forgive me if I single out the Chairman, Professor Charles Townes, for particular mention; he did a mountain of work in drawing up the scientific programme, inviting the speakers, and fitting together the contributed papers, and it is due in no small part to his insight, breadth of knowledge, and tactfulness that the Symposium was the scientific success that most participants seemed to feel that it was. On the social and administrative sides of affairs we managed to avoid disasters, and as Chairman of the Local Committee I can testify to the enthusiasm and devotion to duty of each of the members, without whom all plans would have been as dust in the wind. We enjoyed the financial and moral support of Natural Sciences and Engineering Research Council of Canada, the International Astronomical Union, the Canadian National Committee of the IAU, the National Research Council of Canada, and of Dr. J.L. Locke, Director of the Herzberg Institute of Astrophysics, who placed the considerable resources of the Institute at our disposal.

I must mention the memorable contributions of Drs. Bok and Morimoto to the after-banquet speeches. Their oratory, alas, is recorded only in the minds of their audience. Dr. Morimoto presented me with a bottle of saki large enough to take a bath in, and was later kind enough to give repeated demonstrations of its appropriate disposal. Dr. Bok by his colourful reminiscences added much to the younger generation's appreciation of the early days of radio astronomy. I retain the handwritten notes of his talk, and had some thought of publishing them, but I decided eventually not to risk a libel action. I originally asked him to give the after-dinner speech at "our meeting" while he was attending a conference at the Herzberg Institute. He readily agreed. Unfortunately Bart, dedicated to the task at hand, had but one meeting in mind, the one he was attending, while I, immersed in the organisation of Sympoisum 87, had thought only of that. He retired immediately to his hotel and prepared his talk, then sat through the conference dinner that night without being invited to the microphone. All's well that end's well. The speech, like good wine and Bart himself, must have improved with age, for when it was eventually given, it was more than a success.

Finally, I know that all who attended the Symposium will join in expressing their appreciation to Mrs. Ghislaine DesChênes and Mrs. Mary Saver for their unfailing cheerfulness and efficiency at the information desk. Their ability to cope with every crisis smoothed many a potentially rough moment. They both contributed enormously to the pre-conference preparations. Mary, in addition, shouldered the burden of the typing load created by the editor's megalomania.

                                                        Bryan Andrew

Scientific Organising Committee

Chairman:　　C.H. Townes

Members:

B.H. Andrew	B.J. Robinson
M. Guélin	O.E.H. Rydbeck
D.N.B. Hall	V.I. Slysh
M. Morimoto	W.D. Watson
T. Oka	G. Winnewisser

Local Organising Committee

Chairman:　　B.H. Andrew

Members:

L.H. Doherty	J.W.C. Johns
L.A. Higgs	J.M. MacLeod

IAU Symposium No. 87 was sponsored by
IAU Commissions Nos. 14, 34, and 40.

LIST OF PARTICIPANTS

ADAMS, N.G., Space Res. Dept., Univ. of Birmingham, Birmingham,
    England.
ALLAMANDOLA, L.J., Huygens Lab., Wassenaarseweg 78, Leiden 2300 RA,
    The Netherlands.
ANDREW, B.H., Herzberg Inst. of Astrophysics, NRC, Ottawa K1A OR6,
    Canada.
ANICICH, V.G., Jet Propulsion Lab., 4800 Oak Grove Dr., Pasadena,
    CA 91103, U.S.A.
ARNOLD, F., Max-Planck-Inst. für Kernphysik, 69 Heidelberg,
    F.R. Germany.
ASUNDI, R.K., Spectroscopy Divn., BARC, Trombay, Bombay-85,
    India 400085.
AVERY, L.W., Herzberg Inst. of Astrophysics, NRC, Ottawa K1A OR6,
    Canada.
BALLY, J., Dept. of Astronomy, Univ. of Massachusetts, Amherst, MA
    01003, U.S.A.
BAR-NUN, A., Dept. of Geophysics & Planetary Sc., Tel Aviv Univ.,
    Tel Aviv, Israel.
BARRETT, A.H., Massachusetts Inst. Tech., Cambridge, MA 02139, U.S.A.
BARRETT, J., Physics Dept. State Univ. of New York, Stony Brook,
    NY 11794, U.S.A.
BAUDRY, A., Obs. de Bordeaux, 33270 Floirac, France.
BECKWITH, S., Dept. of Astronomy, Space Sc. Bldg., Cornell Univ.,
    Ithaca, NY 14853, U.S.A.
BEICHMAN, C.A., California Inst. Tech., Pasadena, CA 91125, U.S.A.
BETZ, A.L., Dept. of Physics, Univ. of California, Berkeley, CA 94720,
    U.S.A.
BIEGING, J.H., Radio Astronomy Lab., Univ. of California, Berkeley,
    CA 94720, U.S.A.
BIERMANN, L., Max-Planck-Inst. für Physik, Föhringer Ring 6,
    8000 München 40, F.R. Germany.
BLACK, J.H., Center for Astrophysics, 60 Garden St., Cambridge,
    MA 02138, U.S.A.
BLITZ, L., Radio Astronomy Lab., Univ. of California, Berkeley,
    CA 94720, U.S.A.
BOHME, D.K., Dept. of Chemistry, York Univ., Downsview, Ontario,
    Canada.
BOK, B.J., Steward Obs., Univ. of Arizona, Tucson, AZ 85721, U.S.A.
BOOTH, R.S., Nuffield Radio Astronomy Lab., Jodrell Bank,
    Macclesfield, England.
BROTEN, N.W., Herzberg Inst. of Astrophysics, NRC, Ottawa K1A OR6,
    Canada.
BUHL, D., Code 693, NASA Goddard Space Flight Center, Greenbelt,
    MA 20771. U.S.A.
BURDYUZHA, V.V., Space Res. Inst., USSR Academy of Sc.,
    Profsouznaya 84, Moscow, USSR.
BUXTON, R., Massachusetts Inst. Tech. Cambridge, MA 02139, U.S.A.
CARLSON, W.J., Dept. of Astronomy, Ohio State Univ., Columbus,
    OH 43210, U.S.A.

CARRUTHERS, G.R., Code 7123, Naval Res. Lab., Washington, DC 20375, U.S.A.

CHAFFEE, F.H., Mt. Hopkins Obs., P.O. Box 97, Amado, AZ 85640, U.S.A.

CHIN, G., Code 693, NASA Goddard Space Flight Center, Greenbelt, MD 20771, U.S.A.

CHURCHWELL, E.B., Dept. of Astronomy, Univ. of Wisconsin, Madison, WI 53706, U.S.A.

CLARK, F.O., Univ. of Kentucky, Dept. Physics & Astronomy, Lexington, KY 40506, U.S.A.

COHEN, R., Goddard Inst. for Space Studies, 2800 Broadway, New York, NY 10025, U.S.A.

COLOMB, F.R., Inst. Argentino de Radioastronomia, Casilla 5, 1894 Villa Elisa, Buenos Aires, Argentina.

CONG, H-I., Goddard Inst. for Space Studies, 2880 Broadway, New York, NY 10025, U.S.A.

CROVISIER, J., Dépt. de Radioastronomie, Obs. de Meudon, 92190 Meudon, France.

CRUTCHER, R.M., Astronomy Dept., Univ. of Illinois, Urbana, IL 61801, U.S.A.

CUDABACK, D.D., Radio Astronomy Lab., Univ. of California, Berkeley, CA 94710, U.S.A.

CUMMINS, S.E., Goddard Inst. for Space Studies, 2880 Broadway, New York, NY 10025, U.S.A.

DALGARNO, A., Harvard College Obs., 60 Garden St., Cambridge, MA 02138, U.S.A.

DAME, T.M., Pupin Hall, Columbia University, New York, NY 10027, U.S.A.

DANKS, A.C., European Southern Obs., Casilla 16317, Santiago 9, Chile.

DAVIES, R.D., Nuffield Radio Astronomy Lab., Jodrell Bank, Macclesfield, England.

DE JONG, T., Astronomical Inst., Roeterstraat 15, Amsterdam, The Netherlands.

DE ZAFRA, R., Physics Dept. State Univ. of New York, Stony Brook, NY 11794, U.S.A.

DICKEL, H.R., Astronomy Dept., Univ. of Illinois, Urbana, IL 61801, U.S.A.

DICKINSON, D.F., Jet Propulsion Lab., 4800 Oak Grove Dr., Pasadena, CA 91103, U.S.A.

DIERCKSEN, G.H.F., Max-Planck-Inst. für Physik und Astrophysik, Föhringer Ring 6, D8000 München 40, F.R. Germany.

DINGER, A.S., Jet Propulsion Lab., 4800 Oak Grove Dr., Padadena, CA 91103, U.S.A.

DOHERTY, L.H., Herzberg Inst. of Astrophysics, NRC, Ottawa K1A OR6, Canada.

DOWNES, D., Max-Planck-Inst. für Radioastronomie, Auf dem Hügel 69, D5300 Bonn 1, West F.R. Germany.

DRAPATZ, S., Max-Planck-Inst. für Extraterrestrische Physik, 8046 Garching, F.R. Germany.

DULEY, W.W., Dept. of Physics, York University, Downsview, Ontario M3J 1P3, Canada.

DYMANUS, A., Grameystr 35, Nijmegen, The Netherlands.
ELITZUR, M., Dept. of Physics, Univ. of Illinois, Urbana, IL 61801,
    U.S.A.
ELMEGREEN, B.G., Dept. of Astronomy, Columbia Univ., New York,
    NY 10027, U.S.A.
EVANS, N.J., Dept. of Astronomy, Univ. of Texas, Austin, TX 78721,
    U.S.A.
EVENSON, K.M., National Bureau of Standards, Boulder, CO 80302,
    U.S.A.
FALGARONE, E., Dept. of Radioastronomie, Obs. de Paris,
    F-92190 Meudon, France.
FEDERMAN, S., Dept of Astronomy, Univ. of Texas, Austin, TX 78712,
    U.S.A.
FEHSENFELD, F.C., Aeronomy Lab., NOAA Environmental Res. Lab.,
    Boulder, CO 80302, U.S.A.
FELDMAN, P.A., Herzberg Inst. of Astrophysics, NRC, Ottawa K1A OR6,
    Canada.
FIELD, D., School of Chemistry, Cantocks Close, Univ. of Bristol,
    Bristol, England.
FITTON, B., Astronomy Divn., European Space Agency, 2200 AA Noordwijk,
    The Netherlands.
FORSTER, J.R., Radio Sterrenwacht Dwingeloo, 7990 AA Dwingeloo,
    The Netherlands.
FOX, K., Dept. of Physics & Astronomy, Univ. of Tennessee, Knoxville,
    TN 37916, U.S.A.
FRISCH, P.C., Univ. of Chicago, Enrico Fermi Inst., 933 E 56th St.,
    Chicago, IL 60637, U.S.A.
FUKUI, Y., Tokyo Astronomical Obs., Mitaka, Tokyo, Japan.
GEBALLE, T.R., Hale Obs., 813 Santa Barbara St., Pasadena, CA 91101,
    U.S.A.
GEZARI, D.Y., NASA Goddard Space Flight Center, Code 693.2, Greenbelt,
    MD 20771, U.S.A.
GILLESPIE, A.R., Max-Planck-Inst. für Radioastronomie, Auf dem
    Hügel 69, D5300 Bonn 1, F.R. Germany.
GILMORE, W., Dept. of Astronomy, Univ. of Toronto, Toronto, Canada.
GILRA, D.P., Kapteyn Astronomical Inst., Univ. of Groningen,
    Groningen, The Netherlands.
GLASSGOLD, A.E., Physics Dept., New York Univ., 4 Washington Pl.,
    New York, NY 10063, U.S.A.
GOLD, T., Center for Radiophysics and Space Res., Cornell Univ.,
    Ithaca, N.Y. 14853, U.S.A.
GOLDREICH, P., Divn. of Geological & Planetary Sc., California Inst.
    Tech., Pasadena, CA 91125, U.S.A.
GOLDSMITH, P.F., Dept. of Astronomy, Univ. of Massachusetts, Amherst,
    MA 01003, U.S.A.
GOMEZ-GONZALEZ, J., Obs. Astronomico Nacional, Alfonso XII N3, Madrid,
    Spain.
GORDON, M.A., National Radio Astronomy Obs., 2010 North Forbes Blvd.,
    Tucson, AZ 85705, U.S.A.
GOSS, W.M., Kapteyn Astronomical Inst., Univ. of Groningen, Groningen,
    The Netherlands.

GOTTLIEB, C.A., Goddard Inst. for Space Studies, 2880 Broadway,
New York, NY 10025, U.S.A.
GREENBERG, J.M., Huygens Lab., Wassenaarseweg 78, Leiden 2300 RA,
The Netherlands.
GUELIN, M., Obs. de Paris, Section d'Astrophysique, 92190 Meudon,
France.
HAGEN, W., Huygens Lab., Wassenaarseweg 78, Leiden, 2300 RA,
The Netherlands.
HALL, D.N.B., Kitt Peak Natinal Obs., 950 North Cherry Av., Tucson,
AZ 85726, U.S.A.
HARTLEY, C.H., Electronic Space Systems Corp., Old Powder Mill Rd.,
Concord, MA 01742, U.S.A.
HERBST, E., Dept. of Chemistry, College of William & Mary,
Williamsburg, VA, U.S.A.
HERZBERG, G., Herzberg Inst. of Astrophysics, NRC, Ottawa K1A OR6,
Canada.
HIGGS, L.A., Herzberg Inst. of Astrophysics, NRC, Ottawa K1A OR6,
Canada.
HILLS, R.E., Cavendish Lab., Cambridge Univ., Madingley Rd.,
Cambridge, England.
HJALMARSON, A.G., Onsala Space Obs., S-43034 Onsala, Sweden.
HO, P.T.P., Dept. of Astronomy Univ. of Massachusetts, Amherst,
MA 01003, U.S.A.
HOGLUND, B., Onsala Space Obs., S-43034 Onsala, Sweden.
HOLLENBACH, D.J., NASA Ames Research Center, Moffett Field, CA 94035,
U.S.A.
HOLLIS, J.M., National Radio Astronomy Obs., 2010 North Forbes Blvd.,
Tucson, AZ 85705, U.S.A.
HUEBNER, W.F., Los Alamos Scientific Lab., Los Alamos, NM 87545,
U.S.A.
HUGUENIN, R., Dept. of Astronomy, Univ. of Massachusetts, Amherst,
MA 01003, U.S.A.
HUNTRESS, W.T., Jet Propulsion Lab., 4800 Oak Grove Dr., Pasadena,
CA 91103, U.S.A.
IRVINE, W.M., Onsala Space Obs., S-43034 Onsala, Sweden.
JEFFERS, S., Physics Dept., York Univ., Downsview, Ontario M3J 1P3,
Canada.
JENNINGS, D.E., Code 693, NASA Goddard Space Flight Center, Greenbelt,
MD 10771, U.S.A.
JOHNS, J.W.C., Herzberg Inst. of Astrophysics, NRC, Ottawa K1A OR6,
Canada.
JOHNSON, D.R., National Bureau of Standards, Washington, DC 20234,
U.S.A.
JOHNSTON, K.J., Code 7134, Naval Res. Lab., Washington, DC 20375,
U.S.A.
KEGEL, W.H., Inst. für Theoretische Physik, Univ. of Frankfurt,
Frankfurt, F.R. Germany.
KESTEVEN, M.J., Dept. of Physics, Queen's University, Kingston,
Canada.
KIRBY, K., Harvard-Smithsonian Center for Astrophysics, 60 Garden St.,
Cambridge, MA 02138, U.S.A.

KOLLBERG, E., Onsala Space Obs,, S-43034 Onsala, Sweden.
KRAEMER, W.P., Max-Planck-Inst. für Physik und Astrophysik, Föhringer
    Ring 6, 8 München, F.R. Germany.
KRASSNER, J., Grumman Aerospace Corp. Plant 26, Bethpage, NY 11714,
    U.S.A.
KROTO, H.W., Dept. of Chemistry, Univ. of Sussex, Brighton, BN1 9QT,
    England.
KUIPER, T.B.H., Jet Propulsion Lab., California Inst. of Tech.,
    Pasadena, CA 91103, U.S.A.
KUTNER, M.L., Physics Dept., Rensselaer Polytechnic Inst., Troy,
    NY 12181, U.S.A.
KWOK, S., Herzberg Inst. of Astrophysics, NRC, Ottawa K1A OR6,
    Canada.
LANE, A.P., Dept. of Astronomy, Univ. of Massachusetts, Amherst,
    MA 01003, U.S.A.
LANGER, W.D., Dept. of Physics & Astronomy, Univ. of Massachusetts,
    Amherst, MA 01003, U.S.A.
LEES, R.M., Dept. of Physics, Univ. of New Brunswick, Fredericton,
    E3B 5A3, Canada.
LEPINE, J.R.D., CRAAM, Obs. Nacional, Rua Para 277, Sao Paulo-01243,
    Brazil.
LEUNG, C.M., Dept. of Physics, Renssalaer Polytechnic Inst., Troy,
    NY 12181, U.S.A.
LINKE, R.A., Bell Labs., Crawford Hill Lab., Holmdel, NJ 07733,
    U.S.A.
LISZT, H.S., National Radio Astronomy Obs., Edgemont Rd.,
    Charlottesville, VA 22901, U.S.A.
LITVAK, M.M., Jet Propulsion Lab., 4800 Oak Grove Dr., Pasadena,
    CA 91103, U.S.A.
LO, K.Y., California Inst. Tech., Pasadena, CA 91125, U.S.A.
LOCKE, J.L., Herzberg Inst. of Astrophysics, NRC, Ottawa K1A OR6,
    Canada.
LORTET, M.C., D.A.F., Obs. de Meudon, 92190 Meudon, France.
LOVAS, F.J., Molecular Spectroscopy Divn., National Bureau of
    Standards, Washington, DC 20234, U.S.A.
LUCAS, R., Radioastronomie, Ecole Normale Superieure, 75230 Paris
    Cedex 05, France.
LUTZ, B.L., Lowell Obs., P.O. Box 1269, Flagstaff, AZ 86002, U.S.A.
MACDONALD, G.H., Electronics Lab., Univ. of Kent, Canterbury,
    Kent CT2 7NT, England.
MACKAY, G., Dept. of Chemistry, York Univ., Downsview, Canada.
MacLEOD, J.M., Herzberg Inst. of Astrophysics, NRC, Ottawa K1A OR6,
    Canada.
MARQUES DOS SANTOS, P., Obs. Nacional, CRAAM, Rua Para 277,
    Sao Paulo 01243 Brazil.
MATHESON, D.N., SRC Appleton Lab., Chilbolton Obs., Stockbridge,
    Hants, England.
McCABE, E.M., Astronomy Centre, Univ. of Sussex, Brighton, BN1 9QJ,
    England.
McCARROLL, R., Lab. d'Astrophysique, Univ. de Bordeaux, 33405-Talence,
    France.

McCUTCHEON, W.H., Dept. of Physics, Univ. of British Columbia,
    Vancouver, V6T 1W5, Canada.
McEWAN, M.J., Jet Propulsion Lab., 4800 Oak Grove Dr., Pasadena,
    CA 91103, U.S.A.
McLAREN, R.A., Dept. of Astronomy, Univ. of Toronto, Toronto, M5S 1A7,
    Canada.
MEZGER, P.G., Max-Planck-Inst. für Radioastronomie, Auf dem Hugel 69,
    53 Bonn 1, F.R. Germany.
MILLAR, T.J., Dept. of Theoretical Physics, Univ. of Oxford,
    Oxford OX1 3NP, England.
MINN, Y.K., National Astronomical Obs., Yoksam-dong, Kangnam-Ku,
    Seoul, Korea.
MITCHELL, G.F., Jet Propulsion Lab., 4800 Oak Grove Dr., Pasadena,
    CA 91103, U.S.A.
MORAN, J.M., Center for Astrophysics, 60 Garden St., Cambridge,
    MA 02138, U.S.A.
MORIMOTO, M., Tokyo Astronomical Obs., Mitaka, Tokyo, Japan.
MORRIS, M., Astronomy Dept., Columbia Univ., New York, NY 10027,
    U.S.A.
MOUSCHOVIAS, T.Ch., Dept. of Physics & Astronomy, Univ. of Illinois,
    Urbana, IL 61801, U.S.A.
MYERS, P.C., Massachusetts Inst. Tech., Cambridge, MA 02139, U.S.A.
NADEAU, D., California Inst. Tech., Pasadena, CA 91125, U.S.A.
NAKAGAWA, N., Dept. of Chemistry, Univ. of Electro-communicatins,
    Chofu, Tokyo, Japan.
NORMAN, C.A., Huygens Lab., Wassenaarseweg 78, Leiden 2300 RA,
    The Netherlands.
OKA, T., Herzberg Inst. of Astrophysics, NRC, Ottawa K1A 0R6, Canada.
OMONT, A., CERMO, B.P. 53, 38041 Grenoble-Cedex, France.
PALMER, P., Dept. of Astronomy, Univ. of Chicago, 1100 E 58th St.,
    Chicago, IL 60637, U.S.A.
PANKONIN, V., Divn. of Astronomical Sc., NSF, 1800 G St. N.W.,
    Washington, DC 20550, U.S.A.
PARRISH, A., FCRAO, Univ. of Massachusetts, Amherst, MA 01003, U.S.A.
PAULS, T., Max-Planck-Inst. für Radioastronomie, Auf dem Hügel 69,
    5300 Bonn 1, F.R. Germany.
PENZIAS, A.A., Bell Labs., Crawford Hill Lab., Box 400, Holmdel,
    NJ 07733, U.S.A.
PHILLIPS, T.G., Bell Labs., Murray Hill, NJ 07974, U.S.A.
PLAMBECK, R.L., Radio Astronomy Lab., Univ. of California, Berkeley,
    CA 94720, U.S.A.
POPPEL, W., Inst. Argentino de Radioastronomia, Casilla 5,
    1894 Villa Elisa, Buenos Aires, Argentina.
PRASAD, S.S., Jet Propulsion Lab., California Inst. Tech., Pasadena,
    CA 91103, U.S.A.
RADFORD, H.E., Harvard College Obs., 60 Garden St., Cambridge,
    MA 02138, U.S.A.
RADHAKRISHNAN, V., Raman Research Inst., Bangalore 56006, India.
RAMSAY, D.A., Herzberg Inst. of Astrophysics, NRC, Ottawa K1A 0R6,
    Canada.

RICKARD, L.J., National Radio Astronomy Obs., Edgemont Road,
    Charlottesville, VA 22901, U.S.A.
RIDGWAY, S.T., Kitt Peak National Obs., Tucson, AZ 85726, U.S.A.
RIEU, N.Q., Obs. de Meudon, 92190 Meudon, France.
ROBINSON, B.J., CSIRO Radiophysics, P.O. Box 76, Epping 2121,
    Australia.
RODRIGUEZ-KUIPER, E.N., Ball Aerospace Systems Divn., Western Labs.,
    Gardena, California, U.S.A.
ROUEFF, E., DAPHE, Obs. de Meudon, 92190 Meudon, France.
RUBIN, R.H., Physics Dept., California State Univ. at Fullerton,
    Fullerton, CA 92634, U.S.A.
RYDBECK, G., Onsala Space Obs., S-43034 Onsala, Sweden.
SAKATA, A., The Univ. of Electro-communications, Chofu, Tokyo, Japan.
SANDERS, D.B., Dept. of Earth & Space Sc., State Univ. of New York,
    Stony Brook, NY 11733, U.S.A.
SANDQVIST, Aa., Stockholm Obs., S-13300, Saltsjöbaden, Sweden.
SCALISE, E., Obs. Nacional, CRAAM, Rua Para 277, 01243 Sao Paulo,
    Brazil.
SCHIFF, H.I., Dept. of Chemistry, York Univ., Downsview, Canada.
SCHLOERB, F.P., Astronomy Program, Univ. of Massachusetts, Amherst,
    MA 01003, U.S.A.
SCHWARTZ, P.R., Naval Res. Lab., Code 7138, Washington, DC 20375,
    U.S.A.
SCOTT, P.F., Cavendish Lab., Cambridge Univ., Madingley Road,
    Cambridge, England.
SCOVILLE, N.Z., Dept. of Physics & Astronomy, Univ. of Massachusetts,
    Amherst, MA 01002, U.S.A.
SHERWOOD, W.A., Max-Planck Inst. für Radioastronomie, Auf den
    Hugel 69, 53 Bonn, F.R. Germany.
SHIVANANDAN, K., Code 7122.1 Space Sc. Divn., Naval Res. Lab.,
    Washington, DC, U.S.A.
SHUTER, W.L.H., Dept. of Physics, Univ. of British Columbia, Vancouver
    V6T 1W5, Canada.
SILK, J.I., Dept. of Astronomy, Univ. of California, Berkeley,
    CA 94720, U.S.A.
SLYSH, V.I., Space Res. Inst., Profsoyuznaya 84, Moscow, U.S.S.R.
SMITH, D., Space Res. Dept., Univ. of Birmingham, Birmingham B15 2TT,
    England.
SMITH, P.L., Harvard-Smithsonian Center for Astrophysics, 60 Garden
    St., Cambridge, MA 01138, U.S.A.
SNELL, R.L., Astronomy Dept., Univ. of Massachusetts, Amherst,
    MA 01003, U.S.A.
SNOW, T.P., LASP, Univ. of Colorado, Boulder CO 80309, U.S.A.
SNYDER, L.E., Astronomy Dept., Univ. of Illinois, Urbana, IL 61801,
    U.S.A.
SOLLNER, T.C.L.G., FCRAO, Univ. of Massachusetts, Amherst, MA 01003,
    U.S.A.
SOLOMON, P.M., Astronomy Program, State Univ. of New York,
    Stony Brook, NY 11794, U.S.A.
SOMERVILLE, W.B., Dept. Physics & Astronomy, Univ. College, London,
    England.

STAUDE, H.J., Max-Planck Inst. für Astronomie, 69 Heidelberg
    Königstuhl, F.R. Germany.
STOREY, J.W.V., Physics Dept., Univ. of California, Berkeley,
    CA 94720, U.S.A.
SUZUKI, H., Dept. of Physics, Kyoto Univ., Kyoto, Japan.
TATUM, J.B., Dept. of Physics, Univ. of Victoria, Victoria, V8W 2Y2,
    Canada.
TENORIO-TAGLE, G., ESO/CERN, Ch. 1211 Geneve 23, Switzerland.
THADDEUS, P., Goddard Inst. for Space Studies, 2800 Broadway,
    New York, NY 10025, U.S.A.
TIELENS, A.G.G.M., Sterrewacht, Wassenaarseweg 78, Leiden 2300 RA,
    The Netherlands.
TOWNES, C.H., Dept. of Physics, Univ. of California, Berkeley,
    CA 94720, U.S.A.
VANDEN BOUT, P., Astronomy Dept., Univ. of Texas, Austin, TX 78712,
    U.S.A.
VANYSEK, V., Dept. of Astronomy & Astrophysics, Charles Univ.,
    Svedska 8, 15000 Prague 5, Smichov, Czechoslovakia.
WALMSLEY, M.C., Max-Planck Inst. für Radioastronomie, Auf dem
    Hugel 69, 53 Bonn, F.R. Germany.
WANNIER, P.G., California Inst. Tech., Pasadena, CA 91125, U.S.A.
WATANABE, T., Space Environment Lab., NOAA Environmental Res. Lab.,
    Boulder, CO, U.S.A.
WATSON, J.K.G., Herzberg Inst. of Astrophysics, NRC, Ottawa K1A OR6,
    Canada.
WATSON, W.D., Depts. of Physics & Astronomy, Univ. of Illinois,
    Urbana, IL 61801, U.S.A.
WELCH, W.J., Radio Astronomy Lab., Univ. of California, Berkeley,
    CA 94720, U.S.A.
WHITE, R., Five College Astronomy Dept., Smith College, Northampton,
    MA 01063, U.S.A.
WILLIS, C., Chemistry Divn., National Research Council, Ottawa
    K1A OR6, Canada.
WILLNER, S.P., Physics Dept., Univ. of California, San Diego,
    La Jolla, CA 92093, U.S.A.
WILSON, R.W., Bell Labs., Holmdel, NJ 07733, U.S.A.
WILSON, T.L., Max-Planck Inst. für Radioastronomie, Auf Dem Hügel 69,
    D53 Bonn 1, F.R. Germany.
WINNEWISSER, G., Max-Planck Inst. für Radioastronomie, Auf Dem
    Hügel 69, D53 Bonn 1, F.R. Germany.
WOODS, R.C., Dept. of Chemistry, Univ. of Wisconsin, Madison,
    WI 53706, U.S.A.
WOOTTEN, H.A., Owens Valley Radio Obs., California Inst. Tech.,
    Pasadena, CA 91125, U.S.A.
WRIGHT, J.P., National Sc. Foundation, 1800 E. St. N.W., Washington,
    DC 20550, U.S.A.
WYNN-WILLIAMS, C.G., Inst. for Astronomy, Univ. of Hawaii,
    2680 Woodlawn Dr., Honolulu, HI 96822, U.S.A.
ZUCKERMAN, B., Dept. of Physics & Astronomy, Univ. of Maryland,
    College Park, MD, U.S.A.

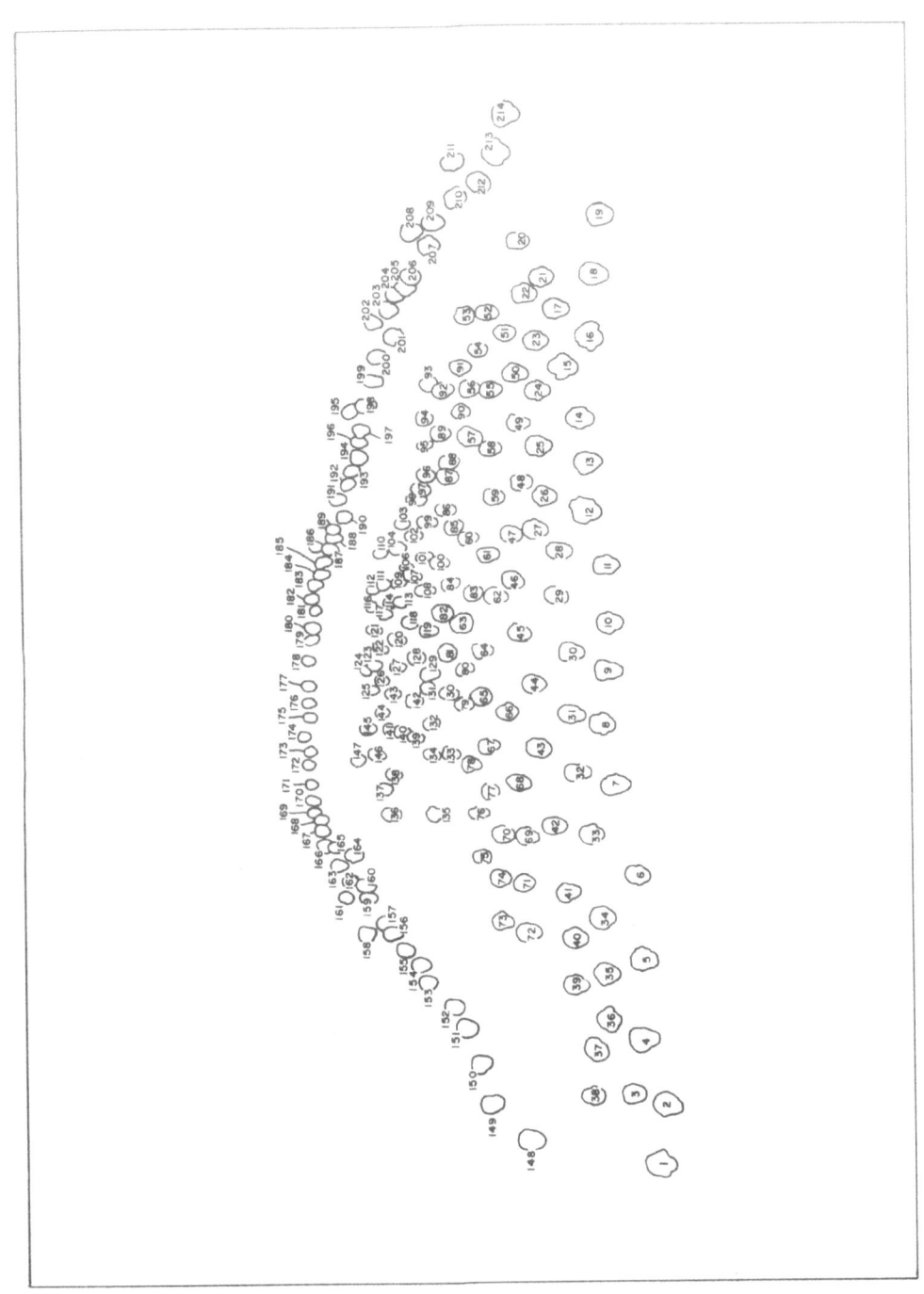

KEY TO PHOTOGRAPH

1.	T.J. Millar	44.	P. Palmer
2.	H.W. Kroto	45.	J.M. Hollis
3.	T. Gold	46.	L.E. Snyder
4.	A.C. Danks	47.	P.R. Schwartz
5.	R. Lucas	48.	W.L.H. Shuter
6.	S.S. Prasad	49.	B. Hoglund
7.	R.A. Linke	50.	J. Gomez-Gonzalez
8.	M. Morimoto	51.	A. Bar-Nun
9.	R.W. Wilson	52.	R.M. Crutcher
10.	B.J. Bok	53.	H.J. Staude
11.	A.A. Penzias	54.	N. Nakagawa
12.	P.G. Wannier	55.	K.J. Johnston
13.	G. Winnewisser	56.	P.F. Goldsmith
14.	B.J. Robinson	57.	T.R. Geballe
15.	R.E. Hills	58.	C.M. Leung
16.	D. Field	59.	H.S. Liszt
17.	B.H. Andrew	60.	A. Dymanus
18.	B. Fitton	61.	D. Smith
19.	G. Herzberg	62.	N.G. Adams
20.	P. Vanden Bout	63.	A.P. Lane
21.	J.P. Wright	64.	J. Silk
22.	T.L. Wilson	65.	C.A. Norman
23.	A. Baudry	66.	G. Rydbeck
24.	T. Watanabe	67.	T. de Jong
25.	Y. Fukui	68.	E. Kollberg
26.	L.W. Avery	69.	J.M. Greenberg
27.	A. Hjalmarson	70.	J.W. Johns
28.	F.J. Lovas	71.	V. Vanýsek
29.	W.M. Irvine	72.	G.R. Huguenin
30.	J.L. Locke	73.	R.A. McLaren
31.	A.H. Barrett	74.	A. Sakata
32.	R. de Zafra	75.	C.A. Gottleib
33.	B.L. Lutz	76.	E. Gottleib
34.	E.M. McCabe	77.	R.E. White
35.	T.B.H. Kuiper	78.	D.D. Cudaback
36.	E.N. Rodriguez Kuiper	79.	L.J. Rickard
37.	A.E. Glassgold	80.	B.G. Elmegreen
38.	W. Pöppel	81.	J. Bally
39.	T.P. Snow	82.	P.C. Frisch
40.	P.L. Smith	83.	R.S. Cohen
41.	J.H. Black	84.	W. Gilmore
42.	L.J. Allamandola	85.	W.J. Carlson
43.	D.Y. Gezari	86.	A.G.G.M. Tielens

87. Y.K. Minn
88. L. Cong
89. D.N.B. Hall
90. Y.P. Viala
91. R.L. Plambeck
92. D.F. Dickinson
93. K. Kirby
94. S. Ridgway
95. T. Oka
96. H.I. Cong
97. M. Elitzur
98. M.A. Gordon
99. M.L. Kutner
100. S.E. Cummins
101. T.M. Dame
102. W.H. McCutcheon
103. S. Beckwith
104. B. Zuckerman
105. J.J. Charfman
106. E. Roueff
107. R.C. Woods
108. J. Krassner
109. J. Tarter
110. A.R. Gillespie
111. J.B. Tatum
112. J.R.D. Lépine
113. H.E. Radford
114. W.J. Welch
115. A. Spare
116. V.I. Slysh
117. M.C. Walmsley
118. G. Chin
119. J.W. Barrett
120. K.Y. Lo
121. P.G. Mezger
122. A. Omont
123. K. Fox
124. A.S. Dinger
125. S. Jeffers
126. V.V. Burdyuzha
127. D. Downes
128. A. Downes
129. A.L. Betz
130. H.A. Wootten
131. P.T.P. Ho
132. S. Drapatz
133. R.L. Snell
134. W.F. Huebner
135. D.E. Jennings
136. W.P. Kraemer
137. F. Arnold

138. M.J. McEwan
139. E.B. Churchwell
140. G.I. Mackay
141. F.H. Chaffee
142. C.G. Wynn-Williams
143. D.P. Gilra
144. R.S. Booth
145. D.K. Bohme
146. H.I. Schiff
147. H. Suzuki
148. W.D. Watson
149. E. Herbst
150. G. Tenorio-Tagle
151. F.R. Colomb
152. E. Scalise
153. T. Ch. Mouschovias
154. P. Marques Dos Santos
155. R.H. Rubin
156. Aa. Sandqvist
157. F.P. Schloerb
158. N.Z. Scoville
159. P.F. Scott
160. P. Goldreich
161. M.M. Litvak
162. W.M. Goss
163. D.N. Matheson
164. V. Radhakrishnan
165. G.H. Macdonald
166. D.R. Johnson
167. P.C. Myers
168. J.M. MacLeod
169. D.A. Ramsay
170. W.W. Duley
171. G.F. Mitchell
172. A. Parrish
173. M.J. Kesteven
174. V. Pankonin
175. J.K.G. Watson
176. C.H. Hartley
177. S. Federman
178. T.C.L.G. Sollner
179. R.M. Lees
180. V.G. Anicich
181. R.B. Buxton
182. W. Hagen
183. S.P. Willner
184. D.B. Sanders
185. F.C. Fehsenfeld
186. J.H. Bieging
187. W.T. Huntress
188. G.R. Carruthers

189.	F.O. Clark	202.	M. Guélin
190.	K. Shivanandan	203.	M. Morris
191.	C.H. Townes	204.	W.H. Kegel
192.	D. Nadeau	205.	W.B. Somerville
193.	W.A. Sherwood	206.	W.D. Langer
194.	C.A. Beichman	207.	P. Thaddeus
195.	D.J. Hollenbach	208.	L. Blitz
196.	L.A. Higgs	209.	P.M. Solomon
197.	T.G. Phillips	210.	E. Falgarone
198.	K.M. Evenson	211.	L.H. Doherty
199.	A. Dalgarno	212.	H.R. Dickel
200.	R. McCarroll	213.	J.W.V. Storey
201.	J.R. Forster	214.	C. Willis

PROGRAM OF EVENTS AT IAU SYMPOSIUM NO. 87

Only the actual speakers are listed here. Consult the Table of
Contents for the complete authorship of the papers presented.

SESSION I   Monday, 6 August 1979   0830 hrs.

Chairman   -   C.H. Townes
Introductory Remarks   -   J.L. Locke

Invited Talk: Molecular Species and Characteristics   P. Thaddeus
   in Cool Dense Clouds
Detection of Submillimeter Lines of CO(0.65 mm) and   T.G. Phillips
   $H_2O(0.79$ mm)
New Detection of Spectral Lines in the Frequency   P. Vanden Bout
   Range 260-285 GHz
Observations of SO in Dark Molecular Clouds   W.M. Irvine
Invited Talk: Optical Observations of Interstellar   T.P. Snow, Jr.
   Molecules
Interstellar Rotational Fine Structure Lines of   F.H. Chaffee, Jr.
   C_2 Towards ζ Persii
Observations of Interstellar Molecules with the   J.H. Black
   International Ultraviolet Explorer

SESSION II   Monday, 6 August 1979   1400 hrs.

Chairman   -   D.A. Ramsay

Invited Talk: Infrared Measurements of Molecules,   D.N.B. Hall
   Particularly in Circumstellar Material
Infrared Heterodyne Spectroscopy of Circumstellar   A.L. Betz
   Molecules
Spectroscopic Studies of IRC+10216 and Related   S.T. Ridgway
   Objects
Invited Talk: Recent Laboratory Work on Molecules   G. Herzberg
   of Possible Importance for Interstellar Studies
Polarized Emission in the Broad SiO Feature from R Leo   F.O. Clark
The 4-8 μm Spectrum of the Infrared Source W33A   S.P. Willner
Chemical Identification of Grain Mantles by Infrared   L.J. Allamandola
   Molecular Fluorescence and Absorption
Interstellar Molecule Production by Grains in   J.M. Greenberg
   Molecular Clouds - Laboratory Analog Results
   and Theoretical Applications

SESSION III   Tuesday, 7 August 1979   0830 hrs.

Chairman   -   N.W. Broten

Invited Talk: Long Chain Molecules in the Inter-   L.W. Avery
   Stellar Medium
Long Carbon Chain Molecules in the Laboratory and   G. Winnewisser
   in Space
Formation of Interstellar Linear Molecules   A. Sakata
Invited Talk: Experimental Measurements of Ion-   F.C. Fehsenfeld
   molecule Reactions
An ICR Study of Ion-molecule Reactions in the   M.J. McEwan
   C_2H_2/HCN System
Gas Phase Synthesis of Cyano-, Amino- and Nitroso-   N.G. Adams
   Compounds in Interstellar Gas Clouds
High Resolution 4.8 GHz Mapping of H_2CO using the   J.R. Forster
   Westerbork Synthesis Radio Telescope
Invited Talk: High Frequency Rotational Spectra   K.M. Evenson
   of Free Radicals with Laser Magnetic Resonance
Invited Talk: The Hydrogen Molecule as a Collision   T. Oka
   Partner

SESSION IV   Tuesday, 7 August 1979   2030 hrs.

Chairman   -   C. Willis

Invited Talk: Molecular Formation in Cool Clouds   W.D. Watson
The Formation of Hydrocarbons and Iron Hydrides on   A. Bar-Nun
   Cold Interstellar Grains--Experimental Studies
Laboratory and Modelling Studies of Chemistry in   W.T. Huntress, Jr.
   Dense Molecular Clouds
The Formation of Complex Interstellar Molecules   E. Herbst
   by Radiative Association
Invited Talk: Molecular Formation and Excitation   A. Dalgarno
   in Hot Diffuse Clouds
Molecular Abundances in IRC+10216   E.M. McCabe
The Photodissociation of Interstellar CH^+ K. Kirby
Ammonia Observations of the Molecular Clouds near   D.N. Matheson
   S68, S140 and OMC2

SESSION V   Wednesday, 8 August 1979   0830 hrs.

Chairman   -   M. Morimoto

Invited Talk: Measurements of Isotopic Abundances   A. Penzias
   in Interstellar Clouds
Detection of Deuterated Formaldehyde in Interstellar   W.D. Langer
   Clouds
High Spatial Resolution Studies of $H^{13}CN$, $H^{12}CN$ and   A. Hjalmarson
   HCO^+ J=1-0 Emissions in Orion A

Formaldehyde in Dense Molecular Clouds: H_2 T.L. Wilson
   Densities and ($^{12}C/^{13}C$) Ratios
The $^{12}C/^{13}C$ Ratio in Interstellar Dark Clouds   W.H. McCutcheon
Isotopic Fractionation in the Interstellar Carbon-   V. Vanýsek
   Bearing Molecules not Related to Carbon Monoxide
CO Abundance and Isotopic Fractionation in Dark   P.F. Goldsmith
   Clouds
Invited Talk: Interpretation of Isotopic Abundances   M. Guélin
   in Interstellar Clouds

SESSION VI   Wednesday, 8 August 1979   2030 hrs.

Chairman   -   G. Winnewisser

Invited Talk: Physical and Dynamical Conditions in   N.J. Evans II
   Interstellar Clouds
Hydrostatic Models of Molecular Clouds   T. de Jong
Clumpy Molecular Clouds: A Dynamic Model Self-   J. Silk
   Consistently Regulated by T-Tauri Star Formation
The Large-Scale Distribution of Molecular Clouds in   M. Morris
   Orion and Monoceros
Invited Talk: Behavior and Significance of Circum-   B. Zuckerman
   stellar Clouds
Observation of Circumstellar Clouds   P.G. Wannier
Atomic Hydrogen in and Around the Giant Molecular   Y. Fukui
   Cloud Near W3 and W4
Molecular Clouds Near Supernova Remnants   V.I. Slysh

SESSION VII   Thursday, 9 August 1979   0830 hrs.

Chairman   -   T. Gold

Invited Talk: The Galactic Distribution, Mass   P. Solomon
   and Age of Molecular Clouds
On the Formation of Giant Molecular Cloud Complexes   B.G. Elmegreen
   and the Appearance of Spiral Structure in
   Galaxies
Columbia CO Survey: Molecular Clouds and Spiral   R.S. Cohen
   Structure
The Evolution of Molecular Clouds   C. Norman
The Rotation Curve of the Galaxy to R = 16 kpc   L. Blitz
Invited Talk: Observations of Shock Waves in   S. Beckwith
   Interstellar Clouds
Observations of the v=0, S(2) Line of Molecular   T.R. Geballe
   Hydrogen at 12.28 Microns in the Orion
   Molecular Cloud
High Velocity Gas in the Kleinmann-Low Nebula   N.Z. Scoville
Invited Talk: Theoretical Considerations of Shock   D. Hollenbach
   Wave Behavior

Thursday, 9 August 1979   1930 hrs.

Conference Banquet
    Chairman - C.H. Townes
    Remarks on behalf of Participants - M. Morimoto
    Invited Address - B.J. Bok

SESSION VIII   Friday, 10 August 1979   0830 hrs.

Chairman - V.I. Slysh

Invited Talk: Observational Characteristics of Masers   L. Snyder
    Associated with Stars
Invited Talk: Observational Characteristics of Masers   D. Downes
    in Regions of Star Formation
Further Observations of Extragalactic H_2O Sources   J.R.D. Lépine
The SiO Maser in Orion   W. Welch
Time Variations of Interstellar Water Masers in HII   G.H. Macdonald
    Regions
Invited Talk: Interpretation of High-velocity   V. Burdjuzha
    H_2O Masers
Invited Talk: Interpretation of Circumstellar Masers   P. Goldreich
Λ-Doublet Population Inversion in Collisions of OH, OD,   D. Field
    CH, CD, and NH^+
Pumping Mechanism of OH Masers   A. Omont

SESSION IX   Friday, 10 August 1979   1400 hrs.

Chairman - V. Radhakrishnan

Carbon Monosulphide Emission and Turbulent Cores in   R.A. Linke
    Molecular Clouds
Observations and Theory of an Expanding Molecular Cloud   J. Bally
    Network Surrounding an Old HII Region: W80
    (The Pelican Nebula)
Molecular Fan of 360 pc Radius in the Galactic Center   Y. Fukui
    Region
Round Table Discussion of New Experimental Possibilities
    and the Future
    Maser Amplifiers   E. Kollberg
    Wideband Spectrometers for Millimetre Wavelengths   B.J. Robinson
    Future Possibilities for Ultraviolet Observations G.R. Carruthers
        of Interstellar Molecules
    Future Spectral Line Research with the VLA   K.J. Johnston
    New Experimental Possibilities and Future Prospects   D.N.B. Hall
        for 1-5 μm Infrared Spectroscopy of Inter-
        stellar Molecules
    New Experimental Possibilities and the Future at   C.H. Townes
        Far IR Wavelengths

PHYSICAL AND DYNAMICAL CONDITIONS IN INTERSTELLAR CLOUDS

Neal J. Evans II
The University of Texas at Austin

1. INTRODUCTION

The most far-reaching result to come from the study of interstellar molecules has been the recognition of a new class of galactic structures - molecular clouds. These clouds appear to contain most of the mass of the interstellar medium and are the objects from which new stars are formed. Thus, a prerequisite for any understanding of the star formation process is a knowledge of the physical and dynamical conditions in molecular clouds.

In discussing the parameters that describe molecular clouds, it is useful to divide them into local parameters, which characterize a given location in a cloud, and global parameters, which characterize the cloud as a whole. Examples of local parameters are gas kinetic temperature (T_K), dust temperature (T_D), total density (n), magnetic field strength (B), abundance of species i ($X_i \equiv n_i/n$), volume cooling and heating rates for gas (Λ_g, Γ_g) and dust (Λ_d, Γ_d), and the thermal (V_{th}) and turbulent (V_{turb}) velocities. Global properties include the size, expressed as a length (L), or area (A), the orientation and shape, the mass (M), and the integrated heating and cooling rates. If the cloud as a whole is collapsing or rotating, then the collapse velocity (V_c) and the rotation velocity (V_r) or angular momentum are global properties. A third class of parameters are intermediate: the column density of species i (N_i), the total column density (N), and the average density ($\langle n \rangle \equiv N/L$) share some features of both local and global properties.

In the following sections, techniques for measuring these parameters will be described. Selected results will be used to illustrate the techniques. No attempt will be made to survey the entire literature of this field, since the space is inadequate. Frequent reference will be made to a recent study by Snell (1979), primarily because it illustrates many of the techniques discussed. For future reference, Snell's sample consists of 9 nearby ($\langle d \rangle$ = 170 pc) clouds selected optically as "dark clouds". Based on his maps of CO and ^{13}CO, Snell has artificially "sphericalized" his clouds by averaging all observations at the same

1

B. H. Andrew (ed.), Interstellar Molecules, 1-19.
Copyright © 1980 by the IAU.

distance from the cloud's density peak. While this method loses some information on cloud irregularities, it allows any underlying regularities to be examined.

2. TEMPERATURES

2.1. Gas Kinetic Temperature

The standard thermometer for measuring the gas kinetic temperature (T_K) is the carbon monoxide (CO) molecule. The absolute intensity of the radiation from transitions between several of its lowest levels may be converted into T_K. This technique assumes that the levels are thermalized ($T_{ex} = T_K$) and that the transition is optically thick, so that T_A^* is uniquely related to T_{ex}. Then,

$$T_K = T_{ex} = \frac{h\nu/k}{\ln\left(1 + \dfrac{h\nu/k}{T_A^*/\eta_p + T_{bg}}\right)} = \frac{5.55}{\ln\left(1 + \dfrac{5.55}{T_A^*/\eta_p + 0.83}\right)} \, ,$$

where the last expression is valid for the $J=1 \rightarrow 0$ line, and η_p is the coupling of the source to the forward antenna pattern.

Let us consider how this technique might fail. First, the density might be too low to thermalize the transition. Densities of $\sim 10^3$ cm^{-3} are sufficient to thermalize the $J=1 \rightarrow 0$ transition; thus $T_{ex} = T_K$ is a good assumption over most regions of a cloud, but the outer regions of clouds may not be thermalized. The high abundance of CO generally insures that $\tau \gg 1$, with the exception of some high velocity flows. But the very large τ found in the normal case raises another question: might regions of higher T_K be hidden by a cool, but optically thick, envelope? The answer to this question depends upon the vexing issue of the proper choice of radiative transport model, and hence is coupled to the dynamics. We will defer the issue of dynamics and note only that, for reasons perhaps not fully understood, the $J=1 \rightarrow 0$ line of CO does generally "see into" the warm cores of clouds. A few exceptions have been found where self-reversed profiles indicate partial absorption by cooler foreground gas (cf. Snell and Loren 1977). Several checks on the T_K derived from the $J=1 \rightarrow 0$ line of CO exist. A partial check that the line "sees into" the cloud is afforded by observations of the $J=2 \rightarrow 1$ and, in a few clouds, the $J=3 \rightarrow 2$ lines of CO. Having higher τ, these lines should give different T_K if thick lines fail to see into the cloud. The general agreement of the T_K's derived from different CO transitions supports the reliability of CO as a thermometer.

Another thermometer is provided by the NH$_3$ molecule. In this case T_K is determined, not from the absolute intensity of a single line, but from relative intensities, and hence populations, in the J,K = 1,1 and 2,2 inversion doublets. From studies of the transitions across both of these inversion doublets, the populations in each of the two doublets $n(J,K)$ can be determined. Then the rotational temperature T_R is given by:

$$T_R = \frac{-41.5}{\ln\left(\frac{3}{5}\frac{n(2,2)}{n(1,1)}\right)}$$

To a first approximation, $T_R = T_K$ because radiative transitions are not allowed between the 2,2 and 1,1 doublets. However, $\Delta K = 1$ collisional transitions to non-metastable states (e.g. $2,2 \rightarrow 2,1$) followed by radiative decay $(2,1 \rightarrow 1,1)$ can cause T_R to be less than T_K. A statistical equilibrium calculation must be performed to obtain T_K. Such calculations (Morris et al. 1973) indicate that $0.8\ T_K \leq T_R \leq T_K$. An analogous procedure may be used for other metastable inversion doublets. Generally speaking, the T_K derived from NH_3 are in reasonable agreement with those derived from CO (see Table 1). Since the NH_3 emission has been interpreted as coming from small clumps deep in the cloud, this agreement is further evidence that the CO sees into the clouds.

Table 1
T_K DETERMINATIONS

Source	$T_K(CO)$	$T_R(NH_3)$	$\dfrac{T_R(NH_3)}{0.8}$	$T_K(SO_2)$	$T_K(CH_3OH)$
S255	33[1]	30[2]	38[2]		
S140	30[3]	23[2]	29[2]		
NGC6334N	47[3]	19[4]	24[4]		
DR21(OH)	31[3]	22[4]	28[4]		
OMC1	94[3]			65[5]	90[6]

[1]Evans et al. 1977, [2]Ho 1977, [3]Loren, private communication, [4]Cheung 1976, [5]Pickett and Davis 1979, [6]Kutner et al. 1973

Other thermometers have been used in the specialized case of OMC1, most notably SO_2 (Pickett and Davis 1979) and CH_3OH (Kutner et al. 1973), but these probes are not useful in other regions. In summary, the accuracy of our knowledge of T_K is often limited by our ability to calibrate the CO measurements. We can determine T_K over large regions in clouds to ±10% or so, making T_K the best determined parameter in a molecular cloud.

Using CO, Dickman (1975) found that most "dark clouds" have $T_K \sim 10$ K. Maps of CO by Snell (1979) confirm that $T_A^*(CO)$ is remarkably uniform over his clouds. A more detailed analysis indicates that near the edge of the clouds, n is insufficient to thermalize CO, and the uniform or slightly declining CO emission actually translates into a rising T_K. Studies of other samples of clouds indicate the frequent presence of "hot spots", where the CO indicates $T_K > 20$ K. (cf. Blair, Peters, and

Vanden Bout 1975). Such hot spots are very often also dense and
associated with recent star formation.

2.2. Dust Temperatures

The temperature of dust grains in the cloud (T_D) can be determined
from far-infrared or sub-millimeter observations. Measurements of the
flux density $S_\nu(\lambda)$ at two different wavelengths yields a color temperature which together with an emissivity law defines the physical grain
temperature,

$$T_D = \frac{\frac{hc}{k}\left(\frac{1}{\lambda_2} - \frac{1}{\lambda_1}\right)}{(3+\beta)\ln\left(\frac{\lambda_1}{\lambda_2}\right) + \ln\left(\frac{S_\nu(\lambda_1)}{S_\nu(\lambda_2)}\right)} \quad,$$

where the grain emissivity is assumed to follow $\varepsilon(\lambda) = \varepsilon_0\lambda^{-\beta}$ and
$e^{hc/kT_D\lambda} \gg 1$ holds for both λ_1 and λ_2. Several factors limit the
accuracy of this method. The need for spatial chopping suppresses any
relatively uniform emission at low T_D and this can bias the T_D (de Muizon
et al. 1979). Also, the exponent in the emissivity law is not accurately
known. The range, $0 \leq \beta \leq 2$, is often used, with $\beta = 1$ or $\beta = 2$ favored
by various groups. The difference between $\beta = 0$ and $\beta = 2$ translates into
uncertainties in T_D of 50% or larger, depending on the exact circumstances.
Once T_D has been determined, the absolute value of the flux density at
either wavelength can be used to get an emissivity or optical depth at
that wavelength from

$$\varepsilon(\lambda) = (1 - e^{-\tau_\lambda}) = \frac{S_\nu(\lambda)\lambda^3}{2hc}\frac{(e^{hc/kT_D\lambda} - 1)}{\Omega} \quad,$$

where Ω is the solid angle.

Mapping of T_D and $\varepsilon(\lambda)$ over molecular clouds has just begun, largely
because far-infrared observers have concentrated on the brighter emission
from H II regions and their immediate vicinity. Such studies have
already shown how molecular clouds adjacent to H II regions may be heated
by the exciting stars. Maps show a decline in T_D and an increase in τ_λ
as the beam moves from the H II region into the molecular cloud (cf.
Gatley et al. 1979). In other cases, such as OMC1, part of the heating
comes from stars or protostars embedded in the molecular cloud (cf. Werner
et al. 1976). Recent observations of S140 by de Muizon et al. (1979)
indicate that such embedded sources are able to raise T_D above 20 K over
regions 20' in extent, comparable to the extent of the $T_K > 20$ K region.

3. MEASURES OF DENSITY

3.1. Column Density

It is often useful to know the total column density of material along some line-of-sight. This quantity is most conveniently expressed as $N(cm^{-2})$, the column density of gas, although a measure of the dust column density A_v, the visual extinction in magnitudes, is often used as well. Studies of diffuse clouds (Jenkins and Savage 1974) indicate that these are related by $N = 2.5 \times 10^{21} A_v$, but we are not assured that this relation will hold in molecular clouds. Indeed, neither of these quantities is directly measurable over most of the extent of molecular clouds and we are forced to use surrogate measures based on trace constituents. The most commonly used measure is $N_{13}(cm^{-2})$, the column density of ^{13}CO. This measure is useful because ^{13}CO is widely detectable, the $J=1 \rightarrow 0$ line is generally optically thin, and N_{13} can be deduced from the observations with uncertainties of no more than a factor of 2. Furthermore, Dickman (1978) showed that N_{13} and A_v were well correlated up to $A_v \sim 6$, where traditional star counting techniques begin to fail. On this basis ^{13}CO has become the standard probe of the column density of matter, using the relation, $N = 5 \times 10^5 N_{13}$ (Dickman 1978). If one also has a measure of the size (L) and assumes spherical symmetry, one may also obtain the average density $\langle n \rangle = N/L$. Over most of the extent of molecular clouds, this is the only available estimate of n.

Snell (1979) has used his sphericalized clouds to study the variation of N_{13} with radius (r). Over most of the cloud (0.2 pc < r < 0.5 – 1.0 pc) he finds that $N_{13} \sim r^{-1}$, indicating that $n \sim r^{-2}$. Inside $r \sim 0.2$ pc he finds a weaker dependence of N_{13} on r. More direct measures of n (see below) inside r = 0.2 pc are consistent with a continued $n \sim r^{-2}$ law, suggesting that N_{13} fails as a probe in dense cloud cores. Snell also extended measures of A_v by using infrared colors of stars which appear to be behind the clouds. He found that the relation between N_{13} and A_v does seem to break down above $A_v \sim 5\text{--}10$.

3.2. Density

The direct determination of the total volume density (n) is much less accurate than the determination of T_K. In principle, one simply requires observations of an optically thick but unthermalized transition. Then T_A^* leads to T_{ex}, and T_{ex} can be related to n via statistical equilibrium calculations. However, the presence of radiative trapping means that T_{ex} is also a function of optical depth, or of the molecular abundance and velocity gradient in the combination $X_i(dv/dr)^{-1}$. Thus more than one line is required; the best situation exists when the two lines used respond in different ways to changes in n and $X_i(dv/dr)^{-1}$.

This situation can be seen more easily in contour plots of equal intensity in the $(n, X_i(dv/dr)^{-1})$ plane. Such plots have been made for the 2 mm and 2 cm transitions of H_2CO (Snell 1979). A given intensity of the 2 mm line alone can be fit by any $(n, X_i(dv/dr)^{-1})$ combination lying along a curve. Measurement of the 2 cm intensity constrains the solution to lie along another curve. If one requires a simultaneous solution, implying that both lines arise under the same conditions, then the solution is given by the intersection of the two lines. The assumption that the two lines arise under the same conditions is a critical one. First one must take care to obtain the two measurements with similar beam sizes. Second, the two transitions should be excited at comparable densities, so that density variations along the line-of-sight do not cause the transitions to arise largely in different regions of the cloud.

The 2 mm and 2 cm transitions of H_2CO satisfy these conditions rather well and we have relied heavily on them to determine densities. Let us use them to explore various uncertainties inherent in this method. First, there is the choice of geometry and radiative transport. Spherical large-velocity gradient (LVG) models were used to construct the contour plots used to find densities. How would they differ for other choices? Snell has also calculated a less extensive set of models for turbulent slabs (TS) and compared the two models. At a given n, the TS model gives a weaker 2 mm line but there is little change in the 2 cm line. Thus a given 2 mm line strength observed from a turbulent cloud would imply a larger n; the difference depends on the conditions but seldom exceeds a factor of 2. Turbulent spheres and LVG slabs would probably give n larger or smaller by factors of 3 or less. Thus if we profess total ignorance of the cloud dynamics and geometry, we must admit uncertainties of a factor of 3-5 in either direction about the results from LVG spheres and turbulent slabs.

Secondly, the construction of these diagrams requires a knowledge of the collisional cross sections. Thanks largely to the work of S. Green and associates, we now have good theoretical collision rates for He collisions with H_2CO (Green et al. 1978) and many other molecules. Collisions with H_2, the more common collision partner, are usually assumed to be the same as those with He. This assumption, and the absence of direct laboratory tests of these rates, introduces an additional uncertainty which is hard to quantify but which should always be borne in mind.

We have used H_2CO to illustrate the technique, but the same basic method and uncertainties apply to several other molecules. Extensive studies of n have been made using H_2CO, HC_3N, CS, NH_3, and ^{13}CO. Table 2 compares the densities derived from these different molecules for the same sources. Only a few studies have been included, in order to achieve uniformity at the expense of extensiveness.

The trend in Table 2 is that the lines giving the highest n are those requiring the highest densities to excite. A measure of the density needed to excite each line (n*) was calculated by setting the collision rate equal to the spontaneous decay rate. Note that a single n* characterizes H_2CO because the 2 cm line strength is controlled by collisions between the same levels that produce the 2 mm line. The

Table 2
DENSITY DETERMINATIONS

Source	H_2CO^a	HC_3N^b	CS^c	NH_3^d	$^{13}CO^f$
L134N	1.2×10^4	2×10^4	$\leq1\times10^4$	$>4\times10^3$	9×10^3
L1529	1.2×10^4	8×10^4	$\leq1\times10^4$	3.8×10^3	
NGC2264	2.5×10^5	7×10^4	6×10^4		3×10^3
DR21(OH)	7×10^5	2×10^4	6×10^4		
S255	6×10^5	4×10^4	1×10^5	1.1×10^{3e}	5×10^3
S140	5×10^5			$>3.2\times10^{3e}$	5×10^3
NGC6334N	8×10^4	3×10^4	6×10^4		
M17SW	7×10^5	4×10^4	6×10^4		2×10^3
OMC2	1.5×10^6	6×10^4	2×10^5		2×10^3
$\bar{n}$	5×10^5	5×10^4	7×10^4	3×10^3	4×10^3
n*(smaller)	1×10^6	7×10^4	8×10^4	2×10^3	3×10^3
$\langle n(H_2CO)/n\rangle$	1.0	12	6	270	230
n*(H_2CO)/n*	1.0	18	17	870	460

a. Wootten, et al. (1978) $J_{K_{-1}K_1} = 2_{12}-1_{11}$ (2mm), ($1\rlap{.}'8$ beam);
$J_{K_{-1}K_1} = 2_{12}-2_{11}$ (2 cm) (2' beam); spherical LVG; n* = 1.3×10^6.

b. Vanden Bout, et al. (1979) J=5 → 4 ($2\rlap{.}'6$ beam); J=9 → 8 ($3\rlap{.}'1$ beam);
$\tau \ll 1$; $n^*_{54} = 7.4 \times 10^4$, $n^*_{98} = 4.5 \times 10^5$.

c. Linke and Goldsmith (1980) J=1 → 0 ($2\rlap{.}'6$ beam); J=2 → 1 ($2\rlap{.}'1$ beam);
spherical LVG; $n^*_{10} = 7.6 \times 10^4$, $n^*_{21} = 4.9 \times 10^5$

d. Ho (1977) (J,K) = (1,1) → (1,1); $1\rlap{.}'3$ beam; 2-level model; $n_* = 1.5\times10^3$.

e. These values are for a filling factor $\Phi = 1$. Ho favors $\Phi = 0.1$,
which gives n = 1.4×10^4 for S255 and 2.0×10^5 for S140.

f. Plambeck, Snell, and Loren (1979), J=1 → 0 ($1\rlap{.}'0$ or $2\rlap{.}'3$ beam); J=2 → 1
($1\rlap{.}'2$ beam); spherical LVG; $n^*_{10} = 2.8\times10^3$, $n^*_{21} = 1.6\times10^4$.

average n measured by a given probe ($\bar{n}$) is always close to the smaller
n* of the two lines used. The ratio of n derived from H_2CO to n derived
from another probe shows the same correlation as the ratio of n*'s. The
most obvious explanation of this situation is that the density varies
along the line-of-sight and that different lines sample different density
regimes. Thus the easy-to-excite ^{13}CO samples much more of the low
density envelope, while the hard-to-excite H_2CO lines arise almost entirely in much denser regions. In such a situation, the use of two lines
of differing n* will bias the result toward lower n. To test this explanation, one must construct a cloud model with density gradients. We
have done this for several clouds with sufficiently detailed H_2CO data,
and find that n ∿ r^{-2} laws appear to match H_2CO data from up to five

different lines reasonably well. However the ^{13}CO lines produced in the model are far too strong if Dickman's (1978) value for $X(^{13}CO) = 2 \times 10^{-6}$ is used. Thus we found that $X(^{13}CO)$ had to be much lower in the dense core where the H_2CO lines are formed, in order to make the ^{13}CO and H_2CO observations consistent (Blair et al. 1978; Wootten et al. 1978).

This conclusion again indicates that N_{13} fails as a probe of N or $\langle n \rangle$ in dense cores. We can now pull together some of these ideas by examining the results for one of Snell's sphericalized clouds, L1407. He has used the 2 mm and 2 cm H_2CO lines to determine densities of $\sim 3 \times 10^4$ cm^{-3} in the dense core. These densities agree with the $n \sim r^{-2}$ law established from N_{13} measurements at larger r, but are more than one would predict from N_{13} measures <u>at</u> the dense core. Moving outward, as n falls below $\sim 10^3$ cm^{-3}, the CO line is no longer thermalized, and T_K must rise to maintain the gradual decline in $T_A^*(CO)$.

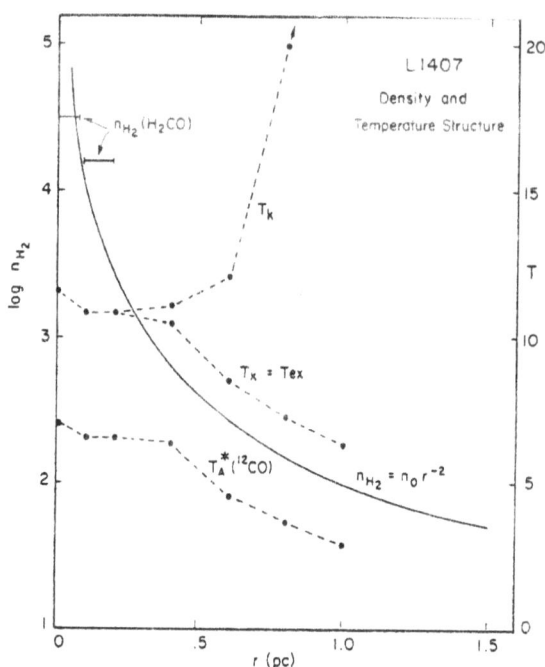

3.3. Chemical Abundance

To test theories of chemical evolution in molecular clouds, observations should provide measures of the chemical abundance of species i. The most useful measure is the density of i <u>relative</u> to the total density, $X_i = n_i/n$. For molecules which are probes of n, the analysis leading to n usually provides as well a measure of $X_i(dv/dr)^{-1}$. Measurement of the line width and extent of emission provide (dv/dr), and hence X_i. This technique has been widely applied to H_2CO (Wootten et al. 1978; Loren, Evans, and Knapp 1979; Snell 1979). The results have shown a strong anti-correlation of $X(H_2CO)$ and n.

If the cloud model derived from H_2CO is assumed in models of the excitation of other molecules, then measurement of a single line of species i is sufficient to determine X_i provided that collision rates are available for species i. Wootten et al. (1978) found that X_i also declines with increasing n for the species HCN, HNC, HCO^+, and ^{13}CO, based on a sample of 13 regions. The results for these other species are more uncertain since the lines could arise in rather different portions of the cloud.

While $X(HCO^+)$ or $X(CO)$ is more directly related to chemical theories, the additional uncertainties make it safer to use $X(H_2CO)$ to study these effects further. Using the data from Wootten et al. (1978), Loren et al. (1979), and Snell (1979), the best fit power law is $X(H_2CO) = 3.34 \times 10^4 n^{-1.26}$. This fit has a correlation coefficient r of 0.94 and is based on 73 positions in 30 clouds. Loren et al. (1979) determined n and $X(H_2CO)$ for eight positions in the Corona Austrina cloud: $X(H_2CO)$ has the same dependence on n, even within a single cloud.

The relation between $X(H_2CO)$ and n expected on the basis of theory is undetermined since the chemistry of H_2CO is poorly known. However, $X(HCO^+) \sim n^{-0.5}$ is predicted by simple steady state chemical models with constant CO abundances, and most molecules formed from HCO^+ would follow a similar law. Time dependent calculations suggest that abundances would decline more slowly or even increase as n increases. Wootten et al. suggested several possible explanations for the more rapid decline which they observe: one suggestion is that higher abundances may exist in lower density clouds because they are much older, and thus more evolved chemically. This explanation now appears unlikely since the anti-correlation exists even for a single cloud. Thus it seems that the best explanation is that the molecules are depleted by sticking to dust grains, a process that operates faster at higher n.

The probability that dust grains in dense molecular clouds have substantial mantles of organic molecules is also suggested by infrared work. The 3.1 μm "ice" band is much stronger relative to the 9.7 μm silicate band in molecular clouds than in the general interstellar medium (Merrill, Russell, and Soifer, 1976). Further, the newly discovered 6.0 and 6.8 μm bands appear only in sources behind substantial molecular material. These bands are consistent with being caused by vibrations of C-H and C-O bonds in organic molecules deposited on the grains (cf. Soifer et al. 1979).

Measurement of suitable chemical abundances may also be used to derive X(e), the electron abundance. The most notable technique here is the study of $X(DCO^+)/X(HCO^+)$ (Guelin et al. 1977) which can be related through the theory of HCO^+ formation and deuteration to X(e) (Watson

1977). Results indicate $X(e) \sim 10^{-8}$ in cool molecular clouds. Wootten, Snell, and Glassgold (1979) have recently devised a method to measure $X(e)$ in warmer clouds and find $X(e) \sim 10^{-7} - 10^{-8}$. For clouds with both kinds of $X(e)$ measures, an upper limit to ζ, the cosmic ray ionization rate is obtained. The resulting limits on ζ of 10^{-18} s^{-1} are consistent with ionization by the high energy portion of the cosmic ray spectrum.

4. ENERGETICS

The gas in molecular clouds cools primarily through line emission by trace elements, emission from H_2 being negligible at the usual cloud temperatures. Goldsmith and Langer (1978) have calculated the cooling due to a number of species under a variety of conditions. They confirm earlier results (Scoville and Solomon 1974) that the dominant coolant is CO, with H_2O becoming important at high T_K and n. Most of the cooling comes from CO transitions with large J, and has not been directly measured. While this is always an uncomfortable situation, our models of CO excitation are probably good enough to predict it correctly, given accurate measures of T_K and n.

The integrated gas cooling rate is $C_g = \int \Lambda_g \, dv$. Studies of several clouds with substantial hot spots have shown that C_g is very modest (5 - 50 $L_\odot$) compared to available energy sources such as embedded and nearby stars (Evans, Blair, and Beckwith 1977; Blair et al. 1978). The difficulty in heating the gas is in coupling radiant energy into kinetic energy of the molecules. The standard mechanism in clouds with hot spots was suggested to be collisions with warmer dust grains (Goldreich and Kwan 1974) which are themselves quite good absorbers of the radiant energy. For this heat source, Γ_{d-g}, to balance Λ_g, T_D must exceed T_K, and n must be rather large ($10^4 - 10^5$ cm^{-3}). These conditions seem to be fulfilled in many dense, hot spots surrounding infrared sources, but the frequent occurrence of extended plateaus at $T_K \sim 15 - 20$ K may be difficult to explain on this basis.

The requirement that $T_D > T_K$ leads in turn to predictions of substantial dust cooling rates, Λ_D and $C_D = \int \Lambda_D dv$. In the clouds with hot spots, the predicted C_D may be $10^4 - 10^5$ $L_\odot$, and thus the dust would clearly dominate the cloud energetics. Observations of these regions have detected the large far-infrared fluxes predicted from the immediate vicinity of the hot spot (Harvey et al. 1978; Rouan et al. 1977). Only recently has very extended emission been detected by de Muizon et al. (1979). The results indicate that $T_D > T_K$ quite far out in the S140 cloud. Nonetheless, estimates of $\langle n \rangle$ in the outer regions suggest that Γ_{d-g} is insufficient to balance Λ_g. The role of other possible heating sources such as collapse and magnetic ion-slip is hard to determine because of our lack of knowledge about the dynamical state and magnetic field strengths in clouds.

In clouds without hot spots $T_K \sim 10$ K, and Λ_g is low enough that heating by cosmic rays appears to be sufficient (Nachman 1979). Recent calculations (Clavel et al. 1978) suggest that chemical heating may be able to raise T_K to $\sim$15 K and that near the outside of a cloud T_K should rise rapidly, in agreement with Snell's (1979) results. Leung (1975) and Clavel et al. (1978) have calculated T_D in a cloud heated only by the interstellar ultraviolet radiation, and predict $T_D \sim 5 - 10$ K, in agreement with the only available observation of such a cloud (Keene et al. 1978). The rough agreement of T_D and T_K in such clouds appears to be coincidental since coupling between gas and dust will be weak until $n > 10^4$ cm^{-3}.

5. MAGNETIC FIELDS

The magnetic field in molecular clouds can potentially play a critical role in energetics and dynamics. The latter area is especially important, since magnetic fields may largely control the collapse, fragmentation, and angular momentum of molecular clouds (Mouschovias 1978). The problem is that magnetic fields have proven essentially impossible to measure in molecular clouds. Zeeman studies of H I clouds have indicated that B increases steadily with increasing n. If this increase is represented by $B \sim n^n$, the exponent, n, would be 2/3 for isotropic collapse and flux freezing. The data may be better fit if $n = 1/3$ (Mouschovias 1978), a value which agrees better with theoretical calculations. In molecular clouds, Zeeman splitting in OH (Chaisson and Vrba 1978) and SO (Clark et al. 1978) has been searched for but never found outside maser regions. Limits as low as 50 μG have been set in two dark clouds (Crutcher et al. 1975) with $n \sim 10^3$ cm^{-3}. Using $B_O = 3$ μG at $n_O = 1$ cm^{-3} as initial values (Mouschovias 1978), we predict $B = 30$ μG if $n = 1/3$ and 300 μG if $n = 2/3$. n larger than $\sim$1/2 would conflict with the limits, but n may vary with position in the cloud, complicating the interpretation.

Zeeman effects are widely suspected of being involved in the polarization of OH masers, but clear patterns seldom emerge from the data. A few promising cases have been interpreted as evidence for $B \sim 10^3$ μG. The density in such regions is poorly known, making the implications of these results unclear.

Studies of polarization in dark clouds do suggest the importance of magnetic fields in the evolution of such clouds. While estimates of field strength based on polarization depend on many poorly known parameters, the alignment of magnetic field <u>direction</u> with some cloud structures suggests that the magnetic field has played a major role (Vrba et al. 1976). Recent evidence suggests that this role may extend to the dense star-forming cores. Dyck and Lonsdale (1979) have compared the direction of infrared polarization of 31 protostars and compact H II regions with the average direction of the surrounding interstellar polarization. For 65% of the sample, the two directions agree to within 30°. The authors conclude that the magnetic field has strongly affected the cloud evolution even down to the very compact scales that determine the infrared polarization.

Thus the role of magnetic fields remains tantalizing. Our ignorance of their strength in molecular clouds may represent one of our most serious unknowns.

6. GLOBAL PROPERTIES

6.1. Cloud Size, Shape, Orientation

Discussions of cloud size often degenerate into disagreements over definitions of "cloud" and "size". One man's cloud is another man's complex of clouds. A given cloud will have a different size as mapped in different molecular lines. One has to decide also on the limit which defines the cloud's size - the half-power point or some arbitrary limit of antenna temperature are generally chosen.

Typical cloud sizes have been estimated from galactic plane CO surveys to be 5-20 pc (Gordon and Burton 1977) and 10-80 pc (Solomon, Sanders, and Scoville 1977). Mapping of eight individual clouds found in such surveys and lying in the 4-8 kpc molecular ring indicated maximum cloud dimensions of 20-100 pc at the 3 K contour level of CO; the clouds were elongated but showed no tendency to align with the galactic plane (Sanders and Solomon 1977). The group at Goddard Institute have mapped a number of molecular cloud complexes near OB associations with $\langle d \rangle \sim 1.4$ kpc, as summarized by Blitz (1979). Using the $T_A^* \Delta V = 1$ K km s^{-1} level of CO to define the size, they find that these complexes are elongated with a mean largest dimension of 90 pc within a range of 60 - 110 pc; the elongation has some tendency to lie along the galactic plane. The more local ($\langle d \rangle = 170$ pc) clouds in Snell's sample have an average diameter of 1.1 pc using the $T_A^* = 1$ K level of ^{13}CO emission. Snell suggests that much more extensive and tenuous envelopes exist and often encompass several apparently separate clouds. Since Blitz (1979) reports that the complexes near OB associations contain 20-50 clouds with sizes of 2 pc and up, it would be useful to determine whether such complexes are more distant versions of nearby "dark cloud" complexes. For such studies, it would be useful to measure higher order structure parameters, such as the size and spacing of regions of enhanced T_K and n.

6.2. Mass

The mass of a cloud may be estimated in several ways. The most common method is to use a column density tracer such as ^{13}CO. Then

$$M = A_i \frac{N_i}{X_i} m_H \mu$$

where A_i is the projected area as mapped in species i, m_H is the hydrogen atom mass, and μ is the mean molecular weight. If N_i varies substantially over the cloud, $A_i N_i$ can be replaced by some suitable integral of N_i over the cloud area. Uncertainties in M so obtained are large. A_i may be poorly defined (cf. Section VIa) and depends on d^2, where d is the distance. If i = ^{13}CO, the variation of X(^{13}CO) from Dickman's value may cause masses to be underestimated, especially in dense cores.

An alternative method is available in dense cores where n has been determined. There,

$$M_c = v_c \, n \, m_H \mu$$

where v_c is the volume of the core. v_c is usually determined from the area by assuming spherical symmetry and also depends on d^3. M_c is obviously extremely uncertain, but often appears to be a significant fraction of the total mass (Snell 1979; Evans et al. 1977; Blair et al. 1978).

Finally, a virial mass can be computed by assuming the cloud is in equilibrium, supported by turbulence, or by assuming free-fall collapse with V $\sim$ r. In both cases, some suitable measure of the line width (ΔV) and cloud radius (R) are used to calculate

$$M_{vir} \sim \frac{R\Delta V^2}{G} \; .$$

The masses of the clouds (or complexes) in the molecular ring have been estimated at $10^4 - 5 \times 10^6$ $M_\odot$ (Solomon et al. 1977), while the cloud complexes accompanying OB associations have an average mass of 10^5 $M_\odot$ (Blitz 1979). Solomon and Sanders (1979) have estimated a mass distribution function from a survey of clouds in the molecular ring and find that most of the mass is contained in the most massive (M $\sim$ 10^6 $M_\odot$) clouds. For the clouds in Snell's sample, the average mass is 70 $M_\odot$ within the $T_A^*(^{13}CO)$ = 1 K contour.

7. DYNAMICAL STATE OF MOLECULAR CLOUDS

This issue has surely provoked more controversy than any other issue in this controversial field. One may begin by noting that, based on the T_K's measured as discussed earlier, the thermal velocity in molecular clouds, V_{th}, is generally much less than the observed linewidths, ΔV. Thus the line-widths must be produced by mass motions. Suggestions for these motions include small-scale turbulence, "macroturbulence", or the random motion of rather large blobs, and systematic motions, primarily collapse. Rotation can be ruled out as the cause of ΔV, as can expansion, with the exception of a few very small regions. The arguments against turbulence of any kind have been summarized by Penzias (1975), while the case against collapse has been presented by Zuckerman and Evans (1974) and by Zuckerman and Palmer (1974). I will consider only some of the recent developments.

One of the strongest arguments against small-scale turbulence has been the absence of self-reversed profiles (cf. Liszt et al. 1974). With the advent of higher sensitivity and spectral resolving power, such profiles have begun to show up rather commonly in nearby dark clouds (Langer et al. 1978), though in high excitation molecules rather than in CO. Thus the dense cores which produce these lines may be largely turbulent. The more distant sources may not show such self-reversals because of superposition of several such cores (a kind of macroturbulence). Further evidence along these lines comes from studies of the J=2→1 and 1→0 CS lines (Linke and Goldsmith 1980), which find a constant ratio of these two lines across the profile. This conflicts with collapse models wherein each part of the line would represent a different part of the cloud, and hence different densities.

On the other side of the fence, Myers et al. (1978) have found a strong correlation between the spatial extent of emission of a molecular line and the linewidth, as predicted by collapse models with $V_c \sim r^\alpha$; $\alpha > 0$. Snell (1979) has found a similar effect and fits his data well with $V_c \sim r^{0.5}$, a retarded collapse. Such a velocity law might join smoothly to a turbulent core. This result suggests a possible resolution of the dilemma. If a substantial fraction of the gravitational potential energy released by contraction of a cloud can be coupled into turbulence, perhaps mediated by the magnetic field, then free fall collapse could be slowed, creating a turbulent core and a contraction velocity $V_c \sim r^{0.5}$ in the outer regions.

The dynamical effect of newly formed stars on molecular clouds is only beginning to be explored. Such stars often have strong stellar winds which may compress, accelerate, or push holes in the molecular cloud. Such effects are suggested by the phenomena of Herbig-Haro objects and high velocity H_2O masers. In addition, expanding H II regions and eventual supernova explosions may have severe effects on molecular clouds. All of these phenomena will result in a shock propagating through the molecular cloud. Observations of H_2 emission (Gautier et al. 1976; Beckwith et al. 1978) have spurred a number of calculations of shocks in molecular clouds. Such shocks may play major roles in initiating further star formation (Elmegreen and Lada 1977), in disrupting the cloud (Wheeler, Mazurek, and Sivaramakrishnan 1979) and in increasing the general kinetic energy content of the cloud.

8. SUMMARY

The discovery of interstellar molecules led to the recognition of molecular clouds as an important new galactic structure. As our understanding of molecular excitation and line formation has improved, we have begun to use observations of the molecular lines to probe the conditions in molecular clouds. Techniques now exist for determining most of the parameters that are needed to characterize the clouds. While discrepancies among different methods and uncertainties in line formation still are serious problems, a consensus on techniques appears to be emerging for at least some parameters. Indeed, the worst problem in trying to characterize clouds from a survey of the literature is the unsystematic nature of many investigations, the failure to publish a standard set of parameters determined in a standard way, and the lack of overlap in the clouds studied by different techniques. Because of this problem, I have relied heavily on a few studies and used them as examples of the techniques. With a few exceptions, the appropriate data do not yet exist to define average values or distributions of parameters.

This research has been supported in part by NSF Grant AST77-28475, by NASA Grants NSG-7381 and NSG-2345, and by the Research Corporation.

REFERENCES

Beckwith, S., Persson, S.E., Neugebauer, G., and Becklin, E.E.: 1978, Astrophys. J. 223, 464.
Blair, G.N., Peters, W.L., and Vanden Bout, P.A.: 1975, Astrophys. J. (Letters) 200, L161.
Blitz, L.: 1979, Proceedings of the Third Gregynog Workshop on Giant Molecular Clouds, P.M. Solomon, ed., in press.
Chaisson, E.J., and Vrba, F.J.: 1978, in Protostars and Planets, ed. T. Gehrels (Tucson: Univ. of Arizona Press) pg. 189.
Cheung, A.C.: 1976, unpublished dissertation, University of California, Berkeley.
Clark, F.O., Johnson, D.R., Heiles, C.E., and Troland, T.H.: 1978, Astrophys. J. 226, 824.
Clavel, J., Viala, Y.P., and Bel, N.: 1978, Astron. and Astrophys. 65, 435.
Crutcher, R.M., Evans, N.J., II, Troland, T., and Heiles, C.: 1975, Astrophys. J. 198, 91.
de Muizon, M., Rouan, D., Lena, P., Nicollier, C., and Wijnbergen, J.: 1979, Astron. and Astrophys., in press.
Dickman, R.L.: 1975, Astrophys. J. 202, 50.
Dickman, R.L.: 1978, Astrophys. J. Suppl. 37, 407.
Dyck, H.M., and Lonsdale, C.J.: 1979, preprint.
Elmegreen, B.G., and Lada, C.J.: 1977, Astrophys. J. 214, 725.
Evans, N.J.II, Blair, G.N., and Beckwith, S.: 1977, Astrophys. J. 217, 448.
Gatley, I., Becklin, E.E., Sellgren, K., and Werner, M.W.: 1979, Astrophys. J. 233, 575.
Gautier, T.N., III, Fink, U., Treffers, R.R., and Larson, H.P.: 1976, Astrophys. J. 207, L129.
Goldreich, P., and Kwan, J.: 1974, Astrophys. J. 189, 441.
Goldsmith, P.F., and Langer, W.D.: 1978, Astrophys. J. 222, 881.
Gordon, M.A., and Burton, W.B.: 1977, Bull. Am. Astr. Soc. 9, 553.
Green, S., Garrison, B.J., Lester, W.A., Jr., and Miller, W.H.: 1978, Astrophys. J. Suppl. 37, 321.
Guelin, M., Langer, W.D., Snell, R.L., and Wootten, H.A.: 1977, Astrophys. J. (Letters) 217, L165.
Harvey, P.M., Campbell, M.F., and Hoffmann, W.F.: 1978, Astrophys. J. 219, 891.

Ho, P.T.P.: 1977, unpublished dissertation, M.I.T.

Jenkins, E.B., and Savage, B.D.: 1974, Astrophys. J. 187, 243.

Keene, J., Harper, D.A., Hildebrand, R.H., Loewenstein, R.F., Moseley, S.H., Whitcomb, S.E., Winston, R., and Steining, R.F.: 1978, Bull. Am. Astr. Soc. 10, 687.

Kutner, M.L., Thaddeus, P., Penzias, A.A., Wilson, R.W., and Jefferts, K.B.: 1973, Astrophys. J. (Letters) 183, L27.

Langer, W.D., Wilson, R.W., Henry, P.S., and Guelin, M.: 1978, Astrophys. J. (Letters) 225, L139.

Leung, C.M.: 1975, Astrophys. J. 199, 340.

Linke, R., and Goldsmith, P.F.: 1980, Astrophys. J., in press.

Liszt, H.S., Wilson, R.W., Penzias, A.A., Jefferts, K.B., Wannier, P.G., and Solomon, P.M.: 1974, Astrophys. J. 190, 557.

Loren, R.B., Evans, N.J., II, and Knapp, G.R.: 1979, Astrophys. J., 234, 932.

Merrill, K.M., Russell, R.W., and Soifer, B.T.: 1976, Astrophys. J. 207, 763.

Morris, M., Zuckerman, B., Palmer, P., and Turner, B.E.: 1973, Astrophys. J. 186, 501.

Mouschovias, T. Ch.: 1978, in Protostars and Planets, ed. T. Gehrels, (Tucson: Univ. of Arizona Press) pg. 209.

Myers, P.C., Ho, P.T.P., Schneps, M.H., Chin, G., Pankonin, V., and Winnberg, A.: 1978, Astrophys. J. 220, 864.

Nachman, P.: 1979, Astrophys. J. Suppl. 39, 103.

Penzias, A.A.: 1975, Atomic and Molecular Physics and the Interstellar Matter, ed. Balian, Encrenaz and Lequeux (North Holland Publ. Co.) pg. 373.

Pickett, H.M., and Davis, J.H.: 1979, Astrophys. J. 227, 446.

Plambeck, R., Snell, R.L., and Loren, R.B.: 1979, in preparation.

Rouan, D., Lena, P.J., Puget, J.L., deBoer, K.S., and Wijnbergen, J.J.: 1977, Astrophys. J. (Letters) 213, L35.

Sanders, D.B., and Solomon, P.M.: 1977, Bull. Am. Astr. Soc. 9, 554.

Scoville, N.Z., and Solomon, P.M.: 1974, Astrophys. J. (Letters) 187, L67.

Snell, R.L.: 1979, unpublished dissertation, The University of Texas at Austin.

Snell, R.L., and Loren, R.B.: 1977, Astrophys. J. 211, 122.

Soifer, B.T., Puetter, R.C., Russell, R.W., Willner, S.P., Harvey, P.M., and Gillett, F.C.: 1979, Astrophys. J. (Letters) 232, L53.

Solomon, P.M., Sanders, D.B., and Scoville, N.Z.: 1977, Bull. Am. Astr. Soc. 9, 554.

Solomon, P.M., and Sanders, D.B.: 1979, Proceedings of the Third Gregynog Workshop on Giant Molecular Clouds, P.M. Solomon ed., in press.

Vanden Bout, P.A., Loren, R.B., Snell, R.L., and Wootten, A.: 1979, in preparation.

Vrba, F.J., Strom, S.E., and Strom, K.M.: 1976, Astron. J. 81, 958.

Watson, W.D.: 1977, in CNO Processes in Astrophysics, ed. J. Audouze (Dordrecht: Reidel).

Werner, M.W., Gatley, I., Harper, D.A., Becklin, E.E., Loewenstein, R.F., Telesco, C.M., and Thronson, H.A.: 1976, Astrophys. J. 204, 420.

Wheeler, J.C., Mazurek, T.J., and Sivaramakrishnan, A.: 1979, preprint.
Wootten, H.A., Evans, N.J., II, Snell, R.L., and Vanden Bout, P.A.:
    1978, Astrophys. J. (Letters) 225, L143.
Wootten, A., Snell, R., and Glassgold, A.E.: 1979, Astrophys. J. 234, 876.
Zuckerman, B., and Evans, N.J. II: 1974, Astrophys. J. (Letters) 192,
    L149.
Zuckerman, B., and Palmer, P.: 1974, Ann. Rev. Astron. and Astrophys.
    12, 279.

DISCUSSION FOLLOWING EVANS

Crutcher: A recent OH experiment by Crutcher, Heiles, and Troland gave 3σ limits to the magnetic field of ~ 25 μG toward dust clouds with $n_H \sim 10^3 cm^{-3}$. A possible detection of OH Zeeman splitting in Taurus gave $B = 12 \pm 4$ μG.

Evans: These new results are beginning to put pressure on the theory.

Biermann: Would it not seem, taking into account the very low relative electron abundances you mentioned, that the coupling between mass motions and magnetic fields becomes so small that no substantial increase of the magnetic field strength by compression is expected?

Evans: The very low electron abundances really apply only to the dense cores of the clouds, so that the magnetic field may still play a major role in most of the volume of the cloud. One might expect that the magnetic field is no longer important in the dense cores for the reason you suggest, but the results of Dyck and Lonsdale give one pause.

Ho: The derivation of X_i depends on dV/dr. What happens if dV/dr varies with radius? Observations indicate that line widths vary between the centers and edges of clouds. In fact "cores" or "condensations" towards the centers of clouds may be very quiescent, in which case dV/dr may be very small. Would X_i then be larger than has been deduced?

Evans: In most cases we have used dV/dr as deduced from H_2CO itself; thus our velocity gradients do apply to the dense core. A calculation of dV/dr for ^{13}CO actually indicated a lower dV/dr on the average, although the difference was very small (Wootten et al. 1978).

Ho: The H_2 densities derived from NH_3 observations seem comparatively low (Table 2) because you have ignored the effects of the clumped distribution of NH_3.

Evans: That explanation works only if NH_3 is more clumpy than the other molecules; otherwise the discrepancy persists. Since NH_3, analyzed with simple assumptions, follows the same trend as ^{13}CO, we may not have to invoke clumpiness after all.

Ho: If different molecules such as NH_3, CS, and H_2CO are really sampling regions of different densities, how do you explain their similar spatial distribution? If $n \propto 1/r$, would a difference in derived n by a factor of 10 imply a difference in spatial extent by a comparable factor? Spatial extents of different molecules should be considered, because a very steep radial decrease of n may be implied.

Evans: We suggested $n \propto r^{-2}$, giving changes of n by factors of 10 for changes of r by factors of 3. More generally, I agree that a completely satisfactory model would also account for the spatial extent of the emission of each line.

Mouschovias: On the important issue of large linewidths in molecular clouds, the theoretical arguments against turbulence are well known, as you said, and one need not repeat them. Have you not also presented in your talk *observational* evidence against turbulence? I am referring, of course, to the results of optical and infrared polarization measurements, which show unambiguously a beautiful ordering of the magnetic field over large length scales. The presence of significant turbulence would have destroyed this ordering, would it not? Incidentally, in 1975 I suggested that linewidths are due to large-scale oscillations at supersonic, but sub-Alfven, speeds of self-gravitating magnetic interstellar clouds about stable equilibrium configurations [(Ph.D. Thesis); see also IAU Symposium No. 75.]

Evans: Your question brings into focus a certain conflict that is just under the surface at this meeting. It seems to me that there is an essential conflict between the evidence for very large scale structure (eg. the paper by Morris et al.) which appears to be ordered by magnetic fields, and the picture of giant molecular clouds growing by collisions of essentially isolated clouds. On the smaller scale of the dense cores, however, there is some observational evidence now for turbulence. I agree that it is difficult to reconcile that with the polarization results

Kwok: One observational parameter that you did not discuss is the line profile of ^{12}CO. The centrally peaked profiles observed in many molecular clouds place severe constraints on the allowable density and velocity laws. For example, there must be a density gradient, and velocity must vary with radius according to a power law with a positive index. What is needed now is a dynamical justification (eg. collapse calculations) for such density and velocity laws.

Evans: I will venture to say that no one in this room really understands why these lines have the shapes that they do, and I would be delighted to be contradicted.

Clark: Do the observations absolutely rule out rotation? A small near-rigid rotation consistent with Mouschovias' calculations could very nicely reproduce your observed variation in linewidth.

Evans: It is pretty clearly ruled out as the *cause* of the linewidth in the clouds I described. Rotation may still play a role, of course.

Penzias: In your talk you enumerated some of the problems encountered in modelling the intra-cloud flows which manifest themselves as large linewidths. I think it is worth emphasizing that the physical constraints which I set down in my Les Houches review lead to contradictions when one attempts to construct self-consistent slowly dissipative macro-turbulent models. If, on the other hand, one could identify an adequately powerful source with which to drive dissipative flows, it seems to me that one could resolve the remaining theoretical difficulties with a judicious combination of ordered motion and macroturbulence.

Evans: I tend to agree that macroturbulent models are probably closest to the actual situation, at least in more distant objects, where the beam could include a number of separate blobs. The suggestions of Silk and Norman for driving the flows with T-Tauri winds can be tested in nearby clouds by infrared searches for the stars. One should also explore non-stellar sources for driving turbulence, including the collapse (or contraction) itself.

Gilmore: You made a comparison between Snell's sample of local dark clouds and more distant cloud complexes associated with HII regions. In no way is either of these sets complete. In fact local clouds have a very diverse nature, ranging from less than 1 pc to several parsecs in size, ranging in opacity, and ranging in degree of association with other clouds. More distant clouds represent an unknown sample, since most of those known are associated with HII regions and star formation. There are local large complexes similar in size to distant ones, yet different in physical characteristics observed via molecular lines at millimeter wavelengths. A most significant indicator of the difference is the presence of star formation. The nature of the clouds being observed, especially with respect to evolutionary state and the presence, for some external or internal reason, of star formation, could strongly bias the nature of the chemistry observed as well as other general deductions one might make concerning any of the parameters you mentioned.

Evans: Your comment supports my call for more systematic studies.

DETECTION OF SUBMILLIMETER LINES OF CO (0.65 mm) AND H_2O (0.79 mm)

T. G. Phillips and John Kwan
Bell Laboratories, Murray Hill, NJ 07974

P. J. Huggins
New York University, New York, NY 10003

The 91.5 cm telescope of NASA's Kuiper Airborne Observatory has been used, in conjunction with an InSb heterodyne receiver, to detect the J = 4-3 submillimeter transition of CO (461 GHz) and the 4_{14}-3_{21} transition of H_2O (380 GHz). The water emission was detected from the Orion "plateau" region.

1. INTRODUCTION

    The submillimeter portion of the spectrum is important for molecular astronomy because it contains the fundamental transitions of many of the simple hydride molecules, and also major cooling lines of abundant molecules such as CO and possibly H_2O. Observations in the submillimeter require low atmospheric water vapor, so that an ideal telescope for initial studies is that of the NASA Kuiper Airborne Observatory. Here we report some line detections with that telescope at wavelengths in the 0.8-0.6 mm range. An InSb heterodyne bolometer receiver was used in conjunction with a diode harmonic generator for local oscillator (LO) power. System temperatures varied from 400-1500 K depending on the amount of LO power available.

2. CO (J=4-3)

    Figure 1 shows a spectrum of CO (J=4-3) in the direction of the Becklin-Neugebauer-Kleinmann-Low (BNKL) region of the Orion molecular cloud. As usual for CO both narrow line and broad line components are observed coming from the extended cloud and compact "plateau" region respectively. The line center is found to be at a velocity of 9 km/sec (LSR) using the calculated frequency of 461.0408 GHz (Lovas and Tiemann, 1974). Since our beam size is about 2.5 arc min and the plateau source size for CO is about 50 arc sec (as determined from J = 2-1 measurements using the OVRO 10 m telescope--to be published separately), the broad line is severely diluted. For the narrow line, probably the most immediately striking point is that there is still no

21

B. H. Andrew (ed.), Interstellar Molecules, 21–24.

self-reversal feature, even with the very large opacity in the 4-3
line. It seems that the large cloud is probably heated from the front
(or by a not too deeply imbedded source) if there is to be so little
foreground cool gas.

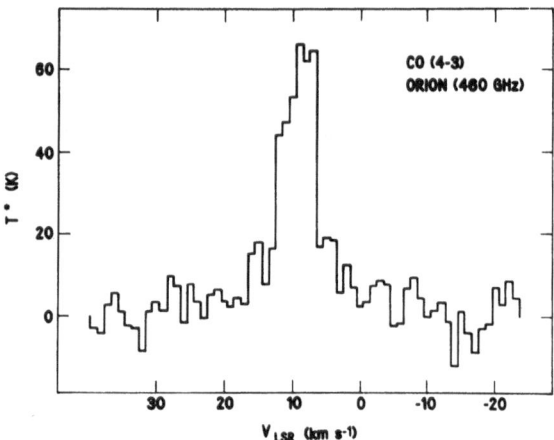

Figure 1. CO (J=4-3) in the direction of BNKL.

3. H_2O ($4_{14}-3_{21}$)

      Interstellar water was detected by the powerful masing $6_{16}-5_{23}$
(22 GHz) transition (Cheung et al, 1968) and has recently been
observed in the $3_{13}-2_{20}$ transition (Waters et al, 1979) and in the
$H_2^{18}O$ line (Phillips et al, 1978). The $3_{13}-2_{20}$ observations have lead
to an abundance estimate relative to H_2 of about 10^{-5} and an explanation of this high value has been given by Elitzur (1979) in terms of
shock chemistry. That explanation may well be appropriate since the
observations were made in the direction of BNKL, where it is currently
thought that the observed "plateau" lines and H_2 vibration-rotation
lines are due to shock waves (e.g. Kwan, 1977; Hollenbach and Shull,
1977; Kwan and Scoville, 1976).

      Figure 2 is a spectrum of the $4_{14}-3_{21}$ line towards BNKL. The
rest frequency is given by DeLucia et al (1972) as 380.1974 GHz. The
shape and width of the line are more typical of the plateau source
rather than the large cloud, so that a discussion of excitation under
the physical conditions of the plateau region is appropriate. To
get a value for the brightness temperature of the source we need to
know the beam dilution factor. The beam size at 380 GHz is 3 arc min

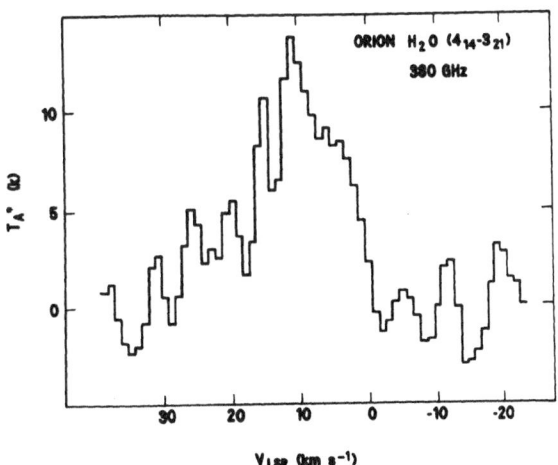

Figure 2. H_2O $(4_{14}-3_{21})$ in the direction of BNKL.

and the source size presumably lies in the range between 50 arc sec
found for CO (see above) and 30 arc sec found for SO_2 (Phillips et al,
1980). Consequently the brightness temperature for H_2O $(4_{14}-3_{21})$
would be between 160 and 430 K. Assuming a source density of
10^7 H_2 molecules per cm^3, a velocity gradient of 50 km s^{-1} per 3×10^{17}
cm and a gas temperature of 100 K, we find from an excitation calcula-
tion a brightness temperature of > 160 K in the abundance range
$[H_2O]/[H_2]$ = 10^{-6}-2×10^{-5}. Within that range the brightness temperature
peaks at $\sim$ 500 K. The observations and calculations are consistent
with peak maser $(-\tau \stackrel{<}{\sim} 3)$ action for both the $4_{14}-3_{21}$ and $3_{13}-2_{20}$ lines
from the plateau source.

REFERENCES

Cheung, A.C., Rank, D.M., Townes, C.H., Thornton, D.D., and Welch, W.J.:
    1969, Nature 221, pp. 626-628.
DeLucia, F.C., Helminger, P., Cook, R.L., and Gordy, W.:   1972, Phys.
    Rev. A 5, pp. 487-490.
Elitzur, M.: 1979, Astrophys. J. 229, pp. 560-566.
Hollenbach, D.J. and Shull, J.M.: 1977, Astrophys. J. 216, pp. 419-426.
Kwan, J. and Scoville, N.: 1976, Astrophys. J. (Letters) 210, pp. L39-
    L43.
Kwan, J.: 1977, Astrophys. J. 216, pp. 713-723.
Lovas, F.J. and Tiemann, E.: 1974, J. Phys. Chem. Ref. Data 3, pp.
    609-770.
Phillips, T.G., Scoville, N., Kwan, J., Huggins, P.J., and Wannier,
    P.G.: 1978, Astrophys. J. (Letters) 222, pp. L59-L62.

Phillips, T.G., Pickett, H.M., Knapp, G.R., Huggins, P.J., and Redman,
    R.: 1980, to be published.
Waters, J., Gustincic, J.J., Kuiper, T.B.H., Roscoe, H.K., Swanson,
    P.N., Kerr, A.F., and Thaddeus, P.: 1979, Astrophys. J., to be
    published.

DISCUSSION FOLLOWING PHILLIPS

Hollenbach: There seems to be some disagreement in the literature
about the theoretical gas-phase abundance of H_2O in *cold* molecular gas.
Is it well established that the H_2O in Orion is necessarily produced
behind the shock in the *hot* gas?

Phillips: In the case of $H_2{}^{16}O$ the smallest beamwidth used is
3 arc min so that there is no direct proof that H_2O is confined to the
hot region. However, the deduced brightness temperatures of >160 K and
the observed linewidths of ∿50 km/sec imply that the H_2O is produced
behind the shock. Attempts to observe $H_2{}^{16}O$ lines in cold gas regions
have been unsuccessful.

Black: What is the ratio of abundances of ortho and para species
of H_2O?

Phillips: Lines of both species have been detected in the Orion
plateau source. The observations seem to be consistent with an ortho
to para ratio of 3:1, but the errors are at least a factor of 2.

Kuiper: The shock chemistry depends largely on the effect of
increased temperature on temperature-sensitive reactions. The SO_2
plateau observations of Pickettt and Davis (Ap.J. 1979, 227, 446), as
well as those reported by you, indicate that SO_2 in the plateau is in
thermodynamic equilibrium at 70 K. Even if the plateau region was
originally shocked, the absence of any residual temperature effects
makes it doubtful whether an H_2O enhancement could be attributed to a
shock.

Phillips: The current temperature of the region is not relevant
to the temperature at the time of the formation of the molecules. Once
formed, the water will obviously remain until chemically converted into
something else. The time scale for the conversion would have to be
worked out.

Kuiper: A search with a typical sensitivity of 0.2 K was made on
the Kuiper Airborne Observatory for the 380 GHz line of water by
de Graouw, Lidholm, van Vliet, Nieuwenhuyzen, van de Stadt (U. of Utrecht),
and myself. The sources included the molecular clouds M17, W51, W49,
Sgr B2, ρ Oph, and L134, the star χ Cygni, and the galaxy M101. The
line was not detected. This result, coupled with the detection reported
here by Phillips et al., suggests that the excitation of this transition
bears some similarity to that of the 183 GHz transition (Waters et al.,
Ap.J., 1980, in press).

NEW DETECTIONS OF SPECTRAL LINES IN THE FREQUENCY RANGE 260-285 GHz

N. Erickson
University of California-Berkeley
J.H. Davis, N.J. Evans II, R.B. Loren, L. Mundy,
W.L. Peters III, M. Scholtes, and P.A. Vanden Bout
University of Texas-Austin

The upper half of the frequency band lying between the atmospheric water lines at 183 and 323 GHz is virtually unexplored by spectroscopists of interstellar matter. The only published observations are those of Huggins et al. (1979) who detected the J=3-2 lines of HCN, HNC, and HCO^+ in the Orion A molecular cloud (OMC-1). We report here the results of extensive observations made in March and April of 1979 with a University of California (Berkeley) receiver on the Texas 4.9-m antenna.

The receiver had a quasi-optical diplexer for local oscillator injection (Erickson 1977) and a klystron and doubler for a local oscillator source. A Schottky-barrier diode at room temperature was used in the mixer and the system temperature was typically 4500 K (SSB). A room temperature FET amplifier was used for the 1.7-GHz IF signal. The aperture efficiency of the antenna at 282 GHz was roughly 25% from observations of Jupiter. The same planet was used to map the main beam at 267 GHz and we estimate the beam efficiency to be about 40%. The weather during the run was generally excellent with atmospheric optical depths typically 0.3 and occasionally as good as 0.1. The line intensities were calibrated by synchronously detecting a chopper wheel (Penzias and Burrus 1973).

The results of these observations are summarized in Table 1. Except for HCN, HNC, and HCO^+, all of the lines are new detections. The SO_2, HC_3N, and unidentified lines were observed only in OMC-1. The U-lines frequencies are: 278.263, 278.306, 278.889, and 281.958 GHz. Sally Cummins of the NASA Institute for Space Studies has suggested the $8_{17}-7_{16}$ transition of H_2CS at 278.8865 GHz as identification for U278.889. The frequencies agree to within one line width, assuming the U-line occurs at $v_{LSR}=9.0$ km/s. It might be expected that all four of these relatively strong U-lines are higher transitions of known, relatively abundant interstellar molecules.

The widths of the SO_2 and HC_3N lines are typical of lines associated with the plateau source in OMC-1. Examples are shown in Figure 1

B. H. Andrew (ed.), Interstellar Molecules, 25–30.
Copyright © 1980 by the IAU.

Table 1

Molecule	Transition	Frequency (GHz)	Comments
SO_2	$11_{3,9} - 11_{2,10}$	262.257	detected in OMC-1 (K-L)
	$13_{3,11} - 13_{2,12}$	267.537	
	$15_{3,13} - 15_{2,14}$	275.240	
	$15_{1,15} - 14_{0,14}$	281.763	
	$6_{2,4} - 5_{1,5}$	282.037	
	$20_{1,19} - 20_{0,20}$	282.293	
HC_3N	30-29	272.885	detected in OMC-1 (K-L)
	31-30	281.977	
H_2CO	$4_{1,4} - 3_{1,3}$	281.523	detected in (12) sources
N_2H^+	3-2	279.513	detected in (10) sources
HCO^+	3-2	267.557	detected in (27) sources
HCN	3-2	265.886	detected in (14) sources
HNC	3-2	271.983	detected in (4) sources

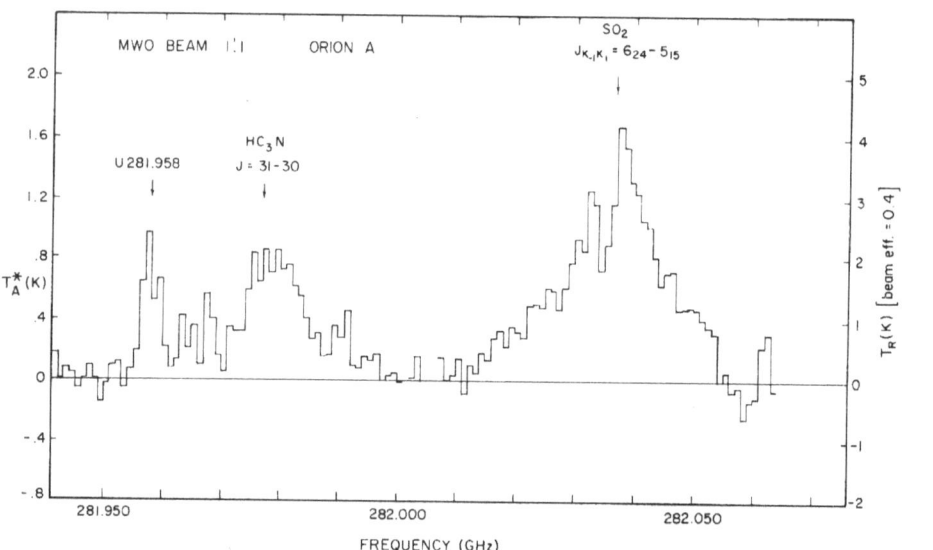

Figure 1. Sample OMC-1 spectra: HC_3N, SO_2, and an unidentified line near 282 GHz.

where very roughly one 10-MHz division on the horizontal axis is 10 km/s. The spike component which can be seen in the 6_{24}–5_{15} SO_2 line is not present in any of the other SO_2 spectra. This is expected because the 5_{15} level lies only 10.9 cm^{-1} above the ground state whereas the other observed SO_2 transitions have lower levels lying 48.8 to 128.8 cm^{-1} above the ground state.

The plateau source apparently has several components. The central velocity of HC_3N line is 7 km/s whereas the spike source has a velocity of 9 km/s and typically the central velocity of other plateau lines lie even farther to the red. This is in agreement with the velocity of 5 km/s found by Wannier and Linke (1978) for broad J=9–8 HC_3N emission and the velocity of the emission from vibrationally excited HC_3N detected by Clark et al. (1976) at 4 and 6 km/s. (Our spectra demonstrate one advantage of observing at higher frequencies: the broad HC_3N lines we observed are ten times stronger ($T_A^* \sim 1K$) than those observed by Wannier and Linke ($T_A^* \sim 0.1K$) as a result of both the narrower antenna beam and greater optical depth of the higher frequency transition.) Further evidence for multiple plateau components can be seen in Figure 2 which shows that the broad HCO^+ emission can be seen 1'N of the K-L position and is, therefore, more extended than the SO_2 emission which is seen only at the K-L position.

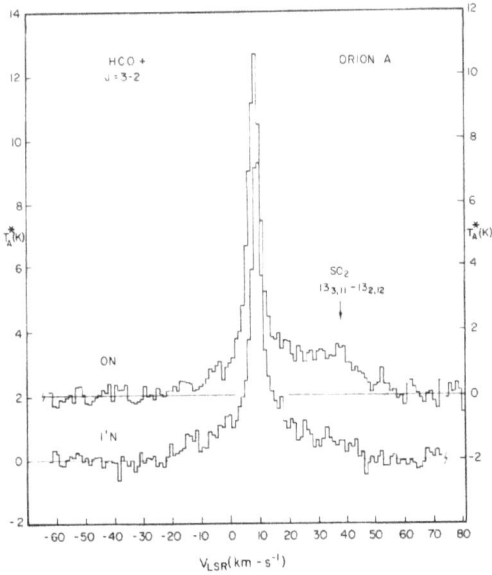

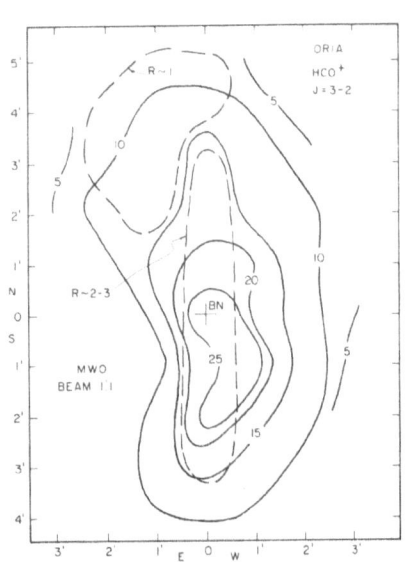

Figure 2. HCO^+ J=3–2 lines observed at the K-L position and 1'N in OMC-1. Both scales are for T_A^*.

Figure 3. A map of HCO^+ J=3–2 emission in OMC-1. Contours are of T_R. $R \equiv T_R(3-2)/T_R(2-1)$.

The most extensive map made during this observing run was of J=3-2 HCO$^+$ emission from OMC-1. Figure 3 shows the map which consists of 36 points with 1' spacings. Our map is similar to the map of J=1-0 emission published by Turner and Thaddeus (1977) in that the familiar N-S ridge can be seen in both maps. However, the J=3-2 map does not show a strong secondary peak at the 4'N 1'E position as is seen in the J=1-0 map. Both lines have comparable strengths at this position whereas the J=3-2 line is 2 to 3 times stronger than the J=1-0 line at the K-L position and along the N-S ridge. The two regions in question are labelled R=T_R(3-2)/T_R(1-0)=1 and R=2→3 in Figure 3.

The strong HCO$^+$ J=3-2 lines by themselves indicate the presence of high density gas in OMC-1. The characteristic density, n, for this transition, defined by $nR_{23} \equiv A_{32}$, is n≈5(+6) cm^{-3}, where the rate coefficient R_{23} for a kinetic temperature of 70K was estimated by extrapolating those given by Green (1975). A more realistic density estimate can be derived from the ratio R. Simple large-velocity-gradient (LVG) calculations show that R=1 corresponds to n≈5(+5) cm^{-3} for a wide range of abundances [n(HCO$^+$)/n=5(-12)→10(-10)]. The ratio has values larger than one only for optically thin lines. For n(HCO$^+$)/n=5(-12)→10(-11), R=2→3 corresponds to n≈10(+6) or greater. Densities of this order have been reported previously for the N-S ridge based on HC$_3$N (Morris et al. 1977) and H$_2$CO (Evans et al. 1979) observations. The validity of this density estimate depends on the accuracy of the line strengths. Our HCO$^+$ line temperature is 2.5 larger than that of Huggins et al. (1979). It is possible that there are significant contributions to the signal from extended HCO$^+$ lying outside the main beam of the 4.9-m antenna. Until these data are more carefully calibrated, the density estimate is tentative.

Figure 4 shows OMC-1 spectra of N$_2$H$^+$, H$_2$CO, HNC, and HCN. Our J=3-2 line intensities for HCN and HNC, after correction for beam efficiency, are in good agreement with those of Huggins et al. (1979). The N$_2$H$^+$ line shown in Figure 4 is for a position 4'N 1'E of K-L; the line is weaker for the K-L position. Again, as for HCO$^+$, the J=3-2 and J=1-0 N$_2$H$^+$ lines are comparable in strength at the 4'N 1'E position and the J=3-2 line is 2 to 3 times stronger than the J=1-0 line at the K-L position. The similar behavior in the OMC-1 maps of the ratio R for both N$_2$H$^+$ and HCO$^+$ is not unexpected. Both molecules have little optical depth in OMC-1 and the ratio R reflects the excitation, which should be similar for the two molecules because their collisional cross sections are similar. The fact that the line strengths themselves for both transitions of N$_2$H$^+$ are stronger at the 4'N 1'E position than at K-L, with the reverse true for HCO$^+$, means the N$_2$H$^+$/HCO$^+$ abundance ratio changes between these two positions in OMC-1, as concluded by Turner and Thaddeus (1977).

The plateau component is just visible at roughly 20% of the spike component strength in the H$_2$CO 4_{14}-3_{13} profile for the K-L position. We have not yet incorporated this line in a model of all the observed H$_2$CO lines in OMC-1. The 4_{14}-3_{13} line strength seen in S140 is roughly consistent with our model for that source.

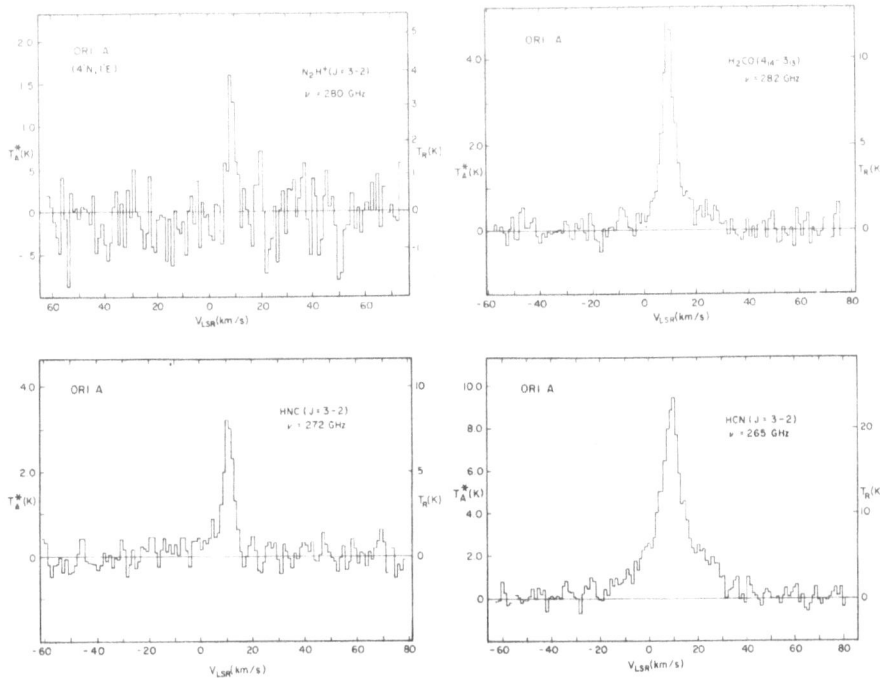

Figure 4. Spectra of N_2H^+, H_2CO, HNC, and HCN emission from the
K-L position in OMC-1 except the N_2H^+ line is for a
position 4'N 1'E of K-L. The T_R scale assumes a beam
efficiency of 40%.

The HCN J=3-2 line profile for the K-L position clearly shows both
spike and plateau components; only the spike component is visible in
the HNC J=3-2 profile. The ratio R is about R=2 for both spike and plateau components of HCN and for the spike component of HNC, using the
J=1-0 data of Gottlieb et al. (1975) for HCN and of Snell and Wootten
(1979) for HNC. Again, a ratio larger than unity is indicative of
small optical depths. The ratio is less than that observed for HCO^+
but this is expected because the characteristic density for excitation
of the HCN J=3-2 line is roughly ten times larger than that for HCO^+.

These initial observations and results indicate both the richness
of the spectrum in this band and the possibilities high-frequency observations offer for determining the properties of galactic molecular
clouds. This work was supported by NSF Grant AST 75-13511 and AST
77-28475.

REFERENCES

Clark, F.O., Brown, R.D., Godfrey, P.D., Storey, J.W.V., and Johnson,
     D.R.: 1976, Astrophys. J. (Letters), 210, pp. L139-L140.

Erickson, N.R.: 1977, IEEE Trans. Microwave Theory and Tech., MTT-25,
    pp. 865-866.
Evans, N.J., II, Plambeck, R.L., and Davis, J.H.: 1979, Astrophys. J.
    (Letters) 227, pp. L25-L28.
Gottlieb, C.A., Lada, C.J., Gottlieb, E.W., Lilley, A.E., and Litvak,
    M.M.: 1975, Astrophys. J. 202, pp. 655-672.
Green, S.: 1975, Astrophys. J. 201, pp. 366-372.
Huggins, P.J., Phillips, T.G., Neugebauer, G., Werner, M.W., Wannier,
    P.G., and Ennis, D.: 1979, Astrophys. J. 227, pp. 441-445.
Morris, M., Snell, R.L., and Vanden Bout, P.A.: 1977, Astrophys. J.
    216, pp. 738-746.
Penzias, A.A., and Burrus, C.A.: 1973, Ann. Rev. Astron. Astrophys. 11,
    pp. 51-72.
Snell, R.L., and Wootten, H.A.: 1979, Astrophys. J. 228, pp. 748-754.
Turner, B.E., and Thaddeus, P.: 1977, Astrophys. J. 211, pp. 755-771.
Wannier, P.G., and Linke, R.A.: 1978, Astrophys. J. 226, pp. 817-823.

DISCUSSION FOLLOWING VANDEN BOUT

Kuiper: Do you have any information on the extent and position of
the "plateau" source in Orion? Besides the one spectrum taken 1' north
of KL, is there any other evidence for extended HCO^+ "plateau" emission?

Vanden Bout: The other positions in our HCO^+ map have sufficient
signal to noise to define the spike component intensity but not to
reveal weak plateau-component emission, so we cannot say if the broad
emission is extended beyond KL and 1' north.

Guelin: An HCN 2-1 strip map in Orion A made by Thaddeus and myself
shows the plateau to fall down very quickly off the KL position ($\leq 2'$).

Goldsmith: High velocity dispersion emission in the $J=3 \rightarrow 2$
transition of CS has been detected with the 14-m FCRAO antenna near the
KL position. The center is between 0" and 30" north of KL, and the
size $\leq 40"$, but signal to noise ratio is the limiting factor.

Winnewisser: Do you have any idea of the optical depth of the
280 GHz lines. How useful are these lines for isotopic substitution?

Vanden Bout: Given the line strengths and kinetic temperature of
OMC-1, it is difficult to see how these lines can be optically thick
unless the excitation is severely sub-thermal. Our LVG calculations
indicate that it is not, and the lines are optically thin, so they
may be suitable for isotopic abundance measurements in Orion; other
HCO^+ sources may well have optically thick J=3-2 lines because several
self-reversed J=1-0 profiles are known.

FURTHER OBSERVATIONAL STUDIES OF THE HIGH VELOCITY
MOLECULAR SOURCE ("PLATEAU") IN ORION

T.B.H. Kuiper
Jet Propulsion Laboratory, California Institute of Technology
E.N. Rodriguez Kuiper
Western Laboratories, Ball Aerospace Systems Divisions
B. Zuckerman
Astronomy Program, University of Maryland

ABSTRACT

Sensitive measurements have been made of the profiles of various molecular lines in the direction of the BN/KL infrared complex in Orion. Similar line shapes are found for the high velocity wings ("plateau") of most lines observed. The position of the high velocity source (HVS) was found in SO and HCO^+ relative to the SiO maser. A search for HV emission in other sources was unsuccessful.

OBSERVATIONS

Low noise spectra of the J=0-1 transitions of ^{12}CO, ^{13}CO, HCN, HCO^+, and N_2H^+, of the 3_2-2_1 transition of SO, the $8_{17}-8_{08}$ transition of SO_2, and the J=7-6 transition of OCS were obtained with the NRAO 11-m telescope for the KL/BN region in Orion. A search was made for the J=5-4 transition of SiS. The 2_2-1_1 transition of SO was mapped simultaneously with the v=1 J=2-1 transition of SiO. Mapping observations of HCO^+ and SiO were interleaved. A search was made for CO J=0-1 emission at large velocities relative to the line center in W3 (IRS5 and OH), NGC1333 (IR and H_2O), Ori MC2, Mon R2 (IRS2 and IRS3), NGC6334 (IRS1), M17 (IRS1), GL2591, DR21(OH), and NGC7538 (IRS1).

LINE SHAPES

We found that ^{12}CO, ^{13}CO, HCN, HCO^+, SO, and SO_2 have emission at $|\Delta V| > 20$ km/s (where $\Delta V=0$ at $V_{lsr}=9$ km/s), and that this emission is centered at $V_{lsr}=9\pm1$ km/s for all these lines. The emission for $|\Delta V| > 20$ km/s has the same dependence on ΔV, namely that the intensity varies as $|\Delta V|^{-2.3}$. Emission for $|\Delta V| < 20$ km/s shows different dependences on $|\Delta V|$ for different lines. Comparison with profiles of the ^{12}CO J=1-2 and J=2-3 lines (Phillips 1978, private communication), of the $3_{13}-2_{02}$ transition of H O (Waters et al. 1980), and the $1_{10}-1_{01}$ transition of H_2S (Kuiper, 1978, private communication), shows the same behaviour for $|\Delta V| > 20$ km/s, although the signal-to-noise ratios are lower. No HV emission was seen in N_2H^+ (<0.1 K at $|\Delta V|$ =20 km/s). OCS showed a broad

31

B. H. Andrew (ed.), Interstellar Molecules, 31–32.
Copyright © 1980 by the IAU.

spectral feature centered at 6 km/s, while a narrow and more intense
component was centered at 4 km/s. The SiS J=5-4 transition was not
seen at all to an upper limit of 0.2 K.

The conclusion that the J=0-1 transition of ^{12}CO is optically thin
in the wings (Wannier and Phillips 1977) is confirmed by the ratio of
the J=1-0 ^{12}CO and ^{13}CO wings, which are in the ratio of ∿40 deduced by
Wannier et al. (1976) for dense clouds. Assuming that the emission in
the other lines comes from the same region and that the emissions are
in thermodynamic equilibrium at 70 K (Pickett and Davis 1979), we find
that HCO$^+$ and HCN are in the ratio found in molecular clouds (Wootten
et al. 1978), that SO_2, SO, and H_2S have ratios typical of molecular
clouds (Gottlieb et al. 1978), but that HCO$^+$ and HCN are overabundant
relative to CO by a factor of at least ten (Wootten et al. 1978).
Statistical equilibrium calculations carried out for the $3_{13}-2_{02}$ line of
H_2O also indicate that H O is overabundant by a similar factor relative
to CO (Waters et al. 1980).

SOURCE POSITIONS

The position of the HVS in SO is indistinguishable from the posi-
tion of the SiO maser source to a 1 sigma accuracy of 2 arcseconds. The
HCO$^+$ centroid was found to be 16 arcsec north and 6 arcsec west of the
SiO maser with an estimated uncertainty of 5 arcseconds.

SEARCH FOR OTHER HIGH VELOCITY EMISSION SOURCES

To a r.m.s. noise level of typically 0.2 K, no emission was seen at
$|\Delta V|$ =20 km/s in any source except M17, in which the presence of multiple
components made it impossible to determine whether broad wing emission
was present. Taking into account the distances of the sources, we conclude that emission of the intensity seen in Orion is not present in
NGC1333 (IR or H_2O), Ori MC2, Mon R2 (IRS2 or IRS3), or NGC6334 (IRS1).
For the other sources, HV emission of the intensity seen in Orion would
not have been detected.

REFERENCES

Gottlieb, C.A., Gottlieb, E.W., Litvak, M.M., Ball, J.A., and Penfield,H.:
    1978, Astrophys. J. 219, pp. 77-94.
Phillips, T.G., Huggins, P.J., Neugebauer, G., and Werner, M.W.: 1977,
    Astrophys. J. (Letters) 217, pp. L161-L164.
Pickett, H., and Davis, J.H.: 1979, Astrophys. J. 227, pp. 446-449.
Wannier, P.G., Penzias, A.A., Linke, R.A., and Wilson, R.W.: 1976,
    Astrophys. J. 204, pp. 26-42.
Wannier, P., and Phillips, T.G.: 1977, Astrophys. J. 215, pp. 796-799.
Waters, J.W., Gustincic, J.J., Kakar, R.K., Kuiper, T.B.H., Roscoe,
    H.K., Swanson, P.N., Rodriguez Kuiper, E.N., Kerr, A.R., and
    Thaddeus, P.: 1980, Astrophys. J. in press.
Wootten, A., Evans, N.J., II, Snell, R., and Vanden Bout, P.: 1978,
    Astrophys. J. (Letters) 225, pp. L143-L148.

HIGH VELOCITY GAS IN THE ORION NEBULA

Nicholas Z. Scoville
Department of Physics and Astronomy
University of Massachusetts

ABSTRACT

Observations at both millimeter and infrared wavelengths reveal energetic activity within the core of the Orion molecular cloud in the vicinity of the KL-BN cluster. New observations of the high velocity CO emission at 2.6-mm with improved angular resolution (HPBW = 44") show that the source diameter averages 4×10^{17} cm and the center of mass is displaced 10-12" north of the Kleinmann-Low nebula to a position close to the Becklin-Neugebauer object. The total mass of high velocity gas in the core region is ~10 $M_\odot$ (assuming 10% of the carbon is in CO); the <u>present</u> kinetic energy is 4×10^{47} ergs. Further evidence that BN may be the ultimate source of this energy is provided by high resolution infrared spectra which show both ionized and high temperature ($T_k \gtrsim 3000$ K) neutral gas in this source. CO bandhead emission ($v = 2 \to 0$, $3 \to 1$, and $4 \to 2$) seen in BN is thought to arise from collisional excitation at high temperatures in a very dense ($n_H > 10^{10}$ cm^{-3}) region only 1 AU in size. And high spectral resolution profiles of the Br α and γ recombination lines show that the HII region previously detected in BN apparently has motions over 100 km s^{-1}.

Over the last few years our impression of the evolutionary status of the Kleinmann-Low nebula at the core of the Orion cloud has radically changed as a result of several entirely independent observational discoveries. Taken together, these new observations suggest that the core region is evolving on timescales as short as a few thousand years rather than 10^4 to 10^5 years as previously contemplated. Observations of millimeter wavelength rotational transitions of carbon monoxide reveal emission far out in the line wings at ± 50 km s^{-1} relative to the cloud center of mass at $V_{LSR} = 9$ km s^{-1} (Zuckerman, Kuiper, and Kuiper 1976; Kwan and Scoville 1976, Phillips et al. 1977). In the near infrared, high spectral resolution observations by Gautier et al. (1976) detect several lines at 2μ identified with decay from the $v = 1$ state of H_2 at the equivalent energy $h\nu/k$ of 6800 K (reviewed here by Beckwith). In the immediate vicinity of the strongest near infrared source (BN), there is also now evidence of both ionized gas (Grasdalen 1976)

33

B. H. Andrew (ed.), Interstellar Molecules, 33–38.
Copyright © 1980 by the IAU.

and hot, high density neutral gas (Scoville, Hall, Kleinmann, and Ridgway 1979).

For several years observers of millimeter wavelength molecular lines in the Orion nebula have recognized that the emission profiles show two distant components: a spike feature at 9 km s^{-1} of width $\Delta V \simeq 5$ km s^{-1} and a "plateau" component centered at about the same velocity but with much larger width $\Delta V \simeq 40$ to 100 km s^{-1}. The narrow component is spatially extended and is therefore emitted from the large Orion cloud; the broad feature is seen only at the infrared nebula. Our most recent millimeter wavelength observations (Solomon, Huguenin, and Scoville 1979) are designed to better define the spatial character-istics of the J = 1 → 0 CO line. We chose this line for the best possi-ble indication of the mass distribution since the excitation require-ments of CO are very modest, CO is least likely to suffer abundance gradients due to shock chemistry, and choosing from amongst the low rotational transitions of CO, the 1 → 0 line is the most optically thin. For these observations the highest possible angular resolution (44") was provided by the 45-foot FCRAO telescope at the University of Massachusetts. Figure 1 shows data taken along a declination strip which crosses both KL (at $\Delta\delta = 0$) and BN (at $\Delta\delta = 0'.2$). The apparent half-power size of 57" for the high velocity emission (measured at 20 km s^{-1} from line center) translates into a true radius of ~2 x 10^{17} cm when corrected for the 44" primary beam of the FCRAO telescope.

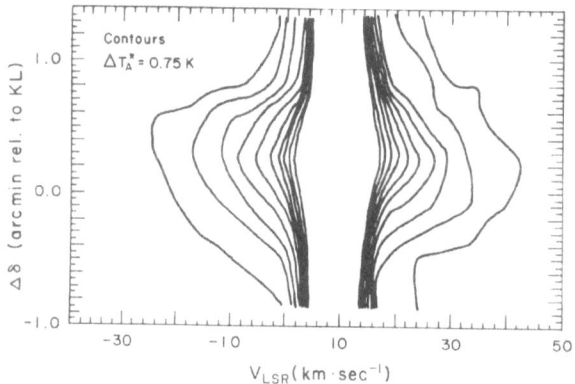

Figure 1. Map of the J = 1 → 0 CO emission along a declination strip crossing KL at $\Delta\delta = 0'$ and BN at $\Delta\delta = 0'.2$. Data was taken with the 45-foot FCRAO telescope for which the HPBW = 44".

For the declination strip shown in Figure 1, a careful comparison of the source size measured at different velocities reveals evidence that the size increases somewhat for greater velocities further out in the line wings (e.g. compare 10 km s^{-1} with 25 km s^{-1} from line center). If verified for the right ascension direction, this behavior would pro-vide strong support of kinematic models in which the largest veloci-ties occur at greatest radius in the source (e.g. v ∝ r).

The smoothness of the CO and ^{13}CO profiles strongly favors dynam-ical models with large scale motions such as rotation, collapse, or expansion to account for the line width. The alternatives, that the broadband emission arises from a superposition of narrow lines from randomly moving clumps of gas each unresolved in the beam, would pro-duce a "spikey" appearance in CO and ^{13}CO profiles unless there were a very large number of such clumps. The absence of observable velocity gradients across source and the high degree of symmetry in all the line profiles make rotation models unlikely. Moreover, if high velocity rotation is to be stabilized gravitationally by the mass of the region itself, then an interior mass of 10^4 $M_\odot$ is required. Such a high mass, if it is contained in the gas, would be expected to produce very opti-cally thick CO lines ($\tau > 400$) and cause the ^{13}CO to be equally bright. The same argument also rules against gravitational collapse models. Collapse and expansion models where the radial motion is all at a con-stant velocity may also be ruled out from the profile shapes.

For a <u>differentially</u> collapsing or expanding envelope the velocity can be either a decreasing or increasing function of radius. The ar-gument against the first case is provided by the observational result that the size of the emission at larger velocities is at least equal to that at the lower velocities. A final argument favoring an expansion model over collapse is the discontinuous change in chemical abundances between the outer cloud, the "spike" feature, and the "plateau" source. If the high-velocity gas is merely a collapsed core pulled in from the larger cloud, then why are the chemical abundances so radically dif-ferent? On the other hand, if the high-velocity gas is ejected from some condensed core where the chemistry is different, an abrupt change in abundances is entirely expected at the interface between the out-flowing material and the ambient medium. Indeed the relatively high abundance of SiO in the high-velocity gas is perhaps indicative that the temperature of the core where this gas was chemically processed was much higher than the outer Orion cloud (e.g. Lada, Oppenheimer, and Hartquist 1978).

The total mass of the high velocity gas may be estimated assuming that the CO is optically thin and has a mean excitation temperature of 150 K throughout the envelope. For a $[CO/H_2]$ abundance ratio of 1.2×10^{-4} (10% of C in CO), the mass estimate is about 10 $M_\odot$. The kinetic energy of the high velocity gas is 4×10^{47} ergs at present but would of course have to have been much greater ($\sim 10^{50}$ ergs) in the past if most of the 10 $M_\odot$ is swept up material rather than original ejecta. If the high velocity gas has been accelerated by radiation pressure over the expansion time scale ($R/v = 1000$ yrs), the rate of momentum transfer must be 3×10^7 $L_\odot/c$. Since the required luminosity is 300 times greater than what is observed in the far infrared from the Kleinmann-Low nebula (Werner <u>et al</u>. 1976), it appears unlikely that the envelope is driven by radiation pressure.

In the above model which was originally proposed by Kwan and Scoville (1977) for the CO emission, a shock heated gas layer must

naturally occur where the expanding gas confronts the stationary molecular cloud. High gas temperatures, possibly giving rise to the observed 2μ H_2 emission, should be expected in a shell with angular extent
on the sky equivalent to the $\geq 40''$ size of the high velocity CO. The
hot shell would be thin ($\Delta r \leqslant 10^{14}$ cm) due to the fast cooling of the
gas (reviewed here by Hollenbach). In view of this proposed connection
between the H_2 emission and the plateau source, it is significant that
the diameter measured here for the mm CO emission is nearly equal to
that of the 2μ H_2 emission, and the 60-90 km s^{-1} widths of the H_2 lines
measured at high resolution (Nadeau and Geballe 1979, Hall et al. 1979)
are comparable with the millimeter line widths.

A second goal of the recent millimeter CO observations was to
identify the source of origin of the energetic phenomena in the infrared cluster. From the data of Figure 1 it is clear that the center of
mass is shifted about 10-12" north of the KL position to a point close
to BN; similar data in right ascension shows no significant displace-
ment from $\Delta\alpha = 0\overset{s}{.}0$. Frequent pointing on planets at the time of these
observations indicates an absolute position accuracy of ±5"; moreover
three repetitions of this experiment at FCRAO and one at the 36-foot
NRAO telescope all gave consistent displacements north of KL. (The 5"
accuracy of this central position determination does not decisively
rule out the source IRc2 which was shown by VLBI between FCRAO and
NEROC to be the source of SiO maser emission (Genzel et al. 1979)).

Other evidence of the very special nature of the BN object has
been contributed by high resolution infrared spectroscopy recently made
possible by a Fourier-Transform-Spectrometer on the 4-m KPNO telescope.
The IR absorption lines seen in the continuum of BN sample the entire
line-of-sight down to radius $r \leqslant 10^{15}$ cm. At 4.6μ Hall et al. (1978)
have reported two CO absorption systems at V_{LSR} = +9 and -20 km s^{-1}
which they ascribe to the large Orion molecular cloud and to the foreground half of the plateau source respectively. More recently a third
CO system (in emission!) has been seen at V_{LSR} = +20 km s^{-1}
(Kleinmann et al. 1979).

Perhaps even closer to BN is the ionized gas responsible for the
near IR recombination lines (Grasdalen 1976; Joyce, Simon, and Simon
1978). The doppler widths of these lines are up to 100 km s^{-1} (Hall
et al. 1978). Finally there now exists data on the neutral gas close
to the surface of BN. At 2.3μ we recently detected the CO overtone
bandheads (v = 2 → 0, 3 → 1, and 4 → 2), the last of which arises from
a level at E/k = 19000 K (Scoville et al. 1979). Analysis of the latter
observations indicates that the overtones must arise from an ultrahigh
density ($n_H > 10^{10}$ cm^{-3}), compact region (~1 AU) near BN. The
temperature here must be > 3000 K.

I have described here only a part of the observational data at
both millimeter and infrared wavelengths which seem to be converging
towards a very exciting picture of the Orion nebula. We find consistent evidence from both sets of observations suggesting the

occurrence of very energetic phenomena within the young star cluster. The total kinetic energy estimated to exist in the core at present is 4×10^{47} ergs or about 0.1% the energy of a supernova. The original energy must have been much greater as much of the kinetic energy could have been liberated earlier as heat. At present BN (as opposed to IRc 2) appears as a prime candidate for the origin of this energy due to the near-IR detection of an HII region and high temperature neutral gas in close proximity. Future near IR observations at high sensitivity will be important to clarify the nature of IRc 2.

REFERENCES

Gautier, T.N., Fink, U., Treffers, R.R., and Larson, H.P.: 1976, Ap.J.(Letters), 207, L29.
Genzel, R., Moran, J.M., Lane, A.P., Predmore, C.R., Ho, P.T.P., Hansen, S.S., and Reid, M.J.: 1979, Ap.J.(Letters), 231, L73.
Grasdalen, G.L.: 1976, Ap.J.(Letters), 205, L83.
Hall, D.N.B., Kleinmann, S.G., Ridgway, S.T., and Gillett, F.C.: 1978, Ap.J.(Letters), 223, L47.
Hall, D.N.B., Ridgway, S.T., Kleinmann, S.G., and Scoville, N.Z.: 1979 (in preparation).
Joyce, R.R., Simon, M., and Simon, T.: 1978, Ap.J., 220, 156.
Kwan, J., and Scoville, N.Z.: 1976, Ap.J.(Letters), 210, L39.
Kwan, J.: 1977, Ap.J., 216, 713.
Lada, C.J., Oppenheimer, M., and Hartquist, T.W.: 1979, Ap.J.(Letters), 226, L163.
Phillips, T.G., Huggins, P.J., Neugebauer, G., and Werner, M.W.: 1977, Ap.J.(Letters), 217, L161.
Ridgway, S.T., Kleinmann, S.G., Gillett, F.C., Hall, D.N.B., and Scoville, N.Z.: 1979 (in preparation).
Scoville, N.Z., Hall, D.N.B., Kleinmann, S.G., and Ridgway, S.T.: 1979, Ap.J.(Letters), 232, L121.
Solomon, P.M., Huguenin, G.R., and Scoville, N.Z.: 1979 (in preparation).
Zuckerman, B., Kuiper, T.B.H., and Rodriguez Kuiper, E.N.: 1976, Ap.J.(Letters), 209, L137.

DISCUSSION FOLLOWING SCOVILLE

Elitzur: If the shock velocity is ∿50 km s^{-1}, the post-shock temperature is ∿10^5K. So far there is no evidence for such high temperatures. Do you have any idea how this can be resolved?
Scoville: Perhaps the highest velocity gas is accelerated in clumps so that the bulk of the material can be maintained at low temperature. I should point out that the rotational lines at millimeter wavelengths are not sensitive indicators of the very hot shocked gas, since the column length of hot gas behind the shock is only 10^{13}-10^{14}cm.
T. Wilson: The maps of NH$_3$, made with the Bonn 100-m telescope, and of HCN, made with the Onsala 20-m dish, both with angular resolution

~40", indicate that the peak of the plateau appears to be south of your peak position. The angular size of the NH3 plateau is consistent with a region of size <20".

Scoville: There are other molecules in the plateau source which peak to the north of our position (e.g. HCO$^+$). It is now becoming evident that different molecules peak at different locations in the Orion core. The reasons may lie in the varying excitation requirements of different molecules, or in the occurrence of shock-chemistry which can cause differential gradients in molecular abundance across the source. In my own opinion the CO emission gives the best estimate of the true center of mass because CO is most easily excited and is rather insensitive to shock processing. Among the various CO lines (J = 3 → 2, 2 → 1, and 1 → 0), the 1 → 0 line is preferable due to its lower optical depth, τ < 1 for this region.

Zuckerman: In distinguishing between stellar-wind models and explosive models, the dependence of the source size on velocity is important. You said your observations suggest that the source size is increasing at the larger velocities. Would Dr. Phillips be willing to comment on source-size measurements he may have made of the 2→1 line of CO, and indicate whether his measurements agree with yours?

Phillips: CO(2→1) observations from the Owens Valley 10-m telescope have 25 arcsec resolution. We do not find evidence that the size of the plateau source increases as velocity shifts from the central velocity. In fact, there may be some evidence for a decrease, at least in right ascension.

Scoville: Our measurements of the putative increase are in declination only. Other possible discrepancies between our J=1 → 0 and your 2 → 1 observations might be ascribed to the higher opacity of the 2 → 1 line (assuming no observational errors!).

Staude: Your sketch of the immediate surroundings of BN suggests axial symmetry, maybe the presence of a dense disk around the central star. Is there any specific evidence for that? In describing the IR polarization data of BN we assumed scattering in an unresolved bipolar distribution of dust (Elsasser and Staude, 1978, Astron. Astrophys. 70, 43) similar to that observed in the spatially resolved case of S106 (Eiroa et al., 1979, Astron. Astrophys. 74, 89).

Scoville: We have no direct evidence of a disk surrounding BN, only indirect evidence that there is both dense neutral gas and ionized gas near BN. The former is implied by the CO Δv = 2 bandhead observations, the latter by the near-IR Brackett α and γ lines. One possible geometry in which both may occur is that of a neutral disk, with ionized gas above and below the disk but still in view of the central UV source.

Elmegreen: The temperature, density, and size of the source, and the fact that neutral gas occurs close to the BN star (as derived from your CO bandhead observations), all suggest that there is a protoplanetary type of disk surrounding the star.

Scoville: That is the model presented in my sketch. I should emphasize again that there is no direct evidence of a disk, only indirect evidence. The precise geometry is not clear.

HIGH SPATIAL RESOLUTION STUDIES OF $H^{13}CN$, $H^{12}CN$ AND HCO^+ J = 1-0 EMISSIONS IN ORION A

O.E.H. Rydbeck, Å. Hjalmarson, G. Rydbeck,
J. Elldér, A. Sume and S. Lidholm
Onsala Space Observatory, S-430 34 Onsala, Sweden

The distributions of the $H^{13}CN$, $H^{12}CN$ and $H^{12}CO^+$ J =1-0 lines have been mapped with 20" spacing towards the Orion A molecular cloud using the new Onsala 20 m millimeter wave telescope equipped with a room temperature mixer. The aperture and main beam efficiencies are about 49 and 65% and the half power beam width is ~ 43". The absolute pointing accuracy is estimated to be better than 5" rms in the Orion elevation range.

The Orion A ridge and pedestal cloud emissions have been separated by gaussian decomposition. The HCN pedestal emission region has been partly resolved. Its position and shape[*] agree with the features of the Becklin-Neugebauer/Kleinmann-Low infrared cluster, suggesting that the broad molecular line emissions are composite results of expanding envelopes around the individual objects in the cluster. The HCO^+ pedestal emission is weak compared to that of HCN while the narrow line intensities of the two species are about the same in the N-S elongated ridge cloud. The latter cloud exhibits notable velocity shifts also on a scale as small as 20" (0.05 pc). The asymmetric map center ridge lines are decomposed into two gaussians at 7.8 and 10.0 km s^{-1}, the latter being strongest. A theoretical model analysis of the data has been performed.

A similar mapping has been done around S140 IR. The HCN/HCO$^+$ emission maximum is very close to the 2μ/100μ source position. No pedestal feature was detected, which may suggest that this object - in the infrared very similar to BN - has not yet reached the BN stage of evolution.

---

[*]Position (1950.0): $5^h32^m46.9^s\pm0.2^s$, $-5°24'25"\pm4"$. The half power size of an assumed gaussian source is ≲19" N-S, ≲13" E-W. These data are almost in complete agreement with Effelsberg NH$_3$ data by Wilson, Downes and Bieging (1979, Astron. Astrophys. 71, 275). Our position also agrees within 1" with that of the SiO (v=1,J=1-0) maser (Moran et al. 1977, Astrophys. J. 217, 434).

B. H. Andrew (ed.), Interstellar Molecules, 39–40.
Copyright © 1980 by the IAU.

DISCUSSION FOLLOWING HJALMARSON

Zuckerman: The Kuipers and I have measured the positions of the HCO^+ and SO high velocity "plateau" sources in Orion with the NRAO 36-ft. telescope. We agree with your position for HCO^+ (i.e. north of BN), but find the SO source to be centered within a few arc seconds of the SiO maser sources. Therefore, those people who consider the chemistry of the plateau source should be aware that plateau emission from different molecules may originate in different regions.

Hjalmarson: It is very interesting and, I think, also convincing that our data agree so well. Our HCN plateau position is almost identical to your SO source. Definitely the different distributions of the plateau molecules must be important for "astrochemists". Shock enhancement could be a very important process.

Kuiper: I should like to amplify the remark by Zuckerman. While we have high confidence in the SO plateau position relative to the SiO maser, the formal errors in the HCO^+ position may well be dominated by systematic errors in interpolating the pointing offsets deduced from the SiO maps. These are hard to estimate but I would guess the uncertainty is about 5 arcseconds.

I also want to draw your attention to another part of our paper which concerns the line shapes. The shapes of the lines reported are the same for relative velocities outside ± 20 km/s, but are very different for different lines inside this limit. In the most sensitive of these data, it is very clear that the plateau line-shape is not Gaussian. One should therefore bear in mind that a Gaussian decomposition may lead to deceptive results about the amount of plateau emission in any given line.

Hjalmarson: I am very happy with the close agreement between your data and Onsala data. We also have a preliminary five point SO_2 ($8_{17}-8_{08}$) map which seems to agree with your SO result. I am well aware of the line-shape problems you mention in the second part of your comment. In the case of HCN the three quadrupole lines of the spike make a meaningful comparison between different plateau line shapes (where there is also a blending of quadrupole lines) very difficult. In the cases of HCO^+, SO, SO_2, CO etc. we definitely will have to work on the line-shape problem when we have cooled mixer high quality profiles.

Kutner: You have mentioned a velocity structure in the ridge feature. Our formaldehyde observations, and NH_3 observations of Ho and co-workers, show this structure to be quite interesting. With your resolution, we can learn about the clumpiness in this structure. Have you made a declination-velocity diagram from your data?

Hjalmarson: The data seem to show a clear velocity structure, ranging from ~ 8.4 km s^{-1} at 80"S of KL to ~ 10 km s^{-1} at 80"N. In the E-W direction we find ~ 8.6 km s^{-1} at 80"W and ~ 10.3 km s^{-1} at 80"E. I therefore have a feeling that Orion A deserves somewhat more extended HCN/HCO^+ Onsala maps with the cooled mixer receiver that is now available.

LUNAR OCCULTATION OBSERVATIONS OF MILLIMETER CO EMISSION IN S255

F. Peter Schloerb and N.Z. Scoville
Department of Physics and Astronomy
University of Massachusetts
Amherst, Massachusetts 01003

Observations of millimeter wavelength emission lines at the highest possible angular resolution are necessary to reveal the energetics and dynamics in compact regions of star formation activity. Lunar occultations of molecular clouds provide a means for obtaining angular resolution that is much better than that possible with a single radio telescope, and make it possible to study in more distant sources features at the same linear scale as those in Orion. We have recently used lunar occultations of the region around two infrared sources embedded in the S255 molecular cloud to determine the angular structure of the CO emission of the surrounding gas, and we have found that this region closely resembles the core of the Orion molecular cloud (Schloerb and Scoville 1980).

During 1978 and 1979 we used the Five College Radio Astronomy Observatory's 14-m antenna to observe three lunar occultations of the S255 molecular cloud in the CO J=1-0 transition. The resolution obtained by the observations was between 4 and 7 arcsec, and enabled us to resolve features that were much smaller than the 44 arcsec telescope beam. In addition to the large-scale structure seen in previous maps of this cloud, the occultation observations reveal two high temperature emission regions in the cloud core that are associated with two compact infrared sources separated by about 20 arcsec. The first of these high temperature regions, which is clearly present in all three occultations, is well modeled by a 40 arcsec (FWHM) gaussian with a peak temperature of about 65 K. The position of this component is close to that of S255 IRS2. The most recent occultation revealed evidence of a smaller (<7 arcsec) and hotter (>200 K) source at the position of S255 IRS1. Due in part to the better signal-to-noise then, this small feature has been observed unambiguously only during the most recent occultation, although there is some evidence for it in the previous two occultations as well. Further observations of this region will be necessary to confirm its existence and clarify its properties

Overall, there exists a remarkable similarity between the scale and characteristics of the phenomena in S255 and those in the Orion molecular cloud. In both cases a grouping of two dominant infrared sources is found in a neutral molecular cloud at the edge of young HII regions. The

41

B. H. Andrew (ed.), Interstellar Molecules, 41–42.

sources are separated by $\sim 10^{17}$ cm. One of the sources in each region (S255 IRS1; and the Becklin-Neugebauer object, BN in Orion) has a relatively high color temperature ($T_C \sim 500$–600 K) and dominates the near infrared emission, while the other, cooler source (S255 IRS2; and the Kleinmann-Low Nebula, KL) becomes bright only at $\lambda \gtrsim 10\mu$. The total infrared luminosities for these regions are also comparable to each other (8 x 10^4 $L_\odot$ for S255 IR, Beichman et al. 1979; and 1.2 x 10^5 $L_\odot$ for KL and BN, Werner et al. 1976). We find that these similarities persist when one compares the millimeter line characteristics. Both the peak temperature (65 K) and half intensity size of 0.5 parsec of the 40 arcsec CO source are in close agreement with the peak intensity and half power size of the Orion cloud core.

Since the infrared characteristics of the two regions are so similar, the similarity of the CO emission properties in the S255 and Orion clouds is easily accounted for if the H2 gas (and CO) is heated by thermal collisions with the warm dust grains which surround the infrared sources. In such a model (Goldreich and Kwan 1974; Scoville and Kwan 1976), when the grains and gas are in thermal equilibrium, the gas temperature surrounding an embedded infrared source should depend mostly on $(L/r^2)^{1/5 \rightarrow 1/6}$. Since the luminosities are so nearly equal for these two regions, we then expect that the temperature structures should also be the same.

In seeking a counterpart to the small, hot CO source associated with S255 IRS1, one might look to the high velocity emission in Orion (Zuckerman et al. 1976; Kwan and Scoville 1976). Solomon et al. (1979) have recently determined that the center of this CO emission is close to BN (the counterpart of S255 IRS1) and its size is similar to that of the unresolved source (~ 2 x 10^{17} cm). A sensitive search for high velocity CO emission at the position of S255 IRS1 would indicate whether the similarity to BN extends to the kinematic properties of the gas.

The Five College Radio Astronomy Observatory is supported by NSF Grant AST76-24610 and is operated with the permission of the Metropolitan District Commission of the Commonwealth of Massachusetts. This is contribution No. 325 of the Five College Astronomy Department.

REFERENCES

Beichman, C.A., Becklin, E.E., and Wynn-Williams, C.G.: 1979, Astrophys. J. (Letters) 232, L47.
Goldreich, P., and Kwan, J.: 1974, Astrophys. J. 217, 448.
Kwan, J., and Scoville, N.Z.: 1976, Astrophys. J. 210, L39.
Schloerb, F.P., and Scoville, N.Z.: 1980, Astrophys. J. (Letters) 235, L33.
Scoville, N.Z., and Kwan, J.: 1976, Astrophys. J. 206, 718.
Solomon, P.M., Huguenin, G.R., and Scoville, N.Z.: 1979, in preparation.
Werner, M.W., Gatley, I., Harper, D.A., Becklin, E.E., Loewenstein, R.F., Telesco, C.M., and Thronson, H.A.: 1976, Astrophys. J. 204, 420.
Zuckerman, B., Kuiper, T.B.H., and Kuiper, E.N.R.: 1976, Astrophys. J. 209, L137.

RADIO-FREQUENCY OBSERVATIONS OF INTERSTELLAR CH_4 AND HC_5N

Kenneth Fox and Donald E. Jennings
Department of Physics and Astronomy, University of Tennessee,
Knoxville, Tennessee 37916, U.S.A.; and
Infrared and Radio Astronomy Branch, NASA Goddard Space Flight
Center, Greenbelt, Maryland 20771, U.S.A.

ABSTRACT. We detected new $^{12}CH_4$ transitions at 21,303 and 94,089 MHz in emission from Orion A, and observed temporal variations of intensity among several velocity components at 76,700 MHz. We also detected an HC_5N transition at 21,301 MHz in emission from IRC+10216.

We reported the first detection of interstellar methane (Fox and Jennings 1978) after observing six radio emission lines from Orion A. Subsequently, infrared absorption by CH_4 was detected in IRC+10216 (Hall and Ridgway 1978), and infrared features in NGC 7027 and other galactic sources were associated with CH_4 fluorescence in interstellar grain mantles (Allamandola and Norman 1978). Methane, a nonpolar molecule by symmetry, may nevertheless have pure rotational vibronic ground-state transitions resulting from vibration-rotation interactions (Fox 1971). Several of these weak lines (Fox 1972) were measured in the laboratory by various techniques (Holt et al. 1975), and molecular parameters were inferred from which accurate transition frequencies have been calculated.

The new J = 19 F2(5)-F(1) CH_4 transition at 94,088.51 MHz was observed in Orion A ($\alpha_{1950}=05^h32^m47\overset{s}{.}0$, $\delta_{1950}=-05°24'21"$) during 1979 March 4-5; a previously detected transition at 76,700.02 MHz was observed several times between 1977 November 20 and 1979 March 4; in both cases the NRAO* 11-m telescope at Kitt Peak was used. In the latest observations the 80-120 GHz receiver was equipped with exceptionally low-noise mixer diodes (Linke et al. 1978). The new J=15 A1(1)-A2(1) transition at 21,303.45 MHz was detected on 1979 February 10 and 11 by the NRAO 43-m telescope at Greenbank equipped with a 18-26 GHz maser receiver. Details are given in Jennings and Fox (1979). At 76,700 MHz, we observed rapid ($\sim$ hours) and slow ($\sim$ months) variability of intensity among several velocity components in Orion A. Figure 1 shows our spectra of these temporal variations in the source.

We also observed the J=8-7 transition of HC_5N at 21,301.247 MHz in IRC+10216 ($\alpha_{1950}=9^h45^m14\overset{s}{.}8$, $\delta_{1950}=13°30'40"$) on 1979 February 14. The detection of HC_5N in this source was reported by Winnewisser and Walmsley

---

NRAO is operated by Associated Universities Inc., under contract with NSF.

B. H. Andrew (ed.), Interstellar Molecules, 43—44.

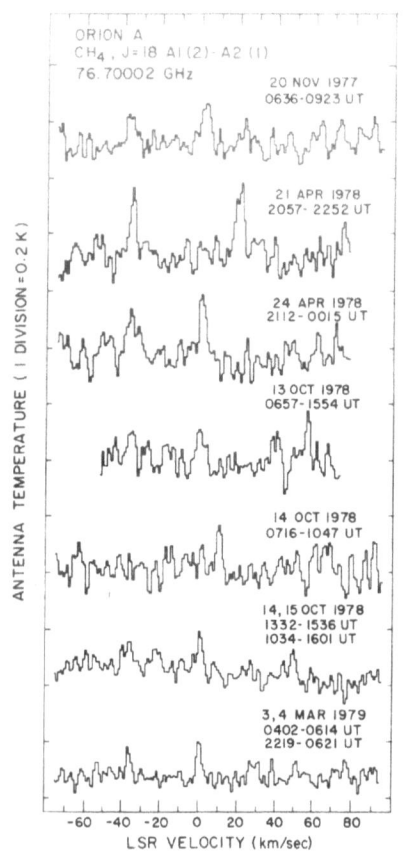

Figure 1. Temporal variations in CH_4 emission from Orion A. The resolution is 1 km s^{-1}; a two-point smoothing has been applied to all data. The spectra show emission components (never all present simultaneously) at -36, 0, $+10$, $+19$, $+49$ and possibly $+29$ km s^{-1}; the two most persistent are at -36 and 0. On 1978 April 21, the 0 component was not visible; it was replaced by a strong line at $+19$. Only 3 days later, the emission structure returned to the original configuration, with the 0 component strongly present and the $+19$ absent. On October 13, the emission was in the most common configuration of components at -36 and 0. During the first 210 min of the session on October 14, however, both seemed absent, being replaced by emission at $+10$. During the final 2 hours of that session, and on the following day, the -36 and 0 features returned, accompanied by one at $+49$. On 1979 March 3 and 4, the -36 and 0 components recurred. Our data also suggest a long-term variation. The spectra seem consistent with a compact, multi-component source, having shell-like structure centered at ~ 8.5 km s^{-1}.

(1978) who observed the J=9-8 transition. Our J=8-7 emission appears as a single line, ~ 7 km s^{-1} wide, and centered near -28 km s^{-1}. This structure is distinctly different from that of the J=9-8 feature which is ~ 28 km s^{-1} wide and flat-topped (characteristic of optically thin emission). The reason for this structural difference is not yet clear.

REFERENCES

Allamandola, L.J., and Norman, C.A.: 1978, Astron. Astrophys. <u>63</u>, L23.
Fox, K.: 1971, Phys. Rev. Letters <u>27</u>, 233.
Fox, K.: 1972, Phys. Rev. A <u>6</u>, 907.
Fox, K., and Jennings, D.E.: 1978, Astrophys. J. (Letters) <u>226</u>, L43.
Hall, D.N.B., and Ridgway, S.T.: 1978, Nature <u>273</u>, 281.
Holt, C.W., Gerry, M.C.L., and Ozier, I.: 1975, Can. J. Phys. <u>53</u>, 1971; and references therein.
Jennings, D.E., and Fox, K.: 1979, in preparation.
Linke, R.A., Schneider, M.V., and Cho, A.Y.: 1978, IEEE Trans. Microwave Theory Tech. MTT-26, 935.
Winnewisser, G., and Walmsley, C.M.: 1978, Astron. Astrophys. <u>70</u>, L37.

ON THE IDENTIFICATION OF MM-WAVELENGTH U-LINES

B.E. Turner
National Radio Astronomy Observatory

During the past five years, a total of about 180 U-lines has been observed by the author and others in the 70-115 GHz region in the prominent molecular clouds Sgr B2 and Ori (KL). A program to identify the carriers of these lines has been undertaken. A computer code, incorporating nine different centrifugal distortion Hamiltonians, generates spectra of asymmetric rotors from given rotational and centrigugal-distortion constants. Spectra have been computed for ~ 550 molecules that are considered potentially interesting astrophysically. Molecules were selected from Landolt-Bornstein Volumes IV (1967) and VI (1974) on Molecular Constants, and from the spectroscopic literature (1973 - present).

A calculated transition frequency is declared "matched" to a U-line if the frequencies agree within ± 3 MHz. The criteria applied in deciding whether a given molecule is a good candidate for interstellar identifications are

(1) that there be at least three matching transitions,
(2) that there be at least two matching transitions having quantum number $K_{-1} \leq 1$ and corresponding to the largest dipole moment component,
(3) that the energies of all matching transitions be sufficiently low (matching transitions with very low J values are weighted highly in deciding the merits of a candidate),
(4) that the microwave spectroscopy be "reliable". All calculations are checked against measured frequencies (typically in the 20-40 GHz region), and agreement is demanded with an accuracy sufficient to justify an extrapolation from the 20-40 GHz region to the 70-115 GHz region.

Molecules that satisfy these criteria and that are therefore judged to be reasonable interstellar candidates appear in the table below. The list in the table has been arbitrarily truncated; many other molecules have statistics as favorable as some that appear, but are subjectively judged less likely (e.g., 1,3,4 thiadiazole, a 5-member ring containing sulfur). Additional comments follow.

B. H. Andrew (ed.), Interstellar Molecules, 45–46.
Copyright © 1980 by the IAU.

(1) "Overlap" refers to two or more transitions having the same frequency within the ± 3 MHz tolerance. When overlap occurs, the group of transitions involved is counted as only one transition. "Yes(n)" means n groups are overlapped, which amounts to a minimum of 2n, but more usually 4n transitions counted as n matches.

(2) The probability of accidental matchups has been estimated, and implies that perhaps 25% of the entries in the table occur due to chance.

(3) Candidates are not rejected on the basis that one, or even two, calculated transitions were not observed to the sensitivity limit (0.1 K) of the survey. "Control" molecules such as ethyl cyanide and methyl formate were not seen in two or more transitions, while their statistics in the format of the table would read 9 6 5 N and 5 3 2 N, respectively.

Molecule		Number of Matches			Overlap
		Total	$K_{-1} \leq 1$	$K_{-1} \leq 1$ & μ_{max}	
nitrosyl cyanide	NOCN	3	3	1	N
1-penten-3-yne	$CH_3-CC-CH=CH_2$	7	4	4	N
formic anhydride	HCO-O-HCO	8	3	3	N
propyl cyanide (gau.)	$CH_3 CH_2 CH_2 CN$	7	5	3	N
(trans)	$CH_3 CH_2 CH_2 CN$	5	3	2	N
crotonitrile (cis)	$CH_3-CH=CHCN$	8	5	3	N
(tr A)	$CH_3-CH=CHCN$	6	3	2	N
(tr E)	$CH_3-CH=CHCN$	4	2	2	N
cyanoallene	$CH_2=C=CHCN$	4	2	2	N
ethoxyethyne	$CH_3 CH_2 OCCH$	6	5	5	N
methyl vinyl ether	$CH_3 OCH=CH_2$	4	3	3	N
glyoxal	HCO-HCO	3	2	2	N
vinyl formate	$HCO-O-CHCH_2$	6	3	3	N
crotonaldehyde (A)	$CH_3 CH=CH-CHO$	5	4	3	N
(E)	$CH_3 CH=CH-CHO$	4	4	2	N
2,5 dihydrofuran		4	4	4	Y(4)
4-pyrone		5	4	4	Y(5)
pyrrole		3	2	2	Y(3)
piperidine		4	3	3	Y(3)

In light of the above comments, the entries in the table should not be regarded as definitive identifications. Additional observations are planned to further test these and other species. The ultimate degree of success of this approach to the identification of U-lines may rest on laboratory verifications. Work is currently underway in this area by several groups.

At a level of 0.1 K there is about one U-line every 170 MHz in the 3-mm wavelength region. Accidental matchups of lines can be expected. Viable identifications of new interstellar molecules must now rest on many more matching transitions than was once considered adequate.

LONG CHAIN CARBON MOLECULES IN THE INTERSTELLAR MEDIUM

L.W. Avery
Herzberg Institute of Astrophysics
National Research Council of Canada, Ottawa

1. INTRODUCTION

The long chain carbon molecules known as the cyanopolyynes (HC_nN, n=3,5,7,9) are becoming increasingly more important in astrophysics. At present, the smallest member of the family, cyanoacetylene (HC_3N), has been observed in at least 32 sources, and cyanodiacetylene (HC_5N) in at least ten. Some 29 transitions of these two molecules have been detected to date, and the number of new sources and new lines is increasing quickly. Although the larger members of the family have not yet been found in sufficient abundance to permit studies in more than a few sources, the fact that they exist at all in detectable amounts is of interest from the standpoint of astrochemistry.

As a family, these molecules share a number of properties that are unique or, at least, unusual among interstellar species. In the following section we shall review briefly some of these properties and how they contribute to the usefulness of the cyanopolyynes as astrophysical probes. Then we shall look at some specific studies in which these molecules have been utilized.

2. CYANOPOLYYNE NOMENCLATURE AND PROPERTIES

The cyanopolyyne molecules consist of a conjugated, unsaturated, carbon chain terminated at one end by an H atom and at the other by the cyano group, CN. So long as the carbon chain is relatively short, the structural resemblance to acetylene is sufficient to justify the commonly used names, cyanoacetylene (HC_3N) and cyanodiacetylene (HC_5N). However, for the larger molecules, it is perhaps better to adopt a more formal terminology as outlined, for example, in (1). The general formula for these molecules can be written $H(C \equiv C)_n CN$, n=1,2,3, ____, and the names deduced from the length of the carbon chains and the bonding. For example, if n=3, we have cyano-hexa-tri-yne. The prefix "cyano" obviously denotes the presence of CN, "hexa" indicates a 6-atom carbon chain, and "tri-yne" indicates there are three (tri) triple bonds (yne)

47

B. H. Andrew (ed.), Interstellar Molecules, 47–58.
Copyright © 1980 by the IAU.

in the chain.   Similarly, for n=4, we have cyano-octa-tetra-yne, and
the rationale for the generic term, cyanopolyynes is clear.

The cyanopolyynes can be described in terms of superlatives.   Their
long carbon chains make them the heaviest known interstellar molecules,
and, to the writer's knowledge, they are the longest linear molecules
known to exist anywhere.   These properties, in turn, result in the
largest moments of inertia and smallest rotational constants of the known
interstellar species.   In addition, they are characterized by very large
dipole moments.

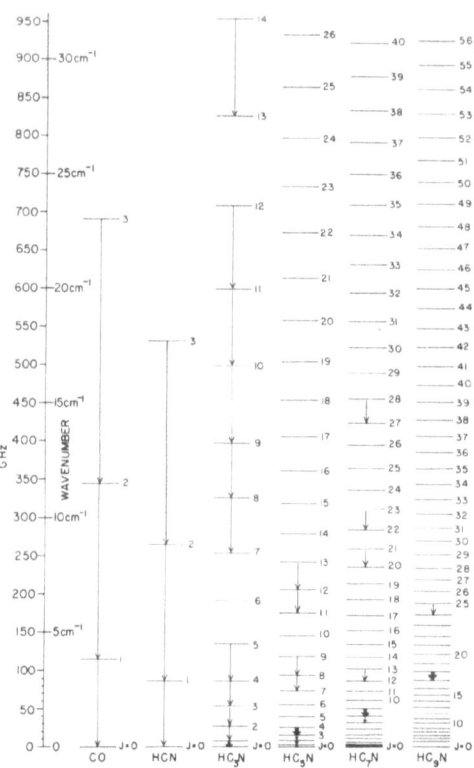

Figure 1:   Ground state
rotational energy levels for
the known cyanopolyynes.   The
light arrows indicate the
observed transitions and the
dark arrows the discovery
lines.

Figure 1   illustrates the large number of uniformly distributed,
rotational spectral lines produced by the known cyanopolyynes.   As a
consequence of the large dipole moments and linear structure, their
spectral lines are relatively strong, with frequencies that are easily
and accurately computed.   A summary of the spectroscopic constants of
the cyanopolyynes, plus two closely related radicals, butadiynyl
(C_4H) and cyanoethynyl (C_3N) is given in Table 1.

Table 1. Molecular Constants of Cyanopolyynes and Related Molecules

	Rotational Constant B_O MHz	Centrifugal Constant D_O kHz	eqQ MHz	Dipole Moment Debyes
HC_3N	4549.0579(4) [2]	0.54311(45) [2]	-4.3187(29) [2]	3.724(30) [3]
HC_5N	1331.3323(5) [4]	0.02826(107) [4]	-4.242(30) [5]	4.33(03) [6]
			-4.275(15)	
HC_7N	564.00074(16) [7]	0.003820(87) [7]	–	5.0 [8]
HC_9N	290.5184(2) [8]	0.00101 [8]	–	5.6 [8]
C_3N	4947.66(10) [9]	1.0(5) [9]	–	2.2 [10]
HC_4	4758.48(10) [9]	<1.0 [9]	–	0.9 [10]

One of the most important attributes of these molecules is the sensitivity with which they reflect density and temperature in molecular clouds. Because of its higher abundance HC_3N is likely to be the most useful for this purpose. As an example, Figure 2a shows how the ratio of brightness temperature for the J=4-3 line (36.4 GHz) and J=9-8 line (81.9 GHz) of HC_3N depends upon H_2 density and kinetic temperature for an optically thin cloud. The ratio varies by orders of magnitude over the range of $n(H_2)$ and T_k encountered in molecular clouds. Note, however, that it exceeds 10 for $n(H_2) \leq 10^4$ cm^{-3} which means that, in practice, the J=9-8 line may be too weak to observe easily. Under such conditions, one need not change molecules; only frequencies. Figure 2b shows the brightness temperature ratio of the J=1-0 (9.1 GHz) and J=3-2 (27.3 GHz) lines. For $n(H_2) \leq 10^4$ cm^{-3} the relative strength of these lines is clearly a useful density indicator.

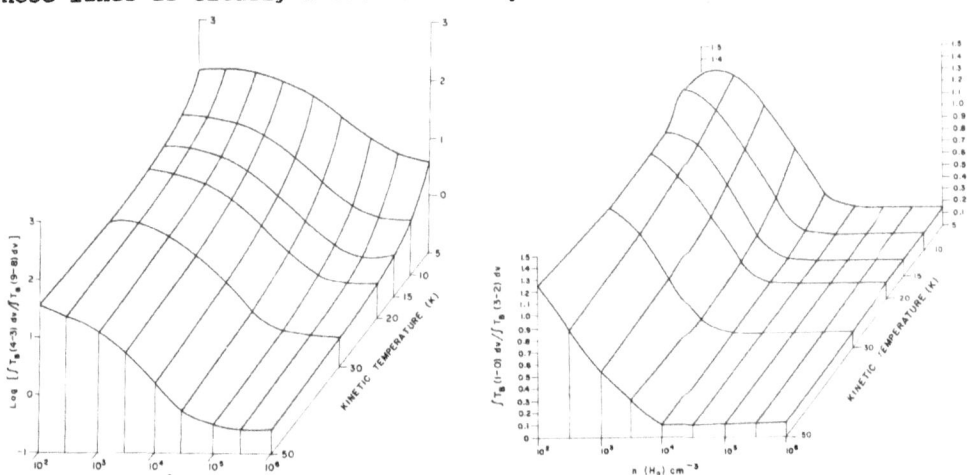

Figure 2. Ratios of integrated brightness temperature of HC_3N lines in in an optically thin cloud. (a) For J=4-3 and 9-8 lines. (b) For J=1-0 and 3-2 lines. Note difference in ordinate scale between (a) and (b).

Another excitation property shared by the cyanopolyynes is a tendency for the J=1-0 line to act as a weak maser. This phenomenon has been discussed by Morris et al. (11) for HC_3N and by Avery et al. (12) for HC_5N. The population inversion in the J=1 level occurs because of the dependence of the downward radiative rate coefficients upon J. For the transition J→J-1, the radiative rate varies as $B_o^2\mu^2[J^3/(2J+1)]$ where μ is the dipole moment and B_o the rotational constant of the molecule. Thus, the downward rate from J=2 is about 5 times greater than that from J=1, resulting in an accumulation of population in J=1 relative to J=0, provided that the collisional rate coefficients, $C_{J,J-1}=n(H_2)\sigma_{J,J-1}V$, are approximately equal to the radiative rates. If the collisional rates significantly exceed the radiative rates, then the inversion is destroyed. This property of HC_5N is illustrated in Figure 3. For $n(H_2)<10^3cm^{-3}$, T_{ex} (1-0)<0. Because of the requirement that $C_{J,J-1} \approx R_{J,J-1}$ for inversion, the density at which maximum inversion occurs for different cyanopolyynes is $\propto B_o^2$. Consequently, if maximum inversion occurs at $n(H_2) \approx 10^2cm^{-3}$ for HC_5N, it will occur at $n(H_2) \approx 10^3cm^{-3}$ for HC_3N.

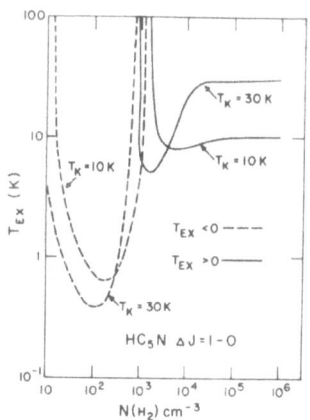

Figure 3. T_{ex} versus $n(H_2)$ for the J=1-0 line of HC_5N. The strongest maser effect occurs when T_{ex} is negative and closest to zero. The broken lines indicate negative T_{ex}.

Broten et al. (13) found evidence that the J=1-0 line of HC_5N in Sgr B_2 is enhanced relative to the J=4-3 and 8-7 lines. Following an analysis based on the principles discussed above, Avery et al. (12) interpreted these observations to indicate the presence of an extended, low density envelope in the Sgr B_2 molecular cloud.

3. SOME INTERESTING RESULTS OF OBSERVATIONS OF LONG CHAIN MOLECULES

Having considered some of the properties of these molecules, let us now turn to some specific studies based on them. It is not feasible to present a complete review of the observational status of the cyanopolyynes, so I shall single out three specific current topics of interest as representative.

3.1 Cyanoacetylene in OMC1

A number of authors have detected various transitions of HC_3N in
OMC1 which have revealed some new properties of this well-studied cloud.
Very recently, Loren et al. (14) have detected the J=31-30 and 30-29
lines at 282.0 GHz and 272.9 GHz. Morris et al. (15) have observed
various lines up to J=16-15 which they interpret as arising in a new,
very dense component ($n(H_2) \approx 10^6 cm^{-3}$) of the north-south ridge in OMC1.
In the same source, Clark et al. (16) observed the ℓ-doublet, J=10-9
transitions of the ν_7 vibrationally excited state of HC_3N. This state
lies 224 cm^{-1} above the ground state, and these lines are among the
highest excitation, non-masering interstellar lines yet found. Deguchi
et al. (17) undertook a statistical equilibrium study of these observations based on a new determination of the rate coefficients involving
the excited state. Their results reveal the presence of a dense, compact
cloud of vibrationally excited HC_3N, some 3" to 8" arc in diameter, with
$n(H_2) > 3 \times 10^7 cm^{-3}$ and a total mass of 10-100 $M_{\odot}$. It is probable that the
recently detected J=31-30 and 30-29 ground state lines also arise from
this cloud. The line widths and LSR velocities of these lines are
$\approx$ 9.0 km. s^{-1} and $\approx +6.5$ km. s^{-1}, close to the values for the vibrational
state lines. Typical values for the lower J lines observed by Morris
et al. (15) are $\Delta V_{FWHM} \approx 4.0$ km s^{-1} and $V_{LSR} + \approx 9.0$ km s^{-1}.

3.2 Studies of the $^{12}C/^{13}C$ ratio and isotope fractionation

The cyanopolyyne molecules, especially HC_3N because of its
relatively strong lines, offer unique opportunities to study ^{13}C
fractionation and abundances. The opacity in the ^{12}C species is generally small so that radiative trapping is not a problem as it is in the
case of isotope studies based on ^{12}CO or ^{12}CS. Churchwell et al. (18)
have observed the J=1-0 transition of the three ^{13}C-substituted isotopic
species of HC_3N in Sgr B_2. They concluded there was strong evidence for
^{13}C fractionation in that $H^{13}CCCN$ was twice as abundant as each of the
other two ^{13}C species. However, Wannier and Linke (19) have observed
the various J=9-8 transitions in Sgr B_2 and OMC1, and find no evidence
for isotope fractionation. They deduced a $^{12}C/^{13}C$ ratio of 22 in Sgr B_2
and 50 in OMC1.

The situation regarding ^{13}C fractionation has been further confused
following a theoretical study by Wolfsberg et al (20). These theoretical findings are at odds with both of the above observational results!
They predict that $H^{13}CCCN$ should be 30% less abundant than $HC^{13}CCN$ and
50% less abundant than $HCC^{13}CN$.

Clearly, this is an area where more work is required, for the
question of whether ^{13}C fractionation occurs and if so to what degree
has an important bearing on determinations of the $^{12}C/^{13}C$ abundance ratio.

3.3 Cyanopolyynes in the Taurus Dark Cloud Complex

In recent years, the collection of dark clouds in the constellation
Taurus has been revealed as a most remarkable location for cyanopolyyne
studies. The premier source within the complex is TMC1, a small, conden-

sation buried within Heiles' Cloud 2. The first indication of the
potential of TMC1 as a cyanopolyyne source was provided by Morris et al.
(11) who detected relatively strong emission in the J=5-4 line of HC_3N.
Subsequently, HC_5N was detected and studied there (21,22,23), and both
HC_7N and HC_9N were first discovered in TMC1 (7,8). Indeed, HC_9N has
not, to date, been detected in any other source. Table 2 shows the
estimated column densities of the various cyanopolyynes at the peak
position in TMC1. Other studies (24,25,26) have revealed HC_5N in three
more locations in the Taurus complex, TMC2 (L1529), L1544 and L1521 B.
In the following, it will not be possible to review all the work relevant
to our topic. I shall restrict the discussion to some of the more
recent developments and implications of the Taurus observations.

Table 2. Cyanopolyyne Column Density in TMC1

Molecule	$NL(10^{13}cm^{-3})$
HC_3N	13.0 †
HC_5N	5.0 *
HC_7N	1.2 *
HC_9N	0.3 *

† This paper.   * Reference (8).

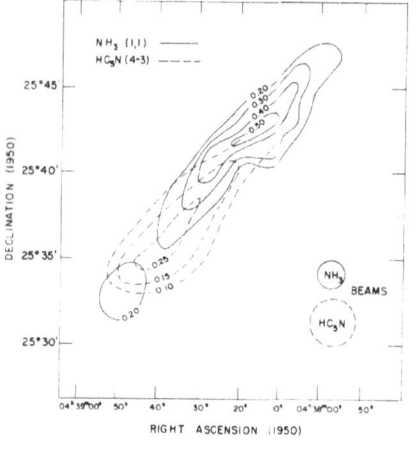

Figure 4. Distributions of
HC_5N, J=4-3 and $NH_3(1,1)$ emission
in TMC1. Contour scale for NH_3 is
T_A of strongest hyperfine component,
and $\int T_A dV$ for HC_5N.

The HC_5N distribution in TMC1 is in the form of a narrow ellipse of
dimensions > 10' x 2' which corresponds to a physical size of >0.4x0.08
pc. The ridge of emission is not distinguished optically within the
much larger area of Heiles' Cloud 2, and represents only 1% of the
Cloud 2 area. Its total mass is 1 $M_\odot$. Figure 4 shows a superposition
of the HC_5N J=4-3 and NH_3 (1,1) line emission observed at ARO[1] (21,27).
Although emission from both molecules overlaps along the same ridge, the
peak of the NH_3 distribution lies some 10' arc northwest of the HC_5N
peak. This has also been observed by Ungerechts et al (28) and Little
et al. (29). Little et al. attribute this effect to an abundance
difference, not an excitation phenomenon. Wootten (30) has mapped TMC1
in C_2H and finds the C_2H contours also form an emission ridge, with two
separate, equal maxima - one at the HC_5N peak, and the other at the NH_3

peak. Apparently, the physical conditions at the southeast end of TMC1 differ in a way not yet understood from those at the northwest end. Conditions at the SE end favour production of the cyanopolyynes; conditions to the NW favour NH_3, and both sets of conditions are kindly disposed to C_2H.

To accurately determine the density and temperature of the cyano-polyyne TMC1 ridge, we have undertaken observations at Kitt Peak[2] of the J=4-3, 5-4, 9-8, 10-9, and 11-10 HC_3N lines at the peak position. The details will be published elsewhere, but preliminary results indicate the observations require a two-component, core-halo model. The model parameters computed for a cylindrical cloud are shown in Table 3. The density and temperature of the core are somewhat higher than earlier estimates, but not exceptional for dark clouds. It thus appears that the enhanced abundance of long chain carbon molecules cannot be attributed to obviously unusual physical conditions. Figure 5 shows the relative contributions of the core and halo regions to the perceived brightness temperatures of different HC_3N transitions at the cloud centre. The cool halo is optically thick over the middle range of J, which accounts for the shape of the core contribution.

Table 3. Model of TMC1 Based on HC_3N Observations

	Dimensions FWHM	T_k K	NL HC_3N	$n(H_2)$	$\dfrac{n(HC_3N)}{n(H_2)}$
Core	1'x3'	20	$2 \times 10^{13} cm^{-2}$	$6 \times 10^4 cm^{-3}$	3×10^{-9}
Halo	2.5'x10'	10	$1.1 \times 10^{14} cm^{-2}$	$9 \times 10^3 cm^{-3}$	4×10^{-8}

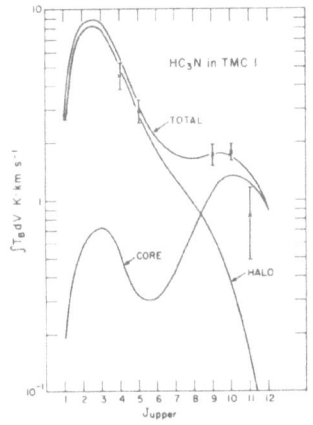

Figure 5. Calculated integrated brightness temperature of HC_3N lines in TMC1 for the model of Table 3. For comparison the observations on which the model is based are also shown. The J=1-0 observation is from G. Winnewisser, private communication.

If we cannot attribute the enhanced abundance of cyanopolyynes in Taurus to unusual densities, temperatures, or grain properties [cf. Elias, (31)], what other factors could be invoked? Myers et al. (26) have put forth a suggestion based on their study of TMC2. TMC2 appears similar to TMC1, except it is round, and the column density of HC_5N is smaller by a factor of three. Myers et al. find $n(H_2) \approx 4 \times 10^4$ cm^{-3}, $T_K \approx 20K$, and the total mass is $1M_\odot$ as for TMC1.

L. W. AVERY

The observed HC_5N line width in TMC2 is less than that expected for a cloud in free-fall contraction, and Myers et al. conclude that TMC2 is a small, dense condensation that is in hydrostatic equilibrium inside a larger, less dense cloud. For a self-gravitating, isothermal spherical cloud surrounded by a medium of $n(H_2)=3\times10^3 cm^{-3}$ and $T_K=10K$, the critical mass above which the cloud collapses is $\simeq 6M_\odot$ [see, for example Jura (32)]. This is consistent with the idea that TMC2 is in hydrostatic equilibrium. Myers et al. argue that TMC2 and the other HC_5N clouds in Taurus are potential star forming regions which are temporarily stable [see also (23)]. As such they may have been at high density - i.e. $n(H_2)>10^4 cm^{-3}$ - for a sufficiently long time that chemical evolution has proceeded to an unusual degree, thereby explaining the high cyanopolyyne abundances.

Quite apart from any conjecture as to why the cyanopolyyne cloudlets exist in Taurus, an important fact is that they do exist. As we have seen, typical parameters are $n(H_2)\sim10^4 cm^{-3}$, $T_K=15-20K$, $M\sim1M_\odot$ and scale sizes ≤ 0.1 pc. It is tempting to identify these clouds as the manifestation of the fragmentation process that has long been discussed by theoreticians. Silk (33) states that detection of clumpiness in large clouds on scales of ~ 0.1 pc, involving $1M_\odot$ or less, at densities $\sim10^4 cm^{-3}$, would provide support for existing theories of fragmentation, and "could provide a vital link between molecular clouds and the star-formation process". It seems that the compact clouds detected in Taurus are exactly what the theoreticians ordered!

1. The Algonquin Radio Observatory is operated by the National Research Council of Canada as a national radio astronomy facility.
2. The National Radio Astronomy Observatory is operated by Associated Universities, Inc., under contract with the National Science Foundation.

REFERENCES

1. Handbook of Chemistry and Physics, 59th edition, p. C-1, ed. Weast, CRC Press, Inc., Florida.
2. Creswell, R.A., Winnewisser, G., and Gerry, M.C.L., 1977, J. Molec. Spectrosc. 65, 420.
3. Lafferty, W.J. and Lovas, F.J., 1978, J. Phys. and Chem. Reference Data, 7, 441.
4. Winnewisser, G., Creswell, R.A., and Winnewisser, M., 1978, Z. Naturforsch. 33a, 1169.
5. Gardner, F.F. and Winnewisser, G., 1978, Mon. Not. Roy. Astron. Soc., 185, 57p.
6. Alexander, A.J., Kroto, H.W., and Walton, D.R.M., 1976, J. Molec. Spectrosc., 62, 175.
7. Kroto, H.W., Kirby, C., Walton, D.R.M., Avery, L.W., Broten, N.W., MacLeod, J.M., and Oka, T., 1976, Astrop. Journ. (Letters), 219, L133.
8. Broten, N.W., Oka, T., Avery, L.W., MacLeod, J.M., and Kroto, H.W., 1978, Astrop. Journ. (Letters), 223, L105.

9. Guélin, M., Green, S., and Thaddeus, P., 1978, Astrop. Journ.
        (Letters), 224, L27.
10. Wilson, S. and Green, S., 1977, Astrop. Journ. (Letters), 212, L87.
11. Morris, M., Turner, B.E., Palmer, P. and Zuckerman, B., 1976,
        Astrop. Journ., 205, 82.
12. Avery, L.W., Oka, T., Broten, N.W., and MacLeod, J.M., 1979, Astrop.
        Journ., 231, 48.
13. Broten, N.W., MacLeod, J.M., Oka, T., Avery, L.W., Brooks, J.W.,
        McGee, R.X., and Newton, L.M., 1976, Astrop. Journ. (Letters),
        209, L143.
14. Loren, R.B., Erickson, N., Mundy, L., and Davis, J.H., 1979,
        private communication.
15. Morris, M., Snell, R.L., and Vanden Bout, P., 1977, Astrop. Journ.,
        216, 738.
16. Clark, F.O., Brown, R.D., Godfrey, P.D., Storey, J.W.V., and
        Johnson, D.R., 1976, Astrop. Journ. (Letters), 210, L139.
17. Deguchi, S., Nakada, Y., Onaka, T., and Uyemara, M., 1979, Pub.
        Astron. Soc. Japan, 31, 105.
18. Churchwell, E., Walmsley, C.M., and Winnewisser, G., 1977, Astron.
        Astrop., 54, 925.
19. Wannier, P.G. and Linke, R.A., 1978, Astrop. Journ., 226, 817.
20. Walfsberg, M., Bopp, P., Heinzinger, K., and Mallinson, P.D., 1979,
        Astron. Astrop. 74, 369.
21. MacLeod, J.M., Avery, L.W., and Broten, N.W., 1979, Astrop. Journ.,
        233, 584.
22. Little, L.T., Riley, P.W., Macdonald, G.H., and Matheson, D.N.,
        1978, Mon. Not. Roy. Astron. Soc., 183, 805.
23. Churchwell, E., Winnewisser, G., and Walmsley, C.M., 1978, Astron.
        Astrop., 67, 139.
24. Walmsley, C.M., Winnewisser, G., and Toelle, F., 1980, Astron.
        Astrop., in press.
25. Little, L.T., Macdonald, G.H., Riley, P.W. and Matheson, D.N.,
        1978, Mon. Not. Roy. Astron. Soc., 183, 45p.
26. Myers, P.C., Ho, P.T.P., and Benson, P.J., 1979, Astrop. Journ.
        (Letters), 233, L141.
27. MacLeod, J.M., Avery, L.W., and Broten, N.W., unpublished.
28. Ungerechts, H., Walmsley, C.M., and Winnewisser, G., 1979, private
        communication.
29. Little, L.T., Macdonald, G.H., Riley, P.W., and Matheson, D.N.,
        in press.
30. Wootten, H.A., 1979, private communication.
31. Elias, J.H., 1978, Astrop. Journ., 224, 857.
32. Jura, M., 1976, Astron. Journ., 81, 178.
33. Silk, J., 1978, Protostars and Planets, p. 172, ed. T. Gehrels,
        Univ. Arizona Press.

DISCUSSION FOLLOWING AVERY

Bok: This was one of the most beautiful papers I have heard, and it was beautifully presented. My congratulations to my Ottawa friends on a marvellous piece of research. Now I have a couple of questions. My first question: how did you arrive at the temperatures and densities shown in your slide?

Avery: We used a multilevel statistical equilibrium programme to compute the excitation temperatures for each of the observed lines, and then used the transfer equation, with appropriate allowance for beam dilution effects, to give calculated antenna temperatures. The kinetic temperature and H_2 density are parameters in the statistical equilibrium programme which are varied until the best agreement between the calculated and observed quantities is obtained. The values quoted in this paper are preliminary as we have not yet allowed for the effects of radiative trapping on T_{ex}. These effects could be important, especially for the J=4-3 and 5-4 lines, which are optically thick in our model.

Bok: Is the search for these molecules being extended to other dark clouds or molecular clouds? Specifically, have you in mind studies of several choice dark clouds of the southern hemisphere?

Avery: A number of investigators, including ourselves, have looked for HC_5N in many dark clouds, but to date it has been found only in four parts of the Taurus complex, and in B 335. However, the search for new cyanopolyyne sources is continuing at a number of observatories. I agree that observations of southern dark clouds would be very interesting, but I am not aware that any systematic searches are under way at present.

Elmegreen: We have made an optical survey of dark filaments, like those in Taurus, that show dark globular condensations along their length. They all show a similar pattern of nearly equally spaced primary condensations, which is typical of the mechanism of self-gravitational fragmentation in magnetic filaments. The catalogue contains some 100 condensations similar to those in Taurus, and it covers both the northern and southern hemispheres. It might be a good sample of objects for your cyanopolyynes search.

Avery: I have seen a preprint of your catalogue, and I believe that it will be a very useful guide for future searches for new carbon chain molecules. Systematic searches in regions other than Taurus will be important to establish the degree to which Taurus is unique with regard to the cyanopolyynes.

Greenberg: Do you have an estimate of the age of the Taurus cloud TMC 2, whose size and density you stated were 0.1 pc and $\sim 10^4 cm^{-3}$ respectively? If it is 10^6 years or more then photolysis of the grains could be substantial, and *if* grains could collide at ~ 0.1 km/sec (is there turbulence supporting this cloud from collapse?) then we have a potential molecule source.

Myers: The cloud age is probably considerably greater than 10^6 years if it is in stable equilibrium.

Feldman: It might not be necessary for TMC 1 and 2 to be very old objects in hydrostatic equilibrium in order to form abundant cyanopolyyne chains. After all, relatively large quantities of HC_5N and HC_7N have been found in the circumstellar envelope of the carbon star IRC+10216.

Avery: The idea of Myers *et al.* is that these clouds may be older than average so that the relatively high abundances of cyanopolyynes can be attributed to advanced chemical evolution. The idea is a speculative one, and it is true that these molecules have also been found in the dynamic envelope of IRC+10216. However the chemistry of the two environments is very probably different in view of the different physical conditions, so I do not think the presence of cyanopolyynes in IRC+10216 necessarily argues against the suggestion of Myers *et al.*

Biermann: Theoretical work done in Munich recently has shown that long-lived stable fragments may be due to an approximate equilibrium between rotations and gravity. Are there any observational indications that rotation plays a role in the clouds you discussed?

Avery: Ungerrechts, Walmsley and Winnewisser at Bonn have made maps of TMC 1 with high angular and velocity resolution. They find that the cloud is very quiescent, and shows little evidence of any systematic rotation.

Ho: A velocity gradient of ~ 1 km s^{-1}pc^{-1} is observed in the TMC 2 region (Ho *et al.* 1977). The problem in dark clouds is that typically size scales are small, ~ 0.1 pc, so that velocity shifts are difficult to detect. Because of the small sizes, the contribution of rotational energy to the dynamical stability is small, despite substantial velocity gradients.

Sherwood: From our study of the visual extinction in Heiles Cloud 2 T. Wilson and I find suggestions of fragmentation; we note some seven fragments with *minimum* extinction in the range 7-8 magnitudes. The age may be crudely estimated in two ways: (i) T Tauri and Hα emission stars in the vicinity imply $\sim 10^7$ years, and (ii) if the gradient in the radial velocity of H_2CO is matched by a transverse velocity, then the separation of components also implies $\sim 10^7$ years. We estimate $n(H_2) > 10^3 cm^{-3}$, and hence a free-fall time of only $\sim 10^5$ years.

Avery: I am interested to learn of these results. I think it would be important to continue these kinds of studies in the infrared, where the fragments might be more sharply defined.

Guelin: Observations of HC_3N, HC_5N, HC_7N and HC_9N with the 36 ft. and 140 ft. NRAO telescopes (Bujarrabal, Guelin, Morris and Thaddeus) also show that TMC 1 is clumpy. LVG statistical equilibrium computations imply densities of $\sim 10^5$ for the clumps and 10^3-10^4 in between. This result has some bearing on the HC_3N/HC_5N, HC_5N/HC_7N and HC_9N/HC_7N abundance ratios, derivations of which depend on the clump model. The 'decrement' of four in the abundances derived by previous studies has to be considered with caution.

Macdonald: Since TMC 1 is elongated along a sharp velocity discontinuity in HI, it is tempting to suggest, as several authors have, that fragmentation in this object has been triggered by a collision between two molecular clouds. Would you care to comment on this idea?

Avery: I think this suggestion was first made by Little et al.; it is an interesting possibility in view of the HI velocity ridge. It would be an unlikely coincidence, given the narrowness of TMC 1 and the velocity discontinuity, if the two features which align so well were unrelated. However, it is unlikely that similar collisions could be invoked to account for the other HC_nN condensations observed, such as TMC 2 or L1544, so fragmentation in this region can apparently proceed without cloud collisions. If TMC 1 is coincident with a collision interface, it may well be flattened in such a way that our line of sight is along a diameter of a disc, resulting in a longer column length of molecules than in the other observed condensations. Thus the enhancement of the carbon chain molecules in TMC 1 relative to TMC 2 and L1544 may be a geometrical effect.

Wootten: The comment that the HC_5N emission region in TMC 1 is not distinguishable in H_2CO maps is, I think, somewhat misleading. It is true if one considers only the 6 cm maps; a 2 cm map we have recently made follows the contours of the C_2H and HC_5N emission very well, with local absorption maxima coinciding with C_2H emission maxima. We attribute this behavior to the fact that the 2 cm line, arising in a more excited level than the 6 cm line, samples a denser region of the cloud.

Avery: I was not aware that a 2 cm H_2CO map of this region existed. My statement, that TMC 1 is not evident in H_2CO observations, referred to the 6 cm map. It may be that the cloud does not appear in the 6 cm line because of the large beamwidth used in the observations.

Gilmore: There exist faint radio continuum sources in the direction of both TMC 1 and TMC 2. The source toward TMC 1 is nonthermal, the one toward TMC 2 has three components, one with a flat spectrum. Both sources may be extragalactic background objects; I mention them because of the unusual chemistry in Taurus.

Thaddeus: I just wanted to make the obvious point that we see the cyanopolyynes only because of their large dipole moments and simple partition functions. Almost the entire table of molecules, certainly all the big ones, would not be visible at the same distances and same column densities as HC_7N or HC_9N. It is therefore almost impossible to prove that the cyanopolyyne clouds are chemically unique - it is possible that all the other substances are there and we just cannot see them, nor are we likely to in the foreseeable future.

LONG CARBON CHAIN MOLECULES IN THE LABORATORY AND IN SPACE

G. Winnewisser, F. Toelle, H.Ungerechts, C.M. Walmsley

Max-Planck-Institut für Radioastronomie, Bonn, Germany
and
Physikalisch-Chemisches Institut, Justus Liebig Universität, Giessen, Germany

The unsaturated long carbon chain molecules of the type $HC_{2n+1}N$
(with n=1,2..) have the remarkable property of being very stable under
a wide variety of different laboratory and interstellar conditions. In
fact, they can be synthesized in the laboratory under the action of a
radio frequency discharge in a mixture of acetylene, HCCH, and hydrogen cyanide, HCN (Creswell et al. 1977 , Winnewisser et al. 1978).
Once they are formed they are for example very stable towards temperature changes as well as saturation of the carbon bonds due to hydrogen
addition. In the laboratory their chief cause of destruction is poly-
merization. These experiments were triggered by our observation that
in the course of producing hydrogen isocyanide, HNC, (Creswell et al.
1976) by reacting CH_3I with N_2 in a gas discharge, also small amounts
of HC_3N were present which clearly requires the joining together of
four heavy atoms from different starting molecules.

Over the last couple of years we have carried out more detailed
experiments of the acetylene/hydrogen cyanide discharges to elucidate
the reaction mechanism and its suspected relevance to the formation of
interstellar cyanopolyynes by using ^{13}C isotopic tracers and different
functional groups, such as CH_3, ... The present state of our laboratory experiments can be summarized by:

$$HCCH + HCN + \text{discharge} \rightarrow HC_3N + H_2; \quad HC_5N + H_2; \quad (HC_7N + H_2?);$$
$$+ \text{ unidentified lines.}$$

If the experiment is repeated with CH_3CCH the appropriate methyl-cyano-
polyynes are obtained. A sample spectrum is shown in Fig.1. If the experiment is performed with carbon-13 enriched hydrogen cyanide the ^{13}C
remains entirely with the cyanide group:

$$HCCH + H^{13}CN + \text{discharge} \rightarrow HCC^{13}CN + H_2$$
$$\rightarrow HCCCC^{13}CN + H_2$$

suggesting that the long carbon chain is essentially being formed from
the acetylene. In this connection it is interesting to mention that we

B. H. Andrew (ed.), Interstellar Molecules, 59–65.
Copyright © 1980 by the IAU.

succeeded in detecting the $J = 1 \to 0$; $F = 2 \to 1$ transition of $HCC^{13}CN$ in
TMC1, yielding an $^{12}C/^{13}C$ ratio of approximately 60. The appropriate
transitions of $H^{13}CCCN$ and $HC^{13}CCN$ were not detected to similar limits,
suggesting ^{13}C fractionation which could be caused by the formation
mechanism. The $HCC^{13}CN$ emission signal was detected from the peak of
the HC_3N cloud in TMC1. This position lies within 1 arc min of the
HC_5N emission peaks. Typical medium velocity resolution ($0.1 \ kms^{-1}$)
spectra of TMC1 are displayed in Fig. 2, indicating that the intrinsic
linewidths in dark clouds can be as low as $0.2 \ kms^{-1}$ which is close
to the thermal limit. Cool dark dust clouds offer a great potential
for studying precisely the physical conditions with high spectral and
spatial resolution.

We have used the high angular resolution of the 100 m radiotelescope to conduct a detailed survey of selected dark dust clouds by employing the cyanopolyyne molecules HC_3N, HC_5N, HC_7N and ammonia NH_3
as molecular probes (see table I). The indication from our surveys of
the core regions of dark dust clouds is that the Taurus region is one
of many dark dust cloud complexes which contain a sizeable number of
small (L < 1 pc) dense ($n(H_2) \sim 10^4 cm^{-3}$) but cool ($T_{kin} \sim 10K$) and low
mass (M < $5M_\odot$) fragments. As judged either from the temperature estimates or from the line widths the studied core regions in TMC1, TMC2,
L1544 and L183 (L134N) all have in common being temporarily in a very
quiescent stage, with no evidence of embedded protostellar objects. Of
the four studied clouds TMC1 shows the highest line intensities in the
cyanopolyyne molecules and encompasses the largest projected area. In
both the cyanopolyyne and ammonia molecular cloud TMC1 exhibits a conspicuous elongated structure forming a common major axis and essentially
two minor axes.

Among the transitions used in this survey (Table 1) the three hyperfine components of the $J = 1 \to 0$ transition of HC_3N can best be used for
high velocity resolution studies of dark clouds, since spectroscopically no additional splitting is expected to be observable. We have,
therefore, used these transitions to study with high spectral resolution ($0.027 \ kms^{-1} = 760$ Hz) the dynamics of TMC1. All three hyperfine

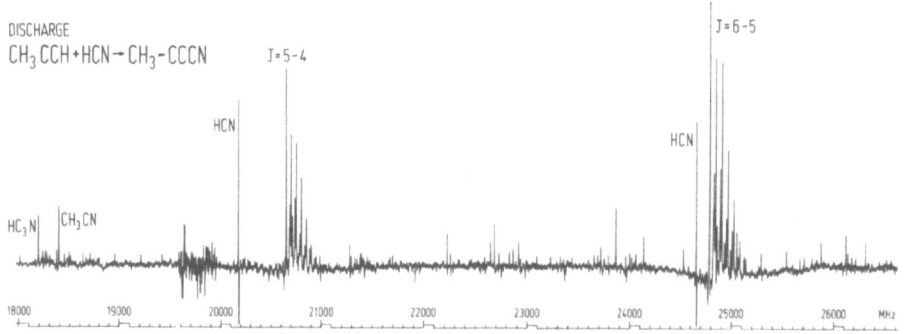

Figure 1. Laboratory Spectrum of CH_3CCCN obtained by discharge.
(from Winnewisser and Ziurys, unpublished results).

components exhibit additional structure. The almost identical line profiles of the $F = 1 \to 1$ and $F = 2 \to 1$ hyperfine components and their relative intensities which are close to the LTE ratio of 3 to 5 resp., show that the lines are optically thin. The line profiles indicate that two or more separate velocity components are present with a velocity spread of ~ 0.3 kms^{-1} and an intrinsic linewidth ~ 0.2 kms^{-1}. This might be caused by a blend of emission from several filaments aligned with the TMC1 main axis and which are rotating slowly (~ 0.1 kms^{-1}) around the minor and major axis. Although the abundances of the cyanopolyyne and ammonia molecules coincide with the predominant molecular ridge (major axis) in TMC1 their distribution is strikingly different. The maximum of the NH$_3$ emission is conspicuously displaced from the cyanopolyyne maxima by more than 400 arc sec to the north-west, (Fig. 4). Simultaneous observation of the NH$_3$(1,1) and (2,2) transitions at three positions have been used to derive a kinetic temperature of (9 ± 1)K along the main ridge. A sample of the NH$_3$ spectra is given in Fig. 5. From statistical equilibrium calculations on HC$_3$N and NH$_3$ we estimate that the gas density is $\sim 4 \times 10^4$ cm^{-3} and nearly constant along the main ridge but declines sharply along the minor axis. A detailed account of the HC$_{2n+1}$N (n = 1,2,3) and NH$_3$ results will be published elsewhere (Toelle, Ungerechts, Walmsley, Winnewisser and Churchwell). We have, therefore, the curious fact that regions of similar physical parameters such as density, and temperature and age show different chemical composition. The laboratory and interstellar data seem to suggest that the formation of cyanopolyynes

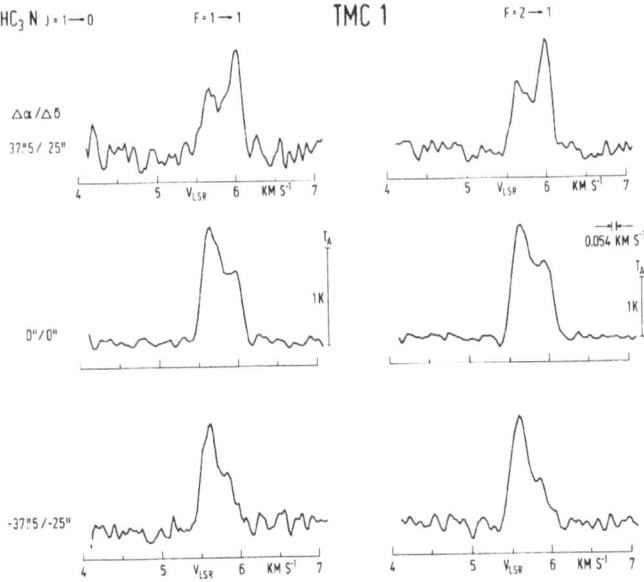

Figure 3. Cyanoacetylene emission in two hyperfine components from three positions along the minor axis of TMC1 taken with high velocity resolution. Crosses (x) in Figure 4 mark the corresponding positions.

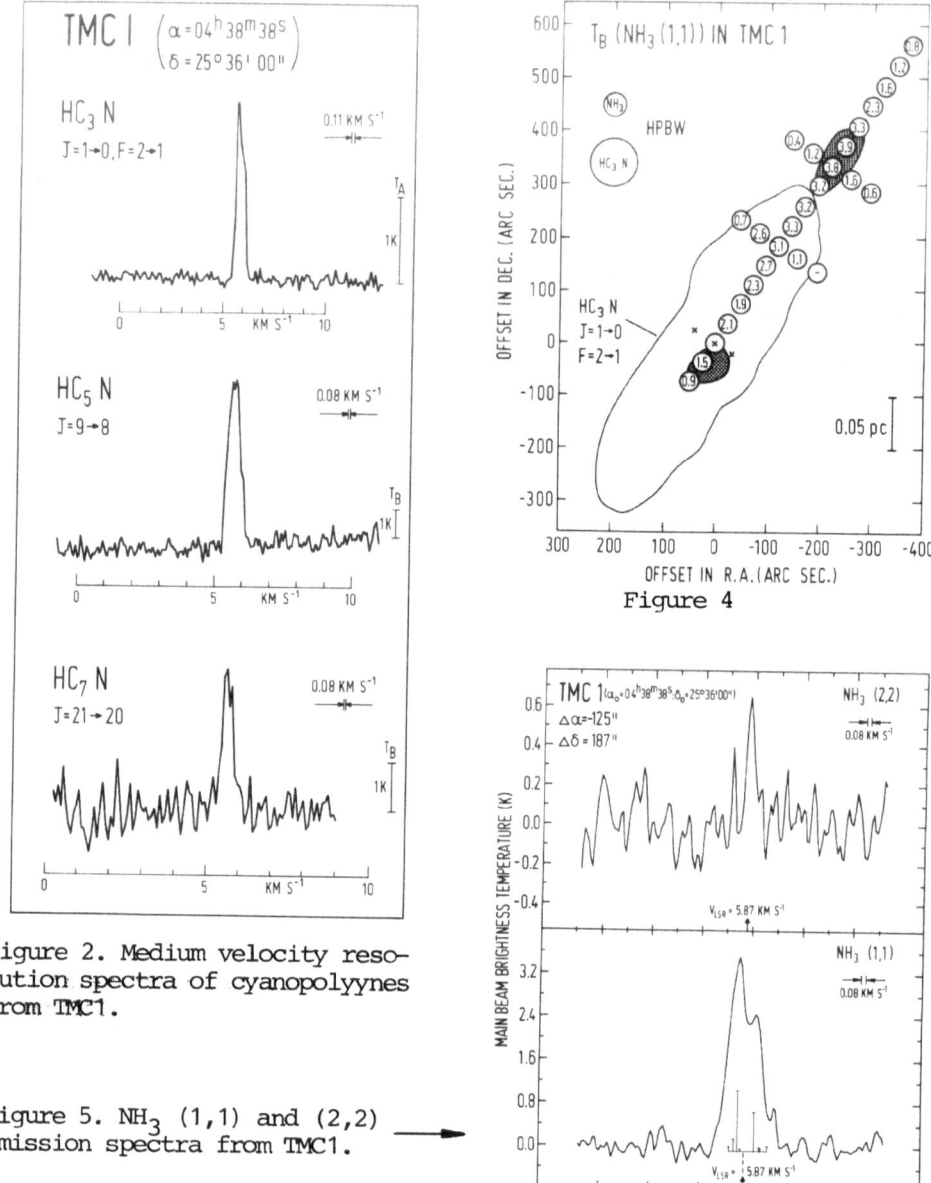

Figure 2. Medium velocity resolution spectra of cyanopolyynes
from TMC1.

Figure 5. NH$_3$ (1,1) and (2,2)
emission spectra from TMC1.

Figure 4. Map of the cyanopolyyne and ammonia clouds in TMC1. The
hatched areas indicate the position of the cyanopolyyne (south-east)
and ammonia (north-west) emission peaks.

Table 1

Molecular Transitions surveyed and mapped in selected sources
(TMC1, TMC2, L1544, L183)

Molecule	Transition	Frequency MHz	Angular Resolution arc sec
HC$_3$N	J = 1 → 0		90
	F = 1 → 1	9 097.0346	
	2 → 1	9 098.3321	
	0 → 1	9 100.2727	
HC$_5$N	J = 9 → 8	23 963.897	40
HC$_7$N	J = 8 → 7	9 024.014	90
	21 → 20	23 687.890	40
NH$_3$	J,K = 1, 1	23 694.495	40
	2, 2	23 722.633	40

is linked with the existence of acetylene. A summary on the distribution of carbon chain molecules has been given (Winnewisser and
Walmsley, 1979). We suggest that the longer cyanopolyyne molecules
are formed via the acetylene "backbone" reaction

$$C_2H_x^+ + C_2H_2 \rightarrow C_4H_y^+ + C_2H_2 \rightarrow C_6H_z^+ + C_2H_2 \rightarrow \ldots$$

the appropriate species will then react with CN, HCN or HCN$^+$ to form
intermediates of the form $H_2C_3N^+$,... which by dissociative recombination form the cyanopolyynes HC_3N,... .We conclude, therefore, that
(1) the abundance of the cyanopolyyne molecules will decrease with
increasing length of the carbon chain, (2) that long chain molecules
with other functional groups such as CH_3, NH_2 should be observable,
(3) that molecules with no permanent dipole moment such as acetylene,
diacetylene,... have to be abundant in space and (4) that unstable
species such as HCCN, H_2CCN may be abundant in interstellar clouds.
Some of the unidentified lines may be caused by them.

Thus our high spectral and spatial resolution data indicate that
TMC1 is presently in a quiescent phase of temporary gravitational equilibrium. This equilibrium might be ended either by slowly accreting
material which eventually causes gravitational collapse or by external
influences such as shock waves which can trigger collapse. The spatial
proximity of low-mass stars seems to support the speculation that its
evolutionary path will lead from its present "pre-protostellar" stage
to the formation of a low-mass star (or stars) in a time scale which is
longer than the free-fall time ($\sim 10^5$ years).

REFERENCES

Creswell, R.A., Pearson, E.F., Winnewisser, M., and Winnewisser, G.,
     (1976), Z. Naturforsch. 31a, 222.
Creswell, R.A., Winnewisser, G., and Gerry, M.C.L. (1977),
     J. Mol. Spectrosc. 65, 420.
Winnewisser, G., Creswell, R.A., and Winnewisser, M. (1978).
     Z. Naturforsch. 33a, 1169.
Winnewisser, G., and Walmsley, C.M., (1979)
     Astrophys. Space Sci. 65, 83.

DISCUSSION FOLLOWING WINNEWISSER

Kutner: How would these chains be destroyed, once they are formed?
Winnewisser: Good question. I don't know.
Thaddeus: For several years now Guelin and I have been trying to
identify a doublet at 85.3 GHz. In TMC 1 the strongest line is remarkably
strong - about 3 K. The best candidate we now have is the triplet
molecule H_2CCN, so we are very interested in any laboratory data that you
have on this radical.
Winnewisser: We have no data on HCCN or H_2CCN. However we do have
some unidentified lines in our HCCH+HCN discharge. Maybe we should
exchange diaries.
Langer: I would like to play the role of archaeologist for a moment
by talking about CO amidst all these modern studies of long chain
molecules. An extensive $C^{18}O$ survey of the Taurus region by Frerking and
myself at Bell Labs. reveals at least five fragments where you see about
two. Clearly there is a great deal of material and fragmentary structure
which is not revealed by the long chain molecules with their large dipole
moments, so one should be careful about interpretations of density and
physical conditions.
Winnewisser: I feel you are correct, but it is very worthwhile trying to check the cyanopolyyne molecules in the positions you have found
$C^{18}O$ emission.
Langer: Is there any evidence for self-absorption effects in the
HC_3N data as is the case with HCO^+ and HNC?
Winnewisser: There is no evidence for self-absorption in HC_3N.
Among other reasons we reach this conclusion from the similarity of the
hyperfine components of HC_3N in the J=1-0 transition. They just look
completely alike.
Clark: We have previously reported an apparent minor axis rotation
in an adjacent region in Taurus, B213 NW. It is superimposed on small
motions about the major axis. Thus your apparent minor axis rotation is
not without precedent, although it is difficult to understand why it
occurs.
Winnewisser: The rotation observed in TMC 1 is only detected by
means of the three hyperfine components of the J=1-0 transition of HC_3N.
Vital to this detection is the high velocity resolution which we have
used ($\Delta V=0.027$ km s^{-1}). At lower velocity resolutions the spatial
difference between the two observed velocity components could have easily

been interpreted as a much larger rotation of TMC 1 than actually is present. I would like to emphasise that very high velocity resolution is necessary to obtain the proper dynamics of these quiescent clouds.

Glassgold: Have deuterated molecules been observed in TMC 1, especially DCO^+, which might give information on the electron abundance, and its possible correlation with long chain molecules?

Winnewisser: We ourselves have made no observations of DCO^+. However, Guélin, R. Wilson et al. have made those measurements. From the data one concludes that the electron abundance is low in TMC 1. But Guélin can comment better on his data than I can. We have not yet made measurements of DC_3N etc.

Guélin: In answer to Dr. Glassgold's question about the DCO^+ abundance, Langer, R. Wilson and myself have completed a survey of HCO^+ isotopic species in TMC 1. The HCO^+ source is exactly like the NH_3 source. It peaks ∿10' to the north-west of the HC_7N maximum. We *do* find DCO^+ everywhere. In other words, there is widespread deuterium enhancement in all the dense regions, which implies that the electron density is very low everywhere the heavy cyanopolyynes (HC_5N, HC_7N....) are observed.

T. Wilson: Henkel, Pankonin and myself have mapped $^{12}C^{16}O$ and $^{13}C^{16}O$ toward TMC 1, and find no variation in the intensities. In H_2CO, the 2_{11}-2_{12} line at 14.5 GHz peaks within 1.3' of the center of TMC 1. The 1_{10}-1_{11} line of H_2CO, at 4.8 GHz, varies more slowly with distance from the center of TMC 1.

Winnewisser: We have defined the center of TMC 1 by the position of the peak of the HC_5N emission. However the NH_3 position of maximum emission in TMC 1 is displaced by ∿7 arc min to the north-west of the HC_3N peak. Our data also show that the HC_5N maximum emission is displaced from the HC_3N position by ∿1 arc min.

Greenberg: What is the visual extinction in these clouds, and what is the photodissociation rate for the chain molecules? The reason for this question is that I would like to know the required production rate.

Winnewisser: The extinction in TMC 1 is A_V ∿ 6. T. Wilson et al. have made star counts and obtained a more precise number. There is nothing known about photodissociation to my knowledge.

T. Wilson: From star counts of POSS red prints, the *minimum* A_V toward TMC 1 is 5 magnitudes.

Irvine: In order to compare different molecular species in regions as quiescent as TMC 1 or TMC 2, it is necessary to know rest frequencies very accurately. How well are these known for the long carbon chains?

Winnewisser: We can supply rest frequencies to a calculated standard deviation of ≲5 KHz. In the case of HC_5N we have measured the rotational spectrum as high as ∿300 GHz. We will publish these laboratory data soon.

THE RELATIVE DISTRIBUTION OF AMMONIA AND CYANOBUTADIYNE EMISSION IN HEILES 2 DUST CLOUD

L.T. Little[1], G.H. Macdonald[2], P.W. Riley[2] and D.N. Matheson[3]
[1]DERAD, Observatoire de Meudon, 92190 Meudon, France
[2]Electronics Laboratory, University of Kent at Canterbury, UK
[3]SRC Appleton Laboratory, Chilbolton Observ., Stockbridge, UK

A map of the NH_3 (1,1) emission in Heiles 2 dust cloud, observed with 2.2 arc min. resolution, is presented and compared with that of the J = 9 → 8 transition of HC_5N. Although the distribution of both molecules lies on a ridge, there is a marked anti-correlation in the observed antenna temperatures of the two molecules along it. This appears to be best explained as a variation in the relative abundances of the two molecules along the ridge by a factor 8-20 in a distance of 10^{18} cm.

Interest has focused on the dark dust cloud Heiles 2 since the discovery within it of numerous cyanopolyynes $HC_{2n}CN$ (e.g. Broten et al. 1978). Using the SRC Appleton Laboratory 25m radio telescope ($\eta_B \approx 0.37$) we have mapped Heiles 2 in the J = 9 → 8 transition of HC_5N at 24 GHz (Little et al. 1978) and also in the NH_3 (1,1) transition at 23.7 GHz first detected there by Rydbeck et al. 1977 (see figure 1).

Although the emission from both molecules lies along a ridge, there is an interesting anti-correlation between the two along it. It is hard to explain this as a subtle excitation effect; a genuine variation in chemical abundance seems more likely.

At the NH_3 peak the optical depth in the NH_3 (1,1) line was found to be $\tau = 1.5 \pm 0.25$ from the main/hyperfine component ratio. Comparing this value with $\tau = 0.4 \pm 0.2$ deduced by Rydbeck et al. at the HC_5N peak shows that the increase in NH_3 antenna temperature going from the HC_5N peak to the NH_3 peak is accompanied by increase in τ. From the measured dimensions of the NH_3 source, the observed antenna temperature, and τ, the excitation temperature of the NH_3 and hence the density of colliding particles (hydrogen molecules) may be deduced, assuming a kinetic temperature 10K. Results in the range $n_{H_2} \sim (1 \text{ to } 5) \times 10^4$ cm^{-3} are obtained. This molecular hydrogen density may then be used to determine the excitation conditions for HC_5N. The result of this procedure is to suggest that, moving along the Heiles 2 ridge from NW to SE, a decrease in NH_3 column density is accompanied by an actual increase in that of HC_5N, with a relative variation by a factor of 8 to 20.

B. H. Andrew (ed.), Interstellar Molecules, 67–68.

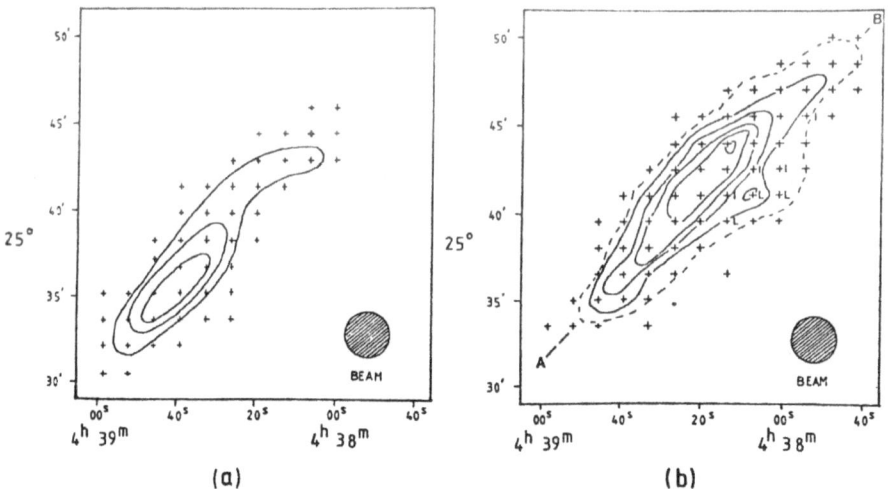

Figure 1. Contours of peak antenna temperature in the lines
(a) J = 9 → 8 HC$_5$N (resolution 0.30 km/s) (b) (1,1) NH$_3$ (resolution 0.43
km/s). Positions lettered I have LSR velocities between 5.4 and 5.7
km/s; those lettered L have velocities between 5.2 and 5.4 km/s, and
represent evidence for a second weak component on the W flank. Otherwise LSR velocities do not differ significantly from 5.8-5.9 km/s. The
line widths suggest Doppler broadening ∿0.4 km s^{-1}. For the HC$_5$N map
the contour interval is 0.2K (lowest contour 0.3K). For the NH$_3$ map the
contour interval is 0.15K (broken contour 0.2K).

The explanation for this behaviour is unknown. In ion-molecule
schemes for interstellar chemistry (e.g. Smith & Adams 1978) CN and HCN
are produced from NH$_3$ by reaction with ions such as C$^+$, CH$_3^+$, and it is
conceivable that the polyynes are formed from these molecules via reactions
with C$_2$H$_3^+$, C$_4$H$_3^+$ etc. (Walmsley et al. 1980). An abundance variation of
some of these ions could lead to an increase in [HC$_5$N] while at the same
time decreasing [NH$_3$]. Variation of the cosmic ray and u.v. ionization
rate with position in the cloud might produce such an effect.

REFERENCES

Broten, N.W., Oka, T., Avery, L.W., MacLeod, J.M., and Kroto, H.W.: 1978,
     Astrophys. J. (Letters) 223, pp. L105-L107.
Little, L.T., Riley, P.W., Macdonald, G.H., and Matheson, D.N.: 1978,
     Mon. Not. R. astr. Soc. 183, pp. 805-811.
Rydbeck, O.E.H., Sume, A., Hjalmarson, A., Elldér, J., Rönnäng, B.O., and
     Kollberg, E.: 1977, Astrophys. J. (Letters) 215, pp. L35-L40.
Smith, D., and Adams, N.G.: 1977, Astrophys. J. 217, pp. 741-748.
Walmsley, C.M., Winnewisser, G., and Toelle, F.: 1980, Astron. Astrophys.,
     in press.

OBSERVATIONS OF HC_5N (J=12-11 AND 13-12) AND HC_7N (J=28-27),
INCLUDING A DETECTION OF B335

Å. Hjalmarson and P. Friberg
Onsala Space Observatory, S-430 34 Onsala, Sweden

The Onsala 20 m telescope, equipped with a travelling wave maser (frequency range ~29-35 GHz; zenith system noise temperature ~90-160 K) has been used in a search for HC_5N (J=12-11 and 13-12) and HC_7N (J = 28-27) to a low detection limit. In addition to observations of the Taurus dust clouds TMC1, TMC2 and L1544, the HC_5N (J =13-12) transition has been detected at a low level ($T_A^* \sim 0.1$ K; p-t-p noise ~ 0.1 K for a velocity resolution of 0.04 km s^{-1}) in the isolated globule B335. The total HC_5N column density estimated from the very narrow line ($\Delta v \sim$ 0.2 km s^{-1}) is $6 \cdot 10^{11}$ cm^{-2}. From our non-detection of HC_5N in L134N a total column density $\lesssim 5 \cdot 10^{11}$ cm^{-2} is deduced.

It is interesting to compare our HC_5N column densities with published NH_3 data. The NH_3/HC_5N abundance ratio is estimated to be ~ 1-10 in the Taurus clouds, ~ 10^3 for B335 and $\gtrsim 2 \cdot 10^3$ for L134N. Although the statistical sample is poor due to the very few detections of HC_5N in dust clouds, it is tempting to suggest that the observed NH_3/HC_5N abundance ratios may give a clue to our understanding of the chemistry of the cyanopolyynes. Perhaps we are observing the competition between formation/destruction of NH_3 and HC_5N - presumably a result of time dependent gas-phase chemistry. The chemical equilibrium time scale may be greater that the gravitational collapse free-fall time. Effelsberg NH_3 and HC_5N observations of TMC1 seem to indicate that the two species have maxima at different positions in the cloud (G. Winnewisser, private communication).

Information on the excitation conditions in the Taurus clouds can be obtained by comparison of our data with published results for lower frequency transitions.

B. H. Andrew (ed.), Interstellar Molecules, 69–70.

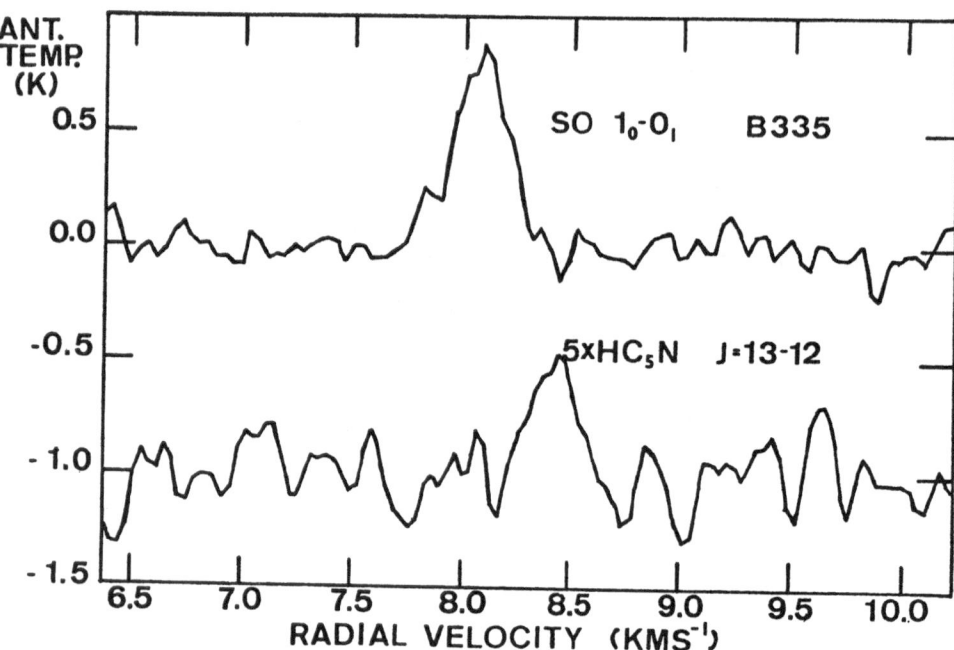

Figure 1. Observed HC₅N and SO spectra towards B335. The antenna temperatures are corrected for radome, atmospheric and antenna losses. The HC₅N intensity is multiplied by five. Spectral resolution is 0.04 km s⁻¹.

SPECTRA OF THE 1_0-0_1 TRANSITION OF SULFUR MONOXIDE IN INTERSTELLAR CLOUDS

O.E.H. Rydbeck, Å. Hjalmarson, G. Rydbeck, J. Elldér,
and E. Kollberg
Onsala Space Observatory, Sweden

W.M. Irvine
University of Massachusetts, USA

Emission from SO towards a number of HII regions and molecular clouds with embedded energy sources has been reported by several authors (e.g., Gottlieb et al. 1978; Clark et al. 1978; Wannier and Phillips 1977; and references therein; cf. also Loren et al., 1974; Loren et al. 1975; Lada et al. 1974). Transitions observed include the $4_5 \rightarrow 4_4$, $4_3 \rightarrow 3_2$, $3_2 \rightarrow 2_1$, $2_3 \rightarrow 1_2$, $2_2 \rightarrow 1_1$, $1_2 \rightarrow 1_1$, and $1_0 \rightarrow 0_1$, with the latter (to the ground state) seen only in Sgr B2. Recently Rydbeck et al. (1980) have detected SO in cold, dark clouds and have made the first astronomical measurements of the $1_0 \rightarrow 0_1$ transition in a variety of sources, including the corresponding ^{34}SO line. The latter authors find that the $1_0 \rightarrow 0_1$ transition of SO is an excellent tracer of structure in dark clouds, and they discuss the fractional abundance $[SO]/[H_2]$ on the basis of column densities derived from observations of the two isotopic species. They also set limits to the magnetic field strength in dark clouds from the absence of observed Zeeman splitting. We shall provide here additional spectra and information on observing procedures, and shall discuss the rest frequencies for the SO and ^{34}SO $1_0 \rightarrow 0_1$ transitions.

The observations were performed between December 1978 and March 1979 with the 20 m millimeter wave telescope of the Onsala Space Observatory equipped with a new traveling-wave maser preamplifier (Kollberg and Lewin 1976). The zenith system noise temperature varied in the range 90-150 K, and the beam-width is about 120". An autocorrelation spectrometer was used to achieve resolutions between 120 and 2.4 kHz, corresponding to 1.2 and 0.024 km s^{-1}. Antenna temperatures T_A^* have been corrected for atmospheric, radome, and antenna losses, using a radome transmission factor of 0.76 and a beam efficiency of 0.72. Observations were performed by position switching in the total power mode, the azimuth offset typically being one degree. The data were calibrated by the use of a noise tube. An absolute intensity scale was obtained from matched loads at ambient and liquid

B. H. Andrew (ed.), Interstellar Molecules, 71–76.

nitrogen temperatures. Rest frequencies employed were, for
^{32}SO: 30 001.523 ± 0.010 MHz as calculated on the basis of many
observed transitions by Clark and DeLucia (1976); for ^{34}SO: the
measured value 29 678.98 ± 0.10 MHz (Tiemann 1974), which was
subsequently revised as listed below.

Because of the relatively small velocity dispersion present,
observations of L 134 N (α,δ = 15^{h}51^{m}30^s, -2°44'0") may be used to
refine the determination of rest frequencies and to resolve hyperfine
structure for molecular species which have not been accurately observed
in the laboratory (e.g., Rydbeck et al. 1974, for CH; Rydbeck et al.
1977, for NH$_3$; Snyder et al. 1977, for N$_2$H$^+$ and N$_2$D$^+$). The core
velocity structure as determined by previous observations is probably
best represented by NH$_3$ (Rydbeck et al. 1977), for which rest
frequencies are very accurately known, hyperfine splitting is resolved,
and core optical depths are small or moderate. According to a
comparison of spectra for L 134 N and the other dust clouds with
available molecular line data (Rydbeck et al. 1980), the predicted
^{32}SO frequency (Clark and DeLucia 1976) appears to be accurate to
within 20 kHz. However, we had to shift the ^{34}SO rest frequency by
about -108 kHz from the measured value quoted by Tiemann (1974) to
bring it into agreement with NH$_3$, C^{18}O and ^{32}SO data. The new rest
frequency (217 kHz lower than Tiemann's <u>calculated</u> as opposed to his
<u>measured</u> value) is $\nu(^3\Sigma; 1_0-0_1; ^{34}SO)$ = 29 678.872 ± 0.020 MHz. This
value is also supported by observations of both SO species at a
nearby position in the cloud (15^{h}51^{m}38^s, -2°47'30") and in TMC-1
(Rydbeck et al. 1980, Fig. 3). As a matter of fact, a ^{34}SO rest
frequency shift was expected from a comparison of the calculated
(using laboratory data) and measured ^{32}SO rest frequencies given by
Tiemann (1974), and the calculated frequency of Clark and DeLucia
(1976). The latter authors were able to measure a number of weak
transitions around 1 mm which are necessary for a direct calculation
of the previously only indirectly estimated electronic spin-spin
interaction constant of SO (in fact dominated by second-order spin-
orbit coupling due to higher electronic states, which in the
Hamiltonian formally contributes in the same fashion as spin-spin
coupling); as a result the difference between the measured and
calculated frequencies for the 1_0-0_1 transition of ^{32}SO changed sign,
consistent with our result for ^{34}SO.

The degree to which SO can trace density and velocity structure
in dark clouds is shown for TMC-1 in Figure 1. That the v = 5.8 km s^{-1}
feature originates in the densest region of the source is confirmed by
the corresponding emission from NH$_3$, HC$_5$N and HC$_7$N at this velocity
(Rydbeck et al. 1977; Hjalmarson and Friberg 1979). In addition, as a
relatively abundant species requiring intermediate densities for
excitation, SO also shows features at v $\approx$ 5.0 and 5.4 km s^{-1} which
presumably arise in less dense regions and which are partially visible
in H$_2$CO (Sume et al. 1975) and are absent for HC$_5$N and HC$_7$N. These
SO "clouds" are all extended relative to our beam size, although
their distributions are not identical. A more detailed mapping is

obviously desirable, since it would contribute to our understanding of this interesting region. It should be borne in mind in this context that the rest frequencies of ^{32}SO and the linear carbon chains HC_nH are all uncertain by at least ~0.1 km s^{-1} at the frequencies of relevance here. Note also that the line-widths ($\lesssim 0.25$ km s^{-1}) of the individual SO features imply very small internal velocity gradients (at 10 K the thermal line-width of SO would be 0.1 km s^{-1}). Winnewisser (1979) reports very narrow lines and considerable velocity structure in TMC-1, as delineated by Effelsberg NH_3 and HC_3N data.

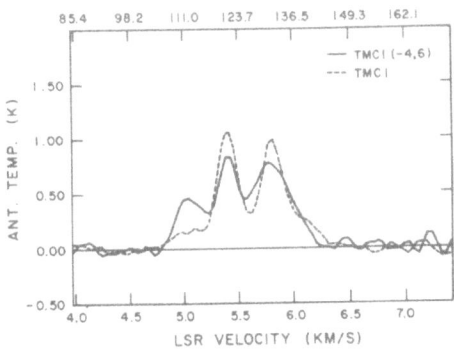

Fig. 1. Position-switched spectra of the l_0-0_1 transition of ^{32}SO (rest frequency = 30,001.523 MHz) towards two positions in TMC-1 (Heiles Cloud 2). Antenna temperature corrected for radome, atmospheric, and antenna losses. Spectral resolution is 4 kHz = 0.04 km s^{-1}. Coordinates (α,δ) = (04^{h}38^{m}20^s, 25°41'45") for ____; (04^{h}38^{m}38.6^s, 25°36'18") for ---.

Figures 2 and 3 illustrate previously unpublished spectra for a number of dark clouds and molecular clouds (HII regions). Measured line parameters are tabulated in Rydbeck et al. (1980). Many of the features are asymmetric, indicating that the cloud velocity structure is not simple. Note also that the antenna temperature of the SO line towards the dark clouds is generally greater than towards the HII regions, either because of beam dilution or lower optical depths for the latter sources in the $l_0 \rightarrow 0_1$ transition.

74

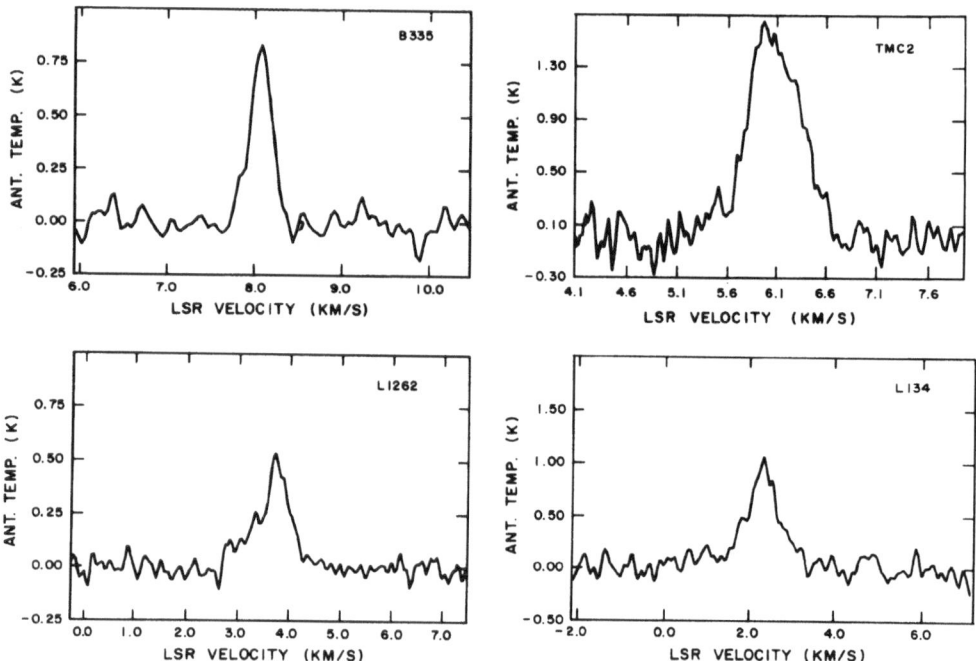

Fig. 2. SO(1_0-0_1) spectra towards several dark clouds. Note differences in scales for both ordinate and abscissa. Coordinates (α,δ) = ($19^h34^m34^s$, $07°27'00''$) for B335; ($04^h29^m43^s$, $24°18'15''$) for TMC-2; ($23^h24^m05^s$, $74°00'00''$) for L1262; ($15^h51^m00^s$, $-04°26'57''$) for L134.

W.I. gratefully acknowledges the support of a University of Massachusetts Faculty Research Grant and of NSF grant AST76-24610. Onsala Space Observatory, Chalmers University of Technology, is operated with financial support from the Swedish Natural Science Research Council. The maser development done by the quantum electronics group at the Research Laboratory of Electronics, Chalmers University of Technology, was supported by the Swedish Board of Technical Development.

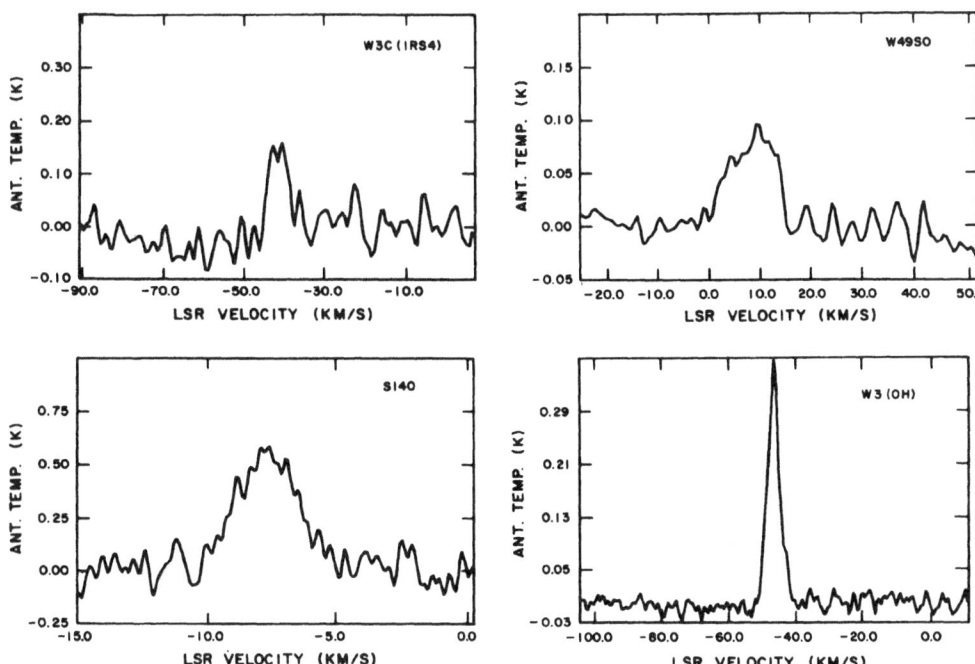

Fig. 3. $SO(1_0-0_1)$ spectra towards several molecular clouds. Note
differences in scales for both ordinate and abscissa. Coordinates
$(\alpha, \delta) = (02^h21^m44^s, 61°52'48'')$ for W3C(IRS4); $(19^h07^m49.8^s, 09°01'17'')$
for W49SO; $(22^h17^m42.0^s, 63°03'45'')$ for S140; $(02^h23^m17^s, 61°39'00'')$
for W3(OH).

REFERENCES

Clark, F.O., Johnson, D.R., Heiles, C.E. and Troland, T.H. 1978,
    Astrophys. J. 226, 824.
Clark, W.W. and DeLucia, F.C. 1976, J. Mol. Spectrosc. 60, 332.
Gottlieb, C.A., Gottlieb, E.W., Litvak, M.M., Ball, J.A. and Penfield,
    H. 1978, Astrophys. J. 219, 77.
Hjalmarson, Å. and Friberg, P. 1979, this volume.
Kollberg, E.L. and Lewin, P.T. 1976, IEEE Trans., MTT-24, 718.
Lada, C., Dickinson, D.F. and Penfield, H. 1974, Astrophys. J. (Letters)
    189, L35.
Loren, R.B., Peters, W.L. and Vanden Bout, P.A. 1974, Astrophys. J.
    (Letters) 194, L103.
Loren, R.B., Peters, W.L. and Vanden Bout, P.A. 1975, Astrophys. J.
    195, 75.
Rydbeck, O.E.H., Elldér, J., Irvine, W.M., Sume, A. and Hjalmarson, Å.
    1974, Astron. Astrophys. 34, 479.

Rydbeck, O.E.H., Sume, A., Hjalmarson, Å., Elldér, J. and Kollberg, E. 1977, Astrophys. J. (Letters) 215, L35.

Rydbeck, O.E.H., Irvine, W.M., Hjalmarson, Å., Rydbeck, G., Elldér, J. and Kollberg, E. 1980, Astrophys. J. (Letters), 235, L171.

Snyder, L.E., Hollis, J.M., Buhl, D. and Watson, W.D. 1977, Astrophys. J. (Letters) 218, L61.

Sume, A., Downes, D. and Wilson, T.L. 1975, Astron. Astrophys. 39, 435.

Tiemann, E. 1974, J. Phys. Chem. Ref. Data 3, 259.

Wannier, P.G. and Phillips, T.G. 1977, Astrophys. J. 215, 796.

Winnewisser, G. 1979, 3rd Nordic Symposium on Atomic and Molecular Physics, "Molecules in the Laboratory and in Space", and private communication.

DISCUSSION FOLLOWING IRVINE

Snyder: Were the different line components in the SO spectrum of Cloud 2 due to different velocities or to self-absorption?

Irvine: That is an important and a difficult question. From a comparison with published spectra for other molecular species, the simplest interpretation of the SO results does involve different velocity components rather than self-absorption. Because the individual features are so narrow, however, small uncertainties in rest frequencies for the different molecular species, or slight systematic errors in velocity determinations between different observatories, might change that interpretation. We hope to clarify the situation during the next observing season by mapping SO in Cloud 2, and by observing other molecules from both Onsala and other observatories in several clouds. We also hope to better define rest frequencies for the 1_0 0_1 transitions of SO and ^{34}SO.

Mouschovias: If the gas density in the clouds in which you searched for magnetic fields is $\sim 10^4 cm^{-3}$, one expects from theoretical calculations that $B \simeq 200-300$ μG. Could you have detected that field?

Irvine: Unfortunately not.

Kuiper: Nearly all the Orion "plateau" spectra are highly symmetrical. The 30 GHz SO line you observed[+]appears to have asymmetrical features. Are these real, or could they be attributed to baseline effects?

Irvine: Baseline effects are possible.

McCutcheon: You assumed a terrestrial value for (SO)/(^{34}SO) and derived $T_{ex} \sim 5$ K for L134N. If you assumed $T_{ex} \sim 10-12$ K (as for CO), would the derived ratio differ significantly from a terrestrial value?

Irvine: You would derive in that case an abundance ratio for the isotopes of about 10, instead of the terrestrial value of 22.5.

[+]The figure referred to is not reproduced in the text but can be found in Rydbeck et al., 1980.

CN OBSERVATIONS IN TAURUS DARK CLOUDLETS

Ed Churchwell
Washburn Observatory
475 N. Charter St.
Madison, WI 53706

ABSTRACT

CN($N = 1 \to 0$, $J = 3/2 \to 1/2$) has been searched for toward 8 locations in the Taurus dark cloud complex where NH_3 and HC_3N and/or HC_5N have been observed. CN was detected in TMC1(NH_3) and TMC2(cont.) and was probably detected in L1533(NH_3) and L1544(NH_3). CN appears to be correlated with NH_3 and anti-correlated with HC_5N and HC_3N. It is postulated that NH_3 is likely a dominant precursor of CN and HCN but that CN and HCN are probably not important precursors of HC_3N and heavier cyanopolyacetylenes.

Several small, low mass ($M \sim$ a few $M_\odot$), cool ($T_K \sim 10K$) cloudlets in the Taurus dark cloud complex have been studied in some detail in the lines of NH_3, HC_3N and HC_5N. The primary results of these studies seem to be that: 1) the NH_3 and HC_3N-HC_5N distributions are different; 2) HC_3N and HC_5N are apparently more abundant than HCN and CN (in TMC1); 3) the cyanopolyacetylene molecules (HC_nN, n=3,5,7,9) are more abundant in Taurus than in other dark cloud complexes with similar densities and temperatures; and 4) the line widths of NH_3, HC_3N and HC_5N are typically $\sim 0.15-0.2$ km s^{-1}, i.e., thermal, so that very quiescent conditions are implied.

The observed CN data are presented in tabular form and spectra are shown. CN was detected in TMC1(NH_3) and TMC(cont.). It may have also been detected at about the 2σ level in L1533(NH_3) and L1544(NH_3), but these detections require independent confirmation.

Allen and Knapp (1978 – hereafter AK) detected CN in L1529 but did not detect it in TMC1(HC_5N); among the four dark clouds detected by AK two are known NH_3 sources (L1529 and B335 – Ho et al. 1978), one (Ori I-2) has a measured upper limit (Ho et al. 1978), and no published NH_3 data could be found for one (IC 1848-1). CN was not detected in any of the positions where HC_5N or HC_3N are strong but was detected where NH_3 is strong and where there is a known compact radio continuum source. From analyses of other molecules it appears that most of the cloudlets where CN was not observed differ little in density and temperature from those where CN was detected.

B. H. Andrew (ed.), Interstellar Molecules, 77–80.

TABLE 1

MEASURED PARAMETERS OF CN TOWARD DARK CLOUDLETS IN TAURUS

SOURCE	OBSERVED POSITION α(1950)	δ(1950)	<ATM>	$F_u \rightarrow F_\ell$	T_L^* K	Δv km s^{-1}	V_{LSR} km s^{-1}	$\int T_L dv$ K km s^{-1}	Notes
TMC-2 (cont)	04 29 33.4	24 14 03	1.26	5/2 − 3/2	≳ 0.60 ± .17	0.6 ± .3	6.0 ± 1	0.37	1
L1529	04 29 43.0	24 16 45	1.12	5/2 − 3/2	≲ 0.70				
L1535	04 32 30.0	23 48 00	1.07	5/2 − 3/2	< 0.50				
L1533 (NH$_3$)	04 32 38.0	24 02 00	1.48	5/2 − 3/2	~ 0.26 ± .14	≲ 0.26	5.8 ± 1	~ 0.18	1,2,2a
TMC-1 (NH$_3$)	04 38 20.3	25 42 00	1.10	3/2 − 1/2 5/2 − 3/2 1/2 − 1/2	0.35 ± .10 0.80 ± .10 0.40 ± .10	0.35 ± .2	5.83 ± .10	~ 0.55	3
TMC-1 (HC$_5$N)	04 38 38.0	25 36 00	1.88	5/2 − 3/2	< 0.39				
L1544 (HC$_5$N)	05 01 08.0	25 07 40	1.73	5/2 − 3/2	< 0.75				
L1544 (NH$_3$)	05 01 14	25 07 00	1.61	5/2 − 3/2	~ 0.4 ± .2	≲ 0.26	7.0 ± .1	~ 0.28	2,2b

Notes

1 The line is probably not fully resolved with 100 kHz (0.26 km s^{-1}) resolution. Unfortunately the 30 kHz filterbank had a large gain step between the 4th and 5th cards of 16 filters which occured very close to where the line was. It was therefore not possible to derive reliable line parameters from the 30 kHz filters.

2 The feature is only 2σ above the noise and is therefore very uncertain. It occurs at the same velocity as NH$_3$ and appears in both the 100 kHz and 250 kHz spectra, which were taken on different days.

2a The feature is only 0.17 ± .10 K in the 250 kHz filters, which, if real, would imply that it is unresolved at this resolution.

2b The feature is only ~ 0.28 ± .13 K in the 250 kHz filters, which, if real, would imply that it is not resolved at this resolution.

3 The lines are probably unresolved and therefore may be more intense. If the relative intensities are roughly correct, then the hf line ratios indicate a mean opacity in the F = 5/2 → 3/2 component of $\langle \tau \rangle_{27} \simeq 1.2^{+.7}_{-.6}$.

The frequency resolution was 100 kHz (0.26 km s^{-1}) and 250 kHz (0.66 km s^{-1}) at each position. At a few positions the 30 kHz filters were used, but there was such a bad instrumental ripple that these were considered unreliable.

CN

Taurus Dark Clouds

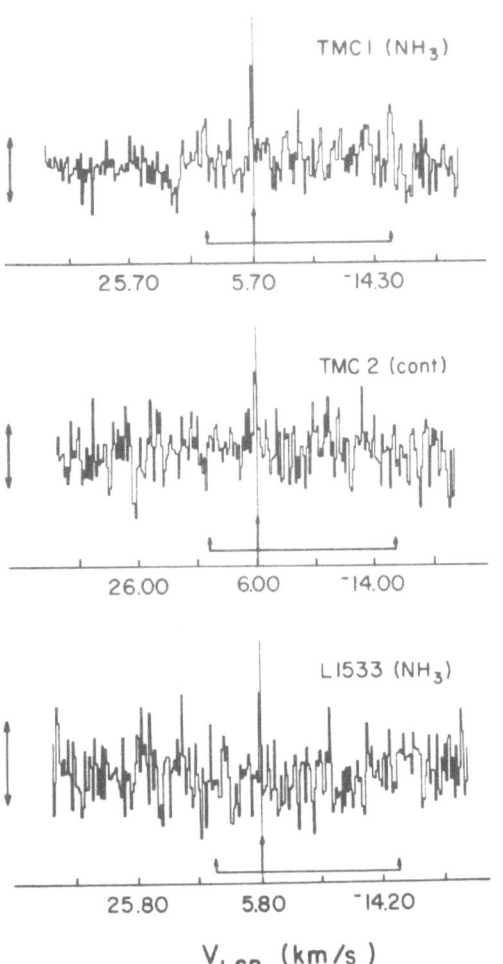

V_{LSR} (km/s)

E. CHURCHWELL

The anti-correlation of CN with HC_3N and HC_5N and its apparent positive correlation with NH_3 suggest that chemistry rather than cloud density or temperature probably plays a dominant role in the observed CN abundance variations. In clouds with $n \geq 10^4$ cm^{-3}, all gas-phase models predict that CN formation is primarily via the reaction $H_2CN^+ + e \rightarrow CN + 2H$, and destruction is via reaction with the ions He^+, H_3^+, and perhaps H^+. Iglesias (1977) suggested that $CN + O_2 \rightarrow NCO^+ + O$ is the primary destruction path in dense clouds ($n \geq 10^4$ cm^{-3}), so that the CN abundance will decrease with cloud age (i.e., increasing density) roughly as $[CN]/[n] \propto n^{-1}$ even if condensation onto grains is not included. NH_3 is typically 10-100 times more abundant than CN or HCN and therefore would not be greatly depleted even if all CN and HCN were formed from NH_3. The apparent correlation of CN with NH_3 would seem to support the idea initially proposed by Herbst & Klemperer (1973) that NH_3 is a primary precursor of CN, probably via the reaction chain $NH_3 + C^+ \rightarrow H_2CN^+ + H$ and $H_2CN^+ + e \rightarrow CN + 2H$. The possible anti-correlation of CN and NH_3 with HC_3N and HC_5N in Taurus is not easy to interpret. Simplistically one might assume that CN, HCN and HNC have mostly been converted to HC_3N and HC_5N, but in this case one would expect also a close correlation of HC_3N and NH_3, which is not observed. A possible ion-molecule formation scheme which does not involve CN, HCN or HNC is: $C_2H_2^+ + CH_4 \rightarrow C_3H_5^+ + H$; $C_3H_5^+ + N \rightarrow C_3H_3N^+ + H_2$; $C_3H_3N^+ + e \rightarrow HC_3N + H_2$.

It is unlikely that these reactions could dominate those involving CN, HCN, and HNC unless CH_4 is overwhelmingly abundant in Taurus.

The observational correlations are not well enough established to rule out or establish one chemical network over another. The correlation of CN with NH_3 and anti-correlation of CN with HC_3N are important clues to CN chemistry, but further observations are required to establish the validity of these. The Taurus complex is probably one of the best regions in which to pursue this problem further because of its small distance and its apparently anomalously high HC_3N abundance.

REFERENCES

Allen, M., Knapp, G.R.: 1978, Astrophys. J. 225, 843.
Herbst, E., Klemperer, W.: 1973, Astrophys. J. 185, 505.
Ho, P.T.P., Martin, R.N., Barrett, A.H.: 1978, Astrophys. J. (Letters) 221, L117.
Iglesias, E.: 1977, Astrophys. J. 218, 697.
Mitchell, G.F., Ginsburg, J.L., Kuntz, P.J.: 1978, Astrophys. J. Suppl. 38, 39.

C_2H AND HC_3N IN INTERSTELLAR CLOUDS

Alwyn Wootten
Owens Valley Radio Observatory, Caltech
Dept. of Astronomy, University of Texas
G.P. Bozyan, D.B. Garrett, R.B. Loren, R.L. Snell,
P. Vanden Bout
Dept. of Astronomy, University of Texas

A survey for the molecules C_2H and HC_3N in a variety of inter-stellar clouds has been completed. Both molecules are very widespread, in cold dark clouds as well as in hot clouds. C_2H emission has been mapped in L1534. In cold clouds the fractional abundance $X(C_2H)$ is found to be $2-6 \times 10^{-9}$. The ratio of abundances $X(C_2H)/X(HC_3N)$ falls in the range 6-10, consistent with some gas-phase reaction schemes for these molecules.

C_2H has not previously been detected in cold dark clouds. Since HC_3N has been detected in abundance in a number of clouds, and since both molecules are thought to derive from the same precursor molecule $C_2H_2^+$ in gas-phase chemistry, we have surveyed a number of clouds for C_2H and HC_3N emission. As a result, both molecules have been found in a variety of clouds: cold dark clouds, clouds near Herbig-Haro objects, and in complexes near HII regions.

The C_2H line was sufficiently strong to permit detailed mapping in L1534, L43, and M17SW. These maps show that C_2H emission is strongly correlated with emission from other molecules excited in dense regions.

In L1534 the C_2H map displays a distinct maximum near the peak HC_5N emission found by Little et al. (1978). An additional peak which does not correspond to an HC_5N emission peak has been found to the northwest. This additional peak has also been found in absorption at 2 cm by formaldehyde.

The column density of C_2H is in general similar to that of HCO^+ or H_2CO. The ratio of collisional to radiative timescales is also similar for these molecules. Therefore C_2H emission probably arises in regions similar to those producing HCO^+ and H_2CO emission. This conclusion is strengthened by the similarity of the maps of these species. We have therefore used a large velocity gradient (LVG) code to determine the abundance of C_2H, $X(C_2H)$, following the procedure detailed in Wootten et al. (1978). We find $X(C_2H) \sim 2-6 \times 10^{-9}$ in cold dark clouds, (L63, L134N, L1534, L1529) but is probably much lower, $X(C_2H) \sim 10^{-10}$, in

B. H. Andrew (ed.), Interstellar Molecules, 81–82.

denser, warmer clouds (M17SW, ρ Oph, NGC2264). Using a similar analysis for the J=5-4 line of HC_3N allows us to estimate the abundance ratio $X(C_2H)/X(HC_3N) \sim 6-10$ for most clouds. In M17SW the ratio appears to be 37.

In a simple chemical reaction scheme, C_2H is created by electron recombination on $C_2H_2^+$ and destroyed by reaction with oxygen. HC_3N is created by reaction of $C_2H_2^+$ and HCN, followed by electron recombination; it is destroyed by reactions with C^+ or He^+. The abundance ratio $X(C_2H)/X(HC_3N) \sim 10$ is consistent with this scheme.

The abundance of C_2H appears to be somewhat higher in the Taurus cloud L1534 than in two otherwise similar clouds L134N and L63. Langer (1976) demonstrated that the CO abundance in a chemically evolving cloud reaches equilibrium only after fairly lengthy timescales; and that before it attains equilibrium the excess of free carbon in the cloud can lead to elevated C_2H abundances. Perhaps equilibrium in carbon chemistry has not yet been reached in the clouds with highest abundances of C_2H.

REFERENCES

Langer, W.D.: 1976, Astrophys. J. 206, 699.
Little, L.T., Riley, P.W., MacDonald, G.H., and Matheson, D.N.: 1978, M.N.R.A.S. 183, 805.
Wootten, A., Evans, N.J., Snell, R., and Vanden Bout, P.: 1978, Astrophys. J. (Letters) 225, L143.

DETECTION OF NEW AMMONIA SOURCES

G.H.Macdonald[1], A.T.Brown[1], L.T.Little[2],D.N.Matheson[3], & M.Felli[4]
[1]Electronics Laboratory, University of Kent at Canterbury, U.K.
[2]DERAD, Observatoire de Meudon, 92190 Meudon, France.
[3]SRC Appleton Laboratory, Chilbolton Observ., Stockbridge, U.K.
[4]Osservatorio Astrofisico di Arcetri, 50125 Firenze, Italy.

Ammonia is a favoured molecule for the study of molecular clouds since several important parameters of the cloud can be deduced from simple observations of the J,K=1,1 and 2,2 inversion doublet transitions and the hyperfine structure in the (1,1) line. With the additional knowledge of the kinetic temperature T_k from observations of CO, for example, it is possible to compute the excitation temperature of the (1,1) line (T_{11}), the rotational temperature between the (1,1) and (2,2) levels (T_{21}), the molecular hydrogen density $n(H_2)$ and ammonia column density $N(NH_3)$ (see, for example, Martin and Barrett, 1978).

We have undertaken a systematic survey of compact HII regions, H_2O masers, Herbig-Haro objects and other protostellar indicators for emission in the (1,1) and (2,2) transitions of NH_3, using the SRC Appleton Laboratory 25 m telescope (η_B=0.37). So far this program has yielded 33 detections which we report here. 5 of these sources were detected independently by Ho (1977) but not published elsewhere, and are included here for completeness.

The objects to be searched were selected from the following lists:

(i)   Compact HII regions (Felli et al., 1978, Israel and
        Felli, 1978);
(ii) H_2O masers (Genzel and Downes, 1977, 1979);
(iii)   Herbig-Haro objects (Gyulbudaghian et al., 1978);
(iv)   Peaks in HCN emission (Tucker, private communication);
(v)   Reflection nebulae (Knapp et al., 1977);
(vi)   CO 'hot spots' (Blair et al., 1975).

The 33 new NH_3 sources are listed in Table 1, which gives the antenna temperatures in each transition $T_A(1,1)$, $T_A(2,2)$, the antenna temperature of the hyperfine structure of the (1,1) line $T_{HFS}(1,1)$, the velocity V_{LSR} and the width ΔV_{LSR} of the main component of the (1,1) line.

B. H. Andrew (ed.), Interstellar Molecules, 83–84.

Table 1. Observational parameters of the (1,1) and (2,2) lines of NH_3 detected in present survey.

SOURCE	α(1950)	δ(1950)	$T_A(1,1)$ (K)	$T_{HFS}(1,1)$ (K)	V_{LSR} (kms^{-1})	ΔV_{LSR} (kms^{-1})	$T_A(2,2)$ (K)
S187	01 20 24	61 38 25	0.15	- ±0.014	-15.1	1.8	0.03
N2-3	03 25 45	30 56 00	0.38	0.1±0.04	7.2	1.1	0.05
N4	05 37 22	23 49 24	0.28	0.16±0.04	2.3	1.9	0.14
S235	05 37 30	35 40 00	0.22	0.11±0.06	-17.0	2.0	-
N7	05 38 24	-8 09 00	0.32	0.16±0.06	5.6	1.3	0.06
G205.11-14.11	05 44 31	00 20 48	0.70	0.22±0.03	8.85	1.7	0.26
Mon R2	06 05 19	-6 22 40	0.46	0.21±0.024	10.3	2.6	0.19
N12-15	06 08 28	-6 10 46	0.35	0.13±0.05	11.7	1.4	0.07
S255	06 10 01	18 01 30	0.25	0.15±0.10	7.0	3.0	-
N16-17	06 10 21	-6 13 00	0.24	0.09±0.05	11.7	1.9	0.18
Rosette IRS	06 31 57	4 15 03	0.22	- ±0.017	12.6	1.3	0.12
S68	18 27 28	1 12 00	0.77	0.30	7.8	1.8	0.22
G23.95+0.15	18 31 41	-7 57 17	0.22	0.12±0.021	80.6	2.1	0.09
G24.49-0.04	18 33 23	-7 33 54	0.24	0.09±0.03	109.6	2.6	0.18
G24.8+0.1	18 33 30	-7 14 27	0.61	0.185±0.028	110.2	3.2	0.31
G28.86+0.07	18 41 8	-3 38 41	0.17	<0.04	99.8	1.9	<0.03
W43S	18 43 27	-2 42 40	0.26	0.09±0.016	97.6	3.2	0.26
G31.4-0.3	18 44 59	-1 16 07	0.26	0.08±0.03	96.2	2.3	0.07
G32.15+0.13	18 46 58	-0 41 30	0.19	≤0.09	94.7	2.1	0.15
G33.9+0.1	18 50 16	0 51 47	0.18	0.07±0.018	107.0	2.6	0.15
G34.3+0.1	18 50 46	1 11 00	0.59	0.18±0.024	58.1	2.8	0.40
W48	18 59 15	1 08 50	0.24	0.17±0.019	42.6	1.9	0.22
G45.49+0.13	19 11 50	11 7 47	0.22	<0.04	59.3	1.9	0.13
S87	19 44 14	24 27 58	0.31	0.1±0.03	24.1	1.7	0.19
S88	19 44 44	25 05 30	0.22	0.01±0.016	21.8	2.1	0.13
ON1	20 08 10	31 22 41	0.67	0.24±0.020	11.0	2.5	0.22
ON2	20 19 50	37 16 30	0.28	0.08±0.021	-0.3	3.0	0.14
R131	20 22 41	42 06 18	0.37	0.15±0.05	5.0	1.7	0.12
S106	20 25 27	37 12 45	0.50	0.24±0.015	1.5	1.7	0.24
R146	21 42 40	65 52 57	0.18	0.08±0.04	-9.8	0.9	0.13
S140	22 17 45	63 04 00	0.69	0.27±0.015	-7.1	1.8	0.38
Cep A	22 54 20	61 45 42	0.29	0.13±0.05	-11.1	3.5	0.22
S156A	23 03 05	59 58 10	0.23	0.12±0.08	-51	3.2	-

REFERENCES

Blair, G.N., Peters, W.L., and Vanden Bout, P.A.: 1975, Astrophys. J. (Letters) 200, L161-L164.

Felli, M., Harten, R.H., Habing, H.J., and Israel, F.P.: 1978, Astron. Astrophys. Suppl. 32, pp. 423-428.

Genzel, R., and Downes, D.: 1977, Astron. Astrophys. Suppl. 30, pp. 145-168.

Genzel, R., and Downes, D.: 1979, Astron. Astrophys. 72, pp. 234-240.

Gyulbudaghian, A.L., Glushkov, Yu. I., and Denisyuk, E.K.: 1978, Astrophys. J. (Letters) 224, L137-L138.

Ho, P.T.P.: 1977, Ph.D. Thesis, Massachusetts Institute of Technology

Israel, F.P., and Felli, M.: 1978, Astron. Astrophys. 63, pp. 325-334.

Knapp, G.R., Kuiper, T.B.H., Knapp, S.L., and Brown, R.L.: 1977, Astrophys. J. 214, pp. 78-85.

Martin, R.N., and Barrett, A.H.: 1978, Astrophys. J. Suppl. 36, pp. 1-51.

AMMONIA OBSERVATIONS OF THE MOLECULAR CLOUDS NEAR S68, S140 AND OMC2

L.T.Little[1], A.T.Brown[2], G.H.Macdonald[2], P.W.Riley[2] & D.N.Matheson[3]
[1]DERAD, Observatoire de Meudon, 92190 Meudon, France.
[2]Electronics Laboratory, University of Kent at Canterbury, UK
[3]SRC Appleton Laboratory, Chilbolton Observ., Stockbridge, UK

Maps of the J=1,K=1 inversion transition of interstellar ammonia are presented and compared with observations of carbon monoxide and formaldehyde.

Recent estimates of the densities within molecular clouds derived from CO observations (Plambeck and Williams, 1979) yield results which are often 10 to 100 times less than those deduced from those of 2mm. H_2CO (Wootten et al., 1978). To shed light on this disagreement we present here maps of the cloud cores in S68, S140 and OMC2, made in the $NH_3(1,1)$ inversion doublet at 23.69 GHz with a 2.2 arc min. beam, using the SRC Appleton Laboratory 25m telescope (see figure 1). Our J=2,K=2 maps are generally similar, but limited by an inferior signal/noise ratio. The ammonia doublet is appropriate for the task, being excited at densities intermediate between those required for CO ($\sim 10^3$ cm^{-3}) and 2mm. H_2CO (10^6 cm^{-3}).

For S140 and OMC2 the dimensions of the NH_3 and H_2CO emission (Blair et al., 1978, Kutner et al., 1976) are rather similar : those of ^{13}CO are noticeably more extended. For S68 the NH_3 and ^{13}CO (Blair et al., 1975) have similar dimensions : no data on H_2CO have yet been published. Near the peak of the NH_3 emission the optical depth τ of the (1,1) line has been deduced from the main/hyperfine component ratios (see, e.g., Schwartz et al., 1977) - results of order unity are obtained. The excitation temperature T_{EX} may be calculated from τ, the source dimensions, and the observed antenna temperatures, assuming that the source is not 'clumped'. The density of molecular hydrogen may then be obtained from T_{EX}, the kinetic temperature T_K (deduced from ^{12}CO observations), and τ. The derived densities $\sim 5 \times 10^3$ cm^{-3} are much less than those deduced from 2mm. H_2CO although the NH_3 emission has similar dimensions. The discrepancy may be resolved by either a "core-halo", or a "clumped" structure for the sources. The latter possibility is more probable, in which case the medium is well modelled as many 'clumps' of density $\sim 10^6$ cm^{-3} and size $<5 \times 10^{-2}$ pc (dominating the NH_3 and H_2CO emission) immersed in a more tenuous medium of density 10^{3-4} cm^{-3} (dominating that of CO).

B. H. Andrew (ed.), Interstellar Molecules, 85–87.

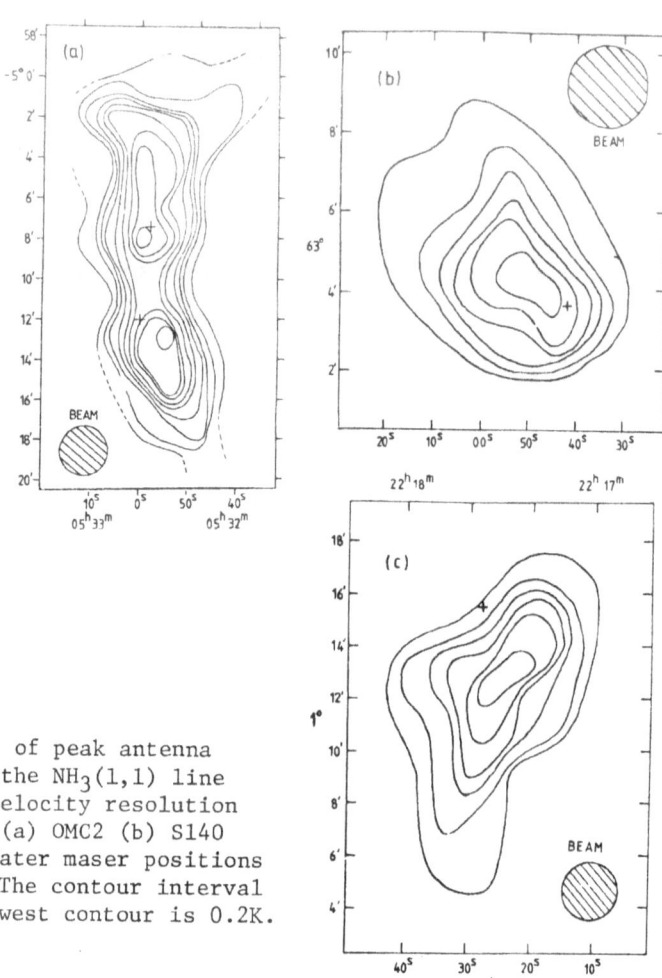

Figure 1. Maps of peak antenna
temperature of the $NH_3(1,1)$ line
observed with velocity resolution
0.86 km s^{-1} in (a) OMC2 (b) S140
and (c) S68. Water maser positions
are marked +. The contour interval
is 0.1K, the lowest contour is 0.2K.

REFERENCES

Blair, G.N., et al.: 1978, Astrophys. J. 219, pp. 896-913.
Blair, G.N., et al.: 1975, Astrophys. J. (Letters) 200, L161-L164.
Kutner, M.L., et al.: 1976, Astrophys. J. 209, pp. 452-461.
Plambeck, R.L. and Williams, D.R.W.: 1979, Astrophys. J. (Letters) 227,
     L43-L47.
Schwartz, P.R., et al.: 1977, Astrophys. J. 218, pp. 671-676.
Wootten, A., et al.: 1978, Astrophys. J. (Letters) 225, L143-L148.

DISCUSSION FOLLOWING MATHESON

Scoville: It might be worthwhile including the effects of far-infrared pumping of NH_3 in your excitation calculations.

Ho: The relevance of this work to this particular session concerns the question of clumping in dark clouds. It is such an important question that great care should be paid to the interpretation of the data. Barrett and I have mapped the same clouds with greater angular and spectral resolution at Haystack. Whereas the general shapes of the emission regions are similar, we find additional structures $\lesssim 2'$ in the cloud associated with S68 (Serpens object), (Ho and Barrett, Ap. J. submitted). This means $T_{ex}(NH_3) \gtrsim 10K$, when beam dilution effects are taken into account. As $T_K \sim 20K$, $n(H_2) \sim few \times 10^4 cm^{-3}$ would account for the excitation of the NH_3 line. If $n(H_2)$ is even higher so that $T_{ex} \sim T_K$, "clumping" is still not severe in the sense that $\sim 50\%$ of the beam would be filled. We would conclude therefore that clumping with many small clumps within the antenna beam is probably incorrect for this case. In addition, mapping results in dark clouds (Ho and Martin, in preparation) reveal that individual fragments are resolved with size scales of ~ 0.1 pc. However these fragments are well isolated and correspond to optical condensations. Hence clumping, such that there are many unresolved clumps within a 2' beam, is probably not well established for cool ($\lesssim 20K$) clouds at the present time.

Matheson: We would agree with the latter comment: for example, observations of the quiescent dark clouds Taurus and Heiles 2, which have very narrow line widths, and appear to lack H_2O masers and internal infrared sources, do not indicate intrinsic filling factors differing substantially from unity (Little et al. MNRAS 1979). However for the sources I have described here, if we invoke a 'clumping' model with the ammonia excited at the kinetic temperature of the source, then from our data it is difficult to avoid the conclusion that our telescope beam remains substantially unfilled.

AMMONIA OBSERVATIONS OF THE MOLECULAR CLOUD NEAR S106

L.T.Little[1],G.H.Macdonald[2],P.W.Riley[2] and D.N.Matheson[3]
[1]DERAD, Observatoire de Meudon, 92190 Meudon, France.
[2]Electronics Laboratory, University of Kent at Canterbury, UK
[3]SRC Appleton Laboratory, Chilbolton Observ., Stockbridge, UK

The region surrounding the optical nebula S106 has been mapped in the J=1,K=1 (1,1) transition of ammonia with a 2.2 arc min. beam. The source is observed as two components with the HII region sandwiched between them. It is likely that the asymmetry of the optical nebula is produced by preferential escape of u.v. radiation from the central source of excitation in directions of reduced density within the molecular condensation. Observations of the (1,1) and (2,2) transitions yield a kinetic temperature for the NH_3 molecules in the range 18-22K. There is evidence for "clumpiness" in the distribution of matter.

Optically the Sharpless HII region S106 has a bipolar appearance, its two halves being separated by a dust cloud. Behind the central dust lies a powerful infra-red source which appears to be heated by a single star (Sibille et al., 1975, Eiroa et al., 1979). Eiroa et al. estimate its distance at 500 pc. HII continuum maps (e.g., Israel and Felli, 1978) reveal an intricate structure for the ionised material. Sibille et al. suggested that the appearance of the nebula arises from its preferential expansion in directions of reduced density perpendicular to the plane of a disc. General models for the radiation transfer in this situation have been developed by Kandel and Sibille (1978), and for the hydrodynamics by Bodenheimer et al. (1979). However, the CO structure in the surrounding 20×25 arc min. molecular cloud appears, if anything, to be extended along the nebulosity (Lucas et al., 1978).

Using the SRC Appleton Laboratory 25m telescope ($\eta_B \approx 0.37$) we have mapped the (1,1) transition of interstellar ammonia at 23.7 GHz in S106 (see figure 1). A tempting interpretation of our map is indeed that of a thick disc- or ring-like structure, in slow rotation, seen nearly edge on. The central 'saddle' could result from absorption of radiation from the background HII region by the ammonia molecules. Observations of the hyperfine ratios of the (1,1) transition at the emission peak allow an estimate of its optical depth $\tau \sim 2$ and, with an observation of the (2,2) transition, the rotation temperature $T_{21} \sim 18K$. H_2CO observations (Lucas et al., 1978) have yielded an estimate of the

89

B. H. Andrew (ed.), Interstellar Molecules, 89–90.
Copyright © 1980 by the IAU.

BEAM

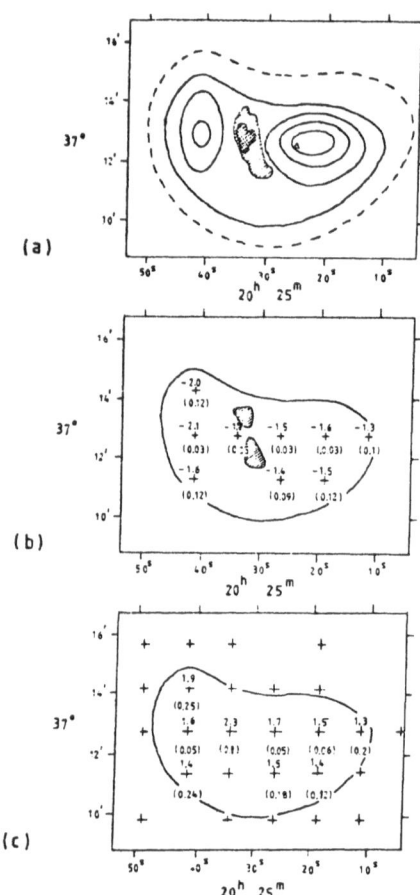

Figure 1.
(a) Map of observed peak antenna temperature for NH3(1,1) line in S106 observed with 2.2 arc min. and 0.43 km s^{-1} resolution. Contour interval 0.1K. Dashed contour 0.1K. Superimposed is a sketch of the HII region derived from the 5 GHz continuum map of Israel and Felli (1978). The triangle marks the position of the H$_2$O maser (Cesarsky et al.,1978).

(b) Observed LSR velocity (km s^{-1}) as a function of position in S106. Superimposed is a sketch of the optical emission derived from a POSS print.

(c) Line widths (FWHM in km s^{-1}) as a function of position across the source. The crosses mark positions observed. The noise is not constant at all positions but it is at worst 0.08K r.m.s. in a single channel (width 0.43 km s^{-1}).

In figures (b) and (c) the 0.2K contour is shown, and estimates of the 1σ errors are given in brackets.

molecular hydrogen density $n_{H_2} \sim 5.10^4$ cm^{-3}. Comparing our apparent excitation temperature (knowing τ) with that predicted from n_{H_2} suggests the existence of clumping within the source, with filling factor ~0.25.

REFERENCES

Bodenheimer, P., Tenorio-Tagle, G., and Yorke, H.W.: 1979, Astrophys J. 233, pp. 85-96.
Cesarsky, C.J., Cesarsky, D.A., Churchwell, E., and Lequeux, J.: 1978, Astron. Astrophys. 68, pp. 33-39.
Eiroa, C., Elsässer, H., and Lahulla, J.F.: 1979, Astron. Astrophys. 74, pp. 89-92.
Kandel, R.S., and Sibille, F.: 1978, Astron. Astrophys., 68, pp. 217-228.
Israel, F.P., and Felli, M.: 1978, Astron. Astrophys. 63, pp. 325-334.
Lucas, R., Le Squéren, A.M., Kazès, I., and Encrenaz, P.J.: 1978, Astron. Astrophys. 66, pp. 155-160.
Sibille, F., Bergeat, J., Lunel, M., and Kandel, R.: 1975, Astron. Astrophys. 40, pp. 441-446.

THE CORE OF A QUIESCENT CLOUD, L183

C.M. Walmsley, H. Ungerechts, G. Winnewisser
Max-Planck-Institut für Radioastronomie, Bonn, Germany

Simultaneous observation of the J,K=1,1 and 2,2 inversion transitions of ammonia (NH_3) with high spatial resolution ($\lesssim 1$ arc min) offers a powerful method of probing the core region of interstellar clouds for evidence of molecular clumping and of prevailing physical conditions which could lead to star formation. We have therefore used the Effelsberg 100-m radiotelescope to make an extensive study of the central region of the nearby dark dust cloud L183 (also known as L134N) in the NH_3 (1,1) transition; the spatial resolution was 40 arcsec. The core region as mapped in the NH_3 (1,1) transition with a velocity resolution of 0.08 km s^{-1} consists of two elongated condensations separated by about 2 arcmin in north-south direction (see Fig. 1). The central part of the NH_3 cloud has an approximate dimension of 6' (N-S) by 2' (E-W) corresponding to a linear extent of 0.17 x 0.06 pc at an assumed distance of 100 pc. The measured velocity structure of the NH_3 cloud seems to reflect the double peaked nature of the cloud in that it increases from 2.30 km s^{-1} in the south to about 2.5 km s^{-1} at the northern end of the southern NH_3 peak, and then decreases again to 2.3 km s^{-1} towards the north. The intrinsic linewidths of NH_3 (corrected for hyperfine blending) do not vary significantly with position and are between 0.2 and 0.3 km s^{-1}. The two ammonia peaks are part of a central molecular ridge from which we have observed NH_3 (2,2) emission at 9 positions (see Fig. 1). The rotation temperature T_{21} as determined from the optical depths of the (1,1) and (2,2) transitions is ~ 9K for all positions, and hence the kinetic temperature T_{kin} seems close to this value as well, i.e. ~ 10K throughout the central part of L183. A more detailed account is being publsihed elsewhere (Ungerechts, Walmsley and Winnewisser).

In addition HC_3N (J=1→0; F=2→1) emission has been detected from the ridge; the emission has a single peak at the northern ammonia condensation (see Fig. 1). The HC_3N velocities (~ 2.4 km s^{-1}) agree within the experimental uncertainties with those of NH_3, suggesting that both molecular species occupy the same volume of space. The linewidths of both molecules (~ 0.2 km s^{-1} for HC_3N) are consistent with a kinetic temperature of ~ 10K. Thus the observations of NH_3 and the detection of

B. H. Andrew (ed.), Interstellar Molecules, 91–92.

HC$_3$N indicate that the core region of L183 is a small (L < 0.2 pc), dense
(n(H$_2$) $\sim$ 5x10^{4}cm^{-3}), and low mass ($\sim$1M$_\odot$) fragment with two NH$_3$ condensations. They exhibit throughout the core region a velocity dispersion
($\sim$0.2 km s^{-1}) which is close to the thermal limit and smaller than
typical free-fall velocities. Thus the cloud is presently in a quiescent
"pre-protostellar" stage which is marked by gravitational equilibrium;
there is no evidence of the presence of a protostellar object embedded
within the cloud. Slowly accreting material or external influences such
as shock-waves are likely causes of eventual collapse.

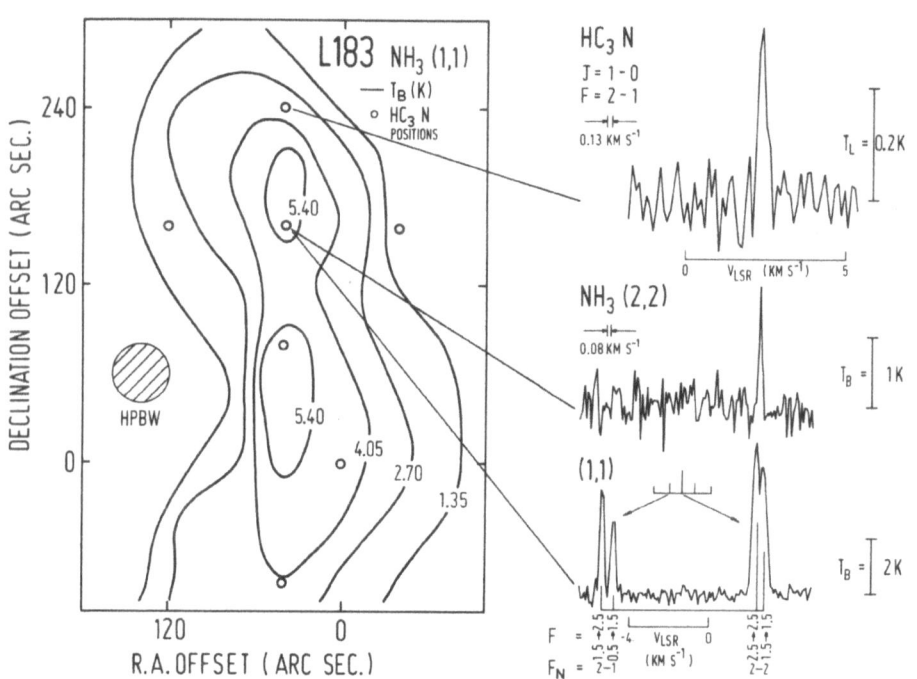

Figure 1. NH$_3$ map of the core region of L183, and sample spectra of
HC$_3$N and NH$_3$ at selected positions. The (0,0) position is
α(1950) = 15^{h}51^{m}30^s, δ(1950) = $-$02°43'31".

OBSERVATIONS OF NH$_3$ TOWARD DR21

T. Pauls and T. L. Wilson
Max-Planck-Institut für Radioastronomie
Auf dem Hügel 69
5300 Bonn 1
Federal Republic of Germany

Figure 1 shows the $(J,K) = (1,1)$, $(2,2)$ and $(3,3)$ spectra of NH$_3$ observed toward the continuum peak of DR21 and four positions offset 50" from the peak. These observations were made with the Effelsberg 100-m radio telescope, which has a half-power beam width of $\sim$43" at these frequencies.

PARA-NH$_3$: The $(1,1)$ and $(2,2)$ lines arise from para-NH$_3$ and both transitions are seen in absorption toward the peak and in emission elsewhere. Our results confirm those of Matsakis et al. (1977) in showing a non-equilibrium distribution of the hyperfine components of the $(1,1)$ line toward the peak and south of the peak. The apparent optical depths of the $(1,1)$ and $(2,2)$ lines are 0.15 and 0.12, respectively; and the rotational temperature between the $(1,1)$ and $(2,2)$ levels is $\sim$26 K.

ORTHO-NH$_3$: The $(3,3)$ line arises from ortho-NH$_3$ and we see that this line is in emission at all positions, including the continuum peak. In addition, the line strength at the peak is $\sim$3 times larger than at the off-peak positions. While the $(3,3)$ line may come from a different region than the $(1,1)$ and $(2,2)$ lines due to formation or past history, we feel our data suggest a population inversion of the $(3,3)$ level.

REFERENCE

Matsakis, D.N., Brandshaft, D., Chui, M.F., Cheung, A.C., Yngvesson, K.S., Cardiasmenos, A.G., Shanley, J.F., and Ho, P.T.P.: 1977, Astrophys. J. (Letters) 214, L67.

B. H. Andrew (ed.), Interstellar Molecules, 93–94.

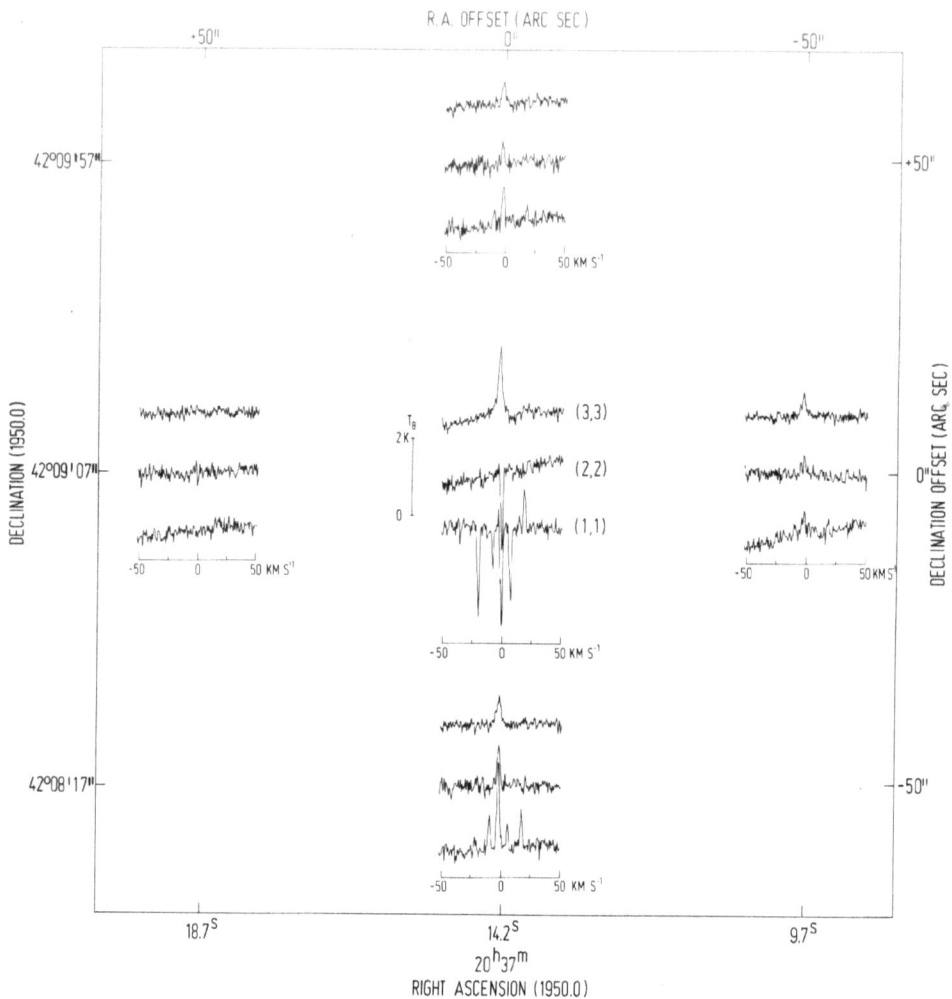

Fig. 1:   NH$_3$ spectra toward DR21.

HIGH RESOLUTION 4.8 GHZ MAPPING OF H_2CO USING THE WESTERBORK SYNTHESIS RADIO TELESCOPE

J.R. Forster[1], W.M. Goss[2], T. de Jong[3], C.A. Norman[4],
H.J. Habing[4], H.R. Dickel[5].

INTRODUCTION

We mapped the distribution of 6 cm H_2CO ($1_{10} \rightarrow 1_{11}$) absorption against the HII regions DR21 and W58 with an angular resolution of $6\overset{''}{.}8$ (RA) and a velocity resolution of 0.73 km s^{-1}. The Westerbork SRT was used with the newly completed 5120 channel digital line receiver. With all 14 telescopes, a maximum baseline of 1440 m, both linear polarizations and a bandwidth per channel of 10 kHz the rms noise in the channel maps was about 7 K. The goal of this work is to measure scale sizes of H_2CO in molecular clouds near HII regions and to study the kinematics of the clouds in the molecular line.

RESULTS

The H_2CO optical depth profile in the direction of maximum intensity in DR21 (Fig. 1a) shows three features at velocities -4.5, -3.1 and 8.6 km s^{-1}. The intensities of the lines change markedly with position. At a distance of 2 kpc (Dickel et al. 1978) the linear extent of the background source is about 0.5 pc and the synthesized beam size is 0.1 pc (RA) by 0.15 pc (DEC). The optical depth distribution of the main feature at the peak of the line (-3.1 km s^{-1}) is mapped in Fig. 1b. The -4.5 km s^{-1} feature is strongest along the western edge and the 8.6 km s^{-1} feature appears inversely correlated with the main feature and is much weaker. The intensity weighted mean velocities of the -3.1 and 8.6 km s^{-1} features vary by about 1.2 km s^{-1} over the source and also appear to be inversely correlated. The linear scale for significant changes in optical depth is about 0.2 pc for all three features.

---

1. Netherlands Foundation for Radio Astronomy, Dwingeloo
2. Kapteyn Astronomical Institute, Groningen
3. Astronomical Institute, Amsterdam
4. Sterrenwacht, Leiden
5. University of Illinois, Urbana, USA.

B. H. Andrew (ed.), Interstellar Molecules, 95–98.
Copyright © 1980 by the IAU.

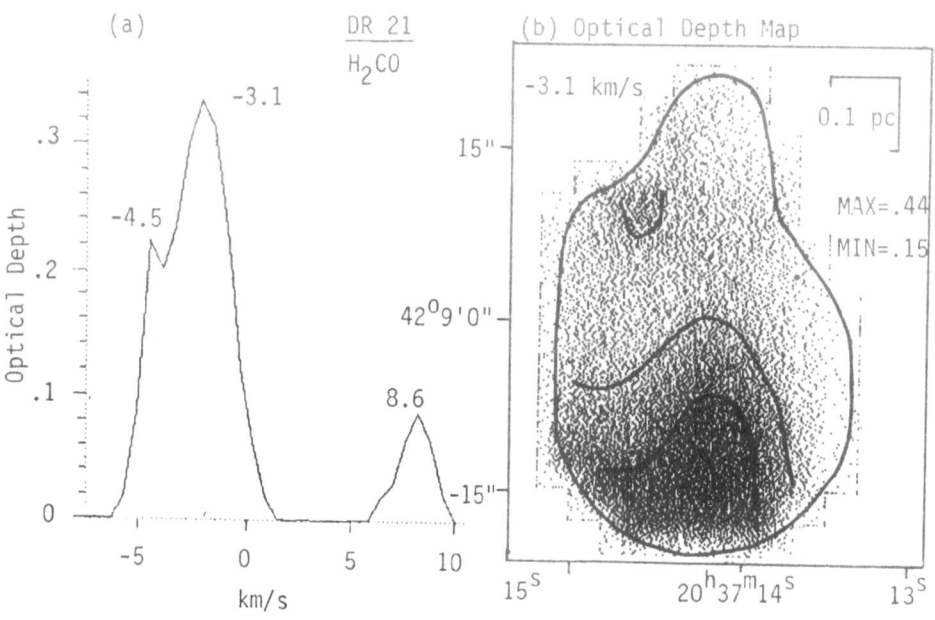

Figure 1. (a) H$_2$CO optical depth profile at the position of maximum
flux in DR 21. (b) Distribution of optical depth at -3.1 km/s.
The contours are at optical depths of 0.3 and 0.4.

Optical depth profiles in the directions of components A (K3-50),
B and C in W58 are given in Figure 2 along with a portion of the 6 cm
continuum map of Israel (1975). The profiles are spatial averages over
the individual components shown. Component C has a much larger H$_2$CO
opacity than the other two components and also shows strong variation
between the two sub-components C1 and C2. The optical depths measured
at the line center (-21.2 km s^{-1}) are listed below for four positions
in W58.

Radio Component	(1950) Position RA	DEC	Optical Depth at -21.2 km s^{-1}
A(K3-50)	19^{h}59^{m}50^s.1	33°24'20"	0.01 ± .01
B	51.9	24 40	0.06 ± .04
C1	58.4	25 51	1.75 ± .01
C2	59.6	25 51	0.13 ± .05

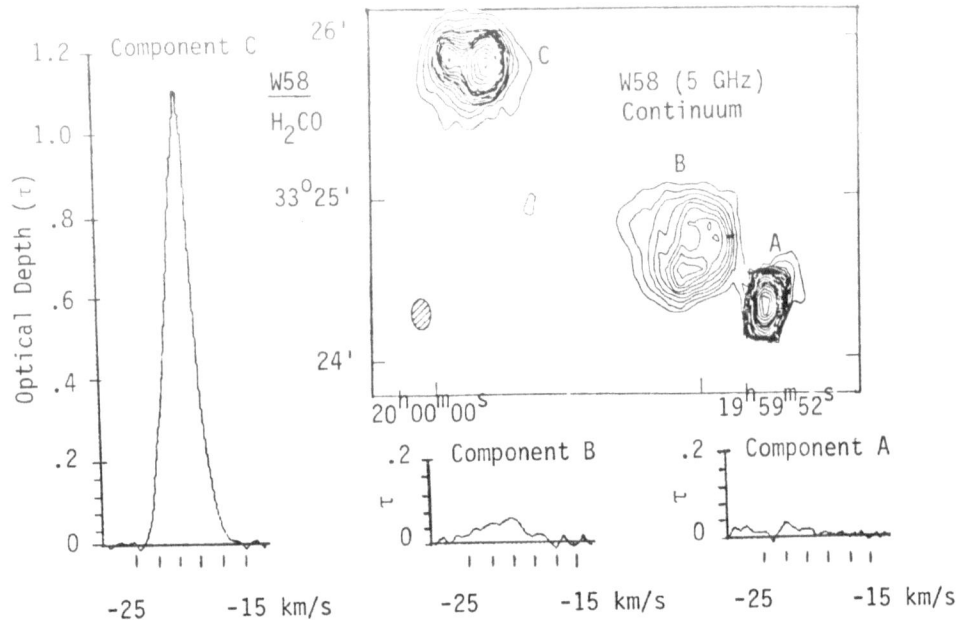

Figure 2. Spatially integrated optical depth profiles in the direction
of components A, B and C in W58. The contimuum map is taken
from Israel (1975).

The most striking aspect of the region is the dramatic change in
H₂CO opacity among the various components. This is not altogether unexpected as component A is associated with the optical object K3-50 and
component C1 has several hundred magnitudes of visual extinction (Wynn-
Williams et al. 1977) and the OH maser ON-3 is associated with it. At a
distance of 9 kpc the linear size of the synthesized beam is 0.46 x
0.84 pc. The H₂CO opacity in front of component C changes by more than
a factor of 10 on an apparent linear scale of about a parsec.

Velocity structure is also observed in component C. The H₂CO velocity field will be compared to recent H109α WSRT data obtained by van
Gorkom et al. (in preparation) in order to help determine the geometry
and kinematics of the region.

REFERENCES

Dickel, J.R., Dickel, H.R. and Wilson, W.J. 1978, Ap. J. 223, 840.
Israel, F.P. 1975, Astron. & Astrophys. 48, 193.
Wynn-Williams, C.G., Becklin, E.E., Matthews, K., Neugebauer, G. and
    Werner, M.W. 1977, MNRAS 179, 255.

DISCUSSION FOLLOWING FORSTER

Goldsmith: Is the variation in H_2CO opacity due to changes in H_2 density or to changes in fractional abundance?

Forster: In W58, at least, the most deeply absorbed components are also optically obscured, suggesting that the H_2CO variations are caused by density changes.

Greenberg: Do you have any way of establishing the exact location of the H_2CO with respect to the HII region. Is it far or close, and can you discriminate between the velocity of the H_2CO and the velocity of the HII region?

Forster: The H_2CO agrees with the CO and other molecular line velocities, and so is assumed to lie within the associated molecular cloud. The velocities of the individual HII region components are given by the recombination line velocities measured with the Westerbork Synthesis Radio Telescope by van Gorkom et al. (in preparation).

SURVEYS OF THE 4.8 GHZ FORMALDEHYDE ABSORPTION LINE IN DARK CLOUDS IN M17 AND NGC 2024

Y.K. Minn
National Astronomical Observatory, Seoul, Korea

The 4.8 GHz H_2CO absorption line was mapped in the directions of the dark clouds in M17 and NGC 2024. The observations were made with a beamwidth of 6.6 arc min and a velocity resolution of 0.12 km s^{-1}.

In M17 two clouds at velocities of about 24 km s^{-1} and 19 km s^{-1} are mapped in antenna temperature T_A. The 24 km s^{-1} cloud, which shows deep and narrow absorption lines, is confined to the immediate vicinity of the radio continuum region. The shape and the peak positions of the cloud are in good agreement with those of the continuum distribution (Mezger and Henderson 1967). The optical depth of this cloud is generally uniform with a mean value of about 0.03. The cloud appears to be located in front of the radio source, as suggested by Lada and Chaisson (1975). The 19 km s^{-1} cloud, which shows wide line halfwidths ($\Delta V \approx 3$ km s^{-1}), has its center at the darkest part of the foreground dust lane. The cloud has an elongated shape which corresponds to that of the dark lane. The equivalent widths in this cloud increase toward the northwest direction and the lines of constant equivalent width run parallel to the dark lane. The principal OH lines at 19 and 24 km s^{-1} also have similar distributions (Gardner and McGee 1971).

The contour maps of the antenna temperature in the space-velocity plane (Fig. 2a and b) show that the two clouds are connected and have a common outer envelope. The 24 km s^{-1} cloud appears to be smooth and narrow in the direction of the radial velocity. The 19 km s^{-1} cloud has an extension to the north where there also appears the peak, as seen in Fig. 2b. In both clouds there are velocity gradients in the directions to the north and to the northeast. The velocity gradients are $dV/d\delta =$ 0.1 km s^{-1} per min of arc for the 24 km s^{-1} cloud and 0.4 km s^{-1} per min of arc for the 19 km s^{-1} cloud. The velocity gradients seem to be the result of rotation of the clouds as there are no indications of cloud collapse. It appears that the 24 km s^{-1} cloud is actually larger than was detected in front of the HII region. This cloud is moving toward the HII region with a velocity of 7 km s^{-1}. The 19 km s^{-1} cloud is within the visible dark cloud in front of the HII region. It is moving toward the HII region with a velocity of 2 km s^{-1}.

B. H. Andrew (ed.), Interstellar Molecules, 99–100.

In NGC 2024 two clouds at velocities of 10 km s^{-1} and 13 km s^{-1} are detected. Both clouds have concentric circular contours with the centers at the positions of the radio continuum peak and the dark bar. It appears that the 10 km s^{-1} cloud is associated with the dark bar in front of the continuum source and the 13 km s^{-1} cloud is located behind the continuum source.

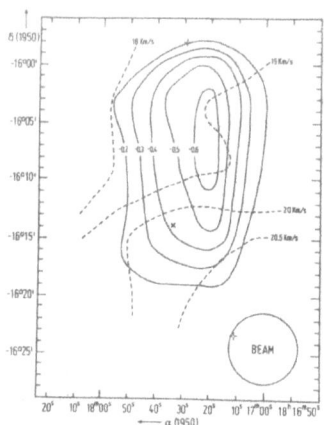

Fig. 1. Distributions of T_A and V_R of 19 km s^{-1} cloud in M17. X is the continuum peak position.

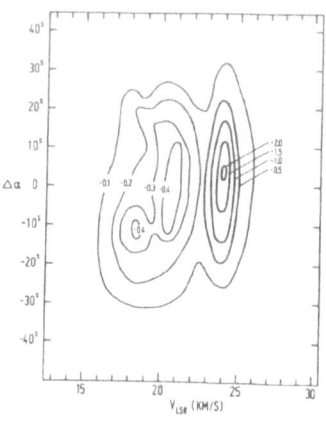

Fig. 2a. The contour map of T_A in the $\Delta\alpha$-V_R plane at $\delta(1950)=-16°15'$ in M17.

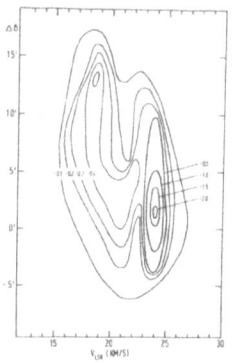

Fig. 2b. The contour map of T_A in the $\Delta\delta$-V_R plane at $\alpha(1950)=18h17m$ 40s in M17.

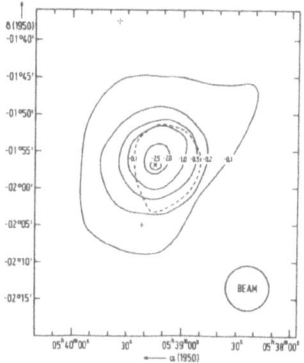

Fig. 3. Contour map of T_A of the 10 km s^{-1} (solid line) and 13 km s^{-1} (dashed line) clouds in NGC 2024.

REFERENCES

Lada, C., and Chaisson, E.J.: 1975, Astrophys. J. 195, 367.
Gardner, F.F., and McGee, R.X.: 1971, Astrophys. Letters 8, 84.
Mezger, P.G., and Henderson, A.P.: 1967, Astrophys. J. 147, 471.

CLOUD-TO-CLOUD VARIATIONS IN H_2CO-TO-H_2 RATIOS

W.A. Sherwood
Max-Planck-Institut für Radioastronomie

The ratio H_2CO/H_2 has been determined for Heiles Cloud 2 (HC2) (Sherwood and Wilson, 1980, and references therein). Rather surprisingly, the ratio was much larger than that found by Kutner (1973) in a cloud KC only 3 degrees away. In both ratios the H_2 density was determined from a relation between star counts and dust column density (see Sherwood and Wilson, 1980). Two other clouds have been studied in a similar manner: Khavtassi 3 (K3) (Myers, 1975) and ρ Oph (Myers et al., 1978). The ratio H_2CO/H_2 varies among the four clouds as does the minimum visual extinction apparently required before H_2CO can form. The data are summarized in the Table. In all cases linear fits were used to determine H_2CO/H_2.

		HC2	KC	K3	ρ Oph
Minimum A_v	(mag)	1.4	0	1.9	0.9
$\log(H_2CO/H_2)$		-7.7	-8.4	-7.3	-8.5
Total extinction A_v	(mag)	≥ 8	≤ 5	4	12.5

The values of total extinction in the four clouds are from Sherwood and Wilson (1980), Batrla (1979), Myers (1975) and Grasdalen et al. (1973) respectively.

There are three points to notice:

1) The production of H_2CO with respect to H_2 appears to increase as the minimum extinction required to produce H_2CO also increases.

2) With the exception of KC, there is an inverse correlation between the minimum extinction needed to produce H_2CO and the total extinction in the cloud. This may mean that in low density dust clouds (N_D cm$^{-2} \propto A_v$) the excitation temperature of H_2CO rises to the background temperature 2.7K and only in the cooler central regions of such clouds does one observe H_2CO. Consequently the zero point for A_v would tend to be larger.

B. H. Andrew (ed.), Interstellar Molecules, 101–102.

3) The depletion of carbon onto grains is likely to increase as the grain density, i.e. A_V, increases. With the exception of KC, the amount of H_2CO in relation to H_2 (which is directly proportional to A_V) decreases as the total number of grains increases. It should be pointed out that the ratio becomes even smaller if A_V has been underestimated.

Kutner's cloud presents a problem. There are definitely more stars visible toward KC than can be attributed to the foreground if KC is optically thick and 135 pc away (Elias,1978). If KC is optically thin one might confirm the $A_V = 5^m$ by finding stars with known colour excess and distance. Inspection of Blanco et al. (1968), Neckel (1967) and Elias (1978) failed to reveal any, although optical selection effects or low IR sensitivity may have critically influenced the results. On the other hand, both Sume et al. (1975) and Elias (1978) report several groups of T-Tauri and Hα emission-line objects indicative of star formation in HC2 and KC. If these numbers were removed from the star counts, KC would have a larger visual extinction and would appear to be a dense cloud, albeit well fragmented by the action of star formation.

ACKNOWLEDGEMENTS

I would like to thank Wolfgang Batrla and Tom Wilson for their cooperation in various aspects of this study. This work was supported by the Deutsche Forschungsgemeinschaft, Sonderforschungsbereich 131, Radioastronomie.

REFERENCES

Batrla, W.: 1979, Ph.D. Thesis, University of Erlangen-Nürnberg.
Blanco, V.M., Demers, S., Douglass, G.G., and FitzGerald, M.P.: 1968, Publ. U.S. Naval Obs., 2nd Series XXI.
Elias, J.H.: 1978, Astrophys. J. 224, 857.
Grasdalen, G.L., Strom, K.M., and Strom, S.E.: 1973, Astrophys. J.(Letters 184, L53.
Kutner, M.L.: 1973, Molecules in the Galactic Environment, ed. M.A. Gordon and L.E. Snyder, J. Wiley, New York, p. 199.
Myers, P.C.: 1975, Astrophys. J. 198, 331.
Myers, P.C., Ho, P.T.P., Schneps, M.H., Chin, G., Pankonin, V, and Winnberg, A.: 1978, Astrophys. J. 220, 864.
Neckel, T.: 1967, Landessternwarte Heidelberg-Königsstuhl 19,
Sherwood, W.A., and Wilson, T.L.: 1980, Astron. Astrophys., to be submitted.
Sume, A., Downes, D., and Wilson, T.L.: 1975, Astron. Astrophys. 39, 435.

FORMALDEHYDE IN L1551, L134 AND THE GALACTIC CENTRE

Aa. Sandqvist and C. Bernes
Stockholm Observatory
S-133 00 Saltsjöbaden, Sweden

The formaldehyde molecule is an excellent probe of physical conditions inside interstellar clouds. We illustrate this by presenting results of our recent series of observations using the MPIfR Effelsberg 100-m radio telescope for the 6-cm transition, the NRAO Green Bank 43-m radio telescope for the 2-cm transition and the NRAO Kitt Peak 11-m mm-wave telescope for the 2-mm transition.

L1551 AND L134

We have mapped the L1551 dark cloud in these three lines (see fig. 1), and a hyperfine analysis (as described by Sandqvist and Lindroos 1976) has been applied to the 6-cm absorption profiles. As a separate approach to the study of this cloud we have constructed a model, fitting the predicted line intensities to the observed intensities in all three lines with the use of a Monte Carlo procedure (Bernes 1978, 1979) for solving the radiative transfer equation. Both analyses indicate decreasing 6-cm excitation temperature towards the centre of the cloud, demanding a corresponding rise in the kinetic temperature. According to the model the kinetic temperature increases from 11 K in the cloud periphery to 20 K in the central regions and the hydrogen number density increases from 10^3 cm^{-3} to slightly above 10^4 cm^{-3} in the centre. Column densities calculated from the hyperfine analysis show a marked increase towards an embedded 2.2 μ infrared object (source no. 5, Strom et al. 1976). A high-density region at this point has been included in the model, which here indicates a hydrogen number density of 4-5 10^4 cm^{-3} but no increase of the kinetic temperature.

A 2-cm H$_2$CO absorption profile of high quality (fig. 2) has been obtained towards L134. The hyperfine analysis of this profile yielded an excitation temperature of 2.0 K, a total optical depth of 1.0 and a velocity dispersion of 0.12 km s^{-1} for the 2-cm absorption line. These results are found to be inconsistent with our upper limit to the 2-mm line intensity and an earlier 6-cm hyperfine analysis by Downes et al. (1976).

B. H. Andrew (ed.), Interstellar Molecules, 103–107.
Copyright © 1980 by the IAU.

A detailed presentation of the above results for L1551 and L134, as well as an analysis of formaldehyde observations of L1317(S187) and NGC 7023 will be published in the near future in Astronomy and Astrophysics.

THE SGR A +40 KM s⁻¹ MOLECULAR CLOUD

The +40 km s^{-1} feature in the galactic centre has been studied in detail in the absorption lines of H I, OH and H$_2$CO (6-cm) seen towards Sgr A (Sandqvist 1970, 1974). Through a series of lunar occultations, observed with the NRAO 43-m radio telescope, the continuum source Sgr A was resolved into several components, now generally known as Sgr A East and Sgr A West. Furthermore, the +40 km s^{-1} cloud was also resolved into several components, with major features having velocities of +50 and +25 km s^{-1}. There is some evidence of a general velocity gradient across the complex. It is noteworthy that the OH observations show a concentration of molecules directly towards Sgr A West that is smaller than in some of the surrounding regions. This effect was also noted in the aperture-synthesis observations of 6-cm H$_2$CO line absorption in the Sgr A complex by Whiteoak et al. (1974); their interpretation was that the molecular region is mainly behind the continuum sources. On the other hand, Liszt et al. (1975) found that there is a decrease of CO emission line intensity towards Sgr A West and hence that the H$_2$CO absorption results do not necessarily imply that the molecular region is on the far side of the galactic nucleus. The HCN emission line results of Fukui et al. (1977), which have a 2' resolution, show a minor decrease of intensity in the direction of Sgr A West, and these authors place the molecular region on the far side of the nucleus.

In the analysis of the occultation and aperture-synthesis absorption data mentioned above, it was assumed that all the continuum sources lay behind the molecular region. If this assumption is abandoned, and a re-analysis is performed and a comparison made with 2-mm H$_2$CO line emission, it may be possible to determine the relative positions of the continuum and the molecular regions; while the absorption line data are affected by the relative arrangement of these regions, the emission line data are not. With this in mind, we have begun mapping a region around Sgr A in the 2-mm H$_2$CO line, and here we present some of the preliminary observations.

The mapping observations so far obtained are presented in fig. 3, where the top small numeral in each square gives the maximum T$_A$-value (K) for each 2-mm profile, and the large numerals represent $\int$line T$_A$ dV (K km s^{-1}). The bottom small numerals give the approximate radial velocity V(km s^{-1}) of the T$_A^*$-peak. The map, consisting of grid points with 1' spacing, is centered on the position of Sgr A West at α(1950.0)= 17^{h}42^{m}29^{s}3, δ(1950.0)= $-28°59'18"$ (the reference "off"-point in the position-switching observing mode was α(1950.0)=17^{h}41^{m}29^{s}3, δ(1950.0)= $-28°59'18"$). The 2-mm H$_2$CO numbers are superimposed on a map of the OH distribution (solid contours) and the 18 cm continuum (dashed contours) as obtained by lunar occultation observations (Sandqvist 1974). The OH

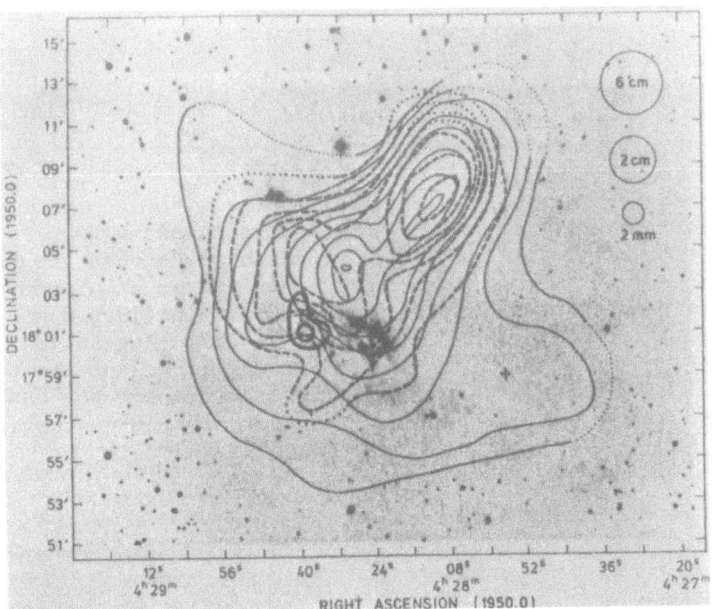

Fig. 1. The dark cloud L1551. H_2CO contours of 6-cm (thin solid lines) and 2-cm (dashed lines) brightness temperatures, and 2-mm (heavy solid line) T_A^*, superimposed on the red print of the National Geographic Society-Palomar Sky Atlas. The crosses mark positions of five 2.2 μ infrared sources (Strom et al. 1976). The large nebulosity is S239. The 6- and 2-cm and the 2-mm beams are shown in the upper right hand corner. Contour intervals are -0.1, -0.03 and +0.2 K for the 6-cm, 2-cm and 2-mm lines, respectively, with the outermost contour having values of -0.2, -0.9 and +0.4 K, respectively.

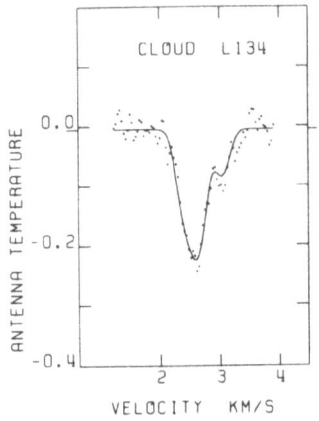

Fig. 2. A 2-cm H_2CO absorption profile (dots) observed towards L134 ($\alpha(1950.0)=15^h50^m54^s$, $\delta(1950.0)=-04°31'00''$). The velocity resolution is 0.08 km s^{-1}, and the integration time was 8 hours. The beamwidth was 2.1 and the beam efficiency 0.5. The profile determined from a hyperfine analysis is shown as a solid line.

Aa. SANDQVIST AND C. BERNES

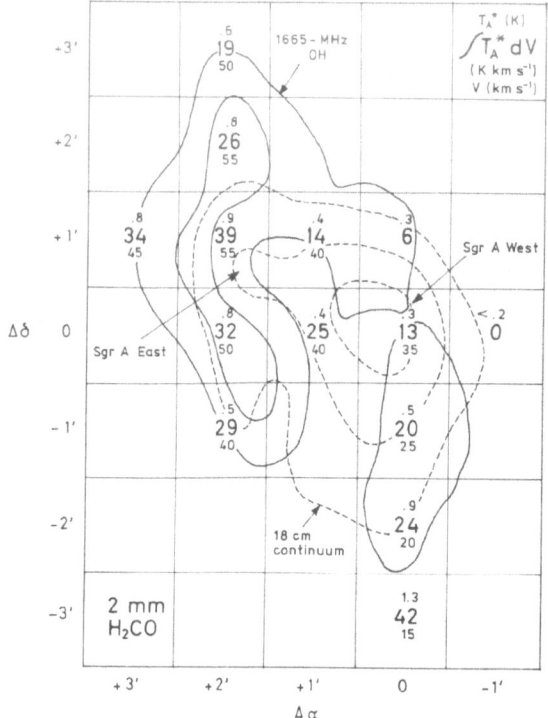

Fig. 3. 2-mm H_2CO line
emission from the Sgr A
+40 km s^{-1} molecular cloud.
Values of T_A^* (K), $\int_{line} T_A^* dV$
(K km s^{-1}) and V(km s^{-1}) are
given at each observed point.
The origin is at $\alpha(1950.0)=$
$17^h42^m29\overset{s}{.}3$, $\delta(1950.0)=$
$-28°59'18''$. The solid line
contours represent the OH
distribution and the dashed
contours the 18-cm continuum
(Sandqvist 1974).

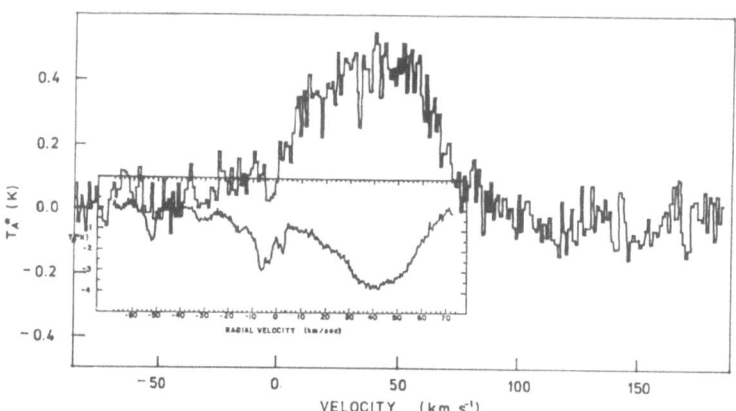

Fig. 4. The averaged profile of all the 2-mm H_2CO observations in
fig. 3. The velocity resolution is 1.1 km s^{-1}. A 6-cm H_2CO absorption
profile (Sandqvist 1970) observed towards Sgr A with a velocity resolution of 0.5 km s^{-1} is inserted for the sake of comparison.

is represented by the distribution of column density divided by excita-
tation temperature for the 1665-MHz OH line. The line profile has been
integrated from -42 to $+102$ km s^{-1}, excluding the -30 and 0 km s^{-1}
features.

The averaged profile of all the 2-mm H$_2$CO observations in fig. 3 is
shown in fig. 4, and can be thought of as having been obtained with an
approximately 3' x 6' beam. There is a remarkable similarity between
this 2-mm H$_2$CO emission profile and the 6-cm H$_2$CO absorption line profile
observed towards Sgr A with a beamwidth of 6.6 (Sandqvist 1970, reproduced in fig. 4), which may be an indication that most of the continuum
sources lie behind the molecular region. The general velocity gradient
across the molecular complex, discovered through the lunar occultation
observations, appears to be confirmed by the 2-mm observations. The
decrease in H$_2$CO column density towards Sgr A West (as indicated by the
decrease of the integrated 2-mm values in fig. 3) still leaves it unde-
cided whether the molecular region is on the near or far side of the
galactic nucleus. It is hoped that the complete mapping of this region
and the above-mentioned analysis may resolve this problem.

REFERENCES

Bernes, C.: 1978, Stockholm Observatory Report No. 15.
Bernes, C.: 1979, Astron. Astrophys. 73, 67.
Downes, D., Wilson, T.L., Bieging, J.: 1976, Astron. Astrophys. 52, 321.
Fukui, Y., Iguchi, T., Kaifu, N., Chikada, Y., Morimoto, M., Nagane, K.,
    Miyazawa, K., Miyaji, T.: 1977, Publ. Astron. Soc. Jpn. 29, 643.
Liszt, H.S., Sanders, R.H., Burton, W.B.: 1975, Astrophys. J. 198, 537.
Sandqvist, Aa.: 1970, Astron. J. 75, 135.
Sandqvist, Aa.: 1974, Astron. Astrophys. 33, 413.
Sandqvist, Aa., Lindroos, K.P.: 1976, Astron. Astrophys. 53, 179.
Strom, K.M., Strom, S.E., Vrba, F.J.: 1976, Astron. J. 81, 320.
Thaddeus, P., Wilson, R.W., Kutner, M., Penzias, A.A., Jefferts, K.B.:
    1971, Astrophys. J. (Letters) 168, L59.
Whiteoak, J.B., Rogstad, D.H., Lockhart, I.A.: 1974, Astron. Astrophys.
    36, 245.

OBSERVATIONS OF THE 3.4-mm HCO$^+$ LINE TOWARD THE GALACTIC CENTER

Yasuo Fukui, Norio Kaifu, Masaki Morimoto, and Takeshi Miyaji
Tokyo Astronomical Observatory, Mitaka, Tokyo, Japan

3.4-mm line of HCO$^+$ has been mapped toward the galactic center. The telescope used was the 6-m mm-wave telescope at the Tokyo Astronomical Observatory, equipped with a GaAs Schottky barrier diode mixer and a 256-ch filter bank of 1-MHz resolution. Fig. 1 shows the observed positions on the antenna temperature contour map of HCN(Fukui et al. 1977). Intensive observations were made of the line SR where the HCN emission is strongest. Fig. 2 shows the HCO$^+$ profile as well as the HCN profile, both of which are averages along SR($0°0 \lesssim \ell \lesssim 0°2$).

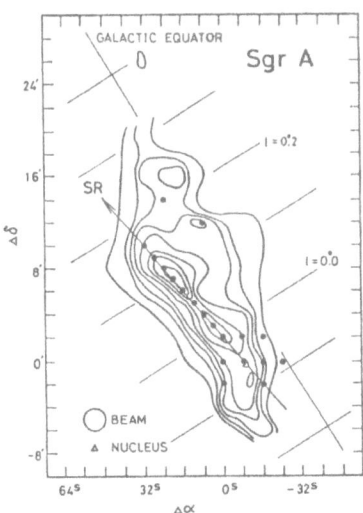

Fig. 1: Observed positions in the 3.4-mm line of HCO$^+$ are shown on the HCN map(Fukui et al. 1977). The origin is R.A. =17h42m40s and Decl.=-28°59' (1950).

The main results can be summarized as follows;
1. As shown in Fig. 2, the HCO$^+$ profile is very broad and asymmetric. The emission ranges at least from -50 km s^{-1} to 110 km s^{-1}.
2. The HCO$^+$ profile has a sharp dip at 0 km s^{-1}. Similar dips were found in CO(Liszt et al. 1975) and HCN(Fukui et al. 1977).
3. The line core of HCO$^+$ resembles well that of HCN, while the line wing found in HCO$^+$ is not recognisable in the HCN profile. The decrease in HCN brightness at the edge of the line core is remarkable.

Fig. 3 shows the correlation of the HCO$^+$ and HCN line shapes. It shows that HCO$^+$ emission is still significant where little HCN emission is detected, which suggests that HCO$^+$ becomes significantly more abundant in the less dense region. This tendency is consistent with theoretical prediction based on the ion-molecule reaction scheme (e.g. Suzuki et al. 1976), and can be interpreted as mainly caused by the increase in the degree of ionization with the decrease of density.

109

B. H. Andrew (ed.), Interstellar Molecules, 109–110.

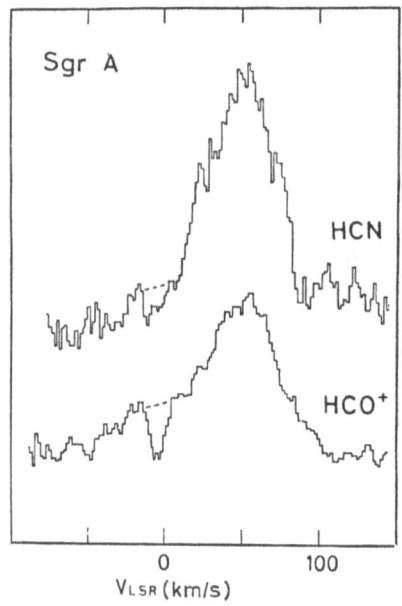

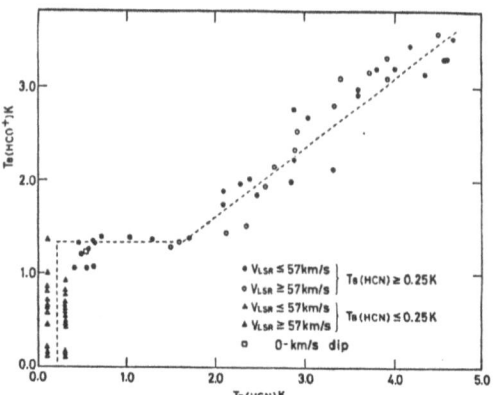

Fig. 2 (left): averaged profiles of HCO^+ and HCN.

Fig. 3 (above): the correlation between HCO^+ and HCN. The profiles in Fig. 2 are sampled at 2 km s^{-1} intervals to provide the data.

Table 1. Molecular abundance in the 0-km s^{-1} cloud

Molecule	Transition(λ mm)	Column density(cm^{-2})	Reference
H^{12}CN	J=1-0(3.4)	$\gtrsim 3 \times 10^{13}$	present work
H^{12}CO$^+$	J=1-0(3.4)	$\gtrsim 3 \times 10^{13}$	present work
^{13}CO	J=1-0(2.7)	$\gtrsim 2 \times 10^{16}$	Liszt et al.(1975)
2H$_2$		3-10×10^{22}*	Scoville and Solomon(1975)

* Estimated from the data of the 2.6-mm ^{12}CO emission.

The 0-km s^{-1} dip is ascribed to self-absorption due to a foreground cold cloud. Table 1 summarizes column densities of HCO^+ and HCN in the 0-km s^{-1} cloud. It is noteworthy that these tri-atomic molecules are still abundant in a diffuse cloud where these lines are not appreciably excited.

This paper is a summary of part of the Doctoral thesis of one of the authors (Y.F.).

References

Fukui,Y.,Iguchi,T.,Kaifu,N.,Chikada,Y.,Morimoto,M.,Nagane,K.,Miyazawa,K., and Miyaji,T.: 1977, Publ. Astron. Soc. Japan 29, 643.
Liszt,H.S.,Sanders,R.H., and Burton,W.B.: 1975, Astrophys. J. 198, 537.
Scoville,N.Z., and Solomon,P.M.: 1975,Astrophys. J. (Letters) 199, L105.
Suzuki,H.,Miki,S.,Sato,K.,Kiguchi,M., and Nakagawa,Y.: 1976, Prog. Theor. Phys. 56, 1111.

CO AND OH IN THE GALACTIC CENTER REGION

Junji Inatani, Nobuharu Ukita,
Department of Astronomy, University of Tokyo
Norio Kaifu,
Tokyo Astronomical Observatory
Shinji Kodaira and Koichi Ishii
Kisarazu Technical College

The two-dimensional distribution of molecular clouds in the galactic center region has been investigated in the CO 115 GHz line and in the OH 1665 and 1667 MHz lines. As the former is an emission line, we can find molecular clouds without the unavoidable bias to continuum sources which is inherent in a survey of OH absorption lines. Because the CO line is usually optically thick, the brightness temperature of the line is directly related to the kinetic temperature of the cloud. On the other hand, the real optical depth of the OH line can be obtained from the intensity ratio between 1665 and 1667 MHz lines (assuming LTE). From this point of view we have compared the CO and OH observational results.

The CO observations were made with the 1.5-meter mm-wave radio telescope of Kisarazu Technical College, which has a half-power beamwidth of $\sim 10'$. During the winter of 1977-78, the region $0°4 \lesssim \ell \lesssim 1°8$ and $|b| \lesssim 0°4$ was surveyed in a grid of $0°2$. The receiver front-end was a room-temperature Schottky-barrier diode mixer with system temperature ~ 4000 K (SSB); a 158-channel 1 MHz filter-bank was used. During the spring of 1979, the region $2°75 \lesssim \ell \lesssim 3°50$ and $|b| \lesssim 0°50$ was surveyed in a grid of $0°25$. At this time a liquid nitrogen cooled front-end was available, and the system temperature was ~ 1200 K (SSB); a 256-channel 1 MHz acousto-optic spectrometer was used as the back-end. The OH observations were made in 1974 with 43-meter telescope of NRAO which has a beamwidth of 18'; the region $-2°8 \lesssim \ell \lesssim 4°0$ and $-50' \lesssim b \lesssim 46'$ was surveyed in a grid of $0°2$. Details of these CO and OH observations are to be published elsewhere.

Here we discuss the four clouds which show the strongest OH absorption in the surveyed region. They are the 80 km/s and −30 km/s clouds located at $\ell = 1°3$, and the 100 km/s and 0 km/s clouds at $\ell = 3°0$. We will designate them as L1.3(80), L1.3(−30), L3.0(100) and L3.0(0).

L1.3(80): The CO distribution (Fig. 1) shows that this is a very massive cloud with a diameter of ~ 140 pc and a hydrogen mass of $\sim 3 \times 10^7$ $M_\odot$ (according to Bania's 1977 formula). The velocity of this cloud changes

B. H. Andrew (ed.), Interstellar Molecules, 111–112.
Copyright © 1980 by the IAU.

from 75-80 km/s at negative latitudes to 100-105 km/s at positive
latitudes.

L1.3(-30): There is no CO counterpart of this strong OH absorption,
but there is an extended plateau of CO emission with velocities from
-50 km/s to 50 km/s.

L3.0(100), L3.0(0): Both the CO emission and the OH absorption of
these clouds show a similar distribution. However their CO brightness
temperatures are considerably lower than those of L1.3(80) despite their
having similar amounts of OH absorption.

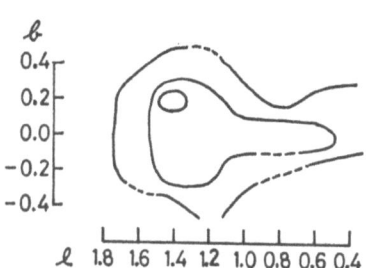

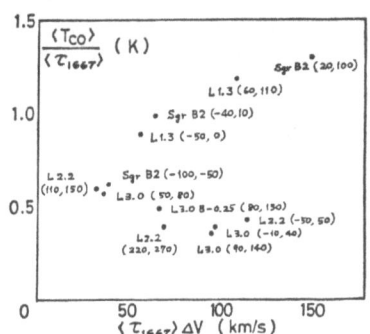

Fig. 1. Spatial distribution of
the integrated CO emission in the
velocity range 42<V<140 km/s.
Contour unit is 100 K·km/s.

Fig. 2. Correlation between the
averaged CO antenna temperature
and the averaged OH real optical
depth.

Fig. 2 shows the variation of the ratio between the CO antenna temperature T_{CO} and the OH real optical depth τ_{1667} from cloud to cloud.
The ordinate indicates the ratio $\langle T_{CO}\rangle / \langle \tau_{1667}\rangle$ and the abscissa
represents the integrated OH optical depth over the velocity width ΔV
in each cloud (given in parenthesis); $\langle\ \rangle$ means a value averaged over
ΔV. The points in Fig. 2 are divided in two groups. L1.3(80) and Sgr B2
form one group, and such clouds as L3.0(100) and L3.0(0) form the other.
The former clouds have a three times greater value of the ratio $\langle T_{CO}\rangle /$
$\langle \tau_{1667}\rangle$ than the latter. This difference may be due to the different
kinetic temperatures of the clouds. The fact that L1.3 and Sgr B2 are
accompanied by HII regions while L3.0 and L2.2 are not (Altenhoff et al.
1978) seems to support this idea.

REFERENCES

Altenhoff, W.J., Downes, D., Pauls, T., and Schraml, J.: 1978, Astron.
      Astrophys. Suppl. 35, p. 23.
Bania, T.M.: 1977, Astrophys. J. 216, p. 381.

IN SEARCH OF ≤ 5 K GALACTIC MOLECULAR GAS

R. H. Rubin*, Neal J. Evans II**, and B. Zuckerman[†]
* California State University, Fullerton.
** Department of Astronomy and Electrical Engineering
  Research Laboratory, the University of Texas at Austin.
[†] University of Maryland.

ABSTRACT

Existing surveys of our galaxy are able to detect CO clouds with an excitation temperature $T_{ex} \gtrsim 5$ K. We have made observations to determine if a substantial molecular component is colder than ~ 5 K. We compared CO emission features with H_2CO absorption against 17 continuum sources near the galactic plane. After elimination of features in the spectra that are probably associated with known H II regions, there were 39 clouds remaining, most of which show an excellent kinematic agreement between H_2CO and CO. The results do <u>not</u> suggest the existence of a large amount of mass in ultra-cold molecular clouds.

INTRODUCTION

Current CO J = 1 → 0 emission surveys are sensitive to molecular clouds of optically thick CO at an excitation temperature $T_{ex} \gtrsim 5$ K (e.g. Gordon and Burton: 1976, Ap. J. <u>208</u>, 346). Extremely cold molecular clouds with $T_{ex} \lesssim 5$ K should show up well in 6 cm absorption spectra against continuum sources, but not in CO emission. The <u>non</u> appearance of CO emission at the position and velocity of formaldehyde absorption would indicate an ultra-cold cloud, since the absence of CO in a region of H_2CO would be unlikely.

OBSERVATIONS

The H_2CO observations made with the 100 m Bonn telescope were kindly provided by Dennis Downes and Tom Wilson, 1978. We used these spectra to select a sample for our CO observations, which were made with the 4.9 m antenna at the Millimeter Wave Observatory (MWO) of the University of Texas at Austin. The use of the MWO telescope for 3 mm CO and the Bonn dish for 6 cm H_2CO gave us a useful combination since the HPBWs of ~ 2!5 were comparable. Observations were first made in

113

B. H. Andrew (ed.), Interstellar Molecules, 113–114.

January 1978, using 128 filters each 250 KHz wide, providing a velocity resolution of 0.65 km s^{-1} over a range of 83 km s^{-1}. Both beam and frequency switching were used, and the reference positions and spectra were checked to be free of lines. Because of poor weather conditions our antenna temperatures were good to only ~ ± 50%. However, additional observations under improved conditions in 1979 of all the weaker (≤ 1.5 K) CO lines, using position switching only, provided calibrations that should be reliable to ± 20%.

RESULTS AND DISCUSSION

Because we intended to look for ultra-cold gas, we separated those clouds that have velocities within ± 10 km s^{-1} of the recombination line velocity, since they presumably are associated with H II regions. The recombination line velocities are from observations of the H110α line kindly provided from the Bonn telescope. The H110α line is the closest H-α line in frequency (and hence resolution) to the 6 cm H_2CO line. The remaining clouds show an excellent velocity agreement between CO and H_2CO, which is strong evidence for the physical association of the two molecular species. In some cases of spectral features associated with the continuum source there is apparent self-absorption. In G9.6+0.2 the most intense H_2CO line occurs at 1.6 km s^{-1}, precisely at a local minimum in the most intense CO emission. This is also close to the H110α line velocity of 3 km s^{-1}. Observations of ^{13}CO would be useful to test whether self-absorption is occurring. Similar situations have been found for G24.8+0.1, V_{LSR} = 108 km s^{-1}, and G34.3+0.14, V_{LSR} = 60 km s^{-1}.

Evans et al. (1979, in preparation) have found lower limits of 4.7 and 4.9 K for T_{ex} for the two coldest features in the study. It is noted that lower temperatures may exist but, in order to ascertain this, better H_2CO data would be needed to allow a more sensitive search program for CO. If these colder clouds do exist, they would correspond to clouds with substantially smaller visual extinctions (A_v ~ 0.3) than those seen in dark clouds. Such ultra-cold clouds would not likely have a sufficient average density to thermalize the CO transition (Snell, 1979, unpublished dissertation, the University of Texas at Austin) which would cause low CO antenna temperatures. Also, clouds with such low extinctions are not likely to contain a significant mass. In a statistical sense, our results do not suggest the existence of a substantial population of cold CO clouds, since none of the definite H_2CO features are lacking a corresponding CO line.

CO (J=2-1) OBSERVATIONS OF SEVERAL GALACTIC HII REGIONS

T. de Graauw[1], S. Lidholm[1], B. Fitton[1], F.P. Israel[2],
A. Sargent[2], T.B.H. Kuiper[3], and H. Nieuwenhuyzen[4]
[1]Astronomy Division, European Space Agency, Noordwijk (NL)
[2]Owens Valley Radio Observatory, C.I.T., Pasadena (USA)
[3]Jet Propulsion Lab., Pasadena (USA)
[4]Astronomical Institute, Utrecht (NL)

We used the new OVRO 10 meter millimeter telescope together with the Estec heterodyne receiver and backend to observe ^{12}CO (J=2-1) emission in the direction of several galactic HII regions. At a frequency of 230 GHz our angular resolution was 30 arcsec and our velocity resolutions were 0.3 and 1.3 km s^{-1}. We observed the cores of 16 molecular clouds associated mainly with compact and subcompact HII regions: S88, S106, S157, S158 (NGC 7538), S159, S187, S228, S255 (IC 2162), S269, ON-1, G45.5-0.1, W1 (center), W58 (K3-50), Cep A and Mon R2. Observations of all sources were in a cross pattern of at least five points. Four sources – Cep A, Mon R2, S158 and W58 – were mapped in more detail. Results for the latter two are shown in Figures 1 and 2.

We also used the OVRO room temperature 115 GHz receiver to obtain ^{12}CO (J=1-0) and ^{13}CO (J=1-0) observations at selected positions with a resolution of 1 arcmin.

Most of the observed line profiles are simple structures having linewidths of about 5 km s^{-1}. Self-absorption features were observed in ON-1, Cep A and Mon R2, and mapped in the latter two sources. Multiple velocity components appear in the line profiles of G45.5-0.1, and possibly S158 and W58 (see Figures 1b and 2b). Considerable line broadening is seen in W58 at the positions of the compact HII region K3-50 and the ultracompact HII region/OH maser source ON-3 (Fig. 1b). The CO map shows these sources to be associated with different CO maxima. The CO cloud near K3-50 and source B is mainly hot, the one near ON-3 is mainly dense.

Although the CO map of S158 (Fig. 2a) shows a smooth distribution extending over several arcminutes around the compact HII regions and maser sources, the line profile broadens noticeably about one arcmin east of the compact HII regions (Fig. 2b).

B. H. Andrew (ed.), Interstellar Molecules, 115–116.

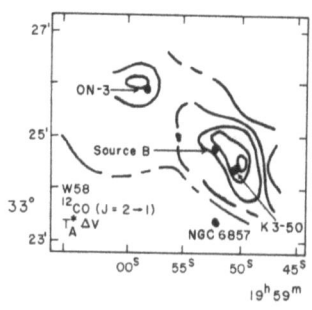

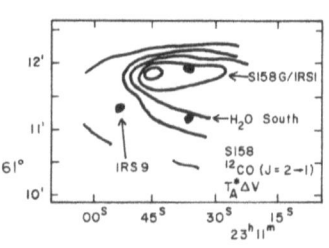

1a   1b

Figure 1. Maps in ^{12}CO (J=2-1) of (a) W58 and (b) S158

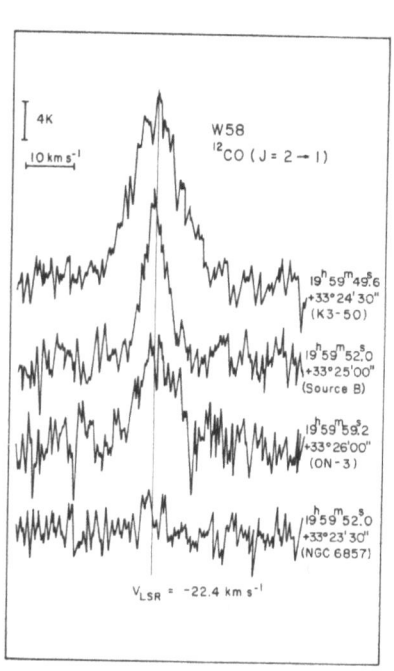

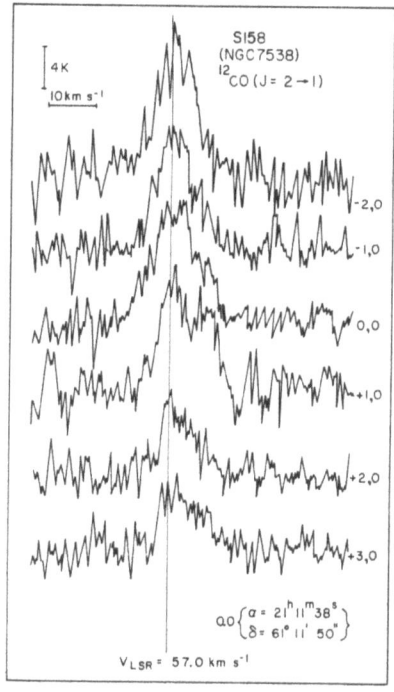

2a   2b

**Figure 2. ^{12}CO (J=2-1) profiles in (a) W58 and (b) S158.
In 2(b) the profiles correspond to positions offset
as indicated from the origin. The offsets are given
in minutes of arc.**

MOLECULAR CLOUD DENSITIES FROM OBSERVATIONS OF CARBON MONOSULFIDE

Richard A. Linke
Bell Laboratories, Holmdel, New Jersey

Paul F. Goldsmith
University of Massachusetts, Amherst, Massachusetts

We have made carefully calibrated measurements of the J=1→0 and J=2→1 transitions of CS in 32 molecular sources in order to obtain density and fractional abundance information from the excitation of this molecule. The antennas used provided beams of nearly equal size. The line intensities, which fall between 1 and 5K, have typically been determined with a (1σ) uncertainty of ~10%. The two transitions show nearly equal intensities and linewidths which are, respectively, about one tenth and one half of the corresponding quantities for CO. Our observations are seen to be in good agreement with the results of excitation calculations involving either a velocity gradient model or a purely micro-turbulent model since in both cases optical depths are small (0.3 to 3.0). Values obtained for n(H$_2$) and X(CS)/dv/dr from the velocity gradient model fall in the range 2x10^4 to 2x10^5 cm^{-3} and 3x10^{-11} to 3x10^{-10} (km s^{-1}pc^{-1})$^{-1}$ respectively.

The accurately calibrated measurement of a number of transitions of a single molecular species provides our only generally applicable method of determining molecular hydrogen densities within molecular clouds. Since carbon monosulfide is a widely observed molecular cloud constituent which has a number of transitions conveniently accessible to millimeter astronomy, we have undertaken a comprehensive study of CS excitation in 32 sources, including both giant molecular clouds and dark clouds. Previous studies of CS (Turner et al 1973; Liszt and Linke 1975; Martin and Barrett 1975) suffered from a number of instrumental limitations and were restricted to a small number of sources.

The data for the CS J=1→0 line at 48991.0 MHz were obtained using a cooled 6mm mixer receiver on the 11m antenna of the National Radio Astronomy Observatory at Kitt Peak with a half power beamwidth of 2.6 arc minutes. The CS (J=2→1) data at 97981.0 MHz were taken at the Bell Laboratories' 7m Antenna in Holmdel, New Jersey with a FWHP beamwidth of 2.1 arc minutes. The receiver used was a cryogenic 60-90 GHz receiver described by Linke, Schneider and Cho (1978). The highest velocity resolution available was 0.2 Kms^{-1} for both

B. H. Andrew (ed.), Interstellar Molecules, 117–121.

transitions. Thus filter dilution, a potential problem only for the narrowest lines, was equal for both transitions. The data are summarized in Table 1. For considerably more detail, including source coordinates, linewidths, velocities and errors as well as line profiles, see Linke and Goldsmith (1980).

TABLE 1   CS CORRECTED ANTENNA TEMPERATURES (K)

SOURCE NAME	J=1→0	J=2→1	SOURCE NAME	J=1→0	J=2→1
W3	0.92	1.41	L134N	1.61	0.82
W3(OH)	2.38	2.71	ρ Oph	2.25	1.65
IC1848	0.56	0.38	G350	2.04	2.27
o PER	1.06	0.66	NGC6334	7.86	8.18
TMC	2.59	1.56	Sgr A	3.47	3.49
"(0,2)	2.47	2.29	Sgr B2	2.92	2.45
Heiles Cl 2	1.70	1.31	W33	5.35	5.32
"(-6,8)	2.52	2.08	M17L	5.56	7.74
Ori A	4.23	7.95	M17A	4.01	4.97
OMC2	2.01	4.24	W43	1.08	0.87
Ori B	4.03	5.53	W49	1.50	1.64
L1622	1.25	0.91	W51	2.51	3.80
B227	0.86	0.67	B335	1.22	0.61
Mon R2	3.34	5.06	DR21	3.24	3.16
S255	2.21	2.87	DR21(OH)	4.41	4.25
NGC2264	4.41	4.24	B361	0.66	0.27
L134	1.04	0.40	NGC7538	3.30	3.96

As demonstrated by Liszt and Linke (1975), the CS sources are significantly beam diluted by a 2.5' beam. In order to determine the sizes of some of our sources we obtained a limited number of maps with a resolution of one arc minute at CS(J=2→1) using the 14m antenna of the Five College Radio Astronomy Observatory. In addition, several of the sources were observed with the 10m antenna of the Owens Valley Radio Observatory at CS(J=4→3) with a 0.6 arc minute beam.

The data presented above include various types of molecular clouds, so we do not expect a single radiative transfer model to accurately describe all of them. Rather, we deal with the problem of radiative transfer only sufficiently accurately to derive characteristic conditions in the regions producing the CS emission. For the small optical depths encountered in CS, line intensities predicted by a microturbulent model (Liszt and Leung 1977) are not very different from results which we obtain from a 10 level large velocity gradient (LVG) model. In view of this fact, as well as the relative computational simplicity of the LVG radiative transfer calculations, we have applied the LVG model to all of our CS data to obtain the results summarized in Figure 1.

The most prominent feature of our results is the relatively narrow range of molecular hydrogen densities which characterize the molecular

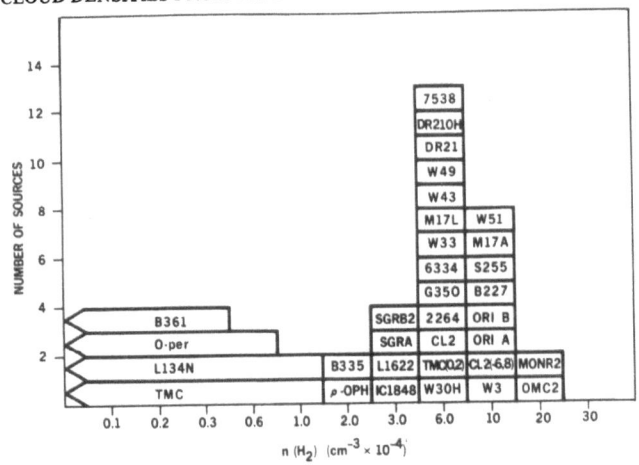

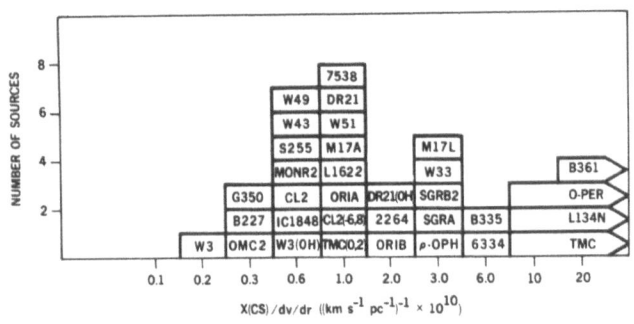

Figure 1   Distribution of molecular hydrogen densities and CS
fractional abundances per unit velocity gradient. Arrows indicate
no unique solution.

clouds. Our CS observations indicate that there is a density
$n(H_2) \simeq 7 \times 10^4$ cm^{-3} characteristic of the central few square arc
minutes of the clouds we have studied. The densities derived from
CS are about 20 times those derived for a number of sources from ^{13}CO
observations (Goldsmith, Plambeck and Chiao 1975; Phillips and Huggins
1977; Plambeck and Williams 1979). Our densities are generally
consistent with those derived from observations of methanol (Gottlieb
et al 1979) and SO (Gottlieb et al 1978) and about one tenth of
those derived from H_2CO observations (Wootten et al 1978). In those
cases where the data are available, there is an inverse correlation
between the spatial extent of the emission from a particular molecule
and the H_2 density derived from the observed line intensities. This
is suggestive of systematic radial density gradients.

The CS abundance per unit velocity gradient shows considerable
source to source variation with most sources having X(CS)/dv/dr
between 3×10^{-11} and 6×10^{-10} (km s^{-1} pc^{-1})$^{-1}$. We can estimate the

velocity gradient and therefore the fractional abundance itself for
those sources for which we have cloud dimensions. These results are
listed in Table 2 together with the total mass implied by the
observations. Half intensity source sizes for Ori-B, M17L and W51
are from Liszt and Linke (1975).

TABLE 2 CS SOURCE PARAMETERS

NAME	SIZE (arc min)	MEAN RADIUS (pc)	X(CS)	H_2 DENSITY (cm^{-3})	TOTAL MASS (m_o)
W3(OH)	1.0x1.6	1.3	$1.8x10^{-10}$	$6x10^4$	$2.5x10^4$
Orion A	1.1x4.4	.35	5.0	10	$8.2x10^2$
Orion B	2x5	.30	9.2	10	$5.1x10^2$
Mon R2	2.5x2.5	.31	4.0	20	$1.1x10^3$
M17L	3x3	1.0	7.5	6	$1.2x10^4$
W51	4x3	4.4	1.8	10	$1.6x10^6$
DR21(OH)	1.5x2.4	1.2	8.0	6	$2.0x10^4$
NGC7538	1.6x2.2	2.7	1.7	6	$2.2x10^5$

We have examined the velocity dependence of the derived quantities
for several sources and find that $n(H_2)$ remains constant while
X(CS)/dv/dr falls off in the line wings. Thus our data are consistent
with either a microtrubulent model or a uniform density LVG model.
In view of the density gradients implied by other observations the
former model may be more appropriate for the CS emission regions.

REFERENCES

Goldsmith, P.F., Plambeck, R.L., and Chiao, R.Y.: 1975, Astrophys. J.
      (Letters) 196, L39.
Gottlieb, C.A., Gottlieb, E.W., Litvak, M.M., Ball, J.A., and Penfield,
      H.: 1978, Astrophys. J. 219, 77.
Gottlieb, C.A., Ball, J.A., Gottlieb, E.W., and Dickinson, D.F.: 1979,
      Astrophys. J. 227, 422.
Linke, R.A., and Goldsmith, P.F.: 1980, Astrophys. J. 235.
Linke, R.A., Schneider, M.V., and Cho, A.Y.: 1978, IEEE, MTT 26, 935.
Liszt, H.S., and Linke, R.A.: 1975, Astrophys. J. 196, 709.
Liszt, H.S., and Leung, C.M.: 1977, Astrophys. J. 218, 396.
Martin, R.N., and Barrett, A.H.: 1975, Astrophys. J. (Letters) 202, L83.
Phillips, T.G., and Huggins, P.J.: 1977, Astrophys. J. 211, 798.
Plambeck, R.L., and Williams, D.R.W.: 1979, Astrophys. J. (Letters)
      227, L43.
Turner, B.E., Zuckerman, B., Palmer, P., and Morris, M.: 1973, Astrophys.
      J. 186, 123.
Wootten, H.A., Evans, N.J. II, Snell, R., and Vanden Bout, P.: 1978,
      Astrophys. J. (Letters) 225, L143.

DISCUSSION FOLLOWING LINKE

T. Wilson: The H_2 density obtained from H_2CO observations of clouds common to the CS survey is $\sim 10^5 cm^{-3}$, only slightly larger than the CS density. Could it be that the CS and H_2CO are preferentially formed at densities of $\sim 10^4$ to $10^5 cm^{-3}$?

Linke: That is very possibly the correct interpretation of the observations. However we must keep in mind that, as Evans pointed out, excitation of CS requires a relatively high density of H_2, so in regions of low density we cannot be sure whether CS is under-abundant, or simply emits at a level below the limit to our sensitivity.

Mouschovias: Finally! I am pleased to make a comment which does not concern magnetic fields. Is not the choice between uniform and non-uniform density models a non-existent problem, in that a non-uniform density distribution is an inevitable consequence of self-gravity? Unless you artificially contrive forces to support the self-gravitating clouds with which you are concerned, a uniform density will violate Newton's Second Law.

Linke: One solution describing the gravitational collapse of a cold spherical cloud of uniform density maintains a uniform density distribution (albeit one that increases with time) and a velocity field of the form $v(r) \propto r$ (c.f. Spitzer, 1968, p. 226). Nevertheless, we conclude that such a model is not appropriate to the clouds we have studied.

Bok: Star counts show that the density of B361 has a steeply negative gradient. A uniform density model is inappropriate.

Glassgold: It was not clear from your description of the observations whether you mapped the clouds or simply made observations at a single position.

Linke: The densities obtained were for the central position (typically the intensity peak) of each cloud. We are currently mapping in the two transitions to obtain the spatial dependence of the derived quantities.

White: Did you assume a constant abundance of CS? One explanation of the apparent conflict could be that (CS/H_2) *decreases* into the cloud as n_{H_2} rises, since the excitation varies approximately as $n_{H_2}^2[CS/H_2]$.

Linke: In the sense that each velocity element is independent of the others in the LVG model, we did not assume any form for the CS abundance. We conclude, however, that $X(CS)$ probably decreases with increasing radius.

Avery: I am concerned about the application of the LVG radiative transfer approximation in some of the very quiescent dark clouds you have considered. Why is this approximation valid for such sources?

Linke: The densities derived from our data using a microturbulent model are not very different from those which we quote. Of course, neither model takes proper account of the foreground absorption which may be present.

CO OBSERVATIONS IN THE SOUTHERN HEMISPHERE

A.R. Gillespie
Max Planck Institut für Radioastronomie, Bonn

ABSTRACT

CO maps of areas around HII regions show many secondary hotspots.
J=2-1 observations of 2 dark clouds are discussed.

INTRODUCTION

Few spectral line observations have been made at high frequencies
at declinations less than -40°. This paper summarises CO observations
with InSb systems, the only others being some by de Graauw (private
communication) using a diode mixer.

THE OBSERVATIONS

The observations at 115 GHz were made with a system from Queen Mary
College, London on the Anglo-Australian-Telescope and at 230 GHz with a
system from the Max Planck Institut on the European Southern Observatory
3.6 m telescope; the beamwidths were 3.2 and 1.5 arcmin respectively,
and the velocity resolutions were 2.6 km s^{-1} at 115 GHz and 1.0, 1.3 or
2.6 km s^{-1} at 230 GHz. The observing techniques are described in
Gillespie et al. (1977 and 1979 respectively).

HII Regions

Spectra were taken of 41 southern Galactic radio sources; the
spatial extent of the CO round the brightest of these was studied by
scanning the source whilst switching the receiver between a line and a
reference frequency. An area of about 1 deg^2 around each source was
completely sampled, usually at several adjacent frequencies. In almost
all the regions mapped there are bright CO peaks near the radio continuum
peak, and also other CO hot-spots embedded in an extensive CO cloud;
examples include RCW38, RCW122, and G316.8-0.1 which have simple radio
continuum structures. Observations of the radio complex near RCW106 at

123

B. H. Andrew (ed.), Interstellar Molecules, 123–124.

115 GHz revealed a giant cloud approximately 100pc by 35pc (at 4.2kpc distance) which is larger than the Orion A complex (Kutner et al. 1977). There is a velocity gradient along it (V_{lsr}=-51 → -55 km s^{-1} North to South) and considerable structure within the cloud, there being four main maxima and an additional peak at the southern end which may be a second cloud with V_{lsr}=-49 km s^{-1}. The correlation of the peaks with the thermal radio sources of Goss and Shaver (1970) suggests a very large rotating cloud containing several well developed HII regions. An area of 1 deg.^{2} centred on the CO peak in Orion was observed at one frequency and shows considerable structure, particularly to the East.

Dark Clouds

The Coalsack is one of the most prominent features of the southern sky, and several (J=1-0) CO spectra were taken at positions in an H_2CO cloud (Brooks et al. 1976) which contains several dust globules. A (J=2-1) CO spectral line at the brightest J=1-0 position (No 4 in Huggins et al. 1977, T_A^*=7.9 K) has T_A^*=10 K and is not significantly resolved at a resolution of 1 km s^{-1}, so that a line width $\lesssim$ 0.5 km s^{-1} is implied. Allowing for resolution effects, the line intensity is consistent with the CO's being optically thick, confirming a mass of several solar masses for this cloud. The more northern dark cloud B68 was observed at 230 GHz; when the data are compared with the 115 GHz work of Martin and Barrett (1978), they imply that the CO is optically thick.

ACKNOWLEDGEMENTS

These sources were observed in collaboration with R.N. Martin of MPI and the authors of Gillespie et al. 1977 and 1979.

REFERENCES

Brooks, J.W., Sinclair, M.W., and Manfield, G.A.: 1976, Mon. Not. R. astr. Soc. 175, pp. 117.
Gillespie, A.R., Huggins, P.J., Sollner, T.C.L.G., Phillips, T.G., Gardner, F.F., and Knowles, S.H.: 1977, Astron. Astrophys. 60, pp. 221.
Gillespie, A.R., White, G.J., and Watt, G.D.: 1979, Mon. Not. R. astr. Soc. 186, pp. 383.
Goss, W.M., and Shaver, P.A.: 1970, Aust. J. Phys., Astrophys. Suppl. No. 14.
Huggins, P.J., Gillespie, A.R., Sollner, T.C.L.G., and Phillips, T.G.: 1977, Astron. Astrophys. 54, pp. 955.
Kutner, M.L., Tucker, K.D., Chin, G., and Thaddeus, P.: 1977, Astrophys. J. 215, pp. 521.
Martin, R.N., and Barrett, A.H.: 1978, Astrophys. J. Suppl. 36, pp. 1.

CO (J=2-1) OBSERVATIONS OF THE CARINA NEBULA AND G 333.6-0.2 AND A
SEARCH FOR CO IN LMC AND SMC

T. de Graauw, S. Lidholm, B. Fitton
Astronomy Division, European Space Agency, Noordwijk (NL)
F.P. Israel
Owens Valley Radio Observatory, California Institute of
Technology, Pasadena (USA)
J. Beckman
Queen Mary College, London (UK)
H. Nieuwenhuyzen, J. Vermue
Astronomical Institute, Utrecht (NL)

We used the ESO 3.6 m telescope at La Silla (Chile) in conjunction
with the Estec heterodyne receiver to observe CO (J=2-1) emission from
southern sources. The telescope HPBW was 2.2 arcmin, the beam efficiency
43 per cent. The receiver used backward-wave oscillators and Schottky
barrier diode mixers. It had a single-sideband noise temperature of
4000 K at 230 GHz. The backend was a 256 channel filterbank of 1 MHz
(1.3 km s^{-1}) bandwidth per channel. Calibration was done in the usual
way.

We mapped the central region of the Carina nebula, and a chain of
sources around longitude 333°, and searched for emission from the
Magellanic Clouds. The CO map of the Carina nebula (Fig. 1) shows two
distinct sources. The northern source consists of a compact component,
not related to any other feature, and an extended, ridgelike component
that follows the ionization front in the optical nebula Car I. The southern source consists of two components in a common envelope. It coincides
with the dark bay southeast of Car II.

From a comparison with OH observations (Dickel and Wall, 1974),
radio continuum observations (Huchtmeier, 1975) and far-IR observations
(Harvey et al., 1979) we deduce that the optical nebulae Car I and Car II
are enveloped by a molecular cloud. The southern CO source is located
in front of the nebula, the southwestern part of the northern source is
situated to the side of Car I, and the remaining part of the northern
source is behind Car I. The far-IR source in Car II is an isolated
object, but the far-IR source in Car I is situated just at the interface
of the ionized region and the molecular cloud.

Around longitude 333° three CO clouds were found to be associated
with seven HII regions (Fig. 2). The appearance of the CO clouds is that

B. H. Andrew (ed.), Interstellar Molecules, 125–126.

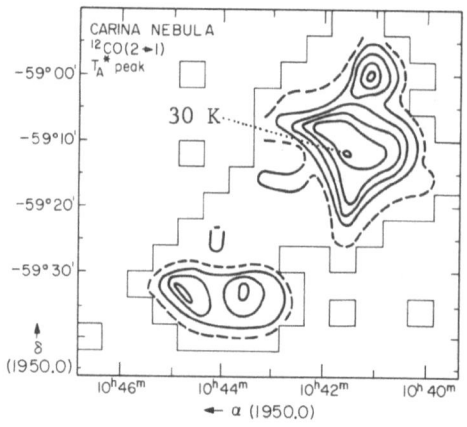

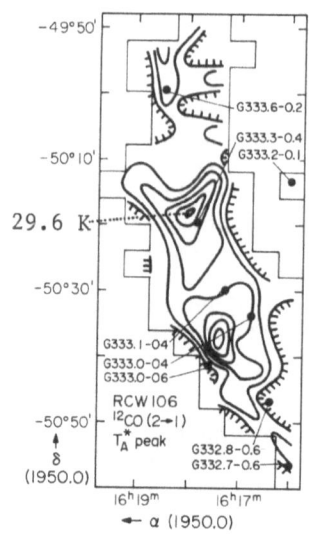

Contour intervals are 2.3 K

Figure 1. CO map of Carina Nebula

Figure 2. CO map of Southern Milky Way around $\ell = 333°$

of a chain of CO maxima in a large, very elongated cloud complex. In this respect, the G 333 CO complex resembles CO cloud complexes observed from the northern hemisphere, such as Orion and M17.

We observed five positions in the Small Magellanic Cloud and seven in the Large Magellanic Cloud, but failed to detect a signal stronger than $T_A^* = 6$ K (three times r.m.s. noise). However, a weak signal may be present at the positions of N90 (SMC), 30 Dor (LMC) and N159 (LMC).

REFERENCES

Dickel, H.R., and Wall, J.V.: 1974, Astron. Astrophys. 31, 5.
Huchtmeier, W.K., and Day, G.A.: 1975, Astron. Astrophys. 41, 153.
Harvey, P.M., Hoffmann, W.F., and Campbell, M.F.: 1979, Astrophys. J. 227, 114.

POPULATION INVERSION AND SUPRATHERMAL EXCITATION IN CARBON MONOXIDE

J. Köppen and W.H. Kegel[*]
Institut für Theoretische Astrophysik
Universität Heidelberg, Germany
[*]Institut für Theoretische Physik,
Universität Frankfurt/Main, Germany

We have investigated under which physical conditions (kinetic temperature, H_2 density, CO column density) the CO molecule shows suprathermal excitation and population inversion. The computations are based on a model in which the excitation is basically due to collisions with H_2 molecules. The collision cross-sections were taken from Green and Thaddeus (1976). The radiative transport in the molecular lines is treated in an on-the-spot approximation (see e.g. Kegel 1979)

$$J = S(1 - e^{-\tau}) + I_{bg}e^{-\tau}$$

in order to keep the computing time low enough to permit the investigation of a wide range of parameters. Our approximation of the radiative transfer is mathematically equivalent to a formalism involving an escape probability $\beta = e^{-\tau}$. This escape probability arises from the fact that the radiative transfer in the line wings is neglected.

For the $J = 1 \to 0$ line we find for $T_{kin} \gtrsim 20$ K suprathermal excitation, and for $T_{kin} \gtrsim 50$ K population inversion for H_2 densities about 10^4 cm^{-3} and not too large column densities. However, only weak masers ($|\tau| < 0.1$) are found.

The calculations also give the parameter range in which the often-used assumption of LTE is justified. For $n_{H_2} \gtrsim 10^4$cm^{-3}, LTE is a good approximation if $I(^{12}CO)/I(^{13}CO) \lesssim 20$. If $I(^{12}CO)/I(^{13}CO) \lesssim 4$, the ^{12}CO is always thermalized.

Most observed clouds seem to have physical conditions which allow the application of LTE analysis. We expect, however, that in clouds of lower density the LTE analysis will overestimate the CO column density by as much as a factor of 100.

REFERENCES

Green, S., Thaddeus, P.: 1976, Astrophys. J. 205, 766.
Kegel, W.H.: 1979, Astron. Astrophys. Suppl. 38, 131.

B. H. Andrew (ed.), Interstellar Molecules, 127.

1.0 mm CONTINUUM OBSERVATIONS OF COOL SOUTHERN CLOUDS

D.Y. Gezari, L. Cheung, and M.G. Hauser
Infrared and Radio Astronomy Branch (Code 693.2)
NASA/Goddard Space Flight Center, Greenbelt, MD. 20771

J.A. Frogel
Cerro Tololo Inter-American Observatory, La Serena, Chile

High surface brightness 1.0 mm continuum emission has been mapped in nine southern hemisphere HII/molecular cloud complexes: NGC 6334, RCW 38, RCW 57, RCW 122, RCW 117, G333.6-0.2, G351.6-1.3, W 33 and W 33A. All of the sources are located in the inner part of the Galaxy near the galactic plane. This paper presents new 1.0 mm continuum mapping results with 65 arc sec resolution. A more detailed discussion of the observations is given by Cheung et al. (1978, 1979).

All known 1 mm continuum dust emission sources are optically thin. The intensity of the 1 mm emission is proportional to the product of temperature and column density of the dust grains in the line of sight (for $T_d > 25K$). Therefore these 1.0 mm continuum observations provide a direct probe of the distribution of matter in dense molecular clouds. The sources presented here have a narrow range of dust density, linear extent, total mass and infrared luminosity. In most cases, the dust clouds are singly peaked and centered about one or more compact near-infrared sources (NGC 6334/I being the notable exception). We find in most sources a fairly steep gradient in the dust distribution around the central peaks consistent with a radial density distribution function of about $\rho(r) \propto r^{-1.5}$ (see Cheung et al. 1979 for a detailed analysis), and comparable to profiles of other known extended 1-mm continuum sources.

The derived average temperature for the region outside the central sub-arc-minute peak may be obtained from a two parameter fit to the observed 40-350 μm to 1 mm flux ratio (Cheung et al. 1978). This quantity is useful since the dust temperature in a centrally heated molecular cloud is expected to fall off slowly with radial distance, $T(r) \propto r^{-0.4}$ (Scoville and Kwan 1976). Not more than one third of the 1.0 mm flux in the central 1 arc min core, or 10% of the total map, can be attributed to free-free emission, as determined by a smooth extrapolation of the radio continuum observations. The derived dust temperatures are generally higher than the observed ^{12}CO antenna temperatures measured with similar beam size (Gillespie et al. 1977), in qualitative agreement with the picture that the molecular gas is collisionally heated by dust grains.

B. H. Andrew (ed.), Interstellar Molecules, 129–132.

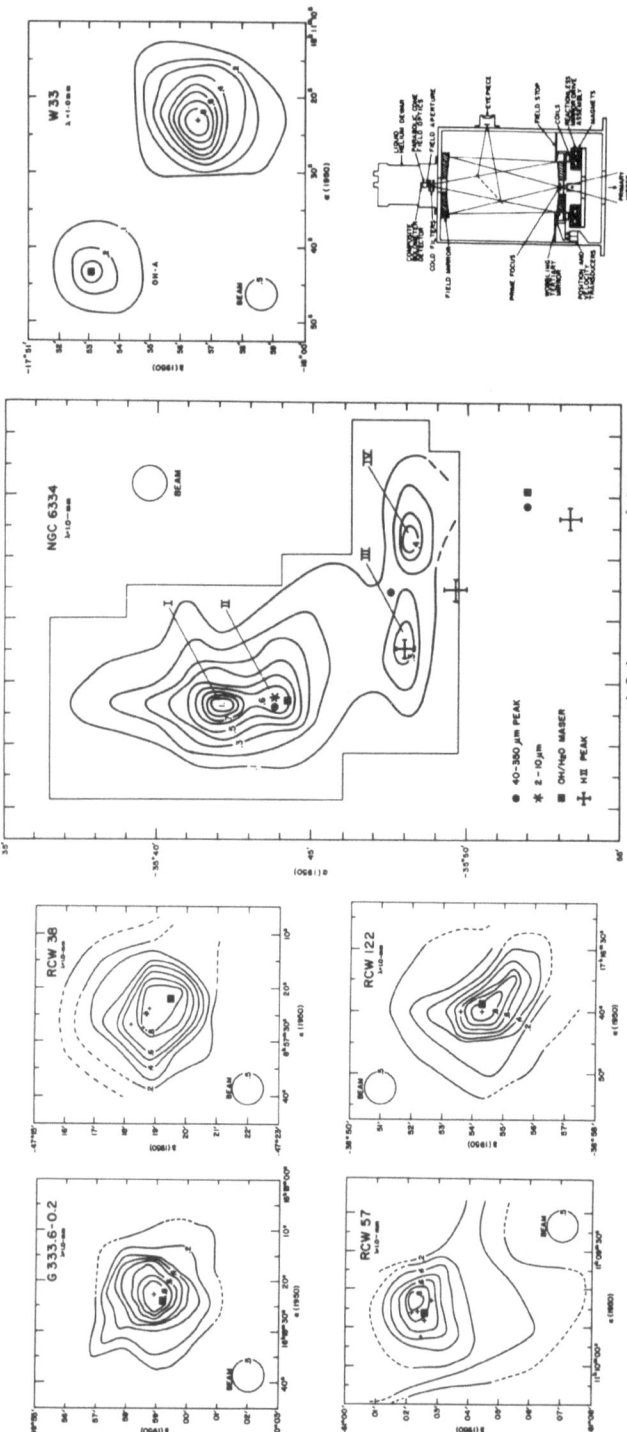

Figures 1-6: 1.0-mm continuum maps of six southern H II/molecular cloud complexes with 65 arc sec (FWHP) resolution. Squares indicate OH/H$_2$O maser sources, small crosses are 2-20μ sources, large crosses are compact H II sources. All six objects are ^{12}CO sources. Each map is normalized to the peak flux density listed in the table. The statistical error is less than 5% of the peak; absolute calibration uncertainty is about 20%.

Figure 7: Remote controlled prime focus photometer (Gezari 1978) used at the CTIO 4-meter telescope. The liquid helium cooled composite bolometer (Hauser and Notarys 1975) was fabricated by our group, and consists of an indium doped germanium thermometer attached to a 2 mm square bismuth coated diamond wafer substrate, with an electrical NEP = 8 x 10^{-15} WHz-1/2 at 1.4K. The square wave driver is essentially vibrationless.

1.0 mm CONTINUUM OBSERVATIONS AND RESULTS

Object	Peak Position		1.0 mm Flux Density		Derived Results				Calculated Cloud Properties	
	$\alpha(1950)$	$\delta(1950)$	Peak§ 65" FWHP (Jy)	Total Map (Jy)	T_{dust}** (K)	D_{dust}§ (gm cm^{-2})	Distance† (kpc)	M_{gas}† ($M_\odot$)	N_{H_2}§ (cm^{-2})	n_{H_2}§ (cm^{-3})
NGC 6334 *	$17^h17^m32.5^s$	$-35°42'00"$	132	2×10^3	25	4×10^{-2}	1.7	4×10^4	1×10^{24}	3×10^6
RCW 38 *	$08^h57^m20.9^s$	$-47°18'50"$	128	2×10^3	30	3×10^{-2}	1.5	3×10^4	8×10^{23}	6×10^5
RCW 57 *	$11^h09^m43.9^s$	$-61°02'09"$	146	1×10^3	-	2×10^{-2}	3.6	9×10^4	8×10^{23}	2×10^5
G333.6-0.2 *	$16^h17^m23.0^s$	$-49°58'54"$	139	7×10^2	35	2×10^{-2}	4.5	1×10^5	6×10^{23}	1×10^5
RCW 117 *	$17^h06^m01.5^s$	$-41°32'20"$	31	-	-	6×10^{-3}	4	-	2×10^{23}	5×10^4
RCW 122 *	$17^h16^m40.1^s$	$-38°54'18"$	53	3×10^2	35	8×10^{-3}	5	5×10^4	2×10^{23}	5×10^4
G351.6-1.3	$17^h25^m53.0^s$	$-36°37'49"$	42	-	-	8×10^{-3}	-	-	2×10^{23}	-
W 33	$18^h11^m18.1^s$	$-17°56'28"$	132	5×10^2	-	2×10^{-2}	4.6	7×10^4	7×10^{23}	2×10^5
W 33A	$18^h11^m43.7^s$	$-17°53'02"$	41	1×10^2	-	8×10^{-3}	4.6	2×10^4	3×10^{23}	6×10^4

* ^{12}CO observations exist for these objects, see Gillespie et al. 1977, Dickel et al. 1977, Scoville and Wannier (1977)

** Dust temperature derived assuming emissivity $\varepsilon_\nu \propto \nu^I$ (Gezari et al. 1973) for those objects with known 100μ fluxes (Furniss et al. 1975).

† Distances from Neckle (1978), Radhakrishnan et al. (1972), Goss et al. (1972).

§ Peak values, in 65 arc sec (FWHP) beam. Values for sources without derived temperatures are calculated assuming $T_{dust} = 40K$.

‡ Total mass within 0.1 contour.

The dust column density D along a line of sight through the dust cloud can be inferred from the relation $D = (4/3)\rho_d\tau/(Q/a)$, where ρ_d is the mass density of dust grains with radius a and 1 mm extinction efficiency Q. τ is the optical depth, which is derived from the observed 1-mm flux density S_ν and the derived grain temperature T_d. To calculate the results in the table we adopt for all the sources the nominal values $Q/a = 1$ cm^{-1} (Aannested 1975), $\rho_d = 1$ gm cm^{-3}, $m_{gas}/m_{dust} = 100$, and spherical source geometry.

The sources W 33A and NGC 6334/I appear to be sites of star formation in its early stages. In contrast to W 33, W 33A is unresolved at 1.0 mm and coincident with an OH emission line source, but shows no radio continuum emission in high resolution interferometric observations (Goss et al. 1978). NGC 6334/I, the strongest peak of 1.0 mm emission, coincides roughly with the extended 40-350μ plateau, but is not associated with any other known compact emission object.

The similarity in mass distribution between these southern clouds and other known 1 mm continuum sources, which differ greatly from our sample in mass, linear extent and infrared luminosity (OMC-1, Sgr B2, W 3, W 49, DR 21, and W 75 - Westbrook et al. 1977) provides further support for the suggestion that extended dust clouds with compact energetic luminosity sources form by a rather general physical process.

REFERENCES

Aannestad, P.A.: 1975, Astrophys. J. 200, 30.
Cheung, L., Frogel, J.A., Gezari, D.Y., and Hauser, M.G.: 1978, Astrophys. J. (Letters) 226, L149.
Cheung, L., Frogel, J.A., Gezari, D.Y., and Hauser, M.G.: 1979, (in preparation).
Dickel, H.R., Dickel, J.R., and Wilson, W.J.: 1977, Astrophys. J. 217, 56.
Furniss, I., Jennings, R.E., and Moorwood, A.F.M.: 1975, Astrophys. J. 202, 400.
Gezari, D.Y.: 1978, (preprint).
Gezari, D.Y., Joyce, R.R., and Simon, M.: 1973, Astrophys. J. (Letters) 179, L67.
Gillespie, A.R., Huggins, P.J., Sollner, T., Phillips, T.G., Gardner, F.F., and Knowles, S.H.: 1977, Astron. Astrophys. 60, 221.
Goss, W.M., Matthews, H.E., and Winnberg, A.: 1978, Astron. Astrophys. 65, 307.
Goss, W.M., Radhakrishnan, V., Brooks, J.W., and Murray, J.D.: 1972, Astrophys. J. Suppl. 24, 123.
Scoville, N.Z., and Kwan, J.: 1976, Astrophys. J. 206, 718.
Scoville, N.Z., and Wannier, P.G.: 1977, Proc. IAU/URSI Symposium (19th COSPAR Plenary) 63, D. Reidel Pub. Co., Holland, 77.
Shaver, P.A., and Goss, W.M.: 1970, Aust. J. Phys. Suppl. 14, 133.
Westbrook, W.E., Werner, M.W., Elias, J.H., Gezari, D.Y., Hauser, M.G., Lo, K.Y., and Neugebauer, G.: 1976, Astrophys. J. 209, 94.

THE EXTINCTION EFFICIENCY OF DUST GRAINS AT 1 MM

W.A. Sherwood, E.M. Arnold, and G.V. Schultz
Max-Planck-Institut für Radioastronomie

We have determined the extinction efficiency factor Q for the objects in the Table by means of maps, scans, or photometry at 1 mm using a composite bolometer at the prime focus of the ESO 3.6m telescope (Arnold et al., 1978; Arnold, 1979). The beam is 2' in diameter.

The flux density F in Janskies (column 2 in Table) is related to the dust temperature T_D and optical depth τ_ν by $F_\nu = B_\nu(T_D)\Omega(1-e^{-\tau\nu})$, where Ω is the beam solid angle. Using the empirical relation found for Sgr B2 by Erickson et al. (1977) between optical depth and frequency, $\tau_\nu = \tau_{\nu_0}(\nu/\nu_0)^m$ with m = 1.5, we have been able to determine dust temperatures (column 3) and optical depths (column 5) by combining 100 μm data from the literature with our 1 mm data appropriately smoothed in resolution.

From the literature we find estimates for the visual extinction (usually converted from infrared values). In the case of Sgr B2 we combined the value from Erickson et al. (1977), $\tau_{100\mu m} = 1.6$, with the observation by Harvey et al. (1979a,b) that $\tau_{.55\mu m}/\tau_{100\mu m} = 100$. In most cases the visual extinction applies to an object at the centre of the dusty region whereas at 1 mm the region is optically thin and we "see" the entire region. Therefore we have increased the value of A_v by a factor of 2 in order to compare relative optical depths with the extinction efficiency factor at 1 mm, Q (column 9). For ρ Oph we adopt a "slab" model that has all the dust in front of the source of heating (Harvey et al., 1979b). $\tau_{1mm} = Q\pi a^2 N_D$ and $A_v = 1.086 Q_v \pi a^2 N_D$, where N_D is the dust column density (cm^{-2}), a is the grain radius 0.15×10^{-4} cm, and $Q_v = 1.5$ is the extinction efficiency in the visual. Q is determined from the above relations; the mean value is 3.1 ± 0.9(m.e.)$\times10^{-4}$. However, errors in T_D, τ_{1mm}, and A_v are large and the true error in Q must be larger than the one derived here. Nevertheless, the mean value of Q agrees well with that predicted by Werner and Salpeter (1969) for a 0.15 μm core-mantle grain.

We note that $<Q>/a = 21\pm6$ cm^{-1}. Using the gas-to-dust mass ratio implied by Bohlin et al. (1978) we find $(1.0\pm0.3) \times 10^{-25}$cm^2 per

B. H. Andrew (ed.), Interstellar Molecules, 133–134.

SOURCE	F_{1mm}	T_{DUST}	REF. T_D	τ_{1mm}	A_v	REF. A_v	A_v/τ_{1mm}	Q_{1mm}
	Jy	K		$\times 10^{-3}$	mag		$\times 10^3$	$\times 10^{-4}$
Sgr A	$\geqslant 30$	50	9	0.9	7	9	7.8	2.1
Sgr B2	463±36	32±4	5	24.6	160	5,13	6.5	2.5
M17/SW	185±27	42±10	11	7.2	20	3	2.8	5.9
267.9-1.1	$\leqslant 101 \pm 3$	69±13	8	$\leqslant 2.2$	21	14	9.5	1.7
287.6-0.6	$\geqslant 52$	55	12	$\geqslant 1.5$	3.0	6	2	8.1
327.3-0.6	94±13	73±15	8	1.9	40	14	21	0.8
333.6-0.2	113±4	84±18	8	2.0	35	14	17.5	0.9
337.9-0.5	52±6	75±15	8	1.0	24	7	24	0.7
ρ Oph	28±4	16.5±1	13	3.7	12.5	10	3.4	4.8

hydrogen atom; i.e. 16 gm cm^{-2}. These values become 2.6×10^{-25} cm^2 and 6 gm cm^{-2} for a gas-to-dust mass ratio of 100:1.

ACKNOWLEDGEMENTS: This work was supported by the Deutsche Forschungs-gemeinschaft, Sonderforschungsbereich 131, Radioastronomie.

REFERENCES

1 Arnold, E.M., Kreysa, E., Schultz, G.V., and Sherwood, W.A.: 1978, Astron. Astrophys. 70, L1.
2 Arnold, E.M.: 1979, PhD Thesis, University of Bonn.
3 Beetz, M., Elsasser, H., and Weinberger, R.: 1974, Astron. Astrophys. 34, 335.
4 Bohlin, R.C., Savage, B.D., and Drake, J.F.: 1978, Astrophys. J. 224, 132.
5 Erickson, E.F., Caroff, L.J., Simpson, J.P., Strecker, D.W., and Goorvitch, D.: 1977, Astrophys. J. 216, 404.
6 Feinstein, A., Marraco, H.G., and Muzzio, J.C.: 1973, Astron. Astrophys. Suppl. 12, 331.
7 Frogel, J.A., Persson, S.E., and Aaronson, M.: 1977, Astrophys. J. 213, 723.
8 Furniss, I., Jennings, R.E., and Moorwood, A.F.M.: 1974, HII Regions and the Galactic Centre, Eslab Symp. No. 8, ed. A.F.M. Moorwood, p.61.
9 Gatley, I., Becklin, E.E., Werner, M.W., and Wynn-Williams, C.G.: 1977, Astrophys. J. 216, 277.
10 Grasdalen, G.L., Strom, K.M., and Strom, S.E.: 1973, Astrophys. J. 184, L53.
11 Harper, D.A., Low, F.J., Rieke, G.H., and Thronson, H.A.: 1976, Astrophys. J. 205, 136.
12 Harvey, P.M., Hoffmann, W.F., and Campbell, M.F.: 1979a, Astrophys. J. 227, 114.
13 Harvey, P.M., Campbell, M.F., and Hoffmann, W.F.: 1979b, Astrophys. J. 228, 445.
14 Persson, S.E., Frogel, J.A., and Aaronson, M.: 1976, Astrophys. J. 208, 753.
15 Werner, M.W., and Salpeter, E.E.: 1969, Mon. Not. R. Astr. Soc. 145, 249.

COMPARISON OF SUBMILLIMETER AND CO BRIGHTNESS IN ORION AND MON R2

David Cudaback and Lawrence Anderson
Radio Astronomy Laboratory, University of California, Berkeley
and
David Lynch and James Smith
Department of Physics, California Institute of Technology

Emission from interstellar dust in L 1641 and Mon R2 was mapped with three arcminute resolution at 400 microns wavelength from White Mountain, California at 3.9 km elevation. Dust brightness distributions and ^{13}CO antenna temperatures are shown below with equal linear scales, emphasizing that the core of Mon R2 is considerably larger and more massive than the core of the Orion region containing OMC-1 and OMC-2. Dust and ^{13}CO brightness distributions for each object are similar, but some of the differences may be real. The differences become accentuated when mass distributions are compared.

Optical depth is determined from 400 micron brightness using roughly estimated temperatures. Temperatures in Orion are estimated from submillimeter colors and temperatures in Mon R2 are estimated from luminosities of previously known sources of heating. Ratios of optical depth to ^{13}CO column density are calculated. The average of these ratios is about 8×10^{-20} square cm for OMC-2 and the eastern part of Mon R2, and is about 30×10^{-20} for the western part of Mon R2. The former is below what would be expected from the distribution over many objects found by Righini-Cohen and Simon (1977), and the latter is above that. This rough procedure for estimating dust optical depth leads to an apparent dust peak in Mon R2 five arcminutes west of the brightness peak, while the CO peak is at the same location as the submillimeter peak. This could be due to the reduction procedure, or it could be due to a real difference in locations of dust peaks and stars as well as a difference in dust and CO distributions. An undetected source of heating would account for the higher brightness of the dust to the west, although this is unlikely as it does not appear in the CO brightness.

These observations support the expectation of dust and CO being closely related, although some deviations in the dust to CO ratio may be visible. Improved measurements of dust color temperature will enable better analysis of these deviations.

REFERENCES
Kutner, M.L., Evans, N.J., II, and Tucker, K.C.: 1976, Ap.J. 209, 452.
Loren, R. B.: 1977, Ap.J. 215, 129.
Righini-Cohen, G., and Simon, M.: 1977, Ap.J. 213, 390.

135

B. H. Andrew (ed.), Interstellar Molecules, 135–136.

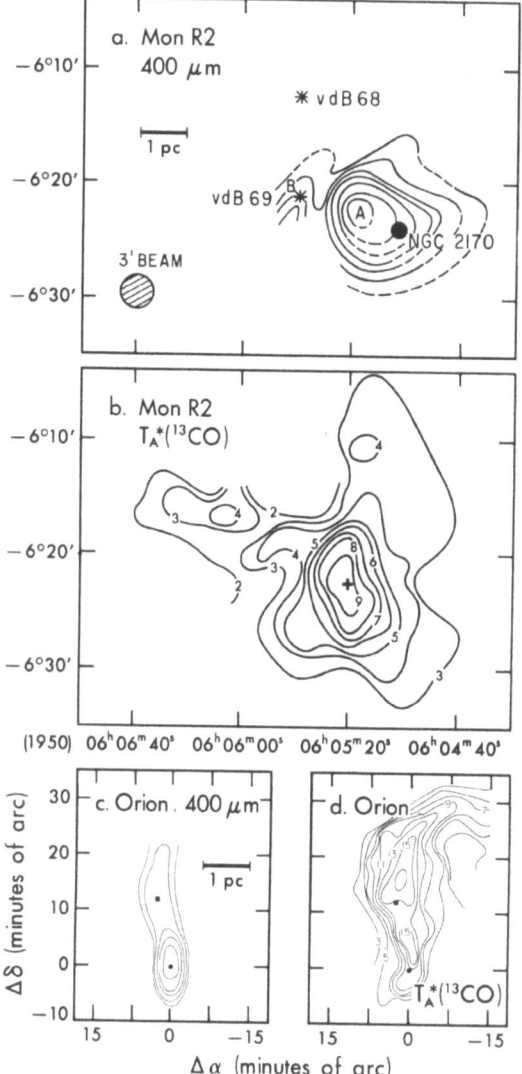

Mon R2, 400 micron contour
intervals are 325 Jy in
3 arcminute beam, objects
A and B are known sources
of infrared heating.
Cross (+) in CO map is at
position of A.   vdB 69 is
assumed to be heated by B.

Orion, L 1641, 400 micron
contour intervals are
logarithmic at 200, 400,
800, 1600 and 3200 Jy in
3 arcminute beam. Solid
circle and square are at
locations of centers of
OMC-1 and OMC-2 respec-
tively. Coordinates are
relative to peak of OMC-1.

Figure 1.   400 micron dust brightness and ^{13}CO antenna temperatures.
400 micron observations do not cover entire area of CO observations.
Dust and CO measurements were made with 3.0 and 2.6 arcminute beamwidths
respectively.   CO data are reproduced by permission of authors and The
Astrophysical Journal.   Orion CO is from Kutner, Evans, and Tucker (1976)
and Mon R2 CO is from Loren (1977).   CO temperatures are corrected for
beam efficiency and atmospheric absorption.

THE EVOLUTION OF GIANT MOLECULAR CLOUDS

Colin Norman
Huygens Laboratory, University of Leiden
and
Joseph Silk
Department of Astronomy, University of California, Berkeley

ABSTRACT

We discuss the origin, lifetime, destruction, spatial distribution and relation to star formation of giant molecular clouds. A coagulation model including the effects of spiral density wave shocks is described. We explore implications for CO observations of external galaxies. The collective effects of OB star winds and supernova remnants in disrupting clouds are considered.

I. INTRODUCTION

Our galaxy contains $\sim 2 \times 10^9$ $M_\odot$ in molecular clouds that have been found in recent CO surveys. Many of these clouds are associated with large molecular complexes and OB associations. A considerable fraction of molecular gas appears to be in complexes with masses of $10^5 - 10^6$ $M_\odot$ (Burton and Gordon 1978; Solomon, Sanders and Scoville 1979). Currently unresolved issues are the lifetimes of these giant molecular complexes, their origin, and destruction, and their relation to ongoing star formation.

In an earlier paper, we have discussed the stability, structure and disruption of molecular clouds in relation to low mass star formation (Norman and Silk 1980). Our goal here is to examine the formation and evolution of the giant molecular complexes or clouds (denoted by GMC) in an attempt to understand many of their physical characteristics and their relation to OB associations.

Strong arguments have recently been presented by Kwan (1979) and Scoville and Hersh (1979) that GMC are built up by the coagulation of many smaller clouds. Further, Solomon and co-workers (Solomon et al. 1979, Solomon and Sanders 1979) have shown that GMC particularly within 4 - 8 kpc of the galactic centre cannot have formed by sweeping up diffuse H I material. This conclusion rests on the value of H_2/H I in this region that is inferred from ^{12}CO and ^{13}CO surveys. A major source of uncertainty in this result lies in the adopted $^{13}CO/H_2$ ratio of 1×10^{-6}. CO appears to become depleted in cloud cores, the correlation

137

B. H. Andrew (ed.), Interstellar Molecules, 137–149.

between A_v and ^{13}CO breaking down at $A_v > 5$. This means that Dickman's (1978) value of $^{13}CO/H_2 = 2 \times 10^{-6}$ will underestimate the H_2 mass by a factor $\gtrsim 2$. We therefore assume that the molecular mass estimated by Solomon et al. (1979) is reliable to within a factor ~ 2.

The coagulation models for GMC indicate long formation times ($\gtrsim 2 \times 10^8$ yr). This is comparable to the rotation time, and the effects of density wave shocks should therefore be included. We show below that inclusion of spiral density wave shocks significantly modifies coagulation models of GMC, and leads to a shorter time scale for GMC formation. The basis of our model is that the ambient interstellar medium can be treated as a fluid on scales greater than or of the order of the mean free path for cloud-cloud collisions, and the existence of a shock is therefore independent of the current controversy about the existence of a hot phase in the interstellar medium.

We argue that the GMC formation rate may be boosted by the magnetic Rayleigh-Taylor instability by a factor ~ 3 (Mouschovias, Shu, and Woodward 1975). Large scale gravitational instability may also play a role in enhancing the GMC formation rate. However, we emphasize that the *coagulation process is inevitable* and suffices to account for the observed properties of GMC whether or not these mechanisms are operative. We point out an observational test of the role of the magnetic Rayleigh Taylor instability in forming GMC.

We present arguments that indicate GMC lifetimes of $\sim 4 \times 10^7$ yr or at least $\leq 10^8$ yr. The mean lifetime of H I at 4 kpc must also be of order $\sim 10^8$ yr, implying that GMC disruption occurs with $\sim 90\%$ of the matter remaining in molecular form. It must however be dispersed into small molecular clouds, which are the building blocks for future GMC upon passage through the spiral arms.

The distribution of OB stars in our Galaxy and other spirals can now be explained if we assume that GMC disruption is triggered by the formation of a giant OB association over the indicated time scale of $\gtrsim 4 \times 10^7$ yr (Bash 1979). The long cloud lifetimes of $\sim 10^8$ yr can be explained if continuing non-disruptive star formation is occurring. For cloud masses of $\sim 10^3 - 10^4$ $M_\odot$ we might expect predominantly less massive O and B stars to be produced. Thus a key feature of our model is the formation of small OB associations throughout the arm and interarm regions, with giant OB associations forming predominantly near the arms.

It is the collective effect of simultaneous formation and evolution of several massive stars that results in GMC disruption.

In the solar vicinity, a substantial fraction of the molecular cloud material will be in small clouds, and we are therefore able to reconcile our star formation model with the observed widespread distribution of OB stars in arm and interarm regions (Mezger 1978).

II. OBSERVATIONAL CHARACTERISTICS

A typical GMC has mass $\sim 2 \times 10^5$ $M_\odot$, maximum dimension ~ 100 pc, projected surface area ~ 2000 pc^2, volume $\sim 10^5$ pc^3 and mean density $\sim 50 - 100$ cm^3 (Blitz 1978). The mass of the molecular material appears to be the largest fraction of mass present, exceeding the H I mass and greatly exceeding the mass in stars and ionized gas. The CO appears to be a good tracer of the sites of active star formation, and there are some 4,000 molecular complexes and OB associations in the Galaxy according to Blitz (1978).

Scoville (1979) has argued that the GMC populate both arm and interarm regions, and Burton and Gordon (1978) estimate that typical cloud sizes lie between 3 and 30 pc. In general, these clouds appear to be grouped into molecular complexes, which show considerable structure, characteristic clump scales being $\sim$ 10 pc with masses $\sim$ 10^3 $M_\odot$ (Blitz 1978).

The association of many GMC with OB associations suggests that their lifetimes may not exceed those of the OB associations. Bash (1979) has developed a kinematical model for the CO distribution which indicates a lifetime of $\sim$ 4 x 10^7 yr on the assumption that the molecular clouds acquire a significant non-circular velocity component in the spiral density wave shock and subsequently move in ballistic orbits. This only constrains the post-shock lifetime of the clouds which could have existed prior to entering the shock.

General constraints on the lifetimes of complexes and clouds come from consideration of star formation efficiency. The present mean rate of star formation is about 3 $M_\odot$ yr^{-1} (Tinsley 1976). Studies of cold molecular clouds indicate a star formation efficiency $\xi \sim$ 10% (Cohen and Kuhi 1979), implying that molecular clouds must be formed and disrupted at an overall rate of 30 $(0.1/\xi)$ $M_\odot$ yr^{-1}. If the total mass of molecular clouds is $\sim$ 2 x 10^9 $M_\odot$ with an additional $\sim$ 2 x 10^9 $M_\odot$ in more diffuse gas, then the mean lifetime of these clouds must be $\lesssim$ $10^8(\xi/0.1)$ yr.

Let us assume that GMC disruption produces predominantly small molecular clouds with a fraction η of the gas dispersed into H I. Within 4 - 6 kpc the ratio of H_2 to H I mass is $\sim$ 10 to 1; a lower bound may be taken to be 5 to 1 (§ I). In a steady state, the mean lifetime for the H I gas must therefore be $\lesssim$ 2 x 10^7 η^{-1} $(\xi/0.1)$ yr. The most plausible mechanism for converting H I to H_2 is associated with passage through the spiral arm, which occurs every $\sim$ 5 x 10^7 yr (at 4 kpc). This therefore provides an estimate of the H I lifetime, and we find $\eta \lesssim$ 0.4 $(\xi/0.1)$. We discuss in § IV a more detailed model for GMC disruption.

III. FORMATION OF MOLECULAR CLOUD COMPLEXES

Oort (1954) originally suggested that cloud growth by coalescence led to gravitational instability and ensuing star formation, and subsequent papers examined this idea in considerable detail (Field and Saslaw 1965, Penston et al. 1969, Field and Hutchins 1968, Taff and Savedoff 1972, 1973). These studies were all of interstellar H I clouds. Formation of giant molecular cloud complexes by a similar aggregation mechanism has been proposed by Kwan (1979) and by Scoville and Hersh (1979). In what follows, we shall describe a simplified model for massive cloud coagulation in which the role of spiral density waves is explored.

a) Kinetic Equation For Cloud Growth

To illustrate the physics of massive cloud growth, we assume that most of the molecular cloud material is in small molecular clouds with mean mass density ρ_{cl}, and that more massive clouds grow predominantly by coalescing with the smaller clouds. We wish to explore the effect of the spiral density wave on cloud growth. In the case of a shocked diffuse intercloud medium, (Shu et al. 1972) find that the density increase amounts to a factor ~ 10. This model may not be relevant if the intercloud medium is pervaded by the hot interiors of old supernova remnants (Cox and Smith 1974; McKee and Ostriker 1977), and a shock will not develop in the hot ($\sim 10^6$ K) component. However, provided that the mean free path for cloud-cloud collisions is sufficiently short, the clouds will simulate the behaviour of a fluid and experience a similar increase in overall density and in collision rate (Shu 1978). Observational evidence for a density enhancement in the H I distribution amounting to a factor ~ 10 in the spiral density wave shock has been found in studies of external galaxies (van der Kruit and Allen 1978). The mean free path for cloud collisions, ℓ, can be inferred from the observed frequency of H I clouds of about 4 per kpc, and consequently $\ell \sim 250$ pc. We can therefore consider the molecular cloud complexes (mean mass M_g and cross-section σ_g) to be dynamically coupled in the spiral density wave via interactions with smaller clouds (mean mass M_s, and cross section σ_s) provided that the spiral density shock width exceeds

$$(\frac{\sigma_s}{\sigma_g})(\frac{M_g}{M_s})(\frac{\ell}{\zeta}) = 600(\frac{10}{\zeta})(\frac{1600 \ pc^2}{\sigma_g})(\frac{\sigma_s}{75 \ pc^2})(\frac{400 \ M_\odot}{M_s})(\frac{M_g}{2 \times 10^5 M_\odot})pc$$

where ζ is the density contrast in cloud material between arm and interarm regions.

If the dominant growth mechanism of molecular cloud complexes is by coagulation with smaller clouds of mean density $\rho_{cl}(t)$, where the time dependence is determined by the dissipative hydrodynamics of spiral density wave theory, the kinetic equation for growth of massive clouds with number density $N(m,t)$ in the mass range $(m, m + dm)$ can be written in the form

$$\frac{\partial N(m,t)}{\partial t} = - \rho_{cl}(t) \frac{\partial}{\partial m} \{N(m,t) \ \sigma(m) \ v_{cl}(m)\} \tag{1}$$

where $\sigma(m)$ is the cross section of a cloud of mass m, and $v_{cl}(m)$ is the mean collision velocity between low mass clouds and clouds of mass m. Note that we have not included a destruction term in (1) since we shall argue later that the destruction mechanism is triggered by an external effect, namely the influence of density waves. The solution of equation (1) is, with subscript o denoting reference quantities,

$$N(m,t) = \int_{-\infty}^{\infty} ds \ A(s)(\frac{\sigma_o v_o}{\sigma v}) \ exp\{s\{\int_o^t \sigma_o v_o \rho_{cl}(t)dt - \int(\frac{m}{\sigma v})dm\}\} \tag{2}$$

where $A(s)$ is chosen to fit suitable initial conditions, namely

$$N(m,o) = \int_{-\infty}^{\infty} ds \; A(s) \left(\frac{\sigma_o v_o}{\sigma v}\right) \exp\left\{ s \int^m \frac{\sigma_o v_o}{\sigma v} \, dm \right\} .$$

Molecular clouds complexes will build up by coagulation of smaller clouds over a time scale $\sim 2 \times 10^8 - 10^9$ yr (Kwan 1979; Scoville and Hersh 1979). We introduce an additional simplification by assuming that the massive clouds move either in a uniform density interarm region or a uniform density arm region. Then, in the interarm region, denoted by subscript i, we have

$$N(m,t) \sim \exp \left(\beta \, \rho_{cl,i} \, \sigma \, v_{cl,i} \, t/\bar{m} \right) ,$$

and in the arm region, denoted by subscript a,

$$N(m,t) \sim \exp \left(\beta \, \rho_{cl,a} \, \sigma \, v_{cl,a} \, t/\bar{m} \right)$$

where β is a coefficient of order unity which can be determined by the complete solution to (2), and $\bar{m}$ is the mean mass of a small cloud.

b) Cloud Coalescence in Arm and Interarm Regions

The ratio of growth rates in arm and interarm regions is $\sim (\rho_{cl,a}/\rho_{cl,i})(v_{cl,a}/v_{cl,i})$. In order for greater growth to be achieved in arms where the cloud spends only ~ 0.1 of a rotation period, we evidently require that the shock strength

$$\frac{\rho_{cl,a}}{\rho_{cl,i}} \gtrsim 10 \; \left(\frac{v_{cl,i}}{v_{cl,a}}\right) \left(\frac{t_i/t_a}{10}\right) \tag{3}$$

where t_i and t_a are the times spent by a cloud in the interarm and arm regions respectively. It appears from equation (3) that the density increase in the arms may only be marginally greater than the total growth in the interarm region in our galaxy, where clouds will experience comparable amounts of growth. Thus coalescence in our galaxy may give only weak correlation of GMC with spiral arms. However, because the relative growth in the arm region is sensitive to shock strength, it is clear that CO observations of external galaxies may give rather different structures. For example M81 and M31 are good candidates for a strong shock, so we expect the GMC distribution to be in the arms. NGC 157 with its weaker and more open arms may have a less marked concentration of GMC in the arms. For filamentary-armed galaxies such as NGC 2841, we expect the GMC distribution to be spread over the entire face of the galaxy disk. An important consideration in selecting galaxies for CO study may be the manner in which the spiral structure is generated (Kormendy and Norman 1979). In particular, theoretical models of barred galaxies give rather strong shock strengths, ~ 20, and thus we may expect a good GMC correlation with the arms.

c) Parker Instability

The magnetic Rayleigh-Taylor instability in the interstellar medium originally proposed by Parker (1967) may lead to a significant enhancement

of the cloud coalescence rate as a consequence of the passage of the
spiral density wave shock. This instability in a uniform medium produces
a density contrast of about a factor of three (Mouschovias 1974), and
may enhance the growth rates (3) by this factor. This could lead to a
significant enhancement of cloud coalescence, and formation of GMC
preferentially in the spiral arms (Mouschovias et al. 1974).

We note here one unique characteristic of cloud coalescence enhanced
by the Parker stability. One would expect both odd and even modes to
develop, leading to a significant velocity dispersion and greater scale
height for the GMC before final equilibrium is attained. Since the time
scale to attain an equilibrium state is comparable to the lifetime of
the complexes after passage through a shock, we would expect many observed complexes to exhibit a significant center-of-mass velocity dispersion. In the absence of significant cloud acceleration, both by shocks
and Parker instability, the GMC that form by coagulation of smaller
clouds attain approximate equipartition of random kinetic energy of their
center-of-mass motions (Penston et al. 1969). The resulting scale height
for the complexes would therefore be relatively small. However acceleration in spiral shocks is inevitable, and will lead to an enhanced
velocity dispersion.

d) Triggering of Star Formation

A key unsolved problem in theories of GMC is the triggering
mechanism by which star formation is initiated in the spiral density
wave. We have argued that cloud collisions and coalescence, possibly
coupled to the Parker instability, could lead to the formation of GMC
preferentially in the spiral arms. However in galaxies with weak spiral
shocks only a moderate contrast in cloud masses can be attained between
arm and interarm regions, and initiation of gravitational instability
will probably not provide an adequate trigger for star formation. In
such systems, a more plausible trigger could be a pressure increase in
diffuse H I clouds in the spiral density wave, initiated either by the
passage of a shock through a warm intercloud medium or by repeated
collisions with other diffuse H I clouds (Shu 1978). This could lead to
the initiation of gravitational instability, fragmentation, and ensuing
OB star formation.

e) Cloud Orbits

If a cloud on entering the spiral density wave sweeps up its own
mass either from the ambient medium or by collisions with other clouds,
it will be strongly momentum-coupled to the swept-up material. Consequently the velocity field inferred from gas dynamical calculations
also applies to clouds in the shocked region. If the cloud is self-
gravitating, it will leave the spiral density wave on a ballistic
trajectory, having acquired a significant non-circular component of
velocity. In this manner the initial conditions required in the model
of Bash (1979) can readily be understood, and it follows that the cloud
subsequently survives for $\sim 4 \times 10^7$ yr before being disrupted by OB
star formation (§ IV).

IV. DISRUPTION OF CLOUD COMPLEXES

The correlation of molecular cloud complexes with OB associations, coupled with the kinematical arguments of Bash (1979), strongly suggests that the lifetimes of cloud complexes are determined by the ages of their OB associations. We emphasize again that molecular clouds may exist in the interarm regions for substantially longer periods; however once prolific massive star formation is triggered, the complexes must be disrupted within $\sim 4 \times 10^7$ yr.

We now consider possible mechanisms for disruption of GMC. These are formations of H II regions by massive stars, supernova explosions, and OB star winds. The energy output in each of these modes is comparable over the lifetime of an OB association, but each possibility must be considered in detail in order to evaluate its overall efficiency for potential disruption.

a) H II Regions

Compact H II regions tend to form preferentially near the outer edges of molecular clouds, leading to a blister-type structure (Habing and Israel 1979). A massive young star is embedded in an open cavity at the edge of the molecular cloud, and ionized gas flows out of the cavity. Whitworth (1979) has studied the erosion and dispersal of massive molecular clouds by this process, and argues that a few O stars can effectively destroy a massive molecular cloud.

However the success of the blister model relies on the maintenance of an extensive (~ 25 pc) low density cavity around a massive star. Now the Stromgren radius for an O6 star is only $1.7\, n_3^{-2/3}$ pc, where $n_3 \equiv 10^3$ cm^3. It seems clear that a realistic model of a molecular cloud complex, which consists of many bound clumps of mass $\sim 10^3\ M_\odot$ and size ~ 10 pc moving with random velocities ~ 5 km s^{-1} (Blitz 1978), will rapidly fill in any extended cavities generated by the blistering process. Thus the Stromgren radius gives an effective measure of the disruption by O stars. Because only early type O stars can play a role, and OB associations are localized within a few pc^3 at formation, we conclude that blistering cannot be a significant erosion mechanism as long as only a small fraction of the surface of the cloud is ionized by O stars. Blistering is ineffective at disruption, and the total volume ionized by an OB association containing, say, 10 O5 stars, will be $\lesssim 150$ pc^3, or less than 0.2 percent of the volume of a typical molecular cloud complex. Note that stars of type later than O5 only contribute ~ 35 percent of the total number of ionizing photons if a Salpeter initial mass function is adopted.

b) Supernovae

A single supernova is incapable of disrupting a molecular cloud complex. To demonstrate this, we note that the maximum radius of the remnant is

$$R_{max} = 4.6\ E_{51}^{0.23}\ n_3^{-0.36}\ T_4^{-0.20}\ \text{pc}.$$

However, it is possible that multiple supernovae may occur, reinforcing one another and thereby disrupting the complex. The relevant criterion for this to occur is similar to that given by Cox and Smith (1974) and McKee and Ostriker (1977) in their studies of the interaction of super-nova remnants with the interstellar medium, namely

$$Q_{SNR} = 10^{-4.61} E_{51}^{1.28} (\frac{S}{10^{-13} pc^{-3} yr^{-1}}) n_3^{-1.44} T_4^{-1.30}$$

where Q_{SNR} is the filling factor of the region with supernova remnants, $E_{51} = E/10^{51}$ ergs, S is the supernova rate per unit volume per unit time, and the <u>effective</u> temperature (including random supersonic veloci-ties) of the ambient medium $T_4 = T/10^4 K$. We have assumed that star formation continues for 10^7 yr. Within this time, ~ 30 massive stars (of spectral type earlier than B1) per OB association of mass $M_* \approx 3 \times 10^3 M_\odot$ have evolved and become potential supernovae (Reeves 1978). In a cloud of mass M_{cl} and volume V_{cl}, we obtain a supernova rate per unit volume per unit time

$$S = 3 \times 10^{-10} (\frac{\xi}{0.1})(\frac{M_{cl}}{3 \times 10^5 M_\odot})(\frac{3 \times 10^3 M_\odot}{M_*})(\frac{10^5 pc^3}{V_{cloud}}) pc^{-3} yr^{-1} .$$

This condition demonstrates that supernovae can sweep out a medium of density

$$n = 10^{2.22} Q_{SNR}^{-0.69} (\frac{\xi}{0.1})^{0.69} (\frac{M_{cl}}{3 \times 10^5 M_\odot})^{0.69} (\frac{3 \times 10^3 M_\odot}{M_*})^{0.69} x$$

$$(\frac{10^5 pc^3}{V_{cl}})^{0.69} E_{51}^{0.88} T_4^{-0.90} cm^{-3} .$$

We conclude that supernovae are probably capable of disrupting molecular cloud complexes. Note that $M_*/M_{cl} \sim 0.01$ is only a lower limit to the mass fraction in stars, since no account is taken here of low mass star formation.

c) OB Stellar Winds

   The discovery that stars of spectral type earlier than B1 (Snow and Morton 1976) undergo extensive high velocity mass outflows provides a significant mode of energy input into the interstellar medium, and in particular into molecular cloud complexes, as we now demonstrate. A theoretical model for the evolution of wind-driven bubbles around massive young stars has been developed by Weaver et al. (1977), who show that the bubble radius at time t is given by

$$R_{bubble} = 6.8 n_3^{-1/5} L_{36}^{1/5} t_6^{3/5} pc$$

where $L \equiv \dot{M} V_w^2$, $\dot{M}$ is the mass loss, V_w is the wind velocity, $L_{36} = L/10^{36} erg s^{-1}$ and $t_6 = t/10^6$ yr. The bubble interior is filled with shocked wind gas at $T \sim 10^6$ K, and the wind is separated by a contact discontinuity from the shell formed by the swept-up ambient medium. It is immediately apparent, because of the weak sensitivity to the ambient density and stellar luminosity, that stellar winds provide a potentially more potent mechanism for cloud disruption than do H II regions. Moreover, since the wind activity is initiated continuously once star formation begins, it can provide the primary mechanism for cloud disruption. To demonstrate that winds are indeed effective in dispersing a molecular cloud complex, we must show that the bubbles can intersect and act coherently in molecular cloud complexes. Following the previous section, we find that, for disruption, the rate of formation of OB star bubbles per unit volume must be

$$S_{OB} = 10^{-8.1} \left(\frac{\bar{v}_{cl}}{5 \text{ km s}^{-1}} \right) n_3^2 L_{36}^{-2} \text{ pc}^{-3} \text{ yr}^{-1}$$

where $\bar{v}_{cl}$ is the internal random velocity dispersion inferred from molecular line widths. Therefore choosing $S_{OB} \sim S$, we find that bubbles can disrupt a cloud of density

$$n \simeq 10^2 \left(\frac{S_{OB}}{10^{-10} \text{ pc}^{-3} \text{yr}^{-1}} \right) \bar{v}_{cl}^{-1/2} L_{36} \text{ cm}^{-3}.$$

We conclude that OB winds and supernova remnants are likely to play comparable roles in the disruption of molecular cloud complexes over a period of $\sim 10^7$ years.

d) Survival of Molecular Clouds

The efficiency arguments of § I indicate that GMC disruption results in the formation of a number of smaller molecular clouds (SMC) with a small fraction of H I produced near 4-6 kpc. In the solar neighbourhood the dominance of H I suggests that molecular cloud disruption may be more complete. This could result because of the longer time spent in the interarm region. The SMC must definitely exist on time scales $\gtrsim 10^8$ yr. They are likely to be bound units and the most plausible stabilizing mechanism would seem to be star formation. A simple statistical model of star formation in which stars first form at low mass and the formation of massive stars is limited by the cloud mass suggests that the SMC will be deficient in their content of very massive stars. Thus, we speculate that the SMC are associated with predominantly low mass stars and occasional small OB associations that provide sufficient momentum and energy to support them in the form of stellar winds and ionizing radiation. This provides us with an interpretation of the hitherto puzzling result that up to $\sim 80\%$ of the Lyman continuum photons inferred from radio continuum studies are produced by OB stars from the interarm regions of the Galaxy. Star formation could also occur in a similar manner in the solar vicinity in the local interarm region where there is evidence for a wide-spread OB star population.

V. FURTHER IMPLICATIONS

We have found that GMC can form in both arm and interarm regions by coagulation of smaller clouds. In many galaxies with strong spiral shocks, GMC's will form preferentially in the spiral arms, and we emphasize that a study of CO in external galaxies of varying spiral structure with different shock strengths and different driving mechanisms (Kormendy and Norman 1979) may resolve many of the issues discussed here. The accumulation process may be further enhanced in the spiral arms by the Parker instability, and a possible observational test of this mechanism has been indicated. Disruption of these complexes preferentially occurs as a consequence of massive star formation after passage through the arms, resulting in coherently interacting stellar winds and supernova remnants. Our estimates of disruption by these mechanisms suggest that only when GMC develop masses of $\gtrsim 3 \times 10^5 \, M_\odot$ (assuming that a fraction $\xi \sim 0.1$ of the GMC has formed predominantly low mass stars with a single OB association) will disruption occur.

We note, finally, several further implications of this model for molecular cloud complexes. Once massive star formation is initiated, the studies of Bash (1979) indicate that disruption will occur on a time scale $\sim 4 \times 10^7$ yr corresponding to the lifetime of OB associations. The galactic rotation period is less than 10^8 yr at a distance $\lesssim 3$ kpc from the Galactic centre. Thus, within this distance it may be impossible for molecular clouds moving on predominantly circular orbits to build up into GMC's. Furthermore, the increase in mean gas density towards the inner regions of the Galaxy implies a coagulation rate that increases with decreasing galactocentric radius. In this manner one might hope to understand the concentration of molecular clouds between 4 - 8 kpc.

The viscosity associated with cloud-cloud collisions leads to an inward radial drift velocity $\sim \frac{1}{3} v_{cl} \, l/r \lesssim 0.1 \, (10 \text{ kpc}/r) \text{ km s}^{-1}$, at radius r. Radial inflow and mixing of gas could, in principle, lead to the development of abundance gradients in spiral disks. However, a radial velocity of several km s^{-1} is required to produce abundance gradients characteristic for disk galaxies (Tinsley and Larson 1978). Radial inflow induced by this mechanism is only significant within ~ 1 kpc of the galactic centre, where it could have interesting implications for feeding active galactic nuclei (cf. Lynden Bell and Pringle 1974) such as Seyfert galaxies.

Observed phenomena that may be related to the disruption of molecular cloud complexes are the large scale features in the H I distribution that have been found in our own Galaxy (Heiles 1977, 1979) in M101 (Allen and Goss 1979) and M31 (Brinks 1979). The large scale (~ 1 kpc) associated with these structures indicate that a coherent phenomenon, similar to that proposed for the disruption of molecular cloud complexes, must be operative.

In summary, we have argued that GMC could be coagulated predominantly in arms in our own Galaxy. The GMC distribution in external galaxies depends on the ratio of arm to interarm growth, and therefore depends on the specific shock strength and relative arm-interarm residence time. Finally we emphasize that only coherent massive star formation can disrupt GMC.

REFERENCES

Allen, R.J., and Goss, W.M. 1979, Astron. and Astrophys. (Suppl.),
    36, 135.
Bash, F. 1979, Astrophys. J. 233, 524.
Blitz, L. 1978, Ph.D. thesis, Columbia University (unpublished).
Brinks, E. 1979, private communication.
Burton, N.B., and Gordon, M.A. 1978, Astron. and Astrophys., 63, 7.
Cohen, M., and Kuhi, L.V. 1979, Astrophys. J. (Suppl.) 41, No. 4.
Cox, P., and Smith, B. 1974, Astrophys. J. (Letters), 189, L105.
Dickman, R.L. 1978, Astrophys. J. (Suppl.), 37, 407.
Field, G.B., and Hutchins, J. 1968, Astrophys. J., 153, 737.
Field, G.B., and Saslaw, W.C. 1965, Astrophys. J., 142, 568.
Habing, H., and Israel, F. 1979, Ann. Revs. Astron. and Astrophys.,
    17 (in press).
Heiles, C. 1976, Astrophys. J. (Letters), 208, L137.
Heiles, C. 1979, Astrophys. J., 229, 553.
Jenkins, E.B., Silk, J., and Wallerstein, G. 1976, Astrophys. J.
    (Suppl.), 32, 681.
Kormendy, J., and Norman, C. 1979, Astrophys. J. 233, 539.
van der Kruit, P., and Allen, R. 1978, Ann. Revs. Astron. and
    Astrophys., 16, 103.
Kwan, J. 1979, Astrophys. J., 229, 567.
Lynden-Bell, D., and Pringle, J.E. 1976, M.N.R.A.S., 168, 603.
McKee, C.F., and Ostriker, J.P. 1977, Astrophys. J., 218, 148.
Mezger, P. 1978, Astron. and Astrophys., 70, 565.
Mouschovias, T.C. 1974, Astrophys. J., 192, 37.
Mouschovias, T.C., Shu, F.H., and Woodward, P. 1974, Astron. and
    Astrophys., 33, 73.
Norman, C., and Silk, J. 1980, Astrophys. J. (in press).
Oort, J. 1954, B.A.N., 12, 177.
Parker, E. 1967, Astrophys. J., 149, 535.
Penston, M., Munday, V., Stickland, D., and Penston, M. 1969,
    M.N.R.A.S., 142, 355.
Reeves, H. 1978, Protostars and Planets, ed. T. Gehrels (Univ. of
    Arizona Press: Tucson), p. 399.
Scoville, N.Z. 1979, in IAU Symposium 84, Large Scale Characteristics
    of the Galaxy, ed. B. Burton (D. Reidel: Dordrecht) p. 277.
Scoville, N.Z., and Hersh, K. 1979, Astrophys. J., 229, 578.
Shu, F.H. 1978, in IAU Symposium 78, Structure and Properties of
    Nearby Galaxies, ed. E.M. Berkhuijsen and R. Wielebinski
    (D. Reidel: Dordrecht), p. 139.
Shu, F.H., Milione, V., Gebel, W., Yuan, C., Goldsmith, D., and
    Roberts, W.W. 1972, Astrophys. J., 173, 557.
Snow, T., and Morton, D. 1976, Astrophys. J. (Suppl.), 32, 429.
Solomon, P.M., Sanders, D.B., and Scoville, N.Z. 1979, in IAU Symposium
    84, Large Scale Characteristics of the Galaxy, ed. B. Burton
    (D. Reidel: Dordrecht) p. 35
Solomon, P.M., and Sanders, D.B. 1979, Giant Molecular Clouds in the
    Galaxy, ed. P. Solomon and M. Edmunds (Pergamon Press: Oxford)
    (in press).

Taff, L., and Savedoff, M. 1972, M.N.R.A.S., 160, 89.
Taff, L., and Savedoff, M. 1973, M.N.R.A.S., 164, 357.
Tinsley, B.M. 1976, Astrophys. J., 208, 797.
Tinsley, B.M., and Larson, R.B. 1978, Astrophys. J., 221, 554.
Weaver, R., McCray, R., Castor, J., Shapiro, P., and Moore, R. 1977,
      Astrophys. J., 218, 377.
Whitworth, A. 1979, M.N.R.A.S., 186, 59.

DISCUSSION FOLLOWING NORMAN

Mouschovias: You made the very strong statement that "coagulation is inevitable." It seems to contain at least three implicit assumptions. First, that some mechanism other than coagulation has formed your building blocks, the $10^3 M_\odot$ self-gravitating clouds. I know of no such mechanism. Thermal instability will not do. (See Protostars and Planets, ed. T. Gehrels, Univ. of Ariz. Press, 1978, pp. 209-242.) Second, that the $10^3 M_\odot$ blobs move randomly with respect to one antoher. Is there any evidence for that, especially since we can only observe radial velocities? Third, that collision between two blobs leads to agglomeration rather than to disruption. The old calculation in 1.5 dimensions, by Stone, although not conclusive, suggests disruptions.

Norman: The calculations of cloud-cloud collisions by Stone were for HI clouds of densities ~ 1 cm^{-3}. I discussed densities ≥ 100 cm^{-3} Thus the condition for inelastic cloud-cloud collisions to result in sticking is well satisfied, namely, that the cooling time is very much shorter than the time taken to cross the cloud by the shock generated in the collision. Thus if these molecular clouds collide, they will certainly coagulate. Do they collide? The estimate of collision rate given by Solomon and co-workers has very little uncertainty since it is based on the observed number of clouds along the line of sight, their observed cross-section, and their observed velocity-dispersion. The velocity-dispersion used in Solomon's calculation is certainly in the range 5-10 km s^{-1} (Blitz, this volume). Thus there is no question that the observed clouds collide and stick and that *"coagulation is inevitable"* for the observed cloud distribution. To make this statement, it is not strictly necessary to discuss how the molecular clouds form, although both coagulation and compression in the density wave have been proposed as possible models.

Elmegreen: In models using collisions to build up giant molecular clouds the time-scales for forming clouds are 10 times longer than the duration of OB-star formation in any one cloud, so there must be 10 times as many quiescent massive clouds as active (OB-star forming) massive clouds. This inference strongly contradicts both the direct observations of giant clouds in the solar neighbourhood and the cloud count for the whole galaxy extrapolated from large surveys.

Norman: From the infrared survey data discussed by Rowan-Robinson (1979, Ap. J., in press), and from the CO survey data presented at this symposium, only $\sim 10\%$ of the giant molecular clouds have $T_{CO} > 20$ K. We associate this warming with the influence of more massive OB-star formation on the cloud. Furthermore, as discussed by Solomon, it is these warm clouds that seem to be associated with the arm region, in good agreement with the model given here.

Gold: What could cause star formation to be restricted to low-mass stars in these clouds? In the presence of turbulence a spectrum of star masses would always be expected. Why should there be a sharp cut-off in some cases?

Norman: That question concerns the details of star formation. Three IAU Symposia from now on this subject I may be able to give you an answer! The model demonstrates that in dark clouds, where there is clearly no OB-star formation occurring, low-mass pre-main-sequence stars provide via their winds sufficient energy to explain the supersonic line widths. Thus, implicit in our model is the assumption that dark clouds make predominantly low mass stars and infrequently some low mass OB-stars. We contend that formation of more massive OB-stars needs a trigger. This trigger seems to be associated with the arm region, since the giant HII regions, which are clearly tracers of the arm regions, are associated with the more massive OB stars. Possible candidates for the trigger are (1) rapid increase in cloud mass $M \gg M_J$ through enhanced coagulation in the arms (although $M > M_J$ anyway, since these clouds appear to be bound), (2) an external trigger such as a nearby supernova explosion or OB-star wind, or (3) an increase in cloud-cloud collision velocity in the arm region, due to acceleration by interaction with the shock. Generally we speculate that OB stars are formed in the process of collision-coagulation. It could be that the mass of the most massive star is correlated with the velocity of collision.

Carruthers: Two of the mechanisms you mentioned, formation of HII regions and OB-star winds, are due to the same OB stars. How do you differentiate between these two effects? It would seem that HII-region formation would be a more important "trigger" mechanism, since it occurs earlier in the life of an OB star.

Norman: The distinction of the OB-star wind is that it can create a bubble structure over a time considerably longer than that of the conventional HII region, of which it is in fact, a modified form (Weaver et al. 1977, Ap. J. 218, 377). The increased time-scale makes it easier for bubbles to intersect, and it is only by coherent interactions of OB-star winds or supernova remnants that a giant molecular cloud can be disrupted.

Gillespie: My observations of large and giant molecular clouds in the southern hemisphere show that in almost all cases there are several sites of increased CO intensity, implying multiple sites of star formation, but not all sites are at the same stage of evolution. What would be the effect on the evolution of the clouds according to your model?

Norman: Different ages at different sites of star formation would be expected if star formation is triggered by cloud-cloud collision, or is triggered in bound smaller clouds before coagulation with the giant cloud. There might also be a sequential process in which the effects of a supernova or OB-star wind trigger star formation in a nearby clump. Detailed observations such as you described may allow us to differentiate between various triggering processes.

THE DISRUPTION OF THE MOLECULAR CLOUD ASSOCIATED WITH THE NORTH AMERICA AND PELICAN NEBULAE

John Bally
Physics and Astronomy Department
University of Massachusetts
Amherst, Massachusetts   01003

Giant molecular clouds can be destroyed in several ways: by conversion of gas into stars; by dissociation of H_2 into HI; by ionization and formation of HII regions; and by dispersal of the cloud into cloudets. The relative importance of these processes can be assessed by observations of <u>evolved</u> HII regions and associated molecular clouds. A number of such complexes have been studied in ^{12}CO at the Five College Radio Astronomy Observatory (FCRAO) as part of an extensive investigation of clouds associated with about 60 Sharpless HII regions. A particularly clear example of cloud disruption is found in the case of W80, the North America and Pelican Nebulae in Cygnus (Bally and Scoville 1980).

The W80 HII region, located at a distance of roughly 1 kpc (Goudis 1976), is highly evolved, as indicated by the low mean electron density of $n_e \sim 9$ cm^{-3} (Wendeker 1968) and its large angular diameter of 50 pc. A fragmented molecular cloud is associated with this complex and is responsible for the foreground obscuration that separates the North America Nebula from the Pelican. The molecular gas in the northern half of the complex exhibits a systematic expansion away from the center of the HII region. All the CO emission from gas near the edge of the complex has a velocity close to $V_{LSR} = 0$ km s^{-1} while emission near the center exhibits line splitting with a peak to peak separation of 10 km s^{-1}. The negative velocity component correlates well with the foreground dust, hence must be gas in front of the HII region. The positive velocity component can be seen in the direction of Hα emission from the Pelican, thus must be emitted by gas behind the nebula. An overall picture of the kinematics of the CO surrounding the northern half of W80 is shown in Fig. 1 where the vertical axis represents distance from the centroid of the free-free emission. Clearly the gas can be characterized as an expanding molecular shell with an expansion velocity of $v = 5$ km s^{-1}. The very opaque gas in front of the southern portion of the N.A. Nebula which has $A_v \gtrsim 10$ mag. does not partake in the general expansion; evidently the column density of this gas is too great to be dynamically affected by the high pressure HII region.

151

B. H. Andrew (ed.), Interstellar Molecules, 151–156.
Copyright © 1980 by the IAU.

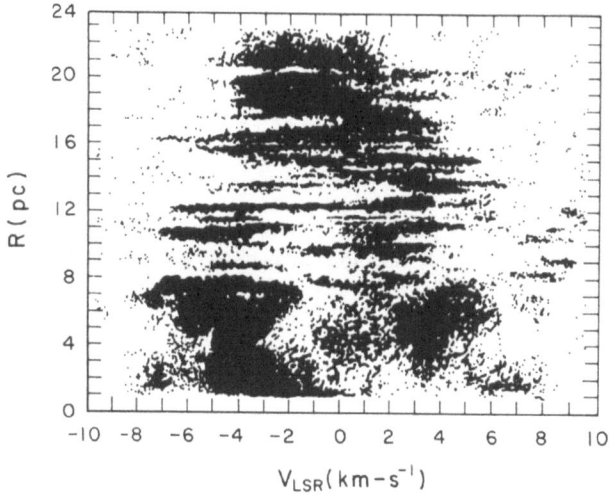

Fig. 1. A radius-velocity diagram of ^{12}CO emission in the northern half of W80 using 200 spectra. The vertical axis represents projected distance from the centroid of the free-free emission to each particular spectrum.

We have modeled the expanding shell surrounding W80 as the last phase in the evolution of the post-shock layer associated with the expansion of the HII region. Once an OB subgroup is born inside a molecular cloud the expansion of the high pressure HII gas sweeps up material behind a weak D-type shock front. Eventually the HII bubble reaches the nearest edge of the molecular cloud and bursts (Bodenheimer et al. 1979). Once the resulting rarefaction wave has reached the ionization front at the inner-edge of the HII cavity, a streaming HII region is established which continues to push the shock front further into the cloud. Eventually the shock runs through the back of the cloud and accretion by the post-shock layer ends. After this, the rocket effect (Oort and Spitzer 1954) can accelerate the shocked gas layer; Fig. 2 shows a schematic diagram of this evolutionary process. Model calculations for a homogeneous molecular cloud indicate that the current stage of evolution is reached between 3 to 6 million years after the onset of 0 & B star formation. The ionizing flux of 5×10^{49} photons s^{-1} has ionized about 2×10^4 $M_\odot$ of the original molecular cloud material. About $3-6 \times 10^4$ $M_\odot$ of gas has been accelerated by the rocket effect and will probably survive in molecular form. A piece of the cloud of $M \sim 3 \times 10^4$ $M_\odot$ in the southern portion of the complex has not yet been significantly affected by the HII region.

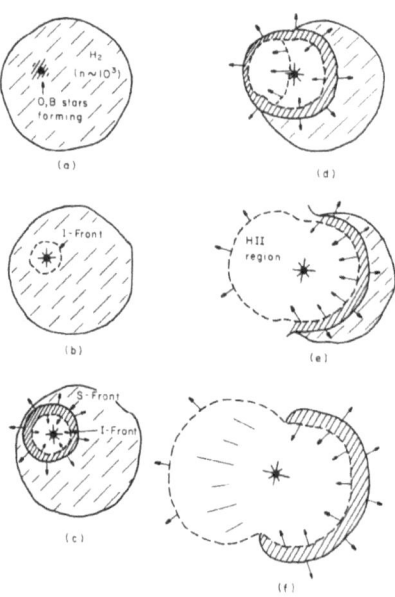

Fig. 2.   Schematic diagram of the evolution of an HII region --
molecular cloud complex.

The various cloud destruction mechanisms can be compared in a
general way. Estimates of the star formation rate indicate that the
conversion of gas into stars proceeds at a rate $\dot{M} = 1$ to 5 $M_\odot$ yr^{-1} in
the entire galaxy (Mezger 1978, Scoville and Herch 1979), corresponding
to a mean rate of $2.5 - 12$ x 10^2 $M_\odot$ in 10^6 yrs per cloud, assuming a
galactic population of 4,000 clouds (Solomon, Sanders, and Scoville
1980). Dissociation, ionization, and dispersal are likely to occur
only after an episode of OB star formation and the creation of an HII
region in the cloud. The effects of supernovae and stellar winds are
minor in comparison with the damage caused by the evolving HII region
(Lada, Blitz, and Elmegreen 1978). The conversion of H_2 into HI is
followed by ionization when the gas flows through the ionization front
at the HII region boundary. The conversion rate $\dot{M}$ is controlled by
the expansion of the HII region since the available flux of ionizing
photons depends on the recombination rate in the ionized gas. From
the equations governing the size of an HII region as a function of
time, and the Strömgren condition (see Spitzer 1968), it follows that

$$\dot{M} = 1.09 \text{ x } 10^{-9} \frac{Q^{2/3}}{n_o^{1/3}} [1 + 1.98 \text{ x } 10^2 \frac{n_o^{2/3}}{Q^{1/3}} t]^{-1/7} \text{ (gm s}^{-1})$$

Table 1:   Parameters of W80 and the expanding molecular shell.

W80		Expanding Shell	
distance	1 kpc	expansion velocity	5 km s^{-1}
ionizing flux	5x10^{49} s^{-1}	mass of shell	3-6x10^4 M$_\odot$
HII mass	1.8x10^4 M$_\odot$	radius of shell	20 pc
diameter	50 pc	shell thickness	2-3 pc
electron density	9 cm^{-3}	expansion center	α = 20^{h}53^{m}15^s
			δ = 43°45°00^s

where equality holds for a Strömgren sphere.   For an HII region that
has burst out of the molecular cloud and is density bounded on one
side $\dot{M}$ can be 2 or 3 times greater since removal of ionized gas from
the region between the star and the molecular cloud increases the flux
of photons available at the I-front.   For an 05 star (Q = 5 x 10^{49}
photons s^{-1}) and a cloud of uniform density n_0 = 500 cm^{-3},
$\dot{M} \gtrsim 2 \times 10^{-3}$ M$_\odot$ yr^{-1}, close to the rate required to generate the W80
HII region if allowance is made for the removal of ionized gas.

A very crude estimate of the conversion rate of H$_2$ into HII on a
galactic scale can be made on the basis of the total ionization rate
in the galaxy (Smith et al. 1978, Mezger 1978).   Assuming that the
flux of Lyman continuum photons responsible for ionization of all
galactic radio HII regions is Q = 4.7 x 10^{52} s^{-1} (corresponding to 16%
of all OB stars in the galaxy), the conversion rate for the galaxy is
$\dot{M}_g \gtrsim$ 2 M$_\odot$ yr^{-1}, comparable to the mean star formation rate.   The mean
rate at which gas is accelerated to escape velocity from clouds by the
rocket effect is difficult to estimate, in general.   In the case of W80,
the mass of the expanding shell is comparable to the mass of the ionized
gas, implying that the mean disruption rate is similar to the
ionization rate.   Assuming that star formation continues at a steady
rate in each cloud, disruption and ionization appear to be the
dominant modes of cloud destruction once OB star formation has occurred.
More detailed studies of the evolved complexes associated with S155,
S184, S125 and other regions are continuing at the FCRAO in order to
quantify the relative importance of various cloud destruction mechanisms.

REFERENCES

Bally, J., and Scoville, N.Z.: 1980, Astrophys. J., in press.
Bodenheimer, P., Tenorio-Tagle, G., and Yorke, H.W.: 1979, Astrophys. J.
    233, 85.
Goudis, C.: 1976, Astrophys. and Space Sci. 39, 173.
Lada, C., Blitz, L., and Elmegreen, B.G.: 1978, in Protostars and
    Planets, ed. T. Gehrels (Tucson: Univ. of Ariz. Press).
Mezger, P.G.: 1978, Astron. and Astrophys. 70, 565.

Oort, J.H., and Spitzer, L.: 1955, Astrophys. J. 121, 6.
Scoville, N.Z., and Herch, K.: 1979, Astrophys. J. 229, 578.
Smith, L.F., Biermann, P., and Mezger, P.G.: 1978, Astron. and
    Astrophys. 66, 65.
Solomon, P.M., Sanders, D.B., and Scoville, N.Z.: 1980, Astrophys. J.,
    in press.
Spitzer, L.: 1968, Diffuse Matter in Space (Interscience)
Wendker, H.: 1968, Zs. f. Ap. 68, 368.

DISCUSSION FOLLOWING BALLY

de Jong: The H_2 shells that you discussed would put a lot of kinetic energy into the interstellar medium. Have you compared this energy input with, for instance, the energy input from supernova explosions?

Bally: Yes. The energy input is quite comparable to that from a supernova if the input due to dissociated gas is included. It is difficult to estimate on a galactic scale the ratio of gas surviving as H_2, to gas that has been dissociated.

Slysh: There is a supernova remnant in the region you mapped. It was in the list presented in my talk.

Bally: The SNR G84.2-0.8, to which you refer, is at a distance of roughly 17 kpc, as determined from the Σ-D relation. It is a background source not related to the W80 complex.

Mouschovias: While a cloud of density $\gtrsim 10^4 cm^{-3}$ is waiting to be churned by an HII region, why is it not collapsing within the free-fall time of about 10^5 yrs? Large linewidths do not necessarily imply support by some unknown mechanism against collapse if they are due to ordered motions.

Bally: Observations show that clouds are *not* forming stars at the rate indicated by the free-fall time scale, so there must be some form of internal support. Turbulence and many forms of ordered motion would damp out too quickly, and there is *no* evidence for all-pervasive *small scale* ordered motion. Some source of pressure, added to the thermal pressure, could support the clouds *and* explain the large observed line widths. For instance, magnetic fields can, in some situations, raise the effective sound speed and act as this source of pressure.

Mouschovias: A word of caution on the rocket effect: since a cloud is not a rocket (i.e. not a solid), a fluid element at one point will not be affected by an expanding HII region until the shock front arrives. Since the length scales that concern you are typically $\gtrsim 20$ pc, and shock velocities are ~ 1 km s^{-1}, a cloud will be affected only in a time $\gtrsim 10^7$ yrs. Is this time-scale compatible with the ideas of churning and dispersing clouds over a few million years?

Bally: Although the shock velocity in our models can reach values as low as 1 km s^{-1}, it is generally larger over most of the lifetime of the region. The numerical models of the W80 region indicate that the *mean* shock velocity is about 5 km s^{-1} between the time of formation of the weak D front and the present. The rocket effect raises the shock velocity above the value it would have for a Strömgren sphere of equivalent radius. The reason is that the post-shock layer stops accumulating matter once it passes the position of the initial cloud boundary and enters the rocket phase.

Mouschovias: Would a dense, massive cloud be dispersed if self-gravity is taken into account (see 1976, Ap. J., <u>207</u>, 141)?

Bally: Except in the case of *extremely* dense molecular clouds, gravity is not capable of halting the expansion of an HII region and the dispersal of the gas accelerated in the post-shock layer.

THE STRUCTURE AND EVOLUTION OF THE W3 MOLECULAR CLOUD

Hélène R. Dickel
Astronomy Department, University of Illinois
and
Sterrewacht, Leiden

A model of the W3 molecular cloud derived from molecular observations is presented and the evolution of the cloud is discussed.

The distributions of the emission from ^{12}CO, ^{13}CO, CS and HCN have been mapped in the W3 molecular cloud (Dickel et al. submitted to Ap.J. 1979). Self-absorption effects are seen in ^{12}CO and probably CS. The abundance of CS is a factor of $\sim$10 lower in the direction of the complex of near-infrared sources designated 5, 6 and 7 than at positions only 0.6 pc away.

From an overall analysis of the data, we conclude that an initial cloud of mass $\sim$5 × 10^4 M$_\odot$ started to fragment and collapse after passage through the spiral density-wave shock of the Perseus arm. The open cluster (Ocl 352) which excites the present W4 HII region formed at the head of the cloud a few million years ago (see figure 1). The ionization front with its associated shock is expanding into the remaining W3 molecular cloud where its effects are seen in the velocities of W3(OH) and the infrared source AFGL 333. The bright dense "core" of the W3 molecular cloud which contains 6 compact HII regions, 8 near-infrared sources, and three H$_2$O masers is collapsing and possibly rotating.

The shock front of the expanding, optical HII region, IC 1795, is moving into the edge of the W3 molecular core at 2 to 3 km s^{-1} as seen by the motion of the CO gas behind the shock front. However, this shocked gas represents only a small fraction of the total material present so the shock can not be responsible for triggering the collapse of the whole W3 core. A full discussion appears in a paper by Dickel (submitted to Ap.J. 1979).

Emission from [OIII] and Hα are seen in a small optical jet which emerges out of the center of the shell radio source W3A close to IRS 2 (see figure 2). This represents the earliest stage of an embedded compact HII region bursting out of its parent cloud to form a "blister" on the edge. Further details will appear in a paper by Dickel and Harten (in preparation).

B. H. Andrew (ed.), Interstellar Molecules, 157–158.

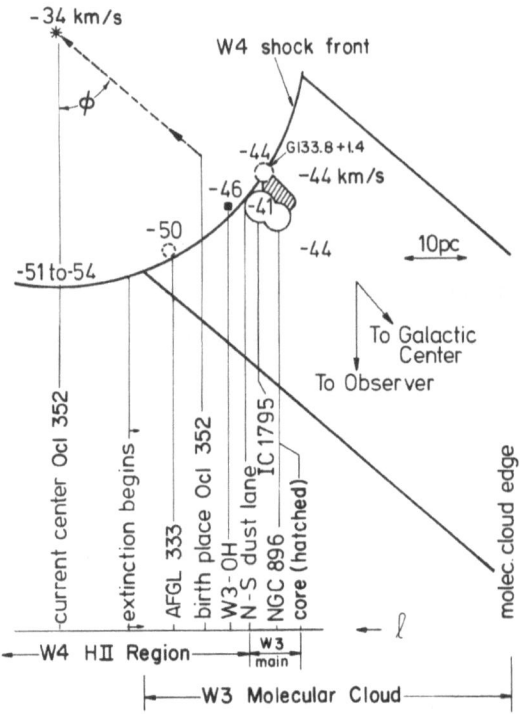

Figure 1. A model of the W3 molecular cloud and adjacent W4 HII region. Observed radial velocities are given at some locations.

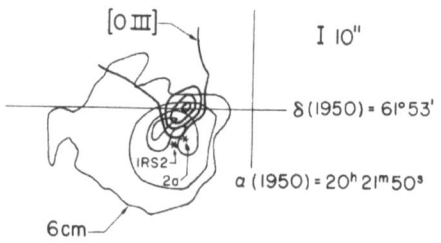

Figure 2. The jet of optical emission emerging from the radio source W3A.

This research was supported in part by the National Science Foundation and by the Sterrewacht, Leiden.

ATOMIC HYDROGEN IN AND AROUND THE GIANT MOLECULAR CLOUD NEAR W3 AND W4

Tetsuo Hasegawa
Department of Astronomy, University of Tokyo, Tokyo, Japan

Fumio Sato
Chiba Prefecture Education Center, Katsuragi-2, Chiba, Japan

Yasuo Fukui
Tokyo Astronomical Observatory, Mitaka, Tokyo, Japan

Abstract: Cold HI gas appears as self-absorption dips in the 21-cm line profiles in and around the giant molecular cloud near W3 and W4. The cold HI cloud is $\sim$150 pc long and extends along the galactic plane. It consists of several fragments, each of which is typically $\sim$25 pc in diameter and $(1 - 4) \times 10^4$ $M_\odot$ in mass. The [H$_2$]/[HI] ratio is estimated to be 15 - 50. The mass of the entire HI cloud amounts to $\sim$10^5 $M_\odot$, which is comparable to that observed in CO emission.

INTRODUCTION

We have analyzed the HI self-absorption features in and around the giant molecular cloud near W3 and W4. The region is a site of massive star formation. HI self-absorption features in dense clouds are caused by residual atomic hydrogen gas whose spin temperature, T_S, is in equilibrium with the gas kinetic temperature, as low as $\sim$20 K. Studies of HI self-absorption are expected to shed a new light on the physical and chemical conditions in giant molecular clouds.

The HI data are from the Maryland-Green Bank Galactic 21-cm Line Survey by Westerhout(1973). The method of data reduction is much the same as that applied to the M17 region by Sato and Fukui(1978), except that the spin temperature of cold HI was assumed to be 20 K uniformly. The radio continuum data are from Rohlfs et al.(1977), and the distance is assumed to be 3.0 kpc.

RESULTS

Near W3 and W4, HI self-absorption features appear at velocities from -50 km s^{-1} to -40 km s^{-1}. The velocities are similar to those of CO emission (e.g.,Lada et al. 1978). Here we concentrate on the features near -50 km s^{-1}, which are the most prominent in the velocity range above. Fig.1 gives the distribution of optical depth, τ(HI), at the velocity of maximum absorption near -50 km s^{-1}, superposed on the photograph taken in

159

B. H. Andrew (ed.), Interstellar Molecules, 159–162.

red light. The cold HI cloud extends nearly along the galactic plane
from the western edge of W4 to ∿150 pc west. The most striking feature
is that it is made up of several fragments. They are designated by the
letters A – F and their peak positions are listed in Table 1. Each of
the fragments is typically ∿25 pc in diameter at the half maximum of
τ(HI). At the edge of W4, the self-absorption dip disappears dramaticall
suggesting that the cold HI cloud is associated with W4. Around W3, on
the other hand, dips are severely blended and determination of expected
profiles is ambiguous, so we excluded this region from our analysis (the
excluded region is shown in Fig.1). The cold HI cloud is distributed
similarly to the region of heavy optical extinction. Lada et al.(1978)
presented CO maps in this region. Fragments C and D coincide well with
the CO emission, while maxima differ a little from each other. As for
fragment E, CO emission is detected only at its center, and fragments
A, B, and F show no detectable CO emission. The HI linewidth is
typically 4 km s^{-1} and shows no systematic change. The velocity at the
absorption maximum is nearly constant around –50 km s^{-1} all over the
cloud, and in each fragment no significant velocity gradient is recognize

PHYSICAL CONDITIONS IN THE HI CLOUD
        The physical conditions inferred are summarized in Table 2. In the
HI cloud with CO detection, the resemblance between the distribution of
cold HI and that of CO suggests that these two species are well-mixed.
T_S is nearly equal to the gas kinetic temperature of 20 K inferred from
CO observations. The [H$_2$]/[HI] ratio is estimated to be 50 from
N(HI)(column density of cold HI gas), N(^{13}CO)(Lada et al. 1978), and
[^{13}CO]/[H$_2$] = 2 × 10^{-6}(Dickman 1978). This value is fairly small compare
with those derived in other molecular clouds; 100 – 350 in bright CO peak
near M17 (Sato et al. 1979), 120 in the ρ Oph dark cloud (Myers et al.
1978), and 500 ± 300 in Heiles' Cloud 2 (Wilson and Minn 1977). Values
as small as 13 – 70 have been derived in the diffuse region of the M17
complex (Sato et al. 1979), and these smaller values may be common among
less dense regions.
        In the HI cloud without CO detection, there is no direct evidence
that T_S is 20 K; but the lack of sharp, narrow HI emission features even
in the region where the HI background emission is weak indicates that the
spin temperature is not very high ($\lesssim$40 K) nor is the optical depth
(τ(HI)$\lesssim$0.5). The upper limit of the total particle density is given by
the lack of detectable CO, and its lower limit is estimated from the
total hydrogen column density inferred from the visual extinction. We
estimate $A_V \gtrsim$ 1.5 mag as a conservative value from eye-inspection of
the Palomar Sky Survey Atlas.

MASSES OF THE FRAGMENTS
        The total masses of the HI fragments, M(total)'s, are estimated from
the physical parameters above and listed in Table 1. The fragments appea
to be in dynamical equilibrium and another independent estimate of the
fragment masses is made from the HI linewidth through the virial theorem.
The virial mass, M(virial) in Table 1, agrees fairly well with M(total),
although M(virial) is systematically larger than M(total) by a factor of
2 – 3. This difference may be caused partly by the assumption of LTE in

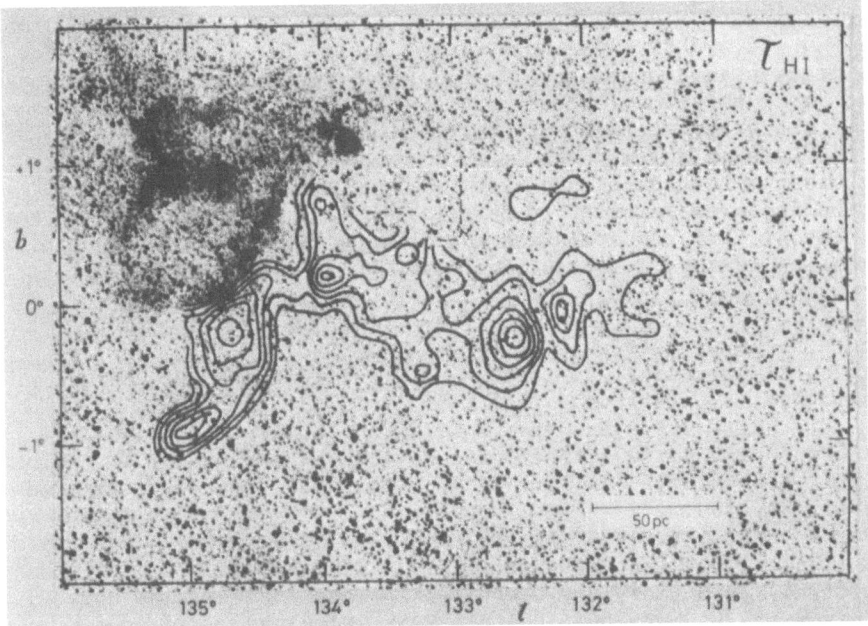

Figure 1.   τ(HI) contours of 0.1, 0.15, 0.2, 0.25, 0.3, and 0.35 for
the velocity of the maximum absorption near -50 km s^{-1} superposed on
the photograph taken in the red light.

Table 1.   Sizes and masses of the fragments.

fragment	(l , b)	diameter(pc)*	M(total)($\times 10^4 M_\odot$)	M(virial)($\times 10^4 M_\odot$)
A	(132°2, -0°1)	20	0.4 - 0.7	2.5
B	(132°5, -0°2)	27	0.9 - 1.8	3.4
C	(133°3, +0°1)	31	2.3	**
D	(134°0, +0°2)	18	0.9	3.1
E	(134°7, -0°2)	30	1.5 - 2.3	4.3
F	(135°0, -0°9)	20	0.4 - 0.8	1.0

*   The diameter at half maximum in the τ(HI) map. D=3.0kpc is assumed.
** The profile is double-dipped and a reliable value cannot be derived.

Table 2.   Physical conditions in the cold HI fragments.

	T_s(K)	τ^{MAX}(HI)	n(HI)(cm^{-3})	[H$_2$]/[HI]	n(total)(cm^{-3})
fragments with CO detection	20	0.36	0.7	50	35
fragments without CO detection	≤40	≤0.47	≤1.8	15 - 50	20 - 35

the calculation of $N(^{13}CO)$ by Lada et al.(1978), or by the assumption of $[^{13}CO]/[H_2] = 2 \times 10^{-6}$. The mass of the entire HI cloud, summed for all the fragments, amounts to $\gtrsim 10^5$ $M_\odot$, which is comparable to the mass observed in CO emission (Lada et al. 1978). Most of the mass is contained in the fragments without CO detection, and this leads us to revise upward the mass of the W3/W4 molecular complex by a factor of 2.

In other molecular clouds also, we have started analyses of HI self-absorption features in order to investigate early stages of cloud evolution. The HI data reduction was done on the FACOM 230-58 computer at the Computing Center, Tokyo Astronomical Observatory.

REFERENCES
Dickman, R.L. 1978, Astrophys. J. Suppl., 37, 407.
Lada, C.J., Elmegreen, B.G., Cong, H.-I., and Thaddeus, P. 1978, Astrophys. J. (Letters), 226, L39.
Myers, P.C., Ho, P.T.P., Schneps, M.H., Chin, G., Pankonin, V., and Winnberg, A. 1978, Astrophys. J., 220, 864.
Rohlfs, K., Braunsfurth, E., and Hills, D.L. 1977, Astron. Astrophys. Suppl. 30, 369.
Sato, F. and Fukui, Y. 1978, Astron. J., 83, 1607.
Sato, F., Fukui, Y., and Hasegawa, T. 1979, this issue.
Westerhout, G. 1973, Maryland-Green Bank Galactic 21-cm Line Survey, (University of Maryland, College Park, Maryland), 3rd ed.
Wilson, T.L. and Minn, Y.K. 1977, Astron. Astrophys., 54, 933.

DISCUSSION FOLLOWING FUKUI

Blitz: The association of giant molecular complexes with HI may be a general property of these complexes. The Rosette molecular complex and the Mon OB1 complex both seem to be embedded in HI clouds, seen in emission, with masses comparable to the molecular masses. Furthermore, examination of Heiles' HI maps of narrow lines shows that there tend to be large HI enhancements coincident with the molecular complexes mapped to date in the second and third quadrants.

Fukui: We are extending the HI-dip studies to the other giant molecular clouds, and hope to examine statistically what fraction of the molecular clouds have such massive less dense components.

Federman: How did you determine the ratio $(H_2)/(HI)$ in regions where no CO emission is seen?

Fukui: The upper limit to the ratio comes from the upper limit to the CO brightness temperature, and the lower limit from the visual extinction. The estimate of the visual extinction ($\gtrsim 1.5$ mag) is crude, and we are now planning a detailed star count in this region with Ohtani and his associates at Kyoto.

Dickel: Why did you adopt 4 kpc for the distance to W3 and W4 when the OB association is at 2 kpc (Ishida, 1969, MNRAS 144, 55; Ishida, 1970, Publ. A.S. Japan 22, 277; Agura and Ishida, 1976, Publ. A.S. Japan 28, 651).

Fukui: We have no strong objection to the distance of 2 kpc. The distance of 3 kpc was just assumed.

HIGH RATE OF DESTRUCTION OF MOLECULAR CLOUDS BY HOT STARS

M. Heydari-Malayeri and M.C. Lortet
Observatoire de Paris, Meudon

L. Deharveng
Observatoire de Marseille

Tenorio-Tagle (1979) first proposed the idea of a third dynamical phase, the champagne phase, following the formation and expansion phases of an HII region; the idea was further explored by Bodenheimer et al. (1979) and Tenorio-Tagle et al. (1979). The champagne phase begins when the high pressure gas of an HII region formed inside a molecular cloud reaches the edge of the cloud and bursts into the lower pressure, low density, intercloud medium. One important implication of the model is the prediction of an enormous enhancement of the rate of erosion of the molecular cloud by the ionising radiation of hot stars, which begins as soon as the process of the decrease of the gas density between the star and the cloud is started. The proportion of hydrogen molecules eroded by ionising photons may reach about 10^{-2}. The mass eroded may exceed the mass of the ionised gas in the case where the ionisation front reaching the edge of the cloud is of D-type. Additional mechanisms (for instance stellar winds), if at work, may even increase the efficiency of the mechanism.

WELL STUDIED HII MOLECULAR COMPLEXES

The examples are very numerous. Disruption of the molecular cloud is found in some cases (for instance Sh2-125=IC 5146, Lada and Elmegreen 1979; Mon OB2, Blitz 1978). Let us focus our attention on molecular clouds containing groups of young objects in different stages of evolution. Four typical instances are K3-50 and the accompanying masers and radio sources (de Graauw et al. 1979); the group Sh2-152 and 153, and an H_2O maser west of Sh2-152 (Israel 1979); the nebulae Sh2-254 to 258 and associated masers and IR sources (Blair 1976); and the molecular cloud containing Sh2-247 and 252 (Baran 1977, Lada and Wooden 1979). In these groups and similar ones it is remarkable that the most diffuse regions (NGC 6857, Sh2-153, Sh2-254, and the diffuse part of Sh2-252) always coincide with the most eroded part of the CO complex or even with the absence of any CO. A more comprehensive analysis will be given by Deharveng (1979).

B. H. Andrew (ed.), Interstellar Molecules, 163–164.
Copyright © 1980 by the IAU.

THE CASE OF SHARPLESS 156

This nebula has been extensively studied by Heydari-Malayeri et al. (1980) and Heydari-Malayeri (1979). In this case the geometry is less favourable than for Sh2-252 or K3-50, where the CO cloud is seen edge on (see Colley and Scott (1977) and Wynn-Williams et al. (1977) for the geometry of K3-50), for the ionising star is seen projected on the bright core. From all available observations and some theoretical predictions on the ionisation structure by Stasinska (1978, 1979), we infer that the ionising star is now located outside the cavity it has eroded in the molecular cloud, so that its radiation reaches obliquely a large area of the surface of the molecular cloud. At this stage of evolution the star may have eroded about 10^3 $M_\odot$ of the molecular cloud.

CONCLUSIONS

A wide variety of objects seen in the same complex may have been formed at the same time, though local conditions (cloud density, mass of the star formed, and its depth in the cloud) have influenced the rapidity of their evolution. The idea of sequential star formation, therefore, would rather apply to large spatial scales, as demonstrated in the case of W3-W4 by Dickel (1979).

It will be very important in the future to obtain local densities and velocities, both for molecular and ionised gas.

REFERENCES
Baran, G.: 1977, private communication.
Blair, G.: 1976, Ph. D. Thesis.
Blitz, L.: 1978, Ph. D. Thesis, N.A.S.A. TM 79 708.
Bodenheimer, P., Tenorio-Tagle, G., Yorke, H.W.: 1979, Astrophys. J. 233, 479.
Colley, D., Scott, P.F.: 1977, M.N.R.A.S. 181, 703.
Deharveng, L.: 1979, Thesis, University of Marseille.
Dickel, H.R.: 1979, this Symposium.
de Graauw, T., Lidholm, S., Fitton, B., Israel, F.P., Sargent, A., Kuiper, T.B.H., Nieuwenhuyzen, H.: 1979, this Symposium.
Heydari-Malayeri, M.: 1979, Ph. D. Thesis, University of Paris VII.
Heydari-Malayeri, M., Testor, G., Lortet, M.C.: 1980 Astron. Astrophys. in press.
Israel, F.P.: 1979, private communication.
Lada, C.J., Elmegreen, B.G.: 1979, Astron. J. 84, 336.
Lada, C.J., Elmegreen, B.G.: 1978, Astrophys. J. (Letters) 226, L39.
Lada, C.J., Wooden, D.: 1979, Astrophys. J. 232, 158.
Stasinska, G.: 1978, Astron. Astrophys. 66, 257; Astron. Astrophys. Suppl. 32, 429.
Stasinska, G.: 1979, to be submitted to Astron. Astrophys.
Tenorio-Tagle, G.: 1978, Astron. Astrophys. 71, 59.
Tenorio-Tagle, G., Yorke, H.W., Bodenheimer, P.: 1979, Astron. Astrophys. 80, 110.
Wynn-Williams, C.G., Becklin, E.E., Mathews, K., Neugebauer, G., Werner, M.W.: 1977, M.N.R.A.S. 179, 255.

THE INTERACTION OF T-TAURI STARS WITH MOLECULAR CLOUDS

Joseph Silk
Department of Astronomy, University of California
Berkeley, California 94720 U.S.A.

Colin Norman
Huygens Laboratorium, Leiden University
Wassenaarseweg 78, 2300 RA Leiden, Netherlands

Winds from T-Tauri stars may provide an important dynamical input into cold molecular clouds. If the frequency of T-Tauri stars exceeds 20 pc^{-3}, wind-driven shells collide and form ram pressure confined clumps. The supersonic clump motions can account for cloud line widths. Clumps collide inelastically, coalescing and eventually becoming Jeans unstable. For characteristic dark cloud temperatures low mass stars form, and we speculate that in this manner clouds can be self-sustaining for $10^7 - 10^8$ yr. Only when either the gas supply is exhausted or an external trigger stimulates massive star formation (for example, by heating the cloud or enhancing the clump collision rate), will the cloud eventually be disrupted. A natural consequence of this model is that dark cloud lifetimes are identified with the duration of low mass star formation, inferred to exceed 10^7 yr from studies of nearby star clusters. Other implications include the prediction of the existence of embedded low mass stars in turbulent cloud cores, the presence of an internal source of radiation in dark clouds, and a clumpy structure for cold molecular clouds.

I. INTRODUCTION

A long-standing problem with molecular clouds lies in understanding their characteristic line widths. On all scales that have hitherto been mapped, the line widths are highly supersonic, although there is some indication that on sufficiently small scales they may not be much broader than thermal (Myers *et al.* 1978). It is evident that simple uniform collapse or expansion models cannot account for observed line widths. Such models imply short lifetimes ($\sim 10^6$ yr) and excessive star formation rates (Zuckerman and Evans 1974). Moreover, recent radiative transfer analyses of molecular line profiles suggest that large scale velocity gradient models are too simplistic (Kwan 1978; Linke and Goldsmith 1979).

165

B. H. Andrew (ed.), Interstellar Molecules, 165–172.
Copyright © 1980 by the IAU.

We wish to present here a new proposal for providing input of stellar energy into these cold, molecular clouds which in general are observed to be bound. It should be noticed that a considerable fraction of the molecular gas ($\sim$ 90% according to Rowan-Robinson 1979) is cold, i.e., $T_{CO} \lesssim 15$ K. We evidently cannot appeal to energy input associated with the interaction of embedded massive stars to account for the dynamical structure of cold molecular clouds.

We note first that the mean lifetime of molecular clouds may be estimated as follows. If Σ is the surface density ($M\odot$ pc^{-2}) and S is the star formation rate ($M\odot$ yr^{-1}), both evaluated at the same galactic radius, and ξ is the efficiency of star formation, the mean cloud lifetime = $\xi\Sigma/S$. In the solar neighborhood we adopt $\Sigma = 2M\odot$ pc^{-2} and $S = 4 \times 10^{-9}M\odot$ yr^{-1} pc^{-2} (Miller and Scalo 1979), and infer a mean cloud lifetime of 5×10^7 (ξ/0.1) yr. A star formation efficiency of $\xi \sim 0.05$-0.1 is indicated by recent studies of dark clouds (Cohen and Kuhi 1979). Comparison of the CO distribution with that of OB associations yields a post-density wave shock lifetime of $\sim (3$-4$) \times 10^7$ yr (Bash, Green, and Peters 1977). These lifetime estimates greatly exceed cloud free-fall times, and show that it is necessary to consider a means of stabilizing clouds against collapse.

Supersonic turbulence will rapidly dissipate. It is evident that in order to maintain the dynamical motions implied by molecular linewidths, either an external or an internal source of momentum must be considered. A review of the problems associated with various possible mechanisms for stabilizing clouds has been given by Field (1978), who suggested that rotation is the principal stabilization mechanism. Mouschovias (1978) has argued in favor of magnetic stabilization. However, detailed observations of molecular clouds provide no evidence for sufficient rotation to stabilize clouds. Also, recent Zeeman-splitting observations strongly constrain the role of magnetic fields in dense clouds. We therefore wish to assert that the most plausible stabilization mechanism is the interaction of embedded stars with molecular clouds. In a later paper, we discuss the interactions of massive stars with warm molecular clouds (Norman and Silk 1979a). Here we consider cold molecular clouds, where only low mass stars can be present. Low mass stars may provide an important dynamical interaction with the surrounding medium during their pre-main sequence convective phase. During much of this time, which lasts $\sim 2 \times 10^7$ yr for a star of 0.8 M$\odot$, these stars are believed to be in the T-Tauri phase. The T-Tauri phenomenon is characterized by extensive mass outflow (and possibly inflow; Ulrich 1976). Characteristic outflow velocities are 100-300 km s^{-1} at mass flow rates of 10^{-7}- 10^{-8} M$\odot$ yr^{-1} (Kuhi 1964).

Let us first ascertain whether the momentum input from T-Tauri winds can provide a significant contribution to molecular cloud dynamics. Since momentum is approximately conserved during the interaction of these radiatively cooled winds with the ambient cloud material, a rough estimate of the resulting velocity dispersion transmitted to an average volume element of cloud material is

$$\langle\Delta v\rangle \sim \left(\frac{M_*}{M_{cl}}\right)\left(\frac{\dot{m}\, t_{\text{T-Tauri}}}{m}\right) V_w$$

where M_* is the stellar mass in the cloud, M_{cl} is the cloud mass, $\dot{m}$ is the mean mass loss rate of an individual T-Tauri star of mass m over time $t_{\text{T-Tauri}}$, and V_w is the wind velocity. Adopting $M_*/M_{cl} \sim 0.1$ (Cohen and Kuhi 1979), we conclude that if a fraction ~ 0.1 of the initial stellar mass is lost in the pre-main sequence phase, then $\langle\Delta v\rangle \sim (1-3)$ km s^{-1}.

This demonstrates that T-Tauri stars could provide significant dynamical input into molecular clouds. To pursue this idea further, we now consider in more detail the nature of this interaction between T-Tauri stellar winds and the ambient molecular cloud.

II. WINDS, BUBBLES, AND SHELLS

Once the mass swept out by the T-Tauri winds exceeds the ejected mass, a strong shock develops which rapidly forms a shell that subsequently snowplows into the molecular cloud. The subsequent evolution of the approximately momentum-conserving shell resembles that of interstellar bubbles (Weaver et $al.$ 1977). One notable difference is that for wind velocities $\lesssim 100$ km s^{-1} the wind will be strongly radiative. Details of T-Tauri bubble evolution are given by Norman and Silk (1979b). The most relevant question that concerns us here is the extent of the bubble and shell radius, R_S. Norman and Silk show that

$$R_S = \left(\frac{\dot{M} V_w}{4\pi\rho V_*^2}\right)^{\frac{1}{2}}$$

$$= 0.1\left(\frac{\dot{M}}{10^{-7}\, M_\odot\, yr^{-1}}\right)^{\frac{1}{2}}\left(\frac{V_w}{100\ km\ s^{-1}}\right)^{\frac{1}{2}}\left(\frac{n}{10^3\ cm^{-2}}\right)^{-\frac{1}{2}}\left(\frac{V_*}{1\ km\ s^{-1}}\right)^{-1}pc.$$

The limiting radius is reached when the ram pressure due to the motion of the stars, V_*, through the cloud balances the wind momentum. The swept-up mass at this stage amounts to ~ 1 M$\odot$.

Consider next the question of how many low mass T-Tauri stars are required in order for the shells to collide. If the T-Tauri star density n_T is sufficient for shell collisions to prevail, then the molecular cloud will effectively develop a 2-phase medium consisting of dense swept-up shells moving in a more diffuse medium. The shell-shell collision time is of order $(n_T\pi R_S^2 V_*)^{-1}$, and the necessary condition is that this be less than $t_{\text{T-Tauri}}$. Adopting $t_{\text{T-Tauri}} = 0.1\ m/\dot{m}$ as before, we infer a critical density of T-Tauri stars of order

$$20\left(\frac{V_*}{1\ km\ s^{-1}}\right)\left(\frac{n}{10^3\ cm^{-3}}\right)\left(\frac{100\ km\ s^{-1}}{V_w}\right)\ pc^{-3}.$$

This is equivalent to requiring the stellar mass fraction in the cloud
to exceed ~ 0.1 $(V_*/1 \text{ km s}^{-1})(100 \text{ km s}^{-1}/V_w)$.

It is of interest to compare the T-Tauri star density with available data. The best studied dark cloud region is the Taurus-Auriga
dark cloud complex. According to Cohen and Kuhi (1979), the considerable number of T-Tauri stars distributed in the outer regions of dark
clouds occur in aggregations that range up to densities of $\sim 30 \text{ pc}^{-3}$.
Jones and Herbig (1979) cite a lower number ($\sim 4 \text{ pc}^{-3}$) for aggregations
in the same region. We should bear in mind that the maximum A_v for the
T-Tauri stars in the Cohen and Kuhi sample is $A_v \sim 4$, and the T-Tauri
star density could be greater in the denser cloud core regions. According to Jones and Herbig (1979), the proper motions of the T-Tauri
stars indicate a velocity dispersion of $2-3 \text{ km s}^{-1}$. Again, we remark
that this value of V_* refers to the outer regions of the dark clouds.

It is evident that these values of V_*, V_w, and n_T are indicative
that T-Tauri stellar winds could indeed drive shells that collide in
dark clouds similar to the clouds in the Taurus region. One consequence of this hypothesis that supersonic line widths especially in
dark cloud cores find a ready explanation in continuously driven turbulence. The interacting shells will build up into clumps of dimen-
sion ~ 0.1 pc and mass $\sim 0.1-1 \text{ M}\odot$. These will only be weakly confined
by ram pressure because of their low Mach numbers, and are likely to
continuously replenish the interclump medium. Thus we envisage that
cloud cores will contain many wind-driven clumps. Clumps of lower
column density will be driven out of the cloud cores by the winds, and
consequently the lower density outer parts of clouds may be undergoing
more systematic large-scale motion. This could either be inflow, if
the cloud cores are accreting material from the larger, more diffuse
molecular cloud complex in which they are embedded, or outflow, if
interacting winds from the core can drive large-scale motions. One
might speculate that cycles of inflow and outflow could alternate,
being regulated, as we now argue, by low mass star formation.

III. LOW MASS STAR FORMATION

We have indicated a means of explaining cold cloud line widths.
However it is also necessary for the clouds to be sufficiently long-
lived. Longevity and stability require continuous low mass star
formation, and we now speculate on a mechanism for achieving this.

For temperatures characteristic of clumps, the Jeans mass equals
$10(T/10K)^{3/2}(10^4 \text{cm}^{-3}/n)^{1/2} \text{ M}\odot$. We argue elsewhere that inelastic clump
collisions and clump coagulation will dominate over competing processes
such as leakage and acceleration. Thus, within a few clump-clump collision times or $\sim 10^7$ yr, the clumps will become Jeans unstable and
form low mass stars. It seems unlikely that massive stars could be
formed unless an external heat source is supplied to raise the Jeans
mass above $\sim 1 \text{ M}\odot$.

This leads us to the following model. Low mass stars will form, and develop winds which sweep up shells. The shells intersect, break up into clumps that are driven together by the winds, gradually coalescing and eventually forming low mass stars. Thus, low mass star formation is self-sustaining, and proceeds continuously either until the gas supply is exhausted or an external trigger stimulates disruptive massive star formation. For example, we envisage that interaction with a nearby OB star association would provide sufficient energy input to raise the Jeans mass and enhance the coagulation rate sufficiently to enable massive stars to form. One important piece of evidence is the fact that studies of star formation indicate that star formation is non-coeval (Herbig 1962). Low mass star formation has evidently been occurring over a timescale in excess of $\sim 10^7$ yr. Then *our star formation model identifies the timescales associated with cold molecular clouds and non-coeval star formation, and incorporates both into a unified model.* Both problems can be simultaneously resolved in our picture.

IV. IMPLICATIONS

If cold molecular clouds are supported by interactions of winds from embedded T-Tauri stars, there are a number of observable consequences. Typical T-Tauri bolometric luminosities are $\sim 1\text{-}10\ L\odot$. Moderate improvements in sensitivity of far IR surveys of dark clouds should be capable of testing our hypothesis, particularly if diagnostics specific to T-Tauri stars can be utilized. Specifically, these could include far infrared line emission from fine-structure excitation of CI and OI, and H_2 vibrational lines associated with the shocked gas.

There are also significant consequences for molecular cloud observations. There will be an internal source of UV due to wind shocks equivalent to the mean interstellar radiation field at a visual extinction of $A_V \sim 3$. This could result in significant modification of dark cloud chemistry by means of photo-dissociation of molecules and photo-ejection both of molecules and electrons from grain surfaces.

Typical density contrasts expected amount to a factor ~ 10, with a volume filling factor ~ 0.1 and a clump scale of ~ 0.1 pc. The ram-pressure confined clumps satisfy $n \propto v^2$, and the gas that contributes to emission line wings may therefore be denser than gas contributing to the line cores. The implications of our model for detailed line profiles remain to be explored. However, the basic feature of our model, namely a turbulent clumpy core surrounded by a more extensive, diffuse halo region sustaining a relatively uniform inflow or outflow, are consistent with recent models for molecular line profiles.

We acknowledge stimulating correspondence with Dr. G.B. Field and helpful discussions with Drs. M. Cohen and L. Kuhi. This research has been supported in part by the NATO Scientific Affairs Division and by NASA under grant NGR-05-003-578.

REFERENCES

Bash, F.M., Green, E., and Peters, W.L.: 1977, Astrophys. J., 217, 464.
Cohen, M., and Kuhi, L.V.: 1979, Astrophys. J. (Suppl.) 41, No. 4.
Field, G.B.: 1978, *Protostars and Planets*, ed. T. Gehrels (University
     of Arizona Press), p. 243.
Herbig, G.: 1962, Astrophys. J., 135, 236.
Jones, B.F., and Herbig, G.H.: 1979 (preprint).
Kuhi, L.V.: 1964, Astrophys. J., 140, 1409.
Kwan, J.: 1978, Astrophys. J., 220, 147.
Linke, R.A., and Goldsmith, P.F.: 1979, *Interstellar Molecules*, IAU
     Symposium 87 (in press).
Miller, G., and Scalo, J.: 1979, Astrophys. J. (Suppl.) 41, 513.
Mouschovias, T.Ch.: 1978, *Protostars and Planets*, ed. T. Gehrels
     (University of Arizona Press), p. 209.
Myers, P.C., Ho, P.T.P., Schneps, M.H., Chen, G., Pankonin, V., and
     Winnberg, A.: 1978, Astrophys. J., 220, 864.
Norman, C., and Silk, J.: 1979a, *Interstellar Molecules*, IAU Symposium
     87 (this volume).
Norman, C., and Silk, J.: 1979b, Astrophys. J. (submitted).
Rowan-Robinson, M.: 1979, Astrophys. J. (in press).
Ulrich, R.K.: 1976, Astrophys. J., 210, 377.
Weaver, R., McCray, R., Castor, J., Shapiro, P., and Moore, A.: 1977,
     Astrophys. J., 218, 377.
Zuckerman, B., and Evans, N.J.: 1974, Astrophys. J. (Letters), 192, L149.

DISCUSSION FOLLOWING SILK

Gilmore: You are using your model to account for the stability of
all clouds. However Zuckerman and Evans estimated that if all clouds
formed stars, then the rate of star formation would be 10-100 times
greater than observed. Can your model be consistent with their estimate?

Silk: Unacceptably high star formation rates arise only if
molecular clouds are assumed to have short lifetimes comparable to the
free-fall time-scales. We are proposing that the clouds are stabilized
by star formation over a time-scale of order a few tens of millions of
years with an efficiency of order 10%. The corresponding mean star
formation rate is then of order 0.1 x (mass in molecular clouds)/cloud
lifetime, or a few solar masses per year, consistent with the observed
rates.

Mezger: There is a severe problem with your proposed model if it
is to work in giant molecular clouds: the star formation rate in the
Galaxy is ~ 3 $M_\odot$ yr^{-1}. An upper limit for the age of the T-Tauri stage
of low mass stars is $\sim 10^6$ yr. It follows that there are some 10^6
T-Tauri stars in the Galaxy, or some 100 per giant molecular cloud.
This number falls orders of magnitude below what your model requires.

Silk: My estimates of T-Tauri stars needed and likely to be present
differ from yours. First, what does the model require? To provide
efficient stirring up of, say, 10^9 $M_\odot$ in cold molecular cloud cores
requires $\sim 10^8$ T-Tauri stars in the galaxy. How many are present?

The observed star formation rate of $\sim 3 M_\odot$ yr^{-1} implies that ~ 4 yr^{-1} is the formation rate of low mass stars (mean mass 0.8 $M_\odot$, say) and therefore is also the approximate formation rate of T-Tauri stars. If the lifetime of the T-Tauri phase is even 10% of the duration of the pre-main sequence convective phase ($\sim 2 \times 10^7$ yr for a star of 0.8 $M_\odot$), there must be $\sim 10^7$ T-Tauri stars present in the galaxy at any time. The stellar winds from these stars suffice to stir up $\sim 10^8$ $M_\odot$ of gas, perhaps enough to account for a significant fraction of the cold molecular cloud cores that show evidence for turbulent line broadening. Because of uncertainties in the mass-range of T-Tauri stars, the duration of their outflow phase, and the initial stellar mass function, it seems not unlikely that up to $\sim 10^9$ $M_\odot$ of molecular cloud gas could be affected by T-Tauri stellar winds.

Bok: Can Herbig-Haro Objects take the place of T-Tauri stars? We now have one globule near the edge of the Gum Nebula which shows two Herbig-Haro objects being ejected at high speed.

Silk: It is indeed possible that Herbig-Haro objects could provide significant heating in dense clouds and globules. Acting much like "interstellar bullets," I believe they could penetrate distances up to ~ 1 parsec.

Zuckerman: Strong stellar winds from T-Tauri stars are required for your model, but their existence has been disputed by Ulrich and Knapp. Also, the IRAS satellite scheduled for launch in 1981 will be capable of detecting, in the infrared, the embedded stars required in your model.

Silk: I agree.

Myers: Earlier in the symposium we heard that there are "clumps" in the Taurus complex which have size about 0.1 pc, mass about 1 $M_\odot$, and which appear to have significant density contrast. These parameters appear consistent with your model. However we have noticed that the linewidths are extremely small, only about 2 or 3 times the thermal width. Are these widths consistent with your model?

Silk: Such narrow linewidths would imply mildly supersonic motions for the clumps. In our model, these motions lead to some degree of ram pressure confinement, with density contrast of order 10.

Kutner: For a simple spherical cloud with a small number of T-Tauri stars inside, how would you expect the observed linewidth to vary as a function of the distance of the line of sight from the centre of the cloud?

Silk: Our model suggests the existence of many clumps, with the more slowly moving clumps confined to the cloud core. Thus I would expect beam-to-beam variations (if a beam corresponding to ~ 0.1 pc resolution is used) in velocity structure, with smaller linewidths contributed from the core material than from the cloud material. However ram pressure confinement means that the gas contributing to the line wings could be denser than the lower velocity gas, so there might be significant opacity in the line wings. This effect complicates predictions of linewidth variations.

Mouschovias: What fraction of molecular clouds contains sufficient T-Tauri stars for your mechanism to work? If polarization observations show large-scale ordering of the magnetic field, would that not be an embarrassment to every model appealing to disordered gas motions in dense clouds?

Silk: Observations of T-Tauri stars have been made only in the outer fringes of a few nearby dark clouds. The observed frequency indicates that T-Tauri stars could be sufficiently common in core regions for our model to work. Polarization observations of embedded infrared sources could indeed provide significant information on the large-scale structure of the field. If large-scale ordering could be unambiguously demonstrated, it would suggest that magnetic forces play a major role in the cloud structure.

OBSERVATIONS OF THE J=1-0 AND J=2-1 LINES OF ^{12}CO IN L1551: EVIDENCE
FOR ANISOTROPIC MASS LOSS

R.L. Snell, R.B. Loren (University of Texas-Austin), and
R.L. Plambeck (University of California-Berkeley)

Observations of the J=1-0 lines of ^{12}CO have been made to determine the extent and nature of the broad secondary velocity feature in
L1551 (S239) first detected by Knapp et al. (1976). L1551 is associated with an extended Herbig-Haro object, HH102 (Strom et al. 1974),
and two compact HH objects, HH28 and HH29 (Herbig 1974). Two nearby
stars, HL Tau and XZ Tau, show extreme T-Tauri characteristics. The
broad CO velocity component is seen toward HH102. Strom et al. (1976)
have detected an infrared source near the HH objects which has a luminosity and infrared colors consistent with a late B star reddened by
roughly 20 magnitudes of visual extinction (Snell 1979). The infrared
source lies in the direction of the core of L1551, which has a density
of 10^5 cm^{-3} (Loren et al. 1979; Snell 1979). The infrared source is
likely to be embedded in this dense core.

Our observations show that the broad CO feature occurs in two locations symmetrically placed about the embedded infrared source. One
region to the SW of the infrared source has broad emission at velocities lower than the narrow primary velocity component of the cloud, and
the other region to the NE has broad emission at higher velocities.
The high and low velocity features are at roughly 11.5 km s^{-1} and
1.2 km s^{-1}, shifted by ±5 km s^{-1} from the narrow cloud-emission at
7 km s^{-1}. The total velocity extent of the broad components is 25 km s^{-1}.

Observations of J=2-1 ^{12}CO show broad emission features that are
enhanced over the J=1-0 emission in many of the positions to the SW,
suggesting that the broad CO emission comes from warm, optically thin
material. The low optical depth is substantiated by the non-detection
of J=1-0 ^{13}CO emission at the velocity of the broad component. The
excitation temperature of the narrow CO component is T_{ex}=10-15K, but
from the ratio of the J=2-1 to J=1-0 emission the broad CO component to
the SW has T_{ex}>25K in some regions.

The broad CO emission likely has a common origin with the HH objects. It has been suggested that the HH emission arises from shocked
regions generated by mass loss from young, luminous stars embedded in

B. H. Andrew (ed.), Interstellar Molecules, 173–174.

the clouds (Schwartz 1978; Dopita 1978; and Raymond 1978). The radial velocities of HH102 and HH29 have been measured by Strom et al. (1974) and these objects are moving toward the observer at much higher velocities than the CO from which the broad lines come. Recently the proper motions of HH28 and HH29 were measured (Cudworth and Herbig 1979). These are directed away from the infrared source with velocities around 140 km s^{-1}. The broad CO emission is likely produced by material moving away from the infrared source with velocities of roughly 5 km s^{-1} directed in two streams, one to the NE and away from the observer, and one to the SW and toward the observer. Shock heating produces the higher excitation temperatures observed in the broad CO features. HH28 and HH29 move with the SW stream but at much higher velocities. The optical emission is seen only in the SW stream because it is directed toward the front face of the cloud.

The high velocities of HH28 and HH29 fit in well with the model of Norman and Silk (1979). The HH objects are ejected at high velocities by the interaction of mass loss from a young, luminous object and the infall of cloud material. The broad CO features may be the result of an interaction between the ejected material and the ambient cloud. The total mass of gas of the broad components is 0.05 $M_{\odot}$ in the NE feature and 0.10 $M_{\odot}$ in the SW feature. The total kinetic energy in the broad components is 3×10^{43} ergs. A rough time scale for mass-loss activity can be estimated from the expansion time for HH28 and HH29, assuming they have been moving at a constant velocity. This time scale is around 1000 years. Over this period of time a mass-loss rate of 10^{-7} $M_{\odot}$/yr. with a wind velocity of 100 km s^{-1} is needed to produce the observed kinetic energy present in the broad features, neglecting radiative cooling losses. This is a modest mass-loss rate; many estimates of the mass-loss rate for T-Tauri stars and Herbig Ae-Be stars are as large as 10^{-5} $M_{\odot}$/yr. This research was supported in part by NSF Grants AST 77-28475 and AST 75-13511.

REFERENCES

Cudworth, K.M., and Herbig, G.: 1979, Astron. J. 84, 548.
Dopita, M.: 1978, Astrophys. J. Suppl. 37, 117.
Herbig, G.: 1974, Lick Obs. Bull. #658.
Knapp, G.R., Kuiper, T.B.H., Knapp, S.L., and Brown, R.L.: 1976,
    Astrophys. J. 206, 443.
Loren, R.B., Evans, N.J., and Knapp, G.R.: 1979, Astrophys. J. 234, 932.
Norman, C., and Silk, J.: 1979, Astrophys. J. 228, 197.
Raymond, J.C.: 1979, Astrophys. J. Suppl. 39, 1.
Schwartz, R.D.: 1978, Astrophys. J. 223, 884.
Snell, R.L.: 1979, Ph.D. thesis, University of Texas (unpublished).
Strom, K.M., Strom, S.E., and Vrba, F.J.: 1976, Astron. J. 81, 320.
Strom, S.E., Grasdalen, G.L., and Strom, K.M.: 1974, Astrophys. J.
    191, 111.

AMMONIA OBSERVATIONS OF DARK CLOUDS CONTAINING HERBIG-HARO OBJECTS

Paul T.P. Ho and Alan H. Barrett
University of Massachusetts and
Massachusetts Institute of Technology

Abstract. A study in NH_3 was conducted towards H-H objects. Cloud fragmentation appears to have occurred in the regions mapped. Rotation is present with velocity gradients of 1-2 km s^{-1} pc^{-1}. We suggest that in the regions containing H-H objects, formation of stars of different spectral types may be taking place.

The Herbig-Haro (H-H) objects are semi-stellar knots or irregular patches of nebulosity whose optical spectra are characterized by emission lines of hydrogen, unusually strong lines of $\boxed{SII}$ and $\boxed{OI}$, and other lines associated with low excitation gas. Numerous models have been proposed to explain the nature of these objects, including *in situ* excitation from an embedded star, reflection nebulae illuminated by obscured but displaced young H-H stars, FU Orionis-type events, shock excitation of interstellar cloudlets and the stellar wind itself, as well as models of H-H objects as expired H_2O masers.

We present here a different approach to the problem of H-H objects, the study of their associated dark clouds. We observed the (J,K)=(1,1) line of NH_3 using the 36.6 m telescope of the Haystack Observatory. The purpose of this study is to confirm the association of H-H objects with high density material, set limits on local densities, search for dynamical interaction between H-H objects and ambient cloud matter, determine the spatial relationship between the optical H-H objects and the density distribution within the clouds, and to study the dynamics of associated dark clouds themselves. We surveyed 25 regions containing H-H objects, detecting NH_3 towards roughly half of them. We also mapped two of the strongest sources, the NGC1333 region and the Serpens object. The survey results combined with the mapping results indicate that although H-H objects are in general associated with dense neutral material with typical density $n(H_2) \gtrsim 5 \times 10^3$ cm^{-3}, some of these regions cannot be detected in NH_3 to very low limits. This is especially clear in the NGC1333 region, where some of the H-H objects are found in the lower-density periphery of the NH_3 cloud. Furthermore, regions containing H-H objects appear to be quiescent with line widths $\lesssim 0.6$ km s^{-1}, so that if

175

B. H. Andrew (ed.), Interstellar Molecules, 175–176.

dynamical interactions with dark cloud matter are present they must occur in the low density region where NH_3 cannot be detected. Although our data appear to argue against the *in situ* model, we note that the present study cannot detect very compact <0.02 pc condensations. The nondetection of compact HII regions and H_2O masers by Haschick et al. (1979) and Rodriguez et al. (1980) in the direction of H-H objects may be more relevant to the *in situ* question by eliminating the possibility of massive main sequence stars. Discrete and continuous outflow phenomena are also considered. We find that our present experiment cannot detect the very small ejected condensations proposed by Norman and Silk (1979) and Rodriguez et al. (1980), although observed hints of outflow from compact HII regions in the Serpens object may be consistent with a strong stellar wind. One major conclusion from our study is that there is poor spatial correlation between the detected dark cloud matter and other signposts of star formation such as the optical H-H objects, H_2O masers, compact HII regions, and IR sources, which themselves are not well correlated in position. Considering the various difficulties of all current models of the H-H phenomena, we suggest the possibility that the aforementioned signposts may not be directly related to each other. Instead, we may be observing the formation of a spectrum of stars with different masses and possibly in different stages of evolution. In this context, detected cloud fragments may be the sites of future star formation.

Mapping results of NGC1333 and Serpens reveal distinct density condensations with masses of 30-200 $M_\odot$. Rotation with a velocity gradient of the order of 1-2 km s^{-1} pc^{-1} is present, which accounts for the large scale velocity structures. Mutual rotation between fragments appears to be sufficient for stable orbits, although rotation by itself does not provide adequate support against the collapse of individual condensations. In particular there is in the NGC1333 region a massive fragment whose gravitational energy greatly exceeds the thermal energy; its apparent instability against collapse is confirmed by an observed region of enhanced line width spatially coincident with the densest position. Our results are also relevant to modeling of CO observations in the NGC1333 region. Snell and Loren (1977) proposed nonhomologous collapse to explain the CO line shapes observed towards NGC1333. We find however that the very narrow NH_3 lines observed in the high density core appear to be inconsistent with such a collapse law. More detailed modeling of the cloud structures in these regions appears to be necessary.

Haystack Observatory is supported in part by NSF Grant AST78-18227. The authors are supported in part by AST76-24610 and AST77-12960.

REFERENCES

Haschick, A.D., Moran, J.M., Rodriguez, L.F., Burke, B.F., Greenfield, P.E., and Garcia-Barreto, J.A.: 1979, in preparation.
Norman, C.N., and Silk, J.: 1979, Ap.J., 228, p.197.
Rodriguez, L.F., Moran, J.M., Ho, P.T.P., and Gottlieb, E.W.: 1980, Ap.J., in press.
Snell, R.L., and Loren, R.B.: 1977, Ap.J. 211, p.122.

HYDROSTATIC MODELS OF MOLECULAR CLOUDS

T.de Jong[*], A.Dalgarno[+] and W.Boland[*+]
[*]Astronomical Institute, 1018 WB Amsterdam, The Netherlands
[+]Center for Astrophysics, Cambridge, MA 02138, USA

1. INTRODUCTION AND SUMMARY

It is customary to convert observed brightness temperatures of molecular lines to molecular column densities by assuming some excitation temperature to describe the population of the energy levels of the molecule. Observers often speak about "abundances" after dividing these column densities by the dimension of the cloud that is being studied. This procedure can be misleading because the level populations are usually not well characterized by one single excitation temperature and because clouds are probably quite centrally condensed. Some molecules are formed preferentially at high densities in the opaque cores of clouds, others reach their maximum abundance in the moderately shielded much less dense outer regions of clouds.

We have constructed a plane-parallel model of the well-studied molecular cloud L134 assuming that it is in hydrostatic equilibrium supported by turbulent pressure. This cloud model is quite centrally condensed. Molecular abundances are calculated by solving the coupled set of equations of chemical equilibrium and thermal balance as a function of depth in the cloud. Depletion of atoms and molecules onto dust grains is taken into account. Column densities of several molecules are predicted and compared with the observations.

2. INGREDIENTS OF THE MODEL

We consider a self-gravitating plane-parallel cloud layer that is in hydrostatic equilibrium supported by turbulent pressure. The turbulence is assumed to be Gaussian and is characterized by the Doppler width δV_D. The chemistry consists of about 150 reactions between about 40 species of the Carbon-Oxygen family (Herbst and Klemperer 1973; Black and Dalgarno 1977). The gas is heated by photoelectrons from dust grains

[+]The research of WB is supported by the Netherlands Organisation for the Advancement of Pure Research (ZWO).

177

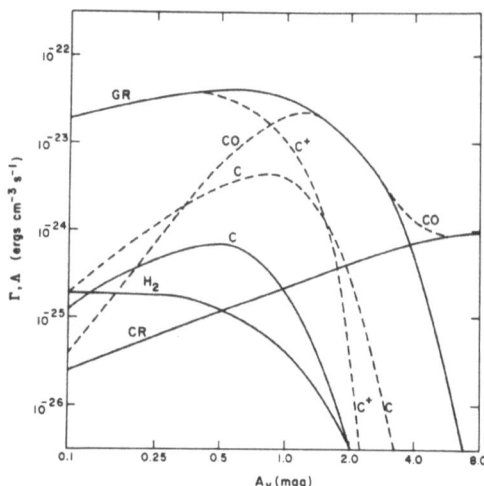

Figure 1. Heating (——) and cooling (- -) rates in the cloud
model of L134.

and by cosmic ray (CR) ionizations. It is cooled by collisional excitation of C^+ions, C^oatoms and CO molecules followed by radiative decay.
Radiation trapping of the cooling photons is very important (in the
core of the cloud). The escape probability of the cooling radiation is
calculated assuming the same turbulent velocity field characterized by
the parameter δV_D. The depletion of Carbon and Oxygen by collisions
with and sticking onto dust grains is approximately included in the
calculations by assuming that depletion occurs through CO collisions. The
depletion depends on the temperature and the density of the gas. The
amount of depletion increases with depth and depends on the age of the
cloud. For more details the reader is referred to de Jong, Dalgarno and
Boland (1979).

TABLE 1. Parameters characterizing the cloud model of L134

A_V	$n_H(cm^{-3})$	$T(K)$	δ
0	35	68	0.250
0.1	840	33	0.209
0.25	2040	27	0.168
0.5	4000	24	0.120
1	7710	21	0.068
2	14400	13.3	0.036
4	24700	6.7	0.023
8	32800	7.0	0.010

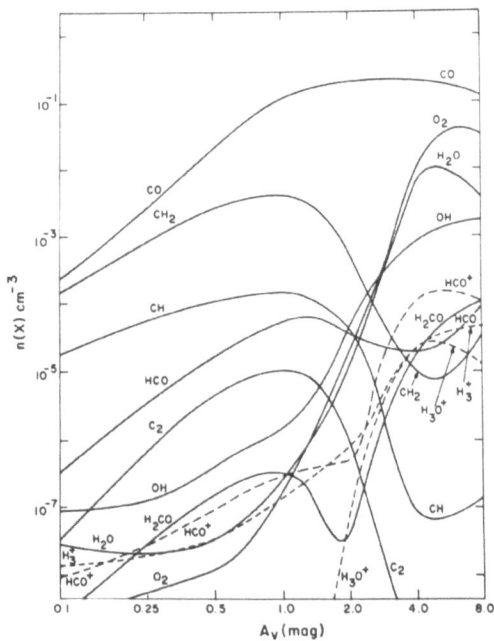

Figure 2. Molecular abundances in the cloud model of L134.

3. A MODEL OF L134

The dark cloud L134 is about 1pc in diameter and has a visual extinction in the centre larger than 7 magnitudes. Molecular line observations of L134 have recently been summarized by Mattila, Winnberg and
Grasshoff (1979). The temperature in the core of the cloud is about
10K (from CO) and the molecular Hydrogen density in the core exceeds
$10^4 cm^{-3}$ (from H_2CO). The full width at half maximum of the molecular
lines is in the range 0.3 -1 km s^{-1}. In the core of the cloud Carbon
and Oxygen are strongly depleted and the degree of ionization is quite
low (Watson, Snyder and Hollis 1978). The "observed" column densities
of several molecules are given in Table 2.

Our model of L134 consists of a symmetrical plane-parallel cloud
layer with a visual extinction of 8 magnitudes to the centre. The ambient gas pressure is $3\times10^3 K$ cm^{-3} and the ambient radiation field is
interstellar. The turbulence parameter δV_D equals 1 km s^{-1} and we have
assumed a CR ionization rate of 2×10^{-18} s^{-1}. The resulting density and
temperature distribution is given in Table 1. The thickness of the cloud
layer is 0.9pc. The depletion factor δ decreases strongly towards the
centre. The temperature distribution is hardly affected by the depletion because the cooling lines are thermalized (cooling independent of
abundance cf. de Jong, Chu and Dalgarno 1975).

TABLE 2. Observed and predicted column densities (cm^{-2}) in L134.

Species	N(H)/T[*]	CH	OH	CO	HCO^+	H_2CO
observed	1×10^{18}	1×10^{14}	5×10^{14}	4×10^{17}	4×10^{13}	1×10^{14}
predicted	1.3×10^{18}	9.7×10^{13}	7.2×10^{14}	2.0×10^{17}	6.7×10^{13}	3.0×10^{13}

Species	CH_2	H_2O	H_3O^+	HCO	C_2	O_2
predicted	2.4×10^{15}	3.2×10^{15}	1.1×10^{13}	6.0×10^{13}	4.7×10^{12}	1.1×10^{16}

[*]quantity proportional to 21 cm optical depth (units cm^{-2} K^{-1})

In figure 1 we show the heating and cooling rates as a function of depth in the cloud. In the outer parts the gas is heated by photoelectrons from dust grains and cooled by C^+ fine structure transitions. In the core CR heating and CO rotational line cooling dominate.

In figure 2 we show the variation with depth of the abundances of several molecular species. There are clearly two regimes. In the outer parts molecules are formed by diffuse cloud chemistry (CH, CH_2, C_2 and HCO) and in the core dark cloud chemistry dominates (OH, H_2O, HCO^+, H_2CO and O_2). All Carbon is in CO over most of the cloud. The use of more up-to-date chemical schemes (cf. Huntress 1979) changes some of these features but the general pattern persists (de Jong et al. 1979). The predicted column densities in Table 2 are in fair agreement with those derived from the observations.

The degree of ionization in the cloud core is very low ($\sim 10^{-8}$). The main ion is HCO^+. This low degree of ionization is of importance for the dynamical evolution because the ambipolar diffusion time of the cloud becomes of the same order of magnitude as the cooling time, the chemical equilibrium time and the free-fall time (all of order few times 10^5yrs). Since these time scales are smaller than the age of the cloud (few times 10^6yrs derived from the amount of depletion) nothing, apart possibly from rotation, seems to prevent fragmentation and star formation in the core of L134.

REFERENCES

Black, J.H. and Dalgarno, A.1977, Astrophys.J.Suppl.Ser.34, 405.
de Jong, T., Chu, S.I. and Dalgarno, A.1975, Astrophys.J.199, 69.
de Jong, T., Dalgarno, A. and Boland, W.1979, submitted to Astron. Astrophys.
Herbst, E. and Klemperer, W.1973, Astrophys.J.185, 505.
Huntress, W.T.1979, these proceedings.
Mattila, K., Winnberg, A. and Grasshoff, M. 1979, Astron. Astrophys. 78, 275.
Watson, W.D., Snyder, L.E. and Hollis, J.M.1978, Astrophys.J.(Letters) 222, L195.

DISCUSSION FOLLOWING DE JONG

Kutner: You compare certain abundances predicted by the model with observed abundances, but the observed abundances are really derived quantities dependent on some cloud model. Would it not be a more meaningful test of any cloud model to synthesize line profiles for various species at various positions, and compare them with the observations?

de Jong: Yes, I think so. We are planning to calculate the brightness temperatures of several molecular lines in L134 to carry out such a comparison. The only observed line whose brightness temperature is calculated in our model is the $J=1 \rightarrow 0$ line of CO at 2.6 mm, because it is one of the cooling lines. Its predicted brightness temperature is about 12K, in excellent agreement with the observations.

Churchwell: L134 and TMC1 are two cloudlets where the line widths are roughly equivalent to the expected thermal width. There is also an indication of a significant rotation rate. Therefore one must depend on other mechanisms to support the clouds. How fast must the core rotate to support itself against gravitational collapse? Is it slow enough that the rotation could have been missed? Or are these clouds perhaps being supported in some other way (e.g. magnetic field)?

Clark: Published data establish reasonably unambiguously that L134 is rotating (Brooks et al.) and has an equatorial bulge. It would seem that your model may be inappropriate.

de Jong: These questions have crossed our minds. We have also constructed models that are supported purely by thermal pressure. These models are even more strongly centrally condensed. However the observations indicate that, though the core of these clouds may be supported by thermal pressure alone, in the outer parts the line widths are superthermal. As far as rotation is concerned, of course rotation affects the density stratification of the cloud. Thermal pressure and rotation could in principle be incorporated in our models. The present approach has been chosen because it is simple and easy.

Mouschovias: How did you obtain your estimate of the time scale for ambipolar diffusion? Are you aware of a recent result that this time scale varies by many orders of magnitude from point to point within a dense cloud (Ap. J. 228, 475, 1979)?

de Jong: I am arguing that the ambipolar diffusion time varies strongly with position in the cloud because it is proportional to the degree of ionization. Only in the core of the cloud does the degree of ionization become sufficiently small that the magnetic field cannot prevent collapse.

CONTAGIOUS B STAR FORMATION IN THE RHO OPHIUCHI DARK CLOUD

E. Falgarone
Département de Radioastronomie, Observatoire de Paris,
Meudon, France

Three young stars are still embedded in the small (0.7 pc wide) and densest region of the Rho Ophiuchi dark cloud. One of these stars coincides with the continuum source Oph 4 (Brown and Zuckerman, 1975). Their existence is indicated by the following: i) thermal continuum sources detected at 21 cm and 6 cm with the Westerbork Synthesis Radiotelescope (Netherlands) which display the presence of compact HII regions surrounding each of three stars; ii) peaks of CO emission (J = 1 $\rightarrow$ 0 transition) detected with the MWO antenna of MacDonald (Texas) which are associated with asymmetrically shaped lines; iii) C 158 α (and S 158 α for Oph 4) line emission mapped with the Nancay Radiotelescope. All these observations imply that the three stars have B3 - B2 spectral types.

A possible interpretation of the simultaneous formation of three similar stars within a small area of the Rho Oph cloud is proposed, based on the contagious process of star formation described by Elmegreen and Lada (1977) for the case of OB associations. A group of nine background stars of types ranging between B2 and B9 (Chini et al., 1977) are thought to be associated with the cloud (Elias, 1978) but not embedded in it. These stars likely excite the evolved HII region detected at Westerbork (Falgarone et al., 1978) because their spectral types are consistent with the continuum flux of the HII region. The dense gas layer (called the cooled post-shock or CPS layer) that is compressed between the ionization front and its preceding shock front may become gravitationally unstable as it enters the densest parts of the cloud, thus leading to star formation. Such an instability is easily reached, as far as hot stars are concerned. The emergent fact is that, even if driven by a group of a few late B stars, the CPS layer can reach a Rayleigh-Taylor instability within $5 \cdot 10^5$ to $5 \cdot 10^6$ years, provided that the neutral parts of the adjacent cloud are dense enough ($n = 5 \cdot 10^3 \mathrm{cm}^{-3}$). The observed parameters, i.e. the size of the HII region, the Lyman continuum flux of the parent stars, and the penetration velocity of the CPS layer into the cloud, which is supplied by the CO self-reversed profiles (Encrenaz et al., 1975), are consistent with the current existence of gravitational instability. In the case of Rho Oph, the

B. H. Andrew (ed.), Interstellar Molecules, 183–184.
Copyright © 1980 by the IAU.

mass of the CPS layer, at the onset of instability, is between 10 $M_{\odot}$ and 70 $M_{\odot}$, a surprisingly low value.

Thus, the following main conclusions can be drawn:

1) if the contagious process is relevant, a group of a few B stars cannot induce the formation of stellar associations including several O stars;

2) on the other hand, the number of low mass stars formed through that mechanism is drastically constrained by the mass of the CPS layer, if the formation rate is supposed to reproduce the slope of the Initial Mass Function (Lequeux, 1979).

A strong selection effect appears if the hypothesis of the second conclusion is correct: if M_o is the typical stellar mass of an association, the later associations generated by the step-by-step process will exhibit stellar masses lying in a narrow interval around M_o.

In the case of Rho Ophiuchi, this effect seems to be supported by the fact that all the visible stars associated with the cloud are B stars or AO stars, with the exception of one M1 star component of a double star.

REFERENCES

Brown, R.L., and Zuckerman, B.: 1975, Astrophys. J. (Letters) 202, L125.
Chini, R., Elsässer, H., Hefele, H., and Weinberger, R.: 1977, Astron. Astrophys. 56, 323.
Elias, J.H.: 1978, Astrophys. J. 224, 453.
Elmegreen, B.G., and Lada, C.J.: 1977, Astrophys. J. 214, 725.
Encrenaz, P.J., Falgarone, E., and Lucas, R.: 1975, Astron. Astrophys. 44, 73.
Falgarone, E., Cesarsky, D.A., Encrenaz, P.J., and Lucas, R.: 1978, Astron. Astrophys. 65, L13.
Lequeux, J.: 1979, Astron. Astrophys. 80, 35.

A NEARBY EXAMPLE OF A GIANT MOLECULAR CLOUD

J.W. Barrett, R.L. deZafra, D.B. Sanders and P.M. Solomon
Astronomy Program, S.U.N.Y. at Stony Brook

We have mapped an extensive molecular cloud in Perseus in the 115 GHz line of ^{12}CO. Observations were made every 10' in right ascension and declination over most of the cloud, and every 2' in the regions of most intense emission, near the open cluster IC 348 and near the reflection nebula NGC 1333. We also obtained 110 GHz ^{13}CO data every 2' in the latter regions, as well as every 10' in several long strips across the cloud. A total of 812 positions were observed in ^{12}CO, and 200 in ^{13}CO. This work was done using the 5 m antenna of the Millimeter Wave Observatory of the University of Texas. The half-power beam size was 2.6.

The cloud is elongated (Fig. 1), with its longest dimension, about 6°, corresponding to 35 parsecs at an assumed distance of 330 parsecs. The width is variable, typically 1° to 2°. The long dimension is inclined to the galactic plane at an angle of about 50°. The mean velocity of the emission shows a smooth gradient from 7 km/s near the western end to 10 km/s near the eastern end (Fig. 2). In the several regions of enhanced emission apparent in this figure $T_A{}^* \gtrsim 9$ K, implying a brightness temperature $\gtrsim 14$ K. This indicates significant heating above typical dark cloud temperatures by imbedded stars, probably recently formed B stars, with the most luminous occurring near NGC 1333. This activity appears scattered throughout the cloud, and there is no evidence of sequential star forma-

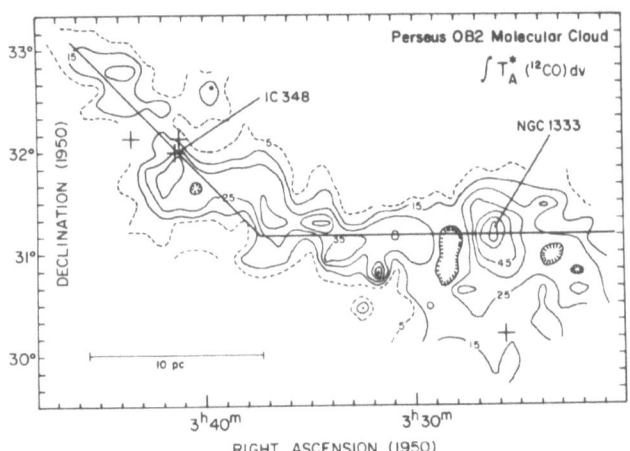

Fig. 1. Integrated intensity in ^{12}CO in units of K km/s. Crosses represent stars belonging to the Perseus OB2 association. o Persei is the northernmost of the two stars near IC 348.

B. H. Andrew (ed.), Interstellar Molecules, 185–186.

tion. The velocities of these 'hot spots' are displaced by as much as
±1.5 km/s from the median velocity of the emission at their positions
along the length of the cloud. The components that appear at velocities
of 1-4 km/s seem to be smaller separate clouds overlapping the main
cloud along the line of sight.

LTE column densities for ^{13}CO range up to $6.5 \times 10^{16} cm^{-2}$ in the 'hot
spots', with $1 \times 10^{16} cm^{-2}$ being a typical value. Assuming $N(^{13}CO)/N(H_2) =$
1×10^{-6}, we estimate the cloud's mass to be 2.5×10^4 solar masses. Sargent
(Ap.J. 1979 233, 163) finds from CO observations a mass consistent with
ours, if adjustment is made for her use of a different $^{13}CO/H_2$ ratio. The
somewhat higher mass found by Baran (Ph.D. Thesis, 1979, Columbia University, in prep.) includes a second molecular cloud to the northeast. These
two clouds, together with several smaller fragments, may form a cloud
complex associated with the Perseus OB2 association, whose center of
expansion lies between them.

A much smaller region near NGC 1333 has been mapped in detail and
analyzed by Loren (Ap.J. 1976 209, 466) in ^{12}CO and ^{13}CO. He presented
evidence that a collision of two clouds furnished the immediate triggering
mechanism for the intense star formation known to be taking place there.
The clouds considered by Loren are seen here to be small sections or wisps
of a much larger structure, whose gross velocity contours (Fig. 2) show
no evidence of separate clouds in collision.

The star o Per, which has a rich ultraviolet and optical interstellar
spectrum (e.g. T.P. Snow, 1976, Ap. J. 204, 759) is projected on the edge
of the Perseus Molecular Cloud. The molecules observed in uv and optical
absorption therefore represent the material on the edge of a true dense
molecular cloud rather than a "diffuse" interstellar cloud. The radiation
from o Per may be dissociating the edge of the molecular cloud.

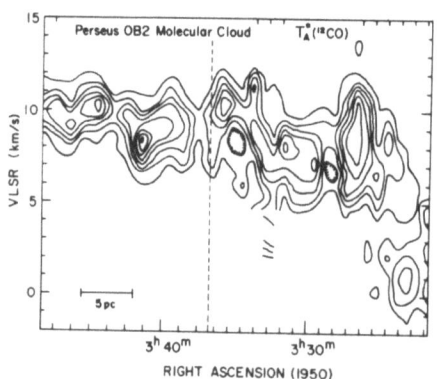

Fig. 2. $T_A^*(^{12}CO)$ along the bent line shown crossing the cloud in Fig. 1.
Horizontal scale is linear in length along the line. Contours every 2 K,
starting at 1 K.

COLD HI GAS IN THE REGION OF THE GIANT MOLECULAR CLOUD NEAR M17

Fumio Sato[1], Yasuo Fukui[2], and Tetsuo Hasegawa[3]
[1]Chiba Prefecture Education Center, Katsuragi-2, Chiba, Japan
[2]Tokyo Astronomical Observatory, Mitaka, Tokyo, Japan
[3]Department of Astronomy, University of Tokyo, Tokyo, Japan

Sato and Fukui (1978, hereafter SF) studied the HI 21-cm line in the giant molecular cloud near M17 and showed that the self-absorption of the 21-cm line is a useful tool for probing star-forming clouds. The giant molecular cloud at V(LSR) $\sim$ 20 km s^{-1} has been found to extend 4° ($\sim$170 pc) southwest of M17 (Elmegreen et al. 1979). To gain a better knowledge of its physical conditions, we used the data from the Maryland-Green Bank Survey (Westerhout 1973) to examine the self-absorption features of the 21-cm line in and around the whole molecular cloud except the two fragments adjoining M17. The HI profiles at every 0°.1 in galactic latitude and longitude in the region were analyzed in almost the same manner as in SF by assuming that the spin temperature of the cold HI gas is equal to 20 K. The distance was taken to be 2.5 kpc.

The main results are as follows:

1. As shown in Fig. 1, the overall distribution of column density N(HI) is somewhat similar to that of the ^{12}CO emission. However, the cold HI gas is not distributed so fragmentarily as the CO gas, and seems to form a single cloud. Moreover, at lower longitudes, the maxima of the N(HI) and of the CO emissions do not coincide with each other. The N(HI) peak at ℓ = 12°.9, b = -0°.2 is located in a region of very weak CO emission, and the CO maximum near ℓ = 12°.8, b = 0°.6 has no remarkable HI counterpart.

2. The velocity at maximum absorption V_c varies between 17 and 23 km s^{-1}. In the region south of the line AB in Fig. 1, V_c is mostly 20 km s^{-1} or higher, and in the northern half it is usually lower than that. Across the line AB, a gradual velocity gradient of $\sim$0.1 km s^{-1} pc^{-1} is found over $\sim$1°. It can be interpreted as a rotation of the HI cloud about the axis AB with a period of $\sim$5 x 10^7 years.

3. The line width at half maximum, ΔV, ranges from 4 to over 10 km s^{-1}. A few regions have double-dipped profiles with $\Delta V \gtrsim$ 8 km s^{-1}. The marginal area of the cloud usually has less wide profiles than the inner region, which may represent a contraction as argued in SF.

B. H. Andrew (ed.), Interstellar Molecules, 187–188.

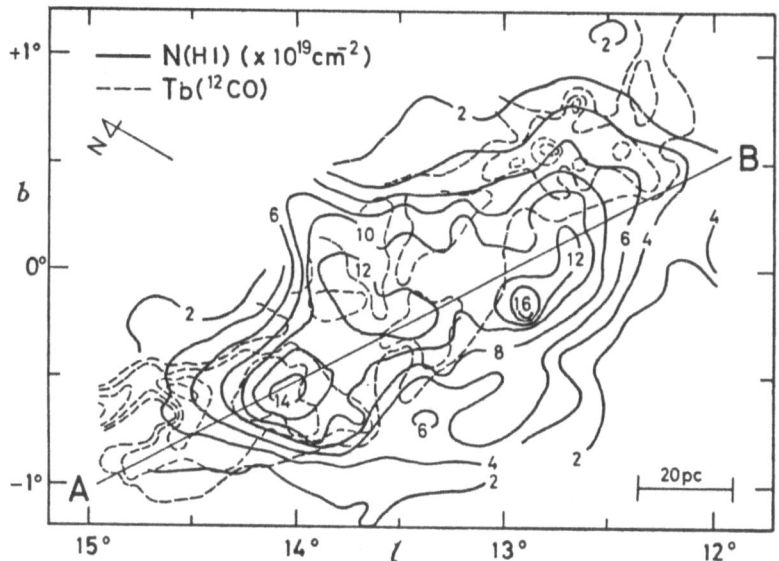

Figure 1. Distribution of the cold HI column density and that of the
^{12}CO brightness temperature (Elmegreen et al. 1979).

    4. Ratios of number densities [H$_2$]/[HI] were estimated for some CO
emission peaks by using our N(HI), N(^{13}CO) (Elmegreen et al. 1979), and
[^{13}CO]/[H$_2$] = 2 x 10^{-6} (Dickman 1978). The results are 350 for fragment
C ($\ell \sim 14°6$, b $\sim -0°6$), 120 for fragment D ($\ell \sim 14°1$, b $\sim -0°6$), and 230
for the peak near ℓ = 12°8, b = 0°6. In the region with extended, weak
CO emission between ℓ = 13° and 14°, the ratio becomes significantly
lower, i.e., from 13 to 70. These low ratios are also found in the W3
and W4 molecular complex (Hasegawa et al. 1979). Such a wide range of
[H$_2$]/[HI] ratios may reflect different physical conditions and evolutionary stages in the various parts of a giant molecular cloud.

    5. In the southern region where there is no detectable CO emission,
the total mass can be estimated to be $\gtrsim 10^4$ M$_\odot$, if [H$_2$]/[HI] $\gtrsim 10$ is assume

    The HI data reduction was made on the FACOM 230-58 computer at the
Computing Center, Tokyo Astronomical Observatory.

REFERENCES

Dickman, R.L.: 1978, Astrophys. J. Suppl. 37, 407.
Elmegreen, B.G., Lada, C.J., and Dickinson, D.F.: 1979, Astrophys. J.
    230, 415.
Hasegawa, T., Sato, F., and Fukui, Y.: 1979, this volume.
Sato, F., and Fukui, Y.: 1978, Astron. J. 83, 1607.
Westerhout, G.: 1973, Maryland-Green Bank Galactic 21-cm Line Survey,
    3rd ed. (University of Maryland, College Park, Maryland).

SEARCH FOR CO IN ATOMIC HYDROGEN CLOUDS

I. Kazès and J. Crovisier
Département de Radioastronomie,
Observatoire de Meudon

The relationship between dense molecular clouds and diffuse clouds, as well as the mechanisms connected with the formation of molecules in diffuse clouds, may be studied using HI 21-cm line observations and molecular line observations in the same directions. For this purpose we previously studied the OH 18-cm main lines (Kazès et al., 1977) and the 2.6-mm CO lines (Crovisier and Kazès, 1977) in directions where strong 21-cm absorption features had been detected in the Nancay survey (Crovisier et al., 1978). Liszt and Burton (1979) also measured CO lines toward 19 directions observed in the Arecibo 21-cm emission/absorption survey (Dickey et al., 1978). This paper presents preliminary results of a more comprehensive search for ^{12}CO in directions previously studied in the Nancay survey.

The observations were made at 2.6 mm with the NRAO 11-m millimetre-wave radio telescope on Kitt Peak (the National Radio Astronomy Observatory is operated by Associated Universities, Inc., under contract with the National Science Foundation). Each direction was observed for 20 minutes with 0.26 and 0.65 km/s velocity resolutions and a position-switching technique (detection limit: $T_A^* = 0.5$ K). Seventy-six directions were investigated, mainly at low and intermediate galactic latitudes toward extragalactic sources showing strong 21-cm absorption in the Nancay survey. The CO line was detected in 22 HI clouds, out of which 16 were new detections.

All the detected CO features correspond in velocity to HI absorption features. However, the antenna temperatures of the detected lines are neither correlated with the corresponding HI optical depths, nor with the HI column densities at the same velocities (see Figure 1). This result does not confirm the strong correlation claimed by Liszt and Burton (1979).

The present investigation shows that in approximately 1/4 of the lines of sight, physical conditions are such that HI to H_2 conversion has occurred in appreciable quantities to allow the formation and hence the detection of CO. The overwhelming presence of HI in a line of sight may result in a blend of several features corresponding to different clouds. The deduced spin temperatures (90 K on the average) are the harmonic means

B. H. Andrew (ed.), Interstellar Molecules, 189–190.

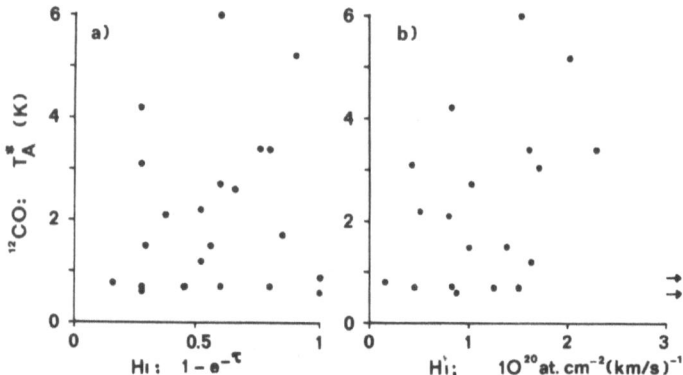

Figure 1. The equivalent antenna temperature of the CO detected spectral features as a function of a) the HI 21-cm absorption, b) the HI column density per unit velocity, at the CO velocity.

of the real spin temperatures along the line of sight. Therefore, the detectable molecular emission cannot be directly related to the deduced spin temperature, but it is plausible that the CO is formed in a re- stricted, cold and well-shielded region of the line of sight. The rela- tively low detection rate obtained from our fairly large sample tends to support the scarcity of these privileged regions.

CO velocities do not coincide with the central velocities of the HI absorption features: the velocity difference has a dispersion of $[(V_{CO} - V_{HI})^2]^{\frac{1}{2}} \sim 1.6$ km/s. While this value is comparable to that of the internal velocity dispersion of the HI features (σ_V = FWHM/2.35 $\sim$ 1.6 km/s), the CO lines are significantly narrower ($\sigma_V \sim 0.7$ km/s). This result adds weight to the argument that the detected CO is associated with only a small fraction of the HI in the line of sight. A more com- prehensive study of the distribution of molecules in diffuse clouds would need mapping in both molecular and HI lines.

REFERENCES

Crovisier, J., and Kazès, I.: 1977, paper presented at the 21st Liege
    International Astrophysical Symposium.
Crovisier, J., Kazès, I., and Aubry, D.: 1978, Astron. Astrophys. Suppl.
    32, pp. 205-282.
Dickey, J.M., Salpeter, E.E., and Terzian, Y.: 1978, Astrophys. J. Suppl.
    36, pp. 77-114.
Kazès, I., Crovisier, J., and Aubry, D.: 1977, Astron. Astrophys. 58,
    pp. 403-410.
Liszt, H.S., and Burton, W.B.: 1979, Astrophys. J. 228, pp. 105-111.

OPTICAL AND THEORETICAL STUDIES OF GIANT CLOUDS IN SPIRAL GALAXIES

Bruce G. Elmegreen

Astronomy Department, Columbia University

Debra Meloy Elmegreen*

Hale Observatories, Carnegie Institution of Washington,
California Institute of Technology

ABSTRACT

   An optical study of four spiral galaxies, combined with radiative
transfer models for transmitted and scattered light, has led to a
determination of the opacities and masses of numerous dark patches and
dust lanes that outline spiral structure. The observed compression
factors for the spiral-like dust lanes are in accord with our expectations from the theory of gas flow in spiral density waves. Several
low density (10^2 cm^{-3}) clouds containing 10^6 to 10^7 $M_\odot$ were also
studied. We discuss these results in terms of recent theoretical
models of cloud and star formation in spiral galaxies. The long-term
evolution of giant molecular clouds is shown to have important consequences for the positions and ages of star formation sites in spiral
arms.

   From the distribution of H II regions in external spiral galaxies,
we may infer that massive star formation is enhanced in spiral arms.
This increase may be the result of a heightened probability for triggering star formation in pre-existing clouds as they enter or cross a
spiral arm, or it may be due to enhanced cloud formation in the arm,
followed by some spiral-independent mechanism for star formation inside these clouds. Both points of view have been extensively investigated theoretically, but observations have been inadequate to choose
between the two alternatives. The problem is that there has been no
significant detection of giant molecular cloud complexes (10^5 $M_\odot$) that
are known to be without massive OB stars. If such massive clouds
were determined to occur with comparable abundance in both the arm and

---

*Visiting Astronomer, Cerro Tololo Inter-American Observatory and
Hale Observatories

B. H. Andrew (ed.), Interstellar Molecules, 191–196.

interarm regions (as defined by the underlying red spiral of stars), then the first alternative given above would seem to apply. On the other hand, if galactic surveys showed molecular clouds to be associated with spiral arms, then the results would not be so conclusive, since they might indicate either that the clouds form in the arms, or that out of a uniform distribution of clouds, the spiral arm clouds are the hottest (perhaps due to star formation). Most likely, both of these processes occur to varying degrees in the same galaxy.

New optical observations have been undertaken that eventually may resolve this indeterminancy (D. Elmegreen 1979). The object was to locate dust cloud complexes in nearby and nearly face-on spiral galaxies and to determine the gaseous column densities and masses of these individual clouds, to see if any are similar to giant molecular clouds in our own galaxy. Observations of objects the size of the giant cloud complexes (e.g., 50 to 100 pc) require the high angular resolution that currently is available with optical techniques. UBVRI plates obtained at the 4-m CTIO telescope and the 1.2-m Palomar Schmidt telescope were analyzed with 1" resolution on a scanning microphotometer and calibrated with photoelectric photometry. Seeing during the plate exposures was estimated to be 2", corresponding to linear sizes of 23 pc, 70 pc, 76 pc, and 93 pc in the four program galaxies NGC 7793, M101, M74, and M51, respectively.

Apparent dust opacities for discrete clouds were estimated from the differences in surface brightness between the clouds and the adjacent stellar backgrounds (chosen to be free of H II regions and bright stars). True dust opacities were obtained by comparing the observed brightness differences at five bandpasses with models of emission from idealized galaxies containing dust lanes and discrete clouds. Radiative transfer calculations were developed which included the effects of scattered light (D. Elmegreen 1979). The underlying stellar population in the model was fit to the observed colors of the comparison regions by assuming only that the relative scale heights for different stellar types was the same as for our own galaxy. In this way, the distinction between a low opacity, high-latitude cloud and a high opacity, low-latitude cloud can be determined, even though these two clouds may have the same degree of darkness in any one bandpass. This distinction is possible because the stellar population reddens with increasing height above the plane, so the surface brightness in blue light is a sensitive indicator of the cloud's height; while the surface brightness in red light is a better probe of the cloud's intrinsic absorption, primarily due to selective extinction.

A typical value for the visual extinction through the galaxies in the intercloud (comparison) region was found to be in the range of 0.6-1.0 magnitudes, while dust lanes (whether they were on the inside or outside of prominent spiral arms), in addition to numerous dust patches throughout the galaxies, showed some 3 magnitudes of internal extinction. (Measured apparent visual magnitude differences are only several tenths of a magnitude; the eye deceptively increases the contrast between bright and dark regions on a photographic plate). These results explain the common appearance of dust patches and lanes in external galaxies: the distributed gas is so close to unit optical depth in the visual wavelengths that any slight compression or convergence of the gas will give the clear appearance of a dark patch or lane.

Evidently, dust configurations are a sensitive probe of disturbances in the interstellar medium. For the spiral dust lanes, compression factors of 2 to 3 relative to the ambient medium were typical. The six cases (in two galaxies) of spiral arm branches with large pitch angles all had higher compression factors of around 10. Both of these results, especially the observed enhancements of the compression factor in regions where the pitch angle was large, are in accord with our expectations for gas compression due to spiral density waves. However, the presence of equally compressed dust lanes in both the inner and outer parts of many arms may point to additional effects. In any case, it is clear that most dust lanes cannot be interpreted simply as long chains of giant molecular clouds like those seen in our galaxy, because (a) the measured dust lane opacities correspond to average densities of only 6 to 20 cm^{-3}, and (b) the dust lane widths of 100 to 200 pc are much larger than the dimensions of the molecular clouds in our galaxy. It is conceivable, however, that some dust lanes contain denser unresolved molecular cores.

Several high latitude clouds with a few magnitudes of extinction were also found; these appear to be similar to high latitude cloud complexes or the top parts of shells seen in our own galaxy, with dimensions of 100 to 200 pc.

Perhaps one of the most interesting results is that in no case was a large (i.e., resolved) dark cloud seen with an intrinsic visual opacity exceeding 12 magnitudes, although such a cloud, if it existed, would have been obvious even in the R and I plates. The three clouds studied that had this intrinsic extinction all had masses in the range of 10^6 to 10^7 $M_\odot$, and their mean gas densities were around 100 cm^{-3}. The implication is that denser dark clouds in this mass range were not observed because either they are illuminated by massive star formation and H II regions, or they are much smaller than the seeing limit. This means that giant cloud complexes that extend for about 100 pc and contain more than 10^6 $M_\odot$ probably initiate star formation before their extinction greatly exceeds some 12 magnitudes.

These results may illustrate some of the aspects of a recent theoretical calculation of self-gravitational cloud formation in spiral density wave shocks (B. Elmegreen 1979). The clouds that are expected to form by large-scale galactic processes are similarly massive H I objects (10^6 to 10^7 $M_\odot$), with large sizes and low mean densities. They should not be identified directly with the observed giant molecular cloud complexes in our own galaxy; they are more like superclouds out of which the smaller molecular clouds (10^5 $M_\odot$) will form by condensation and fragmentation.

Probably star formation will begin in a 10^5 M$_\odot$ fragment at about the same time as molecule formation, because the density thresholds are similar. For the purposes of understanding the details of star and cloud formation in spiral arms, we must consider the long-term consequences of massive star formation in one of the individual molecular clouds (see B. Elmegreen 1979). On a timescale of some 10^7 years, the cloud will be pushed and partially disrupted by the pressures associated with such star formation (H II regions, supernovae, etc.). The sequential formation of OB subgroups may occur at the same time, but on a smaller scale (e.g., every 2 to 3 million years). Since the total mass that becomes ionized may be only several times 10^4 M$_\odot$ at this later stage, (as observed in OB associations), and since only 10^3 to 10^4 M$_\odot$ of stars will form, a considerable amount of neutral material (10^5 M$_\odot$) should remain that simply will be pushed around. If the forces are centralized with respect to the cloud, then a large shell consisting of most of the cloud's original mass may form. Alternatively, if the OB stars are consistently on one side of the cloud (as is often observed to be the case), then the cloud will be pushed as a whole to one side, while a relatively small shell may form toward the other side. In any case, a considerable amount of mass still will be available for re-collection or re-collapse into a second generation molecular cloud complex, and star formation most likely will begin again. The total timescale for this generation-to-generation shuffling will be some 20 to 50 million years, and the cloud's total excursion before star formation resumes may cover 200 to 300 pc. If a spiral density wave passes the cloud before the cloud completely disintegrates by these repetitive bursts of star formation, then the cloud may be shocked by the galaxy into forming new stars (Woodward 1976). On the other hand, if the cloud has had enough time to go through several generations and most of the remaining gas has been converted back into the lower density interstellar medium (e.g., in the outer part of a galaxy), then the gas will be available for the formation of a new cloud complex when the spiral density wave eventually does pass. On the average, a steady-state cloud population could be maintained. Evidently, the details of individual cloud evolution on 30 million year timescales and over regions covering 100 to 300 pc play an important role in generating the observed irregularity in positions and ages of star formation sites inside spiral arms.

REFERENCES

Elmegreen, B.G. (1979). Astrophys. J., 231, pp. 372-383.
Elmegreen, D.M. (1979). Ph.D. Thesis, Harvard University.
Woodward, P.R. (1976). Astrophys. J., 207, pp. 484-501.

DISCUSSION FOLLOWING ELMEGREEN

Goss: Using the Westerbork Synthesis Radio Telescope at 21 cm we have mapped large HI complexes associated with HII regions in M101. The sizes are ∿1 kpc, masses 10^7-5×10^7M$_\odot$, and densities 1-10 cm^{-3}. Are these your superclouds that were observed optically?

Elmegreen: I would expect that the optical superclouds we observed would indeed be prominent in 21 cm line emission. There are several such clouds, however, so we should compare our maps to see if we detect the same ones.

Silk: According to Solomon, molecular clouds fill both arm and inter-arm regions. In order for your mechanism of spiral arm formation and subsequent shuffling to be valid, very long lifetimes for individual molecular clouds are presumably needed. How long a lifetime do you require at a distance of, say, 8 kpc from the center of our galaxy, and how do you think this is achieved?

Elmegreen: Individual cloud lifetimes are an important factor in determining the arm/inter-arm contrast in CO, but only in comparison to the mean time between passages of a spiral density wave. An important point of my work is that the lifetime of an individual cloud is not just the timescale for OB star formation. It may be as large as 3 or 5 times this value due to cloud-shuffling and inefficient destruction. In the inner regions of our galaxy, the spiral density wave timescale may be less than the cloud-shuffling time, so relatively low arm/inter-arm contrasts would be possible. It is true, however, that I still expect some delineation of spiral structure in carbon monoxide, because the primary cloud-formation mechanism that I have proposed will work in phase with a spiral density wave. I believe that this expectation is not in contradiction to the results of the Columbia Sky Survey Telescope.

Townes: For some time it has appeared that the variation of isotopic ratios from cloud to cloud requires the maintenance of some form of material integrity for clouds for at least as long as about 10^9 years. Thus the lifetimes in the range 10^8 years now suggested by Solomon and by the speaker, while going some distance from the approximately 10^6 years which used to be frequently chosen, are still not quite long enough. Can you make any good estimate from your approach as to how long the material of large clouds would be approximately segregated?

Elmegreen: In the inner regions of our galaxy, near the 5 kpc ring, two processes combine to allow clouds to be very long lived, possibly as long as 10^9 years or more. One is that the spiral density wave comes by a cloud before it has time to "shuffle" much more than one generation, so each cloud does not become completely disrupted by internal processes. In addition, the self-gravitational re-collection of cloud pieces that will occur in spiral arms has minimum opposition from galactic tidal forces at 5-6 kpc from the galactic center. Thus cloud formation or re-collection is more efficient, and cloud self-destruction is less effective, at 5-6 kpc than in the solar neighbourhood. The timescales are such that some clouds or cloud-pieces may never be completely destroyed at 5-6 kpc. This also may be true, but to a lesser degree, in the solar neighbourhood.

Mouschovias: Earlier speakers mentioned that observations show that the typical extent of a giant molecular cloud is about 10^2 pc in the galactic plane. If you consider a tube of radius ~ 100 pc, the tube must have a length larger than 1 kpc in order to contain your required $\simeq 5 \times 10$ $M_\odot$. Therefore, you need to move material along the tube over a distance of over 1 kpc, if you are to explain giant molecular clouds in this manner. Even if velocities of about 10 km s^{-1} could be induced by self-gravity at these low densities - which is doubtful - you would still need at least 10^8 years to form these clouds. Is that not too long a time?

Elmegreen: The primary collapse time of a supercloud formed by the self-gravity of a spiral-density-wave shock is calculated to be some 30 million years for our galaxy in the 5 kpc ring. This value was calculated with a shock compression-factor of about 9, and it does account for the elongated geometry. It is fast enough for significant collapse to occur before the gas emerges from the spiral-density-wave shock. After this initial condensation, the final collapse to molecular-cloud densities would presumably occur at a faster rate. I believe that the densities I calculated are slightly larger than your values, and may account for the difference in our time scales. A streaming velocity of 10 km s^{-1}, or slightly larger, does not contradict 21 cm line observations in our own galaxy or in other galaxies. Of course the gas will move from both ends of the perturbation to the center, so the largest excursion of any collapsing element will be only half the wavelength; this immediately reduces your estimate of 10^8 years by a factor of two.

MOLECULAR CLOUDS IN ORION AND MONOCEROS

Mark Morris & J. Montani
Astronomy Department, Columbia University

and

P. Thaddeus
Goddard Institute for Space Studies, New York, NY

ABSTRACT: A 1.2-meter millimeter-wave telescope has been used to survey CO in the constellations of Orion and Monoceros. Many new molecular clouds have been found. The distribution of molecular material shows two striking characteristics: 1) Most of the molecular clouds in this region appear to be connected by continuous extensions and filaments. To judge from continuity in radial velocity, most of these connections appear to be real, and are not merely the result of projection along the line of sight. 2) There are at least two slender filamentary features longer than 10° in angular extent. These filaments may connect the molecular clouds lying well out of the Galactic plane to clouds lying in the plane. Their shape and orientation suggest that magnetic fields may play a role in their evolution. The observed velocity gradients may be explained by accelerated gas flow along the filament.

Previous CO studies of molecular clouds in the second and third galactic quadrants focussed on optically prominent regions containing clear examples of star formation. These studies revealed the presence of a number of quite discrete objects having maximum diameters of about 100 pc and masses on the order of $10^5 M_\odot$ (Kutner et al. 1977, Lada et al. 1978, Blitz 1978).

Do such objects exist which are not accompanied by star formation? The clear bias in the early CO observations towards H II regions and infrared sources meant that very little light was shed on this interesting question.

Here we report the results of a survey of CO at negative galactic latitudes mainly in Orion and Monoceros, which greatly extends the spatial coverage of prior surveys. Several new features emerge, the most striking of which are very long (10-20°) molecular filaments which appear to connect molecular clouds lying far below the galactic plane to molecular features in the plane itself. In addition, many new clouds covering a large range of sizes have been found.

197

B. H. Andrew (ed.), Interstellar Molecules, 197–203.
Copyright © 1980 by the IAU.

The observations were made with a 4-foot millimeter-wave telescope at Columbia University, which has a beamwidth of 8 arc minutes at the frequency of the J = 1→0 CO line. Spectra of CO were taken at over 5000 positions. Corresponding observations of ^{13}CO have not yet been made, so mass estimates for the molecular objects described here are not yet available. We report here only the main findings of the spatial distribution and kinematics of CO.

Maps of peak intensity of CO emission are presented in figures 1 and 2 which together cover about 300 square degrees of sky. The sampling is fairly complete in these regions, and the major molecular features have probably all been found. The extent of molecular gas in these regions is clearly very great. Molecular emission in figure 1 extends from the Orion A and Orion B clouds in Orion's Sword observed by Kutner et al. (1977) north almost to Betelgeuse ($\alpha = 5^h52.5^m$, $\delta = +7°4$), and from there east almost to the molecular cloud observed by Blitz (1978) associated with the Rosette nebula (just off fig. 1 to the east). Figure 2 shows the southern portion of the survey, excluding the southeastern extension of the Orion A cloud, for which $V_{LSR} <$ 8 km s^{-1}. The filamentary structure on the right in figure 2 actually overlaps the Orion A cloud along the line of sight at $\alpha = 5^h35^m - 5^h45^m$, but the two are clearly separated in velocity and hence may not be physically connected. CO emission is nearly continuous from the direction of the Orion A cloud (b = -19°) to the CMa OB1 molecular clouds lying in the galactic plane (Blitz 1978). Contained in the CO complex in figure 2 is the large Mon R2 molecular cloud, the central part of which was previously studied by Kutner and Tucker (1975) and by Loren (1977).

Large Molecular Filaments

Two large features with unprecedented filamentary structures are apparent in CO. They lie at $\delta \sim 5°$ and $\delta \sim -10°$ and are approximately parallel to the celestial equator and to each other. Both intersect the galactic plane at a large angle (~50°). On the Palomar Sky Survey these large-scale molecular filaments can be partially discerned as a sequence of very diffuse patches; with a small amount of additional foreground extinction they would be virtually invisible. The two features have many other remarkable points in common:

1) They are both quite long -- 12° for the one at $\delta \sim 5°$ and 22° for the one at $\delta \sim -10°$ (if it is indeed a single continuous structure). Curiously, their projected separation is about equal to their projected lengths.

2) Relative to their lengths, they are very narrow -- only 15 to 30 arc minutes between one-half peak intensity contours. This continuous and extreme narrowness means that these objects are almost undoubtedly true filaments, and not thin disk-like clouds seen edge-on.

3) Line widths are large in the filaments (4-8 km s^{-1}) compared to quiescent molecular clouds ($\Delta v \sim 1-2$ km s^{-1}) in regions devoid of apparent

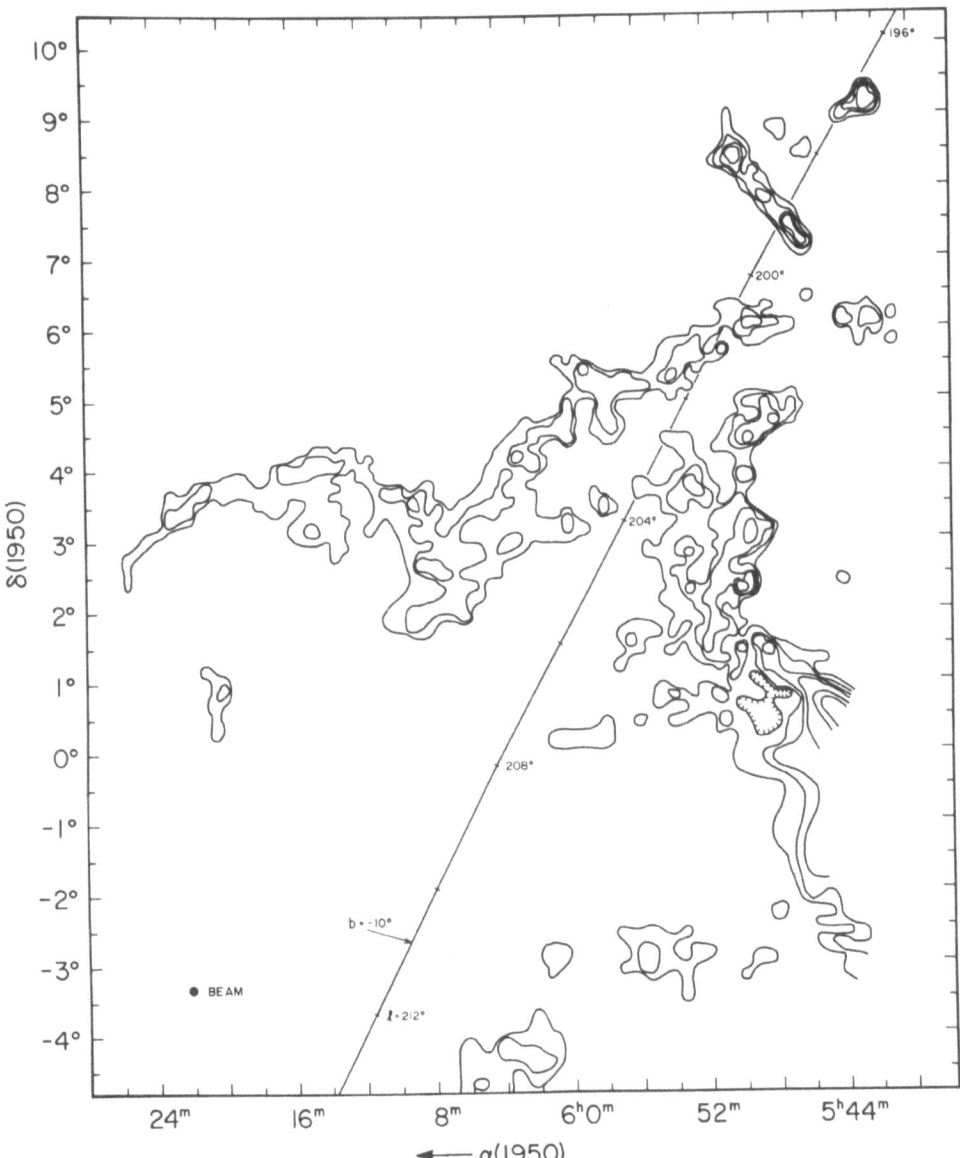

Figure 1: Contours of peak intensity of J = 1→0 CO emission in the
northern portion of the survey field. Units are °K of antenna temperature corrected for atmospheric losses (the correction for telescopic
losses, which amounts to a factor of ∼1.7, is not included and will
be dealt with in a later publication). The lowest contour and the contour interval are 1K. Galactic coordinates are indicated along the line
at b = -10°.

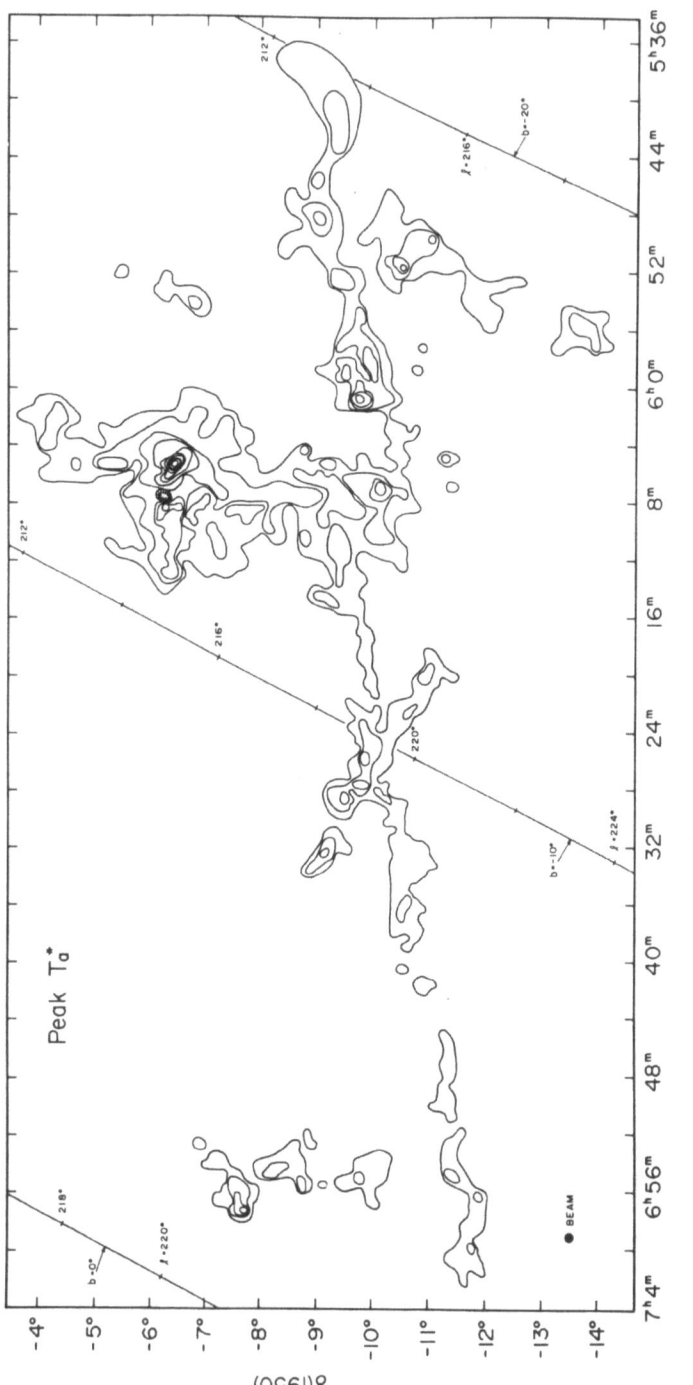

Figure 2: Contours of peak intensity of J = 1→0 CO emission in the southern portion of the survey field. Same contour values as in figure 1.

star formation. Consequently, the central ridges of the filaments are more prominent in a map of integrated intensity than the peak intensities of figures 1 and 2. The increased line broadening does not appear to be associated with the presence of stellar sources of heat as in many large molecular clouds.

4) Velocity gradients exist along both filaments. In the northern one, gradients occur along short (3-5°) segments, and abrupt velocity discontinuities sometimes occur where two such segments overlap. The southern filament is similar except for a more pronounced velocity gradient along a major portion of its length: between $\alpha = 6^h18^m$ and 6^h52^m the velocity changes monotonically from ~8 to ~18 km s^{-1}. Because these filaments are unlikely to be rotating around an axis perpendicular to their length, it is tempting to interpret the velocity field in terms of flow along the filament. If there is a flow along the filament, the gradient may reflect either a changing orientation of the filament with respect to the line of sight as one proceeds along the cloud, or an acceleration of the flow.

The shape and motion of these filaments suggest to us the influence of a magnetic field which is highly ordered on a large scale. The filaments can be interpreted as magnetic flux tubes along which the gas is constrained to flow, although the fattening of the filament at places where the observed velocities are relatively complex suggests that there may exist pools along the filament in which the field orientation and gas motions are relatively chaotic. It is natural to suppose that the direction of flow is from the Orion region towards the plane, but we have as yet no way of proving this. The local scale height for young objects (e.g., OB stars) is about 50 pc, and it is reasonable to assume that this figure holds for molecular clouds as well (at 5 kpc from the galactic center the scale height for molecular clouds is known to be ~50 pc [Cohen and Thaddeus 1977]). Therefore, almost all the CO observed in this study is farther from the galactic plane than a molecular scale height. The gravitational force of the galaxy at such distances is sufficient to provoke an inflow of molecular gas.

The history and structure of the Orion region as a whole have long posed a number of interesting and difficult problems. What has been responsible for elevating so much material -- a mass probably in excess of $10^6 M_\odot$ -- so far above the plane? And what keeps it there?

By suggesting an exchange of matter with the plane along large scale magnetic filaments, the CO data presented here may provide important clues towards resolving these questions.

REFERENCES

Blitz, L.: 1978, Ph.D. Thesis, Astronomy Dept., Columbia University.
Cohen, R.S., and Thaddeus, P.: 1977 Ap.J.(Letters) 217, pp.L155-159.

Kutner, M.L., and Tucker, K.D.: 1975, Ap.J. 199, pp.79-85.
Kutner, M.L., Tucker, K.D., Chin, G., and Thaddeus, P.: 1977, Ap.J. 215, pp.521-528.
Lada, C.J., Elmegreen, B.G., Cong, H.-I., and Thaddeus, P.: 1978, Ap.J. (Letters) 226, pp.L39-42.
Loren, R.B.: 1977, Ap.J. 215, pp129-150.

DISCUSSION FOLLOWING MORRIS

Blitz: Using a best guess for the distance to the filaments, can you estimate the mass and mean density of the filamentary structure?

Morris: Lack of ^{13}CO observations prevents me from making a reliable estimate of the masses or densities. They are conceivably much smaller than in the giant molecular clouds.

Churchwell: The Orion filament which you showed is remarkably straight over ∿200 pc. Since it is almost perpendicular to the galactic plane, it is surprising that galactic rotation has not caused significant bending. Do you have any comment on this?

Morris: I am equally surprised. Perhaps one can infer that these filaments have very little extent along the direction to the galactic center.

Glassgold: What is the relation of your observations to HI studies, especially with regard to the filamentary structure?

Morris: We have taken a preliminary look at the HI studies of this region, and have found no obvious features which correspond to the filaments. Aside from the vague presence of these filaments as absorption features on the Palomar Sky Survey prints, they appear to be definable only by virtue of their CO emission.

Linke: Might there not be other CO features in the large regions of your maps which were not examined?

Morris: The sampling of the regions shown varied from 1° intervals to 1 beamwidth spacing. Most clouds greater than 20'-30' in extent have probably been found.

Crutcher: Is there any information available from optical polarization studies about the possible alignment of the interstellar magnetic field with the molecular filaments?

Morris: The polarization has been measured for only five or six stars in the direction of the two filaments, and these few measurements do indeed indicate a magnetic field aligned along the filament. However, more systematic optical polarization studies are needed.

Gilmore: The two long CO filaments that you showed, the one in the Orion B cloud, and the other south of the Monoceros region, were at a large angle to both the plane of Gould's Belt and the galactic plane. Furthermore, the Parker stability has a size scale of ∿l kpc, not much larger than the size of Gould's Belt. What do you see as the relation of these two CO filaments to Gould's Belt? Would you consider such a large-scale order to the magnetic field at a large angle to Gould's Belt as significant or anomalous?

Morris: I think it is too early to speculate on the relation of these filaments to larger-scale structures, but they are quite provocative because they suggest that magnetic fields may play a significant role in the evolution of 0.1-1 kpc structures.

Sandqvist: We have analysed the local neutral hydrogen gas over the entire sky as well as many of the nearby dark clouds, and have found that these components are kinematically and spatially related to Gould's Belt of bright stars. They form a subset of Gould's Belt, we believe, that does not reach out as far as your Orion clouds, nor attain as high expansion velocities. Could you describe the relation of your clouds to the interstellar matter in Gould's Belt?

Morris: The model for the Gould's Belt system presented by Stothers and Frogel (1974, Astron. J., 79, 456) and Frogel and Stothers (1977, Astron. J., 82, 890) does extend to the distance of the Orion clouds. However the mean velocity of molecular gas in the Orion-Monoceros region, $\sim$10 km s^{-1}, is indeed higher than that usually found for Gould's Belt gas. Also, at 850 pc, the Mon R2 cloud probably lies beyond the Gould's Belt system. Therefore we have no strong evidence that the filaments are related to Gould's Belt.

Kuiper: It would seem that any external input of kinetic energy (such as by supernova, spiral arm passage, or radiation pressure from 0 stars) would give elongated features because of the eventual separation of faster moving material from the slower. One might avoid in this way the problem of not seeing elongated features in HI. On the other hand, if magnetic fields were to control the shape, there would be no reason not to see extended curved shocks. Do you have a reason for ruling out such streaming as the cause for the observed shapes?

Morris: Your suggestion probably applies to the cloud at $\alpha=5^h48^m$, $\delta=+8°$, which borders the HII region surrounding λ Ori. The two long filaments, however, do not appear to be arranged near any obvious sources of kinetic energy.

COLUMBIA CO SURVEY: MOLECULAR CLOUDS AND SPIRAL STRUCTURE

R.S. Cohen, T.M. Dame and P. Thaddeus
Goddard Institute for Space Studies and Columbia University,
New York City

We have recently completed at Columbia University a survey with a
1.2 meter telescope of 2.6 mm CO emission from the first galactic quad-
rant. The beamwidth of this telescope - 7.5' at 2.6 mm - is optimal for
a large-scale galactic survey: it is as small as that of the best 21-cm
surveys but not so small that undersampling is required to finish a survey in a few hundred days of observing. The survey contains more than
3000 spectra between longitudes $12°$ and $60°$ and latitudes $-1°$ and $+1°$. Spec-
tra were taken every beamwidth for $|b| \leq 0.25°$ and every two beamwidths
elsewhere. Integration times, which averaged about 8 minutes, were ad-
justed to give an rms noise level of 0.3 K per 0.5 MHz spectral channel.

With the dense sampling of our survey it is possible for the first
time to improve upon the axisymmetric models used in analyzing previous
undersampled surveys. (See Solomon et al. (1979) for a summary of
previous work.) The so-called molecular ring, visible along the lower
right of the longitude-velocity map in figure 1, is now resolved into
two distinct parallel lanes. As the figure inset shows, these are the
classical Scutum and 4-kpc arms. The other main 21-cm features also
have distinct CO counterparts: A and B are the local arm, and C is the
Sagittarius arm. The arms are separated by regions largely free of molecular clouds. The holes within the loop of the Sagittarius arm, between the Sagittarius and Scutum arms, and between the Scutum and 4-kpc
arms, all represent extensive interarm regions.

The confinement of molecular clouds to spiral arms implies that
they are transient objects with lifetimes less than 10^8 years, the time
it takes interstellar matter to cross a spiral arm. A corollary, based
on simple conservation of mass, is that the molecular clouds cannot
represent much more than half the interstellar gas.

REFERENCES

Burton,W.B. and Shane,W.W. 1970, in IAU Symposium 38, p.397.
Lindblad,P., Grape,K., and Sandqvist,A., Schober,J. 1973, Astr.Ap. 24,309.
Shane,W.W. 1972, Astr.Ap. 16, 118.
Solomon,P., Sanders,D., and Scoville,N. 1979, in IAU Symposium 84, p.35.

205

B. H. Andrew (ed.), Interstellar Molecules, 205–207.
Copyright © 1980 by the IAU.

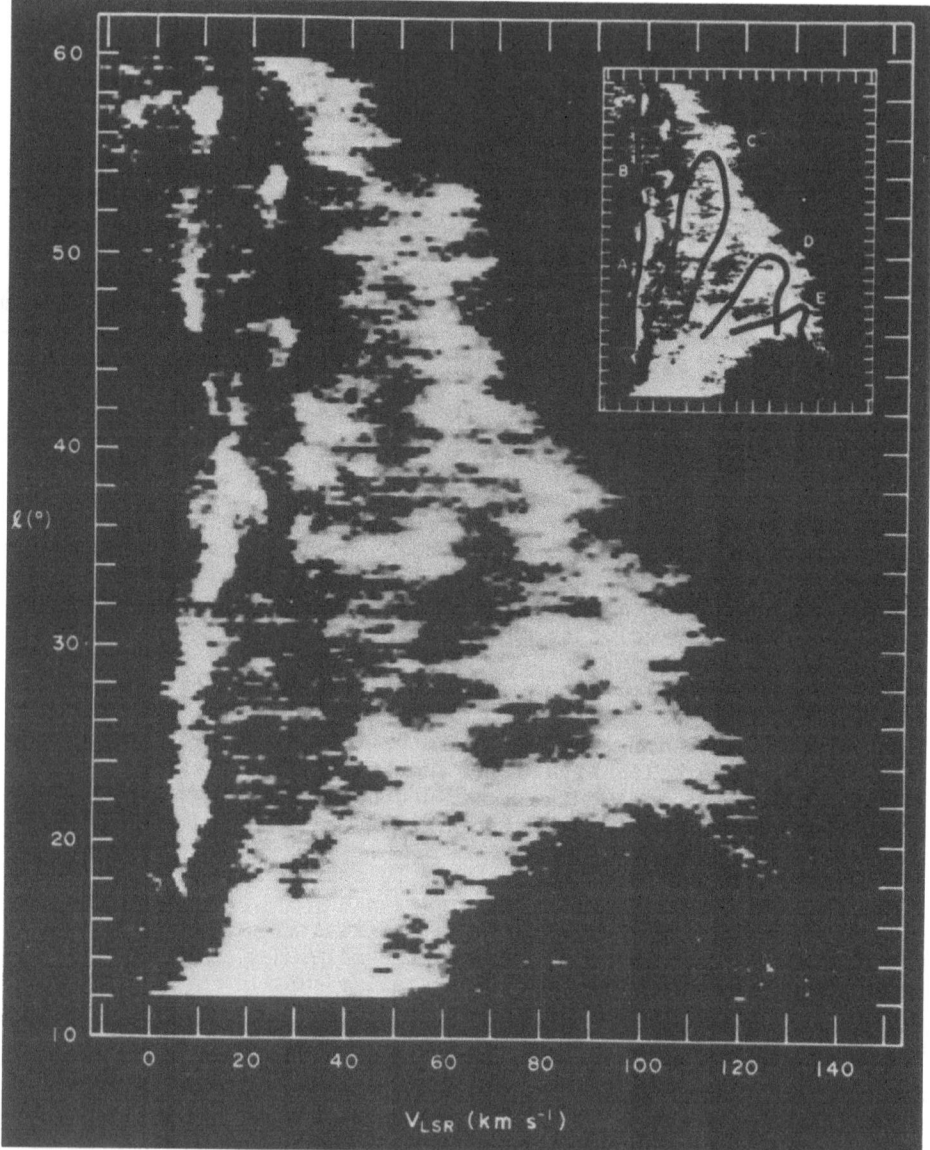

Figure 1. The ℓ, v diagram obtained when our survey is integrated across the galactic plane from $b=-1°$ to $+1°$ and smoothed by $0.25°$ in ℓ. The insert locates the main 21-cm arms with respect to our CO ℓ, v diagram. A and B are Lindblad's local ring (Burton and Shane 1970; Lindblad et al. 1973); C is the Sagittarius arm (Burton and Shane 1970); D and E are the Scutum and 4-kpc arms (Shane 1972).

DISCUSSION FOLLOWING COHEN

Sanders: You have used the term "spiral arms" in describing the large ordered features that seem to be present in your plot of ^{12}CO emission in the longitude-velocity plane in the area $12° \leqslant \ell \leqslant 60°$, $|b| \leqslant 0°.25$. The apparent correlation with HI 21-cm emission-contour ridges was also mentioned as further evidence of "arms", yet no clear spiral structure has emerged from extensive analysis of the HI data. Are you saying that the ^{12}CO data gives a more clear picture of spiral structure?

Cohen: Of course there are many difficulties in going from the ℓ-v diagram to real space. The point is that the large 21-cm objects that have been traditionally called spiral arms are seen as well or better in CO. It remains an open question whether these can be connected into a "grand design" of spiral structure.

Silk: Could the evidence you have found for spiral features correspond to the spiral structure noted by Solomon in the hotter CO clouds? In other words, is your sensitivity lower than his?

Cohen: Our sensitivities are about the same. The real difference is in coverage; we have taken spectra every beamwidth or every other beamwidth over the molecular disk, while the small beam of the NRAO 36-foot telescope forces its users to undersample area by a factor of about 100.

Gordon: Comparison of a 2-arm density wave model with our CO observations shows some agreement (see the side-by-side illustrations by Gordon and Burton, Scientific American, May 1979). However, even more striking is the large discrepancy still remaining. Evidently the arrangement of molecular clouds in our galaxy is much more complicated than simple spirals, either in physical or velocity space. In this regard our computer analysis agrees quite well with the results of Sanders and Solomon.

Cohen: There are certainly some molecular clouds between the arms, but I think the data show that molecules are even more concentrated in the arms than is HI.

MOLECULAR FAN OF 360-pc RADIUS IN THE GALACTIC CENTER REGION

Yasuo Fukui
Tokyo Astronomical Observatory, Mitaka, Tokyo, Japan

Molecular hydrogen of $\gtrsim 10^8 M_\odot$ exists in the galactic center region, as has been revealed by recent observations of molecular emission lines (see e.g. Oort 1977). In the inner region of $\ell \lesssim 3°.0$ most of the dominant emission features concentrate at $0°.0 \lesssim \ell \lesssim 2°.0$ and 0 km s^{-1} $\lesssim v \lesssim 100$ km s^{-1} extremely unevenly with respect to the galactic center (see Fig. 1). As a model of the molecular complex we propose a fan of 360-pc radius whose pivot is at the center. The vertical angle of the fan is about $50°$ and the central line of the fan makes an angle of about $60°$ to the line of sight. Molecules in the fan are flowing out radially from the center with a velocity of 110-140 km s^{-1}. The ℓ-v pattern of the fan model is superposed on the CO map in Fig. 1. The model can explain the whole structure of the molecular complex as well as several fine details such as asymmetry in emission line profiles (Fukui et al. 1979). As for Sgr A and Sgr B2, numerical calculations of molecular line profiles have been made by using the large velocity gradient approximation. The calculations show that the broad and asymmetric line profiles in the complex are well reproduced by the fan model. Further, an isotope effect on line shape is predicted, which will be useful as an observational check of the fan model. Additionally, the carbon isotope ratio $^{12}C/^{13}C$ in HCN and CO was estimated to be 10-20 in the Sgr A +50-km s^{-1} cloud.

The origin of the molecular fan is most directly interpreted as anisotropic mass ejection from the center. Similar fan-like features are found in the external galaxies by optical workers. The total kinetic energy of the complex amounts to $\gtrsim 10^{55}$ erg and the characteristic duration of the phenomenon is estimated to be 3 x 10^6 yr. At $\ell \sim 1°.3$ the complex shows an enormous thickness of about 140 pc (Inatani 1978), and a past active phase of the nucleus may be suggested. Fig. 2 shows a schematic diagram of the fan model.

This paper is a summary of part of the Doctoral thesis of the author (Fukui 1978).

B. H. Andrew (ed.), Interstellar Molecules, 209–211.

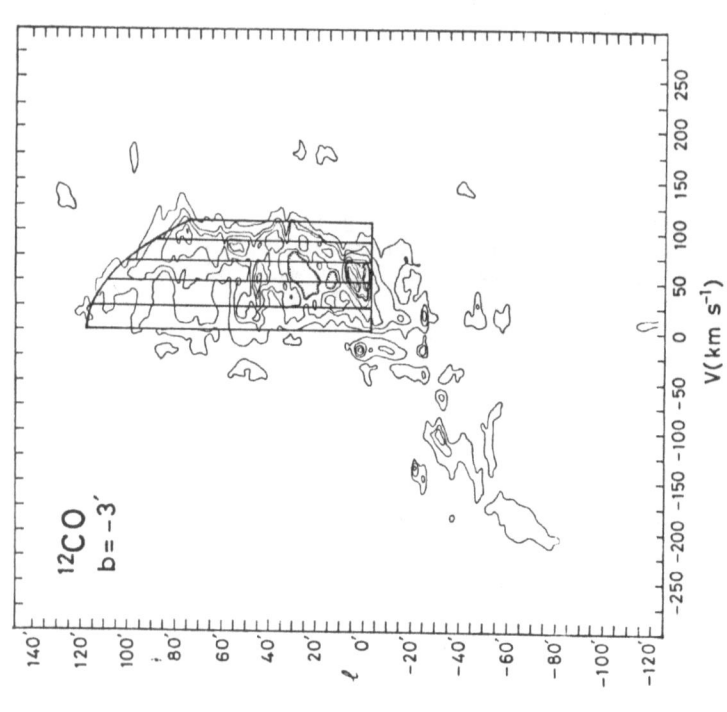

Fig.1(right):The ℓ-v pattern of the fan model(thick line)
superposed on the CO map(Liszt and Burton 1978). Contour
levels less than 5 K are omitted in the CO map.
Fig.2(left):A schematic view of the fan model. The lower
panel shows the top view and the upper the view in the sky.
The dashed line is an HI hole suggested by Sanders and
Wrixon(1973).

REFERENCES

Fukui, Y.: 1978, Doctoral thesis, University of Tokyo.
Fukui, Y., Kaifu, N., Morimoto, M., and Miyaji, T.: 1979, this volume.
Inatani, J.: 1978, Doctoral thesis, University of Tokyo.
Liszt, H.S., and Burton, W.B.: 1978, Astrophys. J. 226, 790.
Oort, J.H.: 1977, Ann. Rev. Astron. Astrophys. 15, 295.
Sanders, R.H., and Wrixon, G.T.: 1973, Astron. Astrophys. 26, 365.

DISCUSSION FOLLOWING FUKUI

Sandqvist: I am in the process of using the 11m telescope at Kitt Peak to map a 10 arcmin region around Sgr A in the 2-mm H_2CO line. The resolution is about 1 arcmin. Preliminary results reveal a velocity gradient of about 5 km s^{-1} per arcmin in the direction of galactic rotation. Did you observe a similar gradient and did you include galactic rotation in your model?

Ho: We also discovered a strong velocity-gradient across the Sgr A cloud when we mapped the Galactic Center region in the (3,3) line of NH_3. The line shapes are relatively narrow, $\sim$20 km s^{-1}, shifting by $\sim$10 km s^{-1} every 1.5.

Fukui: I guess that the strong velocity gradient you observed is in the +30 km s^{-1} cloud, which is located on the negative longitude side of the nucleus. We also observed the same type of velocity gradient there in the HCN emission (P.A.S. Japan, 29, 643, 1977). The +30 km s^{-1} cloud is a minor component of the Sgr A molecular complex, compared to the +50 km s^{-1} cloud at $\ell \gtrsim 0°$, and it is not included in the present model, which aims to explain the predominant emission features. The +50 km s^{-1} cloud shows little sign of rotation according to our HCN and HCO$^+$ results, and the velocities at the emission maxima are almost uniformly $\sim$50 km s^{-1} throughout the cloud (-3' $\gtrsim \ell \gtrsim$ 10'). Therefore, I did not include strong rotational motion in the model.

Pauls: Your previous HCN survey was also interpreted in terms of a jet model. Is your current model of the CO data consistent with that HCN model?

Fukui: Yes. The present model is just a simple extension of the jet model for the HCN data, and they are consistent with each other.

Liszt: Can you suggest a mechanism for ejecting large quantities of purely molecular gas from a small region near Sgr A at velocities $\sim$100 km s^{-1}. Acceleration of pre-existing gas would not be a successful explanation, because there is no central hole in your model.

Fukui: I think the acceleration due to cosmic ray pressure, for example, can push the molecular gas outward. The present state of the nucleus does not show strong signs of activity, and it is not clear that sufficient cosmic ray protons are supplied from the nucleus. The strength of the nuclear activity may have decreased significantly over the last $\sim$10^6 yr, because the anomalous thickness of the +100 km s^{-1} cloud indicates a past active phase of the nucleus, some 10^6 yr ago. The event must be a transient phenomenon. If there were to have been strong acceleration at the earlier phase of the ejection, no additional acceleration would be required.

THE GALACTIC ROTATION CURVE TO R = 18 kpc

Leo Blitz and Michel Fich
Radio Astronomy Laboratory
University of California, Berkeley
and
Antony A. Stark
Princeton University Observatory

The major stumbling block in the determination of a rotation curve beyond the solar circle has been the lack of a suitable set of objects with well defined and independently measured distances and velocities which can be observed to large galactocentric radii. Two things have changed this situation. The first was the realization that essentially all local HII regions have associated molecular material. The second was the acquisition of reliable distances to the stars exciting a sizable number of HII regions at large galactocentric radii (Moffat, FitzGerald, and Jackson 1979). Because the velocity of the associated molecular gas can be measured very accurately by means of radio observations of CO, we have been able to overcome the past difficulties and have measured the rotation curve of the Galaxy to a galactocentric distance of 18 kpc.

We have obtained the velocity of the CO related to 184 HII regions. Most of these are new detections, but we have made extensive use of the published literature and have checked the published values to the degree that it was feasible. The results presented here were obtained with the 5 m telescope at the Millimeter Wave Observatory, the 7 m telescope at Bell Laboratories, and the 11 m telescope at Kitt Peak. We have relied heavily on the distances determined by Moffat, FitzGerald and Jackson (1979) who have obtained spectrophotometric distances to 40 of the most distant HII regions. We have also made use of the distances published by Georgelin, Georgelin and Roux (1973), Georgelin and Georgelin (1976), Crampton, Georgelin and Georgelin (1978) and Humphreys (1979) among others to obtain distances to 92 molecular complexes. These complexes are plotted in Figure 1 which shows their locations projected onto the Galactic plane. It is evident from Figure 1 that there should be no systematic errors due to limited longitude coverage.

We have determined the rotation curve under the assumption that the complexes are in circular rotation about the Galactic center, that R_0 (the solar distance to the Galactic center) is 10 kpc and Θ_0 (the circular velocity of the sun) is 250 km sec^{-1}. HII regions within $\pm$ 15° of the Galactic center and anticenter were omitted from the de-

B. H. Andrew (ed.), Interstellar Molecules, 213–220.

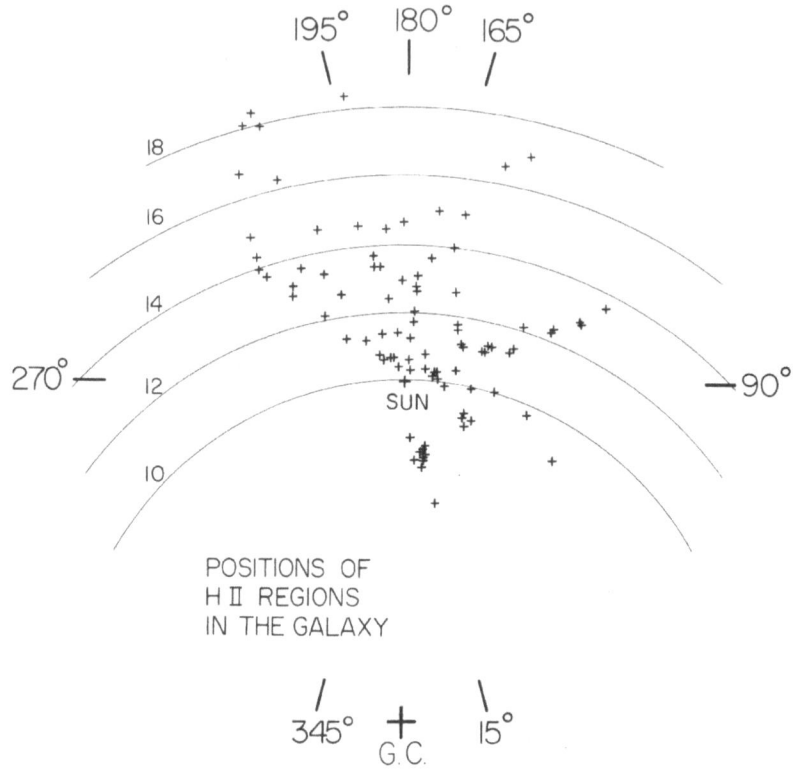

Fig. 1 – The distribution of HII regions used in this investigation projected onto the Galactic plane. Where several HII
regions appear to be related to a single molecular complex,
such as the S254-S258 complex, they are considered as one object. All of the objects plotted have known distances and
CO velocities.

termination of the rotation curve because the radial projection of the
circular velocity in these regions is comparable to the random velocities of the complexes. We have used the omitted HII regions to determine the velocity dispersion of the giant molecular complexes.

Figure 2 shows the rotation curve of the Galaxy from 7 to 18 kpc
determined from the CO velocity data. The curve is a fourth order polynomial fit which excludes the HII regions in the Perseus arm. These are
the nine HII regions near R = 11 kpc with Θ < 240 km sec^{-1}. Inclusion
of these points lowers the polynomial fit by 5 km sec^{-1} near 11 kpc,
but the overall fit is not substantially changed. In the portion of
the Galaxy we have sampled, no region shows as large a systematic deviation from the mean rotation curve as the Perseus arm does.

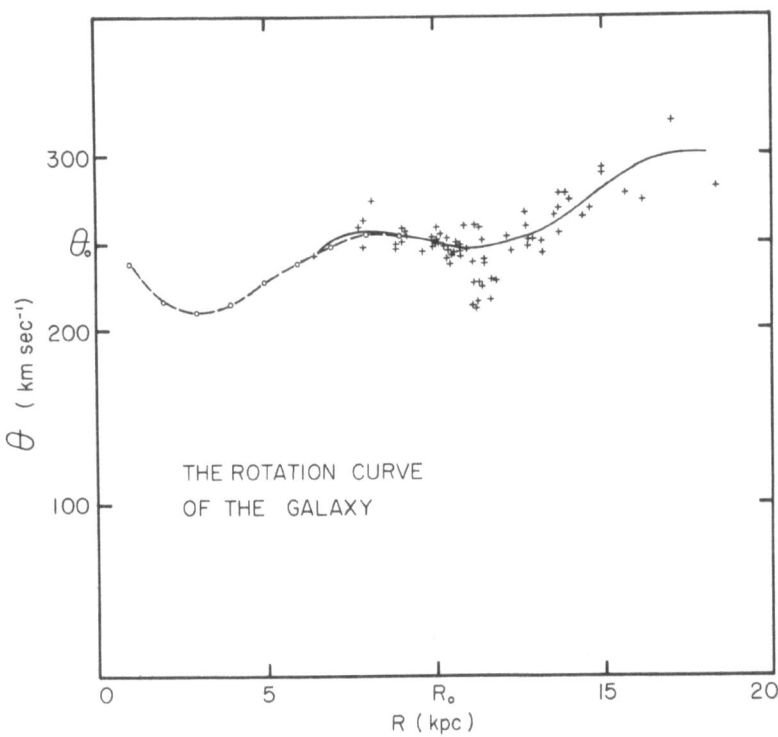

Fig. 2 – The rotation curve of the Galaxy. Dashed line
connecting the circles are from atomic hydrogen data.
The solid curve is a fourth order least squares fit to
the data, which omits the Perseus arm anomolous velocity
observations.

The primary result of our observations is that there is no decrease
(turnover) in the circular velocity to the last measured point. The
Galaxy is therefore typical of the spiral galaxies whose rotation
curves were measured by Rubin, Ford and Thonnard (1978). The rotation
curve rises from about 12 kpc to 18 kpc and attains a circular velocity
of $\sim$ 300 km sec^{-1} at the last measured points. This rise appears to
be real because to make the rotation curve flat requires a change in
the distance scale by a factor of two. Such a change would affect the
points internal to the solar circle causing a marked deviation from the
HI rotation curve, shown by the dotted line connecting the circles in
Figure 2. Our data show the agreement between the HI and CO rotation
curves to be quite good.

We may use the data to find the Oort A constant by plotting Ω vs. R
and performing a least squares polynomial fit to the data, which is

shown in Figure 3. A is given by $-\frac{1}{2} R_o \left(\frac{d\Omega}{dR} \right)_{R_o}$ and has the value

12.9 km sec^{-1} kpc^{-1} if we exclude the Perseus arm and 13.3 km sec^{-1} kpc^{-1} if we include it. These values are considerably lower than the generally accepted value of 15 km sec^{-1} kpc^{-1} (Schmidt 1965), and are reminiscent of the values first suggested by Weaver (1955, 1958). We may also determine the curvature parameter α given by $-\frac{1}{4} R_o \left(\frac{d^2\Omega}{dR^2} \right)_{R_o}$ in the same manner. We find that $\alpha = -1.8$ km sec^{-1} kpc^{-2} and is relatively insensitive to the inclusion or omission of the Perseus arm observations.

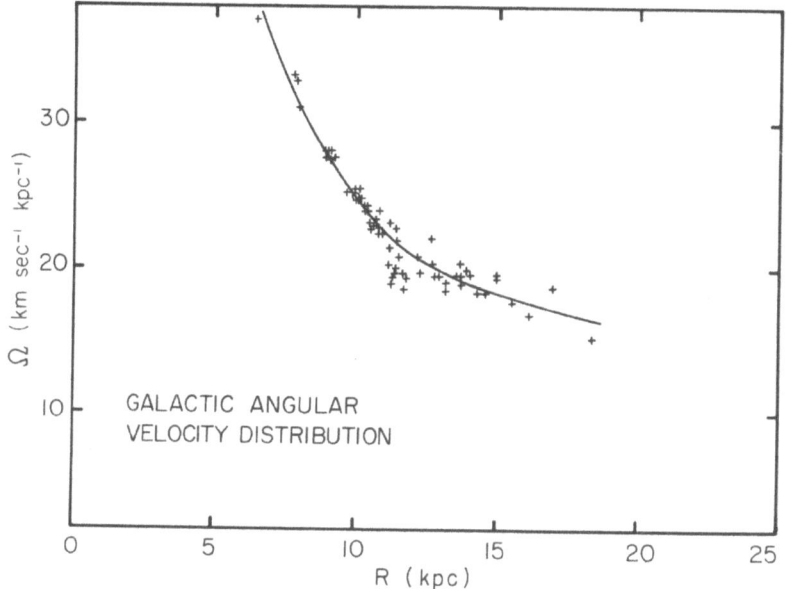

Fig. 3 - Plot of angular velocity Ω vs. R. The solid line is a third order polynomial fit to the data which omits the Perseus arm data. The circle gives the solar values.

The complexes within $\pm$ 15° of ℓ = 180° can be used to determine the velocity dispersion of the giant molecular complexes. We use the spectrophotometric distance and the rotation curve in Figure 2 to obtain a circular velocity for the anticenter complexes and subtract the projection of the circular velocity from the observed radial velocity. The result is just the radial component of the random velocity of each complex. The velocity dispersion of 17 giant complexes in the anticenter is 7 km sec^{-1} which should be contrasted with a dispersion of 8 km sec^{-1} obtained by Stark (1979) for <u>all</u> of the gas toward the galactic anticenter. Stark's observations sample low intensity, low mass molecular clouds. We therefore conclude that the velocity dispersion of high and low mass molecular material is very nearly the same.

The mean velocity of the anticenter complexes is -7 km sec^{-1}. Toward the Galactic center, the mean of four complexes with known distances is $+7$ km sec^{-1}. If we assume that the molecular complexes with unknown distances are at R = 8 kpc, the mean velocity of ten complexes within $\pm$ 15° of ℓ = 0° is $+5$ km sec^{-1}. These values are consistent with a net radial outward motion of the Local Standard of Rest (LSR) of $\sim$6 km sec^{-1}, a suggestion first proposed by Kerr (1961). This conclusion is reinforced by our observations of four high latitude clouds. These objects are probably within $\sim$100 pc of the sun and their motions are presumably tied to the LSR. The radial velocity of each of these clouds is nearly zero. This implies that the observed mean velocities toward the Galactic center and anticenter are due more likely to an outward motion of the entire LSR rather than an error in the measurement of the solar motion relative to the LSR.

Finally, we use the rotation curve to determine the mass of the Galaxy interior to 18 kpc. Using the simplest of mass models, we assume that the mass is distributed in a uniform sphere to the last measured point, and find that the mass of the Galaxy is 3.4 x 10^{11} M$_{\odot}$. Using different models produces only inconsequential changes to the mass estimates. This mass is considerably larger than has been previously thought and is $\sim$50% larger than the value derived for M31 under the same set of assumptions. Our observations indicate, therefore, that the Galaxy is the most massive member of the local group.

This work was partially funded under NSF grant AST 78-21037. A.A.S. acknowledges support by a National Science Foundation Graduate Fellowship. M.F. acknowledges support by an NSERC (Canada) Postgraduate Scholarship.

REFERENCES

Crampton, D., Georgelin, Y.M., and Georgelin, Y.P.: 1978, *Astron. and Ap.* 66, 1.
Georgelin, Y.M., and Georgelin, Y.P.: 1976, *Astron. and Ap.* 49, 57.
Georgelin, Y.M., Georgelin, Y.P., and Roux, S.: 1973, *Astron. and Ap.* 25, 337.
Humphreys, R.M.: 1978, *Ap. J. (Suppl.)*, 38, 309.
Kerr, F.J.: 1961, M.N.R.A.S., 123, 327.
Moffat, A.F.J., FitzGerald, M.P., and Jackson, P.D.: 1979, *Astron. and Ap. (Suppl.)*, 38, 197.
Rubin, V.C., Ford, W.K., and Thonnard, N.: 1978, *Ap. J. (Letters)*, 225, L107.
Schmidt, M.: 1965, in *Galactic Structure*, ed. A. Blaauw and M. Schmidt (Chicago: University of Chicago Press), p. 513.
Stark, A.A.: 1979, Ph.D. Dissertation, Princeton University.
Weaver, H.F.: 1955, *A.J.*, 60, 202.
Weaver, H.F.: 1958, Trans. IAU X, 696.

DISCUSSION FOLLOWING BLITZ

Guelin: There is strong evidence (e.g. W.W. Roberts) that HII, HI gas and stars have large ($\sim$15 km/s) non-rotational velocities in the Perseus arm. CO cloud velocities could also be affected in this region (Yuan, 1976). Since your data partly relies on the Perseus arm, your results could be affected.

Blitz: The CO shows the same velocity anomalies as the other Population I tracers, but the Perseus arm represents only $\sim$10% of the data presented here. The shape of the rotation curve changes only slightly, and the value of A goes from 12.85 to 13.30, when the data from the Perseus arm are included.

Thaddeus: What happens when you compare the ordinary rotation curve obtained from the stellar radial-velocities with that derived from CO? How much does the use of CO improve over the standard method? It would be interesting to see the two side by side.

Blitz: It has not been possible to obtain a reliable *quantitative* rotation curve without the CO data. Jackson, FitzGerald and Moffat, on whose optical data we have heavily relied in the outermost portions of the Galaxy, published a rotation curve last year in the form of a plot of $\Omega - \Omega_0$ vs R (Proceedings of IAU Symposium #84). While their data indicate that the rotation curve deviates markedly from the Schmidt model, the uncertainties in the measurements of radial velocity are so large that no reliable quantitative conclusions can be drawn. For example, if my memory is correct, there are many cases where the CO and stellar data differ by as much as 10 km s^{-1}, although the differences do not appear to be systematic.

Lo: As you pointed out yourself, the derived galactic circular velocities depend critically on the adopted distances of the CO sources. How accurate are the distances to the HII regions associated with the CO sources, and how reliable is the assumption that the CO sources are at the same distance as the HII regions?

Blitz: I believe that the assumption that the CO is at the same distance as the HII is very accurate. We mapped some of the most distant sources and found that the HII regions occurred at or near the CO peaks, as always seems to be the case locally. Also, for most sources in the second and third quadrants, only a single line is detected toward any HII region. Although a few points may be in error, they are likely to be a very small proportion of the total. The errors in the distances are the largest source of uncertainty in the determination of the circular velocity. Most of the largest distances are probably good to $\sim$20-25%, and the corresponding uncertainty in θ from all sources of error is $\leq$20 km s^{-1} for any one point. Not all of the distances are as accurate, however.

Lo: One further comment. The velocity dispersion of outlying globular clusters measured by Hartwick and Sargent suggests that the total mass of the galaxy is many times $10^{11} M_\odot$.

Blitz: If one extends the rotation curve we have presented to $\sim$50 kpc, one derives a mass which agrees, within the errors, with the mass derived by Hartwick and Sargent for an isotropic velocity distribution.

Morris: How does the slope of the rotation curve change if you assume $R_0=8.5$ kpc and $\theta_0=220$ km s^{-1}, values which are suggested by recent studies?

Blitz: For these values the shape is unaffected because θ_0/R_0 is unchanged. The rise in the rotation curve shown here, however, indicates that $\theta_0=250$ km s^{-1}, if one uses the analysis of Knapp, Gunn and Tremaine. We have not yet determined how the shape changes with $R_0=8.5$, $\theta_0=250$ km s^{-1}.

Bok: I hope that everyone realizes that an expansion velocity of 6 km s^{-1} implies that the LSR is moving outward in the Galaxy at a rate of 6 kpc in only one billion years, i.e. four or five galactic revolutions. Before we accept such a fantastic rate of expansion, we should see to what extent we can adjust our distance scale - or impose streaming motions - to obtain a smaller rate of expansion or no expansion. Is our Sun really on the way out?

Blitz: It is not clear why one should expect the 6 km s^{-1} outward motion of the LSR to persist for 10^9 years. The stars which define the LSR would probably disperse on this timescale. Furthermore, we know from the random velocities of the giant molecular complexes that coherent radial motions of the same magnitude as we postulate for the LSR are quite common, even for masses as large as 2×10^5 M$_\odot$. The observed motion of the LSR might also result from the radial component of an elliptical orbit of the LSR. Finally, the motion could also be cyclical or irregular, depending on the perturbations in the gravitational potential of the disk, brought about, say, by the passage of a density wave every $\sim 2 \times 10^8$ yr. In this regard, Lin, Yuan and Roberts have recently argued from an analysis of the observational data that according to the density wave theory, the LSR *should* have an outward component of ~ 6 km s^{-1}.

Radhakrishnan: A recent analysis of the HI absorption spectrum towards Sgr A (Radhakrishnan and Sarma, submitted to Astronomy and Astrophysics) clearly indicates that all of the hydrogen in the direction of the galactic centre - excluding the gas within 3 kpc of the centre - has a mean velocity with respect to the Local Standard of Rest (LSR) of 0 ± 0.25 km s^{-1}. Your conclusion that the LSR has a net radial outward motion of approximately 6 km s^{-1} (as suggested by Kerr (1961)) seems to be untenable unless you are also prepared to conclude that all of the HI gas within 7 kpc of the Sun in the direction of the galactic centre also shares precisely this motion.

Blitz: One cannot use the HI data to refute the hypothesis that the LSR has some radial motion with respect to the molecular gas until one can reconcile the different results implied by your observations and the observations of HI in emission toward the Galactic center. These results (note particularly those already published by Burton and Liszt (1978) and Burton, Gallagher and McGrath (1977)) show unquestionably that the ridge of maximum emission does not occur at a velocity of 0 km sec^{-1} but at about +5 km sec^{-1}, in rough agreement with the CO data. I would be more inclined to trust the emission data since it samples all of the gas along the line of sight and is less likely to be affected by anomalies caused by, say, a single massive cold cloud along the line of sight. Burton has also pointed out in the past that the HI seen in emission toward the anticenter has a net negative velocity, a result which is also in agreement with the CO results. It is important to emphasize that the motion which we claim to exist has an outward component with respect to the *molecular* gas, particularly the giant molecular complexes. If these were to have a preferential inward motion with respect to the center of the Galaxy, as one might expect if they are formed as a result of compression behind a density-wave shock, then the radial motion of the LSR with respect to the Galactic Center could be zero. It would still be necessary to explain, however, the residual motion of the LSR with respect to the HI in emission.

Radhakrishnan: Since when does emission data sample all of the gas along the line of sight and since when is it less likely to be affected by anomalies than absorption data? By definition the optical depth is always added all along the line of sight. On the other hand emission samples all the gas only when it is optically thin. If there is one thing we know and have known for years about the HI gas towards Sgr A, it is its high optical depth at near zero velocities. I have no quarrel with the elegant interpretation of Burton and Liszt of the inner galaxy where high velocities prevail; but for sampling the gas outside the inner galaxy their emission measurements at low velocities and all other such measurements are no match for direct optical depth determinations against Sgr A. The molecular complexes must have a preferential inward motion.

Blitz: It is impossible for me to try to reconcile the HI absorption and emission data without seeing your results in detail. However the depth of the HI absorption profile toward Cas A demonstrates that the absorption toward Sgr A could be dominated by a single cold cloud or a small path length through the Galaxy. Furthermore the mean positive residual of the CO and the peak HI emission appears to be shared by the HI emission which is at low optical depths ($|v| \lesssim 7$ km s^{-1} from the line center; not what one might call high velocity emission) and which thus samples all the gas along the line of sight. The same is true for the HI at high latitudes toward $\ell=0°$. I think more observations or more detailed analyses of existing observations are necessary to decide whether the LSR is moving out or the molecular complexes are moving in.

THE HYDROGEN MOLECULE AS A COLLISION PARTNER

Takeshi Oka
Herzberg Institute of Astrophysics
National Research Council of Canada
Ottawa, Ontario, Canada

Because of its abundance in molecular clouds, the hydrogen molecule is the most important collision partner in a discussion of thermal equilibrium or non-equilibrium in most interstellar molecules. The purpose of this talk is to summarize the properties of the H_2 molecule as a collision partner based on laboratory experiments on NH_3-H_2 collisions.

The basic process to be studied is shown in Figure 1. The NH_3 molecule interacts with the H_2 molecule and changes its inversion-rotation state ($J',K',p' \leftarrow J,K,p$), where the quantum numbers J,K,p correspond to the total angular momentum, its projection along the molecular axis, and the parity, respectively.

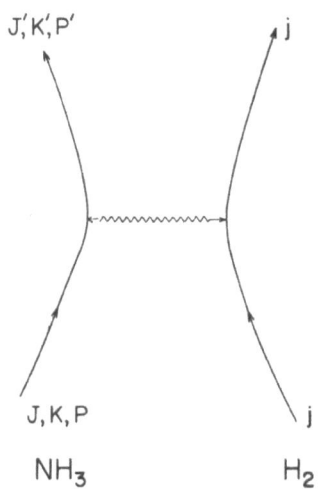

Fig. 1. The elementary process

B. H. Andrew (ed.), Interstellar Molecules, 221–229.

Note that the interaction does not change the rotational quantum number j of the H_2 molecule because of the large energy separation between the levels. (The smallest separation in one spin species is ~ 360 cm^{-1} ~ 520 K between the j=2 and j=0 levels, which is much larger than either the energy change in NH_3 or kT, not only in interstellar space but also in the laboratory. The transition between the j=1 and j=0 levels is forbidden because of the ortho $\leftrightarrow$ para rule). The $\Delta j \neq 0$ transitions of H_2 are neglected in the following discussion because of their small probability.

1.   THE INNER CORE-OUTER SHELL MODEL OF H_2

The potential energy between a polar molecule and a hydrogen molecule can be divided into two parts:

(a) The long range dipole-quadrupole potential,

$$V(\underset{\sim}{r}) = - \frac{1}{r^4} \left[\underset{\sim}{e} \cdot \underset{\approx}{Q} \cdot \underset{\sim}{\mu} - \frac{5}{2}(\underset{\sim}{e} \cdot \underset{\approx}{Q} \cdot \underset{\sim}{e})(\underset{\sim}{e} \cdot \underset{\sim}{\mu}) \right]$$

where $\underset{\sim}{\mu}$ is the dipole moment of the polar molecule, $\underset{\approx}{Q}$ is the quadrupole tensor of the H_2 molecule and $e = \underset{\sim}{r}/r$ is the unit position vector between the two molecules. The higher multipole moments are neglected.

(b) The shorter range potential due to induction, dispersion and valence forces whose radial part is approximated for example by the Lennard-Jones potential. The angular dependent part of such a potential causes rotational transitions in NH_3.

The former potential plays the major role for weaker collisions with large impact parameters, the latter for stronger collisions with small impact parameters. Thus we can consider a model in which the H_2 molecule as a collision partner consists of two parts, a hard core and a soft outer shell. Note that the He atom, which is the next important collision partner, has only the hard core.

2.   FOUR-LEVEL DOUBLE RESONANCES

The existence of the double structure is most clearly revealed by four-level double resonance experiments (Oka 1973). In these experiments two microwave radiation fields ν_p and ν_s and a low pressure (~ 1 torr) mixture of NH_3 and H_2 (or for comparison He) with a mixing ratio of $\sim 1:100$ are used. The first radiation ν_p (pump radiation with a power of ~ 1 W) saturates an inversion transition and thus introduces a non-Boltzmann population in the rotational levels (J,K). This population anomaly is then transferred to other levels by collisions. The second radiation ν_S (signal radiation with a power of $\sim$ 1 µW) monitors the result of the population transfer by detecting a variation in intensity of other inversion lines (J',K'). Using many combinations of (J,K) and (J',K') we can study systematic-ally the collisional process shown in Figure 1.

The effect of the outer shell is most clearly seen when we choose (J' = J-1, K' = K) levels which are connected to the pump levels by the dipole selection rules (Figure 2).

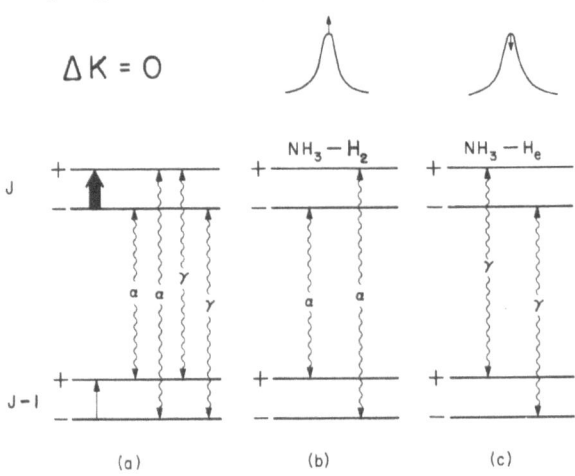

Fig.2 The four-level microwave double resonance experiments, for dipole-allowed transitions. The pump is shown by the bold arrow, the signal by the thin arrow, and the collision-induced transitions by wavy arrows. The H_2 molecule and the He atom act very differently as collision partners.

For the NH_3-H_2 experiment the signal intensity is increased by pumping, indicating that the NH_3 molecule changes its parity by collision. This is expected from the first order treatment of the dipole-quadrupole interaction. On the other hand for the NH_3-He experiment the signal intensity is decreased by pumping, indicating that the parity of the NH_3 molecule is preserved in the collision. These experiments demonstrate the qualitative difference between the core and the outer shell as perturbers.

The effect of the inner core of H_2 is seen when we choose a double resonance system with K'≠K. Such transitions are not dipole-allowed and are caused by stronger collisions. For some reasons which are not completely clear to me, even such stronger collisions have some parity rules and double resonance signals are easily observed. If we compare the relative intensities of such signals for the NH_3-H_2 and the NH_3-He mixtures, we observe a remarkable similarity, as shown in Figure 3. This indicates that in spite of the fact that the H_2 molecule and the He atom have cores of different shapes, they act very similarly for strong collisions.

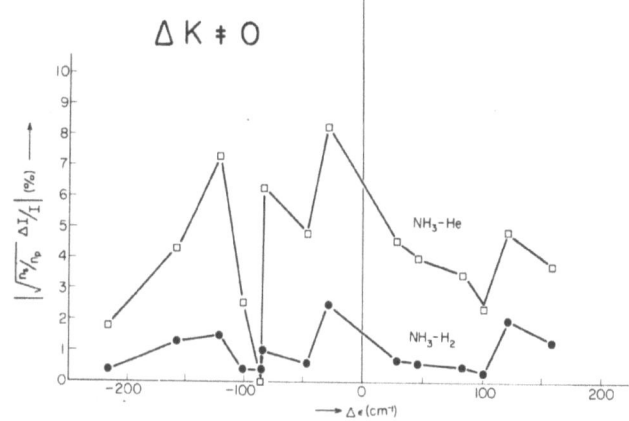

Fig.3. Results of four-level microwave double resonance for dipole-forbidden transitions ($\Delta K = 3n, n \neq 0$). Here the H_2 molecule and the He atom act very similarly on NH_3 as the collision partners.

3. H_2 IN $J=0$

In the collision process shown in Figure 1 in which the H_2 molecule does not change its rotational state, the H_2 molecule in the ground rotational level $J=0$ acts as if it does not have the outer shell. This is expected from the matrix element of the dipole-quadrupole potential $V(\underset{\sim}{r})$ which can be rewritten (Oka 1973),

$$V(\underset{\sim}{r}) = - \frac{8\sqrt{\pi^3}}{r^4} \mu Q \sum_{m_1,m_2,m} \begin{pmatrix} 1 & 2 & 3 \\ m_1 & m_2 & m \end{pmatrix} D^3_{o,m}(\Omega_r) D^1_{o,m_1}(\Omega_\mu) D^2_{o,m_2}(\Omega_Q)$$

where $D^\ell_{n,m}(\Omega)$ are rotation matrices for the spatial orientations Ω_r, Ω_μ, and Ω_Q of the position vector $\underset{\sim}{r}$, the dipole moment $\underset{\sim}{\mu}$, and the quadrupole moment $\underset{\sim}{Q}$, respectively. It is seen from this formula that $\langle j|V(\underset{\sim}{r})|j\rangle = 0$ for $j=0$ because $\langle 0|D^2_{o,m_2}(\Omega_Q)|0\rangle = 0$, that is, the H_2 molecule in the $j=0$ level behaves as a collision partner as if it does not possess the quadrupole moment. Thus we arrive at the models shown in Figure 4.

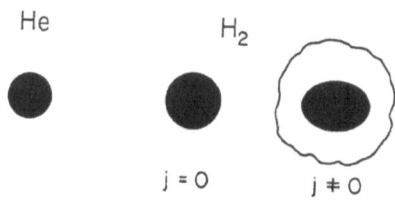

Fig.4. Models of the He atom and the H_2 molecule as collision partners.

4. PRESSURE BROADENING

The special characteristics of the H_2 molecule with $j=0$ can be seen most directly from pressure broadening of spectral lines. There is a large difference in the percentages of the $j=0$ H_2 molecules in normal-H_2 and para-H_2 (18.4% in normal-H_2 and 73.2% in para-H_2 at dry ice temperature; 13.0% and 51.7% at room temperature). This can be used to distinguish the effect of the $j=0$ H_2 molecules from the rest. Figure 5 shows a recent measurement in the infrared region. A difference frequency laser

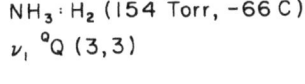

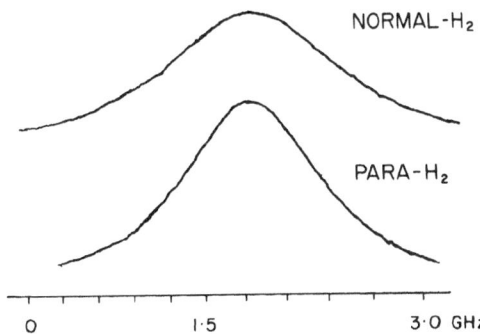

Fig.5. Pressure broadening of the infrared transition line ν_1 $QQ(3,3)$ of NH_3 by normal-H_2 and para-H_2 at dry ice temperature. The total pressure is 154 torr for both traces. (D.R. Rao and T. Oka, 1979).

system was used as a frequency tunable infrared source. The difference between the normal-H_2 and para-H_2 as collision partners is clearly seen. Measurements in the infrared and microwave regions of NH_3 lines showed that the normal-H_2 containing more $J \neq 0$ H_2 is more effective as a perturber than the para-H_2 by a factor of 1.17±0.06 at room temperature and 1.29±0.03 at dry ice temperature. Using the simple assumption that all $j \neq 0$ H_2 molecules have the same collision cross section as partners to NH_3, we determine the ratio of the total collision cross section for the $j \neq 0$ H_2 to that for the $j=0$ H_2 to be

$$\sigma_{j \neq 0} / \sigma_{j=0} = 1.60 .$$

Thus in the model shown in Figure 4, the effect of the outer shell in causing rotation-inversion transitions in NH_3 is 60% of that of the inner core.

It should be noted that although the outer shell is somewhat less effective than the inner core in causing rotation-inversion transitions in NH_3, its effect is concentrated in causing dipole-type transitions and may dominate for such transitions. The results shown in Figure 2 are understood from such a consideration.

5. ASTROPHYSICAL IMPLICATIONS

From the observations given above, it is seen that any discussion on thermal equilibrium or non-equilibrium of interstellar molecules depends critically on whether the H_2 molecule considered as the collision partner is in the $j=0$ level or not. In the former case, the H_2 molecule ($j=0$) acts as the He atom and the results of several theoretical calculations of the (polar molecule-He) system are probably approximately transferrable to the (polar molecule-H_2) system, if allowance is made for a difference between the absolute magnitudes of the cross sections of the two systems (the ratio estimated from pressure broadening is about 1:1.5 for He to H_2 $j=0$). This type of collision is more likely to cause population anomalies which are required for the maser or the anti-maser action.

On the other hand, the H_2 molecules in $j\neq0$ levels act to reduce such population anomalies. This is because (a) the outer shell of such a H_2 molecule causes efficiently the collision-induced transition between the levels observed by microwave radiation and thus tends to thermalize the population and (b) it also tends to make symmetric the transition rates between various levels of K-type, Λ-type or inversion doubling. Thus for example the cooling mechanism proposed by Townes and Cheung (1969) for the $1_{10}\leftarrow1_{11}$ transition of H_2CO is less likely to work for $j\neq0$ H_2 because (a) the collision-induced $1_{10}\longleftrightarrow1_{11}$ transition will reduce the anomaly and (b) the required asymmetry in the rates of the $2_{11}\leftarrow1_{11}$ and the $2_{12}\leftarrow1_{10}$ transitions is less likely.

When the H_2 molecules are initially formed on dust grains and released in space, the ortho-para ratio is thought to be statistical, that is, initially there are more ortho-H_2 than para-H_2 (Black and Dalgarno 1976). The ratio of $j=0$ H_2 to the rest depends on how efficiently the mechanism of $H^{+} + o - H_2 \longleftrightarrow H^{+} + p - H_2$ (Dalgarno, Black and Weisheit 1973) thermalizes the two species in the low temperature molecular cloud.

REFERENCES

Takeshi Oka, "Collision-Induced Transitions Between Rotational Levels", Adv. At. Mol. Phys. 9, 127 (1973).
C.H. Townes and A.C. Cheung, Ap. J. Lett. 157, L103 (1969).
A.H. Black and A. Dalgarno, Ap. J. 203, 132 (1976).
A. Dalgarno, J.H. Black, and J.C. Weisheit, Ap. Lett. 14, 77 (1973).

DISCUSSION FOLLOWING OKA

Carruthers: How did you make pure parahydrogen to use in your experiments?

Oka: We boil off liquid H_2 then check the pureness of para-H_2 by observing the rotational Raman spectrum.

Thaddeus: You can buy pure para-H_2. Presumably it is used by bubble chamber people.

Oka: Yes, but the trouble is that suppliers of liquid H_2 for bubble chambers sell only by tons.

Carruthers: I think that all of the Copernicus observations of interstellar H_2 are consistent with a thermal ortho to para-hydrogen ratio.

Oka: The circumstellar H_2 seems to be all thermalized, but the ratio for dark clouds is not known.

Townes: Have you been able to make any measurement concerning the refrigeration of formaldehyde?

Oka: Yes, Evenson, Freund and I attempted that experiment specifically for the Townes-Cheung four-level system of H_2CO. The result is not conclusive. The main problem is that we cannot simulate the interstellar conditions, especially the temperature.

W. Watson: If, in fact, the ortho to para ratio for H_2 is determined by gas-phase exchange-processes, and is near the thermodynamic equilibrium value at the temperature of the gas, then in cooler clouds there should be a higher fraction in the para state which causes the H_2CO refrigeration. Do you know of any astronomical studies that would suggest such an effect?

Oka: No, I am not aware of any. Whether the ortho-para ratio is indeed thermalized to the low temperature of a cloud depends on the efficiency of the Dalgarno-Black-Weisheit reaction, that is, it depends on the proton density in the cloud. The ortho-para ratio, if thermalized, does not critically depend on temperature as long as the temperature is much lower than 171 K, the excitation level of the j=1 state.

Omont: Could you give a value for the ratio of the collisional rates for transitions inside K doublets to the rates of rotational excitation in NH_3?

Oka: The ratio is about 10 to 1. (Ap. J. Lett. 165, L15 (1971))

Winnewisser: Do collisions of j=0 H_2 and other H_2 affect differently the K=1 and 2 levels and the K=3 levels of NH_3, i.e. para and ortho ammonia? What would be the difference?

Oka: As long as we consider K=0 transitions, there is not much difference between ortho and para NH_3. For K=3 transitions, that is, k = ±1 ↔ ±2 and k = 3 ↔ 0, there is a difference, because the former transition is symmetric for inversion levels while the latter is not.

T. Wilson: Are there large differences between the NH_3-H_2 cross-sections for ortho and para-NH_3?

Oka: No. If you consider the total cross section there is not much difference. There is no detectable difference in pressure broadening parameters of ortho-NH_3 and para-NH_3, apart from the usual K-dependence.

Townes: I am a little puzzled. The conversion from ortho to
para-NH_3 is presumably through the exchange reaction with protons, as
is the case with H_2, although there may also be some other mechanism.
In general it seems that NH_3 is approximately in thermal equilibrium,
and that would seem to have some bearing on whether H_2 is also in
thermal equilibrium.

Oka: I am surprised to hear that from you. Your initial measurement of NH_3 in Sgr B2 showed much higher intensity in the ortho lines
than in the para lines.

Townes: The ratios are, I would say, not far from the thermal
equilibrium values.

Oka: A factor of two?

Townes: Yes, though our measurement was not very precise.

Oka: Two is a large factor. I think other measurements by Morris
et al. and by the Bonn group also show a large difference between ortho
and para-NH_3. One consideration here is that the conversion between
ortho and para-H_2 is expected to be slower than that between ortho and
para-NH_3 or between spin modifications in other polyatomic molecules.

Townes: Is that true? What is the ammonia colliding with?

Oka: Even if NH_3 collides with neutral H_2, conversion can occur.

Townes: Does the collision cause an interchange of hydrogen atoms
between the molecules?

Oka: No, there is just a normal collision without chemical reaction.

Townes: But surely the conversion occurs only by substitution?

Oka: Not necessarily. Ortho-para symmetry in polyatomic molecules
is not as rigorous as in H_2. There is little mixing of ortho and para
species due to nuclear spin-rotation interaction. For example, in
spherical top molecules, the different spin modifications can convert
easily with each other by collision.

Townes: I thought that there is a laboratory experiment indicating
that collisions with molecular hydrogen cannot vary the NH_3 ortho to
para ratio, whereas collisions with atomic hydrogen cause very rapid
conversion.

Oka: I am not aware of that experiment. As far as I know there
has been no observation of transitions between different spin modifications other than for H_2 and D_2. In other words, nobody has yet
produced in the laboratory anomalous spin populations of polyatomic
molecules such as formaldehyde, water or NH_3.

Townes: I believe I was told about the experiment by a chemist in
Berkeley. I think his reasoning was based on the possibility of
reactional substitution. It is not a direct experiment but an experiment
associated with deuterium.

Oka: Then I understand.

Thaddeus: I believe that in the liquid phase of hydrogen, it takes
about 10^{15} collisions to accomplish conversion from ortho to para.
Presumably for NH_3 the number of collisions is comparable.

Oka: It could be much less.

Thaddeus: But the rate of conversion is still very slow?

Oka: Yes, much slower than normal. My double resonance experiments
showed this rather clearly. However, compared with H_2 conversion,
polyatomic molecules can convert much faster. I might mention that

Professor Curl and his collaborators attempted to make CH_4 molecules with anomalous spin modification ratio. They cooled CH_4 to He temperature, then boiled it suddenly, hoping that, in the same way as H_2, the CH_4 molecules would go to one spin modification at low temperature (there are nuclear magnetic susceptibility measurements indicating this) and boil adiabatically to form anomalous spin species. They could not observe any anomaly. A theoretical calculation shows that such conversion is very fast in CH_4, because the rotational energy levels corresponding to different spin modifications are very close and the mixing due to spin-rotation interaction is large. I spent a few months with Retallack doing the same experiment in NH_3 where theory predicts much slower conversion. We could not detect an anomaly.

Field: I should just like to comment that in experiments involving $\tilde{A}\ ^2A_1\ NH_2$ in collisions with H-atoms (and other partners), no ortho-para conversion is observed in rotationally inelastic events.

Oka: That agrees with my own experimental results. However, here we are discussing much slower processes. These slower processes are not likely to be observed in normal laboratory experiments other than those I mentioned earlier.

Dalgarno: I think what is happening in the stellar case is that the interchange of protons will also be effective in the case of NH_3, so NH_3 is thermalized just as H_2 is thermalized. Thus it can be characterized by the same temperature.

Oka: Yes. All I am saying is that there is an extra process which thermalizes NH_3, and therefore NH_3 conversion is at least as fast as H_2 conversion. Now if NH_3 spin modifications are not completely thermalized in interstellar space as observation seems to show, there is more possibility that H_2 spin modifications are not thermalized.

W. Watson: In exchanging the ortho and para forms of NH_3, reactions of the form $NH_3 + H_3^+ \rightarrow NH_4^+ + H_2$, $NH_4^+ + e \rightarrow NH_3 + H$ may be more important than reactions with H^+. If so, one would not expect the two forms to be thermalized.

Thaddeus: One point about your laboratory experiments. Even if you cannot go down to the extremely low temperature of interstellar molecules, your results provide a very valuable check of theory, in that theorists should be able to calculate your results just as easily as they can calculate interstellar cross sections.

Oka: I think Sheldon Green has much more difficulty in calculating the high temperature case. At low temperatures his multi-channel equations are much simpler. For the high temperature case he presumably has to worry much more about how to treat the translational motion.

Thaddeus: Then the comparison should be at intermediate temperatures.

Oka: Yes. His calculations come up to liquid-N_2 temperatures and my experiment goes down to dry ice temperature.

Dalgarno: In their present states the theory and observations are mutually orthogonal.

NRCC   18508

RECENT LABORATORY WORK ON MOLECULES OF POSSIBLE IMPORTANCE FOR INTERSTELLAR STUDIES

G. Herzberg
National Research Council of Canada
Ottawa, Ontario, K1A OR6
Canada

One of the aims of our work at the National Research Council (Ottawa) during the last thirty years has been to contribute laboratory data that are of possible astronomical interest especially for the study of the interstellar medium.

NEUTRAL MOLECULES

The detection of the first two interstellar molecules, CH and CN in the near ultraviolet region, needed no special laboratory work since all the information for this identification had been assembled a long time before. As is well known, the radio-frequency detection of CH presented considerable difficulty. Various attempts to determine the Λ-type doubling in the CH ground state, X $^2\Pi$, from the optical spectrum were required (Douglas and Elliott (1965), Goss (1966), Baird and Bredohl (1971)) before this transition was observed by Rydbeck et al. (1974). Even now the radio lines of CH have not been observed in the laboratory.

A considerable amount of work has been done at NRC on the spectra of H_2, HD and D_2. The earlier work on these molecules, with the exception of that of Wilkinson (1968), was marred by the lack of good standards. This was not a difficulty for the new laboratory measurements of HD (Dabrowski and Herzberg (1976)) and as a result the agreement between the Copernicus observations and the laboratory spectra is now excellent.

For the detection of diatomic carbon, that is, the C_2 molecule, it was first necessary to establish the nature of the ground state of this molecule. This was done in the laboratory work of Ballik and Ramsay (1963) who established that the ground state is $^1\Sigma_g^+$. In Figure 1 the known

B. H. Andrew (ed.), Interstellar Molecules, 231–238.
Copyright © 1980 by the IAU.

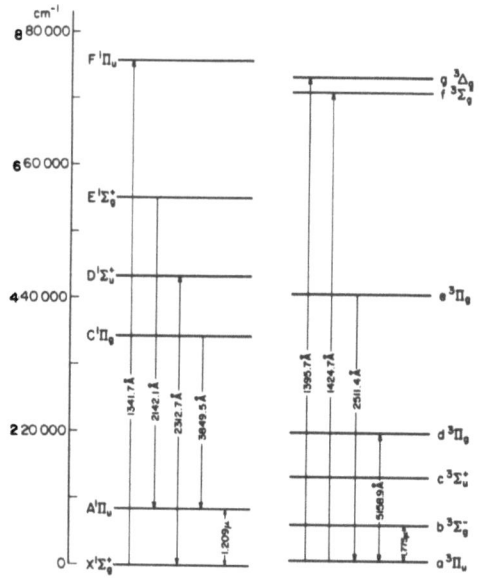

Fig.1. Electronic
energy levels
and observed
transitions in C_2

electronic states of C_2 are presented. It was only last
year that the transition from the ground state to the lowest
excited state (the Phillips bands) was observed in interstellar absorption by Chaffee and Lutz (1978). At the same
time Snow (1978) observed in the ultraviolet the Mulliken
bands in absorption.

It is surprising that the OH molecule, which was the
first molecule identified by radio-astronomy in interstellar
space, was only quite recently observed optically, first in
the vacuum ultraviolet by Snow (1976) using the 1222 Å band
found in the laboratory somewhat earlier by Douglas (1974)
and still more recently in the ordinary ultraviolet near
3080 Å by Crutcher and Watson (1976).

In the 1950's the optical spectra of HCO and HNO were
discovered at NRC (Herzberg and Ramsay 1955, Dalby 1958).
Subsequently microwave spectra were observed in the
laboratory by Saito (1972) and Saito and Takagi (1973) and
on that basis these two radicals were detected in the
interstellar medium by Snyder et al. (1976) and Ulich et al.
(1977). It is somewhat surprising to find HCO present in
spite of the fact that its dissociation energy is rather
small, of the order of 1 eV. It is of interest to mention
that HCO in the visible region (near 5400, 5900 and 6460 Å)
has diffuse absorption bands (in addition to sharp ones).

The R(0) lines of these bands should be observable in the interstellar medium if sufficient concentrations of HCO are present. However they do not agree with the known diffuse interstellar absorption features.

The spectrum of the NH_2 radical in the optical region was discovered in 1952 (Herzberg and Ramsay) and was analysed by Dressler and Ramsay in 1959. The laboratory work on this spectrum is continuing (Johns, Ramsay and Ross 1976). While some microwave lines have been found in the laboratory by microwave optical double resonance and laser magnetic resonance no features have yet been identified in the interstellar medium. The optical spectrum of NH_2 is, however, a prominent feature in the spectra of the heads of comets.

The CH_2 radical has been found by laboratory experiments to have a triplet ground state ($^3B_1$) and a number of low-lying singlet states (Herzberg 1961, Herzberg and Johns 1971). The optical absorption spectrum lies in the vacuum ultraviolet near 1415 Å. In CH_2, unlike CD_2, the individual rotational lines of the absorption band are fairly diffuse. The R(0) and P(1) lines lie at 1415.50 and 1416.05 Å. It is perhaps significant that Snow, York and Resnick (1977) report in a number of stars as the only diffuse feature in the vacuum ultraviolet one somewhat questionable line at 1416 Å.

The CH_3 radical (Herzberg 1961) has diffuse absorption bands with peaks at 2157.6, 2163.6, 1502.4 and 1496.89 Å. It seems likely that they will eventually be found in interstellar absorption.

Recently we have found (Herzberg 1979) in hollow cathode discharges in hydrogen a number of emission bands which have been definitely assigned as belonging to the Rydberg spectrum of triatomic hydrogen (H_3 and D_3). All rotational lines are diffuse, for H_3 much more so than for D_3. The observed intensity alternation confirms the D_{3h} symmetry of this molecule, i.e. that the nuclei form an equilateral triangle with a H-H distance of 0.87 Å as in H_3^+. The new spectra are emitted in the process of dissociative recombination of H_3^+ and D_3^+. This process has been assumed to be the cause of the eventual destruction of H_3^+ in the interstellar medium. It seems possible therefore that the H_3 spectrum will be observable in astronomical spectra.

MOLECULAR IONS

Recently Alberti and Douglas (1975) have made new
measurements of the NO^+ spectrum, first discovered by Baer
and Miescher (1953), with the principal aim of having
accurate predictions for possible identification of interstellar lines of this ion. Table I summarizes their

Table I. NO^+ lines of astronomical interest

Transition	R(0)
$^1\Pi(v=0)-^1\Sigma(v=0)$	1368.2404
$^1\Pi(v=1)-^1\Sigma(v=0)$	1339.6361
$^1\Pi(v=2)-^1\Sigma(v=0)$	1312.9400
$^1\Pi(v=3)-^1\Sigma(v=0)$	1288.0254
$^1\Pi(v=4)-^1\Sigma(v=0)$	1264.7900
$^1\Sigma(v=1)-^1\Sigma(v=0)$	2347.74 (cm^{-1})
$^1\Sigma(J=1-J=0)$	119187±30 MHz

results. Colbourn and Douglas (1977) made new measurements
of N_2^+ and O_2^+ with the same aim. It would seem worthwhile
to make an all-out attempt to observe the N_2^+ spectrum
since the N_2^+ lines have a high f value and are very
conveniently located in the spectrum (3914 Å).

Ms. Dabrowski and I (1977) have used the theoretical
potential function of HeH^+ derived by Wolniewicz (1965)
and Kolos and Peek (1976) in order to predict the infrared
spectrum of this ion. We have also calculated the quasibound levels of this ion which lie above the dissociation
limit at 1.845 eV. The inverse predissociation via these
quasibound levels may lead to a recombination spectrum that
has also been predicted. Attempts in this laboratory to
observe the infrared spectrum of HeH^+ have so far been
unsuccessful.

Discharges through He-Ne mixtures yield two features
near 4200 Å and similar spectra occur near 1400 Å in Ar-He
mixtures. Tanaka, Yoshino and Freeman (1975) first
correctly recognized the nature of the transitions
responsible for these spectra. They are charge transfer
spectra in which the molecule goes from a state that is
essentially He^+ + Ne to a lower state that is essentially
He + Ne^+ and similarly for $HeAr^+$. Figure 2 gives a
potential energy diagram for $HeNe^+$ based on a fairly
complete analysis of the spectrum near 4200 Å (Dabrowski
and Herzberg 1978). The ground state of $HeNe^+$ has a fairly
large dissociation energy (0.6 eV). On account of the

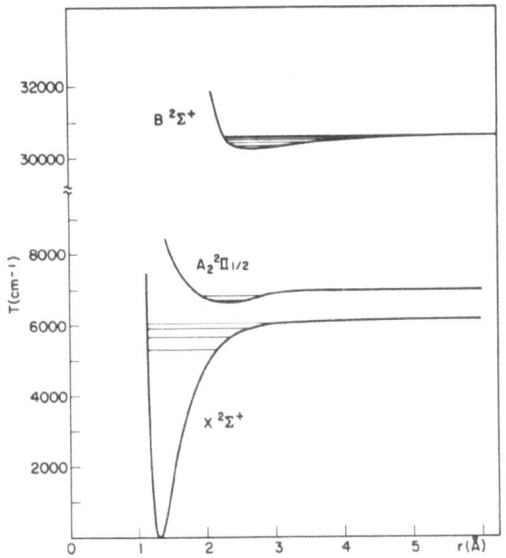

Fig.2. Potential
energy functions
of HeNe[+]

Franck-Condon principle we were unable to observe the
lowest vibrational levels of the ground state but were able
to establish the vibrational numbering on the basis of
isotope shifts in ^{3}HeNe[+]. In the interstellar medium it
seems likely that HeNe[+] if once formed will reach by infrared transition the lowest level of the ground state. A
microwave line at 173550 MHz and an infrared line at
1280 cm^{-1} are predicted.

As is well known, the ions HCO[+] and N$_2$H[+] were first
observed by radio-astronomical methods before they were
observed in the laboratory by Woods et al. (1975) and
Saykally et al. (1976) in absorption in electric discharges
through suitable gas mixtures. We have attempted, so far
unsuccessfully, to observe the ultraviolet spectra of these
ions. It seems possible that some of the unidentified
interstellar lines in the vacuum ultraviolet might be due
to the electronic transition of these molecular ions.

An electronic transition of H$_2$O[+], very similar to that
of NH$_2$, was first observed by H. Lew (1973, 1976) in
emission of hot cathode discharges. In the upper state the
ion is essentially linear while in the lower state it is
strongly bent with an angle of about 110°. The molecular
constants of H$_2$O[+] may be found in Dr. Lew's paper. In
Table II some of the predicted interstellar lines are
listed. None have yet been found.

Table II. Predicted interstellar lines of H_2O^+

a) in the microwave region			b) in the visible region	
Rotational transition	Spin component	Frequency (GHz)	Transition	λ_{lab}(Å)
$2_{20}-3_{13}$	F_1	144.44 ± 0.39	$(0,6,0)-(0,0,0)$	6973.720
	F_2	187.19 ± 0.42		6970.152
$4_{14}-3_{21}$	F_1	134.88 ± 0.40	$(0,8,0)-(0,0,0)$	6146.802
	F_2	103.87 ± 0.39		6147.375
$3_{30}-4_{23}$	F_1	218.61 ± 0.48	$(0,10,0)-(0,0,0)$	5478.548
	F_2	266.56 ± 0.50		5481.217
$4_{22}-3_{31}$	F_1	34.27 ± 0.54	$(0,12,0)-(0,0,0)$	4922.382
	F_2	-13.37 ± 0.54		4917.854
$6_{33}-5_{42}$	F_1	278.89 ± 0.72	$(0,14,0)-(0,0,0)$	4469.358
	F_2	234.41 ± 0.72		4468.823
$6_{52}-7_{43}$	F_1	64.07 ± 1.6		
	F_2	105.63 ± 1.6		

From the photoelectron spectrum of NH_3 it is readily seen that there is an excited state of NH_3^+ at 4.77 eV above the X $^2A_2''$ ground state (assuming D_{3h} symmetry). Therefore an emission spectrum near 2300Å may be expected. Since the excited state is in all probability a $^2E'$ state the transition between it and the ground state is electronically forbidden. However, vibronic interaction will make some of the vibrational transitions in the forbidden electronic transition allowed provided that either in the upper or lower state a degenerate vibration is excited. Using electron bombardment excitation of a NH_3 jet we have observed a number of weak but distinct features in the 2400-2200 Å region. One simple band near 2338 Å appears in which apparently alternate lines are missing. If this were true it would prove that the band is a sub-band of a $\perp$ band in a planar molecule with D_{3h} symmetry as expected for NH_3^+.

The ion-molecule reaction

$$H_2^+ + H_2 \rightarrow H_3^+ + H + 1.75 \text{ eV}$$

is one of the fastest known reactions. As a result one finds that in every discharge through H_2 the H_3^+ ion is the most abundant ion (except at very low pressure). It is believed that also in the interstellar medium H_3^+ is a fairly abundant ion. It seems to play an important role in the processes that lead to molecule formation (see Herbst and Klemperer 1973, Watson 1973, Suzuki 1979).

In order to detect H_3^+ spectroscopically one is dependent on its infrared bands since according to theoretical calculations all excited electronic states seem to be unstable. Up to now the infrared spectrum has not been observed in the laboratory. However, H. Lew, J. Sloan and I at Ottawa have made a number of attempts to observe this spectrum in emission from a hollow cathode discharge cooled to liquid nitrogen temperature. Using a Fourier transform spectrometer we have observed a number of infrared lines in the region between 5000 and 2000 cm^{-1}. Several of these lines are known to be due to H_2 (transitions between various excited electronic states) but a good number of the lines are unidentified. In order to ascertain whether some of these unidentified lines are due to H_3^+ one must either have reliable predictions or differentiate by discharge conditions or by electromagnetic deflection. We find that some of the unidentified lines are much weaker in the anode glow than in the cathode glow. Three years ago Carney and Porter (1976) made some predictions about the theoretically expected infrared spectrum. We find that the strongest predicted lines agree with those observed infrared lines that show considerable difference between the cathode and anode glow. It would be hasty to consider this as proof of the observation of the spectrum of H_3^+ since there are quite a number of H_2 lines also. We either have to observe complete bands allowing the use of the method of combination differences for identification or we have to look at an electronically deflected beam of H_3^+ ions. Such experiments are now in preparation.

References

Alberti, F. and Douglas, A.E.: 1975, Can. J. Phys. 53, 1179.
Baer, P. and Miescher, E.: 1953, Helv. Phys. Acta 26, 91.
Baird, K.M. and Bredohl, H.: 1971, Astrophys. J. 169, L83.
Ballik, E.A. and Ramsay, D.A.: 1963, Astrophys. J. 137, 61.
Carney, G.D. and Porter, R.N.: 1976, J. Chem. Phys. 65, 3547.
Chaffee, F.H. Jr. and Lutz, B.L.: 1978, Astrophys. J. 221, L91.
Colbourn, E.A. and Douglas, A.E.: 1977, J. Mol. Spectrosc. 65, 332.
Crutcher, R.M. and Watson, W.D.: 1976, Astrophys. J. 203, L123.
Dabrowski, I. and Herzberg, G.: 1976, Can. J. Phys. 54, 525.
Dabrowski, I. and Herzberg, G.: 1977, Trans. N.Y. Acad. Sci. 38, II, 14.
Dabrowski, I. and Herzberg. G.: 1978, J. Mol. Spectrosc. 73, 183.
Dalby, F.W.: 1958, Can. J. Phys. 36, 1336.

Douglas, A.E.: 1974, Can. J. Phys. 52, 318.
Douglas, A.E. and Elliott, G.A.: 1965, Can. J. Phys. 43, 496.
Dressler, K. and Ramsay, D.A.: 1959, Phil. Trans. 251A, 553.
Goss, W.M.: 1966, Astrophys. J. 145, 707.
Herbst, E. and Klemperer, W.: 1973, Astrophys. J. 185, 505.
Herzberg, G.: 1961, Proc. Roy. Soc. 262A, 291.
Herzberg, G.: 1979, J. Chem. Phys. 70, 4806.
Herzberg, G. and Johns, J.W.C.: 1971, J. Chem. Phys. 54, 2276.
Herzberg, G. and Ramsay, D.A.: 1952, J. Chem. Phys. 20, 347.
Herzberg, G. and Ramsay, D.A.: 1955, Proc. Roy. Soc. 233A, 34.
Johns, J.W.C., Ramsay, D.A. and Ross, S.C.: 1976, Can. J.
    Phys. 54, 1804.
Kolos, W. and Peek, J.M.: 1976, Chem. Phys. 12, 381.
Lew, H.: 1976, Can. J. Phys. 54, 2028.
Lew, H. and Heiber, I.: 1973, J. Chem. Phys. 58, 1246.
Rydbeck, O.E.H., Elldér, J., Irvine, W.M., Sume, A. and
    Hjalmarson, Å.: 1974, Astron. Astrophys. 33, 315, 34, 479.
Saito, S.: 1972, Astrophys. J. 178, L95.
Saito, S. and Takagi, K.: 1973, J. Mol. Spectrosc. 47, 99.
Saykally, R.J., Dixon, T.A., Anderson, T.G., Szanto, P.G.
    and Woods, R.C.: 1976, Astrophys. J. 205, L101.
Snow, T.P. Jr.: 1976, Astrophys. J. 204, L127.
Snow, T.P. Jr.: 1978, Astrophys. J. 220, L93.
Snow, T.P. Jr., York, D.G. and Resnick, M.: 1977, Publ.
    Astron. Soc. Pacific 89, 758.
Snyder, L.E., Hollis, J.M. Lovas, F.J. and Ulich, B.L.:
    1976, Astrophys. J. 209, 67.
Suzuki, H.: 1979, Prog. Theor. Phys. 62, October.
Tanaka, Y., Yoshino, K. and Freeman, D.E.: 1975, J. Chem.
    Phys. 62, 4484.
Ulich, B.L., Hollis, J.M., Snyder, L.E.: 1977, Astrophys. J.
    217, L105.
Watson, W.D.: 1973, Astrophys. J. 183, L17, 188, 35.
Wilkinson, P.G.: 1968, Can. J. Phys. 46, 1225.
Wolniewicz, L.: 1965, J. Chem. Phys. 43, 1087.
Woods, R.C., Dixon, T.A., Saykally, R.J. and Szanto, P.G.:
    1975, Phys. Rev. Letters 35, 1269.

DISCUSSION FOLLOWING HERZBERG

Hollenbach: Although grain composition may vary from place to place, I would think the correlation of diffuse bands with A_v more likely if the bands were caused by grains; molecular abundances generally are very sensitive to parameters such as gas density and radiation fields along the line of sight.

Herzberg: The references I have quoted, and several others, suggest considerable difficulty for that explanation.

Geballe: What type of H_2 lines might be contaminating the infrared spectrum of H_3^+ ?

Herzberg: The H_2 lines in question represent transitions between several excited states of H_2, for example E-B.

FAR INFRARED LASER MAGNETIC RESONANCE SPECTROSCOPY*

K. M. Evenson and R. J. Saykally
National Bureau of Standards
Time and Frequency Division
Boulder, Colorado 80303 USA

I. INTRODUCTION

Within the last few years, laser magnetic resonance (LMR) has emerged
as a powerful technique for the study of pure rotational spectra of
transient species in the gas phase. The sensitivity of this method is
considerably higher than that of competing techniques, such as conventional optical and microwave spectroscopy and gas-phase EPR, while the
resolution attainable is comparable with that of the latter two methods.
Its domain of applicability presently includes atoms, ground states of
molecules with up to 5 atoms, and most recently, metastable molecular
electronic states in the millisecond lifetime range, and molecular ions.
The only rigorous constraint on this applicability is that the species
of interest must be paramagnetic.

While LMR had its inception and early development in the far-infrared
(FIR) region of the spectrum (50-1000 μm), it has recently been extended to include vibrational transitions in the mid-IR (9-10 μm),
where it similarly has exhibited substantial capabilities. Furthermore, the laser Stark analogue of this experiment has been developed
quite effectively in the mid- and near- IR regions, although not yet
in the far-infrared. In this paper, we focus on the magnetic resonance experiment in the far-IR.

II. DESCRIPTION OF THE METHOD

The LMR experiment itself is intimately related to the technique of
gas-phase electron paramagnetic resonance, (EPR)[1], so successfully
exploited by Radford and by Carrington and his collaborators a decade
ago. In EPR, appropriate paramagnetic energy levels of an absorbing
sample are tuned by a DC magnetic field until their difference frequency equals that of a fixed-frequency source in the microwave region.
The principal difference between these two experiments is that in EPR
the relevant transitions are between different magnetic sublevels (M_J)
of the same angular momentum state (J), typically occuring in the
microwave region for normally accessible laboratory magnetic fields

B. H. Andrew (ed.), Interstellar Molecules, 239–245.

(2 tesla), while the transitions in LMR are between rotational states (in molecules) or fine-structure levels (in atoms), and occur in the far-infrared. In EPR, the transition can in principle be tuned to coincidence with any frequency lower than its maximum tunability (10 GHz, on the average), but in LMR one must rely on a coincidence between the laser frequency and the transition frequency to within about 1%, given the same tunability. In both EPR and LMR the sample is contained inside a resonant cavity. The increased sensitivity of LMR is mainly derived from operating at frequencies roughly 100 times higher than those of the microwaves normally used in EPR, since absorption coefficients normally depend on either the square or the cube of frequency for $h\nu \ll kT$. Also, by placing the absorbing sample inside the laser cavity, additional sensitivity (up to 3 orders-of-magnitude) can result from its interaction with the gain medium of the laser. With a one second time constant, the detection limit for OH radicals by LMR is presently about 1×10^6 cm^{-3}, whereas for EPR it is about 2×10^{12} cm^{-3}, for optical absorption it is about 10^{11} cm^{-3}, and for uv resonance fluorescence with a water vapor discharge source it is 3×10^9 cm^{-3}. High sensitivity for spectroscopic detection of OH radicals is obtained by laser induced fluorescence, 3×10^6 cm^{-3} in ambient air, but probably less than 10^6 cm^{-3} with optimum conditions.

III. FREE RADICALS

Twenty-five free radicals have been detected with LMR techniques in the three laboratories using FIR LMR Spectrometers: atoms: $O(^3P)$[2,3], $C(^3P)$[4]; diatomics: $O_2(^3\Sigma^-)$[5,6,7], OH $(^2\Pi)$[8], $NO(^2\Pi)$[9], $CH(^2\Pi$, v=0, 1, $\cdots$)[10,11], $PH(^3\Sigma^-)$[12], $NH(^3\Sigma^-)$[13,14], $ClO(^2\Pi)$[15], $CF(^2\Pi)$[16]; triatomics: $NO_2(^2A_1)$[17], $HO_2(^2A'')$[18,19,20], $HCO(^2A')$[21], $PH_2(^2B_1)$[22], $NH_2(^2B_1)$[23,24] $CH_2(^3B_1)$[25], $CCH(^2\Sigma^+)$[26]; polyatomics CH_3O[27], CH_2F[28]; metastables: $O_2(^1\Delta_u)$[29], $PH(^1\Delta)$[12], $HO_2(^2A_1)$[18], $CO(A^3\Pi)$[30], and ions: $HBr^+(^2\Pi)$[31], and $DBr^+(^2\Pi)$[31].

We will discuss four of these: two have been discovered in the interstellar medium (CH and CCH); and two more (C and CH_2) will probably be discovered via the frequencies provided by LMR.

Perhaps the single most important development of the LMR method occurred in 1971 with the detection of the J = 5/2 → 7/2, N = 2 → 3, pure rotational transition of the extremely elusive CH radical in a low-pressure oxyacetylene flame using the 118.6 μm H_2 laser. Although EPR had yielded spectra of many similar transient radicals with $^2\Pi$ ground states, all attempts to observe such spectra from CH had failed. Similarly, astronomical searches for the 10 cm lambda-doubling transition had resulted only in frustration. The successful detection of CH clearly demonstrated the high sensitivity of the LMR technique;

CH in the ground vibrational and rotational state yielded signals
260 times noise for an absorption path of 5 cm - roughly 30 times
the S/N reported for an optical absorption experiment! This experiment
used the water vapor laser powered by an electrical discharge. The
course of LMR was dramatically altered by the invention of the optically-[32]
pumped far-infrared laser by Chang and Bridges in 1970. The sensitivity of LMR was quite evident from the experiments performed with the
dozen or so H_2O and HCN laser lines, but because the method relies on
a close coincidence between a cw laser line and a relevant molecular
transition, the general application of LMR as a spectroscopic technique
seemed quite limited. Furthermore, these lines were all shorter than
337 μm (891 GHz), ostensibly limiting applications mainly to paramagnetic
hydrides. With the advent and rapid development of CO_2-pumped FIR lasers,
the number of useful cw laser lines has grown rapidly to the present
total of nearly 1000 - many of these in the wavelength region from 500
to 1000 μm.

In 1977 an optically-pumped LMR spectrometer was constructed using a[33]
transversely-pumped gain cell. This system has the advantage of
accommodating high CO_2 laser powers, and consequently operates on a
large number of FIR lines. With this new system CH spectra were
observed[11] with 8 optically-pumped laser lines with an improved
signal-to-noise ratio, resulting from a more intense source of CH
from a F atom/CH_4 flame. A rather complete analysis of all nine LMR
spectra in this work was obtained deducing values for the 4 hyperfine
constants a, b, c, and d, as well as improved values for lambda doubling intervals in J = 3/2, 5/2, and 7/2 states. A fit of all the data
now available should provide a complete set of improved molecular constants of this important species in the near future.

The important methylene radical (CH_2) was first observed in the LMR
system by Mucha et al.,[25] who reported the detection of several hyperfine triplets. The identity of the carrier of these triplets was
established as the $^3B_1$ ground state of CH_2 from a series of isotopic
substitution experiments. Several other sets of triplets, have
subsequently been found which have been shown to be from CH_2.[34] A tentative assignment of the spectrum observed at 85.3 μm ($^{13}CH_3OH$) has
yielded a structure for the radical, which, however, must be viewed as
extremely tentative because of the obvious perils of using only one
observed transition as the basis for such a determination. Another
interesting result obtained in this study was the observation of a
series of triplets at 171.8 μm ($^{13}CH_3OH$) occurring in stimulated
emission.

The detection of the ethynyl radical (CCH) by LMR was reported in[26]
1978. The N = 6 → 7 pure rotational transition in the $X^2\Sigma^+$ ground
state was found using the 490 μm laser line of CD_3I. The identity of
the carrier was again established through isotopic substitution experiments. This work constitutes the first spectroscopic detection of

gaseous CCH in the laboratory, although the observation of its microwave emission spectrum by radio astronomy has established it as a
ubiquitous constituent of interstellar clouds.[35]

The $2^3P_0 - 2^3P_1$ and $2^3P_1 - 2^3P_2$ fine-structure transitions within the
carbon atom ground state were detected in the same F atom/CH_4 flame
that produced CCH, CH_2, CH, CH_2F and CF spectra. The J=0 → 1 transition was measured with six different laser lines near 610 μm, while
the J=1 → 2 was measured with 4 lines near 370 μm. Mass shifts and
hyperfine splittings were observed for the ^{13}C isotope. Analysis of
these measurements has yielded precise frequencies for the zero-field
fine-structure transitions in both isotopic forms.

In 1978 a new technique for studying transient species in a laser magnetic resonance system was developed.[3] The positive column of a DC
glow discharge was sustained in the sample region of the laser cavity,
and the transverse magnetic field used in the previously described
LMR experiments was replaced by a longitudinal field, provided by a
7.6 cm diameter and 33 cm long liquid nitrogen cooled solenoid magnet.
This magnet was capable of producing a 5 kG field with a 0.1% homo-
geneity over a 15 cm length. The optically-pumped gain cell was
essentially the same. The longitudinal magnetic field configuration
readily accommodates the live intracavity discharge, although producing
some visible plasma constriction, and also provides an increased detection length (~ 15 cm instead of 1.5 cm). This apparatus permitted
the observation of the first laser magnetic resonance spectra of a
molecular ion.[31] Four isotopic forms of HBr^+ ($H^{79}Br^+$, $D^{79}Br^+$, $H^{81}Br^+$,
$D^{81}Br^+$) in its $^2\Pi_{3/2}$ ground state were detected in glow discharges
through a dilute (1%) mixture of HBr (DBr) in helium. For the hydrogen
isotopes, the J=3/2 → 5/2 transition was observed with laser lines at
251.1 μm (CH_3OH) and 253.7 μm (CD_3OH), and the J=5/2 → 7/2 transition in
v=1 was found with the 186.2 μm line of CH_3OH. The J=3/2 → 5/2 transition of DBr^+ was detected with the 496.1 μm line of CH_3F. All of the
spectra showed hyperfine lines from both bromine isotopes and exhibited
small lambda doublings. The proton hyperfine structure was not resolved.
On a single Zeeman component of this transition, a signal-to-noise ratio
of ~ 100 was achieved with a 1 sec time constant on a single scan. While
this was quite a difficult experiment, because of the extremely specific
condition required to produce the spectra, with a potential signal of
this magnitude, other molecular ions should be detectable by the same
method.

Since laser magnetic resonance spectroscopy exhibits a very high sensitivit
with nearly microwave resolution, it has produced detailed spectroscopic
information on a number of interesting transient species which have
eluded detection by other high resolution techniques. In the future,
these same qualities of LMR augmented by the fact that it is much easier
to search for unknown spectra using swept-field rather than swept-
frequency techniques, should make it possible to detect certain astrophysically important radicals for which only theoretical (ab initio)

estimates of parameters are available. Analysis of these spectra will produce precise rest frequencies which can then serve as a basis for astronomical searches for these species in the interstellar medium.

REFERENCES

1. Carrington, Alan (1974), "Microwave Spectroscopy of Free Radicals", Academic Press, New York.
2. P. B. Davies, B. J. Handy, E. K. Murray-Lloyd, and D. K. Russell, J. Chem. Phys. 68, pp. 3377 (1978).
3. R. J. Saykally and K. M. Evenson, "Laser Magnetic Resonance Measurement of the $2^3P - 2^3P_1$ Splitting in Atomic Oxygen", J. Chem. Phys. 71, (1979).
4. R. J. Saykally and K. M. Evenson, "Direct Measurement of Fine-Structure Intervals in the 2^3P Ground State of Carbon Atom by Laser Magnetic Resonance", in preparation.
5. K. M. Evenson, H. P. Broida, J. S. Wells, R. J. Mahler, and M. Mizushima, Phys. Rev. Lett. 21, 1038 (1968).
6. K. M. Evenson and M. Mizushima, Phys. Rev. A 6, pp. 2197 (1972).
7. L. Tomuta, M. Mizushima, C. J. Howard, and K. M. Evenson, Phys. Rev. A 12, pp. 974 (1975).
8. K. M. Evenson, J. S. Wells, and H. E. Radford, Phys. Rev. Lett. 25 pp. 199 (1970).
9. M. Mizushima, K. M. Evenson, and J. S. Wells, Phys. Rev. A 5, pp. 2276 (1972).
10. K. M. Evenson, H. E. Radford, and M. M. Moran, Jr., Ap. Phys. Lett. 18, pp. 426 (1971).
11. J. T. Hougen, J. A. Mucha, D. A. Jennings, and K. M. Evenson, J. Mol. Spectry. 72 pp. 463 (1978).
12. P. B. Davies, D. K. Russell, and B. A. Thrush, Chem. Phys. Lett. 36 pp. 280 (1975).
13. H. E. Radford and M. M. Litvak, Chem. Phys. Lett. 34, pp. 561 (1975).
14. F. D. Wayne and H. E. Radford, Mol. Phys. 32, pp. 1407 (1976).
15. R. M. Stimpfle, R. A. Perry, and C. J. Howard, J. Chem. Phys. in press (1979).
16. R. J. Saykally and K. M. Evenson, "The Far-Infrared Laser Magnetic Resonance Spectrum of CF", 34th Symposium on Molecular Spectroscopy, Columbus, Ohio; June, 1979; Paper TF-4.
17. R. F. Curl, Jr., K. M. Evenson, and J. S. Wells, J. Chem. Phys. 56, pp. 5144 (1972).
18. H. E. Radford, K. M. Evenson, and C. J. Howard, J. Chem. Phys. 60, pp. 3178 (1974).
19. Jon T. Hougen, J. Mol. Spectry. 54, pp. 447 (1975).
20. J. T. Hougen, H. E. Radford, K. M. Evenson, and C. J. Howard, J. Mol. Spectry. 56, pp. 210 (1975).
21. J. M. Cook, K. M. Evenson, C. J. Howard, and R. F. Curl, Jr., J. Chem. Phys. 64, pp. 1381 (1976).
22. P. B. Davies, D. K. Russell, and B. A. Thrush, Chem. Phys. Lett. 37, pp. 43 (1976).

23. P. B. Davies, D. K. Russell, B. A. Thrush, and F. D. Wayne, J. Chem. Phys. 62, pp. 3739 (1975).

24. P. B. Davies, D. K. Russell, B. A. Thrush, and H. E. Radford, Proc. R. Soc. Lond. A 353, pp. 299 (1978).

25. J. A. Mucha, K. M. Evenson, D. A. Jennings, G. B. Ellison, and C. J. Howard, "Laser Magnetic Resonance Detection of Rotational Transitions in CH_2", accepted by Chem. Phys. Lett.

26. R. J. Saykally and K. M. Evenson, "The Ethynyl Radical (CCH) Detected by Laser Magnetic Resonance", 33rd Symposium on Molecular Spectroscopy, Columbus, Ohio; June, 1978; Paper FB-8.

27. H. E. Radford and D, K. Russell, J. Chem. Phys. 66, pp. 2223 (1977).

28. J. A. Mucha, D. A. Jennings, K. M. Evenson, and J. T. Hougen, J. Mol. Spectry. 68, pp. 122 (1977).

29. A. Scalabrin, M. Mizushima, R. J. Saykally, H. E. Radford and K. M. Evenson, "Laser Magnetic Resonance Spectrum of the $a^1\Delta_g$ state of O_2", in preparation.

30. R. J. Saykally and K. M. Evenson, "Far-Infrared Laser Magnetic Resonance Spectra of $a^3\Pi$ CO", 34th Symposium on Molecular Spectroscopy, Columbus, Ohio; June, 1979; Paper TF-6.

31. R. J. Saykally and K. M. Evenson, "Observation of Pure Rotational Transitions in the HBr^+ Molecular Ion by Laser Magnetic Resonance", 1979, Phys. Rev. Lett. 43, pp. 515.

32. Y. Y. Chang and T. J. Bridges, Opt. Commun. 1, pp. 423 (1970).

33. K. M. Evenson, D. A. Jennings, F. R. Petersen, J. A. Mucha, J. J. Jimenez, R. M. Charlton, and C. J. Howard, IEEE J. Quantum Electron. QE-13 pp. 442 (1977).

34. K. M. Evenson, R. J. Saykally, and J. T. Hougen, "Far-Infrared Laser Spectra of Methylene", 34th Symposium on Molecular Spectroscopy, Columbus, Ohio; June, 1979; Paper TF-5.

35. K. D. Tucker, M. L. Kutner, and P. Thaddeus, Ap. J. 193 pp. L115 (1974).

DISCUSSION FOLLOWING EVENSON

Kuiper: In June 1979 de Graauw and Lidholm (ESTEC), van Vliet, Nieuwenhuyzen, and van der Stadt (U. of Utrecht), and myself attempted to detect the 492 GHz CI line with an InSb hot electron bolometer on the Kuiper Airborne Observatory. Instrumental difficulties precluded obtaining any useful data.

Hollenbach: Can you measure the wavelength of the C^+ fine structure line at $\sim$156μ?

Evenson: We are very anxious to try C^+, and we will soon.

Phillips: Would it be easier to examine CHD, which has both a and b transitions, to help in the understanding of the CH_2 experiment?

Also, can you look for HCl^+ as well as HBr^+?

Evenson: That is a very good suggestion. We must eventually look for CHD and CD_2 to arrive at the structure of the molecule. HCl^+ is the next ion on our list of molecular ions to investigate.

Huntress: Were any of the searches for interstellar HO_2 or CH_3O successful?

Evenson: No. However I believe that searches are continuing.

Wootten: Have you been able to measure a frequency for CCD?

Evenson: We have not yet looked at CCD.

Wootten: Can you measure the frequency of the $^3D_{5/2} - {}^3D_{3/2}$ transition of NI at $\sim$250 GHz?

Evenson: No. The lowest frequency our laser will operate at is 340 GHz.

Shivanandan: Has the C^+ line been accurately identified in the laboratory?

Evenson: No, but we shall make a search for it. It will not be easy since we will be looking for magnetic dipole transition of an ion, and obtaining a sufficient number density will be difficult.

OPTICAL OBSERVATIONS OF INTERSTELLAR MOLECULES

Theodore P. Snow, Jr.
Laboratory for Atmospheric and Space Physics and
Department of Astro-Geophysics
University of Colorado
Boulder, Colorado 80309
U.S.A.

ABSTRACT

Recent observational data are summarized on molecular species in diffuse interstellar clouds, and a comprehensive list of all species detected or sought is included. The discussion is focussed on the use of molecular observations to determine physical conditions in clouds, and on important species which provide constraints on cloud chemistry models. Directions for new research are suggested, including intensive work on the unidentified diffuse bands, which may have a molecular origin.

I. INTRODUCTION

This reviewer was asked to summarize current knowledge of molecular species and characteristics in hot diffuse clouds. The title which has been adopted reflects the fact that, to date, observations of molecules in diffuse clouds have been carried out almost exclusively by optical measurements of absorption lines in the spectra of background stars. There have been cases of radio detections of molecular emission lines arising in diffuse clouds, but these detections generally require very long integration times which are only infrequently devoted to this purpose. Hence the present review will be concentrated upon the optical absorption-line data.

The first detections and identifications of molecular species in diffuse clouds (or in any clouds, for that matter), were achieved before 1940, when lines of CH, CH^+, and CN were found in the spectra of reddened stars (Adams 1941; Adams and Dunham 1937). With the possible exceptions of the unidentified carriers of the diffuse interstellar bands (about which more later), these three diatomics remained the only molecules detected through traditional ground-based spectroscopic techniques until well into the 1970's. By the 1960's, ultraviolet spectroscopy of astronomical objects became feasible, at first with sounding rocket instruments, and nearly ten years ago H_2 was detected by Carruthers (1970). Later rocket experiments succeeded

B. H. Andrew (ed.), Interstellar Molecules, 247–256.
Copyright © 1980 by the IAU.

in measuring absorption lines of CO (Smith and Stecher, 1971). Since
the launch of Copernicus in August 1972, a considerable renewal of
activity has occurred, both in ultraviolet and in visible-wavelength
observations of molecules. New species such as OH, C_2, and possibly
H_2O, have been detected and rigorous searches for others have been
carried out. Along with the observations, theoretical work on chemical
reactions in interstellar clouds has provided increasingly detailed
understandings of results already in, and has issued numerous challenges
for new observations to be attempted.

With this tradition of close interaction between observation and
theory in mind, the present review is designed to acquaint scientists
in related areas with the current state of knowledge of molecular
species in diffuse clouds (where the chemistry is far simpler than
that in the dark clouds studied by the radio observers) and to challenge
not only the theorist but also the laboratory spectroscopist, since
one of the major requirements for progress in this field is the
accumulation of basic data on the species sought.

II.   OBSERVATIONAL RESULTS

In this section are briefly described the results of optical
spectroscopic observations of molecular species. Table 1 lists all
those known to the author to have been detected or sought. The
discussion following will be concentrated on a few of these, singled
out for their significance as probes of physical conditions or tests
of chemical models. The table gives references to recent observations
of ·all the listed species. No detailed description of observational
instruments or techniques will be included here; this information may
be obtained from the original references.

a.   Indicators of Physical Conditions

Molecular observations can be useful to the student of the
nature of interstellar clouds as well as to the cloud chemist. Such
parameters as kinetic temperature, density, cosmic ray ionization
rate, and radiation field intensity all can be derived from observations
of molecular species.

Diffuse cloud kinetic temperatures are presumed equal to the
rotational excitation temperature indicated by the ratio of hydrogen
molecules in the first rotational excited state and the ground state,
since H_2 is a homonuclear molecule with no allowed radiative dipole
transitions between adjacent rotational states (Dalgarno, Black, and
Weisheit, 1973). Measurements of this ratio in diffuse clouds have
revealed values of T_{kin} ranging from about 40K to 100K or more (e.g.,
Spitzer and Cochran 1973; Savage et. al. 1977). Another homonuclear
molecule with potential to be an indicator of cloud temperatures is
C_2, recently detected by several observers (Lutz and Souza 1977; Snow
1978; Hobbs 1979; Chaffee et. al. 1979). For the ζ Ophiuchi cloud,
Snow (1978) found $T_{kin} \lesssim 16K$ from ultraviolet observations with Coper-

Table 1. Interstellar Molecules Observed in Optical Absorption

Species	Wavelength	Star	$N(cm^{-2})$	N/N_H	Reference[+]
H_2	912–1108	ζ Oph	4.47×10^{20}	0.32	13,20
HD	912–1108	ζ Oph	1.58×10^{14}	1.12×10^{-7}	20
C_2	8758	ζ Per	1.2×10^{13}	8.51×10^{-9}	4,8
C_3	4050	ζ Oph	$<2 \times 10^{11}f^{-1}$	–	5
CH	4300	ζ Oph	3.39×10^{13}	2.40×10^{-8}	1,7
CH^+	4232	ζ Oph	1.2×10^{13}	8.51×10^{-9}	22
$^{13}CH^+$	4232	ζ Oph	1.32×10^{11}	9.36×10^{-11}	21,22
CN	3874	ζ Oph	4.79×10^{12}	3.40×10^{-9}	7
CN^+	2181	ξ Per	$<1 \times 10^{12}f^{-1}$	–	9
CO	1088	ζ Oph	1.2×10^{15}		14
^{13}CO	1395	ζ Oph	$<5.62 \times 10^{13}$	$<3.99 \times 10^{-9}$	12
CO^+	4251	ζ Oph	$<5.75 \times 10^{12}$	$<4.08 \times 10^{-9}$	2,7
CS	2577	ζ Oph	$<2.55 \times 10^{13}$	$<1.81 \times 10^{-8}$	17
CH_2	1397	ζ Oph	$<1.56 \times 10^{12}f^{-1}$	–	17
CO_2	1089	ζ Oph	$<2.95 \times 10^{13}f^{-1}$	–	17
NH	3358	o Per	$<7 \times 10^{11}$	$<4.49 \times 10^{-10}$	6
NH^+	2890	ζ Oph	$<1.65 \times 10^{11}f^{-1}$	–	17
N_2	958	δ Sco	$<3.8 \times 10^{12}$	$<2.62 \times 10^{-9}$	11
NO	2262	ξ Per	$<1.70 \times 10^{15}$	$<1.21 \times 10^{-6}$	9
NO^+	1313	ζ Oph	$<1.07 \times 10^{14}$	$<7.59 \times 10^{-8}$	12
OH	1222	ζ Oph	5.24×10^{13}	3.72×10^{-8}	17
O_2	1144	o Per	$<3.47 \times 10^{11}f^{-1}$	–	16
H_2O	1114	ζ Oph	$<2.65 \times 10^{13}$	$<1.88 \times 10^{-8}$	15,19
NaH	3991	ζ Oph	$<1.7 \times 10^{11}$	$<1.21 \times 10^{-9}$	18
MgH	5187	ζ Oph	$<2.2 \times 10^{11}$	$<1.56 \times 10^{-10}$	7,10
MgH^+	2806	ξ Per	$<2 \times 10^{11}f^{-1}$	–	9
AlH	2242	ζ Oph	$<2.56 \times 10^{11}f^{-1}$	–	17
SiH	4119	ζ Oph	$<1.41 \times 10^{12}$	$<1.00 \times 10^{-9}$	7
SiO	1310	ζ Oph	$<3.24 \times 10^{11}f^{-1}$	–	12
SH	1257	o Per	$<1.45 \times 10^{11}f^{-1}$	–	16
HCl	1290	ζ Oph	$<1.29 \times 10^{12}$	$<9.15 \times 10^{-10}$	15,23
CaH	2717	ζ Oph	$<9.34 \times 10^{10}f^{-1}$	–	17

[+]References to Table 1.

1. Black and Dalgarno 1973b
2. Bortolot and Thaddeus 1969
3. Chaffee and Lutz 1977
4. Chaffee et. al. 1979
5. Clegg, Lambert, & Snell 1979
6. Crutcher and Watson 1976b
7. Herbig 1968
8. Hobbs 1979
9. Jenkins et. al. 1973
10. Kirby, Saxon, & Liu 1979
11. Lutz, Owen, & Snow 1979
12. Morton 1975
13. Savage et. al. 1977
14. Smith, Krishna Swamy, & Stecher 1978
15. Smith, Yoshino and Parkinson 1979
16. Snow 1975
17. Snow 1976a
18. Snow and Smith 1977
19. Snow and Smith 1979
20. Spitzer, Drake et. al.
21. Vanden Bout 1972
22. Vanden Bout 1979
23. Wright and Morton 1979

nicus. Hobbs (1979), utilizing a near-infrared transition, found
T_{kin} =78 ± 25K for the cloud towards ζ Persei while Chaffee et al.
found T_{kin} = 92 ± 10K for the same cloud. The C_2 result for ζ Oph
showed a much lower T_{kin} than that found for the same cloud from H_2
measurements, whereas good agreement between the H_2 and C_2 results
was found for the ζ Per cloud. It is noteworthy that the detailed
models of both clouds by Black and Dalgarno (1977) and by Black,
Hartquist, and Dalgarno, (1979) invoke a bi-modal structure, with a
cold core and a warmer outer region. Chaffee et. al. (1979) show
that radiative pumping can affect the rotational distribution in C_2
so that the C_2 data may be in harmony with the values of T_{kin} derived
from the H_2 observations.

Cloud densities can be derived from observations of rotational
excitation in CO, which is collisionally excited primarily by H_2
molecules in diffuse clouds. Ultraviolet observations of CO molecules
with Copernicus have resulted in values of $n(H_2)$ ranging from less than
100 cm^{-3} to over 1000 cm^{-3} (Snow 1976b, 1977; Smith, Krishna Swamy, and
Stecher 1978; Snow and Jenkins 1979).

The cosmic ray ionization rate can be derived from observations
of HD or OH abundances, since the reaction sequences leading to these
molecules are initiated by H^+ ions. By observing the abundance of HD
or OH, one can infer the equilibrium abundance of H^+, which in turn
indicates the cosmic ray ionization rate.

It was noted early in the lifetime of the Copernicus satellite
that H_2 often shows excess populations of the high rotational levels
(i.e., J=4-6) as though a second, higher temperature governed their
populations rather than that which controls the low levels. This
excess excitation was first reported by Spitzer and Cochran (1973)
and Spitzer, Cochran, and Hirshfeld (1974), and subsequently attributed
to excess kinetic energy of formation of H_2 molecules as they are
liberated from grains (Spitzer and Zweibel 1974). At about the same
time, Black and Dalgarno (1973a) suggested that pumping via ultraviolet
resonance line absorption was responsible for the high J-level excitation, and later Jura (1975 a, b) was able to use this model to show
how radiation field intensities can be derived from observations of
the excitations of H_2. The basic idea is that electronic transitions
from the ground state via the Lyman and Werner bands leave the molecules
in excited vibrational and rotational states of the first excited
electronic level. As the molecules return radiatively to the ground
state, the equilibrium populations of the high J-levels are determined
by the branching coefficients which indicate the relative probabilities
of the various paths through the vibration-rotation cascade. By
analyzing the high J-level populations, Jura (1975 a, b) and Spitzer
and Morton (1976) found radiation field intensities ranging up to
10-15 times the average interstellar radiation field, revealing
clouds which are in close proximity to their background stars. The
analysis of Spitzer and Morton showed that in several lines of sight,
the highly-excited H_2 (inferred to be close to the star) inhabits a

cloud which has a negative velocity with respect to the star, suggestive of an expanding shell, perhaps produced by a stellar wind or an
old supernova. Very recently, Wright and Morton (1979) were able to
utilize the observed rotational excitation in HD to derive the radiation field intensity of the main cloud seen towards ζ Ophiuchi, thus
determining the cloud's distance from the star. Hence, molecular
observations, through their revelations concerning cloud physical
conditions, can also indirectly yield information on cloud geometries
and kinetics.

b.   Molecular Formation and Tests of Chemical Models

Every detection or new upper limit for a molecular species
provides useful data on the mechanisms which govern cloud chemistry.

For the diffuse clouds, it now appears likely that most species
(excepting H_2) are formed through gas-phase reaction sequences initiated
by cosmic ray or UV ionization, followed by charge-exchange and ion-
molecule reactions. The alternative process, formation of molecules
on grain surfaces, is likely the principal source of H_2 but not of
other species, although some workers (e.g., Allen and Robinson 1977
and Pickles and Williams 1977) contend that formation on grains can
occur rapidly enough to compete with the gas-phase processes. In
this observational review no attempt is made to settle the issue, but
results are described which bear on the question.

Evidence favoring the gas-phase formation hypothesis comes in
two forms: (1) recent detections, at about the predicted level, of
species expected to be abundant under this hypothesis; and (2) failure
to detect, despite intensive searches, species expected to be produced
efficiently by grain surface formations but not by gas-phase reactions.
In the former category fall OH (Crutcher and Watson 1976a; Chaffee
and Lutz 1977; Snow 1976a) and C_2 (Souza and Lutz 1977; Lutz and
Souza 1977; Snow 1978; Hobbs 1979; Chaffee et al. 1979); in the latter
category are NH (Crutcher and Watson 1976 b) and NaH (Snow and Smith 1977).

Important tests which have been attempted but are not yet successfully completed include searches for N_2, HCl, and H_2O, all three of
which are predicted to have substantial abundances in diffuse clouds,
if gas-phase reactions dominate (e.g., Black and Dalgarno 1977). The
search for N_2 (Lutz, Owen, and Snow 1979) is hampered by the fact
that the most favorable transitions are below 1000A, where it has
been feasible to observe only little-reddened stars. In the case of
HCl, predicted to be abundant by Jura (1974) and by Dalgarno et. al.
(1974), the f-value has just been determined (Smith, Yoshino, and
Parkinson, 1979), and the current upper limit (Wright and Morton
1979) is well below the predicted level. The discrepancy between the
observation and theory may be too large to be explained by uncertainties in the photoionization rate for ClI, as proposed by Jura and York
(1978). The chemistry of HCl is discussed further by Black and Smith
(1979).

The triatomic species H_2O, formed through the same reaction sequence as OH, has been marginally seen (Snow and Smith 1979), at about the 2σ level. In this case the line strength has also just been determined (Smith, Yoshino, and Parkinson 1979) yielding a probable column density for H_2O which is larger than that predicted by Black and Dalgarno (1977). A more solid determination will help establish the value of the branching ratio in the reaction

$$H_3O^+ + e \longrightarrow \begin{array}{l} OH + H_2 \\ H_2O + H, \end{array}$$

recently estimated (Herbst 1978) to favor OH by about a factor of 10, helping to make H_2O difficult to detect.

An outstanding problem for theories of molecule formation is posed by the large quantities of CH^+ that are observed. Since the reaction

$$C^+ + H_2 \rightarrow CH^+ + H$$

is endothermic by 0.4ev, and the destruction process is very rapid, it is difficult to understand CH^+ abundances in diffuse clouds. Recent suggestions have invoked CH^+ formation in either shock fronts (Elitzur and Watson 1978) or in the warm outer portions of clouds (de Jong 1979), where the necessary energy can be provided. Careful observations of CH^+ velocities in lines of sight known to contain shocked clouds should help test these suggestions.

IV. CHALLENGES FOR THE FUTURE

A number of important tasks remain. It will be useful to continue to search for new molecular species in diffuse clouds, in order to carry on the testing and refinement of the molecular formation schemes. For example, diatomic species containing silicon may be sufficiently abundant for detection (Turner and Dalgarno 1977), as may some larger molecules such as H_2O (already tentatively identified), C_3, NH_3, CH_2, and CH_3 (Black and Dalgarno 1977). In many of these cases, the laboratory data are incomplete, so that the wavelength at which to search, or the f - value, or both are not known. Further work on N_2 and HCl seems justified, since both are expected to appear at levels above or near the current upper limits. For these species, whose best transitions are in the ultraviolet, new instrumentation will be required, since present UV instruments are either too insensitive at the required wavelengths (as in the case of Copernicus), or have insufficient resolution and photometric accuracy (IUE).

A new class of molecules has been proposed to exist in diffuse clouds. These are carbon chains, which have been detected in the radio spectrum of dark clouds (Broten et al. 1978; Gardner, Whiteoak, and Winnewisser 1978), and which have been proposed as possible carriers of the diffuse interstellar bands (Douglas 1977; Thaddeus 1978). The optical spectra of such species must be determined before a sensible search can be performed, although at least one progenitor to such species, C_3, can be sought now. Very recent work by Clegg, Lambert and Snell (1979) has failed to detect C_3 absorption in the spectra of ζ Oph and the heavily-reddened supergiant HD143183, but the f-value of the 0-0 band at 4050Å is not known, so the significance of this result is unclear.

Regardless of the validity of the carbon-chain hypothesis, the diffuse bands pose an interesting challenge which may well fall into the province of the molecular spectroscopist. Smith, Snow, and York (1977) reviewed the properties of these bands, which appear in the spectra of all reddened stars, and suggest that a molecular origin is more satisfactory in many ways than the chief alternative, absorption resonances in solid grains. The challenge for the observer is to obtain high-resolution profiles of diffuse bands with the highest possible photometric accuracy to search for possible fine structure; the challenge for the spectroscopist is to seek species, preferably composed of cosmically abundant elements, which reproduce these profiles and wavelengths.

Optical observations of interstellar molecules can be a frustrating endeavor, because of the paucity of laboratory data and the difficulties inherent in detecting very weak absorption lines. On the other hand, it can be a very rewarding area in which to work, because of the close interaction between observation and theory, and because significant observations often are concise and readily-defined. It is exciting to contemplate the new developments which are sure to occur in years to come, particularly with the advent of increasingly sophisticated observational techniques.

The preparation of this review has been supported in part by NASA grant NS6-7477 with the University of Colorado.

REFERENCES

Adams, W.S.: 1941, Ap. J. 93, 11.
Allen, M. and Robinson, G.W.: 1977, Ap. J. 195, 81.
Black, J.H. and Dalgarno, A.: 1973a, Ap. J. (Letters) 184, L101.
Black, J.H. and Dalgarno, A.: 1973b, Astrophys. Lett. 15, 79.
Black, J.H. and Dalgarno, A.: 1977, Ap. J. Suppl. 34, 405.
Black, J.H., Hartquist, T.W. and Dalgarno, A.: 1979, Ap. J. 224, 448.
Black, J.H., Smith, P.L.: 1979, this volume.
Bortolot, V.J., and Thaddeus, P.: 1969, Ap. J. (Letters), 155, L17.
Broten, N.W., Oka, T., Avery, L.W., MacLeod, J.M., and Kroto, H.W.:
        1978, Ap. J. (Letters) 223, L105.
Carruthers, G.: 1970, Ap. J. (Letters), 161, L81.
Chaffee, F.H. and Lutz, B.L.: 1977, Ap. J. 213, 394.
Chaffee, F.H. and Lutz, B.L.: 1978, Ap. J. (Letters) 221, L91.

Chaffee, F.H., Lutz, B.L., Black, J.H., Vanden Bout, P. and Snell, R.:
    1979, Ap. J., submitted.
Clegg, R., Lambert, D.L., and Snell, R.: 1979, private communication.
Crutcher, R.M. and Watson, W.D.: 1976a, Ap. J. (Letters) 203, L123.
Crutcher, R.M. and Watson, W.D.: 1976b, Ap. J. 209, 778.
Dalgarno, A., Black, J.H., and Weisheit, J.C.: 1973, Astrophys. Lett.
    14, 77.
de Jong, T.: 1979, preprint.
Douglas, A.E.: 1977, Nature 269, 130.
Dunham, T. and Adams, W.S.: 1937, Pub. A.S.P. 49, 26.
Elitzur, M. and Watson, W.D.: 1978, Ap. J. (Letters) 222, L141.
Gardiner, F.F., Whiteoak, J.B., and Winnewisser, G.: 1978, Astr. Ap.
    67, L23.
Herbig, G.H.: 1968, Z. f. Ap. 68, 243.
Herbst, E.: 1978, Ap. J. 222, 508.
Hobbs, L.M.: 1979, preprint.
Jenkins, E.B., Drake, J.F., Morton, D.C., Rogerson, J.B., Spitzer, L.,
    and York, D.G.: 1973, Ap. J. (Letters) 181, L122.
Jura, M.: 1974, Ap. J. (Letters) 190, L33.
Jura, M.: 1975a, Ap. J. 197, 575.
Jura, M.: 1975b, Ap. J. 197, 581.
Jura, M., and York, D.G.: 1978, Ap. J. 219, 861.
Kirby, K., Saxon, R.P., and Liu, B.: 1979, Ap. J. 231, 637.
Lutz, B.L., Owen, T., and Snow, T.P.: 1979, Ap. J. 227, 159.
Lutz, B.L. and Souza, S.P.: 1977, Ap. J. (Letters) 213, L129.
Morton, D.C.: 1975, Ap. J. 197, 85.
Pickles, J.B. and Williams, D.A.: 1977, Ap. Space Sci. 52, 453.
Savage, B.D., Bohlin, R.C., Drake, J.F., and Budich, W.: 1977, Ap. J.
    216, 291.
Smith, A.M., Krishna Swamy, K.S., and Stecher, J.P.: 1978, Ap. J.
    220, 138.
Smith, A.M., and Stecher, J.P.: 1971, Ap. J. (Letters) 164, 143.
Smith, W.H., Snow, T.P., and York, D.C.: 1977, Ap. J. 218, 124.
Snow, T.P.: 1975, Ap. J. (Letters) 201, L21.
Snow, T.P.: 1976a, Ap. J. (Letters) 204, L127.
Snow, T.P.: 1976b, Ap. J. 204, 759.
Snow, T.P.: 1978, Ap. J. (Letters) 202, L93.
Snow, T.P. and Jenkins, E.B.: 1979, Ap. J., submitted.
Snow, T.P. and Smith, W.H.: 1977, Ap. J. 218, 124.
Snow, T.P. and Smith, W.H.: 1979, in preparation.
Spitzer, L. and Cochran, W.D.: 1973, Ap. J. (Letters) 186, L23.
Spitzer, L., Cochran, W.D. and Hirshfeld, A.: 1974, Ap. J. Suppl. 28, 373
Spitzer, L., Drake, J.F., Jenkins, E.B., Morton, D.C., Rogerson, J.B.
    and York, D.C.: 1973, Ap. J. (Letters) 176, L127.
Spitzer, L. and Zweibel, E.G.: 1974, Ap. J. (Letters) 191, L127.
Thaddeus, P.: 1978, private communication.
Turner, J.L. and Dalgarno, A.: 1977, Ap. J. 213, 386.
Vanden Bout, P.A.: 1972, Ap. J. (Letters) 176, L127.
Vanden Bout, P.A.: 1979, preprint.
Wright, E.L. and Morton, D.C.: 1979, Ap. J. 227, 483.

DISCUSSION FOLLOWING SNOW

Mouschovias: Do you have a favorite number for the cosmic-ray ionization rate and, if so, does this rate correlate with the gas density?

Snow: I do not. Furthermore, molecular data are not available for a sufficiently large number of diffuse clouds to estimate cosmic ray ionization rates or densities, so I don't think an adequate data base exists to permit such a correlation to be sought.

Bok: You have a range of kinetic temperatures for your coolest clouds of 50 K to 75 K. Have you any evidence for clouds with temperatures of 10 K to 20 K - as shown by radio CO observations?

Snow: We have one C_2 result (Snow 1978) that implies a value $T_{kin} \lesssim 16$ K for the ζ Ophiuchi cloud. There are uncertainties in this result, due to low data quality and to apparent inconsistencies in the oscillator strengths. For the same cloud, de Boer and Morton (referred to in Morton 1975) derived a value $T_{kin} = 19$ K for observations of fine-structure excitation in atomic carbon. It is possible that diffuse clouds contain very cold core regions, and that the values of T_{kin} derived from the H_2 results refer to a larger volume containing warmer gas.

Kirby: It's not surprising that NaH has not been observed, as calculations by Kirby and Dalgarno (1978) show that, if it were formed on grains, it would be photo-dissociated very rapidly in the interstellar radiation field. Considering the abundance of Mg relative to Na, it would seem that MgH might be a more likely candidate for a search.

Snow: The search for NaH was based on a hypothesis that it might be formed with great efficiency on grains, then ejected and quickly dissociated, accounting for the low depletion of sodium from the gas. Therefore, we thought it a reasonable candidate for a search because of its suggested high formation rate, despite knowing that it probably also has a high destruction rate. While it's true that Mg is more abundant than Na, it is also true that Mg is highly depleted from the gas and is largely bound up in solid grains. I know of no suggestions that MgH is formed rapidly.

Winnewisser: What do you think are the reasons for the low column densities of molecules such as NaH, MgH, etc. Are they frozen out during formation?

Snow: My guess is that these species are not readily formed in diffuse clouds, although we have hypothesized that NaH would be relatively easily formed on grains, *if* this mechanism is important at all. It may not be. In any event, a more likely destruction mechanism is photo-dissociation, since we are talking about diffuse clouds rather than dark clouds.

Thaddeus: There is no example to my knowledge of optical absorption spectroscopy through a bona fide molecular cloud. What in your opinion is the prospect with new photo-detectors of finding a nova or other bright distant object, with more than, say, 10 magnitudes of extinction?

Snow: Certainly modern detection techniques will enhance our ability to take advantage of such an opportunity, should one arise. I don't know of anyone who has made specific plans to do this, but I think a number of observers would react quickly. I certainly agree that optical measurements of molecular species should have high priority whenever we get a chance to make absorption-line observations of a dense molecular cloud.

Radhakrishnan: Is there any evidence from UV observations of a correlation between temperature and the peculiar velocity of clouds?

Snow: I know of none. It would be very difficult to look for, because peculiar-velocity clouds are usually low column density, secondary components, in lines of sight that are probed optically. Therefore the kinetic temperatures that are derived refer to the dominant, low-velocity clouds, rather than any high-velocity clouds that may be present.

OBSERVATIONS OF INTERSTELLAR MOLECULES WITH THE INTERNATIONAL
ULTRAVIOLET EXPLORER

John H. Black[1]
Harvard-Smithsonian Center for Astrophysics
Cambridge, Massachusetts, U.S.A.

The ultraviolet spectra of 25 early-type stars have been obtained with
the International Ultraviolet Explorer observatory. Bands of the 4th-
positive system of interstellar CO are seen towards 12 of these stars.
Spectra of HD46223 have been examined for interstellar lines of CH, C_2,
CH_2, OH, HCl, and H_2O.

INTRODUCTION

     Electronic transitions of many simple interstellar molecules lie
in the wavelength region $\lambda\lambda 1150$-3200 Å which is accessible to the
spectrographs of the International Ultraviolet Explorer satellite (IUE).
Measurements of the abundances of molecules in diffuse interstellar
clouds permit tests of theories of molecule formation and, in some
instances, provide diagnostic probes of physical conditions inside the
clouds.

     The design and performance of IUE have been described by Boggess
et al. (1978a,b). Although the resolving power (R $\simeq$ 13000) of IUE is
smaller than that of the Princeton ultraviolet spectrometer on the
Copernicus satellite, IUE can be used to observe more highly-reddened
stars. The details of our survey of interstellar absorption lines are
presented elsewhere (Black *et al.* 1979).

A SURVEY OF INTERSTELLAR CO

     Ultraviolet absorption lines of interstellar CO have been
observed previously by Smith and Stecher (1971), Snow (1975, 1977),
Morton (1975), and Jenkins *et al.* (1973). Smith *et al.* (1978)
analyzed *Copernicus* observations of CO in the ζ Oph cloud, and showed
that the rotational excitation of CO is dominated by its interaction
with the cosmic background radiation, indicating that the density is
too low to thermalize the rotational populations at a kinetic temperature much greater than 3 K.

B. H. Andrew (ed.), Interstellar Molecules, 257–260.

J. H. BLACK

As part of a survey of interstellar lines with IUE, the spectra of
25 O and B stars were examined for the presence of the 6-0 through 0-0
bands of the 4th-positive system $(A^1\Pi-X^1\Sigma)$ of CO. Profiles of several
of these bands in the spectrum of HD46223 are shown in Figure 1.

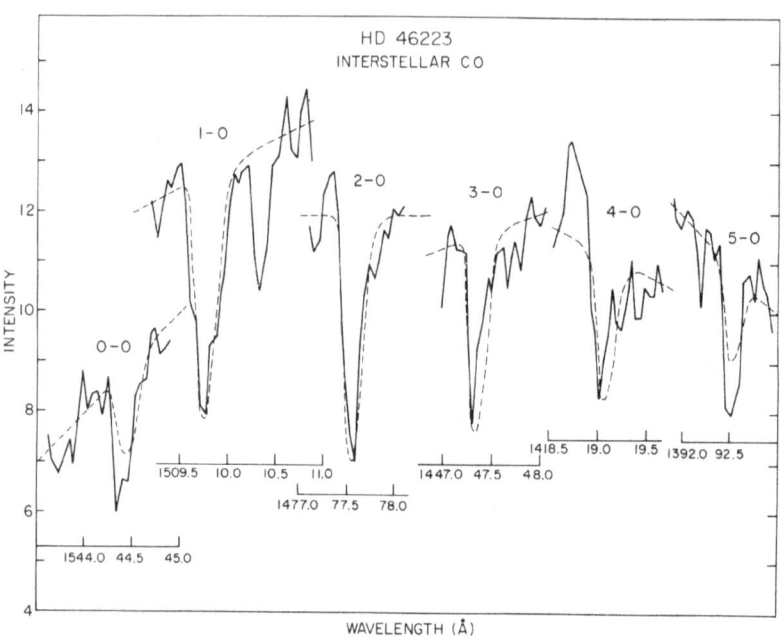

Figure 1. Profiles of CO bands in HD46223.

The bands cannot be resolved with IUE, and the observed band profiles must be compared with theoretical profiles which have been con-
volved with the instrumental response in order to extract column
densities from the data. The widths and central depths of the band
profiles are sensitive to the rotational excitation temperature, T_{rot},
and to the Doppler parameter, b, as well as to column density, $N(CO)$.
The dashed curves in Figure 1 represent theoretical profiles calculated
for b=5 km s^{-1}, T_{rot}=7.5 K, and N(CO)=1.9x10^{14} cm^{-2}. This value of b
is an upper limit determined from a preliminary curve-of-growth analysis
of lines of C I, Mg I, and S I, so that the corresponding value of
N(CO) must be considered a lower limit. The other observations of CO
in our sample are all consistent with T_{rot} < 10 K. Effective curves
of growth have been computed by integrating theoretical band profiles
for the conditions T_{rot}=3 K and b=2 km s^{-1}. Column densities estimated from the equivalent widths and these curves of growth are presented in Table 1 together with the HD catalogue number and color excess,
E(B-V), of the star, the equivalent width of the 1-0 band, W_{1-0}, and the
number of bands, n, upon which the estimate is based.

Table 1. Observations of Interstellar CO

HD	E(B-V)	W_{1-0} mÅ	$N(CO)$ cm^{-2}	n
23180	0.30	57±8	1.5 +14	6
30614	0.34	19±10	3.9 +13	3
36879	0.54	40±9	1.2 +14	4
41117	0.45	80±13	2.1 +14	4
46223	0.54	104±6	5.5 +14	6
147933	0.42	48±9	1.5 +14	6
148937	0.66	89±12	5.0 +14	5
149404	0.74	91±7	6.1 +14	6
151804	0.38	36±6	5.5 +13	3
152424	0.70	66±9	3.5 +14	5
209975	0.39	35±8	1.2 +14	5
213087	0.62	63±14	7.5 +14	5

Upper limits ranging from $1-5 \times 10^{13}$ cm^{-2} can be placed upon the estimated N(CO) for 13 other stars. Noteworthy limits include $N(CO) < 1.5 \times 10^{13}$ cm^{-2} and $N(CO) < 7.5 \times 10^{12}$ cm^{-2} towards 20 Tau and 23 Tau, respectively: these are stars in the Pleiades whose visible spectra show unusually strong lines of interstellar CH^{+} and CH. Although the estimated column densities in Table 1 are of limited quantitative significance individually, they do show a tendency to correlate with E(B-V) according to
$$N(CO) \simeq 5 \times 10^{12} \exp(7 E(B-V)) \text{ } cm^{-2} \text{ .}$$
This differs considerably from the results of studies of millimeter-wavelength rotational lines of CO in dark clouds in which the column densities of CO are found to be much larger and to be directly proportional to the amount of extinction (Tucker *et al.* 1976, Dickman 1978).

SEARCHES FOR OTHER MOLECULES

Four exposures at long wavelengths (2000-3200 Å) and five exposures at short wavelengths (1150-2100 Å) of HD46223 provide the data of highest signal/noise ratio in our survey. The combined spectra have been searched for molecular lines of C_2 (C-X), CH (C-X, D-X, E-X, F-X, and G-X), CH_2 (B̃-X̃ and C̃-X̃), H_2O (C̃-X̃), HCl (C-X), and OH (C-X). With the exception of CH, no other species are detected. An interstellar line appears near 1370 Å in HD23408, 23480, 46223, 149404, and 152424. Using the 1347.240 Å line of Cl I to adjust for radial velocity shifts, we find a rest position for the new line of $\lambda=1369.58\pm.05$ Å and $\tilde{v}=73015\pm2$ cm^{-1}. It is tempting to identify this feature with a line of the G(3d)-X system of CH (cf. Herzberg and Johns 1969). Further laboratory studies of this system are needed in order to resolve its complex structure and to establish its identification in interstellar absorption.

        This research has been supported by NASA grants NSG-5237 to the
University of Minnesota and NSG-5380 to Harvard College Observatory.
I am grateful to Drs. A.K. Dupree, L.W. Hartmann, and J.C. Raymond for
numerous discussions and assistance with the observations.

REFERENCES

Black, J.H., Dupree, A.K., Hartmann, L.W., and Raymond, J.C.: 1979,
        in preparation.
Boggess, A., *et al.*: 1978a, Nature 275, 372-377.
Boggess, A., *et al.*: 1978b, Nature 275, 377-385.
Dickman, R.L.: 1978, Astrophys. J. Supp. Ser. 37, 407-427.
Herzberg, G. and Johns, J.W.C.: 1969, Astrophys. J. 158, 399-418.
Jenkins, E.B., Drake, J.F., Morton, D.C., Rogerson, J.B., Spitzer, L.,
        and York, D.G.: 1973, Astrophys. J. Lett. 181, L122-L127.
Morton, D.C.: 1975, Astrophys. J. 197, 85-115.
Smith, A.M., Krishna Swamy, K.S., and Stecher, T.P.: 1978, Astrophys. J.
        220, 138-148.
Smith, A.M. and Stecher, T.P.: 1971, Astrophys. J. Lett. 164, L43-L47.
Snow, T.P.: 1975, Astrophys. J. Lett. 201, L21-L24.
Snow, T.P.: 1977, Astrophys. J. 216, 724-737.
Tucker, K.D., Dickman, R.L., Encrenaz, P.J., and Kutner, M.L.: 1976,
        Astrophys. J. 210, 679-683.

[1]Guest Investigator with the International Ultraviolet Explorer
satellite, sponsored and operated by NASA, by ESA, and by the SRC
of the United Kingdom.

DISCUSSION FOLLOWING BLACK

Vanden Bout: Are the CO spectra averages of several exposures?
Black: There are repeated exposures for about half the stars in
our sample.
Glassgold: How do your column densities compare with those deter-
mined by Copernicus for the same thickness (A_v)?
Black: The column densities in Table 1 tend to be smaller than
Copernicus results for similar A_v, but the arbitrary selection of b and
T_{rot} in the analysis probably causes underestimated column densities.
W. Watson: What are the current and future prospects for detecting
CH_2?
Black: The ultraviolet bands of CH_2 are quite diffuse, and may be
difficult to distinguish from stellar features. Strong features appear
in HD46223 near 1415 Å, but they are probably stellar. Our data will
be analyzed further by differencing techniques to eliminate some of the
confusion with stellar lines.

INTERSTELLAR LINE SPECTRA OF A DENSE CLOUD: THE VI CYGNI ASSOCIATION

Steven P. Souza*
State University of New York at Stony Brook

Barry L. Lutz*
Lowell Observatory, Flagstaff, Arizona

Spectroscopic observations of the dense interstellar cloud which obscures the VI Cygni association have yielded an optical map of the atomic and molecular constituents in a region which may represent a transition between the diffuse and radio clouds. The velocity of the optical components found is consistent with that of a CO cloud seen in front of No. 12.

INTRODUCTION

The chemistry of interstellar clouds has long been a topic of active astronomical research. Study of the atomic and molecular constituents of such clouds has been primarily limited to extremes of cloud types: diffuse clouds and dark clouds. The midrange clouds, with optical extinctions $\gtrsim 2^m$ or 3^m, may bridge the gap between the two extremes; but they have rarely been probed to any degree of sensitivity, either optically or in the radio regime.

During the past three years we have undertaken to study such midrange clouds. In particular, we have selected a rather extended cloud which is known as VI Cygni. Two of the prime reasons for choosing the VI Cygni cloud were that it is an association of early-type stars in a large region of moderately dense obscuration with a gradient of extinction ranging up to 10 magnitudes, and it contains one of the most highly obscured early-type stars known. The high cloud density that the extinction implies suggested that this region is amenable to optical searches for new molecules. Indeed, this selection proved fruitful, for the 1μ-spectrum and No. 12 provided the first detection of two rotational lines, Q(2) and R(2), of interstellar C_2 (Souza and Lutz, 1977). Since that time we have undertaken an observational program of other members of the association, including Nos. 4, 5, and 9, both for new molecules and to map the optically observable interstellar species.

---

*Visitor, Kitt Peak National Observatory, which is operated by the Association of Universities for Research in Astronomy under contract with the National Science Foundation.

B. H. Andrew (ed.), Interstellar Molecules, 261–262.
Copyright © 1980 by the IAU.

OBSERVATIONS AND RESULTS

Our observations were carried out with the 1.5-meter reflector and echelle spectrograph at Mount Hopkins Observatory and with the echelle and R-C spectrographs on the 4-meter Mayall telescope at Kitt Peak National Observatory. Details of the observations are given by Souza and Lutz (1977, 1979).

Figure 1 shows a new 1μ spectrum of No. 12, which is a composite of all data, both those originally reported and those obtained subsequently. The presence of interstellar C_2 is confirmed, and additional rotational lines unambiguously identified. These new data are consistent with the abundance and rotational temperature of our previous work.

Figure 2 illustrates spectra obtained for Nos. 4, 5, and 9. Curve-of-growth column densities have been determined to provide the first-order distribution of the interstellar material within the cloud, and simple chemical models have been calculated for the limited data available.

Preliminary radio mapping of this region has been carried out in collaboration with M. Simon and G. Righini-Cohen at SUNY at Stony Brook, and CO emission has been found at the same cloud velocity as the optical components exhibit.

This research has been supported by National Science Foundation grants AST 76-04149-A01 and AST 78-20131.

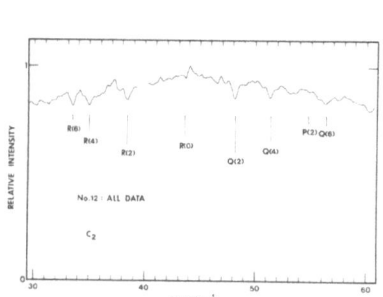

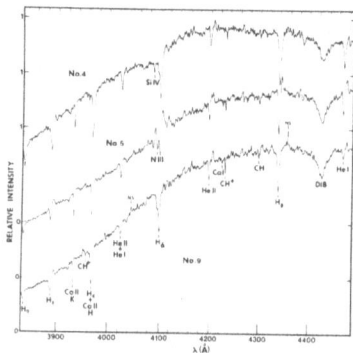

Figure 1. Spectrum of VI Cygni No. 12 showing lines of C_2.

Figure 2. Spectra of VI Cygni Nos. 4, 5, and 9.

References

Souza, S. P., and Lutz, B. L. (1977). *Astrophys. J.* **216**, L49.
Souza, S. P., and Lutz, B. L. (1979). *Astrophys. J. (Lett.)*, in press.

ROTATIONAL FINE STRUCTURE LINES OF INTERSTELLAR C_2 TOWARD ζ PERSEI

Frederic H. Chaffee, Jr.
Smithsonian Astrophysical Observatory, Mt. Hopkins Observatory

Barry L. Lutz
Lowell Observatory

John H. Black
Harvard-Smithsonian Center for Astrophysics

Paul A. Vanden Bout and Ronald L. Snell
University of Texas at Austin

ABSTRACT

We have detected 9 of the rotational fine structure lines of the 2-0 Phillips band of interstellar C_2 toward ζ Persei using the Tull spectrograph and Reticon detector on the 2.7 m telescope at the McDonald Observatory. These data yield a total C_2 column density of 1.2_3 x 10^{13} cm^{-2} and a rotational temperature of 97 K compared to 1.4 x 10^{13} cm^{-2} and 45 K predicted by the detailed model of the cloud by Black, Hartquist and Dalgarno. We suggest that radiative pumping through the Mulliken and Phillips systems has modified the C_2 level populations in such a way as to produce an observed rotational temperature which exceeds that arising in pure thermal equilibrium.

I. INTRODUCTION

In their discussion of the formation of interstellar (IS) molecules through gas phase reactions, Dalgarno and Black (1976) showed that C_2 can be produced through the same chain of reactions which produces CH and CH^+--the two most frequently observed IS molecules in the optical window. They suggested that C_2 is produced by the dissociative recombination of C_2H^+ which itself is formed in a series of reactions beginning with C^+ and H_2.

Recently, detailed models have appeared for the diffuse IS clouds toward ζ Oph (Black and Dalgarno 1977) and ζ Per (Black, Hartquist, and Dalgarno 1978, hereafter BHD). BHD predict column densities of many as yet unobserved molecules of which C_2 seemed the most likely to exhibit lines of sufficient strength to be detected with present techniques.

263

B. H. Andrew (ed.), Interstellar Molecules, 263–267.

The IS R(2) and Q(2) lines of the 1-0 Phillips band were first detected by Souza and Lutz (1977) toward VI Cyg No. 12, a star reddened by nearly 10 magnitudes. They suggested that because C_2 and H_2 have a similar molecular structure, the relative strengths of C_2 fine struc- ture lines arising from different J levels can be used to estimate the cloud temperature just as in the case of H_2.

BHD's model predicts that the R(2) and Q(4) lines of the 2-0 Phillips band of C_2 might be of sufficient strength to be detected optically toward ζ Per, and the present observations were undertaken to search for these lines.

II. OBSERVATIONS AND C_2 ROTATIONAL TEMPERATURE AND COLUMN DENSITY

Spectra were obtained with the Tull spectrograph and the direct illuminated Reticon (Vogt, Tull and Kelton 1978) on the McDonald Obser- vatory 2.7 m telescope. A total of 8 hours of integration over two nights were used to produce the final spectrum. The data were smoothed using the Fourier filtering techniques described by Brault and White (1971), and the smoothed data are displayed below the raw data in figure 1. Q(4) has an equivalent width of 1.49 mA and is the strongest feature in the spectrum. We consider R(10) and R(12) only marginally detected, and we have not used them in our analysis. The weakest line we have used is R(8) with an equivalent width of 0.37 mA.

We have adopted the f-values suggested by Roux, Cerny and d'Incan (1976) to calculate the C_2 column density for each line in figure 1. From the relative line strengths we have determined that the C_2 rota- tional temperature is 97 ± 10 (1σ) K. Using this temperature we have computed the appropriate partition function which allows the total C_2 column density to be calculated from the strength of any line. The 7 C_2 lines in figure 1 yield a mean column density of $N(C_2) = 1.17 ± .08$ (1σ) x 10^{13} cm^{-2}.

III. IMPLICATIONS OF INTERSTELLAR C_2 OBSERVATIONS

The model of the ζ Per diffuse cloud presented by BHD predicts that if C_2 arises in gas phase reactions it should have a column den- sity of 1.4 x 10^{13} cm^{-2}, a prediction in remarkably good agreement with the observed value. Observations of the rotational line strengths of H_2, on the other hand, are best explained by a two component model of the cloud with most of the molecules confined to a cold core having a kinetic temperature of 45 K. A rotational temperature this low is inconsistent with the C_2 observations, but the rotational and kinetic temperatures will be the same only if inelastic collisions with the abundant species (i.e. H and H_2) dominate the statistical equilibrium of rotational levels. Both spontaneous radiative transitions and absorption and fluorescence in electronic systems can alter the rota- tional populations and could explain a disparity between the observed rotational temperature and the actual kinetic temperature of the cloud.

A reasonable estimate of the probability of spontaneous radiative transitions between rotational levels of C_2 suggests that inelastic collisions with H and H_2 are almost certainly more rapid than spontaneous radiation even at IS densities and temperatures. Thus such transitions can not produce a rotational temperature which differs from the kinetic temperature of the cloud.

On the other hand, absorption and fluorescence in the Phillips and Mulliken systems <u>can</u> compete with collisions in redistributing rotational populations, and calculating these populations for C_2 resembles the corresponding analysis for H_2 (Black and Dalgarno 1976). Virtually every absorption in an electronic system is followed by a downward transition to some level of the ground electronic state, and the final rotational quantum number is frequently different from the initial one. If the fluorescent transition is to a vibrationally excited level, the rotational distribution is further modified through quadrupole rotation-vibration transitions.

We have assumed that the level populations of C_2 are in statistical equilibrium and have considered collisional and radiative transitions in our analysis. Cross sections for rotationally inelastic collisions involving C_2 are unknown and can be only estimated, but the effects of absorption and fluorescence can be calculated explicitly.

The results of such an analysis do not permit the exclusive determination of the characteristics of the radiation field, the number density of collision partners and the kinetic temperature from observations of C_2 because of our lack of knowledge of the collisional cross sections and of the details of the radiative cascade. However, under reasonable assumptions of these quantities, a reasonable solution of the statistical equilibrium equation can be found in which the rotational populations are characterized by a <u>single</u> <u>rotational</u> <u>temperature</u> which differs significantly from the kinetic temperature because of radiative pumping.

When better molecular data become available, observations of C_2 can be used to extract information about densities, temperatures and radiation fields in IS clouds. Although the C_2 lines observed so far are very weak, because they lie in the near infrared the potential exists for using them to probe regions of much higher extinction than can be observed in the ultraviolet lines of H_2.

This paper was extracted from a much more detailed one which has been submitted for publication to the <u>Astrophysical Journal</u>.

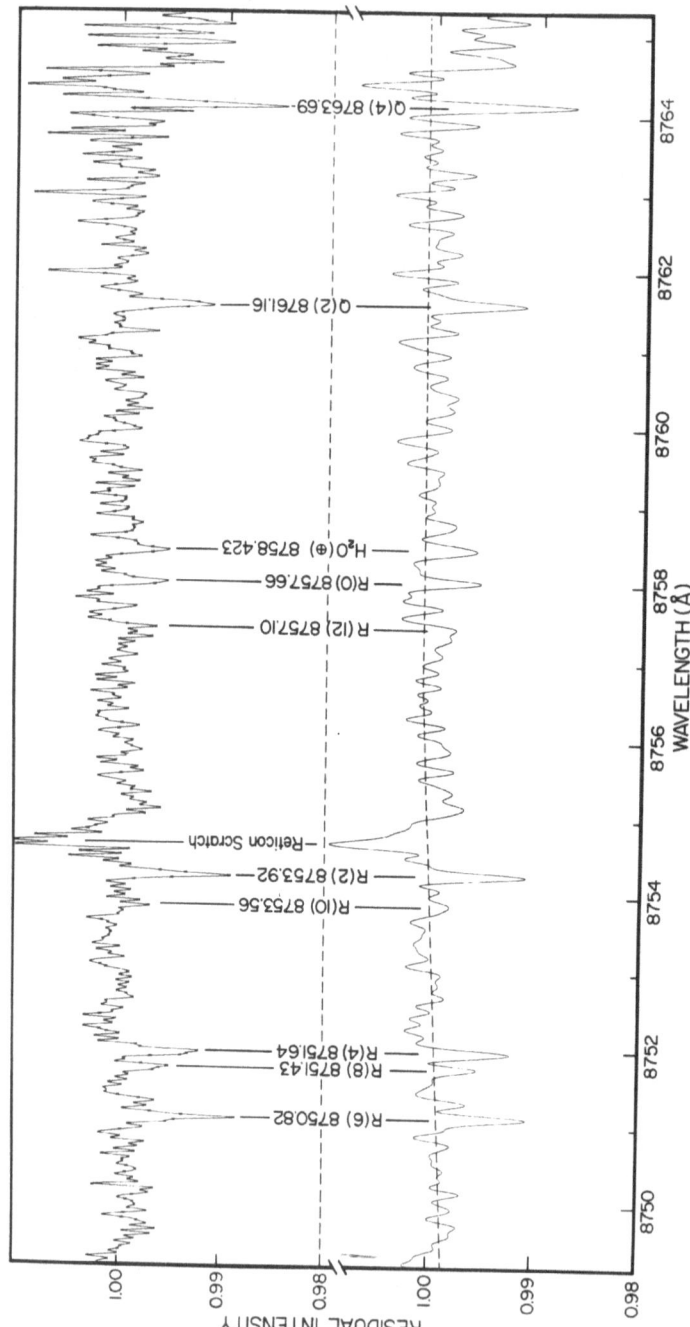

Figure 1. Fine structure lines of the 2-0 Phillips system toward ζ Per: raw data (upper curve) and processed data (lower curve).

REFERENCES

Black, J. H., and Dalgarno, A. 1976, Astrophys. J., 203, 132.
Black, J. H., and Dalgarno, A. 1977, Astrophys. J. Suppl., 34, 405.
Black, J. H., Hartquist, T. W., and Dalgarno, A. 1979, Astrophys. J.,
    224, 448 (BHD).
Brault, J. W., and White, O. R. 1971, Astron. Astrophys., 13, 169.
Dalgarno, A., and Black, J. H. 1976, Rept. Progr. Phys., 39, 573.
Roux, F., Cerny, J., and d'Incan, J. 1976, Astrophys. J., 204, 940.
Souza, S. P., and Lutz, B. L. 1977, Astrophys. J. (Letters), 216, L49.
Vogt, S. S., Tull, R. G., and Kelton, R. 1978, Appl. Optics, 17, 574.

DISCUSSION FOLLOWING CHAFFEE

Thaddeus: Does a plot of log intensity vs rotational energy give a simple linear relationship for your C$_2$ data, characteristic of thermal equilibrium? If so, it would seem to me that the details of the C$_2$ collisions with H or H$_2$ are not an important consideration.

Chaffee: The relationship *is* a simple linear one, but one of the main points of our paper is that it is possible for such a relationship to obtain even when the actual kinetic temperature differs significantly from the measured rotational temperature. In the diffuse cloud toward ζ Persei radiative pumping and collisions are of nearly equal importance in determining the C$_2$ level populations, and the combined effect still produces a linear relationship for log intensity versus rotational energy. The slope of this relationship gives a temperature that is more than a factor of two higher than the actual kinetic temperature of the cloud.

Carruthers: Have you looked for C$_2$ towards ζ Ophiuchi?

Chaffee: Yes. The Q(2) line of the 2-0 Phillips band was detected, and reported in Ap.J. 1978, 221, L91. The column density of C$_2$ is again in good agreement with the Black and Dalgarno model.

Lutz: I want to remark that in addition to this C$_2$ detection in ζ Per and to Chaffee's and my detection in ζ Oph, Souza and I have confirmed the presence of C$_2$ in the VI Cyg cloud and, based on six rotational lines, have found an excitation temperature of 45 K. We have also detected C$_2$ in the ρ Oph cloud - in HD147889 - at a rotational temperature of 60 K. This latter result was based on four lines.

LABORATORY MEASUREMENTS OF OSCILLATOR STRENGTHS OF ULTRAVIOLET MOLECULAR
LINES OF HCl AND H_2O AND COLUMN DENSITIES OF THESE MOLECULES IN THE ZETA
OPHIUCHI CLOUD

Peter L. Smith, K. Yoshino, and W.H. Parkinson
Harvard-Smithsonian Center for Astrophysics
Cambridge, Massachusetts, U.S.A.

The oscillator strengths of two ultraviolet molecular lines expected to
be seen in diffuse interstellar clouds have been measured. For the
1290.257 Å line of HCl, f=0.16±0.06. For the 1114.225 Å line of H_2O,
f=(5.0±2.0)x10^{-3}. These results have been used in conjunction with
observational data to compare measured and predicted column densities
of these molecules in the ζ Ophiuchi cloud.

INTRODUCTION

    The study of molecules in interstellar clouds can provide informa-
tion about the physical conditions prevailing in the interstellar gas
and can be used to test models of the chemical evolution of such clouds.
Quantitative comparisons of observed molecular column densities with
those predicted by the model calculations require oscillator strengths
(f-values) for the lines studied. This paper presents the results of
f-value measurements for the R(0) line of the $C(0)^1\Pi-X(0)^1\Sigma^+$ band of
HCl and of the $1_{11}-0_{00}$ line of the $\tilde{F}-\tilde{X}$ band of H_2O. Rotational analyses
of these bands have been given by Tilford et al. (1970) and by Johns
(1978), respectively. The astrophysical implications of the f-value
data are also discussed; the results of observations and model predictions for the column densities in the ζ Ophiuchi cloud are compared.

MEASUREMENT & DATA ANALYSIS

    The apparatus and method were similar to those used for our measure-
ments of the f-values for the $\tilde{C}(0)^1B_1-X(0)^1A_1$ band of H_2O (Smith and
Parkinson, 1978); some differences in equipment and procedures are discussed in this section. For the measurements on HCl, an absorption cell
8.1 mm long and HCl pressures of 2.1, 2.7, and 3.2 N/m² were used. For
the measurements on H_2O, the entire spectrograph (6.65 m focal length,
2400 line/mm grating) was filled with water vapor and the absorption by
the $\tilde{F}-\tilde{X}$ band was compared to that by the $\tilde{C}-\tilde{X}$ band. Consequently our
results for H_2O are relative to the absolute scale established by Smith

269

B. H. Andrew (ed.), Interstellar Molecules, 269–270.

and Parkinson (1978). Two plates with H_2O pressures in the ratio of approximately 1.5:1 were obtained.

The lines of the C(0)-X(0) band of HCl had a width (FWHM) of more than 70 mÅ whereas the instrumental line width was $\leq$15 mÅ. For the study of the $\tilde{C}$(0)-$\tilde{X}$(0) and $\tilde{F}$-$\tilde{X}$ bands of H_2O, the second order of the spectrograph was used and the line widths were $\geq$30 mÅ and $\geq$15 mÅ respectively whereas the instrumental width was $\leq$6 mÅ. A computer program using Lorentz profiles was used to fit the absorption spectra.

RESULTS AND DISCUSSION

The oscillator strength of the 1290.257 Å line of HCl is 0.16±0.06. This is larger than the value, f=0.05, estimated by Morton (1975). When our value is used to revise the analysis of the observations of absorption by the ζ Oph cloud by Wright and Morton (1979), an upper limit to column density of 8.4×10^{11} cm^2 is obtained. That predicted by Black and Dalgarno (1977) is 2.6×10^{13} cm^2. A discussion of this discrepancy and of chlorine chemistry in diffuse interstellar clouds is given by Black and Smith (1979).

The f-value of the 1114.225 Å line of H_2O is $(5.0±2.0) \times 10^{-3}$. This value has been used by Snow and Smith (1979) to estimate a probable column density of 1.7×10^{13} cm^{-2} for H_2O in the ζ Oph cloud based on a detection at the 1.8 σ level. The value predicted by Black and Dalgarno (1977) is 1.1×10^{12} cm^{-2}.

The oscillator strengths of other lines in the C(0)-X(0) band of HCl and the $\tilde{F}$-$\tilde{X}$ band of H_2O will be published (Smith et al. 1979).

The authors are indebted to J.R. Esmond and H.E. Griesinger for their indefatigable assistance. We also thank J.H. Black and T.P. Snow for their comments and advice and T.P. Snow and Wm. H. Smith for permission to quote their unpublished data. This work was supported in part by NASA Grants NSG-7034 and NSG-7156.

REFERENCES

Black, J.H., and Dalgarno, A.: 1977, Astrophys. J. Supp. Ser. 34, pp. 405-423.
Black, J.H., and Smith, P.L.: 1979, proceedings of this symposium.
Johns, J.W.C.: 1978, personnal communication.
Morton, D.C.: 1975, Astrophys. J. 197, pp. 85-115.
Smith, P.L., and Parkinson, W.H.: 1978, Astrophys. J. Lett. 223, pp. L127-L130.
Smith, P.L., Yoshino, K., and Parkinson, W.H.: 1979, in preparation.
Snow, T.P., and Smith, Wm. H.: 1979, in preparation; (cf. Snow, 1979).
Snow, T.P.: 1979, proceedings of this symposium.
Tilford, S.G., Ginter, M.L., and Vanderslice, J.T.: 1970, J. Mol. Spectrosc. 33, pp. 505-519.
Wright, E.L., and Morton, D.C.: 1979, Astrophys. J. 227, pp. 483-488.

CHLORINE CHEMISTRY IN DIFFUSE INTERSTELLAR CLOUDS

John H. Black and Peter L. Smith
Harvard-Smithsonian Center for Astrophysics
Cambridge, Massachusetts, U.S.A.

The results of recent observations of the column densities of chlorine species in diffuse interstellar clouds are inconsistent with the column densities predicted by model calculations. The use of new atomic and molecular data fails to decrease the discrepancy. We speculate that the exothermic reaction $Cl^+ + H_2 \rightarrow HCl^+ + H$ may not be rapid at the low temperatures prevailing in diffuse interstellar clouds.

INTRODUCTION

The chemistry of chlorine in interstellar clouds has been discussed by Jura (1974) and Dalgarno *et al.* (1974). The abundances of chlorine-bearing species have been predicted for the ζ Oph and ζ Per diffuse clouds by Black and Dalgarno (1977) and Black *et al.* (1978), respectively. Observations of chlorine species have been made by Morton (1975), Jura and York (1978), and Wright and Morton (1979), who used the *Copernicus* satellite, and by one of us (JHB) who used IUE.

ABUNDANCES OF CHLORINE-BEARING SPECIES IN DIFFUSE INTERSTELLAR CLOUDS

Wright and Morton (1979) set an upper limit of $W < 1.9$ mÅ for the equivalent width of the 1290.257 Å line of HCl in the ζ Oph cloud. Smith *et al.* (1979) have measured the oscillator strength of this line, $f = 0.16 \pm 0.05$; thus the column density of HCl is $N(HCl) < 8.4 \times 10^{11}$ cm^{-2}. The observed column densities of Cl and Cl$^+$ in the ζ Oph cloud are $N(Cl) = 1.1 \times 10^{14}$ cm^{-2} and $N(Cl^+) = (1.2$–$4.3) \times 10^{13}$ cm^{-2} (Morton 1975). Black and Dalgarno (1977) predicted $N(HCl) = 2.6 \times 10^{13}$ cm^{-2} for this cloud. Even with an improved cross section for photoionization of Cl (Brown *et al.* 1978) in the model calculations, the predicted abundance, $N(HCl) \approx 1.6 \times 10^{13}$ cm^{-2}, exceeds the observed limit by a factor of 18. The model also underestimates the abundance of Cl$^+$.

Jura and York (1978) predict a ratio of column densities $N(HCl)/N(Cl) = 0.01$ in typical diffuse clouds with high H_2 concentrations. A

B. H. Andrew (ed.), Interstellar Molecules, 271–272.

recent survey of interstellar molecules with the IUE satellite (cf. Black 1979) establishes N(Cl) and limits to N(HCl) for the lines of sight to 13 highly-reddened stars. By neglecting the saturation of the strong 1347.24 Å line of Cl I, it is found that N(HCl)/N(Cl)<<0.1. Toward HD46223, N(HCl)/N(Cl)<0.003. Such a low limit to the HCl/Cl ratio and the relatively large Cl^+/Cl ratio in regions of high H_2 concentration like α Cam, ξ Per, and ζ Oph (cf. Jura and York (1978)) are difficult to explain merely by reducing the rate of photoionization of Cl as suggested by Jura and York (1978).

DISCUSSION

The reaction
$$Cl^+ + H_2 \rightarrow HCl^+ + H$$
both leads to the formation of HCl and reduces the abundance of Cl^+ relative to that of Cl (Jura 1974, Dalgarno *et al.* 1974). Although this reaction has a large rate coefficient, $k=7\times10^{-10}$ cm^3 s^{-1} at T=297 K (Fehsenfeld and Ferguson 1974), it is exothermic by a small amount, 0.22±0.02 eV, and the existence of a modest energy barrier can presumably not be excluded. The presence of an energy barrier would reduce the rate coefficient at interstellar temperatures, T=20-100 K, with the result that the predicted abundance of Cl^+ would increase while that of HCl would decrease. A temperature dependence in the rate of the Cl^++H_2 reaction could remove the discrepancies between the observed and predicted column densities. An experimental test of this conjecture would be useful.

We acknowledge discussions with M. Jura, T.P. Snow, and T.W. Hart-quist. This work was supported in part by NASA Grants NSG-5380 and NSG-7034.

REFERENCES

Black, J.H.: 1979, proceedings of this symposium.
Black, J.H., and Dalgarno, A.: 1977, Astrophys. J. Supp. Ser. 34, pp. 405-423.
Black, J.H., Hartquist, T.W., and Dalgarno, A.: 1978, Astrophys. J. 224, pp. 448-452.
Brown, E.R., Carter, S.L., and Kelly, H.P.: 1978, Phys. Lett. 66A, pp. 290-292.
Dalgarno, A., de Jong, T., Oppenheimer, M., and Black, J.H.: 1974, Astrophys. J. Lett. 192, pp. L37-L39.
Fehsenfeld, F.C., and Ferguson, E.E.: 1974, J. Chem. Phys. 60, p. 5132.
Jura, M.: 1974, Astrophys. J. Lett. 190, pp. L33-L34.
Jura, M., and York, D.G.: 1978, Astrophys. J. 219, pp. 861-869.
Morton, D.C.: 1975, Astrophys. J. 197, pp. 85-115.
Smith, P.L., Yoshino, K., and Parkinson, W.H.: 1979, proceedings of this symposium.
Wright, E.L. and Morton, D.C.: 1979, Astrophys. J. 227, pp. 483-488.

MOLECULAR FORMATION IN HOT DIFFUSE CLOUDS

A. Dalgarno
Harvard-Smithsonian Center for Astrophysics

Abstract. A description is given of the processes of molecular formation and destruction in diffuse interstellar clouds and detailed models
of the clouds lying towards ζ Ophiuchi, ζ Persei and o Persei are used
to assess the validity of gas phase chemistry. Modifications that may
arise from shock-heated regions are discussed.

Diffuse clouds are local concentrations of interstellar gas that
do not obscure entirely the light from the stars which lie behind
them and they may be studied observationally by measuring the absorption of the starlight by the individual constituents of the cloud
(cf. Snow 1979). The systems CO, C^+, CH and OH have been studied also
by measuring their emission in the radio region of the spectrum (cf.
Knapp and Jura 1976, Crutcher 1977, 1979, Lang and Willson 1978, Liszt
1979). Because diffuse clouds are simpler astrophysical entities than
are the dense molecular clouds where star formation occurs the study of
diffuse clouds provides a more stringent assessment of the molecular
formation and destruction processes which occur in the interstellar
gas. However, the chemistry of diffuse clouds is modified substantially
by photodissociation and photoionization processes and is not always
free of the complicating influence of interstellar shocks.

In this brief review I will attempt to summarize recent progress
in our knowledge of the processes involved in diffuse cloud chemistry
and draw attention to areas of significant uncertainty. I will not
discuss the formation of the most abundant molecule, H_2, except to
note that at low temperatures H_2 is formed by association on the surfaces of grains in a reasonably well understood way (cf. Hollenbach
and McKee 1980) although we cannot predict with confidence the initial
rotational and vibrational distribution of the newly created molecules
(cf. Hunter and Watson 1978), an uncertainty which affects the analysis
of the observed rotational populations of H_2 and the calculation of the
associated heating rate.

At high gas temperatures, found in shocked regions and in ionized

273

B. H. Andrew (ed.), Interstellar Molecules, 273–280.

nebulae, associative detachment (cf. Dalgarno and McCray 1973)

$$H + H^- \rightarrow H_2 + e \tag{1}$$

may be a significant source of H_2 (cf. Black 1978) which leads to an emission spectrum containing lines from highly excited vibration - rotation levels (Bieniek and Dalgarno 1979). The spectrum is characteristic and its observation would be a powerful diagnostic probe of the mechanism of associative detachment and of the environment in which it occurs.

Given the formation of H_2 all the other molecules detected in diffuse clouds can be explained by gas phase mechanisms. In the proposed chemical schemes the reactions are initiated by cosmic rays or by ultraviolet photons. Cosmic ray ionization of H and H_2 produces H^+ ions which undergo charge transfer with atomic oxygen:

$$H^+ + O \rightarrow H + O^+ \tag{2}$$

The resulting O^+ ions react with H_2 initiating an abstraction sequence which terminates in the production of H_3O^+. The H_3O^+ ions recombine dissociatively to create OH and H_2O in unknown proportions. In diffuse clouds, H_2O is photodissociated to give OH and essentially every H^+-O charge transfer leads to OH.

The absolute fluxes are based upon a rate coefficient for reaction (2) of $5\times10^{-10} \exp(-232/T)$ cm^{3}s^{-1} inferred from an analysis of OH abundances measured towards ζ Oph (Black and Dalgarno 1977). A study by de Boer (1979) of atomic oxygen towards ζ Oph (Morton 1975) indicates that the abundance of atomic oxygen should be increased by a factor of three. The inferred rate coefficient of reaction (2) is reduced correspondingly to a value of $1.7\times10^{-10} \exp(-232/T)$ cm^{3}s^{-1}, in harmony with calculations (Chambaud et al. 1979) which can be represented approximately by $1.5 \times 10^{-10} \exp(-232/T)$ cm^{3}s^{-1}.

There remain uncertainties in the depletion factor of carbon (cf. Liszt 1979) and in the OH destruction rates but it is unlikely that the derived fluxes are in error by more than a factor of three.

There has been a marginal detection of H_2O (Snow and Smith 1979, Snow 1979) at about the 2σ level. With the oscillator strength measured by Smith, Yoshino and Parkinson (1979), the implied abundance is 2.65×10^{13} cm^{-2}. The calculated abundance is 2×10^{12} f cm^{-2} (Black and Dalgarno 1977) where f is the fraction of dissociative recombinations which lead to H_2O. According to Herbst (1978), f is less than 0.5 and may be as small as 0.1. The discrepancy between theory and observation is large and probably cannot be accommodated by changes in molecular rate coefficients or model parameters. If the detection of H_2O is real, a new source of H_2O must be sought.

Cosmic ray ionization of H^+ also leads to HD through the sequence

$$H^+ + D \rightarrow H + D^+ \tag{3}$$

$$D^+ + H_2 \rightarrow HD + H^+ \tag{4}$$

(Black and Dalgarno 1973a, Watson 1973). Accurate values of the rate coefficient of (3) have been presented by Watson et al. (1978) It is of interest to note that the existence of H_2 is the strongest evidence of the importance of grain chemistry and the existence of HD is the strongest evidence of the importance of gas phase chemistry.

The abundance of HD is directly proportional to the cosmic ray flux and the cosmic (D)/(H) abundance ratio. The cosmic ray fluxes have been inferred from the OH abundances so that (D)/(H) may be inferred from the HD abundances. For ζ Oph, ζ Per and o Per the ratio is 1.4×10^{-5} to within the considerable HD observational uncertainties (Hartquist, Black and Dalgarno 1978), a value in·harmony with the mean for distances up to 200 pc from the Sun (York and Rogerson 1976). The general agreement lends support to the postulated OH and HD chemistries.

The chemistry can be tested also by observations of CO which is formed as a consequence of the C^+-OH and C^+-H_2O reactions and destroyed mainly by photodissociation. A comparison of theory and observation is presented in table 1. A photodissociation rate which reproduces the CO abundance measured for ζ Oph is successful in the ζ Per and o Per clouds also.

Of the constituents ionized by ultraviolet photons, only Cl^+ reacts chemically with H_2 (Jura 1974, Dalgarno et al. 1974). The sequence of

$$Cl^+ + H_2 \rightarrow HCl^+ + H \tag{5}$$

$$HCl^+ + H_2 \rightarrow H_2Cl^+ + H \tag{6}$$

and

$$H_2Cl^+ + e \rightarrow HCl + H \tag{7}$$

leads to HCl. Hydrogen chloride is destroyed by photodissociation and by reaction with C^+ according to

$$HCl + C^+ \rightarrow CCl^+ + H \tag{8}$$

(Dalgarno and Black 1976). For ζ Oph, Black and Dalgarno (1977) predict a column density of 1.0×10^{13} cm^{-2}. An upper limit to the equivalent width of the absorption line at 1290 Å has been obtained by Wright and Morton (1979) which is consistent with a column density of 1.3×10^{12} cm^{-2} when the oscillator strength measured by Smith, Yoshino and Parkinson (1979) is used. The discrepancy is at least an order of magnitude. It could be resolved by a decrease in the interstellar photoionization efficiency of Cl (Jura and York 1978) or by postulating an activation energy for reaction (5) (Black and Smith 1979).

Of greater fundamental importance to interstellar chemistry is the reaction of C^+ with H_2. In a cold gas, C^+ does not react chemically with H_2. Radiative association

$$C^+ + H_2 \rightarrow CH_2^+ + h\nu \qquad (9)$$

may occur (Black and Dalgarno 1973b), initiating a sequence which produces CH and CH^+ and, through a reaction scheme suggested by Watson (1973), also C_2 (Dalgarno and Black 1976), C_2H and $C_2H_2^+$. The molecules C_2H and $C_2H_2^+$ are important in chemical schemes for the formation of the complex molecules found in dense clouds.

The rate coefficient of the initial step, reaction (9), is a critical parameter. The measured abundance of CH towards ζ Oph can be reproduced with a rate coefficient of 5×10^{-16} cm^3 s^{-1}, a value which is successful also for the direction towards ζ Per and o Per, as table 1 illustrates.

Table 1
Column densities log N (cm^{-2}) *

	ζ Oph		ζ Per		o Per	
	Obs	Th	Obs	Th	Obs	Th
CO	15.0–15.2	15.0	14.7–15.0	15.0	14.7–15.0	15.1
CH	13.5–13.6	13.6	13.0–13.4	13.6	13.4–13.6	13.5
CN	12.94	12.95	12.6	12.5	12.3–12.7	11.9
C_2	12.9	12.9	13.1	13.1	–	13.3
CH^+	13.0	11.4	12.2	11.5	12.7	11.5

*References to the observational data are given by Snow (1979)

The interstellar molecule CN is produced by various reaction sequences involving atomic nitrogen and the CH cycle. In table 1, the predicted abundances are compared with measurements. The photo-dissociation rate of CN has been slightly modified to achieve precise agreement between theory and measurement for ζ Oph. There is a small discrepancy for o Per which may be due to an overestimate of the nitrogen depletion; the [N]/[H] ratio, measured by Snow (1976), is unusually small.

Molecular carbon, C_2, is produced by reactions of C^+ with the CH cycle. Its existence in about the correct abundances (cf. table 1) was predicted (Black and Dalgarno 1977, Black et al. 1978) before its detection (Lutz and Souza 1977, Snow 1978, Hobbs 1979, Chaffee et al. 1979). The importance of C_2 as a probe of the interstellar environment has been discussed by Lutz and Souza (1977) and by Chaffee (1979).

The chemical scheme fails to predict the measured abundances of CH^+ (cf. table 1). It can be made to work by the simple expedient of increasing the rate coefficient of reaction (9) to about 10^{-14} cm^3s^{-1} (Black, Dalgarno and Oppenheimer 1975), a value that is consistent with the current estimate based upon a combination of theory (Herbst, Schubert and Certain 1977) and laboratory measurement (cf. Fehsenfeld 1979). The scheme then predicts too much CH, CN and C_2. There is sufficient arbitrariness in the mechanisms forming and destroying CN and C_2 that the predicted overabundances may not be significant. That for CH could be remedied either by assuming that CH_2 preferentially undergoes dissociative ionization rather than photodissociation or that the reaction

$$CH^+ + H_2 \rightarrow CH_2^+ + H$$

becomes slow at low temperatures.

If the postulated rate coefficient of about $5x10^{-16}$ cm^3s^{-1} is correct, an additional source of CH^+ is necessary. In a hot gas, the endothermic reaction

$$C^+ + H_2 \rightarrow CH^+ + H$$

can proceed, and provided the gas remains hot long enough, a substantial amount of CH can be created The high temperature diminishes the destruction by dissociative recombination (Mitchell and McGowan 1978) and the resulting abundance may depend more upon the efficiency of photodissociation (Kirby et al. 1979, Kirby 1979) and, because of the reverse of (11), on the concentration ratio of atomic and molecular hydrogen.

Elitzur and Watson (1978,1979) have suggested that the hot gas is produced by a shock resulting from the expansion of an H II region around the parent star and they have developed a quantitative model which gives fair agreement with the measured abundances with plausible choices of the shock velocity and the pre-shock densities of H_2 and H. The alternative suggestion has been made by de Jong (1979) that the hot gas is located at the exterior boundary of the cloud where it is heated by interstellar shocks driven by supernovae. This model would help to explain some anomalies in the measured populations of high rotational levels of H_2 (Hartquist and Dalgarno 1980). Collision-induced dissociation may not have been taken into account correctly in these shock models (Dalgarno and Roberge 1979).

Gas phase chemistry is radically altered in a hot gas (Iglesias and Silk 1978, Elitzur and de Jong 1978, Hartquist, Oppenheimer and Dalgarno 1980). Endothermic reactions with H_2 play a dominating role and the H_2/H concentration ratio is a critical parameter. Molecular ions such as SiH^+, SH^+, NaH^+, MgH^+, FeH^+ and HCl^+ may be formed and survive long enough to be detectable. An upper limit of $2x10^{11}$ f_a cm^{-2} where f_a is the absorption oscillator strength, has been obtained by

Jenkins et al. (1973) for the column density of the molecular ion MgH^+ towards ζ Oph. A laboratory determination of f_a might be valuable in assessing the significance of shocked gas, though the Mg depletion may be severe in the molecular cloud (Snow and Meyers 1979).

Neutral molecules are readily produced in heated gas. Depending upon the extinction of the radiation field, the density of the heated gas and the cooling rate of the gas, large abundances of OH and H_2O (Iglesias and Silk 1978, Elitzur and de Jong 1978, Elitzur 1979) and of SH and H_2S (Hartquist et al. 1980) can result. If confirmed, the tentative detection of H_2O (Snow 1979) would be strong evidence of a shocked region and the earlier conclusion about gas phase chemistry and cosmic ray ionizations would require revision.

I remarked earlier that the abundances of all the molecular species detected in diffuse clouds can, with the exception of H_2, be attributed to gas phase. The agreement may in fact be illusory stemming in large part from the considerable uncertainties in our knowledge of the multiplicity of molecular processes that occur.

REFERENCES

Black, J.H. 1978, Ap. J. 222, 125.
Black, J.H. and Dalgarno, A. 1973a, Ap. J. (Letters) 184, L101.
Black, J.H. and Dalgarno, A. 1973b, Ap. J. (Letters) 15, 79.
Black, J.H. and Dalgarno, A. 1977, Ap. J. Suppl. 34, 405.
Black, J.H., Dalgarno, A. and Oppenheimer, M. 1975, Ap. J. 199, 633.
Black, J.H. and Smith, P.L. 1979, this volume.
Bieniek, R.J. and Dalgarno, A. 1979, Ap. J. 228, 635.
Chaffee, F. 1979, this volume.   .
Chaffee, F.H., Lutz, B.L., Black, J.H., Vanden Bout, P.A., and Snell,
      R.L. 1979, Ap. J. (Letters) in press.
Chambaud, G., Launay, J.M., Levy, B., Millie, P., Roueff, E. and
      Tran Minh, F. 1979, this volume.
Crutcher, R.M. 1977, Ap. J. (Letters) 217, 109.
Crutcher, R.M. 1979, Ap. J. (Letters) 231, L151.
Dalgarno, A. and Black, J.H. 1976, Rep. Prog. Phys. 39, 573.
Dalgarno, A., de Jong, T., Oppenheimer, M. and Black, J.H. 1974,
      Ap. J. (Letters) 192, L37.
Dalgarno, A. and McCray, R.A. 1973, Ap. J. 181, 95.
Dalgarno, A. and Roberge, W.G. 1979, Ap. J. (Letters) 233, L25.
de Boer, K.A. 1979, Ap. J. 229, 132.
de Jong, T. 1979, Astron. Ap., submitted.
Elitzur, M. 1979, Ap. J. 229, 560.
Elitzur, M. and de Jong, T. 1978, Astron. Ap. 67, 323.
Elitzur, M. and Watson, W.D. 1978, Ap. J. (Letters), 222, L141.
Elitzur, M. and Watson, W.D. 1979, Ap. J., in press.
Fehsenfeld, F.C. 1979, this volume.
Hartquist, T.W., Black, J.H. and Dalgarno, A. 1978, Mon. Not. Roy.
      Astr. Soc. 185, 643.

Hartquist, T.W. and Dalgarno, A. 1980, Mon. Not. Roy. Astron. Soc.,
    in press.
Hartquist, T.W., Oppenheimer, M. and Dalgarno, A. 1980, Ap. J. in
    press.
Herbst, E., Schubert, J.G. and Certain, P.R. 1977, Ap. J. 213, 696.
Hobbs, L.M. 1979, Ap. J. (Letters) in press.
Hollenbach, D.J. and McKee, C.F. 1980, Ap. J. Suppl. in press.
Hunter, D.A. and Watson, W.D. 1978, Ap. J. 226, 477.
Iglesias, E.R. and Silk, J. 1978, Ap. J. 226, 851.
Jura, M. 1974, Ap. J. (Letters) 190, L33.
Jura, M. and York, D.G. 1978, Ap. J. 219, 861.
Knapp, G.R. and Jura, M. 1976, Ap. J. 209, 782.
Kirby, K. 1979, this volume.
Kirby, K., Roberge, W., Saxon, R.P. and Lin, B. 1979, Ap. J., in
    press.
Lang, K.R. and Wilsson, R.F. 1978, Ap. J. 224, 125.
Liszt, H.S. 1979, Ap. J. (Letters), 233, L147.
Lutz, B. and Souza, S.P. 1977, Ap. J. (Letters) 213, L129.
Mitchell, J.B.A. and Mcgowan, J.W. 1978, Ap. J. (Letters) 222, L77.
Morton, D.C. 1975, Ap. J. 197, 85.
Smith, W.H. and Snow, T.P. 1979, Ap. J. 228, 435.
Smith, P.L., Yoshino, K. and Parkinson, W.H. 1979, this volume.
Snow, T.P. 1978, Ap. J. (Letters) 202, L93.
Snow, T.P. 1976, Ap. J. 204, 759.
Snow, T.P. 1979, this volume.
Snow, T.P. and Meyers, K.A. 1979, Ap. J. 229, 545.
Watson, W.D. 1973, Ap. J. (Letters) 102, L69.
Watson, W.D., Christensen, R.B. and Deissler, R.J. 1978, Astron. Ap.
    69, 159.
Wright, E.L. and Morton, D.C. 1979, Ap. J. 227, 483.
York, D.G. and Rogerson, J.B. 1976, Ap. J. 203, 378.

DISCUSSION FOLLOWING DALGARNO

Glassgold: A long-standing problem in chemical modelling of diffuse
clouds is the fairly large uncertainty in the total gaseous abundances
of carbon and oxygen. How do you deal with this situation?

Dalgarno: The uncertainties in the depletion directly affect the
calculated molecular abundances.

Solomon: The division between "dense clouds" and diffuse clouds
may in some cases be arbitrary. The star which is observed optically or
in the UV may be on the edge of a very large dense molecular cloud. For
example Omicron Persei is on the edge of a molecular cloud extending for
at least 25 parsecs (see Barrett et al., this conference), so that what
appears as a "diffuse cloud" is in reality a small part of a dense
molecular cloud. The chemistry observed in the optical or UV may reflect
the recent past of the molecular cloud, and result in part from the
interaction of the OB star with the cloud.

Dalgarno: The cloud models have a high density core, and for ζ Oph and o Per the derived radiation field is substantially larger than the interstellar radiation field, suggesting that the clouds are close to, and probably physically associated with the stars.

Huntress: The radiative association rate constant for $C^+ + H_2$ is strongly dependent upon temperature. What temperature are you using for the clouds where you have used the value $5 \times 10^{-16} cm^3/s$ for this rate constant?

Dalgarno: In the core of one cloud, T=22K and in the cores of the other two T=45K. We used $5 \times 10^{-16} cm^3 s^{-1}$ for all three clouds.

D. Smith: You have stressed that your chemical models of diffuse clouds predict a lower concentration of CH^+ than is observed. We have recently shown in a laboratory study that the atom reactions $C_2H^+ + N$, $C_2H_2^+ + N$ and $C_2H^+ + O$ generate CH^+ ions (rate coefficients $\sim 10^{-10} cm^3 s^{-1}$ at 300K). Could these reactions significantly increase the CH^+ concentration and so narrow the gap between prediction and observation?

Dalgarno: The reactions are of considerable interest. Their inclusion will lessen but not remove the discrepancy.

Herbst: The reactions $CH^+ + H_2 \rightleftarrows CH_2^+ + H$ are nearly thermoneutral. The backward reaction should be studied in the laboratory as an additional CH^+ source.

Dalgarno: The reaction and its inverse are indeed critical to the CH^+ formation scheme.

Huntress: The new value for the heat of formation of CH^+ does considerably reduce the exothermicity of the $CH^+ + H_2$ reaction. It is possible therefore that this reaction may be temperature dependent so the rate constant at 20-40 K could be considerably smaller than the measured value at 300 K. Would a smaller rate constant for this loss process relieve the long-standing problem of the predicted CH^+ densities in diffuse clouds?

Dalgarno: Yes.

Elitzur: If the high H_2O abundance toward ζ Oph is correct, then the pre-shock density for the CH^+ calculation can be increased. H_2O represented the most severe constraint on the pre-shock density since it can be produced in abundance behind the shock.

Dalgarno: I agree.

de Jong: Could you elaborate on the remark you made that collisional destruction rates of molecules in interstellar shocks may be quite a bit smaller than has been assumed up to now?

Dalgarno: The laboratory values are appropriate to high densities. In interstellar shocks, radiative stabilization is a significant process which can drastically decrease the rate of collision-induced dissociation, depending on gas density and on the particular molecule. Preliminary estimates of collision-induced dissociation rates for H_2 and CO, which take radiative stabilization into account, are given by Dalgarno and Roberge (Ap. J. (Letters) **233**, L25, 1979).

FORMATION OF SIMPLE MOLECULES BY C^+ REACTIONS ON OXIDE GRAINS IN
DIFFUSE CLOUDS

W.W. Duley
Physics Department, York University, Toronto, Canada

ABSTRACT

The reaction between C^+ ions and OH^- ions on the surface of oxide
or silicate grains in diffuse clouds is shown to be a source of CH, OH,
CO, HCO and H_2CO molecules.

The analysis of elemental depletions in the ISM by Duley and
Millar (1978) has shown that selective depletion can arise through reactions of gas-phase ions with OH^- ions on the surface of oxide or silicate
grains. The large ionization potential of C together with the fact that
most C is present as C^+ in diffuse clouds ensures that C is <u>not</u> selectively depleted by this mechanism. In the notation of Duley, Millar and
Williams (1978) the reaction between C^+ and surface OH^- can be written

$$C^+ + {}^V OH^-)_s^- \rightarrow (HCO)^*$$

or $$\rightarrow^H (H_2CO)^*$$

where ${}^V OH^-)_s^-$ is a surface OH^- ion adjacent to a cation vacancy. The
complexes $(HCO)^*$ or $(H_2CO)^*$ are highly excited on formation and are
assumed to leave the grain either as a whole or in part. Millar et al.
(1979) have calculated the rate of H_2CO formation via this process and
conclude that it can account for the observed H_2CO abundance in diffuse
clouds (Davies and Matthews, 1972).

The formation rate of HCO or H_2CO on grains is

$$F = n_C v_C s_g n_g \qquad cm^{-3} sec^{-1}$$
$$F = 1.9 \times 10^{-20} n_H^2 \qquad cm^{-3} sec^{-1}$$

for oxide grains with $n_g = 4 \times 10^{-10} n_H$ and 10nm radius. n_H = the hydrogen
space density, n_C = the carbon density, v_C = the thermal velocity of C^+
ions (100 K assumed), s_g = the geometric cross-section of a grain, and

B. H. Andrew (ed.), Interstellar Molecules, 281–282.

n_g = the space density of grains. For a typical diffuse cloud with n_H = 100 cm^{-3} and A_v = 1 the observed density of H_2CO can be shown to require a formation rate F_{H_2CO} = 5 x 10^{-17} cm^{-3}sec^{-1} (Millar et al. 1979). The efficiency of H_2CO formation is then ε_{H_2CO} = 0.26 assuming n(H_2CO)/n_H= 1.5 x 10^{-9}.

The complexes H_2CO^* and HCO^* may also dissociate. Products of the dissociation of HCO^* would be H+CO, O+CH, or C+OH. It is of interest to calculate ε for each of the diatomic molecules formed in this dissociation, as grain reactions may generate appreciable quantities of these species. Using relative molecular abundances for CH, CO and OH as given by Snow (1977) for ζ Per, one obtains, where L_X is the molecular loss or destruction rate,

X	n_X/n_H	L_X(sec^{-1})	F_X(cm^{-3}sec^{-1})	ε_X
H_2CO	1.5 x 10^{-9}	3.5 x 10^{-10}	5 x 10^{-17}	0.26
CH	10^{-8}	10^{-11}	10^{-17}	0.05
CO	5 x 10^{-7}	10^{-12}	5 x 10^{-17}	0.25
OH	5 x 10^{-8}	2 x 10^{-11}	10^{-16}	0.5

We conclude that since ε < 1 for each of these molecules, C^+ -grain reactions occur at a sufficient rate to account for observed abundances. Since N is neutral in diffuse clouds, while C is almost wholly ionized, an analogous reaction scheme is not available for N. Thus NH should be reduced in abundance relative to CH and OH. The low abundance of NH in diffuse clouds has previously been ascribed to the dominance of gas-phase reactions (Crutcher and Watson, 1976).

REFERENCES

Crutcher, R.M., and Watson, W.D.: 1976, Ap. J. 209, 778.
Davies, R.D., and Matthews, H.E.: 1972, M.N.R.A.S. 156, 253.
Duley, W.W., and Millar, T.J.: 1978, Ap. J. 220, 124.
Duley, W.W., Millar, T.J., and Williams, D.A.: 1978, M.N.R.A.S. 185, 915.
Millar, T.J., Duley, W.W., and Williams, D.A.: 1979, M.N.R.A.S. 186, 685.
Snow, T.P.: 1977, Ap. J. 216, 724.

THE PHOTODISSOCIATION OF INTERSTELLAR CH$^+$

K. Kirby
Harvard-Smithsonian Center for Astrophysics

Abstract. Three new excited states of CH$^+$ which can be reached by a dipole transition from the ground state with photon energies less than 13.6 eV have been obtained from accurate theoretical calculations. As these states are unbound with respect to nuclear motion, photon absorption into them results in dissociation of the molecule. Photodissociation cross sections have been calculated, and the CH$^+$ photodissociation rates in the interstellar radiation field as a function of optical depth have been determined. The rates are approximately 500 times larger than previously assumed. We have compared these new photodissociation rates with the rates of other CH$^+$ destruction mechanisms in three different models of the interstellar medium. Photodissociation appears to be a significant destruction mechanism in any model requiring large interstellar radiation fields.

1. DESCRIPTION OF THEORETICAL CALCULATIONS

Using accurate configuration interaction methods Kirby, Saxon and Liu (1979a) have calculated energies and wavefunctions for seven previously undetermined states of CH$^+$ lying within ~17 eV of the ground state. These excited states of $^1\Sigma^+$ and $^1\Pi$ symmetry which can be reached by a dipole transition from the ground $X^1\Sigma^+$ state are shown in Figure 1. The $X^1\Sigma^+$ and $A^1\Pi$ potential curves are known from previous theoretical work (Green et. al. 1972). The new states, labelled by numbers in Figure 1, are unbound with respect to nuclear motion, and electronic transitions to them will result in photodissociation of the CH$^+$. Dipole transition moments between the ground and excited electronic states have been computed as a function of internuclear separation (Kirby et al. 1979a) and used in the calculation of the photodissociation cross sections (Kirby et al. 1979b).

With photon energies $\leq$ 13.6 eV, dissociation from the ground vibrational state, v"=0 of the $X^1\Sigma^+$, takes place through the $2^1\Sigma^+$, $3^1\Sigma^+$ and $2^1\Pi$ states, producing C(^{1}D)+H$^+$ and C(^{1}S)+H$^+$ with maximum cross sections of 6x10^{-19} cm^2, 3x10^{-17}cm^2 and 1.3x10^{-17}cm^2, respectively.

B. H. Andrew (ed.), Interstellar Molecules, 283–286.

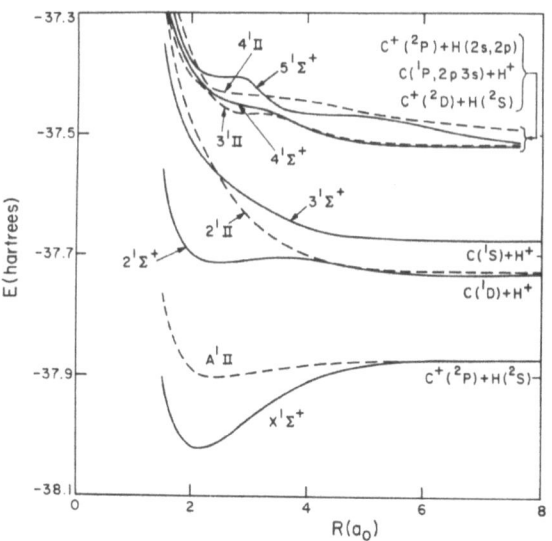

Figure 1.   Potential curves of $^1\Sigma^+$ and $^1\Pi$ states of CH^+.

Previous estimates of the photodissociation rate of CH^+ were based on transitions from $v''=0$ to the $A^1\Pi$ state, producing $C^+(^2P)+H$.   These new cross sections are more than two orders of magnitude larger than the X-A transition cross sections, and the possible ramifications for formation and destruction of CH^+ in the interstellar medium are examined in the next section.

2.   PHOTODISSOCIATION RATES OF CH^+ IN INTERSTELLAR CLOUD MODELS

Photodissociation rates for transitions from $v''=0$ of the $X^1\Sigma^+$ state to the four CH^+ states lying below the Lyman limit have been calculated as a function of visual optical depth, τ_v, in an inter-stellar cloud (Kirby et.al. 1979b).   The interstellar radiation field used was that of Draine (1978) for wavelengths less than 2300 Å, and that of Witt and Johnson (1973) for wavelengths greater than 2300 Å. At the surface of a cloud where $\tau_v=0$, the overall photodissociation rate for CH^+ is 2×10^{-10} s^{-1}.

In order to examine the effect of the large photodissociation rate on the CH^+ chemistry in diffuse interstellar clouds, we have compared the total rate for the photodissociation process:

$$CH^+ + h\nu \rightarrow C + H^+ \qquad\qquad k_1 = 2\times10^{-10}s^{-1} \text{ for } \tau_v=0 \qquad\qquad (1)$$

with the rates of other destruction processes:

Table 1. Parameters and Destruction Rates for CH^+ in the following models: S-W (Stecher and Williams 1974); B-D (Black and Dalgarno 1977); E-W (Elitzur and Watson 1978)

	I	T °K	n(e)	n(H)	n(H$_2$)	Ik$_1$	n(e)k$_2$	n(H$_2$)k$_3$	n(H)k$_4$
	Model Parameters					Destruction Rates			
S-W	1000*		3	100	100	2×10^{-7}	1×10^{-7}	1×10^{-7}	-*
B-D	2.5	110	0.06	300	100	$\sim 1 \times 10^{-11}$	1×10^{-8}	1×10^{-7}	8×10^{-10}
		22	0.25	180	1160	(τ=1)	9×10^{-8}	1×10^{-6}	6×10^{-11}
E-W	10*	4500	$6 \times 10^{-3*}$	15	0.6	2×10^{-9}	1×10^{-10}	6×10^{-10}	4×10^{-9}

$$CH^+ + e \rightarrow C + H \qquad\qquad k_2 = 10^{-7}(300/T)^{1/2} \, cm^3 s^{-1} \qquad (2)$$

$$CH^+ + H_2 \rightarrow CH_2^+ + H \qquad\qquad k_3 = 1 \times 10^{-9} \, cm^3 s^{-1} \qquad (3)$$

$$CH^+ + H \rightarrow C^+ + H_2 \qquad\qquad k_4 = 7.5 \times 10^{-15} T^{5/4} c^{-3} s^{-1} \qquad (4)$$

The values for the rate constants k_2, k_3 and k_4 have been taken from Dalgarno (1976). These rates vary, depending on the electron density, the gas temperature, the molecular and atomic hydrogen densities, and the strength of the interstellar radiation field.

We have examined three model interstellar situations, two of which have been constructed specifically to explain the CH^+ formation and abundance. The first five rows of Table 1 list the parameters characterizing the models which have either been given explicitly by the authors, or which we have deduced (in the latter case, these quantities are starred). The parameter I is the scaling factor multiplying the average interstellar radiation field of Witt and Johnson (1973).

The model of Stecher and Williams (1974) elaborates on the ideas of Bates and Spitzer (1951) proposing evaporation of CH_4 from grain mantles, its subsequent photodissociation and ultimate photoionization of CH to give CH^+. Strong radiation fields generated by the close proximity of a hot star are needed to maintain an ionization rate for CH of $10^{-7}s^{-1}$. The two component model of Black and Dalgarno (1977) was constructed for the ζ Ophiuchi cloud and seriously underestimates the abundance of CH^+. In Table I the values applicable to the warm, diffuse outer zone always appear above those applicable to the cooler, denser inner zone. Elitzur and Watson (1978) propose formation of CH^+ through the endothermic reaction of $C^+ + H_2 \rightleftarrows CH^+ + H$ in the hot gas immediately behind shock fronts.

The destruction rates of CH^+ in s^{-1} are given in the second half of Table 1 for the various models. The photodissociation rates appear in the row labeled Ik$_1$, and the rates for reactions (2), (3), and (4) appear in successive rows. In the Black and Dalgarno (1979) model the primary destruction mechanism for CH^+ is reaction with molecular hydrogen, and photodissociation is comparatively insignificant. In the models of Stecher and Williams (1974) and of Elitzur and Watson (1978) which have enhanced radiation fields, however, the photodissociation rates are comparable to the rates of the other removal mechanisms. Recently de Jong (1979) has suggested that the shocks in which the CH^+ is formed are located at the edges of interstellar clouds (E-W parameters in Table 1 but I = 1). In that case the general interstellar radiation field applies so that photodissociation of CH^+ becomes about one order of magnitude less efficient than destruction by collisions with H atoms.

REFERENCES

Bates, D.R. and Spitzer, L.: 1951, Ap. J. 113, pp. 441-463.
Black, J.H. and Dalgarno, A.: 1979, Ap. J. Suppl. 34, pp. 405-423.
Dalgarno, A.: 1976, *Atomic Processes and Applications*, ed. Burke, P.G. and Moiseiwitsch, B.L. (North Holland: Amsterdam) pp. 109-132.
de Jong, T.: 1979, submitted to Astron. and Astrophys.
Draine, B.T.: 1978, Ap. J. Suppl. 36, pp. 595-619.
Elitzur, M. and Watson, W.D.: 1978, Ap. J. 222, pp. L141-L144.
Green, S., Bagus, P.S., Liu, B., McLean, A.D. and Yoshimine, M.: 1972, Phys. Rev. A 5, pp. 1614-1618.
Kirby, K., Saxon, R.P. and Liu, B.: 1979a, J. Chem. Phys. to be published.
Kirby, K., Saxon, R.P., and Liu, B.: 1979b, Ap. J. 231, pp. 637-641.
Stecher, T.P., and Williams, D.A.: 1974, M.N.R.A.S. 168, pp. 23P-26P.
Uzer, T. and Dalgarno, A.: 1978, Chem. Phys. 32, pp. 301-303.

DISCUSSION FOLLOWING KIRBY

Elitzur: Our model has an extinction of .5 which will further decrease the importance of the CH^+ photodissociation.

Kirby: Nowhere in your published model (Elitzur and Watson, 1978) is there any indication of an extinction of 0.5. However, such an extinction would certainly decrease the relative importance of CH^+ photodissociation as a destruction mechanism.

CHARGE EXCHANGE AND FINE STRUCTURE EXCITATION IN O-H$^+$ COLLISIONS

G. Chambaud[*], J.M. Launay[+], B. Lévy[*], P. Millié[*], E. Roueff[+]
and F. Tran Minh[*]
[*]Groupe de Chimie Quantique ENSJF, 1 rue Maurice Arnoux,
   92120 Montrouge, France; ERA 470 du CNRS
[+]Département d'Astrophysique Fondamentale, Observatoire de
   Meudon, 92190 Meudon, France; GR 024 du CNRS

INTRODUCTION

The charge – transfer reaction between protons and oxygen atoms is critical to the chemistry of the oxygen family; the corresponding rate was evaluated by Field and Steigman (1971) on the assumption of orbiting collisions and statistical probability distribution among the levels. We re-examine this reaction, including the fine-structure excitation process, basing our analysis on a careful description of the different potential curves arising from the O-H$^+$ and O$^+$-H systems and on the evaluation of the coupling responsible for the transitions.

INTERATOMIC POTENTIALS

The charge-transfer reaction is almost resonant; the energy defects between the different exit channels are displayed in Table 1 (below):

Table 1

States	Energy (K)
O ($^3P_2$) + H$^+$	0
O$^+$($^4S_{3/2}$) + H	227
O ($^3P_1$) + H$^+$	228
O ($^3P_0$) + H$^+$	326

Such an accuracy in the relative position of potential energy curves is not available from even the most elaborate ab-initio calculations. In order to avoid this difficulty, we chose to use the experimental values of the energies of the separated atoms limit, use the classical asymptotic behaviours, and join them to the results given by Liu and Verhaegen (1971) at short internuclear distances, since these authors were able to obtain the correct asymptotic energies by an estimation of the correlation energies. For the O-H$^+$ system, when neglecting fine-structure effects, we have two potential curves X $^3\Sigma^-$ and A $^3\Pi$. The asymptotic behaviour is due to a first order charge-quadrupole term varying as

287

B. H. Andrew (ed.), Interstellar Molecules, 287–288.

$1/R^3$ and a second order charge-dipole interaction varying as $1/R^4$. For the O^+-H system, two molecular potentials are involved, $^5\Sigma^-$ and $^3\Sigma^-$. For the $^5\Sigma^-$ curve, the long range interaction is due to the attractive charge-dipole interaction in $1/R^4$ and an additive repulsive term evaluated with valence bond functions and expressed in an exponential form. The $^3\Sigma^-$ potential was determined through the energy difference between the $^5\Sigma^-$ and $^3\Sigma^-$ potentials, which is given in terms of exchange integrals numerically determined for large internuclear distances ($R > 6$ a.u.).

THE COLLISION TREATMENT

At the low energies of the interstellar medium, a full quantum mechanical solution of the scattering problem is required to achieve a satisfactory treatment.

We performed close-coupling calculations, expanding the total wave function in the symmetric-top wave functions, using the formalism described by Mies (1973) for fine-structure transitions. When the charge-exchange channel is taken into account, the couplings between different electronic configurations must also be considered. We neglected the rotational coupling and the fine-structure coupling between different electronic configurations. We kept the fine-structure coupling inside the O (^3P)-H^+ configuration, and by numerical calculations involving bielectronic integrals we determined the electronic coupling between the two $^3\Sigma^-$ states that is essentially responsible for the charge-exchange reaction.

RESULTS

We determined the cross sections for different energies and performed a Maxwell average over the velocities in order to obtain the corresponding rate-constants. The results are given in Table 2 for various temperatures; they are noticeably different from the previous estimates of Field and Steigman (1971). Indeed the orbiting approximation is questionable for this system because of the large energy defect which must be surmounted in inelastic collisions and because of the quadrupolar coupling between fine-structure levels at large distances.

Table 2 - Rate constants in 10^{-9} cm^{3}s^{-1} (k)

	T(K)30	T(K)50	T(K)70	T(K)100	T(K)1000
$k_2 \to 3/2$	$6.30\ 10^{-5}$	$1.38\ 10^{-3}$	$5.31\ 10^{-3}$	$1.52\ 10^{-2}$	$4.77\ 10^{-1}$
$k_2 \to 1$	$9.10\ 10^{-5}$	$2.02\ 10^{-3}$	$8.07\ 10^{-3}$	$2.42\ 10^{-2}$	$9.42\ 10^{-1}$
$k_2 \to 0$	$1.44\ 10^{-7}$	$1.69\ 10^{-5}$	$1.51\ 10^{-4}$	$8.82\ 10^{-4}$	$1.54\ 10^{-1}$

REFERENCES

Field, G.B., and Steigman, G.: 1971, Astrophys. J. 166, 59.
Liu, H.P.D., and Verhaegen, G.: 1971, Intern. J. Quantum Chemistry 5, 103.
Mies, F.H.: 1973, Phys. Rev. A 7, 942.

MOLECULAR SYNTHESIS IN INTERSTELLAR CLOUDS:
THE RADIATIVE ASSOCIATION REACTION H + OH → H_2O + hν

D. Field[†], N.G. Adams[*] and D. Smith[*]
[†] School of Chemistry, University of Bristol, Bristol BS8 1TS U.K.

[*] Dept. of Space Research, University of Birmingham,
Birmingham B15 2TT, U.K.

On the evidence of some recent laboratory data for the association reaction H + OH + He → H_2O + He, and following previously proposed ion-neutral association reaction schemes, we tentatively suggest that the radiative association reaction of H and OH can be a significant mechanism for the production of H_2O in interstellar clouds, when compared with the ion-neutral reaction sequence. The 'H-atom + radical' association process may be of some general importance in the production of other small molecules in interstellar clouds.

INTRODUCTION

    Radiative association is an important, perhaps crucial, first step in gas-phase molecular synthesis. Rate coefficients are immeasurably small at laboratory-attainable temperatures, and methods have been developed for their estimation from third order rate data. It has been suggested that ion-molecule radiative association rate coefficients can be very large at interstellar temperatures[1]. Estimates have shown that rate coefficients for radiative association are generally very much greater than are predicted by statistical theories of unimolecular kinetics[1,2]. It would seem pertinent then to re-ask the question: is radiative association involving neutral radical species a significant mechanism for molecular synthesis in interstellar clouds?

H_2O FORMATION

    In order to examine the question posed above we have chosen to investigate the reaction H + OH → H_2O + hν (k_{rad}). Data exist on the third order association reaction H + OH + He → H_2O + He (k^{3rd}) as a function of temperature between 300K and 230K[3]. The kinetic scheme for this reaction is given by

$$H + OH → H_2O^* \quad (k_1)$$
$$H_2O^* → H + OH \quad (k_{-1})$$
$$H_2O^* + He → H_2O + He \quad (k_2)$$
$$H_2O^* → H_2O + hν \quad (k_3 = 1000 \text{ s}^{-1})$$

289

B. H. Andrew (ed.), Interstellar Molecules, 289–290.
Copyright © 1980 by the IAU.

from which one may deduce

$$k_{rad} = (k_3 k^{3rd}/k_2)$$

and we have derived values for this coefficient down to interstellar
temperatures. Values range from 4(-18) at 300K, 7(-15) at 50K, to
8(-14) at 20K. Lifetimes of H_2O^* ($1/k_{-1}$) that these coefficients
entail have also been calculated, and vary from 0.01 ns to 20 ns to 225
ns over the same temperature range. The temperature variation calculated for the vibrational deactivation of highly excited CO by He [4] has
been employed in estimating the temperature variation of k_2. We have
adopted the working hypothesis that data on the third order association
reaction may be extrapolated to interstellar temperatures from 230K,
and therein lies a major uncertainty. High temperature data do offer
some support to our extrapolation[3] since they indicate that k_{-1} becomes
an increasingly strong function of temperature as temperature drops.
Theoretical considerations would also suggest that at low temperature
shape- and Feshbach- resonances may be dominant in determining lifetimes of such intermediates as H_2O^*, tending to invalidate a basic
approximation of statistical unimolecular theories.

IMPLICATIONS FOR H_2O FORMATION IN INTERSTELLAR CLOUDS

Let us consider the cloud ζ Oph as a representative example. For
species concentrations that hold there[5], it may be deduced that the
rate of production of H_2O by radiative association at 50K is $\sim 1 \times 10^{-16}$
molecules $cm^{-3}s^{-1}$. Ion-molecule schemes suggest a rate roughly 6
times as great, indicating that the neutral-neutral path is at least
significant and might become dominant at lower temperatures or in
regions of low ionisation density.

We should emphasise that our derived values of k_{rad} are conservative and that the lifetimes of H_2O^* fit well with those that emerge
from ion-molecule data. In comparing the importance of neutral mechanisms with ionic the specific environment will in general be decisive.
Other association reactions of interest are $NH_2(H,NH_3)h\nu$ (isoelectronic
with H+OH), $CH(H,CH_2)h\nu$, $CH_2(H,CH_3)h\nu$, $CH_3(H,CH_4)h\nu$, and $NO(H,HNO)h\nu$
amongst others. Further experimental and theoretical work is essential
before radiative association becomes a firmly established mechanism
for neutral systems. It would appear that the mechanism could be a
viable one for many interstellar clouds.

REFERENCES

1. Smith D. and Adams N.G.: 1978, Ap. J. 220, L87.
2. Herbst E.: 1976, Ap. J. 205, 94.
3. Zellner, R., Erler, K., and Field, D.: 1976, Proc. XIV Int. Symp.
      Comb. M.I.T. Cambridge, Mass. pp. 939-948.
4. Verter M.R. and Rabitz M.: 1976, J. Chem. Phys. 64, 2942.
5. Black J.H. and Dalgarno A.: 1977, Ap. J. Suppl. 34, 405.

EXPERIMENTAL MEASUREMENTS OF ION-MOLECULE REACTIONS

F.C. Fehsenfeld
NOAA/ERL, Aeronomy Laboratory, Boulder, Colorado 80303

ABSTRACT

The applicability of ion-molecule reaction rate constants measured at room temperature to simulation of interstellar cloud chemistry is discussed. Three-body association-rate constants of C^+ and NH_3^+ have been measured between 100 K and 300 K. These results give information about the same associations by radiative processes. Possible implications for interstellar molecule production by radiative association and free radical reaction are discussed.

Observations of interstellar clouds made during the past ten years have revealed a variety of complicated molecules (Townes, 1977). It is now generally accepted that positive ion-neutral reactions play an important role in the synthesis of many of these molecules (Black and Dalgarno, 1973; Herbst and Klemperer, 1973; Watson, 1974; Prasad and Huntress, 1979a). Thus far, several hundred positive ion-molecule reactions of astrochemical interest have been measured (Huntress, 1977; Albritton, 1978; Sieck, 1979). Most studies have been carried out at 300 K, a temperature almost an order of magnitude above the temperature characteristic of dense clouds. However, evidence indicates that rate constants for reactions which occur in a large fraction ($\gtrsim \cdot 1$) of the ion-neutral collisions have little if any energy dependence. This encompasses a large majority of the exothermic reactions studied. Although exceptional at low energy, there are also slow exothermic ion-molecule reactions with rate constant less than 10% of the ion-neutral collision rate constant. The reaction

$$S^+ + O_2 \rightarrow SO^+ + O \tag{1}$$

is an example (Dotan, et al., 1979). Reaction (1) has a rate constant, $k_1 = 1.6 \times 10^{-11}$ cm^3s^{-1} at 300 K and is inversely proportional to energy for center-of-mass kinetic energy less than 0.15 eV, a behavior characteristic of many slow reactions at low energy that is attributed to the increasing lifetime of the ion-neutral collision complex with

B. H. Andrew (ed.), Interstellar Molecules, 291–296.
Copyright © 1980 by the IAU.

decreasing energy. For this reason, at interstellar cloud temperatures reaction (1) is expected to have a rate constant several times that at 300 K. Another very slow reaction at low energy is

$$NH_3^+ + H_2 \rightarrow NH_4^+ + H. \qquad (2)$$

The rate constant for this reaction is linear in an Arrhenius plot with $k_2 = 1.7 \times 10^{-11} \exp(-0.09/kT)$ (Fehsenfeld et al., 1975) with $k_2 = 10^{-13}$ cm^{3}s^{-1} at 100 K.

Ion-molecule radiative association in interstellar clouds is potentially a powerful mechanism by which atoms and simple molecules can be combined to form more complex molecules. Theory and indirect experimental evidence suggest these reactions are important. The present understanding of the mechanism is exemplified by the reaction

$$C^+ + H_2 \rightarrow CH_2^+ + h\nu. \qquad (3)$$

Reaction (3) was first proposed by Black and Dalgarno (1973). The rate constant for the reaction has been calculated by Herbst (1976, 77, 79). The reaction begins with the collision of C^+ with H_2 to form an excited complex, CH_2^{+*}, which is unstable against predissociation into the original C^+ and H_2 reactants. The lifetime of the collision complex is determined by the energy which the C^+ and H_2 carry into the complex, the disposition of this energy into the internal modes of the CH_2^{+*}, and the bond dissociation energy of the CH_2^+ states which are formed. The CH_2^{+*} is composed of two electronic states of the molecular ion, the $^2A_1$ and $^2B_1$ states in C_{2v} symmetry. The states are assumed to be completely mixed. The complex is stabilized by the allowed radiative transition from the $^2B_1$ state to the $^2A_1$ state. In the calculation the rate constant for the formation of the excited complex and the lifetime of the complex against predissociation are estimated from the measured rate constant for the analogous three body association (Herbst, 1979)

$$C^+ + H_2 + He \rightarrow CH_2^+ + He. \qquad (4)$$

The data for this reaction are fit by the expression $k_4 = 5.1 \times 10^{-30} (300/T)^{1.3}$. From this data several conclusions concerning reaction (3) are drawn. The three body association rate increases smoothly down to the lowest temperature which implies that there are no energy barriers preventing the formation of CH_2^{+*}. At a pressure of 1.3 torr and temperature of 100 K, the effective binary rate constant for the reaction is 4×10^{-12} cm^{3}s^{-1} and shows no indication of saturation which implies a large fraction of the C^+ collision with H_2 yields CH_2^{+*}. Finally the rate constant of the three body association at 100 K is 2×10^{-29} cm^{6}s^{-1} which implies that the lifetime of CH_2^{+*} is $\gtrsim 10^{-10}$s. These results lend support for the radiative association mechanism.

Another radiative association reaction

$$CH_3^+ + H_2 \rightarrow CH_5^+ + h\nu \qquad (5)$$

is suggested as a key reaction in the formation of CH_4 and ultimately C_2H_2 and C_2H_4 (Prasad and Huntress, 1979b). The analogous three body association

$$CH_3^+ + H_2 + He \rightarrow CH_5^+ + He \tag{6}$$

is fast. Smith and Adams (1979) find $k_6 = 6.3 \times 10^{-27} cm^6 s^{-1}$ at 110 K. This supports the assumption that reaction (5) is fast. Prasad and Huntress (1979b) have also suggested that the radiative association

$$NH_3^+ + H_2 \rightarrow NH_5^+ + h\nu \tag{7}$$

can overcome the bottleneck in the formation of NH_3 caused by the slow rate constants for reaction (2). In this case the analogous three body association

$$NH_3^+ + H_2 + He \rightarrow NH_4^+ + H + He \tag{8}$$

$$\rightarrow NH_5^+ + He$$

is immeasurably slow, $k_8 < 3 \times 10^{-30} cm^6 s^{-1}$, in the variable temperature flowing afterglow apparatus at 100 K and indicates that reaction (7) is also slow.

It seems reasonable to use measured three body association rates at 100 K and the energetics of the association reaction to indicate a crude upper bound on the radiative association rate constant (Herbst, 1979). The limits indicate that radiative association will play an important role in simple molecule (n $\stackrel{<}{\sim}$ 20) building. The largest uncertainty in these estimates is the transition probability for radiative stabilization of the ion-neutral association complex. In the discussion of radiative association, it should be noted that although the radiative association reactions have not been directly observed for molecules of astrophysical interest, the direct attachment of substituted benzenes to Li^+ has been observed at 300 K at low pressure in the ICR and attributed to radiative association (Woodin and Beauchamp, 1978).

Another challenge presented by the interstellar medium is the relatively large concentration of atomic and molecular free radicals. At low cloud temperatures these neutral radicals can be expected to react rapidly with other neutral radicals as well as ions. Experimental difficulty in dealing with these extremely reactive species has limited measurements. Atomic hydrogen reactions have been measured in both ICR (Karpas, et al., 1979) and flowing afterglow reactors (Fehsenfeld and Ferguson, 1971). The study of Karpas et al. (1979) indicates that H_2^+, CO^+, N_2^+ and HCN^+ react with atomic hydrogen while hydrocarbon ions of the type CH_n^+ and $C_2H_n^+$, n = 2, 3 and 4, do not. Atomic oxygen and atomic nitrogen reactions have been studied at 300 K in flowing afterglow and selected ion flow tube reactions (Fehsenfeld and Ferguson, 1972; Fehsenfeld, 1976; Viggiano et al., 1979). Rate constants have been determined for the reactions of H^+, D^+, H_3^+, CH^+, CH_3^+, CH_5^+, C_2^+, C_2H^+,

$C_2H_2^+$, and CO^+ with atomic oxygen and CH^+, CH_2^+, CH_3^+, H_2O^+, C_2^+, $C_2H_2^+$, and O_2^+ with atomic nitrogen. The rate constants obtained for the N and O reactions are generally a substantial fraction of the collision limiting rate constant and characteristically produce C-N and C-O bonded products. The O and N reactions are important steps in the synthesis of molecules observed in interstellar clouds.

REFERENCES

Albritton, D.L.: 1978, Atomic Data Nuclear Data Tables 22, 1-101.
Black, J.H., and Dalgarno, A.: 1973, Astrophys. Lett. 15, 79.
Dotan, I., Fehsenfeld, F.C., and Albritton, D.L.: 1979, J. Chem. Phys. 71, 2728.
Fehsenfeld, F.C., and Ferguson, E.E.: 1971, J. Geophys. Res. 76, 8453.
Fehsenfeld, F.C., and Ferguson, E.E.: 1972, J. Chem. Phys. 56, 3077.
Fehsenfeld, F.C., Lindinger W., Schmeltekopf, A.L., Albritton, D.L., and Ferguson, E.E.: 1975, J. Chem. Phys. 62, 2001.
Fehsenfeld, F.C.: 1976, Astrophys. J. 209, 638.
Herbst, E.: 1976, Astrophys. J. 205, 94.
Herbst, E.: 1977, Astrophys. J. 213, 696.
Herbst, E.: 1979, Astro. and Space Sci. 65, 13.
Herbst, E., and Klemperer, W.: 1973, Astrophys. J. 185, 505.
Huntress, W.T.: 1977, Astrophys. J. 33, 495.
Karpas, Z., Anicich, V., and Huntress, W.T.: 1979, J. Chem. Phys. 70, 2877.
Prasad, S.S. and Huntress, W.T.: 1979a, Astrophys. J. Suppl. (accepted).
Prasad, S.S., and Huntress, W.T.: 1979b, Astrophys. J. (accepted).
Sieck, L.W.: 1979, SNRDS-NBS 64 (Washington, U.S. Government Printing Office).
Smith, D., and Adams, N.G.: 1979, Gas Phase Ion Chemistry, Vol. I, ed. M.T. Bowers, (Academic Press, New York) pp 1-44.
Townes, C.H.: 1977, Observatory, 97, 52.
Viggiano, A.A., Howorka, F., Albritton, D.L., Fehsenfeld, F.C., Adams, N.G., and Smith, D.: 1979, Astrophys. J., (to be submitted).
Watson, W.P.: 1974, Astrophys. J. 188, 35.
Woodin, R.L., and Beauchamp, J.L.: 1978, J. Amer. Chem. Soc. 78, 502.

DISCUSSION FOLLOWING FEHSENFELD

Glassgold: Two long standing questions concerning gas-phase ion molecules have been (a) the reaction of H_3^+ with atomic nitrogen, and (b) charge exchange of molecular ions with neutral metal atoms. Could you comment on the status of measurements of these reactions.

Fehsenfeld: We have not measured the H_3^+ reaction with atomic nitrogen. Charge exchanges between molecular ions and neutral metal atoms are generally rapid. I would expect them to be rapid at interstellar cloud temperatures.

Langer: Can the reaction $H_3^+ + N \rightarrow NH_2^+ + H$ take place rapidly? I believe Huntress has indicated that it may be slow.

Fehsenfeld: We have not studied this reaction, so I do not know.

Huntress: We have made some attempts to measure the rate constant and product distribution for the reaction of H_3^+ ions with N atoms. Preliminary data indicates that the reaction may proceed at a rate on the order of a few times $10^{-10} cm^3/s$, and that the product may indeed be NH_2^+. The present state of the experiment is uncertain, however. The NH^+ ion cannot be produced from this reaction since this channel is endothermic.

Bar-Nun: Could you give us your estimate of the certainty of these complex reaction schemes. The recent findings of strong lightning activity by the Voyageur and Pioneer - Venus space probes has demonstrated the failure of chemical modelling of the atmospheres of Venus and Jupiter.

Fehsenfeld: The only estimate of the certainty of a chemical model is its ability to explain the observations. It is too early to make such a judgement concerning the current models of the gas-phase chemistry of the interstellar clouds.

Allamandola: You notice that the rate constant as the temperature drops from room temperature is higher than that which you would get from an extrapolation of your data. The effect is due to the lower kinetic energy the reacting partners must dissipate, and you suggest that this results in a longer lived transition state which ultimately relaxes into reactants. Do you see a rate constant decrease as you get to much lower temperature due to the dominance of activation energy and the fact that the kinetic energy is lower than the required activation energy?

Fehsenfeld: We have not observed the effect you describe on the ion reactions we have studied, but our measurements do not extend below 100 K.

Allamandola: In our experiment we store atoms, for example oxygen, in solid argon. We find that a very slight rise in temperature (½ K is enough) is sufficient to cause chemiluminescence from O_2, but when the temperature has again dropped the reaction is completely stopped. We believe the effect is due to atoms which are neighbouring, but unable to react because there is insufficient energy to provide the activation energy. The mild warming provides the required energy. The reactions then take place freely until 50-60 K, at which temperature atom mobility is so great that all atoms have apparently encountered a partner or left the system via the gas-phase.

Herbst: Not all polyatomic ion-molecule association reactions will be rapid at $T \leqslant 50$ K. The reasons for this assertion are (1) activation barriers to tightly-held complexes, and (2) the need for significant (>1 eV) bond energies of cluster ions.

Thaddeus: It seems to me obvious that there are pitfalls associated with symmetry if you try to deduce two body radiative association rates from three body rates. You clearly cannot say anything about $H+H \nrightarrow H_2+h\nu$ from $H+H+He \rightarrow H_2+He$.

Fehsenfeld: I agree. The correspondence between three body association and the analogous radiative association, at present, is useful only in establishing an upper limit on the radiative association.

Field: There is some reasonably good experimental evidence to suggest that interaction of *neutral* radicals with H-atoms (e.g. H+OH, H+NH$_2$, H+CH etc.) may involve radiative association proceeding at a significant rate. Such interactions occur on strongly attractive potential energy surfaces (with no energy barriers) and may well be dominated by resonant scattering phenomena.

P. Smith: There are several discrepancies between observations and the predictions of models of chlorine species in diffuse clouds. Black and myself, in a poster paper presented at this symposium, have suggested that these discrepancies can be explained if the reaction $C\ell^+ + H_2 \rightarrow HC\ell^+ + H$ is slow at cloud temperatures. Such a slow rate could be due to the presence of a modest energy barrier in the reaction, which is exothermic by only 0.22 eV. You have stated that the rates for rapid reactions tend to be independent of temperature. Do you have any comment on our conjecture?

Fehsenfeld: There may certainly be exceptions to the generalization that the rate constants of fast reactions are independent of energy. Your suggestion is very interesting.

INTERSTELLAR SULFUR CHEMISTRY

Sheo S. Prasad and Wesley T. Huntress, Jr.
Jet Propulsion Laboratory, California Institute of Technology
4800 Oak Grove Drive, Pasadena, CA 91103, USA

ABSTRACT: This paper summarizes results of a chemical model of SO, CS, and OCS chemistry in dense clouds.

The following results were obtained from a theoretical study of sulfur chemistry in dense interstellar clouds using a large-scale time dependent model[1] of gas-phase chemistry. In the results which follow, $f_x = n(x)/n(H_2) \approx N(X)/N(H_2)$ where $N(X) \equiv$ column abundance of X. Also, $q(x)$ and $L(x)$ will denote net production and loss rates of X.

(a) For the large values $f_{O_2} \approx 5 \times 10^{-5}$ predicted by contemporary models[2,3], the reaction[4] $S + O_2 \rightarrow SO + O$ leads to a large[2] value $f_{SO} = 2 \times 10^{-6}$. Observations[5] indicate $f_{SO} \approx (2\text{-}10) \times 10^{-8}$ in L134 and TMC-1. Consequently, f_{O_2} in dense clouds may be much smaller than predicted.

(b) The large value $f_{SO} \sim 10^{-6}$ for dense clouds predicted by Mitchell, Ginsburg and Kuntz[3] results from the reaction $S + H_3^+ \rightarrow H_2S^+ + H$ (1) in their model. Large H_2S^+ production via reaction (1) leads to correspondingly large HS concentration and large $q(SO)$ through the fast reaction $HS + O \rightarrow SO + H$. Reaction (1) probably does not occur, which is consistent with the lower observed f_{SO}.

(c) Due to activation energy[6], the reaction of CS with O atoms is efficient as a loss mechanism of CS during the early phases of cloud evolution (high temperature), or in hot and oxygen rich sources such as the KL nebula. Reactions of H_3^+, HCO^+, H^+ and C^+ serve merely to recycle CS. Consequently, $L(CS) \sim 4 \times 10^{-15}$ s^{-1} in dense clouds ($n(H_2) = 2 \times 10^5$ cm^{-3}, $T \leq 40K$). On this basis, observed[7] $2 \times 10^{-10} \leq f_{cs} \leq 2 \times 10^{-5}$ for $5 \times 10^4 \leq n(H_2) \leq 2 \times 10^5$ cm^{-3} implies $q(CS) = 1.6 \times 10^{-18}$ cm^{-3} s^{-1} in clouds with $n(H_2) = 2 \times 10^5$ cm^{-3} and low temperature.

(d) If sulfur is not abnormally depleted in dense clouds, then the observed abundances of SO, SO_2, H_2S, CS, OCS, H_2CS and SiS suggest that sulfur is mostly atomic in dense clouds, i.e., $f_s \sim 10^{-5}$. This

B. H. Andrew (ed.), Interstellar Molecules, 297–298.
Copyright © 1980 by the IAU.

value for f_s and the low value for q(CS) deduced above jointly imply
that the reaction CH + S → CS + H has an activation energy and that
$f_{S+} \leq 5 \times 10^{-11}$ in order that contributions to q(CS) via reactions of S^+
with CH_n do not lead to CS in excess of observations. q(CS) could,
however, be substantially higher if condensations onto grains constitute an effective loss mechanism for CS.

(e) In a gas-phase scheme, the reaction chain SO $\xrightarrow{C^+}$ CS^+ $\xrightarrow{H_2}$
HCS^+ $\xrightarrow{e^-}$ CS is the dominant source of CS in dark clouds. L(CS) ≈
4×10^{-15} s^{-1} then implies that n(SO)/n(CS) ≃ $1.6 \times 10^{-5}/n(C^+)$. Using
$n(C^+)$ from our models[2], we obtain n(SO)/n(CS) = 3.4, which agrees well
with observed value of 4 in L134 or TMC-1. Although it is tempting to
interpret this agreement as evidence in favor of gas-phase chemistry,
laboratory measurements of the activation energy in the reaction CH + S
→ CS + H and deduction of upper bounds on f_{S+} from observations are
needed to confirm this inference.

(f) OCS is also stable against reactions with neutral atoms and
radicals in dense clouds. Most ionic reactions serve merely to recycle OCS. Consequently, we assume L(OCS) ≈ 4×10^{-15} s^{-1}. The observations of CS in the absence of OCS in a warm cloud, such as Orion A,
implies that q(OCS) << q(CS) in this cloud. This is consistent with
low f_e in dense clouds[8] and the reaction S^- + CO → OCS + e^- as the
major source[9] of OCS. In relatively colder clouds (T ≤ 20K), such as
Sgr B2, N(OCS) ≳ N(CS) has been reported. This implies that additional
sources of OCS become important at very low temperatures. The reaction
CO + S → OCS + hν, if it occurs, might provide this additional source.

REFERENCES

1. Prasad, S. S., and Huntress, W. T., Jr.: A Model for Gas Phase
    Chemistry in Interstellar Clouds: I., to appear in Ap. J. Suppl.
2. Prasad, S. S., and Huntress, W. T., Jr.: A Model for Gas Phase
    Chemistry in Interstellar Clouds: II., to appear in Ap. J.
3. Mitchell, G. F., Ginsburg, J. L., and Kuntz, P. L.: 1978, Ap. J.
    Suppl. 38, p. 39.
4. Davis, D. D., Klemm, R. B., and Pilling, M. J.: 1972, Int. J. Chem.
    Kinet. 4, p. 367.
5. Rydbeck, O. E. H., Irvine, W. H., Hjalmarson, A., Rydbeck, G.,
    Ellder, J., and Kollberg, E.: Observations of SO in Dark and
    Molecular Clouds, preprint.
6. Lilenfeld, H. V., and Richardson, R. J.: 1977, J. Chem. Phys. 67,
    p. 3991.
7. Liszt, H., and Leung, C. M.: 1977, Ap. J. 218, p. 396.
8. Guélin, M., Langer, W. D., Snell, R. L., Wootten, H. A.: 1977, Ap.
    J. (Letters) 217, p. L165.
9. Oppenheimer, M., and Dalgarno, A.: 1974, Ap. J. 187, p. 231.

AN ICR STUDY OF ION-MOLECULE REACTIONS IN THE C_2H_2/HCN SYSTEM

M. J. McEwan, V. G. Anicich and W. T. Huntress, Jr.,
Jet Propulsion Laboratory, 4800 Oak Grove Drive, Pasadena,
CA 91103

ABSTRACT: Rate constants obtained by the ICR technique are reported for reaction (1), $C_2H_2^+$ + HCN and reaction (2) HCN^+ + C_2H_2 such that $k_1 = 3.6 \times 10^{-10}$ and $k_2 = 6.9 \times 10^{-10}$ cm^3 molecule^{-1} s^{-1}, respectively. Differences between these results and other measurements of reaction (1) are discussed. The relevance of reaction (1) to the formation of HC_3N in interstellar clouds is also briefly assessed.

INTRODUCTION

Three studies[1,2,3] have recently been reported of the rate coefficient for reaction (1). Because of a considerable discrepancy in

$$C_2H_2^+ + HCN \longrightarrow H_2CN^+ + C_2H \tag{1a}$$
$$C_2H_2^+ + HCN \longrightarrow H_2C_3N^+ + H \tag{1b}$$

the reported rate coefficient for this reaction, we decided to reinvestigate the system. In the first study by Huntress[1], the faster reverse charge transfer reaction was neglected. Reaction (2a) is a concurrent reaction in the HCN/C_2H_2 system.

$$HCN^+ + C_2H_2 \rightarrow C_2H_2^+ + HCN \tag{2a}$$

Consequently, in the study of reaction (1), any $C_2H_2^+$ reformed via reaction (2) has the effect of making the observed rate coefficient k_1 of reference 1 (5.3 x 10^{-11} cm^3 molecule^{-1} s^{-1}) too small. Freeman et al[2] in a flowing afterglow study of reaction (1) reported a much larger value, $k_1 = 7.1 \times 10^{-10}$ cm^3 molecule^{-1} s^{-1}, but this result is now known to be too large due to interference from the Penning ionization reaction of HCN with He[*4]. Schiff and Bohme[3] have also reported a selected-ion-flow-tube (SIFT) study of reaction (1) and noted an additional major channel from the three body process (1c), which they estimated accounted for 90% of all products observed in their system.

299

B. H. Andrew (ed.), Interstellar Molecules, 299–303.

$$C_2H_2^+ + HCN + M \rightarrow C_2H_2 \cdot HCN^+ + M \tag{1c}$$

EXPERIMENTAL

The JPL ICR mass spectrometer was used to obtain product distributions and rate constants as described elsewhere[5]. In order to produce $C_2H_2^+$ with a negligible fraction of vibrationally excited ions, the median energy of the ionizing electrons was kept below 12.5eV[5]. Because of the large energy spread in the electron beam, small amounts of HCN^+ were also formed even at these energies, and these were ejected during the ion formation pulse to prevent contamination of the $C_2H_2^+$ decay from the charge transfer reaction (2a). In the measurement of the rate coefficient k_2; an electron energy of 15eV was used and corrections to the observed HCN^+ (m/e=27) decays were made for the natural ^{13}C abundance of $C_2H_2^+$ which also occurs at m/e = 27.

RESULTS AND DISCUSSION

The rate coefficients and product distributions are shown in Table 1.

TABLE 1

Reaction	Rate Coefficient $(10^{-10}$ cm^3 molecule^{-1} s$^{-1})$	Reaction Number in Text
$C_2H_2^+ + HCN$ .66 $\rightarrow H_2CN^+ + C_2H$	} 3.6 ± 1.4	(1a)
.34 $\rightarrow H_2C_3N^+ + H$		(1b)
$HCN^+ + C_2H_2$ .6 $\rightarrow C_2H_2^+ + HCN$	} 6.9 ± 1.3	(2a)
.2 $\rightarrow H_2CN^+$ or $C_2H_3^+$ (+ neutral products)		(2b)
.2 $\rightarrow H_2C_3N^+ + H$		(2c)

Uncertainties shown in the rate coefficient represent the standard deviations of the measurements. We could not distinguish between $C_2H_3^+$ and H_2CN^+ in the products of reaction (2) as their masses coincide with other product ions in the system. Both ions have exothermic pathways leading to their formation. The rate coefficient reported by Schiff and Bohme[3] of 3.9×10^{-10} cm^3 molecule^{-1} s^{-1} for reaction (1) is close to our measured value, but they estimate 90% of the reaction in their system proceeds by a three body pathway. A bimolecular rate for $k_{1(a+b)} \sim 4 \times 10^{-11}$ cm^3 molecule^{-1} s^{-1} is thus implied which is much less than our reported value $k_{1(a+b)} = 3.6 \times 10^{-10}$ cm^3 molecule^{-1} s^{-1}. These results may be reconciled if the $[C_2H_2 \cdot HCN^+]^{**}$ complex can be deactivated at the expense of the bimolecular process.

$$C_2H_2^+ + HCN \rightleftharpoons [C_2H_2 \cdot HCN^+]^{**} \begin{array}{l} \nearrow H_2CN^+, H_2C_3N^+ \\ \searrow_M [C_2H_2 \cdot HCN^+]^* \rightarrow C_2H_2 \cdot HCN^+ \end{array}$$

Reaction (1b) followed by dissociative recombination (3), has been proposed as a source of HC$_3$N in interstellar clouds.[1,2]

$$H_2C_3N^+ + e^- \rightarrow HC_3N + H \tag{3}$$

Recent calculations[6] show that this mechanism can reproduce the observed amount of HC$_3$N in Sgr B2 given a particular scenario for the cloud thermal history, and that the cloud is not in chemical steady-state (equilibrium). Under conditions of chemical equilibrium, reaction (1b) followed by (3) cannot account for the observed abundance of HC$_3$N in Sgr B2. Schiff and Bohme[3] have proposed addition reactions of the type

$$C_nH_2^+ + HCN \rightarrow H_2C_{n+1}N^+ + H \tag{4}$$

which lead not only to HC$_3$N but also to successive members of the cyanopolyacetylene series. However, recent laboratory experiments show proton transfer, not addition, to be the favored process in these reactions.[7] If the chemical steady state prevails in dense clouds such as Sgr B2, then an additional source is required for the cyanopolyacetylene series. Radiative association reactions of the type (5)[8] or (6)[3] have been proposed.

$$H_2CN^+ + C_2H_2 \rightarrow H_2C_2CNH_2^+ + h\nu \tag{5}$$

$$C_2H_2^+ + HCN \rightarrow H_2C_2CNH^+ + h\nu \tag{6}$$

Laboratory experiments[3,9] show that both reactions exhibit collision-stabilized association products at room temperature. Steady-state calculations[8] indicate that reaction (5) is likely to be the stronger source in dense clouds.

ACKNOWLEDGEMENTS

This paper presents the results of one phase of research carried out at the Jet Propulsion Laboratory, California Institute of Technology, under Contract No. NAS7-100, sponsored by the National

Aeronautics and Space Administration. M.J.M. is grateful for the award of a N.R.C. Senior Research Associateship.

REFERENCES

1. Huntress, W. T.: 1976, Ap. J. Suppl. 33, p. 495.
2. Freeman, C. G., Harland, P. W., and McEwan, M. J.: 1978, Astrophys. Letters 19, p. 133.
3. Schiff, H. I., and Bohme, D. K.: 1979, Ap. J. in press.
4. Freeman, C. G., Harland, P. W., and McEwan, M. J.: 1979, private communication.
5. Kim, J. K., Anicich, V. G., and Huntress, W. T.: 1977, J. Phys. Chem. 81, p. 1798.
6. Prasad, S. S., and Huntress, W. T.: 1979, Ap. J. in press.
7. Freeman, C. G., Harland, P. W., and McEwan, M. J.: 1978, Proc. Mon. Not. Roy. Astron. Soc. 187, p. 441.
8. Mitchell, G. F., Huntress, W. T., and Prasad, S. S.: 1979, Ap. J. 233, p. 102.
9. Freeman, C. G., Harland, P. W., Huntress, W. T., and McEwan, M. J.: 1979, unpublished work.

DISCUSSION FOLLOWING McEWAN

Field: What is the kinetic energy of the ions in the I.C.R. experiments? It may influence your derived rate coefficients.

McEwan: The kinetic energy of the ions is small, although it is probably slightly higher than thermal. In most ion-molecule reactions the small amounts of kinetic energy above thermal do not influence the observed rate coefficients. Under typical cell conditions, an ion will experience five to 20 collisions with polyatomic neutrals. Energies <0.1 eV are suggested by a comparison of ICR data to data from other techniques for reactions where kinetic energy does affect the rate.

Glassgold: The abundance of HC_3N implied by the production mechanism $C_2H_2^+ + HCN \rightarrow H_2C_3N^+ + H$ should be very sensitive to the electron abundance. In making your estimate, did you use the low electron fractions $\sim 10^{-8}$ determined for dense clouds?

McEwan: According to the time-dependent model of Prasad and Huntress, the reaction cited can be an important synthetic mechanism for cyanoacetylene if the cloud does not reach steady state. The reaction produces insufficient HC_3N to account for observations if steady state concentrations of $C_2H_2^+$ and HCN are used to predict the HC_3N abundance. The fractional electron abundances used in the model ranged between 10^{-7} and 10^{-8}.

D. Smith: It has been shown that $C_2H_2^+$ has a significant population in v =1 even at 300 K (Buttrill et al). Could you comment on the possible effects of such excitation on your measured association rate for the $C_2H_2^+ + HCN$ reaction?

McEwan: Previous experiments with $C_2H_2^+$ generated under similar conditions of electron impact (ref. 5 above) have shown that vibrational excitation of $C_2H_2^+$, although small, is nevertheless present. In the majority of reactions of $C_2H_2^+$ studied by ICR in the trapped ion mode, good agreement in rate coefficient determinations has generally been observed when compared with other techniques. We would not expect the $v=1$ level to affect this $C_2H_2^+$+HCN rate, as k for this association (ref 3) is too small for the association channel to be observed at the pressures we were using.

Winnewisser: Did you check the analogous reaction in cyanoacetylene and acetylene i.e. $HC_3N+C_2H_2^+$? I expect this reaction to be slow. In our discharge experiments we observe a large amount of polymer formation. Do you observe the same phenomenon at the lower pressures you work at?

McEwan: To answer the first question, we did look at the reaction $C_2H_2^+$+HC_3N, and found no evidence for addition having occurred (see refs. 2 and 7). We do not observe any polymer in our system. We have observed large quantities of polymer in reaction studies of neutral molecules containing the -CN group that have been subjected to a microwave discharge.

Herbst: The reaction $C_2H_3^+$+CN should be investigated for formation of HC_3N.

McEwan: Ion reactions with radicals present more experimental difficulties than conventional ion-molecule reactions. However attempts will shortly be made to measure a series of CN-radical reactions, including the one you have mentioned.

Shiff: Careful search with our system failed to reveal the channel HCN^++$C_2H_2 \rightarrow C_3H_2N^+$+H, which suggests that reactions of this type do not present an attractive route to the cyanoacetylenes. On the other hand, we have recently observed that proton transfer is a minor channel in the case of $C_2H_2^+$+HCN. One should therefore be cautious about rejecting reactions of $C_2H_2^+$ with the cyanoacetylenes to form the next higher member, in spite of observations by McEwan et al. that they proceed exclusively by proton transfer.

McEwan: The reactions of HCN^+ are not significant for interstellar synthesis because of its rapid reaction with H_2 to produce H_2CN^+. The radiative association of H_2CN^+ with C_2H_2 may be relevant for synthesis of cyanoacetylene.

AN ICR STUDY OF AN ASSOCIATION REACTION AT LOW PRESSURE

M.J. McEwan, V.G. Anicich and W.T. Huntress, Jr.
Jet Propulsion Lab., 4800 Oak Grove Dr., Pasadena, CA 91103
P.R. Kemperer and M.T. Bowers
Chemistry Dept., Univ. of California, Santa Barbara, CA 93106

ABSTRACT. An ICR investigation of the association reaction
$$CH_3^+ + HCN \quad CH_3.HCN^+$$
has shown the reaction follows second order kinetics over the pressure range 1×10^{-6} to 3×10^{-4} Torr with a rate coefficient of 2×10^{-10} $cm^3 s^{-1}$. These results can be interpreted in terms of a saturated 3-body or radiative association mechanism.

Investigations of the reactions of the gaseous ion CH_3^+ with a number of different molecules have shown the interesting feature that in several reactions the formation of the association product follows second-order kinetics[1,2]. The two main classes of association reactions that exhibit second order kinetics are (a) saturated 3-body reactions and (b) radiative association reactions. The application of radiative association reactions to interstellar processes provides a convenient means of forming large polyatomic molecules. However, even though radiative association has been invoked in models of interstellar chemistry[3,4], there is still little direct experimental evidence for its participation as an important process. The 3-body association process represented by reaction (1) and radiative association by reaction (2)

$$A^+ + B + M \xrightarrow{k_1} AB^+ + M \tag{1}$$

$$A^+ + B \xrightarrow{k_2} AB^+ + h\nu \tag{2}$$

can be understood in terms of two distinct steps. The first of these steps (reaction (3)) is common to both 3-body and radiative association.

$$A^+ + B \underset{k_{-3}}{\overset{k_3}{\rightleftharpoons}} (AB^+)^* \tag{3}$$

$$(AB^+)^* + M \xrightarrow{k_4} AB^+ + M \tag{4}$$

$$(AB^+)^* \xrightarrow{k_5} AB^+ + h\nu \tag{5}$$

By using the steady state approximation applied to complex $(AB^+)^*$ and assuming that dissociation of $(AB^+)^*$ is rapid compared with stabilization, it is readily shown that the rate of production of the associated product

B. H. Andrew (ed.), Interstellar Molecules, 305–306.
Copyright © 1980 by the IAU.

AB^+ is given by

$$d[AB^+]/dt = (k_3/k_{-3})([A^+][B])(k_4[M] + k_5).$$ (6)

The overall rate coefficients for 3-body and radiative association are therefore

$$k_1 = k_3 k_4 / k_{-3}$$ (7)

and

$$k_2 = k_3 k_5 / k_{-3}$$ (8)

As typical values for the rate coefficients k_4 and k_5 are 10^{-9} cm^3 s^{-1} and 10^2-10^3 s^{-1} respectively[1,5], and $[M] \sim 10^{15}$ molecules cm^{-3} in most flow tube experiments[1,2], then 3-body association should dominate radiative association.

We have applied the ion cyclotron resonance (ICR) technique to investigate the association process

$$CH_3^+ + HCN \longrightarrow CH_3.HCN^+$$ (9)

and have found reaction (9) to follow second order kinetics throughout the entire pressure-range of the experiment, viz. 1×10^{-6} - 2×10^{-4} Torr. Good agreement was found between rate coefficients determined in the trapping mode at the low end of the pressure range and the drift mode at the high end of the pressure range. The observed rate coefficient at room temperature (300 K) was $k_9 = 2 \times 10^{-10}$ cm^3 s^{-1} and was independent of pressure. A most interesting feature of this work is that the observed rate coefficient for k_9 is less than the value of $2 \times 10^{-9} cm^3 s^{-1}$ reported by Schiff and Bohme[2] in their study of the much higher pressure range ~ 0.3 Torr. There are two likely explanations for the difference. In this work the product $k_4[M]$ is a factor $\sim 10^4$ less than in the flow tube of Schiff and Bohme. Consequently we may be in a pressure regime where radiative association is the dominant mechanism for association whereas 3-body association is the major process at ~ 0.3 Torr. Alternatively, as the CH_3^+ ions in our work were formed by the impact of low energy electron on CH_4, they may possess some vibrational excitation. A larger fraction of collisions between CH_3^+ and HCN may then revert back to primary reactants (process k_{-3}) than was the case for thermal ions in the study by Schiff and Bohme.

If the second explanation is true, then the fact that we have observed second-order kinetics implies a value for τ_D, the lifetime of the complex $(CH_3.HCN^+)^*$ to unimolecular decomposition at room temperature, of $\geq 3 \times 10^{-3}$ s.

REFERENCES

1. Smith, D. & Adams, N.G.: 1977, Ap. J. 217, 741; 1978, Ap. J. 220, L87.
2. Schiff, H.I., & Bohme, D.K.: 1979, Ap. J. 232, 740.
3. Black, J.H., & Dalgarno, A.: 1973, Ap. Letters 15, 69.
4. Mitchell, G.F., Huntress, W.T., & Prasad, S.S.: 1979, Ap. J. 233, 102.
5. Herbst, E.: 1976, Ap. J. 205, 94.

LABORATORY STUDIES OF INTERSTELLAR CARBON/NITROGEN ION CHEMISTRY

H.I. Schiff, G.I. Mackay, G.D. Vlachos, and D.K. Bohme
Department of Chemistry, York University, North York,
Ontario, Canada, M3J 1P3

ABSTRACT

Laboratory measurements are reported for ion-molecule reactions involving CN^+, HCN^+, C_2N^+ and HCN, and their implications for interstellar synthesis for C/N compounds are discussed. The reaction of C^+ with HCN was found not to constitute a major loss process for HCN, which is regenerated by reactions of C_2N^+ with NH_3, H_2O, CH_4, C_2H_2 and H_2S. These latter reactions lead to C addition rather than C-N bond formation. Rapid association reactions were observed for CH_3^+ and $C_2H_2^+$ with HCN. These suggest efficient radiative association reactions under interstellar conditions to form ions which may form larger C/N compounds upon neutralization.

INTRODUCTION

Ion-molecule reactions remain an attractive mechanism for the formation of large molecules observed in dense interstellar clouds. Studies of these reactions in our and other laboratories have shown them capable of synthesizing large 'straight-chain' hydrocarbons. In this paper we report the results of our studies involving HCN, HCN^+, CN^+, and C_2N^+.

The reaction channels observed and the rate constants found for these reactions are shown in Table 1. The reactions were studied using a modified form of the selected ion flow tube first developed by Adams and Smith (1976).

REACTIONS WITH HCN

The reactions of C^+ and C_2^+ with HCN proceed exclusively by condensation with H atom elimination, and with a rate constant close to the theoretical limit. In contrast, the reaction of C_2H^+ with HCN proceeds by two almost equal channels, proton transfer from C_2H^+, and H atom transfer from HCN. The combined rate constant is again close to the theoretical value. The somewhat slower reaction between $C_2H_2^+$ and HCN has three primary channels. In the presence of about 0.5 Torr of

307

He, 87% of the reactive collisions lead, by three-body association, to
$C_2H_2^+$.HCN. This suggests a fast rate for radiative recombination under
interstellar conditions. Neutralization of this cluster ion could well
lead to the observed molecules cyanoacetylene and vinyl cyanide. About
8% of the reactive collisions goes by a two-body channel to produce
$H_2C_3N^+$. Neutralization of this ion probably also leads to cyanoacetylene
The third (5%) channel leads to $HCNH^+$ which, upon neutralization,
probably regenerates HCN.

The reaction of CH_3^+ with HCN proceeds exclusively by very rapid
three-body association to form CH_3^+.HCN in the presence of 0.5 Torr He.
This again suggests an efficient radiative association reaction under
interstellar conditions to form the same ion. Neutralization by electron
recombination or proton transfer could give rise to the observed neutral
molecule CH_3CN. For example we have found that the ion does indeed
proton transfer, very rapidly, with NH_3 and $(CH_3)_2O$ to produce methyl
cyanide.

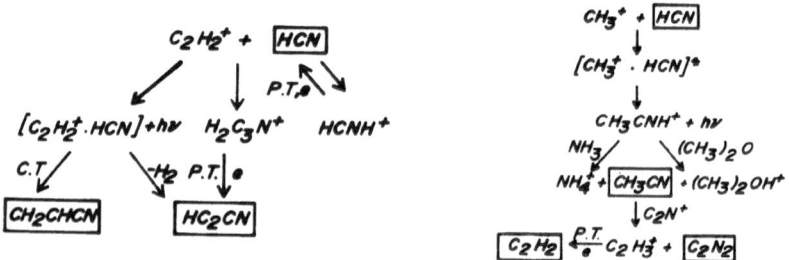

Fig. 1. Possible consequences of the reactions of HCN with $C_2H_2^+$ and
CH_3^+. (P.T. ≡ proton transfer, e ≡ electron recombination, C.T. ≡ charge
transfer)

HCN^+ AND CN^+ REACTIONS

HCN^+ reacts rapidly with C_2H_2 by three channels. The dominant
channel produces HC_3N^+ which can lead to HC_3N by charge transfer. The
channel producing $H_2C_3N^+$ was not observed, which appears to rule out
this route to HC_3N by electron recombination. The other two channels,
charge transfer and proton transfer, do not, of course, lead to C-N bond
formation. Reactions of HCN^+ with CH_4 have two channels. The major
channel (80%) produces H_2CN^+ and therefore, by recombination, HCN. The
minor (20%) channel produces NH_2 and results in C-C bond formation which,
upon neutralization, could yield C_2H_2. The reaction of CN^+ with CH_4 also
has two channels. The major channel is H^- transfer, which produces HCN.
The minor channel is H_2 transfer; recombination of the resulting H_2CN^+
ion probably also leads to HCN. It is of interest to note that there
appears to be little kinetic isotope effect in the reactions of CN^+ with
CD_4, whereas the rate constant for the reaction of HCN^+ with CH_4 is 1.7
times the rate constant with CD_4.

Table 1.   Summary of measured rate constants (in units of 10^{-9} cm^3 molecule^{-1} s^{-1}) at 300 K. [a]

Reactions with HCN

Reaction	
$C^+ + HCN \rightarrow C_2N^+ + H$	3.5
$C_2^+ + HCN \rightarrow C_3N^+ + H$	2.5
$C_2H^+ + HCN \rightarrow HCNH^+ + C_2$	2.5
$\rightarrow C_2H_2^+ + CN$	
$C_2H_2^+ + HCN \rightarrow C_2H_2^+.HCN$	0.38[b]
$\rightarrow H_2C_3N^+ + H$	
$\rightarrow HCNH^+ + C_2H$	
$CH_3^+ + HCN \rightarrow CH_3^+.HCN$	2.0[b]
$CH_3^+.HCN + NH_3 \rightarrow NH_4^+ + CH_3CN$	0.87
$CH_3^+.HCN + (CH_3)_2O \rightarrow (CH_3)_2 OH^+ + CH_3CN$	$\geqslant 1$

Reactions of CN^+ and HCN^+

Reaction	
$CN^+ + CH_4 \rightarrow CH_3^+ + HCN$	0.97
$\rightarrow HCNH^+ + CH_2$	
$CN^+ + CD_4 \rightarrow CD_3^+ + DCN$	1.1
$\rightarrow DCND^+ + CH_2$	
$HCN^+ + CH_4 \rightarrow HCNH^+ + CH_3$	1.3
$\rightarrow C_2H_3^+ + NH_2$	
$HCN^+ + CD_4 \rightarrow HCND^+ + CD_3$	0.77
$\rightarrow C_2D_2H^+ + ND_2$	
$HCN^+ + C_2H_2 \rightarrow C_2H_2^+ + HCN$	1.9
$\rightarrow HC_3N^+ + H_2$	
$\rightarrow C_2H_3^+ + CN$	
$\nrightarrow H_2C_3N^+ + H$	

Reactions of C_2N^+

Reaction	
$C_2N^+ + H_2 \nrightarrow$	$\leqslant 0.0001$
$C_2N^+ + NH_3 \rightarrow HCNH^+ + HCN$	1.8
$C_2N^+ + H_2O \rightarrow HCO^+ + HCN$	0.31
$C_2N^+ + CH_4 \rightarrow C_2H_3^+ + HCN$	0.0044
$H_2C_3N^+ + H_2$	
$C_2N^+ + C_2H_2 \rightarrow C_3H^+ + HCN$	0.89
$\rightarrow HCNH^+ + C_3$	
$C_2N^+ + H_2S \rightarrow HCS^+ + HCN$	1.2
$C_2N^+ + CH_3CN \rightarrow C_2H_3^+ + C_2N_2$	4.1

[a] The branching ratios which were determined are discussed in the text.

[b] At a helium pressure of $\simeq$ 0.5 Torr.

C_2N^+ REACTIONS

No reaction was observed to occur between C_2N^+ and H_2, which suggests that this may be an abundant interstellar ion. It does react with NH_3, H_2O and H_2S by a single channel which can be represented as:

$$C_2N^+ + H_yX \rightarrow H_{y-1}CX^+ + HCN$$

The reaction of C_2N^+ with C_2H_2 has the additional channel giving H_2CN^+ and C_3 as products. However, electron recombination would result in identical neutral products from both channels. The reaction of C_2N^+ with CH_4 is slow - less than 1% of the collision frequency. There are two channels, and a condensation channel giving $H_2C_3N^+$, which therefore could be a minor source of interstellar HC_3N.

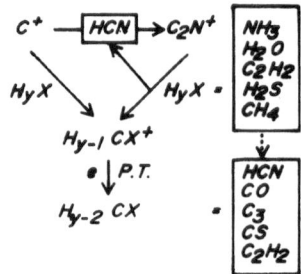

Fig. 2.   C-X bond formation by reaction with C^+ and by C_2N^+ catalysis

There are two important generalizations from the above reactions with C_2N^+. First, they all produce HCN. Thus the reaction of C^+ with HCN to produce C_2N^+ does not constitute a major loss process for HCN since this ion regenerates HCN when it reacts with other major interstellar neutral molecules. Secondly, its reactions do not lead to C-N formation but rather to C addition to the neutral reactant. In fact the C_2N^+ reactions are equivalent to the counterpart reactions with C^+ ions. Thus the reaction of C^+ with HCN followed by C_2N^+ reaction with a neutral simply catalyses the direct reaction of C^+ with the neutral, albeit with a slower effective rate constant than the direct C^+ reaction.

In addition to the slow channel with CH_4, one other, possibly important exception was found to the above generalization, viz. the reaction of C_2N^+ with CH_3CN. This reaction is rapid and has an exclusive channel to form $C_2H_3^+$ and, most likely, the neutral C_2N_2.

REFERENCE

Adams, N.G., and Smith, D.: 1976, Int. J. Mass Spect. and Ion Phys. 21, pp. 349-359.

GAS PHASE SYNTHESIS OF AMINO-, CYANO- AND NITROSO-COMPOUNDS IN INTERSTELLAR CLOUDS

Nigel G. Adams and David Smith,
Department of Space Research, University of Birmingham,
Birmingham, England.

ABSTRACT

   From our laboratory data relating to several hundreds of ion-atom and ion-molecule reactions at thermal energies, we qualitatively describe probable chemical paths to the synthesis of amino-, cyano- and nitroso-compounds in interstellar clouds.

1.   INTRODUCTION

   -NH, -CN, -CO and to a lesser extent -NO groups are common in the observed interstellar molecules, and many attempts have been made during recent years to explain how these molecules are synthesised. Plausible qualitative and semi-quantitative ion-chemical models now exist based on known gas phase ion chemistry and intuitive guesses at likely reactions and their rate coefficients. However, the objective must be to establish the most important routes to the synthesis of the observed molecules by detailed chemical modelling. Before this can be satisfactorily achieved, more data is required on the composition and physical conditions within the gas clouds, as well as a great deal more laboratory data on the rate coefficients and products of ion-neutral and neutral-neutral reactions acquired at appropriately low temperatures.

   Stimulated by the challenge presented to laboratory experimenters by the ion-chemical problems identified in the research papers on interstellar molecular synthesis, we conceived the Selected Ion Flow Tube (SIFT) technique (Adams and Smith, 1976) which we have developed and exploited during the last two years to determine the rate coefficients and product ion distributions of several hundreds of ion-neutral reactions. Most of the reactions studied have been selected because of their potential interest to interstellar chemistry. Thus, we have carried out detailed surveys of the reactions of the ion series CH_n^+ (n = 0 to 4) (Smith and Adams, 1977a,b; Adams and Smith, 1977,1978), $C_2H_n^+$ (n = 0 to 4) (Adams and Smith, 1977) H_nCO^+ (n = 0 to 3) (Adams et al 1978), NH_n^+ (n = 0 to 4) (Adams et al 1979a) and reactions of the

B. H. Andrew (ed.), Interstellar Molecules, 311–315.
Copyright © 1980 by the IAU.

ions N^+, N_2^+, O^+, O_2^+ and NO^+ (Smith et al 1978). In collaboration with the NOAA Group at Boulder, Colorado, we have also studied the reactions of several active ion species with nitrogen and oxygen atoms (Viggiano et al,1979). The measurements have largely been carried out at 300 K but some of the ternary association reactions have been studied at temperatures as low as $\sim$ 100 K (Adams et al, 1979b).

2.   DEDUCTIONS BASED ON SIFT DATA

2.1 Synthesis of Amino- and Cyano-Compounds in Interstellar Clouds.

The laboratory data reveals many ion-neutral routes to the formation of these compounds if it is accepted that N atoms and N^+ ions are present in the gas clouds. Thus the formation of ionized nitrogen hydrides will proceed as follows:

$$N^+ \xrightarrow{+H_2} NH^+ \xrightarrow{+H_2} NH_2^+ \xrightarrow{+H_2} NH_3^+ \qquad (1)$$

However, the binary reaction $NH_3^+ + H_2 \longrightarrow NH_4^+$ does not proceed at a measurable rate at low temperatures and so NH_3^+ is likely to be a relatively long-lived species in the clouds. Our data however indicates the great propensity for NH_3^+ to abstract an hydrogen atom from almost any hydrogen-bearing molecule especially hydrocarbons, e.g. $NH_3^+ + CH_4 \longrightarrow NH_4^+$, and so the synthesis of NH_3 is achieved following the electronic recombination, $NH_4^+ + e \longrightarrow NH_2, NH_3$ although the branching ratio of the recombination is unknown. The synthesis of the hydrocarbons has been discussed previously by several authors and probably proceeds thus:

$$C^+ \xrightarrow{H_2} CH_2^+ \xrightarrow{H_2} CH_3^+ \qquad (2)$$

where the first stage of (2) is the much discussed radiative association reaction. Like NH_3^+, the binary reaction of CH_2^+ with H_2 is immeasurable at gas cloud temperatures, but following our detailed study of the three-body association reactions of CH_2^+ (see Smith and Adams in these Proceedings) we have suggested that the radiative association reaction $CH_3^+ + H_2 \longrightarrow CH_5^+ + h\upsilon$, will proceed at an appreciable rate in molecular clouds after recombination generating CH_4 in relatively high abundance. Reactions of the ions CH_n^+ with CH_4 then generate higher hydrocarbons e.g.

$$C^+ + CH_4 \longrightarrow C_2H_2^+, C_2H_3^+ \xrightarrow{e} C_2H, C_2H_2 \qquad (3)$$

The production of NH_3 and hydrocarbons are important steps in the synthesis of cyano-compounds, since from their reactions with the ions CH_n^+ and NH_n^+ respectively, -CN containing ions are inevitably generated. Examples are:

$$C^+ + NH_3 \longrightarrow H_2CN^+, HCN^+; \quad CH_3^+ + NH_3 \longrightarrow CH_2NH_2^+, CH_3NH_3^+ \qquad (4)$$

and

$$N^+ + CH_4 \longrightarrow H_2CN^+, HCN^+; NH_2^+ + C_2H_4 \longrightarrow H_4CN^+, H_5C_2N^+ \tag{5}$$

which thus can lead to the generation of the observed species CN, HCN, CH$_2$NH and CH$_3$NH$_2$.

The above _ion-molecule_ reactions are all very fast with rate coefficients near to the collisional limit. Positive _ion-atom_ reactions are generally somewhat slower but nevertheless could play an important role in interstellar chemistry. Our recent study of the reactions of N atoms with several important interstellar ions (e.g. CH$^+$, CH$_3^+$, C$_2$H$_2^+$ etc.) has again demonstrated the effectiveness of production of -CN bearing compounds via ion-neutral reactions:

$$CH^+ + N \longrightarrow CN^+, H^+ ; CH_3^+ + N \longrightarrow H_2CN^+, HCN^+ \tag{6}$$

The reaction C$_2$H$_2^+$ + N $\longrightarrow$ HC$_2$N$^+$, C$_2$N$^+$, CH$^+$ is interesting since it indicates how readily the N atoms are incorporated into the hydrocarbon ions generating other reactive ions. The C$_2$N$^+$ is unreactive with H$_2$ and so is available for reaction with more minor species. For example the reaction C$_2$N$^+$ + H$_2$O generates the widely observed HCO$^+$ and HCN. So the more elementary N-atom reactions could probably contribute significantly to the synthesis of the observed cyano-compounds. Indeed most important could be the H$_3^+$ + N reaction which could compete with (1) as the most important first stage in NH$_3$ synthesis (although this has not been studied in the laboratory).

2.2 Synthesis of Nitroso-Compounds in Interstellar Clouds.

Our experiments have shown that few reactions between oxygen and nitrogen bearing ions and neutrals lead to the production of -NO bearing compounds and it is perhaps therefore not surprising that few such compounds have been detected in interstellar clouds. Thus whilst N$^+$ ions react with most observed oxygen bearing molecules as well as with O$_2$ (which must surely be present), the most common ion product bearing an NO bond is NO$^+$ and unless the NO$^+$ can charge transfer with a species of lower ionization energy (and few such species exist, metal atoms being notable exceptions), then the inevitable fate of the NO$^+$ ions is dissociative recombination with electrons producing N and O atoms. In contrast, the C$^+$ + O$_2$ reaction rapidly generates the widely observed CO as well as CO$^+$ which rapidly reacts with H$_2$ (unlike NO$^+$) producing the observed HCO$^+$. The latter ion does not react with N atoms (and indeed is unreactive with most molecules except those of high proton affinity like NH$_3$ and CH$_2$NH$_2$ to which it transfers a proton) and so is not a potential source of -NO containing compounds. We have, however, identified a few potentially important reactions which could lead to NO and HNO in interstellar clouds. For example, the reaction of NH$^+$ with CO$_2$ (which must surely be present) generates HNO$^+$ which on electronic recombination can lead to NO. Similarly the reaction

$NH_2^+ + O_2 \longrightarrow H_2NO^+$, HNO^+ can lead to both NO and HNO production, both of which have recently been detected in Sgr B2.

Nitrogen and oxygen atom reactions with positive ions are very likely to be sources of -NO bearing compounds, although unfortunately few such reactions have been studied in the laboratory. One such relevant reaction which we have studied, however, is $H_2O^+ + N \longrightarrow HNO^+, NO^+$. The reactions of NH_2^+ and NH_3^+ with O atoms may be potential sources of HNO^+ and H_2NO^+.

3. CONCLUSIONS

Many ion-neutral reactions have been identified as sources of amino- and cyano-compounds. Clearly the most important reactions amongst those quoted above are those involving ions which do not react rapidly with H_2. such as C^+, CH_3^+, $C_2H_2^+$ and C_2N^+, but ions such as N^+, NH^+, NH_2^+, H_2O^+ which do react rapidly with H_2 cannot be ignored, since their reactions with relatively abundant neutrals such as CO, CH_4 (and presumably O_2 and CO_2) can clearly generate amino-, and cyano-compounds in their observed relatively small abundances. That nitroso-compounds are much less abundant in interstellar clouds is quite consistent with our laboratory observations of a very large number of ion-neutral reactions in which such products are relatively rare.

REFERENCES

Adams, N.G., and Smith, D.: 1976, Int.J.Mass Spectrom.Ion Phys. 21, pp.349-359
Adams, N.G., and Smith,D.: 1977,Chem.Phys.Letts. 47, pp.383-387
Adams, N.G.,and Smith,D.: 1978, Chem.Phys.Letts. 54,pp.530-534
Adams, N.G., Smith,D.,and Grief,D.:1978, Int.J.Mass Spectrom.Ion Phys. 26,pp.405-415
Adams,N.G.,Smith,D.,and Paulson,J.F.:1979a,J.Chem.Phys.(submitted)
Adams,N.G.,Smith,D.,Lister,D.G.Rakshit,A.B.,Tichy,M.,and Twiddy,N.D.: 1979b,Chem.Phys.Letts.63,pp.166-170
Smith,D.,and Adams,N.G.:1977a,Int.J.Mass Spectrom.Ion.Phys. 23,pp.123-135
Smith,D.,and Adams,N.G.:1977b, Chem.Phys.Letts.47,pp.145-149
Smith,D.,Adams,N.G.,and Miller,T.M.:1978,J.Chem.Phys.69,pp.308-318
Viggiano,A.A.,Albritton,D.L.,Fehsenfeld,F.C.,Adams,N.G.,and Smith,D.: 1979,Ap.J. (submitted)

DISCUSSION FOLLOWING ADAMS

Hollenbach: The impression I have from your remarks and those of Dr. Fehsenfeld is that the radiative-association rate-coefficient for C^+ with H_2 is about $10^{-14} cm^3 s^{-1}$ at low temperatures. What is the probable uncertainty in this coefficient?

Adams: The greatest uncertainty lies in the value of the radiative lifetime for $(CH_2^+)^*$ formed in the C^++H_2 radiative-association reaction (see the papers by Herbst, and by Smith and Adams in these Proceedings).

Glassgold: Despite the excellent new results from laboratory measurements of reaction rates, I believe there remains a fundamental difficulty in explaining nitrogen-bearing molecules with gas-phase chemistry. The reason is that the progenitor ion N^+ you proposed has much too low an abundance in interstellar clouds. (It is presumed to arise from cosmic ray ionization of N; a similar problem arises if most of the nitrogen is in the form of N_2.)

Langer: Direct cosmic ray ionization of nitrogen is not important in initiating nitrogen chemistry. Reactions of N with carbon and oxygen radicals are dominant.

Adams: N^+ is rapidly produced by the reaction of N_2 with He^+, which is assumed to be formed by cosmic ray ionization of the abundant helium. So if most of the nitrogen in the clouds is in the form of N_2, then a ready source of N^+ is available. The reaction $He^++CO \rightarrow C^++O+He$ competes for the He^+, but even if $[CO]>>[N_2]$, sufficient N^+ would still be produced to explain the presence of the NH_3 via the sequence $N^++H_2 \rightarrow NH^++H_2 \rightarrow NH_2^++H_2 \rightarrow NH_3^++H_2 \rightarrow NH_4^++e \rightarrow NH_3$, especially since we have recently shown that the reaction of NH_3^+ with H_2 proceeds at a significant rate at low temperatures.

Langer: Could Dr. Fehsenfeld give a range of values for the radiative association rate of $C^++H_2 \rightarrow CH_2^++h\nu$?

Fehsenfeld: I would expect about $10^{-15}cm^3s^{-1}$ based on Herbst's calculations, perhaps one order of magnitude more or less.

Herbst: The calculated rate coefficient of the $C^++H_2 \rightarrow CH_2^++h\nu$ radiative association reaction could be made more precise by careful calculation of the radiative decay rate.

Winnewisser: You have shown that the fewer chemical pathways leading to N=O bearing molecules such as NO and HNO could be an explanation of their low interstellar abundance. Do you feel that this could also be a likely explanation for the conspicuous absence of larger NO bearing interstellar molecules, for example HNCO versus HCNO?

Adams: Yes. From our studies of many hundreds of ion-neutral reactions, the only reactions leading to HNCO are $NH_2^++COS \rightarrow H_2NCO^+$ ($+e \rightarrow$ HNCO?) and the association reaction $NH_2^++CO \rightarrow NH_2^+\cdot CO$ ($+e \rightarrow$ HNCO?), either or both of which may be responsible for the observed HNCO in interstellar clouds.

THE FORMATION OF COMPLEX INTERSTELLAR MOLECULES BY RADIATIVE ASSOCIATION

E. Herbst
College of William and Mary, Williamsburg, Virginia, U.S.A.

A new statistical theory of ion-molecule association reaction rate coefficients has been formulated and found to give good agreement with three-body association rate coefficients studied in the laboratory in the temperature range 100-300 K (Herbst 1979a). The theory indicates that certain radiative association reactions proceed rapidly at low interstellar temperatures to produce complex interstellar molecules, as suggested by Smith and Adams (1978).

Radiative association is a process in which two gaseous molecules A and B react to form the species AB and a photon:

$$A + B \longrightarrow AB + h\nu \tag{1}$$

The process has not been generally studied in the laboratory because, at laboratory pressures, three-body association

$$A + B + C \longrightarrow AB + C \tag{2}$$

normally dominates (Herbst 1979a,b). Some detailed theoretical work on the radiative association of small species such as CH^+ and CH_2^+ has been undertaken. Herbst (1976) has formulated a less detailed, statistical theory for the rate coefficients of radiative association reactions between molecular ions and neutrals. The motivation behind his work was the feeling that normal ion-molecule pathways could not produce complex interstellar molecules. In the theory, radiative association between the species A^+ and B is modeled as the series of processes

$$A^+ + B \underset{k_{-1}}{\overset{k_1}{\rightleftharpoons}} (AB^+)^* \overset{k_r}{\longrightarrow} AB^+ + h\nu \tag{3}$$

where $(AB^+)^*$ is a "complex" that is formed with a collision frequency k_1 ($cm^3 s^{-1}$), decays randomly with rate coefficient k_{-1} (s^{-1}), or can be stabilized via emission of an infra-red photon with rate coefficient $k_r(s^{-1})$. The overall rate coefficient for radiative association

317

B. H. Andrew (ed.), Interstellar Molecules, 317–321.
Copyright © 1980 by the IAU.

k_{RA} $(cm^3 s^{-1})$ is given by the formula

$$k_{RA} = (k_1/k_{-1})k_r \qquad (4)$$

if $k_{-1} \gg k_r$. Using this equation, Herbst (1976) calculated the rate coefficients of several radiative association processes. He concluded that radiative association is in general much slower than normal ion-molecule reactions and that only if the dominant species H_2 were a reaction partner could radiative association even be competitive in dense interstellar clouds.

More recently, Smith and Adams (1978) investigated three-body association rate coefficients for CH_3^+ and various neutrals using helium as the third body C. The overall rate coefficient k_{3B} $(cm^6 s^{-1})$ can be modeled by the relation

$$k_{3B} = (k_1/k_{-1})k_2 \qquad (5)$$

at sufficiently low pressures (Herbst 1979a), where k_2 is the <u>collisional</u> de-excitation frequency of the collision complex and k_1 and k_{-1} are as defined above. Smith and Adams (1978) found k_{3B} to possess typically a strong inverse temperature dependence of the type T^{-n} where $n > 3$ in the range 225-300 K. They then estimated k_{RA} for the analogous radiative association reactions involving CH_3^+ and various neutrals by estimating k_2 and k_r (Herbst 1976) and using relation (4). Boldly extrapolating to $T \leq 50$ K, Smith and Adams (1978) deduced k_{RA} at these temperatures to be at or near the collision frequency for some, but not all of the reactions involving CH_3^+ and neutrals. Although their procedure is open to criticism (Herbst 1979c), discrepancies between their work and theory (Herbst 1976) were sufficient to provoke a reinvestigation of the statistical theory of radiative association. It was ascertained that the theory violated the principle of detailed balancing (Herbst 1979a) which relates k_1 and k_{-1} via the equation

$$k_1 \, q_{A^+} q_B = k_{-1} \, q_{AB^{+*}} \qquad (6)$$

where q_{A^+}, q_B, and $q_{AB^{+*}}$ are molecular partition functions per unit volume. Use of equation (6) in relations (4) and (5) results in statistical theories for k_{RA} and k_{3B} that do not violate detailed balancing. Herbst (1979a) utilized such a statistical theory to calculate values for k_{3B} that agree well in temperature dependence and magnitude with the work of Smith and Adams (1978) and more recent experimental work in the temperature range 100-300 K. Thus, a proper statistical theory can reproduce laboratory three-body results.

The corrected statistical theory has now been applied to radiative association at low interstellar temperatures (Herbst 1979c). In general, the low temperature extrapolations of Smith and Adams (1978) have been found to be somewhat optimistic, but in no case have there been discrepancies of more than two orders of magnitude with our calculated results. The statistical theory thus indicates that in favorable cases (e.g., some CH_3^+ reactions) radiative association rate coefficients can approach collision frequencies at temperatures under 50 K. Favorable cases are those in which no activation energy barriers exist to block facile formation of a tightly-held complex, no normal exothermic channels are readily available to the tightly-held complex, and the density of vib-rotational states of the complex is large. The last factor depends on both the dissociation energy D_0 of the product molecule and its size. To illustrate these effects, Table I contains some calculated values of $10^9 k_{RA}$ at 10 K and 50 K for reactions of possible interstellar importance. (Note that a typical ion-molecule collision frequency is 10^{-9} $cm^3 s^{-1}$ so that $10^9 k_{RA}$ can be considered a sticking probability.). As can be seen, radiative association reactions of CH_3^+ with CO, H_2O, NH_3 are facile for $T \leqslant 50$ K

TABLE I   Some Calculated Reaction Rates

Reaction	D_0(eV)	$10^9 k_{RA}$(10 K, 50 K)
$CH_3^+ + H_2 \longrightarrow CH_5^+$	1.7	2×10^{-3}, 2×10^{-4}
$CH_3^+ + CO \longrightarrow C_2H_3O^+$	3.6	6×10^{-1}, 2×10^{-2}
$CH_3^+ + H_2O \longrightarrow CH_5O^+$	2.8	6×10^{-1}, 9×10^{-3}
$CH_3^+ + NH_3 \longrightarrow CH_6N^+$	4.5	~ 1, ~ 1
$NH_3^+ + H_2 \longrightarrow NH_5^+$	0.4	8×10^{-8}, 1×10^{-8}
$NH_3^+ + CO \longrightarrow NH_3CO^+$	0.3	1×10^{-8}, 4×10^{-10}
$H_2CN^+ + C_2H_2 \longrightarrow C_3H_4N^+$	1.5 (?)	3×10^{-2}, 5×10^{-4}

whereas reactions with small D_0 (< 1 eV) are very slow even if eight-atom molecules are produced.

Huntress and Mitchell (1979) and Mitchell, Huntress, and Prasad (1979) have estimated that radiative association reactions, if rapid, can make most complex interstellar molecules. In Table II are compared a small selection of their required k_{RA} values with our calculated ones at 50 K. The neutral molecules resulting from subsequent ion-electron reactions are also listed.

TABLE II   Comparison of Results With Model

Reaction	Eventual Product	Req. k_{RA}/Cal. k_{RA} (50 K)
$C^+ + H_2$	CH	1
$CH_3^+ + H_2$	CH_4	1
$CH_3^+ + CO$	CH_2CO	$>10^{-1}$
$CH_3^+ + H_2O$	CH_3OH	10^2
$NH_3^+ + H_2$	NH_3	$1-10^2$
$NH_3^+ + CO$	HNCO	10^7
$H_2CN^+ + C_2H_2$	HC_3N	$>10^3$

It can be seen that their proposed syntheses for methane, ketene, and
possibly methanol and ammonia are not contradicted by our results,
whereas their syntheses of HNCO and maybe HC_3N are too rapid if our
work is valid. Generalizing from these results, we emphasize that
radiative association reactions likely are important in interstellar
chemistry but that each reaction must be examined individually. The
safest method for determining whether a given ion-molecule reaction
will undergo facile low temperature radiative association is to
determine whether or not a rapid three-body channel leading to a stable
species occurs in the laboratory. If so, then a long-lived, tightly-
held complex is indicated and our statistical theory can be utilized
at low temperatures with some degree of confidence.

ACKNOWLEDGEMENTS

I wish to acknowledge support from the Petroleum Research
Fund, the Alfred P. Sloan Foundation, and the National Science
Foundation.

REFERENCES

Herbst, E. : 1976, Astrophys. J. 205, pp. 94-102.
Herbst, E. : 1979a, J. Chem. Phys. 70, pp. 2201-2204.
Herbst, E. : 1979b, Astrophys. and Sp. Sci. 65, pp. 13-20.
Herbst, E. : 1979c, In preparation
Huntress, W.T., Jr., and Mitchell, G.F. : 1979, Astrophys. J. 231,
    pp. 456-467.
Mitchell, G.F., Huntress, W.T., Jr., and Prasad, S.S. : 1979, Astrophys.
    J. 233, pp. 102-108.
Smith, D. and Adams, N.G. : 1978, Astrophys. J. Letters 220,
    pp. L87-L92.

DISCUSSION FOLLOWING HERBST

Hagen: Your calculations indicate that HNCO cannot be formed by
the radiative association of NH_3^+ and CO; it is your "most negative"
result. Our experiments show that this molecule is formed in large
quantities by ultraviolet irradiation of low temperature (10 K) grain
mantle analogs composed of solid CO and NH_3.
   Herbst: The fact that the radiative association of NH_3^++CO is too
slow to produce the observed HNCO abundance does not mean that there do
not exist other, more efficient, such gas-phase syntheses.
   D. Smith: The collisional association reaction NH_3^++CO+He→$NH_3^+ \cdot$CO+He
does not proceed at a measurable rate at 300 K, whereas the NH_2^++CO+He
association reaction proceeds at a significant rate at 300 K.
   Allamandola: You list rate constants as applicable at 10 K and 50 K.
Do you take different activation energy effects into account at these

two different temperatures? If so, how do you do this? If not, while your rate constants at 50 K are probably very reliable I think that they may be significantly lower in the 10 K limit, because at this low temperature the thermal energies are now comparable to activation energies for many low activation energy reactions.

Herbst: The question of whether or not small activation energies exist in ion-molecule systems has not yet been addressed by low temperature experiments.

Prasad: A rate coefficient $k=3\times10^{-10}\,cm^3s^{-1}$ for the reaction $H_2CN^{+}+C_2H_2 \rightarrow H_4C_3N^{+}+h\nu$ was required by Mitchell, Huntress and Prasad on the assumption that HC_3N was totally destroyed by reactions with $-C^{+}$. It is possible that reactions of HC_3N with $-C^{+}$ merely recycle the former species. In this case HC_3N may have a long lifetime ($\geq 10^{15}s$). The postulated radiative association mechanism would then become relevant to HC_3N synthesis even with the smaller k you predict.

THE FORMATION OF INTERSTELLAR MOLECULES VIA RADIATIVE ASSOCIATION REACTIONS

David Smith and Nigel G. Adams
Department of Space Research, University of Birmingham,
Birmingham, England

ABSTRACT

The radiative association rate coefficients and their temperature dependences have been estimated for several likely interstellar ion-molecule reactions from laboratory collisional association rate data. They include the $CH_3^+ + H_2$ and $CH_3^+ + H_2O$ reactions, which we suggest lead to CH_4 and CH_3OH respectively, and the critical association reaction $C^+ + H_2$.

RADIATIVE ASSOCIATION RATES FROM LABORATORY COLLISIONAL RATE DATA

Our initial studies of the ternary association reactions of CH_3^+ ions with the molecules X = H_2, N_2, O_2, CO and CO_2 showed that their rate coefficients, k_3, varied with temperature according to the relation $k_3 = AT^{-n}$, where $n \sim 4$ between 300 K and 225 K (Smith and Adams, 1978a). From this data we have deduced the lifetimes against unimolecular decomposition of the excited intermediate complexes $(CH_3X^+)^*$ and, on the assumption that $(CH_3X^+)^*$ can be stabilised via the emission of an infra-red photon (radiative lifetime $\sim 10^{-3}$s), we have shown that the corresponding binary radiative association reactions (rate coefficients k_r) will proceed at significant rates at the temperatures of interstellar clouds (Smith and Adams, 1977, 1978b). Especially interesting is the $CH_3^+(H_2, CH_5^+)h\nu$ reaction for which we deduce a value of k_r at 50K of $4.0 \times 10^{-13} cm^3 s^{-1}$ and which we suggest generates CH_4 in dense clouds in a relative abundance of $\sim 10^{-4}$.

Subsequent to these studies we have confirmed the T^{-n} variation of k_3 for the above reactions down to a temperature of ~ 100 K (Adams et al., 1979), and now we have extended the temperature range upwards to ~ 500K for the H_2 and CO reactions, firmly establishing the power law temperature dependence over the range ~ 100K to ~ 500 K for these reactions.

The collisional association reactions of CH_3^+ ions with the polar

323

B. H. Andrew (ed.), Interstellar Molecules, 323–324.
Copyright © 1980 by the IAU.

molecules H_2O, NH_3, CH_3OH and H_2CO are especially rapid, k_3 for these being appreciable fractions of their collisional rate coefficients even at 300 K, and consequently their k_r in interstellar clouds should be large. Thus we have suggested that these reactions will contribute significantly to the synthesis of the interstellar CH_3OH and H_2CO via the reaction sequence

$$CH_3^+ \xrightarrow{\quad H_2O \quad} CH_3H_2O^+ \xrightarrow{\quad e \quad} CH_3OH$$

$$CH_3^+ \xrightarrow{\quad CH_3OH \quad} H_3CO^+, CH_3CH_3OH^+ \xrightarrow{\quad e \quad} H_2CO, C_2H_5OH/(CH_3)_2O$$

This suggested scheme is not inconsistent with the astronomical observations of Gottlieb et al. (1979).

It seems likely that many other radiative association reactions contribute to molecular synthesis in interstellar clouds. Therefore we are continuing our laboratory studies over the wider temperature range now attainable in our experiment. Special attention is being given to the reactions of those interstellar ions which, like CH_3^+, do not undergo significant binary reaction with H (e.g. C^+, $C_2H_2^+$, $C_2H_3^+$, NH_4^+, HCO^+ and H_3CO^+), and are thus available to undergo the generally slower radiative association reactions. To date we have studied the reaction $C^+ + H_2 + He \longrightarrow CH_2^+ + He$, for which $k_3 \approx 7.5 \times 10^{-27}T^{-1 \cdot 2} cm^6 s^{-1}$ over the temperature range ~ 100 K to ~ 500 K. The importance of the corresponding radiative association reaction, $C^+(H_2, CH_2^+)h\nu$ is well known, and several theoretical estimates of k_r have been made. These have varied from 10^{-17} to $10^{-14} cm^3 s^{-1}$, depending on the radiation emission process envisaged. Using the analysis adopted for the CH_3^+ reactions we deduce a value for k_r which lies within the range 10^{-16} to $10^{-15} cm^3 s^{-1}$ at 50 K, assuming, as before, that the $(CH_2^+)^*$ excited complex decays via the emission of an infra-red photon. Other collisional association reactions which we have as yet only studied briefly include those between $C_2H_2^+$, $C_2H_3^+$, $C_2H_4^+$ and CO, which proceed at very appreciable rates ($k_3 \sim 10^{-29} cm^6 s^{-1}$ at 300 K).

REFERENCES

Adams, N.G., Smith, D., Lister, D.G., Rakshit, A.B., Tichy, M., and
     Twiddy, N.D.: 1979, Chem. Phys. Letts. 63, pp. 166-170.
Gottlieb, C.A., Ball, J.A., Gottlieb, E.W., and Dickinson, D.F.: 1979,
     Ap. J. 227, pp. 422-432.
Smith, D., and Adams, N.G.: 1977, Ap. J. 217, pp. 741-748.
Smith, D., and Adams, N.G.: 1978a, Chem. Phys. Letts. 54, pp. 535-540.
Smith, D., and Adams, N.G.: 1978b, Ap. J. 220, pp. L87-L92.

ON THE FORMATION OF INTERSTELLAR LINEAR MOLECULES

Akira Sakata
The University of Electro-Communications
Chofu, Tokyo
Japan

ABSTRACT

A possible mechanism for the formation of interstellar linear molecules is studied experimentally by means of synthesis apparatus. C_2 and
CN radicals are abundantly formed from plasmas containing C, N and H
atoms. The C_2 and CN radicals survive electron bombardment. They
collide with each other and recombine to form linear molecules.

INTRODUCTION

It has been shown that unsaturated bonds survive the bombardment
of excited electrons (Sakata et al., 1978). Especially prominent among
unsaturated interstellar molecules are molecules which have linear-
chained structure and conjugated triple bonds. Cyanoacetylene (HC_3N)
(Turner, 1971), cyanobutadiyne (HC_5N) (Avery et al., 1976), cyanohexatri-
yne (HC_7N) (Kroto et al., 1978) and cyanooctatetrayne (HC_9N) (Broten et
al., 1978) have been discovered in interstellar space. It is important
to clarify how these molecules are formed.

APPARATUS AND EXPERIMENTAL DETAILS

Figure 1 shows a schematic diagram of our apparatus. The apparatus
is composed of three parts: a plasma generator at the right of the
diagram, high-vacuum reaction chambers near the center, and at the left
a product detector that employs a quadrupole mass-spectrometer. Initial
gases containing C, N, O and H atoms are introduced from a source gas
reservoir. The gases are heated and decomposed into ions and radicals
by electron bombardment using microwave power generated by a magnetron.
The high energy gases are injected into the vacuum chamber. In these
injected gases ions and radicals recombine to form new molecules. Two
types of experiments have been performed: one detects directly the
products from the plasmas and the other detects trapped molecules on a
cooled copper block at a temperature of 25K.

325

B. H. Andrew (ed.), Interstellar Molecules, 325–329.
Copyright © 1980 by the IAU.

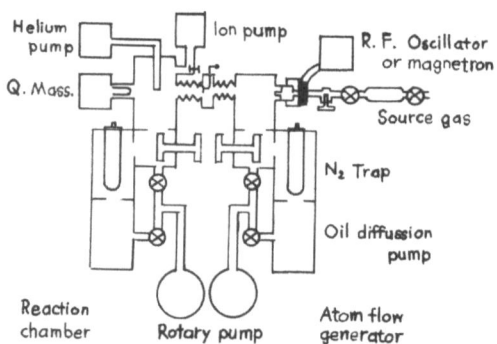

Fig. 1 Molecular Formation Apparatus

RESULTS

When bombarded, both a mixture of CH_4 and NH_3 and a gas of CH_3NH_2, all of which are single-bond molecules, were converted to gas mixtures of N_2, HCN, and C_2H_2, which have triple bonds. A similar pattern of time dependence was observed in both cases. Accordingly it was supposed that there is at some point a stage where the gases in the discharge tube are decomposed into monoatomic or diatomic ions and radicals.

From a gas mixture of methane and hydrogen, acetylene is abundantly formed at the time of discharge. When a mixture of CH_4, NH_3 and H_2 is bombarded, HCN, C_2H_2 and N_2 are produced. HCN and C_2H_2 are increasingly formed as the time of the discharge is increased. When four kinds of gas mixtures containing C, N, O and H atoms in the ratio 1:1:1:10, $C_2H_2 + O_2 + N_2 + 9H_2$, $2CH_4 + O_2 + N_2 + 6H_2$, $CO + NH_3 + 7/2H_2$ and $CH_4 + NO + 3H_2$, are bombarded, the products are as follows: from the gas mixture containing acetylene, CO, N_2, HCN and C_2H_2 are formed, and from the three gas mixtures without acetylene, CO, N_2 and HCN are formed. These results indicate that C_2 radicals survive oxidation. As a consequence of these experiments we think that C_2 and CN radicals are very stable in a radiation field.

Figure 2 and Tables 1 and 2 show the products derived from the bombardment of a gas mixture of C_2H_2, NH_3 and H_2 (2:1:2). They were trapped on a copper block cooled by a helium pump to 20K, and then gradually evaporated. Two types of products are seen: linear molecules such as $H(C\equiv C)_nH$ and $H(C\equiv C)_n \cdot CN$, and ring molecules such as cyclopentadiene, pyrrole, benzene, toluene, indole, indene and naphthalene.

Figure 3 shows typical patterns of the formation of linear molecules and ring molecules. M/e 74 and 75 indicate the linear molecules HC_6H and HC_5N, and m/e 78 indicates benzene (C_6H_6), a ring molecule. Linear molecules and ring molecules behave oppositely. At the beginning of the discharge ring molecules are formed, but with time they form less abundantly. On the other hand linear molecules are formed increasingly with time.

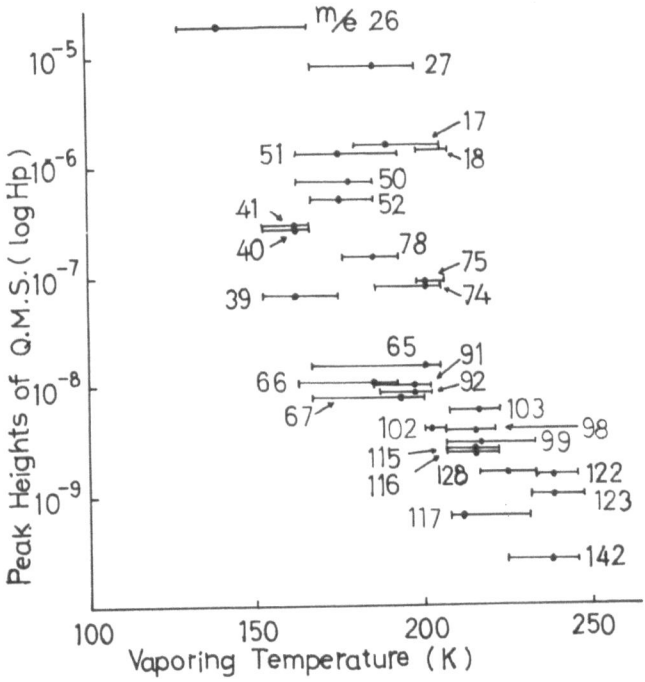

Fig. 2 Discharge products from the gas mixture of C_2H_2, NH_3 and H_2 (2:1:2)

Table 1   Produced molecules which are composed of C and H

m/e		m/e	
16	CH_4	91,92	H_3C-◯
26	C_2H_2	98	HC_8H
40	$H_2C=C=CH_2$	102	$HC\equiv C$-◯
40	$HC\equiv C-CH_3$	116	◯◯
50	HC_4H	116	$CH_3-C\equiv C$-◯
66	◯	122	$HC_{10}H$
72	HC_6H	128	◯◯◯
78	◯	142	◯◯-CH_3

Table 2   Produced molecules which are composed of C, N and H

m/e		m/e	
17	NH_3	67	◯
27	HCN	75	HC_5N
28	N_2	99	HC_7N
41	H_3C-CN	103	◯-CN
51	HC_3N	117	◯◯
52	$NC-CN$	123	HC_9N
65	$H_3C-C\equiv C-CN$		

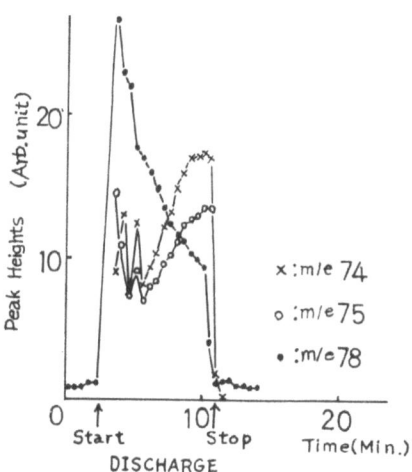

Fig. 3 Time variation of the mass-spectrum of the
products from a gas mixture of C_2H_2, NH_3 and H_2 (2:1:2)

SUMMARY

A formation mechanism of interstellar linear molecules was investigated experimentally. The experiments were done with plasmas containing
C, N, and H atoms. After condensation onto a cooled substrate, unsaturated organic molecules were abundantly formed. $-C\equiv C-$ and $-C\equiv N$ bonds are
very strong and survive electron bombardment. $-C\equiv C-$ and $-C\equiv N$ radicals
recombine each other to form $-C\equiv C-C\equiv C-$, $-C\equiv C-C\equiv N$, $-C\equiv C-C\equiv C-C\equiv C-$, and
$-C\equiv C-C\equiv C-C\equiv N$ radicals. They form a series of cyanopolyyne molecules.

REFERENCES

Sakata, A. and Nakagawa, N.: 1978, in Proceedings of the Second ISSOL
    Meeting and the Fifth ICOL Meeting, ORIGIN OF LIFE, ed. Noda, H.,
    (Japan Scientific Soc. Press), p. 51.
Turner, B.E.: 1971, Astrophys. J. (Letters) 163, L135.
Avery, L.W., Broten, N.W., MacLeod, J.M., Oka, T., and Kroto, H.W.: 1976,
    Astrophys. J. (Letters) 205, L173.
Kroto, H.W., Kirby, C., Walton, D.R.M., Avery, L.W., Broten, N.W.,
    MacLeod, J.M. and Oka, T.: 1978, Astrophys. J. (Letters) 219, L133.
Broten, N.W., Oka, T., Avery, L.W., MacLeod, J.M. and Kroto, H.W.: 1978,
    Astrophys. J. (Letters) 223, L105.

DISCUSSION FOLLOWING SAKATA

Field: In a number of your experiments you consider the reactions of the products of a microwave discharge of mixtures containing H_2, for example. Many molecular species on emerging from microwave discharges will be vibrationally excited - indeed this is a standard method of forming H_2 (V=1). The chemistry which subsequently emerges may be strongly influenced by the presence of such vibrationally excited species, which are presumably not important in the interstellar medium.

Sakata: We have done our experiments supposing mainly highly excited conditions such as those found in the circumstellar clouds, in the shocked region in Orion, and so on.

Willner: Is there a simple physical reason for the formation of completely unsaturated molecules in preference to molecules having some degree of saturation?

Sakata: Saturated bonds which have one shared electron tend to be destroyed more easily by electron bombardment.

Feldman: Regarding the experimentally demonstrated production of ring molecules in microwave discharges, I wish to point out that Andrew and I have searched for pyrrole, pyridine, and benzonitrile at the peak HC_5N position in TMC 1 using the 46 m telescope at the Algonquin Radio Observatory. Our results are negative at the $T_A \sim 10$ mK level.

Allamandola: In what part of the interstellar medium do you think there takes place this process of the formation of diatomic molecules in a plasma, and their subsequent condensation into larger molecules?

Sakata: Diatomic molecules will be formed in such high energy fields as strong radiation and strong mutual collisions among atoms, molecules and radicals. Their subsequent condensation into larger molecules takes place under conditions milder than those in the field in which the diatomic molecules are formed. I think one might expect these processes in circumstellar atmospheres, in clouds irradiated by stars, and in shocked regions in interstellar clouds.

LABORATORY AND MODELING STUDIES OF CHEMISTRY IN DENSE MOLECULAR CLOUDS

W. T. Huntress, Jr., S. S. Prasad and G. F. Mitchell,
Jet Propulsion Laboratory, Pasadena, CA 91103.

Abstract: A chemical evolutionary model with a large number of species and a large chemical library is used to examine the principal chemical processes in interstellar clouds. Simple chemical equilibrium arguments show the potential for synthesis of very complex organic species by ion-molecule radiative association reactions.

Theoretical models are used as key tools in obtaining an understanding of the chemistry occurring in interstellar clouds. These models have two major requirements for input: 1) the available observational abundances of the major elements, and the radiation field, and 2) both laboratory and estimated data on the physical and chemical processes which may occur in these clouds. The output of these models are then compared with observations and iterated to match observations as closely as possible in order to bound the possible chemistry. At this stage in the history of interstellar cloud chemistry, there is insufficient laboratory and observational data to produce models which are as realistic, accurate, or comprehensive as are (for example) the chemical models of the earth's upper atmosphere. The times are exciting nonetheless because it is precisely at this early stage that the learning rate is very high and discoveries tend to be major ones. In this spirit we have conducted two types of theoretical studies. In the first, we have constructed a large, time-dependent chemical model to complement the laboratory work which we have conducted in the past several years on ion-molecule reactions in interstellar clouds. The purpose of the large chemical evolutionary model is not to attempt simulations of the thermal, dynamic or chemical history of particular interstellar clouds. The major purpose is to examine in detail, in as comprehensive a manner as possible, the chemistry in interstellar clouds under a variety of conditions in order to determine what chemistry is of major importance in understanding observations of molecular abundances. In the second approach, we have used simple analytical chemical equilibrium arguments to examine the potential importance of new synthetic mechanisms for producing complex interstellar molecules. If found to be potentially important, then these mechanisms are

331

B. H. Andrew (ed.), Interstellar Molecules, 331–336.
Copyright © 1980 by the IAU.

included in the large model. The large model then identifies the most important production and loss mechanisms for any particular molecule under any set of cloud conditions. These major processes can then be used in dynamical and thermal models where the chemical reaction network must be limited. New directions are also obtained for laboratory measurements and suggestions can be made for observations.

In the large model, the chemical reaction network is not fixed and can change with time as the chemical environment changes. This is achieved by specifying an extraordinarily large library of chemical reactions and photoprocesses without any assumptions as to the importance of any single reaction for formation or destruction of a particular species. At present this library consists of over 1400 reactions amongst 137 species. The computer scans the library at suitable time intervals and selects which reactions in the library are significant under the prevalent conditions at the local time in cloud chemical evolution. The advantages of this approach are: 1) it is a time-dependent, chemical evolutionary model, 2) a large number of species can be managed and there are no a-priori assumptions concerning the reaction network other than the choice of reactions in the library, 3) the chemistry data base is large and can be readily changed and updated, and 4) the coupling of the chemistry between various families of species is not neglected.

The results which have been obtained using this model have been published elsewhere.[1,2] A major product of these models has been diagrams such as Fig. 1, which illustrates the major chemical processes in interstellar clouds and the chemical coupling between chemical families. Temperature, density, thermal history, chemical activation energies, and non-equilibrium effects are shown to be important in the case of particular interstellar molecules. The abundance of O_2 is critically dependent on the activation energies for the neutral reactions involved in its formation and destruction. In turn, the abundance of CN depends critically on the activation energies of the major loss reactions with O and O_2. The molecules C_2H and HC_2CN may be created quite early in the history of a dark cloud when the $C^+ \rightarrow CO$ conversion is still in progress. High abundances of these two species out of chemical equilibrium may be maintained at later times by a combination of lower cloud temperature and activation energies in major neutral loss processes. The NH_3 abundance may be very dependent on temperature, with higher abundances both at very low temperatures and at high temperatures. "Soft" recombination via radiative association of molecular ions (such as NH_3^+) with H_2 before recombination with electrons may lead to more efficient formation of NH_3, H_2CO, HCO and CH_3.

Radiative association other than the C^+-H_2 and $CH_3^+-H_2$ reactions were excluded from the large model. However, it was shown in discussion that the inclusion of certain radiative association reactions would relieve a number of problems with the model. A temperature dependent rate constant for the C^+-H_2 reaction is required in order to

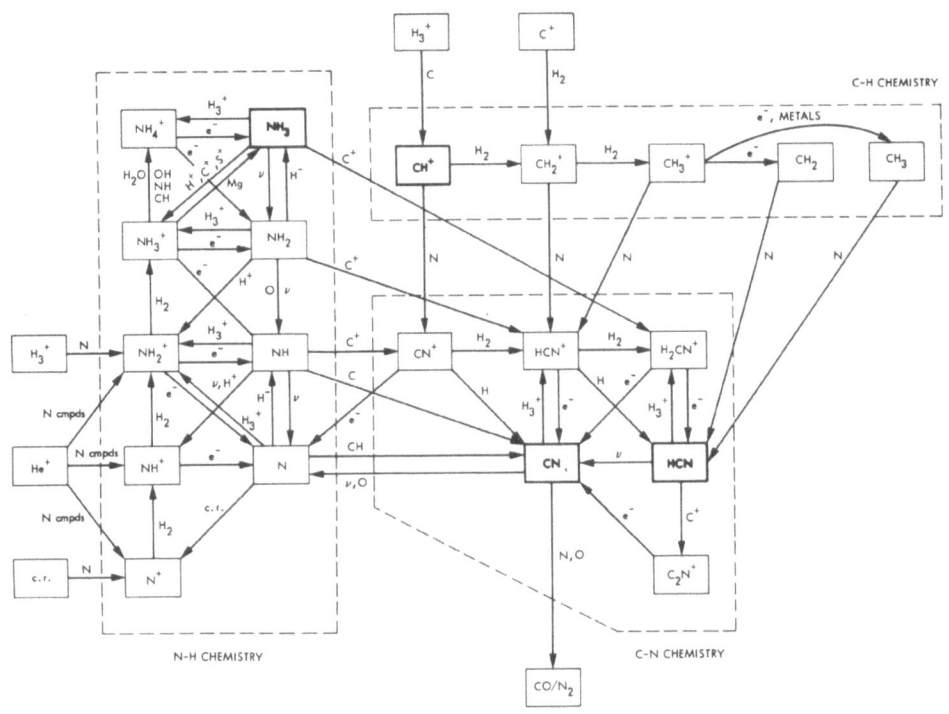

Figure 1

account for CH abundances in both diffuse and dense clouds, and large
rate constants for both the C^+-H_2 and CH_3^+-H_2 reactions at low temperatures are required in order to reproduce the lower CO abundance in
dark, low temperature clouds in comparison to Orion. The observed
amount of HC_2CN and CH_3C_2H in dense clouds requires effective competition by the C^+-H_2 radiative association reaction (compared to the
C^+-H_2O reaction) in order to provide sufficient precursor CH_n compounds
(vs. CO) for formation of these species.

In addition to the large model, simple chemical equilibrium arguments have been applied which show the potential for the gas phase
synthesis of very complex interstellar molecules by ion-molecule radiative association reactions. In addition to the radiative association
reactions suggested by Smith and Adams[3], we have suggested[4] a larger
set of reactions, not yet measured in the laboratory, which can account
for the observed abundances of molecules such as CH_3OOCH, CH_3CHO,
CH_3CH_2OH, CH_3OCH_3 and all of the presently observed complex molecules.
The absolute and relative abundances of the cyanopolyynes, $HC_{2n}CN$, also
can be explained[5] assuming certain ion-molecule radiative association
reactions (Fig. 2). In this case, a laboratory measurement[6] has been
made of the collisional analog of the initiating reaction:

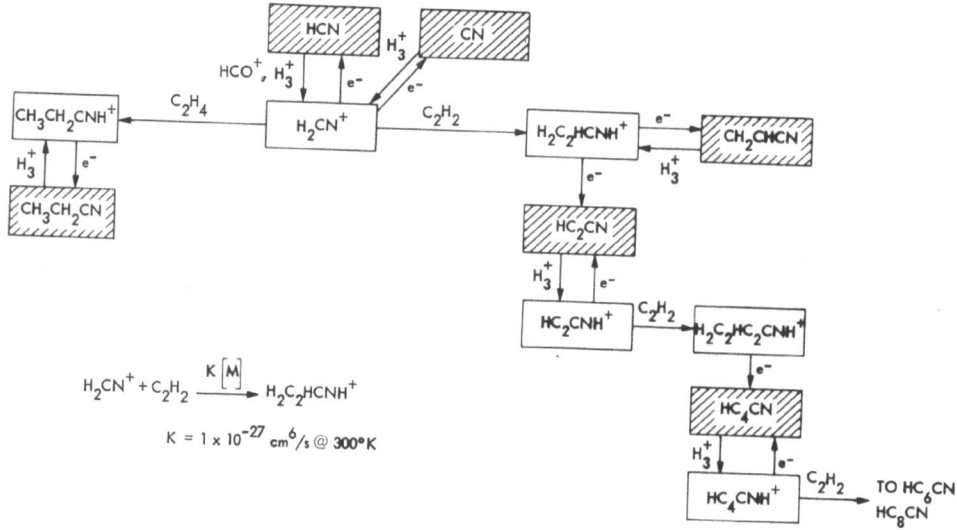

Figure 2

$H_2CN^+ + C_2H_2$. The laboratory and theoretical results thus far lend an optimistic view that gas phase ion chemistry may indeed be capable of synthesizing even the most complex of observed interstellar molecules.

REFERENCES

1. Prasad, S. S. and Huntress, W. T.: 1979, Astrophys. J. Suppl. Ser. in press.
2. Prasad, S. S. and Huntress, W. T.: 1979, Astrophys. J. in press.
3. Smith, D. and Adams, N. G.: 1978, Astrophys. J. (Letters) 220, p. L87.
4. Huntress, W. T. and Mitchell, G. F.: 1979, Astrophys. J. 231, p. 456.
5. Mitchell, G. F., Huntress, W. T. and Prasad, S. S.: 1979, Astrophys. J. 233, p. 102.
6. McEwan, M.: private communication.

DISCUSSION FOLLOWING HUNTRESS

Tatum: How well are reaction rates known for such a large number of reactions, and whence are they obtained?

Huntress: For ion-molecule reactions, a number of groups are very active in producing the required data. There is flowing afterglow work at NOAA Boulder, the University of Birmingham and York University and ion cyclotron resonance work at the Jet Propulsion Laboratory. A great deal of the required data has been, and is being, obtained in these

studies. For neutral-neutral reactions, there are no studies being conducted specifically for interstellar applications, but there is a large
body of data available from a number of laboratories all over the world
working mainly on Earth stratospheric chemistry. The lowest temperature
data is not available, the data set is smaller and the activation
energies are not well enough known, but the general features of this
class of chemistry can be deduced. Dissociative recombination rates have
been measured in microwave discharge work at the University of Pittsburg
and in merged beam work at Western Ontario, but the product distributions
for these reactions are a problem. Photodissociation and photoionization
rates are also a problem, but John Black has compiled much of this data
in his thesis. About 1/3 of the reactions in our chemistry library have
been measured in the laboratory, another 1/3 are derived on the basis of
analogies to known reactions, and another 1/3 are guesses. There is
every expectation that the situation will constantly improve, given the
large amount of laboratory and theoretical activity presently being conducted. For most of the simpler molecules in dense molecular clouds, say
those with about five atoms or less, the rate constants for the major
gas-phase production reactions are fairly confidently known.

Mouschovias: The goal of studies of interstellar chemistry is to
reveal the physical conditions in interstellar clouds, and therefore to
provide us with at least clues to the dynamical processes leading to star
formation.

A question persistently recurs to me as I listen to detailed pre-
sentations (often to the point of saturation of the mind) on interstellar
chemistry. Suppose that we have built the ultimate detectors, and
carried out the most complete observational programs and associated calculations, so that we have detected every detectable molecular transition
of every existing molecular species. We will then have a complete map of
the physical conditions in dense clouds, such as the dependence of gas
density and temperature on position within a cloud, the masses and shapes
of clouds, etc. However, the question of what gave rise to these physical
conditions will still remain. They are, certainly, determined by the
dynamical processes of cloud formation, equilibrium, and collapse.

When these dynamical processes are considered, it becomes imperative
to study in detail the effect of the interstellar magnetic field, whose
energy density is larger than (or at least comparable to) any other
interstellar energy density. It is disappointing to see that of the
seventy or so papers selected for presentation at this symposium, not a
single one deals with new observational information or new theoretical
predictions concerning magnetic fields in dense clouds, and how observations of molecules can test these predictions. Apparently, the
relation between dynamical processes (directly related to star formation)
in magnetic clouds, and observable parameters, such as the velocity
structure of a cloud, is not recognized or appreciated.

For example, we have predicted recently (1979, Ap. J., 230, 204) by
exact analytical calculations that certain fragments within dense clouds
should rotate in a *retrograde* sense with respect to the sense of their
revolution about the axis of rotation of the cloud as a whole. (The
evident significance of this prediction lies in the fact that it offers
a natural explanation of possible retrograde rotation in stellar and

planetary systems as a purely magnetic phenomenon.) The range of
fragment densities in which this effect takes place is from somewhat
less than $10^4 cm^{-3}$ to somewhat greater than $10^6 cm^{-3}$. Since these are
typical densities found in molecular clouds, one may look for retrograde
spin of dense fragments by using optically thin molecular lines (such
as ^{13}CO or an excited state of OH). Such evidence will reveal itself
at the location of a fragment either as a "shoulder" on a plot of
radial-velocity versus distance-from-the-rotation-axis, or as a reversal
of the sign of the slope of such a curve.

A second new result of relevance to this meeting is an as yet
unpublished time-dependent solution for ambipolar diffusion. We find
that supersonic ion-neutral drift can be attained easily and very
rapidly. If a significant fraction of grains are charged, inelastic
collisions between charged and neutral grains can release molecules into
the gas phase, as desired by Dr. Greenberg.

In summary, the interstellar magnetic field can largely determine
the dynamics of dense clouds and, therefore, the physical conditions,
which in turn determine the cloud chemistry. Molecular observations,
on the other hand, can test the importance of magnetic effects in dense
clouds.

Huntress: The goal of studies of interstellar chemistry is not just
the derivation of the physical conditions in interstellar clouds.
Another goal is to obtain an understanding of the evolution of organic
compounds, from atoms and a radiation field in interstellar clouds, to
prebiotic compounds in a primordial life-evolving ocean. How does the
chemistry evolve in a collapsing interstellar cloud, and what is the
resulting composition of a presolar nebula? How is that composition
reflected in the early primitive atmospheres of newly-born planets?
Studies of interstellar chemistry can yield clues to physical conditions
in interstellar clouds because the chemistry is so sensitive to these
physical conditions. Thus it is one goal of such studies, but it is not
the one and only reason these studies are conducted.

MOLECULAR EVOLUTION IN DENSE CLOUDS

H. Suzuki
Department of Physics, Kyoto University, Kyoto, Japan

Molecular abundances were calculated time-dependently using a chemical scheme which was carefully constructed to represent the molecular evolution in dense clouds. All ion-neutral reactions of species containing up to 2 heavy atoms (C, N, O, S and Si) were surveyed and a set of possible reactions was selected from numerous exothermic reactions. As molecules containing 3 to 4 heavy atoms are produced by condensation reactions, the scheme was extended. As a result, 234 species, their 2884 gas-phase reactions, and H recombination on grain surfaces were included in our chemical scheme.

A numerical result based on this chemical scheme is shown in Figure 1. The cloud model is: $\zeta_H/n_H=10^{-23}$ (ζ_H is the cosmic-ray ionization rate of an H atom), T=30K, and all elements are initially atomic and have cosmic abundances. As there are numerous species, Figure 1 is divided into four. Observed molecules which are included in our scheme are produced successfully whether small or large. Most molecules reach their peak abundances at $t_c \sim 10^{15}(10^5/n_H)$ s, when most C atoms have been exhausted in forming molecules (mainly CO), and then attain a steady state at $t_s \sim 2 \times 10^{16}(10^5/n_H)$ s. These time scales are approximately proportional to $1/\zeta_H^{0.6}$ and $1/\zeta_H$ for t_c and t_s respectively, and are very insensitive to n_H. Since these time scales are comparable to the evolutional time scale of dense clouds, time-dependent models are essential. The C/CO ratio is a good indicator of molecular evolution in dense clouds.

Some molecules, especially radicals such as CH, CN and HCO, are overabundant compared with the observations. This should be due to ignoring the effects of neutral-neutral reactions (because they may have activation energies) and condensation onto grain surfaces. Though these processes as well as ejection of molecules from grain surfaces may influence the molecular abundances, the most important part of molecular evolution in dense clouds should be attributed to the ion chemistry in the gas phase.

B. H. Andrew (ed.), Interstellar Molecules, 337–338.

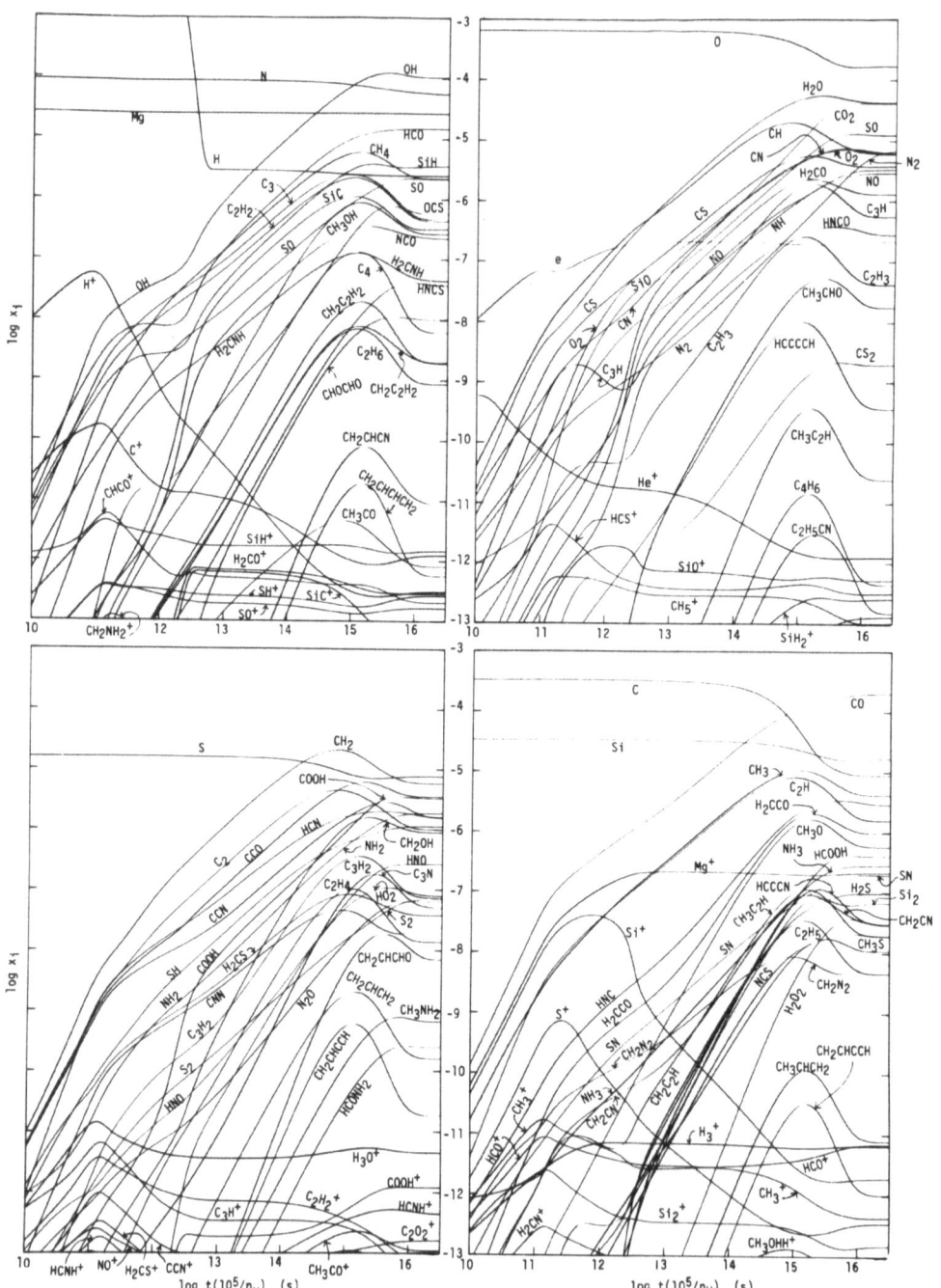

Figure 1.   Relative abundances $x_i = n_i / n_H$ versus time t.

THE DETERMINATION OF ELECTRON ABUNDANCES IN INTERSTELLAR CLOUDS

Alwyn Wootten
Dept. of Astronomy, University of Texas
Owens Valley Radio Observatory, Caltech
Ronald Snell
Dept. of Astronomy, University of Texas
A. E. Glassgold
Dept. of Physics, New York University

A new method for estimating electron fractions in shielded molecular clouds is proposed on the basis of gas phase ion-molecule reactions which involves measuring the quantity $B = n(H_2)n(HCO^+)/n(CO)$. Applied to existing data, it yields upper limits to X_e in the range from 10^{-8} to 10^{-7} for a variety of clouds, warm as well as cool. An upper bound to the cosmic ray ionization rate is also obtained.

We propose a new method for the determination of electron abundances in dense clouds based upon the abundance ratio of two important interstellar molecules, HCO^+ and CO. In common with Watson's (1977) method, ours is derived from a simple application of gas phase, ion-molecule interstellar chemistry. Unlike the fractionation of deuterated molecules, it applies to warm as well as to cool clouds. Both procedures depend upon abundance determinations of interstellar molecules, which are still very uncertain. We shall illustrate this new method primarily with the results of the recent abundance survey of Wootten et al. (1978). In cases where deuterium enhancement is measured, we can also obtain an upper limit to the cosmic ray ionization rate.

Standard application of ion-molecule chemistry leads to the following relation for the HCO^+/CO ratio

$$B \equiv n(H_2) \frac{X(HCO^+)}{X(CO)} = \frac{2\zeta K_1}{(\beta X_e+\delta)(\beta^1 X_e+\delta^1)} \tag{1}$$

where K_1, β, and β^1 are measured rate constants; ζ is essentially the primary cosmic ray ionization rate; and δ and δ^1 are respectively neutral destruction rates for H_3^+ and HCO^+. Because δ and δ^1 are uncertain, the most useful form of (1) is the upper bound

$$X_e \leq (\frac{2\zeta K_1}{\beta\beta^1 B})^{1/2} \tag{2}$$

339

B. H. Andrew (ed.), Interstellar Molecules, 339–340.
Copyright © 1980 by the IAU.

When an upper bound to X_e is already known from Watson's deuteration analysis, (1) can be converted into an upper bound on ζ:

$$\zeta \leq \frac{\beta\beta^1}{2K_1} \; [\text{Max} \; (X_e)]^2 \mathfrak{Z} \tag{3}$$

Estimates of $\mathfrak{Z}$ can be obtained from the analysis of molecular line emission. We illustrate the application of the above results with the data of Wootten et al. (1978) on the dense cores of molecular clouds. By placing the estimates of $\mathfrak{Z}$ on the theoretical curve for the observed temperature, values for X_e can be determined. With a value for ζ of $5.10^{-18}s^{-1}$ upper limits for X_e from (2) lie in the range $10^{-8} - 10^{-7}$. These limits agree with bounds already determined for L134 and L134N from analysis of DCO^+/HCO^+ ratios (Langer et al. 1978, Watson et al. 1978). For these two clouds, we deduce using (3) $\zeta \leq 5.10^{-18}s^{-1}$, in accord with estimates based on the observed cosmic ray spectrum above 1 GeV.

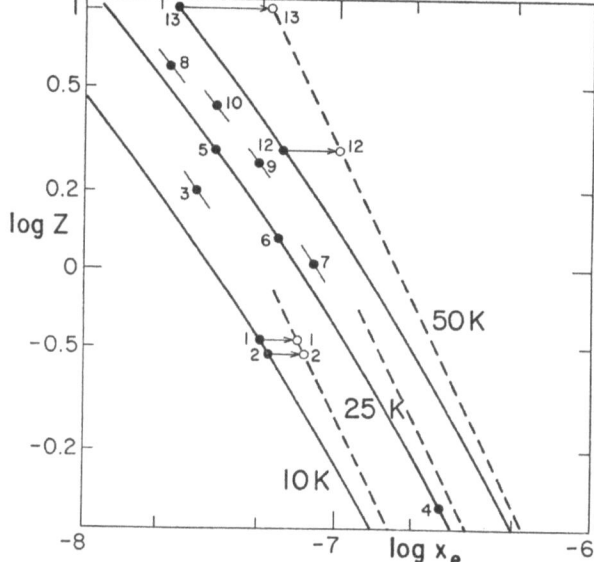

$\mathfrak{Z}$ vs. X_e plot; filled circles are observed values numbered as in Wootten et al. (1978); solid curves are theoretical for several temperatures. Dashed curves are for $\delta=\delta^1=0$ for determining upper bounds.

References

Langer, W.D., Wilson, R.W., Henry, P.S., and Guelin, M., 1978, Astrophys. J. (Letters) 225, L139.

Watson, W.D., 1977, CNO Isotopes in Astrophysics, ed. J. Audouze (D. Reidel, Dordrecht).

Watson, W.D., Snyder, L.E., and Hollis, J.M., 1978, Astrophys. J. (Letters) 222, 145.

Wootten, A., Evans, N.J., Snell, R., and Vanden Bout, P., 1978, Astrophys. J. (Letters) 225, L143.

MOLECULE FORMATION IN COOL, DENSE INTERSTELLAR CLOUDS

William D. Watson
Departments of Physics and Astronomy
University of Illinois at Urbana-Champaign
Urbana, IL 61801

ABSTRACT

A discussion is given of the general processes and considerations that arise in attempting to understand molecular reactions in cool, dense interstellar clouds. Basic elements of the gas phase, "ion-molecule" scheme are given explicitly before surveying topics in which there is considerable current activity. These topics include: (i) refined comparisons of prediction and observation for species of "intermediate" complexity, (ii) numerical computations of cloud models which include numerous chemical reactions and molecular species, (iii) formation of the complex molecules, (iv) isotope fractionation in interstellar molecules and (v) possible contributions from chemistry in shocks.

I. INTRODUCTION

I will discuss reaction processes for molecules under physical conditions which are thought likely to be a satisfactory first approximation in much of the gas of most "dense" interstellar clouds. By dense clouds, I just mean those with extinctions $A_V \gtrsim 10$ for which the number density of the gas is $n \gtrsim 10^4 cm^{-3}$ and the kinetic temperature T is less than about 50 K. Clearly this is not a homogeneous group of clouds. However, it seems most efficient for discussion of the chemistry to consider the impact of their differences as refinements or particular cases within a general outline.

The chief approximation for "dense clouds" is that the intensity of ultraviolet radiation is sufficiently low, as suggested by the large extinction, that photoionization and photodissociation can be ignored. In contrast, these processes have a major if not dominant effect in the chemistry of the classical "diffuse" clouds which will be discussed subsequently by A. Dalgarno. As a result, the gas in dense clouds is mainly in molecular form so that reactions which rearrange molecules are much more important than in diffuse clouds where the gas is

341

B. H. Andrew (ed.), Interstellar Molecules, 341–353.

primarily in atomic form. Hence, the chemical reaction mechanisms tend
to be qualitatively different. A further difference in approach is due
to the generally poor knowledge of physical conditions in dense clouds
and to the substantial uncertainties in many cases in relating the
observed line intensities to molecular abundances in dense clouds.
Tests of proposed reaction schemes that are insensitive to detailed
cloud models are thus preferred. In diffuse clouds, the physical conditions are relatively well known and the number of relevant molecular
species is limited. Hence, detailed models that include the "complete"
set of chemical reactions can be more reliably compared with observations.

II. GENERAL CONSIDERATIONS

Beginning with the earliest quantitative investigation of interstellar chemistry by Bates and Spitzer in 1951, there has been considerable controversy about the relative importance of gas phase versus
catalytic processes on the surfaces of dust grains. Because of the clear
success in understanding the formation of molecular hydrogen as due to
surface chemistry, along with the absence of specific proposals for gas
phase mechanisms except in a limited number of cases, most astronomers
believed before 1973 that surface chemistry is somehow primarily
responsible for interstellar molecule formation. A key problem was
(and still is) how the molecules are ejected from the surface. Most
species, though not H_2, freeze at the temperatures expected for interstellar dust grains. In addition, the lack of predictability of surface
reactions (especially in view of the unknown composition of the surface)
made it seem unlikely that studies of interstellar chemistry could be
put on a quantitative footing. However, recognition of the importance
of reactions between positive ions and neutrals in dense clouds (Herbst
and Klemperer 1973; Watson 1973, 1974) provided the basis for a quantitative, gas phase chemistry that has been explored in considerable
detail in recent years. Reactions involving positive ions ordinarily
are much more effective than those involving only neutrals because the
former usually have no activation energy barriers when the reaction is
exothermic. In addition, possible reactions that arise in dense clouds
tend to be exothermic more often when a positive ion is involved. A
cornerstone prediction of this "ion-molecule" scheme was that the ion
HCO^+ should be abundant. At the time, its microwave frequencies were
unknown. However, a strong, unidentified emission line ("X-ogen") near
the estimated $J = 1-0$ frequency of HCO^+ had been widely observed. First,
detection of another emission line near the frequency expected for
$J = 1-0$ for $H^{13}CO^+$ (see Snyder et al. 1976) was convincing evidence that
"X-ogen" should be identified with HCO^+. Then, measurement of the
frequency for HCO^+ in the laboratory (Woods et al. 1975) provided the
definitive data for the identification.

Evidence that a gas phase chemistry oriented around positive ions
dominates in dense clouds is by no means definitive, but is highly
suggestive as a result of, for example:
    i) the occurrence and the abundance of HCO^+ and N_2H^+,

  ii) the degree and the variation of the enhancement in the deuterium
      to hydrogen ratio in molecules,
 iii) the apparent fractionation of carbon isotopes in molecules in
      certain clouds,
  iv) the presence and abundance of HNC relative to HCN,
   v) the agreement between the predicted and deduced electron
      densities in dense clouds.

To see how it is that reactions involving ions can dominate for a
gas in which the fractional ionization is only $\approx 10^{-8}$, consider the
chracteristic time scale associated with reactions on a grain surface.
Based on the surface area of interstellar dust grains, the mean time
for a <u>specified particle</u> in the gas to strike a grain -- the minimum
time for it to be involved in a surface reaction -- is roughly $t_g \approx$
$(10^{17}/n)$ sec. The minimum time for a <u>specified gas particle</u> to have
its chemical form changed through ion-molecule reactions might be expected to be comparable to the time for it to be ionized directly by
primary ionizing agent. In dense clouds, the agent presumably is high
energy cosmic ray particles. The associated time scale $t_{di} \approx 10^{17}-10^{18}$ sec
is much longer than t_g and probably even longer than the lifetime of
the cloud. On this basis, ion-molecule reactions would ordinarily be
unimportant. The key to the ion-molecule scheme, however, is the
recognition that the direct ionization of hydrogen and helium can be
transferred to the less abundant elements (e.g., C,N,O) with high
efficiency. This then reduces the effective time scale t_i for ionization of C,N,O, etc. from t_{di} by a factor in the neighborhood of the
abundance ratio ($\approx 10^4$). Thus $t_i \approx t_g$ when $n \approx 10^4 cm^{-3}$ and an ion-
stimulated chemistry might reasonably be competitive.

III. BASIC ASPECTS OF THE "ION-MOLECULE" SCHEME

     Essentials of the ion-molecule scheme in dense clouds have been
discussed in a number of reviews (e.g., Herbst and Klemperer 1976;
Herbst 1978a; Watson 1976, 1978), and will not be duplicated here.
The basic chain for the transfer of ionization is illustrated in Figure
3 of Watson (1978). Almost all ionizations of the chief component of
the gas (H_2) produce H_3^+, which frequently reacts with CO because of the
high CO abundance to produce HCO^+. Chemical reactions with CO and other
species containing heavy elements ordinarily compete with electron
recombination in the destruction of H_3^+ so that in fact an appreciable
fraction of the ionizations of hydrogen are transferred to the less
abundant, heavy elements. Ionization of helium plays a key role because the rate coefficients for reaction of He^+ with H and H_2 are
anomalously small. To be neutralized, He^+ must then ionize molecules
containing heavy elements and thus transfer its ionization. Even more
important, the neutralization process of He^+ with diatomics frequently
dissociates the diatomics. Free, reactive atoms and ions are thus
produced from CO, N_2, etc. which ordinarily are unreactive. This may be
the means by which elements remain available for chemical reactions
despite the general tendency toward depletion onto grains. Both CO and
N_2 have vapor pressures which may be sufficient to prevent them from

freezing onto the grains. They then serve as "reservoirs" for atoms which are slowly being converted into other, less abundant species.

The electron density that is predicted as a result of this ionization (e.g., Watson 1978) is considerably smaller than would be the case if the gas were purely atomic. In an atomic gas, neutralization is due primarily to radiative electron recombination. In a molecular gas, molecular ions are likely to be the primary positive ions so that dissociative electron recombination will dominate the neutralization. This process has rate coefficients that typically are a factor of 10^4-10^5 larger than for radiative electron recombination with atomic ions. Even in a molecular gas the neutralization might, in principle, be controlled by radiative recombination if atoms of low ionization energy (e.g., Na, Ca, Si) are present in the gas at the level given by cosmic abundances (Oppenheimer and Dalgarno 1974). Positive ions are, however, neutralized upon collision with the negatively charged dust grains at a rate near $10^{-16}n$ s^{-1} (Watson 1974). This rate ordinarily exceeds that for radiative recombination (rate coefficient $\approx 10^{-11} cm^3 s^{-1}$) in dense clouds.

Representative examples of schemes to form simple molecules via ion-molecule processes include the reactions of H_3^+ with atomic O or N. In the first case, the product OH^+ will react successively with two H_2 molecules, capturing an H-atom in each reaction, to produce OH_3^+. The OH_3^+ will then recombine dissociatively with an electron to produce H_2O and OH. The analogous reaction sequence for nitrogen illustrates the errors that can result from extrapolating information from measurements at 300 K to interstellar temperatures. Reaction of NH_3^+ with H_2 has a negligibly small rate coefficient at 300 K ($\approx 10^{-13} cm^3 s^{-1}$) and it has been assumed the value would be even smaller at T < 100 K. Hence NH_4^+, which could recombine dissociatively with an electron to produce ammonia, would not be produced in this manner. Alternative routes to ammonia are neutralization of the NH_3^+ by charge exchange with metal atoms and radiative association of NH_3^+ with H_2 to produce NH_5^+ which can then produce NH_3 through dissociative electron recombination. Very recently, however, the rate coefficient for $NH_3^+ + H_2 \rightarrow NH_4^+ + H$ was found to remain at about $10^{-13} cm^3 s^{-1}$ at 100 K. The alternative routes are then slower. Another interesting example is the chemistry of C^+. Although it cannot undergo an exchange reaction with H_2, radiative association may be "unusually" rapid with a rate coefficient near $10^{-14} cm^3 s^{-1}$ (see article by E. Herbst in this Symposium). The CH_2^+ then reacts with H_2 to produce CH_3^+ which cannot capture an H-atom from H_2. As discussed for NH_3^+, charge exchange with a metal atom or radiative association with H_2 followed by electron recombination may then produce CH_3 and CH_4.

Specific proposals have been made within the "ion-molecule" scheme for the formation of most molecules of "intermediate" complexity. Various meaningful tests can be performed. A quantitative understanding of these formation processes is clearly desirable as a basis for examining the abundances of the most complex molecules. Examples of proposals for species of particular note are the reaction mechanisms involving C_2H, H_2CO and HCN/HNC, all of which are summarized in Watson (1976)

among other locations. The relatively high abundance of C_2H was predicted in advance of its detection and the reactions that are involved
may be essential to producing the carbon chain molecules. Formaldehyde
is widely observed in interstellar clouds under nearly all physical
conditions. Understanding its formation processes has seemed especially
important as an indication of the ability of gas phase schemes to produce the more complex molecules. In attempting to formulate quantitative
tests of proposed reaction schemes, we (Watson, Crutcher and Dickel
1975) argued that all proposed formation schemes for H_2CO should also
produce a high abundance of HDCO which probably would be in conflict
with the failure to detect the $1_{11}-1_{10}$ transition of this species. A
considerable amount of interpretation, along with a slightly erroneous
laboratory frequency, was involved. The detection of HDCO at millimeter
wavelengths (Langer, Frerking, Linke and Wilson 1979) shows that the
abundance of HDCO actually is not in conflict with the proposed reaction
mechanisms for H_2CO (esp. $CH_3 + O \rightarrow H_2CO + H$). The high abundance of HNC
was immediately recognized as a likely consequence of specific ion-
molecule reactions and more generally as a strong indication that interstellar molecules are produced by non-(thermodynamic) equilibrium
processes in the gas phase. HCN and HNC are expected to be produced by
the dissociative electron recombination of $HCNH^+$ (see Herbst 1978b for
a detailed discussion of the dissociation) which is a result of $C^+ + NH_3$.
Since HCN and HNC are also destroyed through reaction with C^+, a model-
independent (though steady-state) result is $([HCN] + [HNC])/NH_3 \lesssim 0.6$
(e.g., Watson 1976). Though there are uncertainties in deducing the
observed abundances, observations are considered to be compatible with
this ratio. This predicted ratio will be altered if the fractional
abundances of NH_3 and H_2O, which react with C_2N^+ to produce HCN/HNC
(Schiff and Bohme 1979), exceed about 10^{-5}. Recently there has been
some debate about the likelihood that HCN and HNC should both be produced in the recombination of the molecular ion that results from
$C^+ + NH_3$. R. D. Brown (1977) noted that the structure H_2NC^+ may reasonably be expected to be metastable and that it should be formed preferentially because to do otherwise would involve breaking the N-H bonds
already present in the reactant ammonia. In this case, little HCN
might be expected to result from the electron recombination. However,
subsequent molecular orbital calculations (Conrad and Schaefer 1978)
indicate that in fact there is no energy barrier for H_2NC^+ to rearrange
into the configuration of lower energy $HNCH^+$.

IV. DETAILED NUMERICAL MODELS

Since the idealized goal of interstellar chemistry is to understand
all molecular abundances, it is certainly appropriate to attempt to construct detailed numerical models that include the formation and destruction processes for as many species as possible. Most efforts have considered only the small molecules ($\lesssim 4$ atoms). Herbst and Klemperer (1973)
constructed the first such model for dense clouds that includes ion-
molecule reactions. A number of investigators have refined and extended
the original model. More extensive reaction networks have been considered (Mitchell, Ginsburg and Kuntz 1978), along with the time.

dependence involved for the reactions to reach steady-state under fixed
physical conditions (Prasad and Huntress 1979; Suzuki 1979) and including the effects of depletion onto dust grains in a specific approximation (Iglesias 1977). The time dependent calculations do illustrate
that steady-state for the chemical reactions alone is achieved in about
10^6 years under representative conditions for dense clouds. As one
might expect from the comparison of t_g with t_i in Section II, the effects
of depletion are likely to become important in approximately the same time
scale (see Figure 4 of Iglesias 1977). Observations (Wootten et al. 1978)
indicate that fractional molecular abundances do decrease (by a factor
of ≈ 100 in the sample considered) as gas density increases and hence as
t_g decreases. Model calculations that also attempt to simulate changes
in the cloud (especially collapse) involve a time dependence in the
external physical conditions (density, temperature) are also available
(Suzuki et al. 1976; Gerola and Glassgold 1978). The differences between
these calculations already indicate some of the uncertainties that arise
in applying the models to interstellar clouds. In addition, for models
containing only the simplest molecules ($\lesssim 3$ atoms) perhaps 2/3 of the
rate coefficients have not been measured in the laboratory. Only a few
have been measured at temperatures below 300 K. Due to depletion in
even the diffuse phase of the cloud, the initial abundances of the
elements in gaseous form (especially for the metals) are also a major
problem.

V.   FORMATION OF COMPLEX MOLECULES

Understanding the formation of the complex molecules is certainly
the greatest challenge in interstellar chemistry. It is also fraught
with uncertainty because a number of steps are needed to convert individual atoms into molecules. The rate coefficients and branching ratios
for these are even more poorly known than for the smaller molecules
which were discussed in the previous sections.

In the past year or so there have been a number of studies which
indicate that at least a qualitative understanding of the chemistry of
the complex molecules may be near. This optimism is due primarily to
progress in two general areas. (i) Extensive studies have been performed for reactions of the type,

$$AX^+ + BY \rightarrow AB^+ + XY \qquad\qquad (1)$$

Here AB^+ ordinarily will be more complex than AX^+ or BY so that the
overall effect tends to be the step-by-step conversion of simpler to
more complex species. There has been speculation since the earliest
studies of ion-molecule chemistry that such processes might lead to the
more complex molecules. Neutralization by charge exchange with metals
or by dissociative electron recombination might be expected to convert
the ions into the observed neutrals. In the latter case the products
are uncertain and the neutral moelcular complex might break up into two
smaller species rather than into a large molecule. At the time of the

earlier speculations, almost no laboratory data were available on re-actions of the type in equation (1) when larger molecules are involved. Laboratory investigations have now begun to provide information on these reactions (Huntress 1977; Smith and Adams 1977; Freeman, Harland and McEwan 1979; Schiff and Bohme 1979). (ii) The second major area of progress is in the radiative association of larger molecules. Studies (Arnold 1977; Smith and Adams 1978a) show that the rate coefficients for 3-body association involving an ion,

$$A^+ + B + I \rightarrow AB^+ + I \tag{2}$$

increase rapidly (frequently by a factor of 100 or so) as the temperature is decreased from 300 K to 100 K. The increase is interpreted as due to the increased lifetime of the intermediate, excited complex $(AB^+)^*$. The lifetime of this complex can be related to the probability that the complex $(AB^+)^*$ will de-excite in the interstellar medium by emission of radiation. Though there is some uncertainty in converting the laboratory data into radiative association rates and it is unclear whether the temperature dependence observed above 100 K is completely accurate down to 20 K, it does seem clear that at least some radiative association rates are much more rapid than was previously expected. For example, Smith and Adams (1978a) deduce the following rate co-efficients $(cm^3 s^{-1})$ at 50 K: $CH_3^+ + H_2^+$ (4×10^{-13}), $CH_3^+ + CO$ (3.5×10^{-12}), $CH_3^+ + H_2O$ $(\gtrsim 2 \times 10^{-9})$ and $CH_3^+ + NH_3$ $(\gtrsim 4 \times 10^{-10})$. That is, the derived radiative association rates are in some cases near the kinetic collision rate. Herbst (1976) studied radiative association involving molecules from a theoretical viewpoint and will present refinements of his work in light of the laboratory data at this Symposium.

Formation of complex molecules has been discussed within the context of the recent laboratory data by a number of investigators (e.g., Smith and Adams 1978b; Walmsley, Winnewisser and Toelle 1979; Schiff and Bohme 1979; Huntress and Mitchell 1979). Because this work is very recent and much of it is being presented at this Symposium by the authors, I will not go into further detail.

VI. ISOTOPE FRACTIONATION

Studies of the fractionation of deuterium in molecules is perhaps the area where ion-molecule chemistry in dense clouds has been utilized most quantitatively to obtain information of general astronomical interest. Consideration of very different reactions also suggests the $^{13}C/^{12}C$ ratio in molecules may be influenced by chemical fractionation.

In the case of deuterium, reactions of the type,

$$AH^+ + HD \rightarrow AD^+ + H_2 \tag{3}$$

tend to take deuterium from the HD "reservoir" and concentrate it in other molecules. Reaction (3) is known to have a large rate coefficient when $AH^+ = CH_3^+$ and H_3^+, but apparently not for N_2H^+ or HCO^+. If reaction

(3) is to be effective in enhancing the (D/H) ratio in AH^+, it must compete with destruction of AH^+ by other processes. Specifically, it must compete with dissociative electron recombination and with reactions involving other molecular species. This provides an upper limit to the abundance of electrons and of the other species from the observed (D/H) of the molecule. Detection of DCO^+ (Hollis et al. 1976) and N_2D^+ (Snyder et al. 1977) made possible the most reliable application of this diagnostic technique because the relevant reactions are best understood for these species. From the initial detection of DCO^+, upper limits near 10^{-8} were deduced for the electron densities in cool clouds (Watson 1977). More extensive surveys and more refined studies of the radiative transfer confirm this evidence for a low electron density in cool interstellar clouds (Guelin et al. 1977; Watson, Snyder and Hollis 1978; Langer et al. 1978).

Laboratory data on the rate coefficients for dissociative electron recombination of H_3^+ at low energies (≈ 0.01 eV; Auerbach et al. 1977), as well as on isotopic forms of H_3^+ (McGowan et al. 1979) and of N_2H^+ (Mul and McGowan 1979) have been especially valuable in reducing substantially the uncertainties in the chemistry.

The isotope exchange of equation (3), is "driven" by zero point energy differences which give an exothermicity of $\Delta E/k \simeq 200-400$ K depending upon the particular reaction. Although other factors such as electron density also contribute to determining the degree of fractionation, the temperature is not negligible in comparison with $\Delta E/k$ and one might therefore expect a temperature dependence to be present in the observational data. A recent study of the (DNC/HNC) ratio for 18 clouds finds a clear temperature dependence with a derived value $\Delta E/k = 240$ K ± 60 for this species (Snell and Wootten 1979).

Fractionation of carbon isotopes in molecules can occur directly or indirectly as a result of the exchange (Watson, Anicich and Huntress 1976),

$$^{13}C^+ + {}^{12}CO \; \overset{\leftarrow}{\rightarrow} \; {}^{12}C^+ + {}^{13}CO \qquad\qquad (4)$$

It was shown analytically that the $^{13}CO/^{12}CO$ ratio will be enhanced over the $^{13}C/^{12}C$ ratio of the gas if most carbon is in atomic form, but not when most carbon is in molecular form (primarily CO). Secondly, it was shown analytically that preferential depletion of carbon-bearing molecules other than CO onto grains might lead to an overall enhancement of ^{13}C in the gas. The effect of equation (4) has been investigated further in detailed cloud models by Langer (1977) and Liszt (1978). The latter study finds that the overall enhancement due to depletion is not great for the particular assumptions of the models that are involved.

A striking observational example of the first fractionation effect for carbon isotopes apparently occurs in the cloud L134 (Dickman et al. 1977). This interpretation is supported by more detailed studies of the uncertainties due to radiative transfer effects (Dickman, McCutcheon and Shuter 1979; Langer, Goldsmith, Carlson and Wilson 1980).

VII. SHOCKS

There is evidence, especially from the infrared emission by H_2 molecules, that shocks are present to at least some degree in certain dense clouds. Broad molecular lines and "unusually" high abundances have led to proposals that in the Orion cloud the observed SiO and OSC (Lada, Oppenheimer and Hartquist 1978) as well as the H_2O (Elitzur 1979) are formed in the hot gas behind a shock front. These investigators, as well as Iglesias and Silk (1978), have performed detailed computations to assess the contribution by shocks to molecular abundances in dense clouds. Endothermic chemical reactions, which ordinarily are unimportant at the low temperatures in dense clouds, play a key role. Shocks may also contribute appreciably to molecular abundances in diffuse clouds (Elitzur and Watson 1978).

The author's research is supported in part by the U.S. National Science Foundation, Grant AST-7823648.

REFERENCES

Arnold, F.: 1977, Proc. 21st Liege Int. Astrophys. Symp. (in press).
Auerbach, D., Cacak, R., Caudano, R., Gaily, T. D., Keyser, J., McGowan, J. W., Mitchell, J. B. A. and Wilk, S. J.: 1977, J. Phys. B10, 3797.
Bates, D. and Spitzer, L.: 1951, Astrophys. J. 113, 441.
Brown, R. D.: 1977, Nature 270, 39.
Conrad, M. P. and Schaefer, H. F.: 1978, Nature 274, 456.
Dickman, R. L., Langer, W. D., McCutcheon, W. H. and Shuter, W. L. H.: 1977, in "CNO Isotopes in Astrophysics", ed. J. Audouze (D. Reidel, Dordrecht), p. 95.
Dickman, R. L., McCutcheon, W. H. and Shuter, W. L. H.: 1979, Astrophys. J. 234, 100.
Elitzur, M.: 1979, Astrophys. J. 229, 560.
Elitzur, M. and Watson, W. D.: 1978, Astrophys. J. Letters 222, L141.
Freeman, C. G., Harland, P. W. and McEwan, M. J.: 1979, M.N.R.A.S. 187, 441.
Gerola, H. and Glassgold, A. E.: 1978, Astrophys. J. Suppl. Ser. 37, 1.
Guelin, M., Langer, W. D., Snell, R. L. and Wootten, H. A.: 1977, Astrophys. J. Letters 217, L165.
Herbst, E.: 1976, Astrophys. J. 205, 94.
Herbst, E.: 1978a, in "Protostars and Planets", ed. T. Gehrels (Univ. of Arizona Press, Tucson), p. 88.
Herbst, E.: 1978b, Astrophys. J. 222, 508.
Herbst, E. and Klemperer, W.: 1973, Astrophys. J. 185, 505.
Herbst, E. and Klemperer, W.: 1976, Physics Today 29, 32.
Hollis, J. M., Snyder, L. E., Lovas, F. J. and Buhl, D.: 1976, Astrophys. J. Letters 209, L83.
Huntress, W. T.: 1977, Astrophys. J. Suppl. Ser. 33, 495.
Huntress, W. T. and Mitchell, G. F.: 1979 Astrophys. J. 231, 456.

Iglesias, E.: 1977, Astrophys. J. 218, 697.
Iglesias, E. and Silk, J.: 1978, Astrophys. J. 226, 851.
Lada, C. J., Oppenheimer, M. and Hartquist, T. W.: 1978, Astrophys. J.
    Letters 226, L153.
Langer, W. D.: 1977, Astrophys. J. Letters 212, L39.
Langer, W. D., Frerking, M. A., Linke, R. A. and Wilson, R. W.: 1979,
    Astrophys. J. Letters 232, L65.
Langer, W. D., Goldsmith, P. F., Carlson, E. R. and Wilson, R. W.: 1980,
    Astrophys. J. Letters (in press).
Langer, W. D., Wilson, R. W., Henry, P. S. and Guelin, M.: 1978,
    Astrophys. J. Letters 225, L139.
Liszt, H.: 1978, Astrophys. J. 222, 484.
McGowan, J. W., Mul, P. M., D'Angelo, V. S., Mitchell, J. B. A.,
    Defrance, P. and Froelich, H. R.: 1979, Phys. Rev. Letters 42, 373.
Mitchell, G. F., Ginsburg, J. L. and Kuntz, P. J.: 1978, Astrophys. J.
    Suppl. Ser. 38, 39.
Mul, P. M. and McGowan, J. W.: 1979, Astrophys. J. Letters 227, L157.
Oppenheimer, M. and Dalgarno, A.: 1974, Astrophys. J. 192, 29.
Prasad, S. S. and Huntress, W. T.: 1979, Astrophys. J. (in press).
Schiff, H. I. and Bohme, D. K.: 1979, Astrophys. J. 232, 740.
Smith, D. and Adams, N. G.: 1977, Astrophys. J. 217, 741.
Smith, D. and Adams, N. G.: 1978a, Astrophys. J. Letters 220, L87.
Smith, D. and Adams, N. G.: 1978b, "Kinetics of Ion-Molecule Reactions",
    ed. P. Ausloos (Plenum, New York), p. 345.
Snell, R. and Wootten, A.: 1979, Astrophys. J. (submitted for publica-
    tion).
Snyder, L. E., Hollis, J. M., Buhl, D. and Watson, W. D.: 1977,
    Astrophys. J. Letters 218, L61.
Snyder, L., Hollis, J. M., Lovas, F. J. and Ulich, B. L.: 1976,
    Astrophys. J. 209, 67.
Suzuki, H.: 1979, Prog. Theoret. Phys. 62, 936.
Suzuki, H., Miki, S., Sato, K., Kiguchi, M. and Nakagawa, Y.: 1976,
    Prog. Theoret. Phys. 56, 1111.
Walmsley, M., Winnewisser, G. and Toelle, F.: 1979, Astron. Astrophys.
    (in press).
Watson, W. D.: 1973, Astrophys. J. Letters 183, L17.
Watson, W. D.: 1974, Astrophys. J. 188, 35.
Watson, W. D.: 1976, Rev. Mod. Phys. 48, 513.
Watson, W. D.: 1977, in "CNO Isotopes in Astrophysics", ed. J. Audouze
    (D. Reidel, Dordrecht), p. 105.
Watson, W. D.: 1978, Ann. Rev. Astron. Astrophys. 16, 585.
Watson, W. D., Anicich, V. G. and Huntress, W. T.: 1976, Astrophys. J.
    Letters 205, L165.
Watson, W. D., Crutcher, R. M. and Dickel, J. R.: 1975, Astrophys. J.
    201, 102.
Watson, W. D., Snyder, L. E. and Hollis, J. M.: 1978, Astrophys. J.
    Letters 222, L145.
Woods, R. C., Dixon, T. A., Saykally, R. J. and Szanto, P. G.: 1975,
    Phys. Rev. Letters 35, 1269.
Wootten, A., Evans, N. J., Snell, R. and van den Bout, P.: 1978,
    Astrophys. J. Letters 225, L143.

DISCUSSION FOLLOWING WATSON

Glassgold: I was tremendously impressed by the success of the ion-molecule chemistry which you so well described despite the fact that there are relatively few abundance determinations in dense clouds. The recent results from Texas point up the usefulness of absolute abundance determinations, and I urge the observers here to devote some effort in this direction. I would like to cite two important applications requiring better molecular abundance determinations. The first is the estimation of cloud masses, which requires an accurate value for the $^{13}CO/H_2$ ratio. The second is the problem of understanding the thermal properties of dust clouds. Until the abundances of the major molecular coolants have been determined, the temperatures of these clouds cannot be understood.

Kutner: Absolute abundances are still hard to measure, but are coming within the realm of possibility. In the meantime much quantitative information can be obtained from studies of certain relative abundances, and how they vary with physical conditions such as temperature, density and ionization level.

Scoville: Are there any areas in which there is an apparent contradiction between the observed interstellar chemistry and ion-molecule gas-phase theory?

Watson: I know of none in the "dense cloud" regime, though I expect that Dalgarno will mention one in "diffuse cloud" chemistry.

Scoville: Is that due to lack of quantitative data?

Watson: It helps.

Kutner: There is a large range of objects that we would classify as cool, dense and dark, but into which the UV will still penetrate, depending on the grain albedo and phase function. Is there an effort being made to model the chemistry in such intermediate regions?

Watson: There are calculations by Langer and Glassgold in Ap. J.

Elmegreen: It is possible that charged grains play a fairly important role in determining the ionization fractions in dense clouds because they will neutralize the metallic ions at a faster rate than radiative recombination for electron fractions less than about 10^{-7}. Charged grains also may add considerably to the viscous coupling between a magnetic field and the neutral molecules, and may dominate this coupling at electron fractions less than about 5×10^{-8}. These numbers depend on the grain size and abundance, but they should be representative of typical conditions.

Watson: The contribution of grains to the recombination has been noted several times in the literature.

Gold: Gas-phase build-up of molecules would lead to a very different dependence of abundance on atomic number from the process of association on grains. Do we not have enough statistical information of abundances to decide which process is more important?

Watson: It is not clear to me that there would necessarily be a very different dependence of abundance on atomic number, given our lack of knowledge of grain processes and of the physical conditions under which gas-phase reactions proceed.

Allamandola: You make the point that any serious consideration of the contribution of grain reactions to the overall cloud chemistry must first overcome the problem of getting the molecules which have stuck to the grains back off into the gas-phase. In our experiments at Leiden, we are duplicating, as far as possible, molecular mantles at 10 K and subjecting them to ultra-violet radiation. We have found that radicals are generated and stored in samples which have undergone irradiation. When we allow an irradiated sample to undergo a very small rise in temperature of 1 to 5 K, we see visible emission due to the radicals reacting in the solid in a process similar to radiative recombination. This implies that molecules are formed *in the mantle* in excited states (5ev is typical) and are thus an internal energy source. Some of this energy is apparently converted to heat because we observe a substantial vapor pressure increase in the range 15-22 K in irradiated samples as compared to non-irradiated ones. Thus we feel that a mild warming event of only a few degrees can vaporize a small portion of the mantle and inject those molecules into the gas-phase.

Watson: One must keep in mind that a high efficiency is required in order that this process eject a significant fraction of the atoms that stick to a grain in dense clouds. At the *edge* of a cloud, the flux of UV photons due to the galactic background is in the neighbourhood of $10^7 cm^{-2} s^{-1}$. The flux of C,N,O, etc. atoms is $10^4 cm^{-2} s^{-1}$ at a density of 10^4 particles cm^{-3}. However, the UV radiation presumably is reduced by many orders of magnitude in the interiors of dense clouds.

Gold: Can cosmic ray evaporation from grains not account for an adequate re-supply to the gas of molecules formed on grains?

Watson: Various cosmic ray evaporation processes have been investigated in great detail (some by me). At best, they might do the job in the lower density clouds (≤ 100 particles cm^{-3}).

Greenberg: The evidence for accretion on grains has been available for a long time in the *observational* fact that the wavelength dependence of both extinction and polarization in dense clouds shows larger than normal grains.

Watson: True, but data from the study of grains is not sensitive to fractional depletions exceeding perhaps 2/3, whereas the radio observations suggest that less than one percent of the "heavy" elements remain in the gas at high densities.

Greenberg: If you believe that grain accretion takes place, as you implied when you presented results based on gas-phase reactions *including* grain accretion, then I do not see how you can justify the ion-molecule reaction schemes you suggested. The accretion time scale at $n=10^4 cm^{-3}$ is $\sim 10^6$ years, i.e. about equal to the free fall time scale, and if the free fall time scale is a substantial underestimate of cloud lifetimes, then there should be a *very* small number of molecules in 10^4 density clouds. The answer which I have proposed is that, as a result of rather small energy impulses to the grains (an increase in temperature by about 10-20 K), the energy stored in them as radicals can be released, and lead to partial or complete evaporation of the grain mantle, which includes all sorts of molecules.

Watson: I readily acknowledge (even pointed out in my lecture) that the gas-grain interaction is a major problem. However, the two time scales are roughly equal at densities of $10^4 cm^{-3}$, so it is not unreasonable to proceed in the expectation that order-of-magnitude results can be achieved from gas-phase chemistry (the time scale for gas-phase chemistry is reduced by depletion, also). If the effect of grains is non-selective depletion, one might reasonably expect (very roughly) just an overall reduction of absolute abundances, as suggested by the calculations of Iglesias (1977). The comparison between observations and the predictions of gas-phase chemistry for small molecules and molecular ions certainly indicates that the present approach is reasonable.

LABORATORY AND THEORETICAL RESULTS ON INTERSTELLAR MOLECULE PRODUCTION BY GRAINS IN MOLECULAR CLOUDS

J.M. Greenberg, L.J. Allamandola, W. Hagen, C.E.P. van de
Bult and F. Baas.
Laboratory Astrophysics, Leiden University, The Netherlands.

ABSTRACT

   Laboratory and theoretical studies have been made of the effects
of ultraviolet photolysis of interstellar grain mantles. It has been
shown that grain photolysis should be important even in dense clouds.
A large number of molecules and radicals observed in the interstellar
gas appear in the irradiated ices of CO, H_2O, NH_3 and CH_4 which are
deposited at 10 K. Energy released during warm-up is seen from visible
and infrared luminescence and inferred from vapor pressure enhancement
relative to unirradiated samples. Grains are pictured as a source as
well as a sink (capture) of molecules. The photolysis of an individual
grain provides the stored chemical energy which is sporadically released by relatively mild triggering events (such as low velocity
grain-grain collisions in turbulent molecular clouds) to produce the
impulsive heating needed to eject or evaporate a portion of the grain
mantle. An extremely complex and non-volatile substance possessing the
infrared signatures of amino and carboxylic acid groups and having a
mass of 514 amu has been produced at a rate corresponding to a mass
conversion rate of interstellar grains of between 2% and 20% in 10^7
years.

1. INTRODUCTION

   At the Astrophysics Laboratory of Leiden University we have
developed an experimental system which creates for the first time the
relevant interstellar analog conditions leading to the formation of a
wide range of molecules and their injection into interstellar space as
a consequence of the evolution and photoprocessing of grain mantles.
The results on ultraviolet produced changes in chemical composition,
storage of radicals and vapor pressure enhancement due to the heat released by the combination of reactive species stored in the complex
mixtures all lend support to the picture that the interstellar grain
mantles are dynamic centers of activity in which gas phase molecules
and radicals are modified and regenerated by reactions within the grains.

355

B. H. Andrew (ed.), Interstellar Molecules, 355–363.
Copyright © 1980 by the IAU.

We are thus able to demonstrate that the prevailing theories of molecule formation which consider only gas phase and grain surface reactions
are not complete because they have not included the role played by the
interiors of interstellar dust grains in the formation of molecules.

Experiments in the laboratory are aimed at understanding:
(1) the photoprocessing of grains,
(2) grains as a source of molecules up to a very high order of
    complexity and their injection into the interstellar medium
    subsequent to photoprocessing,
(3) grain emission and absorption characteristics from the far
    infrared to the ultraviolet.

2. PHOTOPROCESSING OF GRAINS

Just as gas phase molecules are photodissociated by photons of
sufficient energy, so are molecules imbedded in a solid. A major difference in the solid is that some of the break-up products may more
readily recombine (the cage effect) while others (say H) may readily
diffuse through the solid and combine with another molecule or with a
previously produced radical (see Bass and Broida, 1960, particularly
chapter 4; Meyer, 1971; Hallam, 1973). This means that not every "bond-
breaking" photon which is absorbed by the grain necessarily produces a
net photodissociation. Indicative of the significance of grain mantle
photolysis as an important interstellar process is that the time required for the interaction of every molecular bond in the mantle with
one high energy photon in an ultraviolet flux of 10^8 cm^{-2} s^{-1} (the mean
interstellar field) is only $\sim$ 200 yr (Greenberg, 1973). Thus even if
the photolysis efficiency per photon is taken to be as small as 10^{-3}
(a very conservative estimate), the time required to leave a net 1% of
radicals (a rough upper limit, see Jackson, 1959a,b; or see chapter 10
in Bass and Broida, 1960) in the mantle is still only about 1000 yrs
which is quite small compared with the mantle life-cycle time ($\sim 10^6$ to
10^7 yr). The photolysis time scale for the deepest molecules in the
grain is greater by only about a factor of 10 (Greenberg, 1979).

It may be inferred from earlier calculations (Hagen, Allamandola
and Greenberg 1979, hereafter HAG) that grain photolysis is substantial
during the $\sim$ 3 x 10^8 yr lifetime of a grain mantle as well as the lifetime ($\sim 10^8$ yrs) of a 1 pc cloud whose total visual extinction is about
10 mag.

We thus picture a typical grain mantle as spending a major part
of its lifetime consisting of a mixture of complex and simple molecules
with embedded frozen radicals for which the stored energy may occasionally be released so completely as to totally evaporate the material
(see Fig. 1).

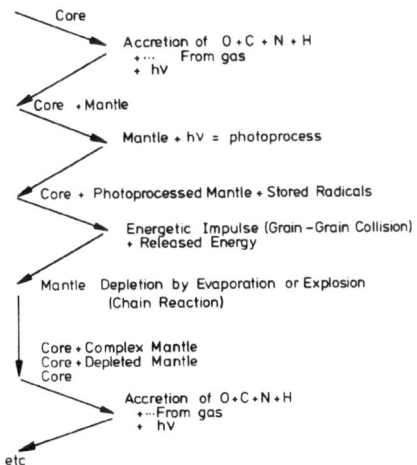

Fig. 1. Schematic of chemical evolution of a grain mantle. Grain mantle cycle time $\approx 10^7$ years.

More frequently there may be reactions which produce local hot spots in a mantle, some of which eject some surface molecules, but many will merely lead to local modifying of the molecular composition. It is interesting to note that any of the stored radicals located at or near the surface of the grain provide the possibility for chemical inter-action with accreting atoms and molecules which should be included in a consideration of the growth as well as the chemical composition of grains.

Since all of the time scales in space are impossibly long for laboratory measurements we have scaled the laboratory photoprocessing as shown schematically in Table 1.

TABLE 1

Comparison between laboratory and interstellar conditions

		Lab	ISM
Mantle {	"Initial" composition	(simple molecules) $CO, H_2O, NH_3, CH_4 \ldots$	All interstellar condensible species
	Thickness	> 0.1 μm	~ 0.15 μm
	Temperature	$\gtrsim 10$ K	$\gtrsim 10$ K
Gas {	Pressure of condensibles	8×10^{-8} mbar	3×10^{-17} mbar
	Number of condensible species	2.4×10^9 cm^{-3}	1 cm^{-3} ($n_H = 10^3$cm^{-3})
Ultra-violet {	$\phi_{uv} \equiv$ Flux (E > 6 eV, $\lambda < 2000$ Å)	10^{15} cm^{-2} s^{-1}	10^8 cm^{-2} s^{-1}
	Equivalent time scales	1 hr	10^3 yr

3. LABORATORY RESULTS

We summarize here a few results which bear on key phases in the life of a grain mantle (see HAG for instrumental details).

3.1 Photoproduction of molecules and radicals at 10 K

A series of infrared absorption spectra of a laboratory sample taken before and after several periods of irradiation is shown in Figure 2.

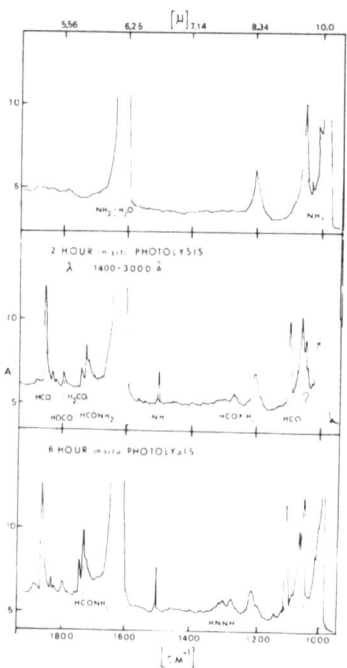

Fig. 2. Infrared absorption spectra of a molecular mixture $CO:NH_3:H_2O:CO_2$ (50:1:1:0.09) at 10 K. Upper, before irradiation; middle, after 2 hrs irradiation; lower, after 6 hrs irradiation.

The formation and growth of new radicals and molecules is clearly demonstrated. The newly formed species which are also known to exist in the interstellar medium are HCO, HNCO, H_2CO, $HCONH_2$, and HCOOH. The infrared spectral changes which take place after the sample is warmed up to 40 K and recooled (not shown) shows a reduction in absorption by the radicals NH_2 and HCO with simultaneous growth of formaldehyde (H_2CO) and formamide ($\widetilde{N}H_2HCO$). The disappearance of the radicals and the growth of the more stable molecules is expected since mild warming up allows diffusion to take place and permits the unstable species to react to produce stable product molecules.

3.2 Warm-up and energy release

Subsequent to the photolysis of the above mixture the system was allowed to warm up slowly. Almost immediately there appeared a blue-green luminescence apparently produced as the radicals combine. This interpretation is borne out by the fact that the visible absorption spectrum of the HCO radical is gradually reduced during warm-up. The sample was first warmed up to about 27 K then cooled back to 10 K (see Fig. 3). Upon subsequent warm-up there was practically no luminescence

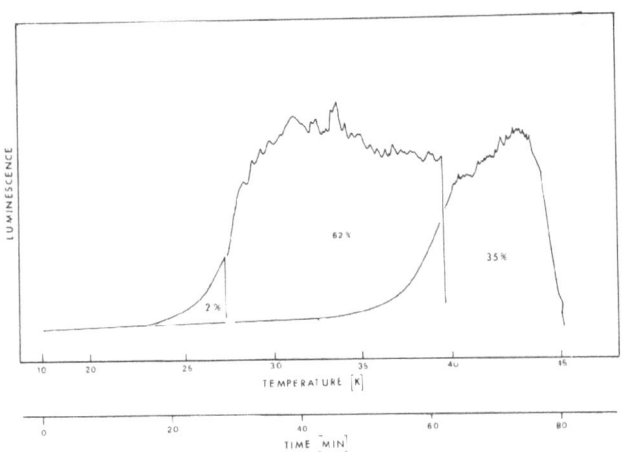

Fig. 3. Luminescence during warm-up of photolyzed mixture $CO:NH_2:CO_2$ (50:1:1:0.09), from 10 K to 45 K with intermediate recooling from 27 K and 40 K.

until the sudden onset at 27 K. The same thing was done up to 40 K with similar results. It appears that only certain frozen radicals are released up to each increase in temperature and that, in cooling and reheating, the radicals which have not been released remain trapped until the appropriate minimum temperature is reached. It is obvious that if the sample had been brought quickly to the temperature of 45 K all the energy seen in Fig. 3 would have been instantaneously released. By implication, since the radical reactions release energy, some of which goes into heating the sample, even a sudden increase to some temperature substantially less than 45 K could also lead to the complete release of stored energy. This is the type of process which we picture as occurring when a slight impulsive heating of an interstellar grain with frozen radicals triggers the release of the stored energy thereby heating the grain further and causing evaporation of the grain. The fact that at least some of the released energy appears as heat has been shown by the fact that the vapor pressure enhancement of various samples increases with the photolysis time (HAG).

The absorption spectrum between 3500 and 10,000 Å has been measured for a number of mixtures. For all cases the region between 4000 and 7000 Å shows the most structure. These absorptions appear only in photolyzed samples. The strong broad absorptions (80 - 100 Å at half height) grow weaker on warm up and are barely detectable at 35 K. It appears that these absorptions are due to HCO (Ewing et al, 1960; Herzberg and Ramsey, 1955; Johns et al, 1963).

3.3 High molecular mass studies

Upon completion of a number of experiments starting with mixtures of NH_3/CO, a yellow non-volatile residue was left on the cold finger. The yellow "stuff" generally appears as a viscous liquid (in one case it seemed to be crystallized) which is water soluble. In one such experiment, 200 μg of the material was recovered from the substrate. This represents a yield of between 0.02% and 0.2% (depending on whether or not we saved all the available residue). Based on the total time of photolysis we infer (see Table 1) that between 2% and 20% of the interstellar dust mantle material is converted into such non-volatile complex molecules in 10^7 years.

The infrared absorption spectrum of the liquid residue is shown in Figure 4. The principal features are readily attributable to amino and carboxylic acid subgroups, but are inconsistent with hydrocarbon or

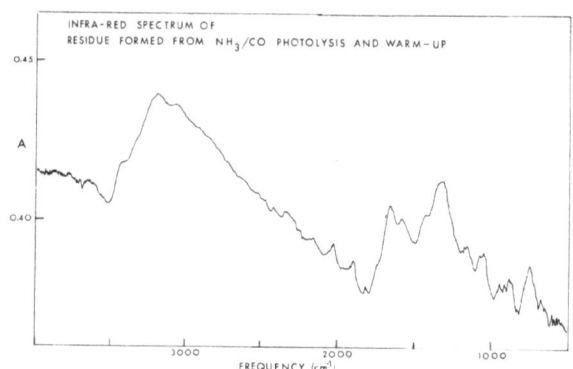

Fig. 4. Infrared spectrum of yellow residue formed from NH_3/CO photolysis and warm-up.

aromatic compounds (Pouchert, 1975). The liquid evaporates between 400 and 500 K (the "crystalline" substance pyrolyzed at 600 K). Both laser and field desorption mass spectrometric techniques yield a mass of 514 amu. Surprisingly this was the only mass. Only the ultimate residue has been studied here. It appears that the more volatile molecules with lower mass numbers must have evaporated. The detection and study of these molecules is now under investigation with an in situ mass spectrometer.

REFERENCES

Bass, A.M., and Broida, H.P.(eds.): 1960, *Formation and Trapping of Free Radicals*, Academic Press, New York.

Ewing, G.E., Thompson, W.E., and Pimentel, G.C.: 1960, J. Chem. Phys. 32, pp. 927.

Greenberg, J.M.: 1979, in B. Westerlund (ed.), *Stars and Star Systems*, IAU 4th European Regional Meeting in Astronomy, Reidel, Dordrecht, pp. 173.

Hagen, W., Allamandola, L.J., and Greenberg, J.M.: 1979, Astrophys. and Space Science, 65, pp. 215.

Hallam, H.E.: 1973, *Vibrational Spectroscopy of Trapped Species*, John Wiley and Sons, New York.

Herzberg, G., and Ramsay, D.A.: 1955, Proc. Roy. Soc. (Lond.) A233, pp. 34.

Jackson, J.L.: 1959a, J. Chem. Phys. 31, pp. 154.

Jackson, J.L.: 1959b, J. Chem. Phys. 31, pp. 722.

Johns, J.W.C., Priddle, S.H., and Ramsay, D.A.: 1963, Discussions Faraday Soc. 35, pp. 90.

Meyer, B.: 1971, *Low Temperature Spectroscopy*, Elsevier Press, New York.

Pouchert, C.J.: 1975, *The Aldrich Library of Infrared Spectra*, 2nd. edition.

DISCUSSION FOLLOWING GREENBERG

Snow: You indicated that many molecules likely to exist in grain mantles would absorb photons in the 4.5-5 eV energy range, which coincides with the 2200 Å interstellar extinction feature. Have you measured the ultraviolet spectrum of your simulated grain mantle material, and if so, did a 2200 Å absorption feature appear?

Greenberg: We do not think that the 2200 Å feature can be produced in our mantle materials. It may arise from our non-volatile residue, but a preliminary UV absorption spectrum of this substance did not seem to show it.

Vanden Bout: Your scheme seems to require both high cloud densities for reasonable rates of accretion onto grains, and low cloud densities for reasonable UV fluxes for photolysis. Can you tell us the density of the molecular clouds in which your proposed process operates?

Greenberg: In clouds with n_{H_2} several times $10^4 cm^{-3}$ there is a rough balance between (1) production of molecules by triggered energy release within previously *adequately* irradiated grain mantles and (2) accretion of molecules on the grains. Without some additional source of energetic photons (over that coming from outside), at higher densities the accretion rate should exceed the production rate, thus leading to a depletion of molecules.

Irvine: The spectrum of comets is due to radicals which, it is usually stated, could not exist for long periods of time in the cometary nucleus. There has been traditionally a problem in time scales if the radicals are produced by photodissociation. Do you think there could be a significant number of such radicals stored in an icy conglomerate nucleus?

Greenberg: I had thought that the photodissociation production of radicals in comets was not a problem. What is a problem, as I understand it, is the identification of the parent molecules. The latter, by my dust conglomerate model of the comet nucleus, would include *some* radicals stored along with the other molecules, so *some* of the radicals seen would come directly from the comet. What fraction of the total would depend largely on the distance of the comet from the sun at the time of observation - the more distant the comet, the larger would be the number of radicals coming directly from it relative to the number being produced by photodissociation.

Avery: The carbon chain molecules are conspicuous by their absence in the products produced in your grain mantles. Are they being made but not detected, or are they not being made at all?

Greenberg: I cannot preclude the production of chain molecules in our present experiments. However, we will not be able to make a definite statement until we have a means of identifying them by their infrared absorption spectra in a solid matrix. We do know that if we start with some CH_4 in our solid, we can photolyze it to produce C_2H_2 so I believe that we are on the way to producing chain molecules.

Thaddeus: It does seem to me that it is a bit of a tease just to compare a list of the molecules that you get in your frost with the molecules that are found in dense pockets of interstellar gas. In that sense your experiment is a variation of the classic Miller-Urey experiment. It has been known for 30 years now that you do get the interstellar molecules out in a number of different ways if you take these simple compounds, put them in a flask, and give it a good kick. The real problem is the time scale of getting the molecules back into the gas, and the quantity of material that you can produce in this way. In spite of what you say, the clouds where we see the big molecules are as black as Satan's snout in many ways. The limit to the rate of production must be the time scale for getting the grain out where the photolysis can take place, then back into this black thing that's as dark as photographer's cloth, then getting the stuff off the grain again.

Greenberg: What we are trying to do is to estimate how much material is ejected by a dust grain when it is hit, and what kinds of molecules are ejected. In a way, it is indeed a bit of a tease to have shown evidence for production in solids of molecules observed in the gas phase in space. On the other hand, it is quite reasonable to expect that these molecules, or at least some combinations of them, will be ejected from the solid provided the right conditions exist to kick them out. I agree that the molecules we see in the solid are not necessarily present in the same relative abundances as those that are dumped off when the solid is evaporated or exploded or whatever. As I have already said in response to Dr. Vanden Bout there must be some mechanism to cause this to happen. One mechanism which I have proposed is that of grain-grain collisions, even at speeds as slow as 0.1 km s^{-1}. This might occur as a result of cloud turbulence, for example. In an earlier paper (Stars and Star System, ed. B. Westerlund, Reidel, 1979, p. 173) I have shown that collision velocities of the order of 0.1 km s^{-1} provide enough of an increase in grain temperature to trigger reactions among the frozen radicals, which then release enough energy to heat the grain and further cause evaporation of some portion of the grain mantle. I have also estimated that an ultraviolet flux large enough to keep the grains in a state of sufficient excitation may be found even in the greater volume of clouds in which the total visual extinction is as high as 20 and the only ultraviolet radiation is that from the diffuse interstellar medium. The irradiation process is slow but steady while the collision process is sporadic. But statistically the two together may provide a quasi steady state. One can put in reasonable parameters, and end up with reasonable number densities of molecules in these clouds, where the rate of production is related to the rate of ejection from grains, and the rate of destruction is related to the rate of accretion onto grains. Now clearly that explanation leaves out all the gas-phase reactions that are also taking place. It is as bad a solution as considering only gas-phase reactions and ignoring the grain-accretion problem, but at least it gives some upper limits. If the molecules are made from the grains and absorbed also on the grains, then some sort of reasonable rate process has to be attained, and that, presumably, we can investigate in the laboratory.

   De Zafra: Do you have any information on rates of free-radical relaxation back to parent molecules? High photo-flux laboratory results could fail to mimic very low-level irradiation processes when relaxation is considered.

   Greenberg: We have left an irradiated sample on the cold finger (without warm-up) for as long as 4 or 5 days without seeing any apparent loss of radical concentration. Although a more quantitative experiment will be necessary, I predict that the rate of radical relaxation will be exceedingly low at 10 K. We have considered a more serious question regarding photon flux. It turns out that there is a negligible possibility that a photon will strike a particular molecule while it is in an excited state produced by a previous photon. For a sample of thickness >0.1 μm, $\tau \sim (\Phi_{uv} d^3)^{-1} \sim 1000$ s is the time between two photons interacting with the same molecule, where Φ is the photon flux $(10^{15} cm^{-2} s^{-1})$, and d is the molecule diameter $\sim 2 \times 10^{-8}$ cm. Molecular relaxation times are generally many orders of magnitude less than this.

   Feldman: Have you used HCN/C_2H_2 mixtures to deposit grain-mantle analogues in any of your experiments? I would be interested to learn what kind of chain and ring molecules are produced.

   Greenberg: We have not used HCN/C_2H_2 as initial material.

INTERSTELLAR MOLECULES ON DUST MANTLES

Naoya Nakagawa
The University of Electro-communications
Tyoohu si, Tokyo, Japan

Condensation temperatures of various interstellar molecules in dark clouds are calculated. Chemical reactions in dust mantles are discussed.

1. CONDENSATION TEMPERATURES IN DARK CLOUDS

Gaseous molecules in dark clouds show more variety and are more concentrated than those in diffuse clouds. Under these conditions the condensation of interstellar molecules on the dust surfaces can occur. We have calculated the condensation temperature of various interstellar molecules (Table 1) using the Clapeyron-Clausius equation. S or L means extrapolation from the data at vapor pressures of 1 Torr and 10 Torr in the solid state(S) or the liquid state(L); other values were calculated using an evaporation heat assumed from Trouton's rule. As the calculated values are given for pure substances, condensation must occur at higher temperatures due to the low vapor pressure of solid solutions.

In Table 1 the molecules can be separated into two groups at $20-30$ K. One group is condensed in the dust mantles and the other group containing CH_4, CO, N_2, and H_2 is gaseous.

2. REACTIONS IN DUST MANTLES

At temperatures below 100 K in dust mantles the usual chemical reactions having activation energies about 50 kJ/mol will have rates with time scales of 10^8 years. However during star formation the temperature becomes high and chemical reactions in dust mantles may occur. Two types of reactions attract notice. One type is the reaction of the main components of dust mantles such as H_2, HCN and NH_3. This mixture is famous among researchers into the origin of life because it produced glycine, alanine, other amino acids, bases of nucleic acids and urea.[3,4]

Another type of reaction is that between water ice and polyynes.

B. H. Andrew (ed.), Interstellar Molecules, 365–366.

Polyynes react with water to form various compounds having $-\overset{|}{C}-O$, $>C=O$, $-\overset{|}{\underset{|}{C}}-H$, $-\overset{|}{\underset{|}{C}}\underset{\diagdown O \diagup}{}\overset{|}{\underset{|}{C}}\diagdown$, as demonstrated by NMR and IR spectra in the laboratory. There is therefore a possibility of the production of sugars from interstellar molecules.

TABLE 1

Molecule	Abundance[1] 10^{-x}	Condensation Temperature (K)		
		10^3/cc	10^5/cc	10^7/cc
H_2NCHO	10	93.4	98.8	105.0 L
H_2O [2]	5	84.8	90.1	97.2 S
HCOOH	10	81.3	85.8	90.7 S
CH_3COOH	(10)	79.7	84.1	89.0
H_2CS	10	73.0	78.0	84.0
CH_3OH	7	65.1	69.4	74.2 L
HCN	6	64.5	68.9	73.9 S
H_2NCN	9	63.0	68.0	73.6
CH_3CH_2OH	10	63.4	67.2	71.4 L
CH_3CH_2CN	10	57.9	61.4	65.4 L
SO_2	7	55.2	58.7	62.6 S
CH_3CN	(9)	55.0	58.5	62.5 L
NH_3	6	52.6	56.1	60.0 S
CH_3OCHO	10	49.8	52.8	56.2 L
CH_3CHO	10	47.9	50.8	54.1 L
(CO_2)	(6)	45.5	48.6	52.1 S
CH_3NH_2	10	45.2	47.9	51.1 L
H_2CO	8	41.6	44.5	47.7 L
(C_2H_2)	(6)	41.2	44.1	47.3 S
CH_3CCH	9	37.9	40.7	43.8
OCS	8	37.2	39.7	42.6 L
CH_3OCH_3	10	36.8	39.1	41.7 L
H_2S	8	36.6	39.0	41.7 S
CH_4	(6)	16.9	18.0	19.5 S
CO	4	14.8	16.1	17.5 S
(N_2)	(6)	12.1	12.9	13.9 S
H_2	0	2.6	2.9	3.2

REFERENCES

1. Watson, A.: 1977, Acc. Chem. Rev. 10, 71.
2. Phillips, T.G., Scoville, N.Z., Kwan, J., Huggins, P.J., and
       Wannier, P.G.: 1978, Ap. J. 222, L59.
3. Yuasa, S., and Isigami, M.: 1977, Geochem. J. 11, 263.
4. Oró, J.: 1961, Nature 190, 326, 389.

THE FORMATION OF HYDROCARBONS AND IRON-HYDRIDES ON COLD INTERSTELLAR GRAINS-EXPERIMENTAL STUDIES

A. Bar-Nun[1], M. Litman[1], M. Pasternak[2] and M.L. Rappaport[2]
[1]Department of Geophysics and Planetary Sciences and
[2]Department of Physics and Astronomy,
Tel Aviv University, Tel Aviv, Israel

1. ABSTRACT

Cold hydrogen atoms at $T \geqslant$ 7K were shown experimentally to react with graphite grains at the same temperature to produce CH_4 and smaller amounts of C_2H_6, C_2H_4 and C_2H_2. At T < 20K the hydrocarbon mantle could polymerize to form carbonaceous substances, similar to those found in carbonaceous chondrites. Further encounters with H-atoms would result in their recombination on the hydrocarbon mantle around the grains. At higher grain temperatures, the hydrocarbons formed could be ejected into the gas phase.

Cold iron atoms at T < 5K were shown experimentally to react with molecular hydrogen in a T < 5K matrix. Mössbauer studies with ^{57}Fe demonstrated the formation of an $Fe-H_2$ bond. FeH_2 and FeH molecules could be formed on grains by encounters of iron atoms with either H-atoms or H_2 molecules.

2. INTRODUCTION

The recombination of H-atoms on grains to form H_2 molecules is by now well established. The two experimental studies which are reported here suggest that H-atoms and H_2 molecules could participate in additional reactions on grains' surfaces. The applications of these studies to the chemistry of interstellar clouds are discussed.

3. EXPERIMENTAL PROCEDURES AND RESULTS

3.1. The Hydrogen-graphite reaction

The experiments at 4.6 - 300K were performed in essentially the same way as those at 78 - 300K, which were reported earlier (Bar-Nun, 1975). The reaction vessel consisted of a quartz tubing 50 cm long and 2 cm I.D. with a side arm 10 cm long and 0.5 cm I.D. attached at the top.

367

B. H. Andrew (ed.), Interstellar Molecules, 367–371.

Several small pieces of pyrolytic graphite (General Electric) were placed at the bottom of the reaction vessel and outgassed at 1200K and 8×10^{-6} Torr for several hours. Hydrogen and helium (Matheson Research Grade) at pressures of 6 Torr each were introduced into the vessel, which was then lowered into a liquid helium dewar until the desired temperature was reached. A cavity of a 100W Kiva model MPG-4 microwave generator was then placed on the side arm and the discharge which generated the H-atoms was continued for 60 min. After the discharge was switched off the vessel was kept at the same low temperature for five more minutes and was then warmed up to room temperature before the products were analyzed by gas chromatography. The experimental results are shown in Table 1. The number of surface carbon atoms was determined by adsorption of argon at 77K and found to be $(1.0 \pm 0.5) \times 10^{17}$. Thus, the effective surface area of the graphite is six times larger than its geometric area. This study was described in detail by Bar-Nun et al (1979).

Table 1. Product Distribution in the Various Experiments

Run	Reactant			Temperature	Products, nmole			
	H_2	He	graphite	K	CH_4	C_2H_6	C_2H_4	C_2H_2
21	+			300	2.35	–	–	–
22	+	+		300	2.38	0.21	0.21	–
4 runs		+	+	300	0.44	–	–	–
35	+	+	+	300	10.37	–	–	–
46	+	+	+	77	29.46	0.59	0.22	–
25	+	+	+	20	45.77	0.59	0.09	0.23
37	+	+	+	12	62.20	0.75	0.58	–
38	+	+	+	12	65.65	0.52	0.32	–
44	+	+	+	9	7.62	0.26	–	–
47	+	+	+	9	7.44	0.07	0.52	–
41	+	+	+	7	5.20	–	–	–
42	+	+	+	7	4.60	0.05	0.48	–
40	+	+	+	4.6	1.60	trace	0.27	–

3.2. The Iron-Hydrogen Reaction

A stream of molecular hydrogen or 10% hydrogen in argon and a stream of iron atoms from a crucible at ∿1000K were allowed to reach a cold substrate (1.5 or 4.2K), freeze there and react. The iron mixing ratio in the matrix was 2×10^{-3}. Mossbauer spectroscopy was used for analyzing the product while still at T < 5K. From the observed isomer shift ($\alpha = 0.42 \pm 0.03$ mm sec^{-1} with respect to metallic iron) and the isomer shift of FeF_2, it can be concluded that one of the iron's 4s electrons is transferred to the molecular orbitals of the hydrogen. A possible configuration of the molecule thus formed is $H_{Fe}H$. It is worth noting that all the iron atoms reacted with hydrogen and that at T > 5K the iron started to agglomerate into metallic grains because of the softening of the hydrogen matrix. This study was described in detail by Pasternak and Barrett (1978).

4. DISCUSSION

4.1. Hydrocarbon Formation

The results of the blank runs show that both the vessel and the graphite were free of organic contamination. The major product is methane but some ethane, ethylene and acetylene are also formed. At 4.6K practically all the hydrogen was adsorbed on the graphite and H-atoms were not available for the reaction. Yet, even at this temperature enough helium was left in the reaction vessel to cause thermalization of the H-atoms, which diffused along 50 cm from the side arm to the graphite.

In the context of reactions in cold interstellar clouds it is important to establish that the H-atom graphite reaction takes place at low temperature and not during the warming up to room temperature. H-atoms can be adsorbed on surface carbon atoms and possibly also between the carbon sheets in the graphite crystal which are separated by 3.4A. The small surface area measured shows that Ar atoms cannot penetrate between the sheets and therefore, even if H-atoms do penetrate there, the products which are larger than Ar would not be able to diffuse out. Thus the observed reaction takes place at the surface. At 12K, 65 nmole of CH_4 were produced, consuming 1.6×10^{17} H-atoms, or on every surface carbon atom should be adsorbed an H-atom. For the reaction to take place during the warming up process, these atoms would have to remain adsorbed on the surface without recombining for at least 300 sec. Barlow and Silk (1976) estimated the time for an H-atom to hop from one site to the other by quantum-mechanical tunnelling as 10^{-5} sec. Even if the time is five orders of magnitude longer, which is very unlikely and would cause severe difficulties for H-atom recombination on grains in the clouds, the time for recombination would have been short in comparison with 300 sec. It can therefore be concluded that the hydrocarbon forming reaction indeed takes place at low temperature. From Barlow and Silk's (1976) probability for recombination and assuming quantum-mechanical tunnelling by hydrogen in the reaction with graphite, with an energy barrier of 0.23 ev (Wood and Wise, 1969) and a width of the barrier of $\sim 1\text{\AA}$, the ratio of probabilities of recombination vs reaction is $1.3 \times 10^{-10} n_H$, where n_H is the H-atom number density. Thus, even in cold interstellar clouds H-atoms would react with bare graphite grains rather than recombine on them.

4.2. Iron Hydrides formation

The major problem in this experiment is the high initial temperature of the iron atoms which are ejected from a crucible at $\sim 1000K$.

However, in the reaction in the matrix which consisted of 10% hydrogen in argon, the hot iron atoms had encountered several argon atoms before encountering a hydrogen molecule. During these encounters they were thermalized to the matrix temperature and, since all the iron atoms formed FeH molecules, it is apparent that the reaction takes place when all the reactants are at T < 5K. This could occur by quantum-mechanical tunnelling of the hydrogen molecules. The reaction $Fe+H_2 \rightarrow FeH + H$ is endothermic by ~ 2.04 ev since the bond energy in H_2 is 4.47 ev and that of FeH according to Walker et al (1972), is 2.43 ev. Thus, FeH would not be formed from hydrogen molecules but could be formed by encounters of H-atoms and Fe-atoms on grains.

5. IMPLICATIONS FOR COLD INTERSTELLAR CLOUDS

In dense and cold clouds where $n_H \sim 10$ cm^{-3} _all_ the H-atoms which remain on the _bare_ graphite grains long enough would react with the surface carbon atoms to form methane and some other C_2 hydrocarbons. At T < 20K the hydrocarbons would be strongly bound to the grains (Watson and Salpeter, 1972) and form a monolayer around them. Further encounters with H-atoms would result in their recombination on the hydrocarbon monolayer. The hydrocarbon mantle could evolve chemically by irradiation with short uv and low energy cosmic rays (Watson and Salpeter, 1972) and could eventually polymerize. Such polymers are indicated by several IR emission bands in clouds (Knacke, 1977). Mantle formation could be inhibited by ejection of the hydrocarbons in very small grains or by photoejection or sputtering during cloud-cloud collisions, and in clouds with grain temperatures greater than 77K by thermal ejection (Bar-Nun, 1975).

Encounters of iron atoms with H-atoms on grains would result in the formation of FeH molecules, with absorption bands at 5900, 6400, 15600, 17100 and 33000 cm^{-1} (Walker et al, 1972). Similar hydrides of Cd, Mn, Co, Ni and Cu, all of which were formed experimentally (Walker et al, 1972), could be responsible in part for the depletion of these species as well as iron in the clouds. Encounters of Fe atoms with H_2 molecules on grains could result in the formation of FeH_2 species. The formation of these hydrides would not prevent metal grains formation, since their encounters with each other result in agglomeration, even at 5K, as demonstrated by the experiments.

REFERENCES

Barlow, M.J., and Silk, J.: 1976, Ap. J. 207, pp. 131.
Bar-Nun, A.: 1975, Ap. J. 197, pp. 341.
Bar-Nun, A., Litman, M., and Rappaport, M.L.: 1979. Submitted to
    Astron. & Astrophys.
Knacke, R.F.: 1977, Nature 269, pp. 132.
Pasternak, M., and Barrett, P.H.: 1978. Proceedings of "Workshop in
    Applications of Mössbauer Spectroscopy" Seeheim/Bergstrasse,
    W. Germany, April, 1978.

Walker, J.H., Walker, T.E.H.,and Kelly, H.P.: 1972, J. Chem. Phys. 57, pp. 2094.
Watson, W.D., and Salpeter, E.E.: 1972, Ap. J. 174, pp. 321.
Wood, B.J., and Wise, H.: 1969, J. Chem. Phys. 73, pp. 1348.

DISCUSSION FOLLOWING BAR-NUN

Carruthers: The conclusion of your paper, then, is that H_2 cannot be formed by pure graphite grains, only coated graphite, or silicates, etc.? If correct, this conclusion may be of significance in the interpretation of the far-UV extinction curve. Perhaps hydrocarbon coating may be in part responsible for the steep rise below 1600 Å. In the directions showing both the steep rise and the 2200 Å "graphite" feature, such as ζ Oph, there is a great deal of H_2, but toward θ Ori, which has the 2200 Å bump but little far-UV rise, there is little or no H_2, although E(B-V) is the same.

Bar-Nun: Yes indeed, it seems that on bare graphite grains at low temperatures, H-atoms would react to form hydrocarbons rather than recombine to form H_2 molecules. I would like to suggest that, since at low temperatures a hydrocarbon mantle would form around the grains, H-atom recombination should be studied on frozen hydrocarbons rather than on bare graphite.

THE CHEMICAL IDENTIFICATION OF GRAIN MANTLES BY INFRARED SPECTROSCOPY

L.J. Allamandola, J.M. Greenberg, C.A. Norman, and W. Hagen
Laboratory Astrophysics, Leiden University, Leiden, Holland.

The composition and physical properties of interstellar grain mantles continues to be an important problem in astrophysics. Part of this importance comes from the fact that grain mantle composition, photochemistry and photophysics are involved in interstellar chemistry (Greenberg et al. 1972, Greenberg, 1979, Greenberg et al. this volume). Because most molecules have a number of fundamental modes of vibration which possess activity between 2.5 and 25 µm (the middle infrared), spectroscopic measurement in this region can provide a direct probe of the molecules making up grain mantles. In addition to molecular composition, under favorable conditions, such measurements can yield molecular abundances, the solid/gas ratio for specific molecules and give an indication of such physical grain properties as temperature and thermal history.

In order to exploit fully the infrared spectrum it must be realised that there are certain spectral characteristics which permit one to distinguish between molecules in low temperature solids, and molecules in the gas phase. Two of the most striking differences are the following: 1) the absence of P and R rotational structure, see Figure 1 (because molecules are no longer free to rotate) and 2) line shifting and broadening which occurs due to interactions within the solid.

The extent to which line broadening and shifting occurs is a sensitive function of the particular molecules involved. While this cannot be quantified on theoretical grounds alone it can be studied empirically (for an example of such an application see the paper by Hagen, Tielens and Greenberg presented at this symposium).

In this paper we discuss the infrared spectra associated with astrophysical objects which show spectral characteristics indicative of molecular mantles.

Objects showing emission.

To the best of our knowledge there are 16 objects which are known to exhibit prominent emission lines. They are listed in Table 1. Inspection of Table 1 shows that the emission features originate

373

B. H. Andrew (ed.), Interstellar Molecules, 373–380.

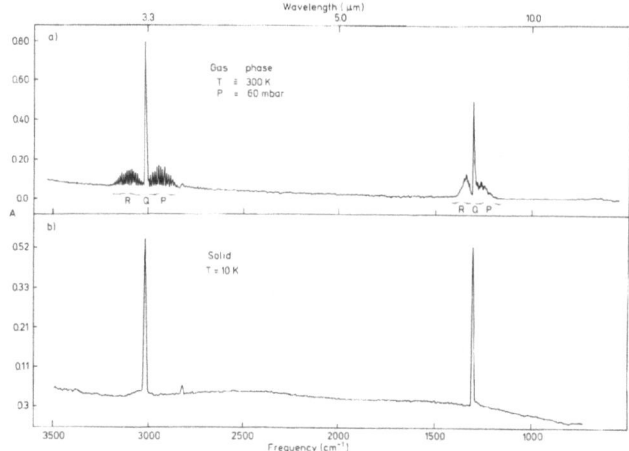

Figure 1. The infrared spectrum of a) gaseous and b) solid methane showing the absence of the P and R branches in the solid.

Table 1.   Infrared objects which show prominent emission features.

Object	Type
NGC 7027	High excitation planetary nebula
BD+30°3639	Low excitation planetary nebula
IC 418	Low excitation planetary nebula
Orion	H-II
M17A	H-II
M17B	H-II
NGC 7538	H-II
AGL 3053	H-II
W51-IRS2	Compact H-II
K3-50	Compact H-II
M 82	Galaxy
NGC 253	Galaxy
HD-44179 (AGL 915)	Nebulosity with imbedded source (late B or early A binary)
AFGL 437	Nebulosity with imbedded source
CRL 2132 (MWC 922)	Nebulosity with imbedded source (Early type star)
CRL 2688	Nebulosity with imbedded source (F?)

in objects which contain substantial amounts of dust in proximity to an ultraviolet source.

NGC 7027 will be used for this discussion because it has one of the richest and best studied spectra of all the objects listed. The spectrum is shown in Figure 2.

Inspection of Figure 2 shows that, superimposed on the peculiar background continuum, there are many prominent emission lines. Those indicated by a dotted line are significantly broader than the resolution

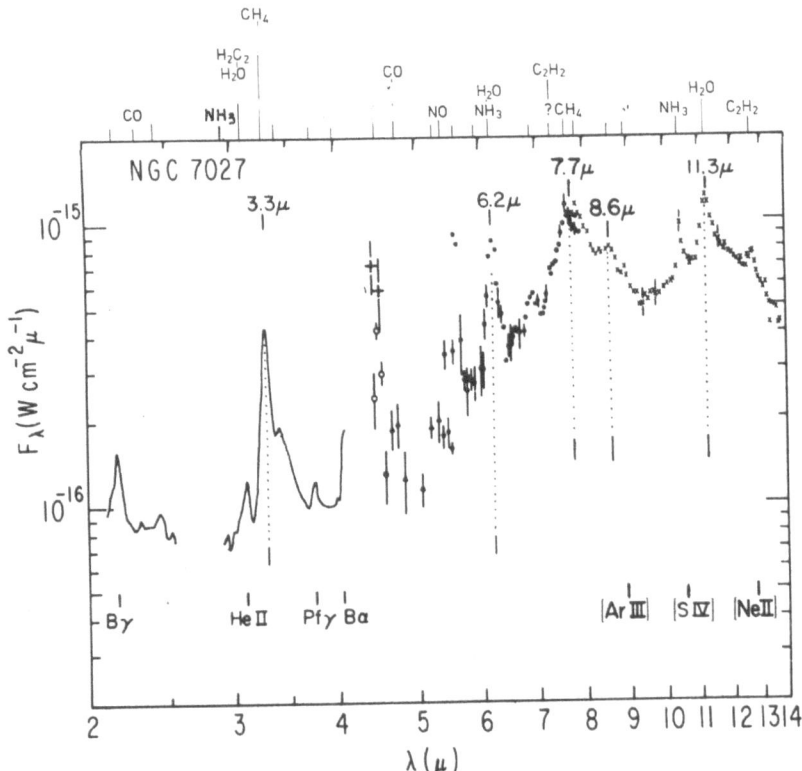

Figure 2. 2-14 μm spectrum of NGC 7027. The top
horizontal axis shows the Class I assignments from
Allamandola and Norman, 1978 and the dotted lines
indicate the features which are broader than the
spectral resolution. The spectrum is reprinted from
a letter of Russell,Soifer and Willner (1977) which
appeared in the Astrophysical Journal, University
of Chicago Press.

of the spectrometer. Another relevant aspect of these lines is the
complete lack of P and R branch-like structure. The broad lines and the
absence of evidence for rotation indicate that these features originate
from molecules in grain mantles. We have assigned a number of these
emission lines in NGC 7027 to molecules in grain mantles based on the
comparison of those lines to absorption lines obtained from molecules
in low temperature solids (Allamandola and Norman (1978), Allamandola,
Greenberg and Norman (1979)).

The very prominent 3.3 μm emission which we have assigned to the ν_3
mode of methane has been measured with 7 cm^{-1} resolution by Grasdalen
and Joyce (1976). We obtain an 18 cm^{-1} line width at half height from

their spectrum. We have measured the temperature dependence of the line
width of the 3.3 μm methane line in two different solids, one pure
methane and the other a mixture of CH_4: H_2O : CO : NH_3 : (2:1:1:1).
For pure methane the line width remains constant at 10 cm^{-1} up to 20 K,
from 20 to 30 K it broadens to about 22 cm^{-1} and at 45 K it is 31 cm^{-1}
wide. In the case of the mixture the linewidth is 14 cm^{-1} up to 20 K
and undergoes slight broadening upon warm-up to 16 cm^{-1} at about 30 K
and 18 cm^{-1} at 38 K. *In addition, the linewidth is a reversible function
of the temperature*. Thus using this model of infrared emitting grain
mantles and the high resolution measurements of Grasdalen and Joyce one
is led to conclude that the grain temperatures lie in the range 25-40 K
in the fluorescing regions.

Investigation of the low temperature spectroscopic literature shows
that some lines due to vibrational transitions remain sharp (< 0.5 cm^{-1})
while others undergo substantial broadening (10-20 cm^{-1} is common, and
in some extreme cases such as H_2O ice as much as 200 cm^{-1} is possible)
(Hallam, 1972). In this light, the features in Figure 2 which are
indicated as being broader than the resolution element are extremely
important. The width of the emission at 3.3 μm has been discussed as
due to CH_4. The extreme breadth of that at 7.7 μm cannot be due to CH_4
or C_2H_2 transitions alone, nor can that of the 8.6 μm features be due
solely to Ar [III]. As pointed out by the original observers the features
at 6.2 and 11.3 μm are also broad (Aitken and Jones, 1973; Gillett et
al., 1973 and 1975; Russell et al., 1977).

We have assigned these to the broad water ice bands, the only
species (with the exception of NH_3) in our Class I assignment category
which normally possesses features broader than the 0.015 resolution
element which holds for most of the spectrum. The ν_4 mode of NH_3, which
overlaps with ν_2 in H_2O at about 6.2 microns also undergoes extensive
broadening when in a multimeric form.

Objects showing absorption lines.

It is also be expected that the infrared spectrum obtained from
dark clouds not associated with a local UV source and lying in front of
a continuous I.R. source should show absorption due to dust molecules
in the cloud. For most molecules these lines should be weak. However,
H_2O molecules acquire anomalously large absorption cross sections when
they are in the ice form (10 x or more, depending on the crystalline
form). This explains why the so-called ice band at 3.1 μm dominates
many spectra and may mislead one into thinking that there are no other
features in the spectrum which are due to molecules.

In order to study the applicability of infrared absorption spectroscopy to this problem of grain mantle analysis we measure spectra of
laboratory prepared samples which are designed to simulate interstellar
grain mantles. The spectrum reduced to 30 cm^{-1} resolution, so as to be
comparable to that of the other spectra, of an $H_2O:CO:CH_3OH:NH_3$ (6:3:3:2)
mixture condensed onto a 10 K substrate is shown in Figure 3-d. This
ratio provides an O:C:N ratio of 6:3:1 which was choosen to approximate
the cosmic abundance ratio (O:C:N:5.2:3.3:1). The experimental techniques

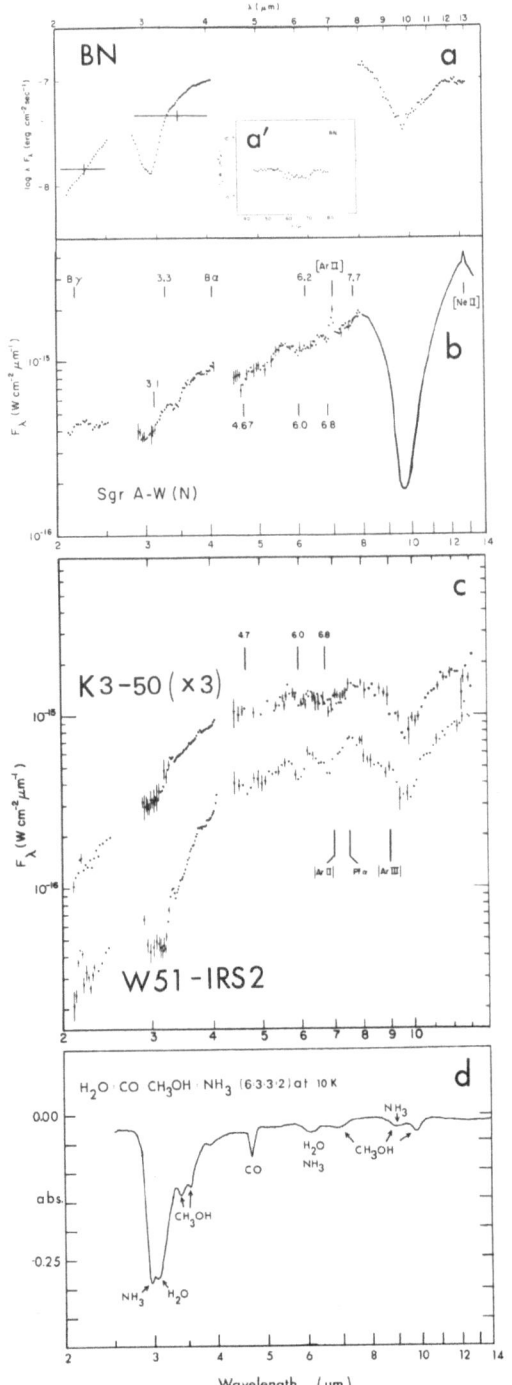

Figure 3. A 2-14 μm spectra
of a, a') The BN object reprinted courtesy of Gillet,
Jones, Merrill and Stein, 1975
and Russell, Soifer and Puetter,
1977; b) The Galactic Center
reprinted from Willner, Russell,
Puetter, Soifer and Harvey,1979;
c) K3-50 and W51-IRS2 reprinted
from Puetter, Russell, Soifer
and Willner, 1979 and; d) The
grain mantle analog.

are described by Hagen et al.,
1979.

This spectrum is to be
compared with the recently
published 2-14 μm spectrum of
the compact H-II regions
W51-IRS2 and K3-50 (Puetter et
al., 1979) which are shown in
Figure 3C, the galactic center
(Willner et al., 1979) shown
in Figure 3b and that of the
BN object shown in Figure 3a.
Examination of this figure
shows that the major features
of the spectrum obtained in
the laboratory agree with those
evident in the other astrophysical objects. Unfortunately
the spectral and spatial
resolution are not yet adequate
to determine the extent to
which some of the features are
contributed to by molecules
which are in the solid phase
and by molecules which are in
the gas phase. However, the
general agreement between the
laboratory spectrum and those
shown in Figure 3 constitutes
strong evidence for molecular
mantles associated with these
objects. The spectra of the
individual objects show differences which are apparent even
at this low resolution. These
differences are probably due
to variations in specific and

relative molecular abundances, solid/gas ratios for particular molecules, grain temperatures and thermal history from object to object (Allamandola, Greenberg, Hagen, 1979). This variation stresses the importance of proper exploitation of the infrared spectrum as a diagnostic tool. When better spectra (higher spectral and spatial resolution) are available one can start to unravel the actual photochemistry and photophysics taking place within a particular cloud and better understand the difference between specific clouds.

In this light, objects like W51-IRS2 and K3-50 are especially interesting since they appear to possess both emission and absorption features and may thus represent the transition case from the BN, and related objects which do not show the emission features at all, to the optically visible H II regions, planetary nebulae and diffuse nebulae which show only the emission features.

References.

Aitken, D.K., and Jones, B.: 1973, M.N.R.A.S. 165, pp. 363.
Allamandola, L.J., and Norman, C.A.: 1978, Astron. Astrophys. 63, pp. L23.
Allamandola, L.J., Greenberg, J.M., and Hagen, W.: 1979, Astron. Astrophys., manuscript in preparation.
Allamandola, L.J., Greenberg, J.M., and Norman, C.A.: 1979, Astron. Astrophys., 77, pp. 66.
Gillett, F.C., Forrest, W.J., and Merrill, K.M.: 1973, Astrophys.J. 183, pp. 87.
Gillett, F.C., Jones, T.W., Merrill, K.M., and Stein, W.A.: 1975, Astron. Astrophys. 45, pp. 77.
Gillett, F.C., Kleinmann, D.E., Wright, E.L. and Capps, R.W.: 1975, Astrophys. J. 198, pp.L65.
Greenberg, J.M.: 1979, Stars and Star Systems, ed. B.Westerlund, Reidel Publishers, pp. 173.
Greenberg, J.M., Yencha, A.J., Corbett, J.W., and Frisch, H.L.: 1972, Mem.Soc.Roy.Sci. de Liège, 6e série, tome III, pp. 425.
Grasdalen, G.L., and Joyce, R.R.: 1976, Astrophys. J. 205, pp.L11.
Hagen, W., Allamandola, L.J., Greenberg, J.M.: 1979, Astrophys. and Space Science, 65, pp. 215.
Hagen, W., Tielens, A.G.G.M., and Greenberg, J.M.: This volume.

Hallam,H.E. (ed.): 1972, Vibrational Spectroscopy of Trapped Species, John Wiley and Sons, New York.
Puetter, R.C., Russel, R.W., Soifer, B.T., and Willner, S.P.: 1979, Astrophys. J. 228, pp. 118.
Russell, R.W., Soifer, B.T., and Puetter, R.C.: 1977, Astron. Astrophys. 54, pp. 959.
Russell, R.W., Soifer, B.T., and Willner, S.P.: 1977, Astrophys. J. 217, pp.L149.
Willner, S.P., Russell, R.W., Puetter, R.C., Soifer, B.T. and Harvey, P.N.: 1979, Astrophys. J. 229, pp.L65.

DISCUSSION FOLLOWING ALLAMANDOLA

Irvine: Is there not a problem getting a sufficiently high quantum efficiency to produce the fluorescent emission lines by the mechanism you suggest?

Allamandola: No, because the quantum efficiency, that is the total number of IR photons emitted per UV photon absorbed, can be less than one. We estimated 0.1, a pessimistic value because one UV photon can contain many vibrational quanta. In addition, we have the broad range of possible UV pumping sources, ranging from the very UV-rich O stars to UV-poor F stars, which can excite this fluorescence if the stellar-cloud geometry is favorable. We have calculated that, when all these things are considered, the effective quantum efficiency can be as much as 100 times lower under favorable conditions.

Herzberg: Have any vibrational transitions been observed in emission in a solid matrix in the laboratory? It seems to me that the lifetimes of vibrational transitions are so long (of the order of milliseconds) that relaxation will occur before emission can take place.

Allamandola: Infra-red emission corresponding to vibrational transitions *has* been observed from inert, low temperature solid matrices, although the radiative lifetime is longer than one would expect for non-radiative decay into the lattice. The mechanism we have proposed here is based on this phenomenon also occurring in low temperature molecular mixtures. Experiments by Prof. Legay and his co-workers at the Molecular Photophysics Institute in Orsay are relevant. They irradiate a thin slab of solid CO with a low power CO laser beam tuned to excite the $V'' = 0$ to $V'' = 1$ transition. When the irradiation is terminated they measure emission not only from $V'' = 1$ to $V'' = 0$, but from much higher V'' levels as well, as high as $V'' = 23$, I believe. Here we have an example of infra-red emission corresponding to vibrational relaxation occurring in a solid molecular matrix. We also have some very recent results from our laboratory. Two weeks ago Fred Baas measured, for the first time, vacuum ultraviolet induced infra-red emission from a 10K CO/NH_3 mixture. These are early results, and the signal detected by the cooled InSb detector was quite weak, but evident.

Millar: The 3.1 μm absorption band does not appear until $A_V \gtrsim 20$ magnitudes. In terms of your grain model, is there a physical reason why this should be so?

Allamandola: The nature of the absorption feature depends critically on the nature of the ice. Unless the history of the mantle is known, it is very dangerous to draw from the observations at 3.1 μm conclusions regarding the presence or absence of H_2O. The absorption cross section of the OH stretch in the H_2O molecule is a very sensitive function of local environment. This cross section can undergo an enhancement by as much as a factor of 100 as H_2O molecules form larger and larger aggregates. There is also a very significant effect on the position and width of the line. Monomeric water absorbs at about 3700 cm^{-1} with a 1-10 cm^{-1} width, complexed "ice" at about 3200 cm^{-1}, with a breadth of 300-400 cm^{-1}.

Willner: There are emission features that occur together as a set in nearly all sources where any of the individual features occurs. These are at 3.3-3.4 μm, 6.2 μm, 7.7 μm, 8.6 μm, and 11.3 μm. Is the 3.4 μm band of CH_4 active or forbidden in your ice samples.

Allamandola: We have found this band to be infra-red active in our complex mixtures. This break-down of "forbiddeness" has something to do with the degree of polarizability of the solid environment. If solid CH_4 is studied, the band *is not* infra-red active, presumably because the lattice is made up of non-polar molecules.

Gilra: Dr. Willner talked about *absorption* features in the previous talk. You have discussed the *emission* features. Since in at least some cases the "dust" producing the emission features is in the same neighbourhood as the dust producing the absorption features, would it be unreasonable to conclude that the dust is the same in the two cases? Some spectral differences between the absorption and emission may be caused by the mode of excitation. For example, the emission spectrum due to molecules in a matrix is somewhat different from the absorption spectrum due to the same molecules in the same matrix.

Willner: The absorption features, with the exception of the silicate feature, have never been seen in emission. The emission features have never been seen in absorption, unless the 3.3-3.4 μm feature turns out to be this wing of the 3.1 μm absorption. If we are talking about ices, we would not expect to see 3.1 μm in emission, because a high temperature would be required, and presumably the ice would evaporate.

Allamandola: It is certainly reasonable to expect absorption if there is a continuum IR source behind the cloud. In this respect the compact HII region W51 IRS2 has some spectral features which resemble emission and some which resemble absorption. Whether the emission and absorption positions occur at the same frequency must await confirmation. My guess is that the frequencies will not be very different, based on the CO laser-excited emission frequency which is within one wavenumber of the absorption position, as shown by Dubost. It is an open question whether one can find the conditions in which the *same* grains both emit and absorb.

INFRARED MOLECULAR ABSORPTION FEATURES

S.P. Willner and R.C. Puetter
University of California, San Diego

Ray W. Russell
Cornell University

B.T. Soifer
California Institute of Technology

Spectra of infrared sources associated with molecular clouds have shown absorption features at wavelengths of 6.0 and 6.8 μm. We suggest that the 6.0 μm feature can be identified with the stretching vibration of C=O and the 6.8 μm feature with the bending vibrations of CH_2 and CH_3. The amount of carbon in the form of hydrocarbon molecules may be comparable to the amount in CO. This abundance of hydrocarbons is probably too large to be consistent with radio observations if the molecules are gaseous, but large abundances of hydrocarbons on the surfaces of grains may explain the infrared features, yet be unobservable in the radio.

OBSERVATIONS

Infrared sources associated with molecular clouds include compact HII regions and sources with much smaller or nonexistent thermal radio emission, often called "protostars". Spectra (typically with resolution $\lambda/\Delta\lambda \sim 60$) in the ground-based 2-4 μm and 8-13 μm windows of many examples of both classes of objects are available in the literature. The 8-13 μm spectra show evidence of silicate dust grains. The silicate feature is seen in emission in the Orion Trapezium, but most objects exhibit various amounts of silicate absorption, believed due to extensive overlying cold dust. In the 2-4 μm spectrum there is often an absorption feature having maximum depth at 3.1 μm; this feature is usually attributed to water ice. The shape of the feature is, however, not entirely consistent with water ice absorption, in that there is significant optical depth between 3.3 and 3.5 μm (Merrill et al. 1976). Neither water nor ammonia ices would be expected to absorb sufficiently at these wavelengths to explain the observed shape of the feature. There is no correlation between the ice and silicate optical depths, except that if ice absorption is present, there is also finite silicate absorption present.

381

B. H. Andrew (ed.), Interstellar Molecules, 381–386.

The Kuiper Airborne Observatory has recently made possible observations in the 4.5 to 8 μm spectral region. A total of 11 sources
obscured by molecular cloud material have now been observed spectroscopically at these wavelengths (Russell et al. 1977, Puetter et al.
1979, Soifer et al. 1979, Puetter et al. 1980). All have depressions
in the spectrum between 6 and 7 μm, and in most cases it can clearly be
seen that there are two dips, centered near 6.0 and 6.8 μm. By far
the strongest features were observed in the source W33A (Soifer et al.
1979), which also has the strongest silicate and 3.1 μm absorptions
known (Capps et al. 1978). The spectrum of this source is shown in
Figure 1. The 6.0 and 6.8 μm features are spectrally resolved and have

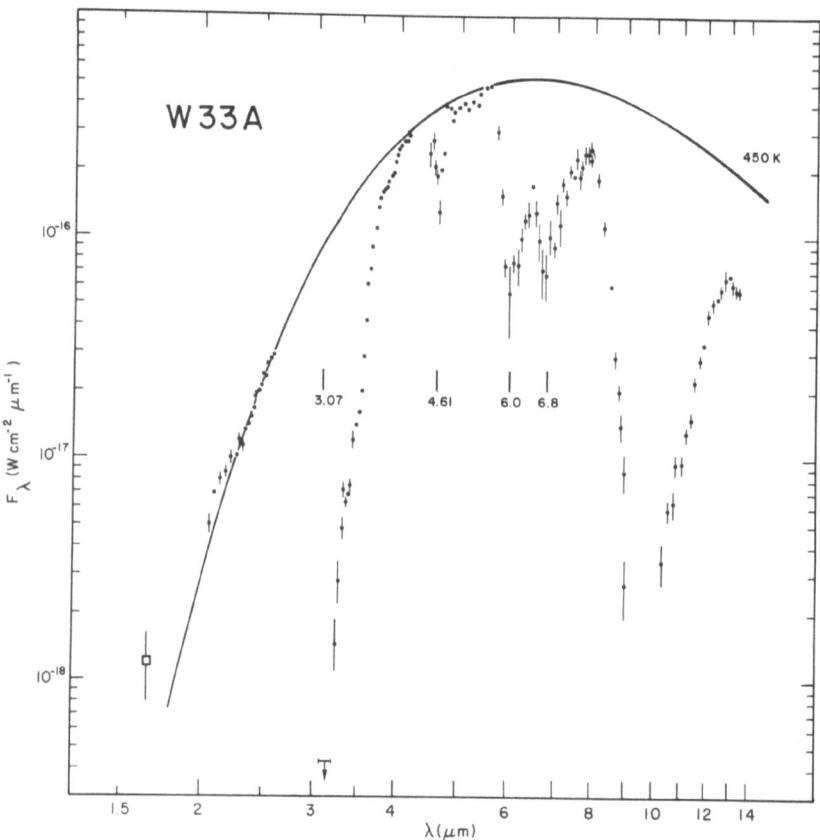

Fig. 1. The 2 to 13 μm spectrum of W33A. The 1.6-4 and 8-13 μm data
are from Capps et al. (1978), and the 4-8 μm data are from Soifer et al.
(1979). Error bars are shown when the statistical uncertainty exceeds
5%. The solid line represents the spectrum of a 450 K blackbody.
Central wavelengths of 4 absorption features are marked.

widths of about 0.6 and 0.7 μm, respectively. The widths and positions
of similar features are less easily determined in the spectra of other
sources, but they are consistent with those in W33A. Table 1 gives the
approximate peak optical depths of the features measured in 11 sources;
the observations will be reported in more detail elsewhere. These
numbers are, of course, strongly dependent on the choice of continuum
level, which has generally been chosen as a blackbody fit through the
spectrum at 4.5-5.5 μm and at 8.0 μm.

TABLE 1

PEAK ABSORPTION OPTICAL DEPTHS

Object	6.0 μm	6.8 μm
Becklin-Neugebauer	0.3	0.3
Sharpless 255	0.2	0.2
AFGL 989	0.3	0.4
W33A	1.9	1.5
AFGL 2136	0.3	0.3
AFGL 2591	0.2	0.2
AFGL 2884	0.5	0.3
NGC7538/IRS1	0.5	0.4
NGC7538/IRS9	0.6	0.4
W51/IRS2	0.5	0.5
K3-50	0.3	0.3

The 6.0 and 6.8 μm features were not seen in the line of sight to
the galactic center (Willner et al. 1979). The features must therefore
be characteristic of material in molecular clouds, rather than low-
density interstellar material. The strengths of the features are not
correlated with the strength of the silicate absorption and are only
weakly correlated with the strength of the ice absorption.

DISCUSSION

It was first proposed that the 6.0 and 6.8 μm features were characteristic of hydrated or otherwise processed silicate minerals. Such
features have been seen between 6 and 7 μm in the laboratory (Duley and
McCullough 1977, Stephens and Russell 1979, Day 1978), although not at
the exact wavelengths of the astronomical features. However, the features have not been seen to be as deep relative to the 10 μm band as in
W33A. Moreover, the water of hydration band in silicates is usually at
6.1-6.2 μm rather than 6.0 μm and is only about half as wide as the
astronomical absorption bands (Stephens and Russell 1979, Day 1978).
Finally, the low temperatures of molecular clouds would seem more con-

ducive to the formation of ice mantles than hydrated grains, whereas in the envelope of an OH-IR star, where hydrated grains might be expected, no 6.0 or 6.8 μm features were found (Forrest et al. 1978). These arguments suggest that silicates are unlikely to be responsible for the 6.0 and 6.8 μm features.

Hydrocarbon bonds presently appear to be the most likely candidates for explaining the 6.0 and 6.8 μm features. The 6.0 μm band would be identified with the stretching vibration of the carbonyl group (C=O) with possibly some contribution from the stretch vibration of the C=C group. The 6.8 μm band would be identified with the scissors vibration in the methyl (CH$_3$) or methylene (CH$_2$) group. It is doubtful that any single molecule would be responsible for the observed absorptions, but rather, a large variety of molecules containing these groups would contribute to the observed bands. This is consistent with the broad absorptions that are observed, with the characteristic signatures of the functional groups in specific molecules being lost.

The column densities of the appropriate absorbers can be crudely estimated using observed equivalent widths and laboratory integrated band intensities (Puetter et al. 1979, Wexler 1967). These column densities, compared either to silicate or to radio CO column densities, generally indicate an amount of carbon comparable to but less than that in the form of CO.

The stretching vibration of the C-H bond is centered at 3.1-3.4 μm, depending on the specific group involved. The absorption is expected to be somewhat greater than that of the 6.8 μm band, and it could even be strong enough to explain all of the observed absorption between 2.9 and 3.5 μm. It is possible that some of the absorption is actually due to ice, but a major portion must be due to hydrocarbons. There will probably be no difficulty explaining the observed shape of the absorption with an appropriate mixture of molecules, either with or without ice.

Radio observations have shown the presence of quite complex hydrocarbons in molecular clouds (e.g., Kroto et al. 1978, Winnewisser and Walmsley 1978), but the derived abundance of known hydrocarbons is considerably smaller than that indicated here (Allen and Robinson 1977 and references therein). The lack of radio detection of such large abundances of hydrocarbons could be explained if most of the molecules are coated onto grains, where the lack of freedom to rotate would prevent radio emission. Such a situation would not be surprising at the low temperatures characteristic of molecular clouds (Watson and Salpeter 1972).

If large abundances of hydrocarbons in molecular clouds can be confirmed, it will probably require that their formation rates be higher than presently believed. Most of the available calculations include only gas-phase reactions, but for a full understanding of molecular clouds, it may be necessary to consider the far more complex subject of surface reactions.

This work was supported by NASA grant NGR 05-005-055 and by NSF grant AST 76-82890.

REFERENCES

Allen, M., and Robinson, G.W.: 1977, Astrophys. J. 212, pp. 396-415.
Capps, R.W., Gillett, F.C., and Knacke, R.F.: 1978, Astrophys. J. 226, pp. 863-868.
Day, K.L.: 1978, private communication.
Duley, W.W., and McCullough, J.D.: 1977, Astrophys. J. (Letters) 211, pp. L145-L148.
Forrest, W.J., Gillett, F.C., Houck, J.R., McCarthy, J.F., Merrill, K.M., Pipher, J.L., Puetter, R.C., Russell, R.W., Soifer, B.T., and Willner, S.P.: 1978, Astrophys. J. 219, pp. 114-120.
Kroto, H.W., Kirby, C., Walton, D.R.M., Avery, L.W., Broten, N.W., MacLeod, J.M., and Oka, T.: 1978, Astrophys. J. (Letters) 219, pp. L133-L137.
Merrill, K.M., Russell, R.W., and Soifer, B.T.: 1976, Astrophys. J. 207, pp. 763-769.
Puetter, R.C., Russell, R.W., Soifer, B.T., and Willner, S.P.: 1979, Astrophys. J. 228, pp. 118-122.
Puetter, R.C., Russell, R.W., Soifer, B.T., Willner, S.P.: 1980, in preparation. For partial abstract see Bull. Amer. Astron. Soc. 9, p. 571.
Russell, R.W., Soifer, B.T., and Puetter, R.C.: 1977, Astron. Astrophys. 54, pp. 959-960.
Soifer, B.T., Puetter, R.C., Russell, R.W., Willner, S.P., Harvey, P.M., and Gillett, F.C.: 1979, Astrophys. J. (Letters) 232, pp. 53-57.
Stephens, J.R., and Russell, R.W.: 1979, Astrophys. J. 228, pp. 780-786.
Watson, W.D., and Salpeter, E.E.: 1972, Astrophys. J. 174, pp. 321-340.
Wexler, A.S.: 1967, Appl. Spectrosc. Rev. 1, pp. 29-98.
Willner, S.P., Russell, R.W., Puetter, R.C., Soifer, B.T., and Harvey, P.M.: 1979, Astrophys. J. (Letters) 229, pp. L65-L68.
Winnewisser, G., and Walmsley, C.M.: 1978, Astron. Astrophys. 70, pp. L37-L39.

DISCUSSION FOLLOWING WILLNER

Thaddeus: It does not seem to me that you are talking about an unreasonable amount of carbon in C=O bonds. We only see a fraction of the total carbon in molecules like CO, and the rest has to be somewhere.

Willner: Yes. There must be a family of molecules with C=O bonds because the spectral features are resolved even with our low resolution, and no one particular molecule would make features as broad as that. If ∿10% of the carbon is in the form of CO, another 10% in the form of hydrocarbons would be more than enough to account for the observations. The implied abundance is, however, much higher than suggested for individual hydrocarbons in the radio.

Greenberg: Earlier the point was made that there is as much oxygen in H_2O as in CO, as if this were an awful lot of H_2O. I figured out that this would mean that at most 10% of all the oxygen is in H_2O. If what you say is true about CO relative to H_2O, then there is a further reduction by a factor of 4, so there is not such a large amount of H_2O compared to what we find in dust.

Willner: The data show considerable scatter, but the 3.1 μm absorption is roughly 10 times as strong as the 6.0 μm absorption. If the absorption efficiency per bond for OH stretch is as great as for C=O stretch, half as much oxygen in OH bonds as in CO would account for the 3.1 μm feature. If I have interpreted them correctly, Hagen's data suggest that OH stretch is a more efficient absorber than C=O, so an even smaller number of OH bonds would suffice.

Allamandola: I will take the bait early. What are your reasons for attributing the 6.0 μm feature to hydrocarbons and not, say, water of hydration or the H-O-H bend or H-N-H bend in the classic dirty ice? This vibration shifts considerably from its uncomplexed position at 1600 cm^{-1} toward 1700 cm^{-1} (6.0-6.2 μm), and undergoes substantial broadening when it complexes.

Willner: We are not aware of the possible shift in wavelength or the suppression of the 11.3 μm ice feature. There are a few sources, such as OH 0739-14, where the 3.1 μm absorption has returned to the continuum level by 3.3 μm, and the 11.3 μm absorption is present, and we felt that these characteristics indicated the presence of ice. Most sources do not look like this, and it seemed natural to suggest something other than ice. Also, the 3.1, 6.0, and 6.8 μm features occur together, and would probably be the strongest absorptions for hydrocarbon molecules.

REPRODUCTION OF THE INTERSTELLAR ICE BAND BY GRAIN MANTLE ANALOGS

W. Hagen, A.G.G.M. Tielens and J.M. Greenberg
Laboratory Astrophysics Group, Huygens Laboratory,
Wassenaarseweg 78, 2300 RA Leiden, Netherlands.

The near-infrared spectrum of many sources associated with molecular clouds shows a broad absorption feature at 3.08 μm (e.g. Merrill et al., 1976; Harris et al., 1978). This feature has usually been attributed to absorption by H_2O ice frozen on grains, but it has been impossible to satisfactorily reproduce the observed band shape (Merrill et al., 1976; Mukai et al., 1978). We have been able to obtain a complete fit of this absorption feature in the laboratory using very low temperature mixtures of H_2O with other polar molecules. The preparation of these interstellar dust grain-mantle analogs has been described elsewhere (Greenberg, 1979; Hagen et al., 1979). They are prepared by allowing a gas mixture of simple molecules (e.g. CO, H_2O, NH_3, CH_4 etc.) to condense on a low temperature (10 K) substrate. This frozen mixture can be heated and recooled. The samples are analyzed with an infrared spectrometer.

With the aim of studying the 3.08 μm band we have examined, in detail, the middle infrared spectra of mixtures of H_2O with other molecules. These samples show absorption lines centered around 3 μm corresponding to the OH stretching vibrations of H_2O monomers, dimers, polymers and complexes between H_2O and other molecules present. Warming up of the sample to about 40 K permits diffusion of the H_2O molecules and the broad band characteristic of amorphous ice appears. Its shape and peak position are dependent on the temperature and on the concentration of the other molecules present. If there are other polar molecules present this band will show a low frequency wing due to further complexing between H_2O ice and these molecules. The strength of this wing is a function of temperature and of concentration as well as polarity of the other molecules present. As an example, our laboratory spectrum of a mixture of CO/H_2O (10/1), deposited at 10 K, heated to 40 K and subsequently recooled to 10 K, is compared in Fig. 1 with the observed 3.08 μm feature associated with the BN object (Gillett et al., 1975). Also shown is the spectrum obtained at 80 K of pure amorphous H_2O ice, Ia (Buontempo, 1972). The agreement between our spectrum and the observed interstellar band with respect to peak position, width *and low frequency wing* is quite good. The satisfactory

B. H. Andrew (ed.), Interstellar Molecules, 387–388.

fit of the band, especially the low frequency wing, suggests that this is part of the long sought evidence for the existence of complex molecular mantles in interstellar grains (Greenberg et al., 1972). We are currently studying this absorption band in more mixtures and incorporating the spectrum into a detailed analysis of the infrared extinction and polarization by interstellar grains in dense clouds.

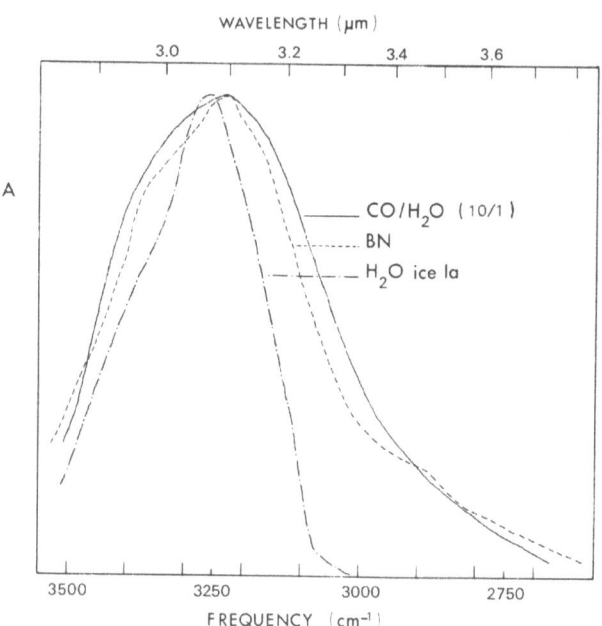

Fig. 1. The infrared spectrum around 3000 cm^{-1};   —— CO:H$_2$O (10:1) mixture, deposited at 10 K, heated to 40 K and recooled to 10 K; --- the BN object (Gillett et al., 1975); -·-·- amorphous H$_2$O ice at 80 K (Buontempo, 1972).

REFERENCES

Buontempo, U.: 1972, Phys. Letters 42A, 17.
Gillett, F.C., Jones, T.W., Merrill, K.M. and Stein, W.A.: 1975,
      Astron. Astrophys. 45, 77. .
Greenberg, J.M.: 1979, The Moon and the Planets 20, 15.
Greenberg, J.M., Yencha, A.J., Corbett, J.W. and Frisch, H.L.: 1972,
      Mem. Soc. Roy. Sciences de Liège, 6e série, Tome III, 425.
Hagen, W., Allamandola, L.J. and Greenberg, J.M.: 1979, Astrophys.
      Space Sci. 65, 215.
Harris, D.H., Woolf, N.J. and Rieke, G.H.: 1978, Astrophys. J. 226, 829.
Merrill, K.M., Russell, R.W. and Soifer, B.T.: 1976, Astrophys. J.
      207, 763.
Mukai, T., Mukai, S. and Noguchi, K.: 1978, Astrophys. Space Sci. 53, 77.

CORRELATIONS BETWEEN THE $\lambda 2200$ FEATURE, THE DIFFUSE $\lambda 4430$ BAND AND E_{B-V}

A.C. Danks[*]
Kapteyn Astronomical Institute

[*]Present address: c/o E.S.O., Casilla 16317, Santiago 9, Chile

ABSTRACT

The $\lambda 2200$ extinction feature has been measured from ANS observations of 30 stars. For each star, the depth of the $\lambda 2200$ feature is compared to E_{B-V} and the equivalent width of the diffuse band $\lambda 4430$.

A good correlation appears out to $E_{B-V} = 1.13$. The various mechanisms for producing the diffuse and $\lambda 2200$ features are discussed, and a discriminatory test is put forward based on observations of the Magellanic Clouds.

INTRODUCTION

Both the $\lambda 2200$ extinction feature and the diffuse interstellar lines have defied identification. Comparisons of the relative strengths of the two bands may help to identify or distinguish the interstellar carrier responsible for the features. Past correlations of the diffuse bands with the $\lambda 2200$ feature have been made by Wu (1972), Nandy et al. (1975), Nandy and Thompson (1975), and Schmidt (1978).

The present study makes use of the high sensitivity of ANS for good photometry of faint (i.e. high E_{B-V}) stars, and of the homogeneous set of photoelectric observations of $\lambda 4430$ given by Gammelgaard (1975). Both the $\lambda 4430$ diffuse band and the $\lambda 2200$ feature pose similar problems of measurement. Due to the large half-width of the $\lambda 2200$ feature, the continuum is difficult to define and so, therefore, is the equivalent width.

Differences in the definitions of satellite systems create problems in comparing the $\lambda 2200$ widths quoted by other authors, e.g. Nandy and Thompson (1975), Savage (1975). Similarly, $\lambda 4430$ and other diffuse features have extended wings and are superimposed on stellar features, so that there are differences in measurement between authors. The 30 stars presented here were all chosen from Gammelgaard's (1975) list.

B. H. Andrew (ed.), Interstellar Molecules, 389–394.

(Some of these stars were chosen by Gammelgaard for measurement of the 6180 diffuse band.)

OBSERVATIONS AND REDUCTION

The ANS photometric system is described in detail by Van Duinen et al. (1975). Basically it is a 5 channel photometer with wavelengths centered at $\lambda 1550$, $\lambda 1800$, $\lambda 2200$, $\lambda 2500$, and $\lambda 3300$. The counts in each channel were inspected for overflow, and the dark current was subtracted. In many cases, data for individual objects was available for several orbits, so a mean value was taken. The counts were converted to flux following the procedures outlined in internal notes (Wesselius 1975, and Aalders 1975). The fluxes were normalized to $\lambda 3300$. Each program star was divided by the normalized flux of a star closely matching the program star in spectral type and luminosity class. These stars were taken from Wu (1975) and corrected for interstellar extinction. The depth of the $\lambda 2200$ (D_{2200}) was derived from the resulting data using the formula $D_{2200} = m_{2200} - (m_{2500} + m_{1800})/2$. Errors in D_{2200} arise from slight spectral and luminosity-class mismatches between program stars and standard stars. Errors in the photometry were also taken into account and are represented in most cases by error bars in Figs. 1 and 2. The 30 stars are listed in Table 1 with D_{2200} and J_c, which is a parameter related to A_c (4430) given in Gammelgaard (1975). J_c is independent of spectral type.

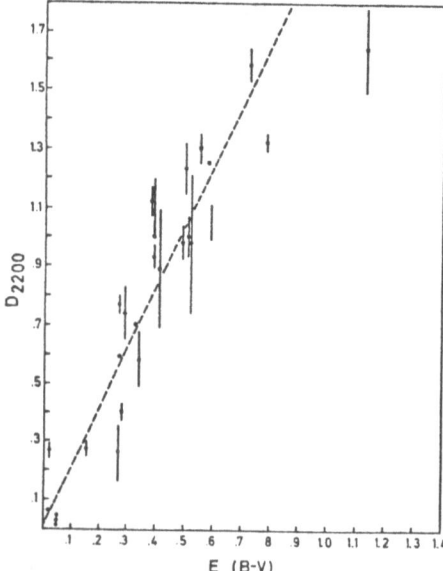 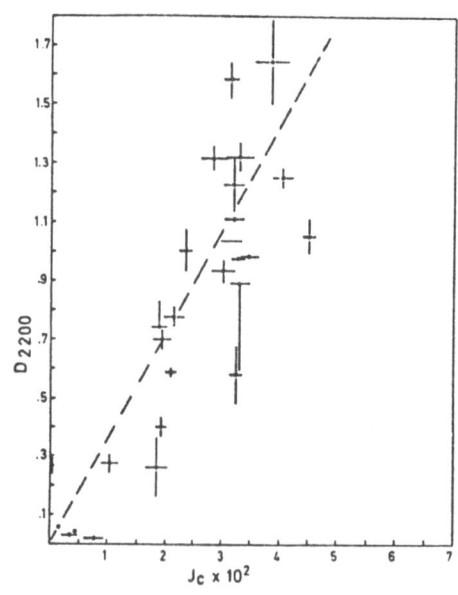

Fig. 1: A plot of D_{2200}, the strength of the $\lambda 2200$ feature, against E_{B-V}.

Fig. 2: A plot of D_{2200}, the strength of the $\lambda 2200$ feature, against $J_c(4430)$, the strength of the diffuse band $\lambda 4430$.

TABLE 1

HD	SpT	V	E$_{(B-V)}$	D$_{2200}$	Jc	kpc
39970	AOIa	6.02	.38	1.124	321	2.5
42088	O6	7.55	.39	.926	303	2.4
44965	B3II	7.82	.49	.978	342	1.5
46149	O8	7.61	.5	1.226	320	1.8
46769	B8Ib	5.79	.02	.265	2	1.8
47240	B1Ib	6.15	.33	.704	196	1.5
48099	O6	6.38	.27	.774	214	1.6
53138	B3Ia	3.04	.05	.021	73	.9
53244	B8II	4.1	−.05	−.045	−44	.4
60308	B2Iab	8.2	.59	1.054	449	3.5
64760	B1Ib	4.24	−.04	.065	26	.9
68450	BOII	6.43	.27	.593	212	1.6
75211	O8V	7.5	.72	1.582	314	1.3
83183	B5II	4.08	.15	.275	103	.4
86440	B5II	3.53	.05	.029	34	.4
87737	AOIb	3.48	−.03	−.030	−72	.6
91316	B1Ib	3.85	.05	.039	48	.8
93843	O6.5V	7.34	.34	.582	324	2.5
97319	O9.5Ib	8.5	.52	.975	333	3.9
97966	O7.5V	8.8	.41	.887	333	3.8
102878	A2Ia	5.69	.27	.256	185	3.0
106068	B9Ia	5.92	.29	.742	189	2.7
109867	BO.5Iab	6.24	.28	.402	180	2.1
111775	AOII	6.32	.02	.058	14	.7
115842	BO.5Iab	6.02	.51	1.023	233	1.4
153919	O6f	6.55	.58	1.246	406	1.2
154368	O9.5Iab	6.11	.78	1.322	287	1.0
165052	O7V	6.86	.39	1.032	314	1.7
167838	B5Ia	6.73	.55	1.324	329	2.6
169454	B1Ia	6.61	1.13	1.638	384	1.9

DISCUSSION

The values of D_{2200} plotted against E_{B-V} are shown in Fig. 1, and D_{2200} against J_c (4430) in Fig. 2. In both figures the dotted line is a weighted least squares fit (LSF) which has been forced through zero. If the fit had not been forced through zero, the lines would have passed through small positive values of J_c and E_{B-V} for zero values of D_{2200}. Both Herbig (1975) and Smith et al. (1976) claimed that the diffuse bands had a finite strength at $E_{B-V} = 0.0$ that has been attributed to the diffuse bands' originating in clouds of low column density. However Somerville (1979) points out that many of the anomalously strong $\lambda 4430$ features reported are due to measurements at low spectral resolution that are contaminated by numerous weak stellar features; higher resolution measurements would result in better correlations passing through the origin. Blades and Somerville (1977) illustrate the effect for ρ Leo.

The correlation of D_{2200} vs. E_{B-V} appears linear and quite tight up to $E_{B-V} = 0.9$. The star showing poor correlation is HD 169454 (B1Ia, $E_{B-V} = 1.13$, d = 1.0 kpc), which is close to M8. Some nebulosity is included in the satellite diaphragm, resulting in the dilution of the $\lambda 2200$ feature (Gilra 1979). Correction for this effect would bring HD 169454 on to the LSF line.

The scatter of observations from the LSF in Fig. 2 (D_{2200} vs. J_c) is predictably larger due to errors in measuring both D_{2200} and J_c. Deviations in the plot of $m_{2100}-V$ index against Q [(U-B) - S(B-V)] were noted by Malaise et al. (1974) at values of $E_{B-V} = 0.6$. Malaise et al. attributed them to a two-component reddening, one component being responsible for the $\lambda 2200$ feature and the other giving rise to a pure reddening with no absorption at 2200 Å. Malaise et al. also demonstrated the problems of defining the continuum in the UV and the effects that stellar lines can have on the effective equivalent width of $\lambda 2200$.

The scatter of individual stars from the linear fit in Fig. 2 is large, but there seems to be a weakening at high J_c (4430) of the relationship between D_{2200} and J_c (4430). However this weakening is not convincingly demonstrated because of the large error bars; high resolution observations of both $\lambda 2200$ and $\lambda 4430$ are needed to define the relationship better.

Despite the large amount of work in the field, the carrier or carriers for these features have not been identified. The most convincing explanation for the $\lambda 2200$ feature has been given by Gilra (1972), who attributes it to small graphite particles. These particles may also produce the visible reddening, but are an unlikely cause of the high extinction at wavelengths shorter than 2200 Å. It has not been established if the carrier of the $\lambda 2200$ feature also produces the visible diffuse bands, or if the diffuse band carriers are simply well mixed with the dust responsible for the visible reddening. As early as 1939, Swings and Ohman proposed that small molecules could cause the diffuse

features. Following the observations at mm wavelengths of large
molecules, Danks and Lambert (1975, 1976) pointed out that relatively
small polyatomic molecules of 7 to 14 atoms could be excited to produce
diffuse bands such as λ5780 and λ5797, although it would need a larger
molecule to produce λ4430.

Duley (1977) suggests MgO and CaO as possible carriers for the
diffuse lines. Millar and Duley (1979) show that λ2200 decreases in
strength in a cloud as carbon depletion increases, contrary to what is
expected if depletion is due to the forming of grains by carbon.
Alternatively Rudkjobing (1978) points out that autoionization of O^-
could also produce the diffuse features.

Large organic molecules have been proposed to explain the λ2200
feature by Wickramasinghe et al. (1977); more specifically, $C_8 H_6 N_2$
has been proposed by Hoyle and Wickramasinghe (1977).

The range and type of suggested carriers is large, and a discriminatory test is needed. Abundances in the Large and Small Magellanic
Clouds may offer a solution. Although the reddening is relatively
small, for instance in the 30 Doradus region (Borgman and Danks 1977),
the λ2200 feature has been measured (Borgman et al. 1975) and is weaker
than expected compared to E_{B-V}. Abundances have been determined from
measurements of HII regions in the Clouds (Dufour and Harlow 1977,
Peimbert and Peimbert 1976, Pagel 1978), and some elements, notably N,
O, Ne, Ar, are found to be underabundant with respect to Orion. If the
carrier of λ2200 or the diffuse bands is dependent on one of these atoms
we may see this dependency reflected in the strength of either λ2200 or
λ4430. However, an accurate relationship between λ2200 vs. λ4430 is
first needed for the Galaxy. Initial steps have been taken by Nandy
and Morgan (1978), who observed λ2200 with the IUE at high resolution in
two stars in the LMC, and by Blades and Madore (1979) and Danks (1979)
who detected λ4430 in several stars in the LMC. If the deficiency
reported by Borgman et al. (1975) proves to be correct for individual
stars, it may be possible to determine whether λ2200 and λ4430 have the
same carrier and which elements are responsible.

Many of the proposed carriers are based on carbon, and it is
important, therefore, to determine the abundance of C as accurately as
possible. With the introduction of unintensified reticons, Dennefeld
(private communication) has been able to measure the red carbon line in
supernova remnants. The carbon abundances from these observations are
essential in identifying the carrier.

Finally, there appears to be no substitute for observations of high
resolution and high signal-to-noise ratio for a range of E_{B-V} and in
particular for high E_{B-V}.

ACKNOWLEDGEMENTS

It is a pleasure to thank Drs. R. van Duinen and P. Wesselius for
useful discussions and V. Pippin for help with the data reduction.

REFERENCES

Aalders, J.W.G.: 1975, internal note 75-23, Correction Table for Bright
    Stars.
Blades, J.C., and Madore, B.F.: 1979, Astron. & Astrophys. 71, 359.
Blades, J.C., and Somerville, W.B.: 1977, M.N.R.A.S. 181, 769.
Borgman, J., and Danks, A.C.: 1977, Astron. & Astrophys. 54, 41.
Borgman, J., van Duinen, R.J., and Koorneef, J.: 1975, Astron. &
    Astrophys. 40, 461.
Danks, A.C.: 1979, preprint.
Danks, A.C., and Lambert, D.L.: 1975, Astron. & Astrophys. 41, 455.
Danks, A.C., and Lambert, D.L.: 1976, M.N.R.A.S. 174, 571.
Douglas, A.C.: 1977, Nature 136, 269.
Dufour, R.J., and Harlow, W.V.: 1977, Astrophys. J. 216, 706.
Duley, W.W.: 1977, Astrophys. & Space Science 47, 185.
Gammelgaard, P.: 1975, Astron. & Astrophys. 43, 85.
Gilra, D.P.: 1979, private communication.
Gilra, D.P.: 1972, in Scientific Results for OAO, ed. D. Code
    (NASA Sp-310), p. 295.
Herbig, G.H.: 1975, Astrophys. J. 196, 129.
Hoyle, F., and Wickramasinghe, N.C.: 1977, Nature 270, 323.
Lynds, B.T.: 1962, Astrophys. J. Suppl. 7, 1.
Malaise, D., Beeckmans, F., and Jamar, C.: 1974, Estrato del Memorie
    Della Societa Astronomica Italiana 45, 233.
Millar, T.J., and Duley, W.W.: 1979, M.N.R.A.S. 187, 379.
Mitchell, G.F., and Huntress, W.T.: 1979, Nature 278, 722.
Nandy, K., Thompson, G.I., Jamar, C., Monfils, A., and Wilson, K.: 1975,
    Astron. & Astrophys. 44, 195.
Nandy, K., and Morgan, D.H.: 1978, Nature 276, 478.
Nandy, K., and Thompson, G.I.: 1975, M.N.R.A.S. 173, 237.
Pagel, B.: 1978, M.N.R.A.S. 183, 1.
Peimbert, M., and Torres-Peimbert, S.: 1976, Astrophys. J. 203, 501.
Rudkjobing, M.: 1978, Astron. & Astrophys. 63, 189.
Savage, B.D.: 1975, Astrophys. J. 199, 92.
Schmidt, E.G.: 1978, Astrophys. J. 223, 458.
Smith, W.H., Snow, T.P., and York, D.G.: 1977, Astrophys. J. 218, 124.
Somerville, W.B.: 1979, IAU Symposium No. 87, this volume.
Snow, T.P., and Cohen, J.G.: 1974, Astrophys. J. 194, 313.
Van Duinen, R.J., Aalders, J.W.G., Wesselius, P.R., Wildeman, K.J.,
    Wu, C.C., Luinge, W., and Snel, D.: 1975, Astron. & Astrophys.
    39, 159.
Van Duinen, R.J., Wu, C.C., and Kester, D.: 1976, internal note 76-4,
    ANS Extinction Curve II.
Wampler, E.J.: 1966, Astrophys. J. 144, 921.
Wesselius, P.R.: 1975, internal note 75-1, Absolute Calibration of
    ANS UVX.
Wickramasinge, N.C., Hoyle, F., and Nandy, K.: 1977, Astrophys. & Space
    Science, 47, 9.
Wu, C.C.: 1975, internal note 75-36, ANS Observations of Early Type Stars.
Wu, C.C.: Luinge, W., and Snel, D.: 1975, Astron. & Astrophys. 39, 159.
Wu, C.C.: 1972, Astrophys. J. 178, 681.

CORRELATIONS FOR INTERSTELLAR MOLECULES AND DIFFUSE BANDS

W.B. Somerville
Department of Physics and Astronomy
University College London

ABSTRACT

Results are presented from a programme of optical spectroscopy and related studies which has the double purpose of investigating the structure of diffuse molecular clouds and of establishing tighter correlations for the unidentified interstellar diffuse bands.

MOLECULAR STUDIES

There has been a tendency for each study of interstellar molecules to be concentrated on one spectral region, neglecting the advantages of relating different observations along the same line of sight. In University College London, D. McNally has developed ideas of analyzing interstellar cloud structures by combining optical and ultraviolet molecular observations, using IUE and other satellites, and comparing them with microwave observations in the same directions made by R.L. Dickman of Aerospace Corporation, Los Angeles. Whittet et al. (1979a) have discussed in this way the abundances of CN and CO in diffuse interstellar clouds. Results are well correlated and lead to the relative abundance

$$N(CN) \ / \ N(CO) \sim 4.5 \times 10^{-3}.$$

This value is consistent with an extrapolation of theoretical results based on ion-molecule reactions in the gas, and is not consistent with predictions involving grain surface reactions.

DIFFUSE INTERSTELLAR BANDS

It is important to establish how precise are the correlations of the unidentified diffuse band strengths with each other and with other quantities such as the colour excess E(B-V) and the strength of the ultraviolet $\lambda2200$ band, and to scrutinize all cases where anomalies may be present. We have found that several reported anomalies cannot be confirmed, and have formed the opinion that all putative anomalies should be approached with caution.

395

B. H. Andrew (ed.), Interstellar Molecules, 395–396.
Copyright © 1980 by the IAU.

From low-resolution observations, the star ρ Leo appeared to have anomalously-strong λ4430; it was frequently cited as an outstanding example of this. However, by high-resolution spectroscopy Blades and Somerville (1977) established that the apparent anomaly is caused by a combination of weak lines in the spectrum of the star itself. Similar results have been found for other stars. This suggests an explanation of the effect in correlation studies of a non-zero λ4430 strength for zero reddening: this has been a puzzle for, whatever the carrier of the band, it is hard to imagine it present where there is no dust. Our work suggests that the λ4430 - E(B-V) correlation line should be lowered, by about 3% in the central intensity $A_c(4430)$, and pass through the origin.

A statistical analysis of the scatter about the mean line in the correlation diagram of Snow et al. (1977) for $A_c(4430)$ with E(B-V) (Somerville, in preparation 1979) indicates that the major part of the scatter is consistent with observational error. There is, however, evidence for some physical departure from a linear relation. This does not in itself prove that the bands are not produced by the grains responsible for the reddening; they could be associated with some property such as impurity content which varies from place to place. The distribution in the sky appears random apart from a tendency in a few regions for stars of weak λ4430 to cluster together, in accord with the well-established result that the diffuse bands are below strength relative to the reddening in certain regions. The statistical analysis predicts that there should be similar numbers of reddened stars where the bands are anomalously strong.

From TD-1 observations, Willis and Wilson (1975) found that the ultraviolet feature λ2200 was anomalously very strong in HD 192163, a Wolf-Rayet star of class WN6. Optical observations (Whittet et al., 1979b) show that interstellar atomic lines and diffuse bands in this star are of normal strength. A quick look at recent observations by Somerville and Whittet suggest that the same conclusion will be reached in a second case, HD 156385, class WC7. If the large anomaly in λ2200 is confirmed, it will thus indicate that the carrier of λ2200 is different from the carriers of all these optical features, and from the dust grains which produce the optical extinction.

REFERENCES

Blades, J.C., and Somerville, W.B., 1977, Mon. Not. R. Astr. Soc. 181, pp 769-776.
Snow, T.P., York, D.G., and Welty, D.E., 1977. Astr. J. 82, pp 113-128.
Whittet, D.C.B., McNally, D., and Dickman, R., 1979a. "The First Year of IUE" (University College London), in press.
Whittet, D.C.B., Somerville, W.B., McNally, D., and Blades, J.C., 1979b. Mon. Not. R. Astr. Soc. 189, pp. 519-525.
Willis, A.J., and Wilson, R., 1975, Astr. Astrophys. 44, pp 205-207.

MEASUREMENTS OF ISOTOPIC ABUNDANCES IN INTERSTELLAR CLOUDS

Arno A. Penzias
Bell Telephone Laboratories, Holmdel, N. J.

While an examination of the available data reveals some seemingly contradictory results, a general framework having the following outlines can be put forward:

1. With the exception of the two galactic center sources SgrA and SgrB, the relative isotopic abundances exhibited by the giant molecular clouds in our Galaxy exhibit few, if any, significant variations from the values obtained by averaging the data from all these sources.

2. The $^{13}C/^{12}C$ and $^{14}N/^{15}N$ abundance ratios are ~130% and ~150%, respectively, of their terrestrial values throughout the galactic plane and somewhat higher, ~300%, near the galactic center.

3. The $^{16}O/^{18}O$ and $^{17}O/^{18}O$ abundance ratios are ~130% and ~160%, respectively, of their terrestrial values throughout the Galaxy, although the former may be somewhat lower near the galactic center.

4. The S and Si isotopes have generally terrestrial abundances.

The data upon which these tentative conclusions are based will be discussed, together with some apparent counter-examples and unresolved questions.

Isotopic abundance data have been obtained for seven of the most abundant elements in interstellar space; hydrogen, helium, carbon, nitrogen, oxygen, sulphur and silicon. This group of elements represents the three fundamental processes of element build-up: cosmological production (hydrogen/deuterium, and helium); the CNO processes, (carbon, nitrogen and oxygen); and explosive nucleosynthesis (sulphur and silicon). Since the light elements will be treated in a separate lecture, they will receive only passing mention here, with the bulk of our attention devoted to the CNO isotopes.

B. H. Andrew (ed.), Interstellar Molecules, 397–404.
Copyright © 1980 by the IAU.

SOURCE	R (kPc)	$\frac{H_2CO}{H^{13}CO}$	$\frac{C^{18}O}{^{13}C^{18}O}$	$\frac{HCO^+}{H^{13}CO^+}$	$\frac{HC_3N}{H^{13}CC_2N}$	$\frac{NH_2CHO}{NH_2^{13}CHO}$	$\frac{OCS}{O^{13}CS}$	$\frac{C^{34}S}{^{13}CS}$ (×23)	$\frac{C^{18}O}{^{13}CO}$ (×500)	$\frac{H_2C^{18}O}{H_2^{13}CO}$ (×500)	$\frac{HC^{18}O^+}{H^{13}CO^+}$ (×500)	$\frac{OH}{^{18}OH}$	$\frac{C^{18}O}{C^{17}O}$	$\frac{^{28}SiO}{^{29}SiO}$	$\frac{^{29}SiO}{^{30}SiO}$	$\frac{C^{34}S}{C^{33}S}$	$\frac{HC^{15}N}{H^{13}CN}$ (×272)
Sgr A		20±10	26±5	21±2				33±2	43±5	60±9		275					2±5
Sgr B	0.1	25±10	23±3	26±4				36±3	32±11	77±4		240					5±5
W43	5.5	42±6			22±1	24±3	14±2	43±4	61±8		>45		3.4±.4	8.4±.5	1.3±.1		
W33	5.7	74±11	42±4					66±3	52±40	79±10			3.2±.2	10.8±.6	1.6±.2	5.3±1	
W51	7.6		32±4					68±5	62±1	96±15			3.6±.2				
W31	8(?)	37±6							54±25				2.8±.1	11.8±1.5			56±5
M17	8.0							59±3	49±7								
NGC 6334	9.3		91±20					48±3	74±3		91±10		3.5±.3				
W49	9.4	53±8	91±28					77±8	39±18		63±7						35±7
DR21	9.9	73±11	143±43					102±12	34±9								
Ori A	10.9		>74	83±24	50±5			66±3	50±4		62±6		3.4±.3				
NGC 2024	11.0	72±11	56±8					69±7	27±4	<75	24±5		3.6±.3	9.0±.5	1.4±.1	4.8±.5	68±5
NGC 2264	11.1		83±31					69±6	60±3				3.5±.2				
W3	12.2	86±13	111±40						37±3			900	3.2±.3				
NGC 7538	12.7		77±21					45±5	42±7				3.5±.3				
Average*		68±18	66±11					70±5	49±13	89	89	500	5.5	20	1.5	5.5	44±8
Solar Sys	10	89	89	89	89	89	89	89	89	89	89	500	5.5	20	1.5	5.5	89
Ref.		(1)	(2)	(3)	(4)	(5)	(6)	(7)	(8)	(9)	(10)	(11)	(12)	(13)	(14)	(15)	(16)

(1) Henkel et al. (1979A) - Data for SgrA and SgrB are from Gardner and Whiteoak (1979) and Wilson et al. (1979) respectively.

(2) Linke (1979).

(3) Stark (1979) - Measurements were made at more than one location in each source. The largest ratio obtained for each source has been tabulated to minimize the effects of possible saturation.

(4) Wannier and Linke (1978).

(5) Lazareff et al. (1978).

(6) Goldsmith and Linke (1979).

(7) Frerking et al. (1979) - The SgrB result is the weighted average of the data at two locations in the source.

(8) Frerking (1979) - The average was obtained giving equal weighting to each source (SgrA and B were excluded) since the dominant errors are probably systematic.

TABLE REFERENCES

(9) Kutner et al. (1979) - These results are an extension of earlier measurements reported by Tucker et al (1979). Similar results have been obtained by Henkel et al (1979B).

(10) Stark (1979) - The SgrB value is from Guelin and Thaddeus (1979).

(11) Whiteoak and Gardner (1975, 1978).

(12) Penzias (1979).

(13) Wolff (1979) - The SgrB values are weighted averages of several positions in the source.

(14) Wolff (1979).

(15) Wilson et al. (1976).

(16) Linke et al. (1977).

Of the seven stable CNO isotopic species (^{12}C, ^{13}C, ^{14}N, ^{15}N, ^{16}O, ^{17}O and ^{18}O*) the isotopes of carbon have traditionally received the most attention. In his review paper at the 1976 IAU Symposium, Peter Wannier (1977) presented data which indicated a relative $C/^{13}C$ abundance ratio of about 50, or ~1/2 the terrestrial ratio, throughout the galactic plane, with a somewhat smaller ratio in the center of the Galaxy**. This conclusion was largely based upon the results of two surveys of giant molecular clouds in our Galaxy. The first of these surveys was a comparison of the common and ^{13}C isotopic species of formaldehyde; the second was a double comparison of the ^{13}C and ^{18}O species of carbon monoxide. More recent work has permitted the interpretation of these results to be refined and has yielded more accurate data leading to somewhat modified abundance values.

In the case of formaldehyde, C. Henkel et al (1979A), have shown that the rotation level population distributions are different in the two isotopic species. This circumstance is due to the fact that the rotation transitions involved in the excitation are themselves optically thick in the more abundant species leading to radiative trapping effects which enhance the collisional excitation. This trapping has the effect of diminishing the intensity of the observed 6 cm K-doubling absorption in the more abundant species and thus serves to diminish its apparent abundance. When appropriate corrections are made for this effect, the resulting $H_2CO/H_2^{13}CO$ abundance ratios are substantially increased but are still somewhat below the terrestrial value. (The results of this work are summarized in Column 1 of the Table.)

In the carbon monoxide study referred to above, comparisons were made between the ^{13}CO and $C^{18}O$ species in order to avoid the use of the heavily saturated spectra of the common CO species. This method suffers, however, from the requirement for a separate determination of the $O/^{18}O$ abundance. An investigation which avoids this requirement has been completed by R. A. Linke (1979). In this measurement the rare $C^{18}O$ isotopic species of carbon monoxide was compared with the yet rarer $^{13}C^{18}O$ species. The results of this work yield a galactic plane $C/^{13}C$ abundance ratio of about 66 (Col. 2) which agrees with the corrected formaldehyde results (Col. 1). It therefore seems reasonable to adopt a value of ~67±10, as more appropriate to the galactic plane than either the terrestrial value of 89 or the value of ~50 referred to above. (In the two galactic center sources, however, the $C/^{13}C$ ratio is considerably lower, a result which is supported by data from other molecules (Col. 1-6).

---

*Following the usual convention the most common isotope of each atomic species will have its atomic weight omitted. Thus $^{13}C^{16}O$ and $^1H^{12}C^{14}N$ will be written ^{13}CO and HCN respectively.

**For the purposes of this discussion, we will use the unmodified term "galactic plane" as excluding the galactic center region.

Returning to the $^{13}CO/C^{18}O$ work, a survey in these species has been recently carried out by M. A. Frerking (1979) with more sensitive equipment, yielding results (Col. 8) in good agreement with the earlier work, but in disagreement with the new $C/^{13}C$ value suggested above if a terrestrial ^{18}O abundance is assumed. This indicates that we must abandon the notion of a terrestrial $O/^{18}O$ abundance in the galaxy. Instead, the new $C/^{13}C$ value can be combined with the results of the $^{13}CO/C^{18}O$ double ratio work (Col. 8) to yield an $O/^{18}O$ abundance in the galactic plane which is about 1.3 times the terrestrial value. While this suggested underabundance of ^{18}O is supported by the $H^{13}CO^+/HC^{18}O^+$ data (Col. 10), the $H_2{}^{13}CO/H_2C^{18}O$ data appear to be more consistent with a terrestrial ^{18}O abundance. It is unlikely that this discrepancy is due to chemical fractionation because of the agreement between the single ratio CO and H_2CO results (Col. 1 and 2). The present state of observational data cannot provide certainty, but an underabundance (relative to terrestrial) of ^{18}O in the galactic plane seems the best fit to the data that we have.

The ^{18}O abundance in the galactic center sources is uncertain, but the data, especially H_2CO (Col. 9) suggest that the $O/^{18}O$ ratio may be appreciably smaller in this region than in the galactic plane. On the other hand, an $^{17}O/^{18}O$ determination (Col. 12) yielded a constant value (~1.6×terrestrial) for this latter ratio over both the galactic center and galactic plane. While one result does not preclude the other, an equal enhancement of ^{17}O and ^{18}O in the galactic center seems unlikely. Earlier OH work, using the 18 cm. λ-doubling transitions, had indicated a substantial ^{18}O enhancement in the galactic center sources (Col. 11). The measurements upon which the OH results were based involved comparisons of optical depths which differed by a factor of several hundred; the conversion of these optical depths into column density ratios assumed equal excitation temperatures in the two species. Since the common species has heavily saturated rotation spectra, its excitation will be different from that of the rarer species; this effect could lead to an underestimate of the relative abundance of the common species by as much as a factor of two (Cernicharo and Guelin 1979). While the weight of evidence seems to be on the side of a lower than galactic $O/^{18}O$ ratio in the galactic center, the present state of our knowledge is far from satisfactory on this point.

Turning now to the other elements, a comparison of the CS data (Col. 7) with the $C/^{13}C$ results shows good agreement indicating a terrestrial abundance of ^{34}S. Other data (Col. 15) suggests that the ^{33}S isotope has a terrestrial abundance as well. In the case of the silicon isotopes on the other hand, while the two rare species seem to have terrestrial abundances relative to each other (Col. 14), their abundances relative to the common species are indicated to be about twice the terrestrial value (Col. 13). The apparent underabundance of the common species does not seem to be an artifact of line saturation, although this possibility cannot be totally ruled out.

Finally, the HCN data in the galactic plane show $^{14}N/^{15}N$ to be enhanced, relative to the terrestrial value by a factor of ~1.5, with a considerably larger enhancement (a factor of ~3) in the galactic center region.

DISCUSSION

With the exception of the galactic center sources no clearly consistent variation of the nuclear abundances between individual sources is evident in the tabulated data. Occasionally the value of one isotope ratio or another seems to differ from those in neighboring sources by a statistically significant amount. However, these variations show little, if any, correlation between one isotopic species and another, or between one set of measurements and another. (Evidence for some source-to-source correlation between the CS and HCN data has been put forward by Frerking et al (1979) however.) It therefore seems premature to interpret any of these anomalies in terms of nuclear differences in the galactic disc. This absence of clear differences suggests that we can obtain representative abundance values by averaging among sources.

Some nuclear differences do exist, of course. The most notable ones are associated with the earth itself. As indicated above, the presolar nebula apparently had only about two-thirds as much ^{13}C, ^{14}N and ^{16}O relative to their respective counterparts ^{12}C, ^{15}N and ^{18}O as does the galactic plane at present. In addition, regions associated with evolving stars such as the envelope around IRC 10216 have been shown to possess nuclear abundances which differ from those of general interstellar space (Wannier and Linke 1977). For the great bulk of material in the Galaxy, however, we seem to see a uniformity which is characteristic of an efficient mixing, or of a common nuclear history, or a combination of the two. The galactic center region appears to be the only exception, a not unsurprising circumstance in view of the markedly greater amount of stellar processing that has taken place there. The much smaller effects due to the decrease in processing with galactic radius in the rest of the Galaxy are largely hidden by the uncertainties in the available data.

The above treatment appears able to obtain agreement between data from different molecules without invoking chemical fractionation effects. It should be emphasized, in this regard, that the tabulated data have been obtained entirely from giant molecular clouds. In the cooler dark clouds, molecular isotopic abundances can be affected by fractionation. In a recent paper, Langer et al (1980) reported studies of three such clouds in which the isotope ratios in the diffuse ($10^3/cm^3$) exteriors showed considerable carbon fractionation. The opaque core regions of these clouds, however, exhibited abundances in agreement with the giant cloud data, indicating little if any fractionation therein.

The increasing clarity of the emerging picture of isotopic abundances in the Galaxy is in considerable part due to the success of

a two-pronged attack on the line formation problem. Improvements in
sensitivity permit observations of very low optical depth transitions
while, at the same time, better analytical treatment has been employed
to deal with the saturation problems characteristic of high optical
depth. While it seems reasonable that some surprises lie hidden by the
uncertainties in our data, a comprehensive observational framework upon
which to base our theoretical understanding appears to be in hand.

REFERENCES

Cernicharo, J., and Guelin, M.: 1979, in preparation.
Frerking, M.A.: 1979, in preparation.
Frerking, M.A., Wilson, R.W., Linke, R.A., Wannier, P.G.: 1979,
       submitted to Ap. J.
Gardner, F.F., and Whiteoak, J.B.: 1979, MNRAS 188, p. 331.
Goldsmith, P.F., and Linke, R.A.: 1979, to be published.
Guelin, M., and Thaddeus, P.: 1979, Ap. J. (Letters) 227, p. L139.
Henkel, C., Walmsley, C.M., and Wilson, T.L.: 1979A, Astron. and
       Astrophys., in press.
Henkel, C., Wilson, T.L., and Downes, D.: 1979B, Astron. and
       Astrophys. 73, p. L13.
Kutner, M.L., Machnik, D.E., Tucker, K.D., and Massano, W.: 1979,
       to be published.
Langer, W.D., Goldsmith, P.F., Carlson, E.R., and Wilson, R.W.: 1980,
       Ap. J., in press.
Lazareff, B., Lucas, R., and Encrenaz, P.: 1978, Astron. and Astrophys.
       70, p. L77.
Linke, R.A., Goldsmith, P.F., Wannier, P.G., Wilson, R.W., and
       Penzias, A.A.: 1977, Ap. J. 214, 50.
Linke, R.A.: 1979, in preparation.
Penzias, A.A.: 1979, in preparation.
Stark, A.A.: 1979, to be published.
Tucker, K.D., Kutner, M.L., and Massano, W.: 1979, Ap. J. (Letters)
       227, L143.
Wannier, P.G.: 1977, CNO Isotopes in Astrophysics, J. Audouze, Ed.
       D. Reidel, p. 71.
Wannier, P.G., and Linke, R.A.: 1977, Ap. J. 214, p. 50.
Wannier, P.G., and Linke, R.A.: 1978, Ap. J. 226, p. 817.
Whiteoak, J.B., and Gardner, F.F.: 1975, Proc. Astr. Soc. Aust. 2, 360.
Whiteoak, J.B., and Gardner, F.F.: 1978, MNRAS 183, p. 67p.
Wilson, R.W., Penzias, A.A., Wannier, P.G., and Linke, R.A.: 1976,
       Ap. J. (Letters) 204, p. L135.
Wilson, T.L., Walmsley, C.M., Henkel, C., Pauls, T., Mattes, H.: 1979,
       Astron. and Astrophys., in press.
Wolff, R.S.: 1979, Ap. J., in press.

DISCUSSION FOLLOWING PENZIAS

Kutner: As listed in column (9) of your table our latest $H_2^{13}CO$ and $H_2C^{18}O$ observations give double ratios of 8.3±1.2, 6.5±0.3, 6.3±0.8, 5.2±0.8 for Sgr A, Sgr B2, W33, W51 (vs the terrestrial value of 5.6), corresponding to carbon ratios of 60±9, 77±4, 79±10, 96±15. We noted in our Jan. 79 Ap. J. Letter that the formaldehyde results are quite consistent with a carbon ratio in the 70's. It still appears that there are some significant discrepancies between the double ratios determined from CO and H_2CO.

Penzias: I quite agree. It may be that the agreement between the single ratio H_2CO and CO data is fortuitous, and that the actual $C^{18}O/^{13}C^{18}O$ ratio is closer to the value one obtains from the double ratio, $C^{18}O/^{13}CO$, and a terrestrial oxygen abundance. In that case, the higher ^{13}C abundance in CO would be due to fractionation. However, I regard this explanation as unlikely, because the differences between the two sets of CO data referred to above seem well established, and the role of chemical fractionation in CO has now been observed to be limited to rather diffuse regions.

Townes: This excellent analysis and the newly obtained results have surely given us a better value of the average $^{12}C/^{13}C$ ratio, a value which is much easier to understand in terms of galactic history than were earlier smaller ratios. In addition to the broad results which Penzias has emphasized, there appears to be very significant information, for example, concerning the variation of isotopic ratios from source to source. The very valuable new measurements of $^{18}O/^{17}O$ ratios show rather striking uniformity. However the values found differ substantially from the terrestrial one, which raises the question whether the earth is really a representative sample of isotopic ratios in our galaxy at its formation about 5×10^9 years ago. In addition, one of the more reliable ratios would seem to be that of $^{13}C^{16}O/^{12}C^{18}O$. However, this varies from source to source by far more than the probable errors listed. This suggests, as has been noted before, that the large molecular clouds have developed over a long period of time in a partially isolated state.

Penzias: Your point is certainly well taken. In attempting to infer broad galactic isotope values from averages, I have pretty much neglected source-to-source variations. For a study of the individual clouds themselves, isotopic abundance differences play a far more central role. Whether the tabulated differences are due to line formation or actual abundance variations seems to be an unresolved issue. For example, NGC 2024 looks low in col. (8) but normal in the others. Similarly W51 looks out of place in col. (2) but not in (7) or (8). Checking this out with appropriate additional observations is clearly the next order of business.

Kutner: I think that within each source the H_2CO isotope data is quite self-consistent. The radiative trapping corrections in Sgr B2 are probably so uncertain that a value of 25 for the $H_2CO/H_2^{13}CO$ must be regarded as tentative.

Penzias: Wilson et al. (1979) have made optical depth measurements in the $1_{10}-1_{11}$, $2_{11}-2_{12}$ and $3_{12}-3_{13}$ transitions of H_2CO in this source. If, as you suggest, they were to have underestimated the optical depth of the lowest lying transition relative to the other two, their model would have to have yielded too high a density. Time does not permit a detailed discussion of their results, but the density they derive is on the low side already, and is unlikely to be a gross overestimate.

Vanden Bout: The $^{12}C/^{13}C$ ratio from optical observations of CH^+ toward ζ Oph, 20 Tau, and ξ Per yield values ranging from 50 to 75, with a mean value close to those in your table for material outside the galactic center.

Penzias: While the agreement between your data and the ratio suggested in my talk is gratifying, I have avoided considering diffuse regions in deriving broad isotope values because of possible fractionation effects.

Vanden Bout: CH^+ is unlikely to be affected by fractionation according to Watson, Anicich, and Huntress (1976, Ap. J. 205, L165).

ISOTOPIC ABUNDANCE RATIOS FROM MICROWAVE OBSERVATIONS OF FORMALDEHYDE

T.L. Wilson, C. Henkel, C.M. Walmsley, T. Pauls
Max-Planck-Institut für Radioastronomie
Auf dem Hügel 69, 5300 Bonn 1, Federal Republic of Germany

This report is a summary of the determination of the ratio of the column density of $H_2^{12}C^{16}O$ (hereafter H_2CO) or $H_2^{12}C^{18}O$ (hereafter $H_2C^{18}O$) to that of $H_2^{13}C^{16}O$ (hereafter $H_2^{13}CO$). With one exception, all of the published ratios have been determined from measurements of the $1_{10}-1_{11}$ lines. The exception is the Orion Molecular Cloud (see discussion of Kutner et al. 1976). The most complete surveys of $H_2^{13}CO$ are those of Wilson et al. (1976) and Gardner and Whiteoak (1979). The ratios obtained from $1_{10}-1_{11}$ data are lower limits because of the effect of photon trapping in H_2CO (see the discussion by Henkel et al. 1979b). From measurements of the $1_{10}-1_{11}$ and $2_{11}-2_{12}$ lines and model calculations, Henkel et al. estimate that the corrections for 10 clouds outside the Galactic center region is 1.6±0.7. This correction gives an average $(H_2CO/H_2^{13}CO)$ ratio of 64±17. The models of Henkel et al. also predict the optical depth of the $3_{12}-3_{13}$ line and the brightness temperature of the $2_{12}-1_{11}$ line. Preliminary measurements of Wilson et al. (1979) show that the $3_{12}-3_{13}$ results are consistent with the predictions; but more sensitive measurements are required to test the model fully. Unpublished $2_{12}-1_{11}$ emission-line spectra are weaker than predicted by the model, but this might be caused by beam dilution, since the observations were made so as to cover a 1.5' uniformly weighted aperture. The lineshapes of the $1_{10}-1_{11}$ and $2_{11}-2_{12}$ absorption lines agree poorly with the $2_{12}-1_{11}$ emission lines, indicating either that self-absorption in the millimeter lines of H_2CO is important, or that there are H_2CO clouds behind the continuum sources. Because of their position relative to the continuum sources these clouds would contribute to the millimeter emission lines but not to the centimeter absorption lines.

In a few sources such as the molecular clouds toward the high brightness HII regions W33, W51, Sgr B2 and Sgr A it is possible to determine the ratio of $H_2C^{18}O$ to $H_2^{13}CO$. Because both of these rare isotopes are optically thin, no trapping corrections are required. The average ratios obtained by Tucker et al. (1979) and Henkel et al. (1979a) are about twice the ratio of the column density of $^{12}C^{18}O$ to $^{13}C^{16}O$ measured by Wannier et al. (1976) for 14 molecular clouds. The difference in these double ratios is an indication that ^{13}CO is enhanced by a fractionation

405

B. H. Andrew (ed.), Interstellar Molecules, 405–408.

process (see discussion by Watson and R.W. Wilson et al., in this volume).
The Sgr B_2 molecular cloud was not analyzed by Henkel et al. (1979b)
because of the large optical depth in the $1_{10}-1_{11}$ line. However, measurements of the $3_{12}-3_{13}$ line, and model calculations indicate that the correction for photon trapping is about a factor of 2. From this factor and
measurements of the $1_{10}-1_{11}$ line, the ($H_2CO/H_2^{13}CO$) ratio is about 25.
Combining this result with the ($H_2^{12}C^{18}O/H_2^{13}C^{16}O$) average ratio from
Tucker et al. (1979) and Henkel et al. (1979a), we find ($^{16}O/^{18}O$) = 160,
which is significantly below the terrestrial value of 489 (Heath 1976).

A partial map of the Orion Molecular Cloud in the $3_{12}-3_{13}$ line with
a $\sim 30''$ angular resolution shows that structure on the order of 20'' exists
in the source. This measurement is consistent with the model of Evans
et al. (1979) which predicts an H_2 density of $\sim 10^6$ cm^{-3}. From the analysi
of the Sgr B_2 cloud, the density is $\sim 10^{4.2}$ cm^{-3}; and for the 10 clouds
studied by Henkel et al. (1979b) the density is $\sim 10^5$ cm^{-3}. These densities
are 10-100 times those obtained from measurements of CO by Plambeck and
Williams (1979) but agree roughly with the CS results of Linke and
Goldsmith (1979). It is possible that different molecules are located
mainly in different regions of the clouds, where different H_2 densities
are present.

REFERENCES

Evans, N.J., Plambeck, R.L., Davis, J.H.: 1979, Astrophys. J. Lett. 227,
    L25
Gardner, F.F., Whiteoak, J.B.: 1979, Monthly Notices Roy. Astron. Soc.
    188, 331
Heath, R.L.: 1976 in "Handbook of Chemistry and Physics", 57th Edition,
    ed. R.C. Weast, CRC Press, Cleveland, p. B-270
Henkel, C., Wilson, T.L., Downes, D.: 1979a, Astron. Astrophys. 73, L13
Henkel, C., Walmsley, C.M., Wilson, T.L.: 1979b, Astron. Astrophys., in
    press
Kutner, M.L., Evans, N.J., Tucker, K.D.: 1976, Astrophys. J. 209, 452
Linke, R.A., Goldsmith, P.: 1979, this volume
Plambeck, R.L., Williams, D.R.W.: 1979, Astrophys. J. 227, L43
Tucker, K.D., Kutner, M.L., Massano, W.: 1979, Astrophys. J. Lett. 227,
    L143
Wannier, P.G., Penzias, A.A., Linke, R.A., Wilson, R.W.: 1976, Astrophys.
    J. 204, 26
Wilson, T.L., Bieging, J., Downes, D., Gardner, F.F.: 1976, Astron.
    Astrophys. 51, 303
Wilson, T.L., Walmsley, C.M., Henkel, C., Pauls, T., Mattes, H.: 1979,
    Astron. Astrophys., in press

DISCUSSION FOLLOWING WILSON

Field: The density you derive for H_2 depends on the rotational-excitation rate-coefficients. The cross-sections given by Sheldon Green are for He-HCHO collisions and may be rather different for H_2-HCHO collisions. Even if H_2 is in the J=0 state, the anisotropy of the potential may be significantly different from that of He. A further point is that very recent calculations by Gerratt and Wilson in Bristol (U.K.) show that cross-sections for rotational excitation are sharply resonant with kinetic energy of He in collisions with HCHO. With a knowledge of the temperatures at which these resonances occur we may be able in the future to characterize cloud temperatures by anomalously high populations of excited rotational states.

Wilson: There are still problems in that we are always making line-of-sight averages in astronomy and in that the density of H_2 is a free parameter. Our density estimates depend on the collisional cross-sections, so estimates of the effect of substituting H_2 for He as a collision partner would be valuable.

Guélin: There is evidence that a cold, low density (10^2-10^3cm^{-3}) foreground cloud lies in the line of sight to the Sgr B2 source at V $\sim$ 62 km/s (e.g. the HCO$^+$ data of Guélin and Thaddeus, 1979). How will the presence of this low density cloud affect your results?

Wilson: The line shapes of the 1_{10}-1_{11} lines of $H_2^{12}C^{18}O$ and $H_2^{12}C^{16}O$ agree, yet $H_2^{12}C^{18}O$ is optically thin, while $H_2^{12}C^{16}O$ has a measured optical depth of 1.3-1.8. Hence we do not believe that self-absorption affects our 1_{10}-1_{11} line. It may be that the H_2CO is in some way related to the low density foreground cloud that you observe. However, the absorption is associated with the region in front of the compact continuum sources, a fact which may explain our higher density (n(H_2) $\sim 10^{4.2}$cm^{-3}) and kinetic temperature ($T_K \sim$ 60K). Similar parameters for n(H_2) and T_K were obtained by Winnewisser, Walmsley and Churchwell (1978) from NH_3 absorption toward Sgr B2.

Greenberg: I know that density alone is not the best way to distinguish the clouds you considered, but did you find any significant difference in the carbon isotopic ratio as a function of density? It appeared to me that Sgr B2, which has the lowest density, also had the lowest $^{13}C/^{12}C$ ratio.

Wilson: Unfortunately, the Sgr B2 cloud is the only one in our sample in which the density and the ($^{12}C/^{13}C$) ratio are low. The ($^{12}C/^{13}C$) ratios of the other 10 sources are scattered, whereas n(H_2) is, in all cases, close to 10^5cm^{-3}.

de Jong: Should your results be interpreted to mean that H_2CO can only form in high density regions or do they mean only that high densities are required to excite the H_2CO molecules sufficiently?

Wilson: Because we make use of measurements of the 1_{10}-1_{11} line, which is the lowest K-doublet in the K_a=1 ladder, the excitation should not be dependent on high densities. The H_2CO is located mainly in regions where n(H_2)$\approx 10^5$cm^{-3}, though you should note that we assume that clouds are spheres or slabs of uniform density and uniform temperature.

Evans: I have to disagree with your conclusion that H_2CO exists only in dense regions. We find that the H_2CO abundance actually decreases in dense regions.

Wilson: "Dense" is a vague term. Our results show that the H_2CO occurs where $n(H_2) \sim 10^5 cm^{-3}$. Plambeck and Williams (1979) derive $n(H_2) \sim 10^2 - 10^3 cm^{-3}$ from a similar study of CO. The H_2CO is located in a denser region than that. We know that regions where $n(H_2) \sim 10^6 cm^{-3}$ (e.g. Orion) exist, but we obtain (except for Orion) no such high density. So in this sense we agree. Finally our results apply to molecular clouds near HII regions, and you are comparing them to results obtained for dark dust clouds.

Evans: Your result in OMC 1 agrees with our model, which requires the presence of small dense clumps, and results in a $H_2^{12}CO/H_2^{13}CO$ ratio higher than 100.

Wilson: Our map of Orion in the $3_{12}-3_{13}$ line of H_2CO should give us a direct check of your angular size predictions.

ISOTOPE RATIOS IN INTERSTELLAR FORMALDEHYDE

Marc L. Kutner and Dennis E. Machnik
Physics Dept., Rensselaer Polytechnic Institute

Kenneth D. Tucker and William Massano
Physics Dept., Fordham University

Formaldehyde is an excellent molecule for studying isotopic abundances. It is easily observed in several connected transitions, allowing an analysis of excitation differences among species. With formaldehyde, one can also measure the same ratio using millimeter emission lines and centimeter absorption lines, which can often be interpreted more directly. A comparison of the results from the millimeter and centimeter lines can give an idea of the uncertainties in the analysis of millimeter emission lines.

There have been direct comparisons of 6-cm absorption of continuum sources by H_2CO and by $H_2^{13}CO$, but such measurements are subject to error because of photon trapping in the H_2CO, resulting in different excitations for the two species. This problem does not exist when comparing $H_2^{13}CO$ with $H_2C^{18}O$, for which the millimeter lines are optically thin; a direct comparison of 6-cm absorptions should give a good abundance ratio. In the first such measurement, Gardner et al. (1971) found a value of about 10 for this double ratio, compared to the terrestrial value of 5.6. We have reported (Tucker et al. 1979) more extensive $H_2C^{18}O$ observations, which, when compared with published $H_2^{13}CO$ spectra, suggested a ratio of about 6.9 for Sgr A, Sgr B2 and W33, with no significant source-to-source variation. Our results, using the NRAO[1] 43-m telescope, are supported by the observations of Henkel et al. (1979b) on the 100-m telescope.

To extend our observations, we have used the 43-m telescope to observe the ^{13}C species in all sources for which we have ^{18}O data. This provides us with data taken on the same system for the comparison. (We have also reobserved the ^{18}O species in Sgr B2 as a check.) The combined results of our observations are summarized in the table below, which gives the $H_2^{13}CO/H_2C^{18}O$ abundance ratio computed in two ways: (1) just using peak line optical depths, correcting for the effect of the large hyperfine splitting in the ^{13}C species (in the narrow line sources) and (2) using the integrated line optical depths. (Uncertainties are 1 σ and pointing is in the direction of the continuum peak in each source.)

B. H. Andrew (ed.), Interstellar Molecules, 409–410.

| Source | $H_2{}^{13}CO/H_2C^{18}O$ Abundance Ratio | | $^{12}C/^{13}C$ |
	Peak	Integrated	
Sgr A	8.1 ± 1.5	8.3 ± 1.2	60 ± 9
Sgr B2	6.5 ± 0.6	6.5 ± 0.3	77 ± 4
W 33	7.5 ± 1.5	6.3 ± 0.8	79 ± 10
W 51	5.6 ± 1.8	5.2 ± 0.8	96 ± 15
NGC 2024	> 6.6	-	< 75

We believe that the integrated profiles are more meaningful despite the fact that they are somewhat more susceptible to baseline uncertainties. Besides the obvious statistical improvement from utilizing the whole line, using integrated intensities eliminates large corrections for the hyperfine structure in the ^{13}C species. Such corrections can be uncertain in sources with multiple peaks. The last column of the table gives the carbon isotope ratio, assuming a terrestrial oxygen ratio and no fractionation. These values are only slightly lower than the terrestrial value of 89, and are consistent with results obtained from 6-cm studies of H_2CO and $H_2{}^{13}CO$, when corrected for trapping, as described by Henkel et al. (1979a), and with the 2-mm observations of OMC-1 by Kutner et al. (1976).

Independent of assumptions of the oxygen ratio, our results yield a double ratio significantly closer to the terrestrial value than that obtained from CO (e.g. Wannier et al. 1976), but in agreement with the results for HCO^+ (Guélin and Thaddeus 1979; Langer et al. 1978). More analysis is required to see the extent to which molecule-to-molecule differences can be explained by fractionation. Until this is done, interpretation of the CO results as being representative of the interstellar medium may be premature.

REFERENCES

Gardner, F.F., Ribes, J.C., and Cooper, B.F.C.: 1971, Ap. Letters, 9, p. 81.
Guélin, M., and Thaddeus, P.: 1979, Ap. J. (Letters), 227, p. L139.
Henkel, C., Walmsley, C.M., and Wilson, T.L.: 1979a, (preprint).
Henkel, C., Wilson, T.L., and Downes, D.: 1979b, Astron. & Ap. 73, p. L13.
Kutner, M.L., Evans, N.J., II, and Tucker, K.D.: 1976, Ap. J. 209, p. 452.
Langer, W.D., Wilson, R.W., Henry, P.S., and Guélin, M.: 1978, Ap. J. (Letters) 225, p. L139.
Tucker, K.D., Kutner, M.L., and Massano, W.: 1979, Ap. J. (Letters) 227, p. L143.
Wannier, P.G., Penzias, A.A., Linke, R.A., and Wilson, R.W.: 1976, Ap. J. 204, p. 26.

[1]The National Radio Astronomy Observatory is operated by Associated Universities, Inc. under contract with the National Science Foundation.

THE $^{12}C/^{13}C$ RATIO IN INTERSTELLAR DARK CLOUDS

W.H. McCutcheon
University of British Columbia

R.L. Dickman
Aerospace Corporation

W.L.H. Shuter
University of British Columbia

R.S. Roger
Dominion Radio Astrophysical Observatory

Since ^{13}C is believed to be produced by non-equilibrium CNO processing in stellar evolution (Truran 1977), measurements of the carbon ratio $R_C \equiv [^{12}C]/[^{13}C]$ in the interstellar medium may provide important information on nucleo-synthesis. Commonly, the ratio $(N_{13}/N_{18})_{LTE} \equiv [^{13}CO/C^{18}O]_{LTE}$ is measured and from this $R_{LTE} \equiv [^{12}CO/^{13}CO]_{LTE}$ is deduced and these values are often identified with R_C. However, this line of reasoning can be misleading for two reasons (Dickman et al. 1979):
(1) The difficulty of determining accurate column densities, $[^{13}C^{16}O]$ and $[^{12}C^{18}O]$, because of the complexity of the radiative transfer problem;
(2) The possible role of fractionation, whereby $R_{CO} \equiv [^{12}CO]/[^{13}CO]$ does not necessarily reflect the initial atomic abundance ratio R_C (Watson et al. 1976, Langer 1977, Liszt 1978).

Mahoney et al. (1976) and Dickman et al. (1977) presented measurements of $(N_{13}/N_{18})_{LTE}$ in the dark cloud L134 which suggested that fractionation was occurring. An analysis by Dickman et al. (1979) showed that $(N_{13}/N_{18})_{LTE}$ in L134 did in fact represent the local abundance ratio $[^{13}CO]/[C^{18}O]$ in which case R_{CO}, obtained by using a terrestrial oxygen isotope ratio, was found to vary systematically with visual extinction A_v. This result was attributed to isotopic fractionation of ^{13}CO. The values of R_{CO} found in regions where $A_v \gtrsim 4$ mag. are close to the terrestrial value and this was interpreted as evidence for a terrestrial value for R_C. These results are in contrast to those of Wannier et al. (1976) who measured $(N_{13}/N_{18})_{LTE}$ in 14 HII regions and deduced R_C to be about 40. Other recent measurements supporting a terrestrial value of R_C have been made by Langer et al. (1978), Guélin and Thaddeus (1979) and Tucker et al. (1979).

B. H. Andrew (ed.), Interstellar Molecules, 411–416.
Copyright © 1980 by the IAU.

Here we present further observations of CO, ^{13}CO and C^{18}O in four additional dark clouds. Data were obtained using the 4.6 m telescope at the Aerospace Corporation with a 2.6 arcmin beam width and a 0.65 km s^{-1} velocity resolution at 115.3 GHz. Table 1 lists the positions observed.

TABLE 1

L183 α_o (1950) = 15^{h}51^{m}18^s, δ_o (1950) = -1° 00'00".

Position ($\Delta\alpha,\Delta\delta$)	T_{12}^* (K) (a)	T_{13}^* (K)	T_{18}^* (K)	$\frac{^{13}C^{18}O}{^{12}C^{18}O}$ (b)	Av(mag)
0.0, 8.0	9.0	3.8	$\leq$0.9	$\geq$6.1	2.4
0.0, 16.0	8.3	4.1	1.4	5.3(+2.2,-1.4)	$\geq$5.7
0.0, 20.0	20.0	3.2	$\leq$0.6	$\geq$4.4	$\geq$5.7
15.0, 8.0	8.6	4.1	0.6	5.5(+4.4,-2.1)	$\geq$8.2

L1524 α_o (1950) = 04^{h}25^{m}00^s, δ_o (1950) = 24°30'00"

0.0, 8.0	6.6	3.8	0.7	5.9(+2.7,-1.6)	1.8
-20.0, 0.0	6.9	4.5	0.9	9.8(+3.1,-2.4)	2.7
20.0, 0.0	7.5	4.3	0.8	6.6(+4.2,-2.2)	3.8
0.0, 4.0	7.6	4.8	1.1	5.9(+2.4,-1.6)	3.8
-20.0, 4.0	(7.6)	5.1	1.1	8.0(+2.6,-1.8)	4.0
20.0, 4.0	(7.6)	4.3	0.8	5.3(+2.8,-1.6)	2.9

L204 α_o (1950) = 16^{h}45^{m}00^s, δ_o (1950) = -12°00'00"

0.0, 0.0	9.0	5.1	1.1	6.9(+1.4,-1.1)	$\geq$9.2
0.0, 5.0	9.6	5.7	1.2	6.6(+2.3,-1.1)	$\geq$9.2
0.0, 10.0	6.2	5.1	1.2	5.5(+3.4,-1.7)	$\geq$9.2
-20.0, 5.0	12.5	5.2	0.6	14.8(+8.6,-4.9)	1.5
30.0, 60.0	11.0	5.6	0.9	12.5(+3.3,-2.5)	$\sim$ 3 (c)
30.0, 75.0	10.4	5.1	0.4	10.8(+6.6,-3.7)	0.6

IC5146 α_o (1950) = 21^{h}51^{m}37^s, δ_o (1950) = 47°02'07"

-61.'2,14.'5	(5.7)	2.3	0.5	6.6(+2.9,-1.9)	$\geq$7.0 (d)

NOTES TO TABLE 1

(a) Values in brackets are assumed values.

(b) Values in brackets are the errors in the double ratio obtained from the 1σ errors in each column density.

(c) Very sharp gradient in A$_v$ at this position; value uncertain.

(d) Stars at this location assumed to be foreground stars and hence were ignored in computing extinction

$(N_{13}/N_{18})_{LTE}$ is tabulated in Table 1 for 17 of the 22 positions observed (five other positions observed in IC5146 will be discussed later). R_{LTE} was obtained by assuming a terrestrial oxygen isotope ratio, an assumption commonly made in the absence of conflicting evidence, particularly for dark nebulae, and values range from approximately terrestrial, 89, down to $\sim$33. Based on the results of Dickman et al. (1979), we assume that the ratios R_{LTE} adequately reflect the values R_{CO} in regions of moderate density and low kinetic temperature and evaluate these results in the framework of fractionation chemistry.

In regions of a low temperature cloud where the extinction is low (and the interstellar UV flux high) [^{13}CO] may be enhanced by large factors (Langer 1977, Liszt 1978). In regions of high extinction $R_{CO} \rightarrow R_C$ after sufficiently long times. In order to evaluate our results extinction was determined at each position from star counts in a reseau size of 2.6 arcmin, taking into account the probability of foreground stars. Fig. 1 shows R_{LTE} plotted against A_v for the four clouds in Table 1.

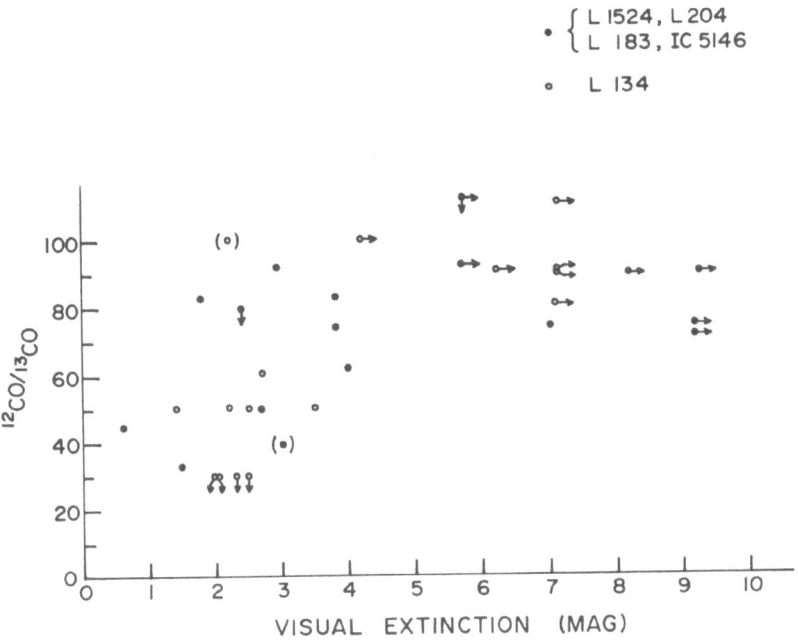

Fig. 1 Ratio of inferred CO column densities vs. visual extinction in magnitudes for the 4 clouds in Table 1 and for L134. For the 4 clouds in this study, the ordinate is R_{LTE}. For L134, the ordinate is R_{CO} and the values are from Fig. 3 of Dickman, McCutcheon and Shuter (1979). R_{LTE} has been derived from the double ratio $[^{13}CO/C^{18}O]$ at each position by assuming $[^{16}O]/[^{18}O]$ is terrestrial. Vertical arrows indicate upper limits to R_{LTE} and R_{CO} because of undetected $C^{18}O$ emission. Uncertainties in R_{LTE} are obtained from the maximum and minimum values of the double ratio in Table 1. Uncertainties in R_{CO} are similar. Horizontal arrows denote lower limits to extinction. The brackets indicate that the extinction for that position is uncertain because a sharp extinction gradient exists over a region comparable to the beam size.

R_{LTE} is generally in agreement with the terrestrial value 89 at positions where $A_v \gtrsim 4$ mag., but at lower extinctions significantly lower values occur. These results are qualitatively consistent with the expected behaviour of carbon monoxide isotope ratios in regions of varying A_v where fractionation is expected to be important. This conclusion is based on the assumption that the ratios R_{LTE} reasonably indicate the values R_{CO} in the dark clouds studied. Considerable confidence is given to this assumption by the close agreement of our R_{LTE} vs. A_V plot with the variation of <u>non-LTE</u> values of R_{CO} vs. A_V for L134 also shown in Fig. 1.

LTE isotope ratios were obtained at six positions in the vicinity of IC5146, a young stellar cluster embedded in the eastern edge of an elongated complex of dark clouds. Fig. 2 shows the values of R_{LTE} placed on a photograph reproduced from the Palomar Sky Survey Print.

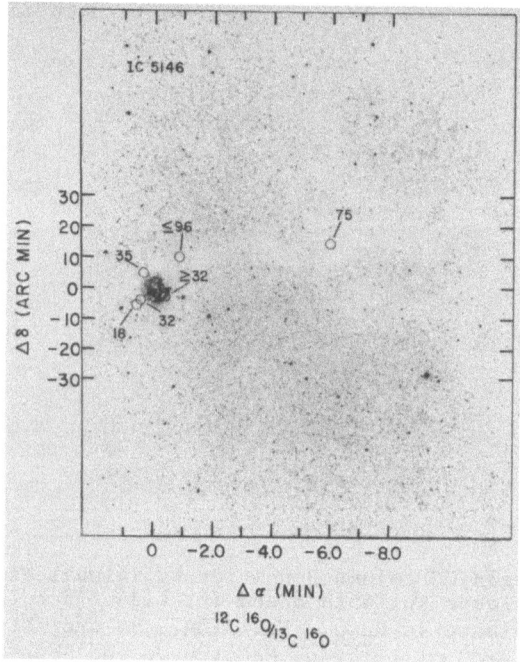

Fig. 2 Positions observed in IC5146 are indicated on a reproduction from the Palomar Sky Survey Prints. The circles represent the beam size and are centered on the observed positions. The numbers are the values of R_{LTE}.

Because of the increased flux of ionizing radiation close to the nebula, these results would appear to be further evidence for fractionation. However, there are two reasons why this effect cannot be interpreted unambiguously. First, the use of LTE may not apply to the warmer and probably denser areas near the ionized nebula and second, at the three positions where R_{LTE} is lowest the kinetic temperature is well in excess of 20 K, and fractionation should be inhibited (Langer 1977, Liszt 1978). However, the trends embodied in Fig. 1 of this paper   and

Fig. 3 of Dickman et al. (1979) do not agree in detail with theory and it may be that theoretical models must be modified.

In summary, the average value for R_{LTE} from twelve data points in Fig. 1 where $A_v > 4$ is 88 ± 3, with the quoted error being the standard error of the mean. We conclude that R_{CO} is statistically consistent with the terrestrial value 89 in the clouds studied here where $A_v > 4$ mag. These results imply a terrestrial value for R_C in dark clouds.

REFERENCES

Dickman, R.L., McCutcheon, W.H., and Shuter, W.L.H. 1979, Ap. J. 234, 100.
Dickman, R.L., Langer, W.D., McCutcheon, W.H., and Shuter, W.L.H. 1977, in "CNO Isotopes in Astrophysics" ed. J. Audouze (Reidel:Dordrecht), p. 95.
Guélin, H., and Thaddeus, P. 1979, Ap. J. (Letters) 227, L139.
Langer, W.D. 1977, Ap. J. (Letters) 212, L39.
Langer, W.D., Wilson, R.W., Henry, P.S., and Guélin, M. 1979, Ap. J. (Letters) 225, L139.
Liszt, H.S. 1978, Ap. J. 222, 484.
Mahoney, M.J., McCutcheon, W.H., and Shuter, W.L.H. 1976, A. J. 81, 508.
Truran, J.W. 1977, in "CNO Isotopes in Astrophysics" ed. J. Audouze (Reidel:Dordrecht), p. 95.
Tucker, K.D., Kutner, M.L., and Massano, W. 1979, Ap. J. (Letters) 227, L143.
Wannier, P.G., Penzias, A.A., Linke, R.A., and Wilson, R.W. 1976, Ap. J. 204, 26.
Watson, W.D., Anicich, V.G., and Huntress, W.T. 1976, Ap. J. (Letters) 204, L165.

DISCUSSION FOLLOWING McCUTCHEON

Penzias: Since a quite similar result could be obtained from a partial saturation of the central ^{13}CO lines, one needs to take possible saturation into account. The excitation of CO in the diffuse exterior should be greater than the excitation in the core, where the ^{13}CO line originates. This excitation difference will lead to an underestimate of the saturation of the ^{13}CO line, a possibility which could be checked by observing the J=2-J=1 transitions.

McCutcheon: The non-LTE analysis using an LVG model has taken account of possible ^{13}CO saturation, and the answer still remains the same. A uniform kinetic temperature (12K) was used in the calculations, and a reasonable variation of this parameter had no sizeable effect upon the results. However, this parameter was not varied across the cloud.

Langer: When Dickman, Langer, McCutcheon, and Shuter (1977) suggested that fractionation explained the CO data in L134, the data were criticized as being under-resolved or filter-diluted because of the velocity resolution of 0.65 km s^{-1}. Could you comment on this?

McCutcheon: The LTE results were obtained by integrating the line optical depths and are not affected by resolution. The non-LTE analysis, however, uses peak temperatures and the results can be affected by filter dilution. In fact, depending upon the intrinsic isotopic line-shapes, dilution by 0.65 km s^{-1} filters can skew the resultant ratios either up or down in a non-linear fashion which depends upon where in the cloud emergent model spectra are computed. In general, though, the trend indicating fractionation was preserved, although the absolute values of the ratios could change by up to 30%.

It is encouraging that Myers, Buxton, and Ho (this conference) have CO results from three dark clouds using a velocity resolution of 0.08 km s^{-1} that agree closely with our results.

Glassgold: In Langer's explicit calculation of Watson's fractionation theory, he finds that the fractionation depends on depth into the cloud as well as on temperature. The fractionation first increases and then decreases on a scale of about $A_v=1$. Do you have sufficient angular resolution to observe this variation?

McCutcheon: In some cases, perhaps. However, variation in extinction is often not uniform. In the centres of many dark clouds, there are large opaque areas where lower limits only may be placed on the visual extinction. These regions may be several of our (2.6 arcmin) beamwidths in diameter. Towards the edges of the clouds, there are sometimes very steep gradients in extinction, over dimensions much smaller than our beamwidth. In other instances, the variation in extinction is more gradual and unit changes of A_v occur over angular extents which we can resolve. Our plot of ratio vs A_v (Fig. 1) does not agree in detail with Langer's theory.

Wannier: Your quoted $^{12}C/^{13}C$ ratio, which you call "consistent with the terrestrial value" is also consistent with the value given earlier by Penzias to within 30% or so. What is the correction that you calculate using the non-LTE calculations? It surprises me that such a correction for the ^{13}CO line intensity at the cold cloud centers could produce ratios accurate to within 30%.

McCutcheon: The non-LTE calculation shows ^{13}CO optical depths to be typically 1.5, which is somewhat higher than found from an LTE calculation. An individual ratio is not claimed to be accurate to 30%. The value quoted, 88±3, is an average of 12 values.

Scoville: One important implication of ^{13}CO fractionation in low extinction dark clouds is that many estimates of giant cloud masses (based upon Dickman's measurements of $^{13}CO/A_v$) must be revised upward. This is because the inferred $^{13}CO/H_2$ ratio obtained for the low A_v regions is anomalously high compared to that for the bulk of giant molecular clouds which have $A_v \gtrsim 10$ mag. If the fractionation in Dickman's original data is a factor 2, then the mass estimates of the giant clouds increase by a factor 2.

McCutcheon: Dickman found $[^{13}CO]/A_v$ to be linear with A_v to $A_v \sim 5$ mag where fractionation is no longer effective. Thus, Dickman's empirical result applies to regions where fractionation is present and where it is not, i.e. his result is independent of fractionation; so it is not obvious that the masses of giant clouds must be multiplied by a factor to account for fractionation. The question, perhaps, is whether Dickman's result is applicable to the hotter, more massive giant molecular clouds.

CO ABUNDANCE AND ISOTOPIC FRACTIONATION IN DARK CLOUDS

Paul F. Goldsmith, William D. Langer
Department of Physics and Astronomy
University of Massachusetts, Amherst

Eric R. Carlson, Robert W. Wilson
Bell Laboratories, Holmdel, New Jersey

In order to determine molecular fractional abundances, both the molecular density and hydrogen density must be known. In this study we have determined these parameters by fitting the observed intensities of $J = 2 \rightarrow 1$ and $J = 1 \rightarrow 0$ ^{13}CO and $C^{18}O$ transitions using a spherical cloud LVG radiative transfer model. The kinetic temperature is determined by observations of ^{12}CO and is found to be between 9 K and 13 K for our sources. The fractional abundance of CO is expected to rise rapidly between regions of low extinction and those with $Av \geq 4$ mag, due primarily to the decrease in photodestruction rates (Langer 1976). The $C^{18}O$ fractional abundance data plotted as function of Av support a nonlinear relationship between $X(C^{18}O)$ and Av for $Av \leq 4$ mag, with an indication of an asymptotoic value $X(C^{18}O) = 2.2 \times 10(-7)$ in highly obscured regions. For $^{16}O/^{18}O$ ratios of 250 (suggested by our data although possible uncorrected saturation of ^{13}CO makes this a lower limit) and 700 (A. Penzias, private communication) the fractions of carbon in CO in well-shielded regions are .08 and .23, respectively.

SOURCE	$n(H_2)$	A_v	$^{13}CO/C^{18}O$	$C^{18}O/^{13}C^{18}O$	$C^{18}O/C^{17}O$
B5	3150	>7.6	4.1	50	3.0
B335	3150	>5.2	2.5	49	2.2
L1262	3150	>4.3	5.3	44	3.3
B5(-8,0)	2150	1.5	17.9		
B5(0,-14)	2150	3.0	25.8		
B5(0,-16)	3150	3.0	29.1		
B335 (-4,2)	2150	1.0	33.1		
L1262(12,0)	1450	1.5	36.0		
L1262(0,-6)	700	3.0	29.4		

The transition region in which the $C^{18}O$ fractional abundance is rising is characterized by an elevated $^{13}CO/C^{18}O$ abundance ratio, which reaches a value 3 to 7 times greater than in the cloud cores. This effect is consistent with the presence of the isotopic exchange reaction discussed by Watson, Anicich, and Huntress 1976 taking place in the

B. H. Andrew (ed.), Interstellar Molecules, 417–420.
Copyright © 1980 by the IAU.

partially shielded cloud edges (Langer 1977). These results are discussed more fully by Langer et al. (1980).

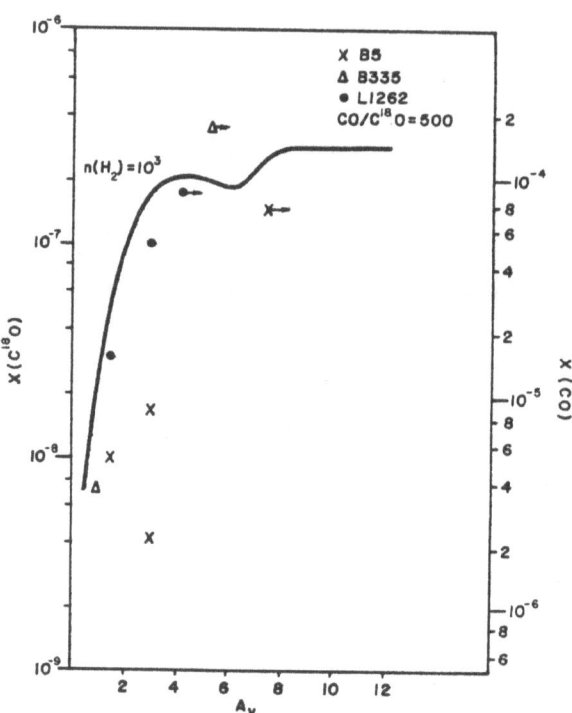

Measured $C^{18}O$ fractional abundance as a function of A_V
with corresponding CO abundances for $^{16}O/^{18}O$ = 500.
The theoretical curve is from Langer 1976.

REFERENCES

Langer, W. D. 1976, Ap. J., 206, 699.
Langer, W. D. 1977, Ap. J., 212, L39.
Langer, W. D., Goldsmith, P. F., Carlson, E. R., and Wilson, R. W.
    1980, Ap. J. (Letters), in press.
Watson, W. D., Anicich, V. G., and Huntress, W. T., Jr. 1976, Ap. J.,
    205, L165.

The research was supported in part by NSF grant AST76-24610 and is
contribution number 324 of the Five College Observatories.

DISCUSSION FOLLOWING GOLDSMITH

Dickinson: Your last slide shows the varying abundance ratio for the two isotopically substituted CO species. The errors are clearly bad in the wings of the line; over how great a velocity range do you consider the ratio reliable?

Goldsmith: The double ratio (of $^{13}C^{16}O/^{12}C^{18}O$ antenna temperatures) increases from ∿4 at line centers to a value of 12-20 where the signal to noise prevents further analysis. This occurs at velocities differing from the central velocity by 1 km s^{-1}, with our present sensitivity. If large scale mass motions of the form $V(r)=r^{\alpha}, \alpha>0$ dominate line formation, then as we look further in the wings we are seeing material further removed from the cloud centers, and the high ratios confirm that the chemical isotopic fractionation process is indeed occurring there.

Plambeck: The large-velocity-gradient model gives you the CO fractional abundance divided by the velocity gradient, not the fractional abundance directly. How did you determine the velocity gradients, and what uncertainty does this introduce into your fractional abundances?

Goldsmith: A good point! We have mapped the clouds, and used the linewidth divided by the linear size for the velocity gradient. This procedure introduces a formal error of approximately 50%, but a bigger uncertainty is in knowing how well this radiative transfer model applies.

de Jong: Your derived CO abundance of $X(CO) \approx 10^{-4}$ seems to be inconsistent with the upper limit derived from the observed DCO^+/HCO^+ ratios in dark clouds. It probably means that the largest contribution to the CO signal comes from the outer parts of the cloud ($A \leq 4$), where the CO abundance may indeed be of the order of magnitude that you mentioned.

Goldsmith: More recent observations of the DCO^+/HCO^+ ratio by Langer and collaborators show that, due to severe self-absorption of HCO^+ (indicated by the study of $H^{13}CO^+$ and other isotopes), the deuterium fractionation is not as large as previously thought. This relaxes the upper limit on $X(CO)$ so that it is now consistent with our derived value.

Kutner: Regarding the so-called "Dickman" ratio, two points should be made:

(1) A point that is often missed is that Dickman's result is essentially an observational prescription for converting apparent ^{13}CO column density into H_2 column density. As such, the details of how that result comes about do not necessarily affect its validity.

(2) If the relation does break down, the effect on derived cloud mass depends on how that ratio is used, so you can get an overestimate or an underestimate.

Goldsmith: Our data are more restricted than Dickman's but have the advantage of higher signal-to-noise ratio, better velocity resolution, and the J=2-1 lines. I feel that our analysis is probably somewhat more accurate, but our results do NOT differ violently from Dickman's: although $X(C^{18}O)$ drops radically at low extinctions, the isotopic fractionation enhances the relative abundance of ^{13}CO, tending to give a more linear relationship between column density and A_v for this molecule.

Bok: Dickman's formula for going from ^{13}CO numbers to H_2 numbers referred originally (thesis) only to very dense globules for which the zone with $A_v < 5$ mag. was too thin to be significant for mass estimates.

Goldsmith: An accurate determination of cloud mass from CO observations requires a knowledge both of the fraction of the mass corresponding to regions of various extinctions and of the variation of the fractional abundance of CO with extinction.

CO ISOTOPE LINE SHAPES IN DARK CLOUDS

P.C. Myers,[*] R.B. Buxton,[*] and P.T.P. Ho[+]
[*]M.I.T. Department of Physics
[+]Five College Radio Observatory

The ratio of ground-state densities $R_O \equiv N(^{13}CO)/N(C^{18}O)$ has been used to infer physical and chemical conditions in giant molecular clouds (Wannier et al. 1976) and dark clouds (Mahoney et al. 1976; Langer et al. 1979). In dark clouds R_O is found to vary from values near the terrestrial ratio $[^{13}C][^{16}O]/[^{12}C][^{18}O] \sim 5$ at positions of high extinction to values ~ 20 at positions of low extinction. In this paper we present high-resolution $J = 1 \rightarrow 0$ spectra of CO, ^{13}CO, and $C^{18}O$ at positions of high extinction in TMC-2, L134, and L134N. The $C^{18}O$ lines have non-Gaussian wings and are $\sim$ half as wide as the ^{13}CO lines. We find that R_O must vary across the line, from a minimum of $R_O \sim 4$ at the peak of the $C^{18}O$ line to a maximum of $R_O \sim 10$ in the wings, unless the ^{13}CO line has peak opacity $\gtrsim 5$. The variation of R_O with position and with velocity is consistent with models of clouds which have a dense core with low velocity-dispersion and low fractionation, and a rarefied envelope with high velocity-dispersion and high fractionation.

The observations were made with the NRAO 36-foot telescope[+] on Kitt Peak, Arizona. Five minute total-power on-off switching was used. The angular and spectral resolutions were 1.1' and 30 kHz (0.08 km s^{-1}). The antenna temperatures $T_A{}^*$ were corrected for atmospheric absorption by chopper-wheel calibration. The observed positions $[\alpha(1950); \delta(1950)]$ were TMC-2, $[04^h29m43s; 24°18'54'']$; L134, $[15^h50m59s; -04°26'58'']$; L134N, $[15^h51^m28s; -02°45'00'']$.

The $C^{18}O$ lines are extremely similar, being strong ($T_A{}^* \sim 2$ K) and narrow ($\Delta v(FWHM) \sim 0.6$ km s^{-1}) with distinct "wings". The ^{13}CO lines have FWHM wider than the $C^{18}O$ lines by a factor ~ 2. Figure 1 shows CO spectra in TMC-2.

We assume plane-parallel radiative transfer, and that the CO line is optically thick and thermalized. We find that R_O is unlikely to be independent of velocity. Only channels in the $C^{18}O$ line core are consistent with $R_O = 5$, while R_O, if ≥ 10, can be constant with respect to line velocity only if the ^{13}CO line has a peak opacity ~ 5 and $T_{ex} \sim 8$ K, which condition requires densities $n \lesssim 10^3 cm^{-3}$ (Kwok 1978). For TMC-2, CS observations indicate $n \sim 6 \times 10^4 cm^{-3}$ (Linke and Goldsmith 1979). Thus (1) R_O cannot be constant with velocity if terrestrial; and (2) R_O can be constant if ≥ 10, but only for a highly saturated, subthermal ^{13}CO line at a relatively low density.

[+]The National Radio Astronomy Observatory is operated by Associated Universities, Inc., under contract with NSF.

B. H. Andrew (ed.), Interstellar Molecules, 421–422.

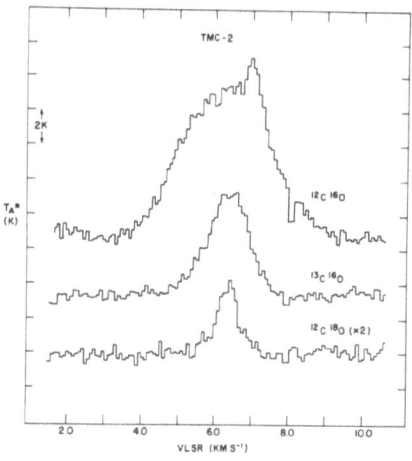

Figure 1

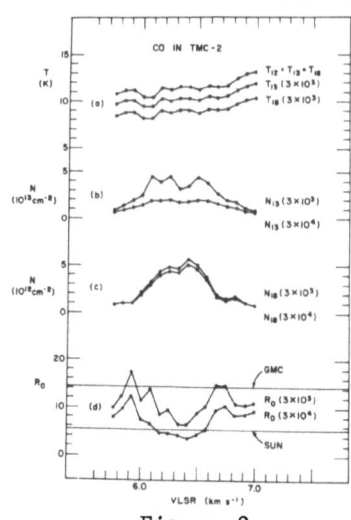

Figure 2

For higher densities 3×10^3 cm$^{-3} \leq$ n $\leq 3 \times 10^4$ cm^{-3} we use the model of Kwok (1978) to predict T_{ex} for each line. We then compute N(^{13}CO) and N(C^{18}O) for each channel where the C^{18}O line has signal-to-noise ratio $\geq$ 4, and calculate the corresponding R_0. The results are shown in Figure 2.

The variation of R_0 from ∿4 to ∿10 is seen in all three clouds. Random errors are too small to obscure the trend. Systematic errors of 15% in each spectrum would change the absolute values of R_0, but the relative variation of R_0 would be unchanged. The variation of R_0 with velocity may be consistent with the decrease of R_0 with extinction found by Langer et al. (1979), if the line core is formed in a dense interior region with low fractionation and if the line wings are formed in a rarefied envelope with high fractionation. If so, then cloud cores have low Δv while cloud envelopes have high Δv, as suggested by the correlation between Δv and spatial extent in the ρ Oph cloud (Myers et al. 1978).

This work was supported by the NSF.

REFERENCES

Kwok, S.: 1978, Astrophys. J. 225, 107.
Langer, W.D.: 1976, Astrophys. J. 210, 328.
Langer, W.D., Goldsmith, P.F., Carlson, E., and Wilson, R.W.: 1979, preprint.
Linke, R.A., and Goldsmith, P.F.: 1979, preprint.
Mahoney, M.J., McCutcheon, W.H., and Shuter, W.L.H.: 1976, Astron. J. 81, 508.
Myers, P.C., Ho, P.T.P., Schneps, M.H., Chin, G., Pankonin, V., and Winnberg, A.: 1978, Astrophys. J. 220, 864.
Wannier, P.G., Penzias, A.A., Linke, R.A., and Wilson, R.W.: 1976, Astrophys. J. 204, 26.

ISOTOPIC FRACTIONATION IN INTERSTELLAR CARBON-BEARING MOLECULES
UNRELATED TO CARBON MONOXIDE

V. Vanýsek
Department of Astronomy and Astrophysics,
Charles University, Prague, Czechoslovakia

ABSTRACT

The isotopic abundance ratio $^{12}C/^{13}C$ in some carbon-bearing
molecules is discussed in the context of chemical fractionation via
ion-molecule and exchange reactions in dense interstellar clouds.
These processes can lead to enhancement of ^{12}C in molecules *not*
related to carbon monoxide. The effect is transient and takes place
preferentially outside the cores of the interstellar clouds. However,
some enhancement of ^{12}C should remain in gases frozen onto dust grains
and may be reflected in larger objects such as cometary nuclei.

The somewhat low $^{12}C/^{13}C$ interstellar ratios in the carbon-bearing
molecules have been discussed by Watson (1977) on the basis of the
fractionation reaction in interstellar gas-phase chemistry (Watson,
Anicich and Huntress 1976). The fractionation mechanism involving the
carbon isotopes is based on the exchange reaction

$$^{13}C^+ + {}^{12}CO \rightleftarrows {}^{13}CO + {}^{12}C^+ + \Delta E.$$

This reaction should enrich CO with ^{13}C when temperature in an interstellar cloud is $T \lesssim \Delta E/k \sim 35$ K. At the same time free C^+ (and all
molecules not closely related to CO) should be enriched by ^{12}C. The
rate for the isotopic exchange reaction is $2 \times 10^{-10} \exp(-\Delta E/kT) \mathrm{cm}^3 \mathrm{s}^{-1}$,
which is higher than the rate of the most significant competing
neutral-neutral reactions.

Langer 1977 has shown that the fractionation mechanism results in
time-dependent isotopic anomalies in diffuse as well as in dense interstellar molecular clouds at low temperatures. Although the effect is
transient, and is without significant influence on the isotopic ratio
in the long term, the abundance of ^{12}C in molecules *not* related to CO,
such as CN, HCN, CH etc., could be considerably enhanced in comparison
with the average local ratio of the carbon isotopes.

B. H. Andrew (ed.), Interstellar Molecules, 423–426.
Copyright © 1980 by the IAU.

We discuss briefly here the numerical results obtained from a calculation in which the most relevant gas-phase reactions are considered (to be published in Bull. of Astr. Inst. of Czech.). The treatment used here is similar to that of Gerola and Glassgold (1978) and of Liszt (1978). Molecules unrelated to CO are formally denoted as species MC, and include molecules such as CN, HCN, and CH_2.

Although the formation mechanisms of these species are not exactly known, it seems reasonable to assume that the molecules are produced by ion-molecule reactions in the gas-phase. These reactions seem to be important not only for smaller polyatomic species but also for the synthesis of larger molecules, especially of those containing the -CN group linked to acetylene-type hydrocarbons (Freeman et al. 1979). The rate coefficient for the formation of the "substituted" molecule MC by reaction with C^+ is assumed to be $2 \times 10^{-9} cm^3 s^{-1}$. The most significant competing reaction $C^+(H_2, h\nu)$ CH_2^+ probably has a rate coefficient no higher than $10^{-16} cm^3 s^{-1}$. The relative abundances and isotopic ratios are $[^{12}C] = 8 \times 10^{-5}$, $[O] = 2 \times 10^{-4}$, $[N] = 10^{-4}$ and $^{12}C/^{13}C = 89$. The model of the cloud is isothermal. A temperature of 10K, an initial hydrogen number density of $10^3 cm^{-3}$, and a radius of 10^{19} cm are adopted. Depletion of light elements including carbon is assumed not to occur during the chemical evolution. The cloud is gravitationally unstable with a free-fall time $\sim 10^6$ years. The numerical results of the "averaged" $^{12}C/^{13}C$ ratio for species unrelated to CO are summarized in Table I.

Table I

$^{12}C/^{13}C$ ratio in a dense cloud for species unrelated to CO. (see text)				
time (10^6 year) / optical depth in visual	0.001	0.01	0.1	1
1	89	90	190	250
5	89	95	300	250
10	89	105	260	150

The $^{12}C/^{13}C$ ratios in Table I serve only to demonstrate qualitatively that isotopic fractionation works efficiently during the early evolution of a dense cloud. The actual $^{12}C/^{13}C$ ratio would be modified by gas depletion via gas-grain coupling. Furthermore the fractionation becomes ineffective due to heating of the gas in the late stages of evolution. Therefore the numerical results for elapsed times of 10^6 years from initial formation are probably unrealistic. However, in the regions outside of cloud cores the enhancement of ^{12}C could be preserved in molecules unrelated to CO which freeze onto dust grains during the fractionation phase. Hence there may be some detectable anomalies in the $^{12}C/^{13}C$ ratio in material which has not been processed through the inner part of the pre-solar nebula. If the carbon-bearing molecules were confined in dust grains with mantles, and were later incorporated in comet-like bodies, then the anomalous $^{12}C/^{13}C$ ratio could be preserved in such objects for a very long time.

Since comets are relics of the pre-solar nebula, some enhancement of ^{12}C can be expected in cometary radicals which are products of parent molecules not related to carbon monoxide. The currently available data for $^{12}C/^{13}C$ in comets do indicate that this ratio is higher than 100 i.e. significantly larger than the terrestrial value (Vanysek and Rahe 1978). It must be noted, however that the *average* $^{12}C/^{13}C$ ratio in comets may not be anomalous and could be terrestrial. Unfortunately the available data for comets relates to the isotope ratio in gaseous C_2 only. The $^{12}C/^{13}C$ ratio in comets may exhibit large differences in the various gaseous and solid cometary species. Confirmation of the effect would be significant proof that chemical fractionation takes place in interstellar dense clouds.

REFERENCES

Freeman, C.G., Harland, P.W. and Mc Ewan, M.J.: 1979, Mon. Not. RAS 187, 441.
Gerola, H. and Glassgold, A.E.: 1978, Astrophys. J. Suppl. 37.
Langer, W.D.: 1976, Astrophys. J. 210, 328.
Liszt, H.S.: 1978, Astrophys. J. 222, 484.
Vanysek, V. and Rahe, J.: 1978, The Moon and the Planets 18, 441.
Watson, W.D., Anicich, V.G. and Huntress, W.T.: 1976, Astrophys. J. (Letters) 205, L165.
Watson, W.D.: 1977 in CNO Isotopes in Astrophysics, ed. J. Audouze (Dordrecht, Reidel) p. 105.

DISCUSSION FOLLOWING VANYSEK

Tatum: There is a difficulty in estimating the ^{13}C abundance in comets. The $^{12}C_2$ bands are strong because $^{12}C_2$ is homonuclear; but $^{12}C^{13}C$ is not homonuclear, and rotational transitions are possible even if the dipole moment is small. Therefore the ^{13}C abundance might be underestimated, and the $^{12}C/^{13}C$ ratio might be much smaller than 120.
Vanysek: So far as I know, this effect in the isotopically shifted C_2 band plays only a secondary role.
Danks: Your suggestion that $^{12}C/^{13}C > 120$ seems unlikely. $^{12}C/^{13}C$ measurements exist for only a few comets, Ikeya 1963 I (Stawikowski and Greenstein), Tago-Sato-Kosaka 1969 IX (Owen), and Kohoutek 1973 XII (Danks, Lambert and Arpigny) which had values of 70±15, 100±20 and 135^{+63}_{-45} respectively. Owen took account of the blending of NH_2 with the $^{12}C^{13}C$ (1-0) bandhead at 4745 Å while Danks et al. resolved

the NH$_2$ lines. More recently Danks and Lambert have observed comets Kobayashi-Berger-Milon 1975h and West 1975a. Their measurements suggest ratios equal to or less than the terrestrial value of 89±4.

Any changes in $^{12}C/^{13}C$ seen in comets may indicate the comets' origin. In theory the ratio can be a few hundred down to the equilibrium value ~ 5. Values of 89 suggest a comet formed in the solar neighbourhood. Lower values suggest a comet formed later than the sun, and therefore elsewhere in the interstellar medium.

The change in the dipole moment from $^{12}C_2$ to $^{12}C^{13}C$ is unlikely to be significant. The most extreme example is the change from H$_2$ to HD, yet it is known to be very small. The temperature distribution in the two (1-0) bandheads of $^{12}C_2$ and $^{12}C^{13}C$ is probably not affected.

Vanýsek: The ratio $^{12}C/^{13}C \sim 70$ obtained by Stawikowski and Greenstein in comet Ikeya 1963 I is obviously underestimated owing to NH$_2$ blending. Although few data exist for comets, the $^{12}C/^{13}C$ values tend to be higher than terrestrial, and the lower limit of the measured values (i.e. ~ 90) seems to be exceeded only by Comet West.

Silk: In principle it is possible to test whether the isotopic composition of grains reveals a different fractionation history from cold molecular gas. Grain destruction occurs in interstellar shock fronts, both by sputtering and by grain-grain collisions, and leads to an enhancement of the abundances of several atomic species observed in diffuse clouds with LSR velocities as low as 20-30 km s^{-1}. It would be interesting to measure the isotopic composition of gas that may have been shocked, for example in the molecular cloud associated with the supernova remnant IC 443.

Irvine: Could accurate measurements of isotopic ratios help solve the fundamental question of whether comets were formed in the Oort Cloud or were formed in the region of the planets and then thrown into their present orbits by planetary perturbations?

Vanýsek: One can predict (if the suggestions in my paper are correct) that the varied history of the material in a cometary conglom- erate may lead to large differences in the $^{12}C/^{13}C$ ratio obtained either from the gaseous constituents or solid material of a cometary nucleus formed in the periphery zone (i.e. outside the core of a cloud). On the other hand cometary nuclei formed in the chemically and physically pro- cessed material of the pre-solar nebula (in the zone of planetary formation) would be isotopically "homogenized". Therefore the cometary isotopic ratios may shed light on the early history of the solar system. They may also indicate the temperature and time at which very primitive large conglomerates can be formed in dense molecular clouds.

Biermann: Some authors have suggested that comets originated during a relatively recent encounter of the solar system with a dense interstellar cloud (say $\sim 10^7$ years ago). A recent origin would imply an isotopic ratio like that of nearby interstellar clouds, whereas an origin along with the solar system would lead us to expect something close to the value 89.

Vanýsek: Yes, I believe that the $^{12}C/^{13}C \gtrsim 100$ ratio in comets is one of the strong arguments against their interstellar origin.

INTERPRETATION OF ISOTOPIC ABUNDANCES IN INTERSTELLAR CLOUDS

Michel Guélin and James Lequeux
Département de Radioastronomie
Observatoire de Paris

1. INTRODUCTION

The abundances of elements in the interstellar medium (ISM) result from a complex sequence of nucleosynthetic processes which started some ten billion years ago in the Big Bang and are still going on. Their study and in particular their comparison with the stellar and the solar system abundances may give clues to: i) the properties of the early Universe, ii) the evolution of galaxies (rate of star formation, Initial Mass Function of stars, stellar nucleosynthesis and rate of ejection of matter by stars), and iii) the flux of low-energy cosmic rays.(See e.g. Reeves 1974, Audouze and Tinsley 1976).

Abundance determinations of a number of atomic species in the ISM can be made directly by studying their optical and radio lines. These determinations however require a precise knowledge of the physical conditions and are often insecure. In the cold ISM, moreover, absolute elemental abundances are difficult to derive from such studies, since a large fraction of the atoms is likely to be depleted into grains and molecules.

The difficulties with the derivation of elemental abundances are partly avoided by using ratios of closely related isotopic species which are usually much less affected by depletion and excitation effects. In return, since the atomic lines of isotopic species cannot in general be resolved in astronomical objects (except, mainly, for hydrogen and helium), the study of isotopic abundances in the ISM has to be carried out essentially through observations of molecular lines, mainly in the millimeter to decimeter range. While optical or UV observations of atomic and molecular isotopic interstellar lines are presently restricted to the lines of sight of bright nearby stars, radio observations can reach all the dense ISM and distant regions in the Galaxy, opening the possibility of addressing the problem of chemical evolution on a truly galactic scale.

The derivation of isotopic elemental abundances from molecular observations, and their interpretation in terms of nucleosynthesis and galactic evolution, have been discussed at length in the literature, and in

427

B. H. Andrew (ed.), Interstellar Molecules, 427–438.

particular at the special session of the IAU in Grenoble three years ago (Audouze 1977). Since that time, the major change in the modelling of chemical evolution has been the recognition that the abundance of an element depends more upon the lifetime of the stars responsible for its synthesis than upon its primary or secondary nature (see e.g. Tinsley 1979, Vigroux 1979). It has also been fully realized that mass exchanges between the different parts of the Galaxy cannot be neglected and introduce extra free parameters in the computations.

On the observational side (see the reviews by Penzias and Watson in this symposium), although improved technical means allow the line intensity ratios of a growing number of molecular species to be measured very accurately, it has become clear that optical depth effects, line confusion and isotopic fractionation often prevent the derivation of the actual isotopic abundance ratios. In the past three years, not only have observations confirmed that deuterium is currently enhanced by several orders of magnitude in the dense IS clouds, but they have provided evidence that ^{13}C also may be strongly enhanced in parts of the cool gas -- a result which forces a reconsideration of the meaning of isotopic ratios involving this species.

2. NUCLEOSYNTHESIS OF OBSERVED INTERSTELLAR ISOTOPES

In this section, we will briefly review the most important processes of nucleosynthesis of the isotopes observed in the interstellar medium (isotopes of H, He, C, N, O, Si and S), and tentatively ascribe the most probable astrophysical sites for their production.

Synthesis of elements is still going on in stars (through thermonuclear reactions) and in the interstellar medium (through spallation). Evidence that element abundance changes are still taking place comes from differences in chemical composition of the atmospheres of stars of various places of origin, abundance gradients in galaxies, and abundance anomalies in probable sites of nucleosynthesis: red giants, novae and supernovae (cf the different chemical composition of the two kinds of filaments in Cas A). It should be noted at this point that there is no reason why the ISM abundances, which result from nucleosynthesis up to the present time, should be the same as the solar system abundances which rather reflect local ISM abundances 4.6 10^9 years ago. Models of the chemical evolution of the Galaxy indeed predict local variations in isotopic ratios up to upper limits of about a factor 2 since that epoch (e.g. Vigroux et al. 1976).

2.1. Equilibrium nucleosynthesis

This type of nucleosynthesis corresponds to the quiet phases of stellar evolution. The most important stages are helium-burning, which produces ^{12}C and ^{16}O in the cores of red giants, and the CNO cycles which transform these species into other C, N and O isotopes.

The CNO cycles consist of a network of proton-capture and beta-decay reactions. While the rates for β-decay are nearly constant, those for p-capture depend drastically on temperature and density. (For example, the rate of the $^{12}C(p,\gamma)$ ^{13}N reaction varies as T^{15} near 2.5 10^7K). At the temperatures and densities which prevail in the H-burning zone of red giants ($T<<10^8K$), the rates for p-capture (k_p) are several orders of magnitude smaller than those for β-decay and control the production of CNO isotopes. For typical temperature and density of 2.5 10^7K and 10^{-2} g cm^{-3}, one has at equilibrium: $^{15}N/^{14}N = k_p(^{14}N)/k_p(^{15}N) = 4\ 10^{-5}$, $^{12}C/^{14}N = 10^{-2}$ and $^{13}C/^{12}C = 1/3$. Since for these conditions equilibrium is reached in some 10^6 years, a time short with respect to stellar lifetimes, most of the pre-existing CNO isotopes are transformed into ^{14}N, and in particular the pre-existing ^{15}N is destroyed rather than new ^{15}N produced; as is well known, ^{15}N has to be produced in drastically different conditions (explosive nucleosynthesis). Substantial amounts of ^{13}C and ^{17}O are synthesized in the "cold" CNO cycles but their observed abundances are difficult to achieve simultaneously at equilibrium. For this, one has to invoke incomplete cycles resulting from a partial mixing of hydrogen from the envelope with the material from the helium shell (Dearborn et al. 1976). During this mixing incomplete p-p reactions may also produce ^3He in zones of low temperature - i.e. $T<10^7K$ (Rood et al. 1976).

2.2. Explosive nucleosynthesis

In explosive nucleosynthesis, matter is compressed and raised to very high temperatures, then cooled and diluted rapidly. It is necessary that the cooling be fast enough to prevent proton- or α-capture of the fragile or unstable nuclei formed at high temperaturess.

p-p explosive nucleosynthesis occuring in the Big Bang, supernova (SN) shells, and possibly supermassive objects (SMOs) may produce not only the stable ^4He but also D, ^3He and ^7Li. Deuterium is so fragile that an extremely fast cooling and dilution is required to save any appreciable amount of it. These conditions seem hard to realize in SN blast waves and SMOs, and the Big-Bang is believed to be the main producer of D as well as of a large part of ^3He and ^7Li (Reeves 1974).

When the temperature in the H-burning zone in a star exceeds 2 10^8K (Nova, SN, SMO) the p-capture reaction rates become comparable to or even larger than the β-decay rates (k_e) in the CNO cycles. Then, large quantities of ^{17}F, ^{15}O, ^{14}O and ^{13}N are quickly built up. If the cooling which follows the explosion is fast enough, these unstable species will decay into ^{17}O, ^{15}N, ^{14}N and ^{13}C respectively without further processing. In this way, one can achieve a $^{15}N/^{14}N$ abundance ratio as high as $k_e(^{14}O)/k_e(^{15}O) = 1.7$ (Lazareff et al. 1979).

Explosive He-burning episodes, like those occuring during the SN blast and in He-rich novae, produce ^{18}O from ^{14}N by the reaction $^{14}N(\alpha,\gamma)^{18}F$ followed by β-decay of ^{18}F. The same reactions also yield some ^{18}O in non-explosive conditions, but this ^{18}O is probably destroyed by subsequent α-capture reactions before getting a chance to reach the surface

(Truran, in Audouze 1977, p.145); negligible amounts of it will thus be released in space, unless the star evolves into a supernova (Dearborn et al. 1978).

The isotopes of Si and S are products of explosive nucleosynthesis in stars. According to Arnett (1973, 1978) ^{29}Si, ^{30}Si and ^{33}S are probably produced in explosive carbon-burning, while their neighbours ^{28}Si, ^{32}S and ^{34}S result from neon and oxygen-burning. Observation of variations in the ratios of silicon and sulfur isotopes would tell us much about the importance of these processes during the successive stages of galactic evolution.

2.3. Possible sites of nucleosynthesis of interstellar isotopes

Deuterium is generally believed to be produced in the Big Bang. It is completely destroyed in the interior of practically all stars, so that D/H must decrease with increasing degree of evolution (i.e. for example be smaller in the galactic center than in the solar neighbourhood), at least if most D is of cosmological origin.

^{3}He can be synthesized in the Big Bang, in SN and SMOs, and in the H-burning shell of red giants. In this latter case, convective mixing may carry it into the stellar envelope from which it will eventually be expelled into the ISM via stellar winds or at the planetary nebula (PN) or SN stage. Unlike D, ^{3}He is not much destroyed by further processing in stars.

^{12}C and ^{16}O are produced by He-burning in the deep interior of stars of mass >2 $M_{\odot}$. They are certainly expelled into the ISM in large quanti--ties during the final explosion of the most massive of these stars. At the present time, however, it is not clear whether they can also be released by the longer-lived lower-mass stars (M < 6 $M_{\odot}$). It is believed that, at least in carbon stars, deep mixing significantly enriches the stellar envelope with ^{12}C; since these stars are known to experience important mass losses, they could be an effective source of interstellar ^{12}C. There is some evidence that carbon stars end their life as PN, and abundance measurements in these objects should in principle allow us to assess the amount of ^{12}C released by this process into the ISM. So far, unfortunately, the results are discrepant: while visible permitted lines of multiply-ionized carbon tend to indicate a large overabundance of this element in several PN, UV observations tend to show in the very same objects a nearly solar abundance (e.g. Natta et al. 1979). It should be noted that the abundances observed in the hot gas may not reflect those of the expelled matter since some carbon may be in the form of dust grains or hidden inside dense molecular envelopes, such as the ones which are known to surround some of the PN. In summary, it cannot be ruled out that a major fraction of ^{12}C is produced by low mass long-lived stars, a possibility which would strongly affect the time behaviour of the ^{13}C/^{12}C ratio in the Galaxy (Tinsley 1979).

Release of ^{16}O by low mass stars seems more difficult than for ^{12}C. In fact ^{16}O abundance measurements in PN, which are more reliable than those of ^{12}C, never show any enrichment in this element. Thus, it is thought that the interstellar ^{16}O comes mainly from high mass stars.

^{13}C is mainly a secondary product of the incomplete CNO cycles in red giants, where the enrichment in ^{13}C can be seen directly because convection has mixed it into the envelope (Dearborn et al. 1976). It is difficult to know the relative contribution of stars of different masses, the present consensus being that most ^{13}C comes from low mass stars ($M < 4\ M_\odot$). Consequently, it is not easy to predict the exact rate of increase of the $^{13}C/^{12}C$ ratio with galactic evolution, although there should be an increase. However, an increase by a factor 2 of this ratio since the birth of the sun, such as the one suggested by the early ^{13}C observations, seems difficult to achieve even with favorable assumptions (Vigroux et al. 1976, Vigroux 1979). A gradient with galactocentric distance is also predicted, but quantitative estimates are uncertain.

^{14}N was believed to be only a secondary product of the CNO cycles in red giants (mainly of low masses), and its overabundance in many PN is well observed (Peimbert 1978). There is increasing evidence however, from the study of the N/O gradients in external galaxies, that this secondary process is not the only one: the N/O ratio does not decrease with O/H as expected, indicating production of ^{14}N by massive stars (Alloin et al. 1979, Talent and Dufour 1979). "Primary" ^{14}N could be synthesized in these stars through the dredge-up mechanism, which incorporates fresh ^{12}C from the core into the H-burning zone where it is converted into ^{14}N .

^{15}N is produced almost entirely by explosive CNO nucleosynthesis in novae and (less likely) in SN. It should be possible to decide between these two sources by studying its abundance gradient with galactocentric distance, since novae and supernovae do not appear to have the same distribution in time and space (Audouze et al., in Audouze 1977, p.155).

^{17}O, a secondary product, can be synthesized both by the cold CNO cycles in massive red giants and the explosive CNO cycles in novae and perhaps SN . Models (Dearborn and Schramm 1974) as well as the few existing observations of ^{17}O enrichment in red giants (Rank et al. 1974, Maillard 1974) seem to indicate only a modest total production in these objects. Thus the usual consensus is that novae are the main producers of ^{17}O . One expects the $^{17}O/^{16}O$ ratio to increase with galactic evolution.

^{18}O appears to be mainly released by SN (no enrichment in this isotope has been observed in red giants), but it could also be produced in large quantities by He-rich novae. Its evolutionary behaviour is not easy to predict, although the $^{18}O/^{16}O$ ratio is expected to increase with galactic evolution since ^{18}O is a secondary (or even tertiary) product. The behaviour of the $^{17}O/^{18}O$ ratio obviously depends on the nature of the sites of synthesis: if ^{17}O is produced mainly in novae and ^{18}O in supernovae, this ratio will increase with the degree of evolution.

3. ISOTOPE FRACTIONATION IN MOLECULES

Isotopic molecular ratios may differ from the true ISM ratios due to chemical fractionation. This effect has been thoroughly investigated only for hydrogen and carbon (Watson, this symposium).

Substitution of hydrogen by deuterium in molecules results in relatively large differences in abundance, weight and zero-vibrational energy which in turn may change the rates of formation and destruction. This is particularly the case for the light ions H_3^+ and CH_3^+ which, in the dense clouds, are believed to be at the origin of most hydrides, including HCN, HCO^+, HNC, N_2H^+, NH_3 and H_2CO, all of which are observed to exhibit large overabundances of the deuterated species. As an example, the deuterium enrichment of H_3^+ is thought to result from the competition between the intrinsically slow formation reaction:

$$HD + H_3^+ \rightleftarrows H_2 + H_2D^+ + \Delta E , \qquad (1)$$

and faster H_2D^+ destruction reactions such as dissociative electron recombination and Langevin-rate reactions with other molecules. Thus, the observation of large deuterium enhancements in HCN, HCO^+, etc. sets stringent upper limits on the abundances of the free electrons and of the trace molecules reacting with H_2D^+ and CH_2D^+ (i.e. N_2, C_2H_2, etc.).

In diffuse clouds, the abundance of HD results from the balance between its formation through $H_2 + D^+ \rightarrow HD + H^+$, and its destruction by selective photodissociation. The measure of the HD/H_2 ratio can be used to set limits to the ionization rate of hydrogen (Watson 1973).

Watson et al. (1976) have suggested that CO may be enriched in ^{13}C through the reaction:

$$^{13}C^+ + ^{12}CO \rightleftarrows ^{12}C^+ + ^{13}CO + \Delta E . \qquad (2)$$

The rate constants of reactions (2) and (1) are about equal (2 and 3 10^{-10} cm^3s^{-1} respectively). However, the excess energy ΔE is much smaller for (2) than for (1), and C^+,the reservoir of ^{13}C,is seldom more abundant than CO in molecular clouds: both factors limit the efficiency of CO fractionation. A detailed analysis (Langer 1977) shows that only the outer regions of the clouds, where the visual extinction A_v is < 4, have enough C^+ to allow a CO enhancement in ^{13}C by more than a factor 3. In the inner regions, there is little C^+ and fractionation becomes ineffective in changing the ^{13}CO abundance; it can however efficiently deplete C^+ in ^{13}C so that molecules formed from this ion could also be depleted in ^{13}C . Observational evidence for such CO fractionation has just been reported by Goldsmith et al. at this symposium: the $^{13}CO/^{12}CO$ ratio increases by factors 6-10 from the center to the edge of three dark clouds, while the CO enrichment in ^{13}C appears to be smaller than a factor 1.5 at the center.

While the lack of an abundant reservoir of ^{13}C limits the fractionation of CO, no such restriction exists for much less abundant molecules. An interesting example is that of HCO^+ for which CO itself acts as a ^{13}C reservoir. One has (Langer et al. 1978) :

$$^{13}CO + H^{12}CO^+ \rightleftharpoons H^{13}CO^+ + {}^{12}CO + \Delta E \quad . \tag{3}$$

Adams and Smith (private communication) have measured a rate constant $\alpha = 3 \ 10^{-10} \ cm^3 s^{-1}$ at 100 K for this reaction, similar to those of reactions (1) and (2). We estimate $\Delta E/k = 17$ K from the ab-initio calculations on HCO^+ of Henning et al. (1977) with an uncertainty of ± 1 K (Kraemer, private communication). α and ΔE are large enough for a substantial enrichment of HCO^+ in ^{13}C in the center of very cold clouds. However no direct evidence exists for such an enhancement : in TMC1, the only cloud for which a good $H^{13}CO^+/HC^{18}O^+$ ratio has been determined (Langer et al. 1978), any enhancement larger than a factor 1.5 seems to be ruled out.

Oxygen and nitrogen isotopic fractionations are not well known. The reaction which would seem the most promising in the case of oxygen is analogous to reaction (3) and involves $C^{18}O$ (or $C^{17}O$) and HCO^+. Its rate for $C^{18}O$ is similar to that of (3) (Adams and Smith, private communication) but its $\Delta E/k$ is only $7 \pm 1K$, again using the HCO^+ vibrational constants of Henning et al.. It is therefore unlikely that ^{18}O is significantly fractionated in this way ; ^{17}O fractionation is expected to be even smaller. As for nitrogen, N_2 and perhaps atomic N appear interesting reservoirs of ^{15}N. The proton-exchange reaction analogous to (3):

$$^{14}N^{15}N + {}^{14}N_2H^+ \rightleftharpoons {}^{14}N^{15}NH^+ + {}^{14}N_2 + \Delta E \tag{4}$$

probably has a similar rate constant. Again, from the calculations of Henning et al., we find a $\Delta E/k$ probably too low for appreciable fractionation to occur via this reaction ($\Delta E/k = 10 \pm 1$ K). In view of the importance of ^{15}N as a test of galactic evolution, laboratory work on nitrogen-bearing ions would be very valuable.

Finally, in the case of sulfur, the reaction of $^{34}S^+$ on $C^{32}S$ might have a large cross section (Watson 1977, ibid). S^+ is likely to be abundant in the outer parts of IS clouds since sulfur is readily ionized by interstellar UV radiation, but whether enough CS remains in these unshielded regions to be detected is questionable.

4. DISCUSSION

The above considerations, coupled to the simplest possible closed-box model of galactic evolution, lead to the predictions on IS isotope abundances listed in Table 1. It should be understood that in this simple model, the "solar system" and "disk ISM" conditions represent only two stages on the same evolutionary track which are separated by 5 billion years, and do not include local inhomogeneities. The "galactic center" conditions, on the other hand, refer to a much more evolved stage where low mass stars dominate strongly the gas production.

TABLE 1.: Predicted versus observed ISM relative isotopic abundances.

		D/H	$^{13}C/^{12}C$	$^{15}N/^{14}N$	$^{18}O/^{16}O$	$^{17}O/^{18}O$
solar system		1	1	1	1	1
disk ISM	predicted	$\gtrsim 0.5$	1 to 1.5	< 1?	> 1	$\gtrsim 1$
	observed	≈ 0.5	1.3	0.7	0.8	1.6
galactic center	predicted	0	1 to 3	< 1	> 1	> 1
	observed	some?	3?	$\lesssim 0.3$	1?	1.6

Let us now discuss the observational results just presented by Penzias (or derived from Copernicus data) in the light of these predictions.

The amount of deuterium produced in the Big Bang is extremely sensitive to the density of matter in the early stages of the Universe, so the measure of the $\underline{d}=D/H$ abundance ratio provides an unique way of determining this important parameter. For this purpose, the value of $\underline{d}$ at the time of the Big Bang should be estimated from the present one by correcting for deuterium processing in stars. It is fortunate, since D is strongly fractionated in molecules, that the atomic D/H ratio can be directly observed in the solar neighbourhood. Copernicus observations in the line of sight to 10 nearby stars indicate an average $\underline{d}$ of 1.4×10^{-5}, with apparently real variations by a factor 2 (Laurent et al. 1979). When the assumption is made that most D is of cosmological origin and the astration correction is applied (a decrease of $\underline{d}$ since the Big Bang by a factor of two), a density much smaller than the "critical" one needed to close the Universe is found (see e.g. Reeves 1974). It has been argued (Ostriker and Tinsley 1975) that SN and possibly SMOs may produce deuterium, in which case this result could be erroneous; were this true, however, the astration correction should be positive and one should observe an increase of $\underline{d}$ towards the inner parts of the Galaxy. Unfortunately, at these large distances from the Sun, the only estimates of $\underline{d}$ result so far from molecular observations (Penzias 1979) and are too uncertain for this prediction to be checked.

Astration is much more severe in the galactic center itself, so that any primordial deuterium has essentially disappeared, unless there is an infall of external gas. Audouze et al. (1976) have shown that the mere detection of D implies either such an infall, or a local production by SN and SMOs, or spallation by a large flux of low-energy cosmic rays (an unlikely possibility). Marginal detections of DCN, DCO^+ and NH_2D in Sgr A and/or Sgr B have been reported (Penzias 1979, Turner et al. 1978). Confirmation of these detections, or detection of any other deuterated molecule in the galactic center would be important.

The probable detection of the 3.5 cm hyperfine line of $^3He^+$ in W51 (Rood et al. 1979) seems to imply a $^3He/H$ abundance ratio in this giant HII region roughly comparable to the protosolar ratio. If confirmed, this result would probably exclude substantial ^3He production in red giants.

As derived from giant molecular clouds, the $^{13}C/^{12}C$ abundance ratio (about 1/70 in the disk sources, according to Penzias's review) shows only a moderate enhancement in ^{13}C with respect to the solar system ratio of 1/89. Such a moderate difference, in contrast with the larger one ($\sim$ 1/40) proposed by Wannier (in Audouze 1977, p.71) is easily explainable by chemical evolution models, but could also at least partly result from fractionation effects. We conclude that: first, the relatively small amplitude of this difference, as it is observed, second, the difficulties due to fractionation in deriving the true isotopic ratio, and third, the present ambiguities in the interpretation of this ratio in terms of stellar nucleosynthesis, make the $^{13}C/^{12}C$ ratio much less interesting than has been thought. It must be recalled that ^{13}C observations of molecules less abundant than CO should be considered with caution since even a moderate ^{13}C enhancement in CO can result in large ^{13}C depletions in molecules which form from C^+ (e.g. Langer 1977). Rather than serving as a dubious test for evolution models, these observations in fact may turn out as valuable tools for studying molecule formation.

As discussed in section 2, ^{15}N is probably the best tracer of explosive nucleosynthetic events either in novae, or in SN . The difference in the $HC^{15}N/H^{13}CN$ ratio observed between the galactic center and the disk IS clouds seems too large to be accounted for by fractionation alone. Taken at its face value, it would rather imply ^{15}N production in high mass stars, i.e. SN or high mass novae (Audouze et al., in Audouze 1977, p.155). Observation of this ratio in other molecules than HCN would be of great interest. Already, the observation of $^{15}NH_3$ in Orion (Wilson and Pauls 1979) might imply a $^{15}N/^{14}N$ ratio smaller than the solar system one.

Although oxygen-bearing molecules such as CO, H_2CO, HCO^+ and OH are widely observed in IS clouds, and although no significant fractionation of oxygen isotopes is expected, a precise determination of the $\underline{a}= {}^{18}O/^{16}O$ and $\underline{b}= {}^{17}O/^{16}O$ abundance ratios remains an arduous problem. By themselves, the very small values of $\underline{a}$ and $\underline{b}$ (1/490 and 1/2670) make them difficult to measure directly. $\underline{a}$ and $\underline{b}$ alternatively can be derived from double isotopic ratios, such as the $^{12}C^{18}O/^{13}C^{16}O$ ratio, but these usually involve the $^{13}C/^{12}C$ ratio which is not too well known; in fact, even the double isotopic ratios are hard to measure since the best suited lines for their studies (those of ^{13}CO and $H^{13}CO^+$ for example) are often optically thick. Thus it appears premature to worry about the mild discrepancy between the predicted and observed values of $\underline{a}$ in Table 1 .

More promising seems to be the $^{17}O/^{18}O$ ratio. The consistency observed from source to source for this ratio, which seems to indicate that line opacity does not pose an insurmountable problem for $C^{18}O$, and the very fact that one would expect oxygen fractionation to decrease rather than

to increase this ratio, seem to give confidence that ^{17}O is slightly
overabundant with respect to ^{18}O in the disk clouds. This would imply a
marked difference of origin for these two isotopes and in particular
would agree with a production of ^{17}O predominantly in novae and of ^{18}O
in supernovae.

The $^{29}Si/^{30}Si$ and $^{33}S/^{34}S$ ratios are relatively easy to determine since
the abundance of each isotope in the pair is not much different. Observations are consistent with solar system ratios, a not unexpected result
since only small variations are forseen. As said before, fractionation
is possible and may hide chemical evolution effects.

5. CONCLUSION

To conclude this session, let us emphasize the few but important astrophysical problems which can be efficiently handled by the present (or
near future) observational and theoretical means.

The first of these is the geometry of the Universe as derived from the
cosmological D/H ratio: some theoretical considerations, and the moderate
yet disturbing variations in the D/H ratio in local diffuse clouds,
still raise the possibility of some deuterium production after the
Big Bang. Direct measurements of this ratio in diffuse clouds at different galactocentric distances and in the Magellanic Clouds are of fundamental importance in this respect. Of lesser interest, since they can
hardly give clear-cut answers, are molecular D/H observations; these
however are the only ones that can reach the innermost regions of the
Galaxy (inside the 4 kpc ring). One undisputable detection of a deuterated molecule in the galactic center would be important.

The second problem which has been and should be further addressed by isotopic measurements is the formation of molecules, in particular of species
which could be formed as well on grains as by gas phase reactions. The
detection of HDCO reported in this symposium illustrates the kind of
work to do. It would be of high interest to observe the deuterated forms
of C_2H, HC_3N and other simple carbon-chain molecules. ^{13}C measurements
can help also in studying the abundance of C^+ and may serve to point out
those molecules formed through this ion.

An important insight into physical conditions inside molecular clouds, in
particular the degree of ionization, can be obtained through D/H measurements in molecules. Analysis of DCO^+ and HCO^+ observations in dense
clouds, for example, indicate a surprisingly low fractional electron
abundance (see e.g. Guélin et al. 1977, Langer et al. 1978). Much can be
said from these studies about the flux of low-energy cosmic rays.

Finally, $^{15}N/^{14}N$, $^{17}O/^{16}O$ and $^{18}O/^{16}O$ observations can still be interpreted in terms of nucleosynthesis and chemical evolution, provided that
they are coupled with studies of the abundance gradients of ^{14}N and ^{16}O
obtained from observations of HII regions, for example. Good galactic

gradients as well as a study of possible cloud-to-cloud variations are quite important for advanced studies of galactic evolution.

We thank J. Audouze for discussions and a critical reading of the manuscript.

REFERENCES

Alloin,D., Collin-Souffrin, S., Joly, M., Vigroux,L. : 1979,Astron. Astrophys. 78, p.200.
Arnett, W.D.: 1973, Ann. Rev. Astron. Astrophys. 11, p.73.
Arnett, W.D.: 1978, Ap. J. 219, p. 1008.
Audouze, J., Tinsley, B.M.:1976, Ann. Rev. Astron.Astrophys. 14, p.43.
Audouze, J., Lequeux, J., Reeves,H., Vigroux, L.: 1976, Ap. J. (Letters) 208, p.L51.
Audouze, J.: 1977 (editor) CNO Isotopes in Astrophysics, D. Reidel, Dordrecht.
Dearborn, D.S., Schramm, D.N.: 1974, Ap. J. (Letters) 194, p.L67.
Dearborn, D.S., Eggleton, P.P., Schramm, D.N.: 1976, Ap. J. 203, p.455.
Dearborn, D.S., Tinsley, B.M., Schramm, D.N.: 1978, Ap. J. 223, p.557.
Guélin, M., Langer, W.D., Snell, R.L., Wootten, H.A.: 1977, Ap. J. (Letters) 217, p.L165.
Henning,P., Kraemer,W.P., Diercksen,G.H.F.:1977, M.P.I. internal report.
Langer, W.D.: 1977, Ap. J. (Letters) 212, p.L39.
Langer, W.D., Wilson, R.W., Henry, P.S., Guélin, M.: 1978, Ap. J. (Letters) 225, p.L139.
Laurent, C., Vidal-Madjar, A., York, D.G.: 1979, Ap. J. 229, p.923.
Lazareff,B., Audouze,J., Starrfield,S., Truran, J.W.:1979, Ap.J. 228,p875.
Maillard, J.P.: 1974, in Highlights in Astronomy 3, p.269, ed. G. Contopoulos, D. Reidel, Dordrecht.
Natta, A., Pottasch, S.R., Preite-Martinez, A.: 1979, Astron. Astrophys. in press.
Ostriker, J.P., Tinsley, B.M.: 1975, Ap.J. (Letters) 201, p.L51.
Peimbert, M.: 1978, in Planetary Nebulae, IAU Symp. 76, p.215, ed. Y. Terzian, D. Reidel, Dordrecht.
Penzias, A.A.: 1979, Ap. J. 228, p.430.
Rank, D.M., Geballe, T.R., Wollman,E.R.: 1974, Ap.J.(Letters) 187,p.L111.
Reeves, H.: 1974, Ann. Rev. Astron. Astrophys. 12, p.437.
Rood, R.T., Steigman,G., Tinsley, B.M.: 1976, Ap.J.(Letters) 207,p.L57.
Rood, R.T., Wilson,T.L., Steigman,G.: 1979, Ap.J.(Letters) 227, p.L97.
Talent, D.L., Dufour, R.J.: 1979, preprint.
Tinsley, B.M.: 1979, Ap. J. 229, p.1046.
Turner, B.E., Zuckerman, B., Morris, M., Palmer, P.: Ap. J. (Letters) 219, p.L43.
Vigroux, L., Audouze,J;, Lequeux,J.: 1976, Astron. Astrophys. 52, p.1.
Vigroux, L.: 1979, Thèse de Doctorat, Université de Paris XI, Orsay.
Wilson, T.L., Pauls, T.: 1979, Astron. Astrophys. 73, p.L10.
Watson, W.D.: 1973, Ap. J. (Letters) 182, p.L73.
Watson, W.D., Anicich, V.G., Huntress, W.T.: 1976, Ap. J. (Letters) 205, p.L165.

DISCUSSION FOLLOWING GUÉLIN

Mouschovias: Have you assumed that explosive nucleosynthesis does not change the isotope ratios established by the CNO cycle prior to the explosion?

Guélin: No. What we refer to as "explosive nucleosynthesis" explicitly takes into account a part of these changes.

Townes: It is not clear how you consider variability from source to source. Are you allowing each large cloud to be quite old, which I believe is the case, so that it develops individually, or are you allowing only a variation with radial distance from the galactic center? What is the significance of your prediction of "probably small" variations of $^{12}C/^{13}C$ from cloud to cloud when the experimental data seems to indicate the opposite?

Guélin: In view of the modest differences observed between the abundance in the solar system and the average abundance in the galactic disk, marked abundance variations from one cloud to another dependent on the age of the clouds should occur only in cases where clouds are at least 5 billion years old. So far, except for Sgr A and B2, no source seems to exhibit in all molecules a consistently high or low $^{12}C/^{13}C$ abundance ratio with respect to the average disk value of 68, at least in the data presented by Penzias this morning. If I were asked to predict the relative importance of such variations in the case of extremely old clouds, I would describe the $^{12}C/^{13}C$ variations as relatively small because of the small difference both observed and predicted between the solar system and the disk for this ratio.

W. Watson: It seems to me that the reactions (3) and (4) are more likely to be important than you indicated. An approximate criterion is that the forward reaction should compete with dissociative electron recombination. Admittedly, application of "currently accepted" data would indicate an effect of only a few per-cent. However the coefficient for dissociative recombination may not continue to increase below 100 K as rapidly as suggested by higher temperature data.

Guélin: ^{13}C exchange between CO and HCO^+ is attractive since it may provide a way to enhance this isotope in the very dense parts of the clouds. Yet, as I said, the observed $H^{13}CO^+/HC^{18}O^+$ ratio (in only one cloud, it is true!) seems to rule out a strong and common effect. More observations have to be made to settle this interesting question.

Scoville: Why do you assume that the gas clouds seen in the galactic center have always been there?

Guélin: Taking into account an infall of fresh gas into the galactic centre introduces extra free parameters and makes it difficult to compare predicted and observed ratios. With the exception of deuterium whose presence in this region has yet to be proven, the observed ratios can be easily explained without infall. In fact, the ^{15}N data even set an upper limit on the importance of such an infall (Audouze et al. in Audouze 1977, p. 155).

DETECTION OF DEUTERATED FORMALDEHYDE IN INTERSTELLAR CLOUDS

William D. Langer
Department of Physics and Astronomy
University of Massachusetts, Amherst, MA   01003

Margaret A. Frerking, Richard A. Linke and Robert W. Wilson
Crawford Hill Laboratory
Bell Telphone Laboratories, Holmdel, NJ   07733

Deuterated formaldehyde has been detected for the first time in interstellar clouds; the observed ratio HDCO/H_2CO implies formation by gas phase chemistry.

Deuterated formaldehyde, HDCO, has been detected for the first time in interstellar clouds from observations of emission from mm radiation in the $2_{02} \rightarrow 1_{01}$ transition at 128.81291 GHz (Langer, Frerking, Linke, and Wilson 1979 henceforth LFLW). The extent to which deuterium is fractionated can be used as a constraint on the chemistry of formaldehyde in interstellar clouds (Watson, Crutcher, and Dickel 1975). To determine the deuterium enhancement the $1_{01} \rightarrow 0_{00}$, $2_{02} \rightarrow 1_{01}$, and $2_{12} \rightarrow 1_{11}$ transitions of H_2CO were observed and upper limits were determined for these transitions of the isotope H_2^{13}CO. The observed deuterium enhancement strongly suggests that formaldehyde is produced primarily by gas phase ion-molecule reactions rather than on grains.

The observation of HDCO and H_2CO were made on the Bell Telephone Laboratories 7m antenna in Holmdel, NJ (further details can be found in LFLW). The detections of HDCO and H_2CO emission in the two dark clouds L134N and Helies Cloud 2 (CLD2) are shown in Figures 1 and 2, respectively (the predicted hyperfine structure is indicated in each figure). For H_2^{13}CO our search yielded only upper limits in these sources and the 1σ limits on emission are 0.05K for the $1_{01} \rightarrow 0_{00}$, $2_{01} \rightarrow 2_{01}$, and $2_{12} \rightarrow 1_{11}$ transitions in CLD2 and 0.07K for the $2_{02} \rightarrow 1_{01}$, transition in L134N.

From the shape of the formaldehyde emission spectra observed for L134N and CLD2 it is apparent that these lines are absorbed by low excitation foreground material in the same manner as HCO^+ (Langer et al. 1978) and HNC (Frerking et al. 1979). The mm emission lines of H_2CO in these clouds must be used cautiously to determine the degree of deuterium enhancement in formaldehyde. Furthermore, the cm absorption lines (Evans et al. 1975) and mm emission lines probably

B. H. Andrew (ed.), Interstellar Molecules, 439–443.

arise from different regions in these clouds and cannot be used together
to determine the physical conditions in the clouds.

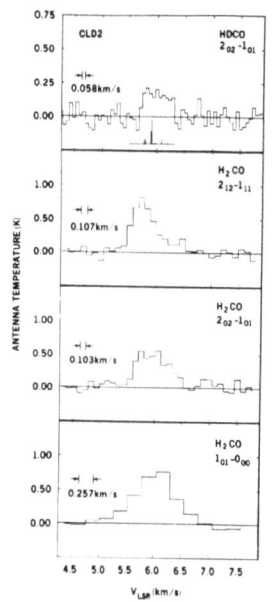
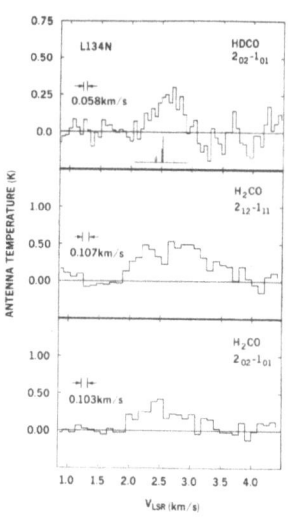

Figures 1 and 2:   The observed emission spectra from L134N
[RA(1950)=15^{h}51^{m}30^s, DEC(1950)=-02°43'30", VLSR=2.5 km/s] and
CLD2 [RA(1950)=4^{h}38^{m}21^s, DEC(1950)=-25°41'00", VLSR=5.9 km/s], for
HDCO and H$_2$CO.   The velocity resolution is indicated in each figure.

     To determine abundances and abundance limits from the measured
antenna temperatures an LVG radiative transfer model was used as a
guide (cf LFLW).   The observed HDCO mm emission from $2_{02} \rightarrow 1_{01}$ and the
limits set by Angerhofer et al. (1978) on the antenna temperature of the
5cm $1_{01} \rightarrow 1_{11}$ line are consistent with hydrogen densities, $n(H_2)$ < 3x10^5
cm^{-3}.   The densities in the regions of L134N and CLD2 where HDCO has
been detected have been estimated, from DCO$^+$ measurements, to be in the
range, $n(H_2)$ = 3x10^4 to 10^5 cm^{-3}, (Guelin et al. 1979).   Within this
range a comparison of HDCO and H$_2$CO antenna temperatures yields a ratio
HDCO/H$_2$CO $\sim$ 0.4.   As has been noted before for DCO$^+$/HCO$^+$ (Langer et al.
1978) self-absorption of H$_2$CO emission by low excitation foreground gas
can result in an HDCO/H$_2$CO intensity ratio which is enhanced with respect
to their abundance ratio.   To circumvent this difficulty we have
determined lower limits on this ratio from our detection limits on
H$_2$^{13}CO by adopting a ^{12}C/^{13}C ratio of 60 (Langer et al. 1979).   At a
density $n(H_2)$ = 3x10^4 cm^{-3} the ratio HDCO/H$_2$CO > 0.08; the corresponding
deuterium enhancement in HCO$^+$, as determined from DCO$^+$ and H^{13}CO$^+$
measurements, is DCO$^+$/HCO$^+$ $\simeq$ 0.04 (Guelin et al 1979).   As discussed

below a rate of $(HDCO/H_2CO) \sim 2(DCO^+/HCO^+)$ is consistent with ion-molecule mechanisms for production of formaldehyde.

In interstellar clouds several reactions have been suggested to inititate production of formaldehyde by gas phase chemistry. By one path or another (cf. Prasad and Huntress 1979) these lead to CH_3 or H_3CO^+. These then produce formaldehyde by the reactions,

$$CH_3 + O \rightarrow H_2CO + H \tag{1}$$

$$H_3CO^+ + e \rightarrow H_2CO + H \tag{2a}$$

in competition with, $\quad H_3CO^+ + e \rightarrow HCO + H_2 \tag{2b}$

$$\rightarrow CO + H + H_2 \tag{2c}$$

In the ion molecule scheme the HDCO is produced by the deuterated form of the same molecules which produce H_2CO, and deuterium enhancement occurs in the precursor ions either through $H_3^+ + HD \rightarrow H_2D^+ + H_2$ or $CH_3^+ + HD \rightarrow CH_2D^+ + H_2$ (cf Guelin et al 1978). Regardless of which reaction dominates production it can be shown that,

$$(HDCO/H_2CO)_{im} = 2 \frac{\Delta(H_2CO)}{\Delta(HDCO)} (DCO^+/HCO^+) \tag{3}$$

where Δ is the destruction rate for each species and where it is assumed that the recombination $H_2DCO^+ + e \rightarrow HDCO + H$ is twice as probable as that to H_2CO+D, and similarly for $CH_2D + O \rightarrow HDCO + H$ versus $\rightarrow H_2CO + D$. The destruction rate $\Delta = \Sigma a_i k_i n_i$, where for each reacting species i, n is the density, k the reaction rate constant (from Prasad and Huntress 1979) and a_i a renormalization term to account for destruction processes which lead back to HDCO or H_2CO.

Two possible mechanisms by which grains could play a role in forming HDCO are: 1) H_2CO is produced by the association of atoms on grain surfaces (Watson and Salpeter 1972) and HDCO is formed by ion-molecular reactions, notably by $H_2D^+ + H_2CO \rightarrow H_2DCO^+ + H_2$, (rate constant k_4) followed by dissociative recombinations; 2) HDCO comes directly from DCO^+ reacting at a negatively charged grain surface by $DCO^+ + H_2 + e \rightarrow HDCO + H$ (Watson et al. 1975). In the first case,

$$(HDCO/H_2CO)_{g+im} = \frac{2}{3} \frac{\alpha k_4 H_3^+}{\Delta(HDCO)} (DCO^+/HCO^+) \tag{4}$$

where $\alpha = k_{2a}/k_{2,tot}$, and in the second case,

$$(HDCO/H_2CO)_{g+DCO^+} = \frac{2}{3} \frac{\Delta(H_2CO)}{\Delta(HDCO)} (DCO^+/HCO^+) \tag{5}$$

assuming the DCO^+ reaction on the grain leads to HDCO twice as often as H_2CO.

The major uncertainty in evaluating the deuterium enhancement is the relative efficiency with which H_3CO^+ and H_2DCO^+ recombine to produce H_2CO and HDCO rather than other products (Eq. 2b and 2c). While laboratory data for the total recombination rate constant for deuterated and non-deuterated ions are similar (Mul and McGowan 1979), there are no measurements of branching ratios. Theoretical calculations by Herbst (1978) suggest that it is unlikely that only one hydrogen will be ejected. We adopt an equal branching ratio for all likely products (Prasad and Huntress 1979) corresponding to $\alpha = 1/3$. For this choice it can be shown, in general, that for ion-molecule production $HDCO/H_2CO \simeq 2\ (DCO^+/HCO^+)$. For production via DCO^+ on grains the ratio is $\sim 2/3\ (DCO^+/HCO^+)$ and for the combined grain plus ion-molecule scheme (the first grain mechanism) the ratio is $< 1/3\ (DCO^+/HCO^+)$ and, when estimates of H_3^+ and the other ion abundances are included, it is closer to $1/10\ (DCO^+/HCO^+)$.

Our observations of HDCO indicate that $(HDCO/H_2CO) \sim 2(DCO^+/HCO^+)$ which strongly suggests that formaldehyde is produced primarily by ion-molecule reactions. Laboratory measurements of the branching ratios for dissociative recombination of H_3CO^+ and H_2DCO^+ are needed to be certain that grains do not play some role.

REFERENCES:

Angerhofer, P., Rossano, G., Vestrand, W., 1978, Astron. J. 83, p. 1417.
Evans, N. J. II, Zuckerman, B., Morris, G., Sato, T. 1975, Ap. J. 196, p. 433.
Frerking, M., Langer, W., Wilson, R., 1979, Ap. J. Letters, 232, L65.
Guelin, M., Langer, W., Wilson, R. 1979, in preparation.
Herbst, E. 1978, Ap. J., 222, p. 508.
Huntress, W. T. Jr., 1977, Ap. J. Suppl. 33, p. 495.
Langer, W., Wilson, R., Henry, P., Guelin, M. 1978, Ap. J. Letters, 225, p. L139.
Langer, W., Goldsmith, P., Carlson, E., Wilson, R., 1980, Ap. J. Letters, in press.
Langer, W., Frerking, M., Linke, R., Wilson, R., 1979, Ap. J. Letters 232, p. L169.
Prasad, S., Huntress, W.T. Jr., 1979, Ap. J., in press.
Mul, P., McGowan, J., 1979, Ap. J. Letters 227, p. L157.
Watson, W., Salpeter, E., 1972, Ap. J. 174, p. 321 and 175, p. 659.
Watson, W., Crutcher, R., Dickel, J., 1975, Ap. J. 201, p. 102.

DISCUSSION FOLLOWING LANGER

Allamandola: You argue that HCO^+/DCO^+ is more consistent with ion-molecule reactions in the gas-phase than with interactions with grains. Would you expect similar behavior for HCO/DCO neutral? Have any such observations been made? Since this molecule is easily made by neutral species combining in the grain, this may be an important test.

Langer: Observations of DCO and HCO could conceivably discriminate between the different production mechanisms.

Winnewisser: I want to congratulate you on finding interstellar HDCO. The transition you have used is the one we measured several years back when I was at the National Research Council. However the frequency accuracy was at best ±150 kHz, so the velocity determination may be in error. How closely do the velocities agree with those of other molecules in TMC1?

Langer: The HDCO peak velocity agrees with that of $H^{13}CO^+$ to within 0.1 km/s, and suggests an uncertainty in frequency of ∿80 kHz, of which half is due to the uncertainty in the determination of the $H^{13}CO^+$ frequency.

W. Watson: If the laboratory rate for $CH_3^+ + H_2 \rightarrow CH_5^+ + h\nu$ is as rapid as suggested by the laboratory people, the channel $CH_3^+ + HD \rightleftarrows CH_2D^+ + H_2$ for fractionation would appear to be much less important than previously thought. Would this affect your analysis?

Langer: It conceivably could if the rate constant were $>> 10^{-13} cm^3 s^{-1}$ and if the CH_2D^+ were formed only by the reaction $CH_3 + HD$. If deuterium fractionation proceeds primarily through $H_3 + HD$, however, then the analysis will not be affected.

Fukui: Watson et al. (1975) obtained an upper limit for HDCO which is significantly smaller than your detection. Could you explain the discrepancy?

Langer: Their search was for the 5 cm line of HDCO, and the interpretation of column density is very sensitive to collision rates and temperature. Recent excitation calculations show that their limits are consistent with the mm emission.

THEORETICAL CONSIDERATIONS OF SHOCK WAVE BEHAVIOR

David Hollenbach
NASA-Ames Research Center, Moffett Field, CA 94035

Interstellar shock waves have a significant influence on the
structure and dynamics of interstellar matter and probably trigger
star formation in suitably dense regions. In this paper the overall
structure of regions near shock waves is reviewed. In addition we
discuss the main observational effects of shocks on interstellar
molecules, including: (i) acceleration to velocities 1 km s^{-1} < v_s
< 100 km s^{-1} relative to the ambient gas, (ii) excitation of infrared
lines in the heated postshock gas, and (iii) production of high abundances of certain molecular species such as H, OH, H_2O, CH^+, OCS, and
SiO through high temperature chemical reactions in the T $\gtrsim$ 1000 K
postshock gas. The molecular region around the BN infrared source in
Orion and the high velocity molecules in IC443 are discussed as
possible examples of shocked molecular gas.

1. INTRODUCTION

    Shock waves in the interstellar medium may induce star formation
(e.g., Shu, Milione and Roberts 1973, Elmegreen and Lada 1977, Herbst
and Assousa 1977, and Loren and Vrba 1979) and may play a significant
role in determining the density, temperature and velocity structure of
the interstellar medium (e.g., Heiles 1979, McKee and Ostriker 1977,
Norman and Silk 1979). Besides their presence in spiral density waves
and cloud-cloud collisions, interstellar shocks frequently accompany
the birth and death of stars, and they may persist throughout the life
of massive stars: stellar winds, expanding HII regions, and supernova
remnants are all capable of driving shock waves.

    The purpose of this paper is to review the overall structure near
interstellar shock waves and to focus on the interaction of shocks
with interstellar molecules. Shock waves change the relative chemical
abundances of molecules (primarily through enhancements in both dissociation and formation rates at the elevated postshock temperatures),
collisionally excite rotational and vibrational transitions of the
molecules in the warm postshock gas, and accelerate molecules to

445

B. H. Andrew (ed.), Interstellar Molecules, 445–453.
Copyright © 1980 by the IAU.

supersonic velocities. The high excitation and the velocity shift of the postshock molecules not only serve as shock signatures but reduce the absorption in the lines caused by any surrounding cold molecular gas, making detection much simpler.

2. LARGE-SCALE STRUCTURE OF SHOCKED SHELLS

The time dependent large-scale structure of the interstellar medium interacting with supernova remnants, stellar winds, or HII regions can be analytically formulated if the approximation of an initial constant hydrogen nucleus density n_o is assumed. A shock front (SF) is established behind which, once it has slowed to $v_s < v_c$, the shocked ambient gas radiatively cools and forms a neutral shell, bounded on the inside by an ionization front (IF). (See Figure 1.) The hydrogen nucleus column density N in the shell, the radius of the shell R, and the timescale t are functions of the velocity of the shell; solutions have been obtained for HII regions (Spitzer 1968), supernova remnants (Woltjer 1973) and early-type stellar winds (Castor, McCray and Weaver 1975) and are summarized in Figure 1 along with the critical

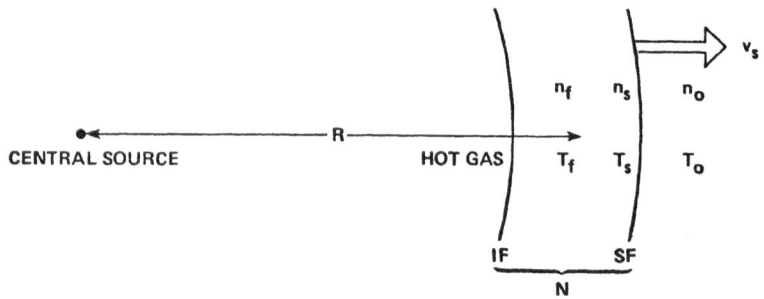

	HII REGIONS	WINDS	SN
v_{cr} (km s^{-1})	14	$76\, n_3^{1/5}\, L_W^{3/5}\, L_{48}^{-2/5}$	$516\, n_3^{2/17}\, E_{51}^{1/17}$
t (yr)	$3.1 \times 10^4\, v_{s6}^{-7/3}\, n_3^{-2/3}\, L_{48}^{1/3}$	$9.7 \times 10^4\, v_{s6}^{-5/2}\, n_3^{-1/2}\, L_W^{1/2}$	$1.1 \times 10^5\, v_{s6}^{-4/3}\, n_3^{-19/51}\, E_{51}^{16/51}$
R (pc)	$0.36\, v_{s6}^{-4/3}\, n_3^{-2/3}\, L_{48}^{1/3}$	$1.6\, v_{s6}^{-3/2}\, n_3^{-1/2}\, L_W^{1/2}$	$4.5\, v_{s6}^{-1/3}\, n_3^{-19/51}\, E_{51}^{16/51}$
N (cm^{-2})	$3.6 \times 10^{20}\, v_{s6}^{-4/3}\, n_3^{1/3}\, L_{48}^{1/3}$	$1.6 \times 10^{21}\, v_{s6}^{-3/2}\, n_3^{1/2}\, L_W^{1/2}$	$4.5 \times 10^{21}\, v_{s6}^{-1/3}\, n_3^{32/51}\, E_{51}^{16/51}$

$n_3 = n_o/10^3\ cm^{-3}$ $L_{48} = L_i/10^{48}\ s^{-1}$

$v_{s6} = v_s/10\ km\ s^{-1}$ L_W = WIND LUMINOSITY IN $10^{36}\ erg\ s^{-1}$

E_{51} = INJECTION ENERGY IN $10^{51}\ erg$

Figure 1. Large-scale structure near shock waves

velocity v_c. We note that Hill and Hollenbach (1978) have shown that around typical expanding HII regions, the FUV photon flux from the central stars dissociates molecules ahead of the shock front until it has slowed to $v_s \lesssim 3$ km s^{-1}.

3.   THE EFFECT OF SHOCK WAVES ON INTERSTELLAR MOLECULES

3.1 Dissociative and non-dissociative shocks

Shocks can be divided into two regimes for the purposes of study-ing the effect of shocks on molecules: (i) non-dissociative shocks ($v_s \ll 30$ km s^{-1}) which mainly heat the ambient molecules and create new molecular abundance ratios by the effects of increased postshock density and temperature on the formation rates of key molecules, and (ii) dissociative shocks ($v_s \gg 30$ km s^{-1}) which fully dissociate any pre-existing molecules, but which reform molecules in the cooling T < 5000 K postshock gas. The exact velocity v_d at which molecules are dissociated has been thrown into question recently by Dalgarno and Roberge (1979), who show the dissociation rate coefficients for mole-cules to be quite density sensitive. Until more quantitative calcu-lations are performed, v_d can be bracketed by its high density ($n_o \gtrsim 10^{10}$ cm^{-3}) value of approximately 25 km s^{-1} (Kwan 1977) and a low density ($n_o \to 0$ cm^{-3}) value of probably less than 50 km s^{-1}, since the postshock collisional ionization will produce a sufficient fractional abundance of electrons, $x(e) \sim 0.2$ (Shutt and McKee 1979), to colli-sionally dissociate H_2.

3.2 Overview of postshock structure

Dissociative and non-dissociative shocks have certain overall similarities. Shocked gas is initially heated and later cooled and compressed as it moves downstream from the shock front. The immediate postshock kinetic temperature T_s is

$$T_s = 3.2 \times 10^5 \; x_{ts}^{-1} \; v_{s7}^2 \; K \, , \tag{1}$$

where $v_{s7} = v_s/100$ km s^{-1} and x_{ts} is the number of gas particles per proton; $x_{ts} = 2.3$ for ionizing shocks with $v_{s7} \gtrsim 1$, $x_{ts} = 0.6$ for non-dissociative shocks into molecular ambient gas. Downstream the gas cools and compresses ($nT \sim$ constant) until magnetic pressure dominates and compression ceases at density n_f.

$$n_f = 77 \; v_{s7} \; B_{\mu G}^{-1} \; n_o^{1/2} \; cm^{-3}, \tag{2}$$

where $B_{\mu G}$ is the ambient magnetic field in μGauss and n_o is in cm^{-3}. Cooling continues to equilibrium $T_f \sim 10 - 100$ K.

However, despite these overall similarities, the detailed post-shock structure behind dissociative shocks is quite different from non-dissociative shocks. The high postshock temperatures ($T \sim 10^5$ K)

behind fast shocks leads to high levels of ionization and UV cooling
to T $\sim 10^4$ K in a column density $N_{cool} \sim 2.1 \times 10^{17} v_{s7}^{4.2}$ cm^{-2} (Shull
and McKee 1979, Hollenbach and McKee 1979a). Initial calculations by
Hollenbach and McKee (1979b) suggest that for $n_o \lesssim 10^3$ cm^{-3} the gas
cools from 10^4 K to 100 K primarily by the CII(156 μ) and OI(63 μ)
lines in a column density $N_h \sim 10^{19}$ cm^{-2}; the column density of warm
(T > 100 K) H_2 is of order $N_2 \sim 10^{16}$ cm^{-2}. Column densities of warm
CO, H_2O, and OH are less than 10^{12} cm^{-2}. Complete reformation of the
molecules occurs in a column density $N_m \sim 10^{21}$ cm^{-2} in the cool (T
< 100 K) gas. UV absorption or radio measurements would generally
detect this cold postshock gas; infrared emission measurements could
not detect the warm postshock gas at present sensitivity. On the
other hand, when $n_o \gtrsim 10^4$ cm^{-3}, the gas behind fast shocks cools to
100 K in a column density $N_h \sim 10^{21}$ to 10^{23} cm^{-2}, the greater column
density being in part due to reduced cooling rates caused by collisional
suppression of atomic and molecular lines and in part due to significant heating by H_2 formation. Molecular hydrogen formation on grains
is less efficient at higher gas and grain temperatures so that the
column density for complete molecular reformation is also $N_m \sim 10^{21}$
to 10^{23} cm^{-3}. The postshock structure of fast, dissociative shocks is
summarized in Figure 2.

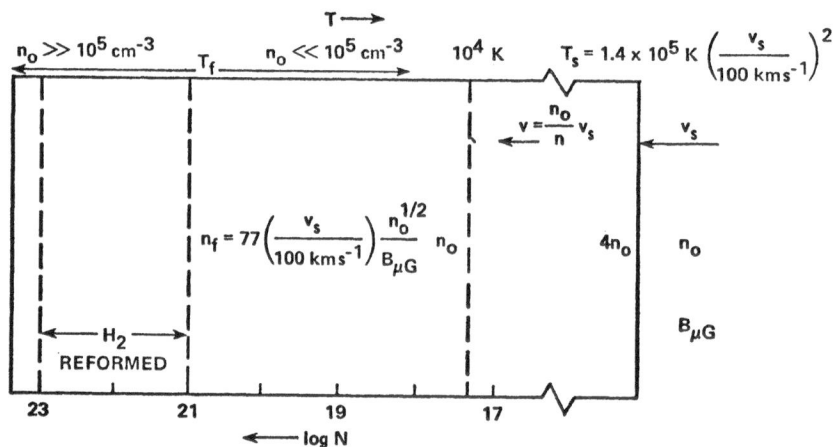

Figure 2. Dissociative shocks ($v_s \gg 30$ km s^{-1})

Slow, non-dissociative shocks initially heat the ambient gas to
temperatures $T_s \lesssim 10^{4.5}$ K. For low ambient densities ($n_o \lesssim 100$ cm^{-3},
mainly atomic) the gas rapidly cools to T < 100 K by H_2 rotational-
vibrational transitions and the 63 μ OI line in a column density
$N_h \sim 10^{19}$ cm^{-2}. Elitzur and Watson (1978) point to the only observational consequence of the hot postshock gas: CH^+ may be produced with
sufficient column density $N(CH^+) \sim 10^{13}$ cm^{-2} to explain the relatively
high amounts of CH^+ seen in absorption toward early-type stars.
Hollenbach, Chu, and McCray (1976), show that H_2 may be formed in the
dense, cool (T < 100 K), wind-driven postshock gas with rotational
populations like those observed by Copernicus toward early-type stars.

For high ambient densities ($n_0 \gtrsim 1000$ cm^{-3}, mainly molecular) the $T \gtrsim 1000$ K postshock molecular gas temperatures initiate the chemical reactions to be discussed in the next section; cooling proceeds via molecular rotational and vibrational transitions. Several authors (Hollenbach and Shull 1977, Kwan 1977, London, McCray and Chu 1977) explain the observed H_2 2 μ 1-0 vibrational transitions in Orion as arising behind such shocked gas. In their calculations, which assumed that H_2 dominated the cooling to $T \sim 1000$ K, the cooling column density N_{1000} to 1000 K was of order

$$N_{1000} \sim 2 \times 10^{13} \, n_0 \text{ cm}^{-2} \tag{3}$$

for $n_0 \gtrsim 10^5$ cm^{-3}. To explain the H_2 2 μ intensity from Orion, $n_0 \sim 10^7$ cm^{-3} is required.

Table 1

Important Chemical Reactions in Hot Neutral Gas

Reaction	Rate Coefficient (cm^{-3}s^{-1})	(1)
$H_2 + 0 \rightleftharpoons OH + H$	$3 \times 10^{-11} \, T_3 \, \exp(-4.48/T_3)$	(2)
	$1.4 \times 10^{-11} \, T_3 \, \exp(-3.5/T_3)$	(2)
$Si + OH \rightleftharpoons SiO + H$	3×10^{-11}	(3)
	$3 \times 10^{-11} \, \exp(-39.5/T_3)$	(3)
$H_2 + OH \rightleftharpoons H_2O + H$	$3.6 \times 10^{-11} \, \exp(-2.59/T_3)$	(2)
	$1.5 \times 10^{-10} \, \exp(-10.25/T_3)$	(2)
$H_2 + C \rightleftharpoons CH + H$	$2.1 \times 10^{-9} \, T_3^{1/2} \, \exp(-14.10/T_3)$	(2)
	$2.1 \times 10^{-9} \, T_3^{1/2} \, \exp(-2.2/T_3)$	(2)
$H_2 + C^+ \rightleftharpoons CH^+ + H$	$4.2 \times 10^{-11} \, T_3^{5/4} \, \exp(-4.64/T_3)$	(4)
	$4.2 \times 10^{-11} \, T_3^{5/4}$	(4)
$H_2 + S \rightleftharpoons HS + H$	$3 \times 10^{-11} \, \exp(-10.0/T_3)$	(3)
	3×10^{-11}	(3)
$HS + CO \rightleftharpoons OCS + H$	$3 \times 10^{-11} \, \exp(-5.36/T_3)$	(3)
	3×10^{-11}	(3)

(1) $T_3 = T/1000$ K, lower rate coefficient for reverse reaction; (2) Iglesias and Silk (1978); (3) Estimated from Lada, Oppenheimer and Hartquist (1978); (4) Solomon and Klemperer (1972).

3.3 Hot chemistry

High temperature ($T \sim 10^3$ to 10^4 K) neutral gas exists behind both slow and fast shocks. The major high temperature gas phase reactions associated with such regions are listed in Table 1, along with their rate coefficients. These reactions, initiated by H_2, proceed rapidly because the rate coefficients are high and because the reactants are generally plentiful. Hence, relatively large quantities of OH, H_2O, SiO, HS, OCS, CH, and CH^+ are produced in the hot postshock gas if H_2 is present. Behind fast dissociative shocks, molecular hydrogen formation must proceed before this "hot chemistry" can commence. Hollenbach and McKee (1979a) have shown that production of H_2 occurs primarily on grain surfaces as long as the grain temperatures are less than about 100 K. Grains are warmed by the radiation (energy flux $\propto n_o v_s^3$) from the cooling postshock gas and the grain temperature condition for molecular hydrogen formation reduces to

$$n_o v_{s7}^3 \stackrel{<}{\sim} 3\times10^3 \text{ cm}^{-3} \quad N \stackrel{<}{\sim} 10^{20} \text{ cm}^{-2}$$

$$(4)$$

$$n_o v_{s7}^3 \stackrel{<}{\sim} 2\times10^6 \text{ cm}^{-3} \quad N \stackrel{>}{\sim} 10^{21} \text{ cm}^{-2}$$

for 0.1 µ silicate or graphite grains. The production of H_2 by the H^- process also produces a small fraction ($x(H_2) \stackrel{<}{\sim} 10^{-3}$) of H_2 in the hot recombining postshock gas (Hollenbach and McKee 1979b).

3.4 Molecular cooling behind shocks

Behind both slow and fast shocks the cooling in the hot neutral regions will be dominated by molecular vibrational-rotational transitions if the gas is substantially molecular. As discussed in Section 3.2, early work focussed on H_2 cooling behind slow shocks, although Kwan (1977) included CO cooling as well. However, Hollenbach and McKee (1979a) have shown that OH and H_2O dominate the cooling when the postshock density exceeds about 10 cm^{-3} in warm molecular gas with solar abundances. This result is due, first of all, to the high fractional abundances ($x(OH) \sim x(H_2O) \sim 10^{-4}$) of OH and H_2O in hot molecular gas. Secondly, densities exceeding 10^5 cm^{-3} collisionally quench CO and H_2 relative to OH and H_2O.

4. APPLICATIONS

Recently, the vibrationally and rotationally excited H_2 in Orion has been observed to have high velocity dispersions (FWHM $\sim$ 50 km s^{-1}) (e.g., Nadeau and Geballe 1979, Beck, Geballe and Lacy 1979). The high preshock densities ($n_o \sim 10^7$ cm^{-3}) probably rule out non-dissociative ($v_s \sim 25$-50 km s^{-1}) shocks. Furthermore, high velocity non-dissociative shocks will produce more 2-1 S(1) H_2 line intensity than is observed (Shull and Hollenbach 1978). Therefore, we are led to postulate: (i), slow ($v_s \sim 10$ km s^{-1}) shocks traversing clumps of gas with velocity dispersions of order 50 km s^{-1}, or (ii), dissociative shocks with $v_s \sim 50$ km s^{-1} with H_2 being reformed at high (T ~ 2000 K) temperatures. The cooling by OH will be important in the second case; preliminary calculations indicate probable fluxes of order 10^{-17} watts cm^{-2} from far infrared OH rotational transitions. These fluxes are within present detection capabilities (Hollenbach and McKee 1979b).

The observation of accelerated HI, OH, and CO (DeNoyer 1978, 1979a, 1979b, and Giovanelli and Haynes 1979) as well as vibrationally excited H_2 (Treffers 1979) in IC443 has led to a picture of shock waves of velocity $v_s \sim 10$ to 100 km s^{-1} incident upon gas of density $10^2 - 10^3$ cm^{-3}. The H_2 probes the hot postshock gas in the non-dissociative shocks while the HI, CO and OH radio observations measure the cold, accelerated, postshock material. Calculations by Hollenbach and McKee (1979b) suggest that the high OH abundance in the hot postshock gas rapidly declines as the gas cools. Therefore, the high observed abundance of cold OH may be an indication of unusual conditions in the cold gas, e.g., a high ionization rate by soft X-rays from the supernova interior.

Finally, as a future application, we note that the simultaneous observation of infrared lines from, for example, H_2, CI, CO, CII, OI, OH, and H_2O in the hot molecular gas near KL and BN will provide, for the first time, a measure of the H:C:O ratio in a dense molecular cloud.

The author acknowledges partial support by NSF Grant AST 77 23069, and by the National Research Council in the form of a Senior Associate-ship.

REFERENCES

Beck, S., Geballe, T.R., and Lacy, J.: 1979, preprint.
Castor, J., McCray, R., and Weaver, R.: 1975, Ap. J. (Letters) 200, pp. L107-L110.
Dalgarno, A. and Roberge, W.G.: 1979, Ap. J. (Letters) 233, pp. L25-L27.
DeNoyer, L.K.: 1978, M.N.R.A.S. 183, pp. 187-194.
DeNoyer, L.K.: 1979a, Ap. J. (Letters) 228, pp. L41-L44.
DeNoyer, L.K.: 1979b, Ap. J. (Letters) 232, pp. L165-L168.
Elitzur, M. and Watson, W.D.: 1978, Ap. J. (Letters) 222, pp. L141-L144.
Elmegreen, B.G. and Lada, C.J.: 1977, Ap. J. 214, pp. 725-741.
Giovanelli, R. and Haynes, M.D.: 1979, Ap. J. 230, pp. 404-414.
Heiles, C.: 1979, Ap. J. 229, pp. 533-544.
Herbst, W. and Assousa, G.E.: 1977, Ap. J. 217, pp. 473-487.
Hill, J.K. and Hollenbach, D.: 1978, Ap. J. 225, pp. 390-404.
Hollenbach, D., Chu, S-I, and McCray, R.: 1976, Ap. J. 208, pp. 458-467.
Hollenbach, D. and McKee, C.F.: 1979a, Ap. J. Suppl., 41, pp. 555-592.
Hollenbach, D. and McKee, C.F.: 1979b, work in progress.
Hollenbach, D. and Shull, J.M.: 1977, Ap. J. 216, pp. 419-426.
Iglesias, E.R. and Silk, J.: 1978, Ap. J. 226, pp. 851-857.
Kwan, J.: 1977, Ap. J. 216, pp. 713-723.
Lada, C.J., Oppenheimer, M. and Hartquist, T.W.: 1978, Ap. J. (Letters) 226, pp. L153-L156.

London, R., McCray, R. and Chu, S.I.: 1977, Ap. J. 217, pp. 442-447.
Loren, R.B. and Vrba, F.J.: 1979, Sky and Telescope 57, pp. 521-526.
McKee, C.F. and Ostriker, J.P.: 1977, Ap. J. 218, pp. 148-169.
Nadeau, D. and Geballe, T.R.: 1979, Ap. J. (Letters) 230, pp. L169-
    L174.
Norman, C. and Silk, J.: 1979, preprint.
Shu, F.H., Milione, V. and Roberts, W.W.: 1977, Ap. J. 183, pp. 819-842.
Shull, J.M. and McKee, C.F.: 1979, Ap. J. 227, pp. 131-149.
Solomon, P.M. and Klemperer, W.: 1972, Ap. J. 178, pp. 389-422.
Spitzer, L.: 1968, Diffuse Matter in Space, Interscience, New York,
    pp. 188-193.
Treffers, R.R.: 1979, Ap. J. (Letters) 233, pp. L17-L19.
Woltjer, L.: 1973, Ann. Rev. of Astron. and Astrophys. 10, pp. 143-145.

DISCUSSION FOLLOWING HOLLENBACH

Field: The rate of reaction of H_2(v=1) with OH has recently been measured to be about 300 times faster than with H_2(v=0). Behind the shock the temperature is sufficiently high that a significant proportion of H_2 (v=1) may be available to react rapidly with OH. This may well influence your chemical modelling.

Hollenbach: It would be very interesting to know whether the reaction of H_2 with O behaves similarly. Both these reactions of vibrationally excited H_2 will affect the chemistry for gas with $n \gtrsim 10^6 cm^{-3}$ and $T \gtrsim 1000$ K.

Slysh: What time is needed for OH to cool from a high "shocked" temperature to the cold, ~ 100K, post-shock temperature?

Hollenbach: The time is approximately 300 years for a 10 km s^{-1} shock incident upon a $10^3 cm^{-3}$ cloud. The time is inversely proportional to the shock velocity and cloud density, for densities up to about $10^4 cm^{-3}$.

Carruthers: Has the (2-1) vibrational H_2 transition been observed toward IC 443? If so, what temperature has been inferred from a comparison with the (1-0) transition?

Hollenbach: It has not yet been observed, but is currently being looked for. If the shock is a 40 km s^{-1} non-dissociative one, we expect the 2-1S(1) line to be almost as intense as the 1-0S(1) line.

Chaffee: Could the A-X OH lines be observed in absorption against a continuum source through the shocks you describe? That is, does the increase in OH production by 10^4 more than offset the increased excitation?

Hollenbach: The OH will be similar to H_2O and CH^+; column densities of $10^{12} cm^{-2}$ to $10^{13} cm^{-2}$, most of which is in the ground rotational state because of rapid radiative decays, will be present for shocks incident upon low (n < $10^3 cm^{-3}$) density gas.

Zuckerman: Rickard, Palmer and myself have observed OH in Orion with the Bonn telescope (to be submitted to Ap. J.). The beam was 2.6 arcmin, i.e. not much larger than the high velocity source in Orion. We saw no evidence of wide OH lines even at low levels of intensity.

Elitzur: The OH/H_2O ratio behind a shock varies with the H_2/H ratio.

The oxygen will be channelled into water and not into OH if the shock runs into dense clouds where the hydrogen is essentially molecular.

Hollenbach: Yes, an absence of hot OH in the KL-BN region would imply that the gas has a high H_2/H ratio. Then if shocks were present, they would be slow non-dissociative shocks as opposed to fast dissociative shocks followed by re-association of H_2 in the hot post-shock gas.

Wootten: I offer observational support of your suggestion of enhanced ionization in cool material between the interior of a supernova remnant and the shock in the surrounding molecular cloud. In the cloud near the W28 remnant there is a small region of very broad HCO^+ emission. At the velocity of the most blueshifted components of HCO^+ the only other molecular emission detected is OH. My study of the HCO^+ indicated that the ionization, in the region in which the blueshifted emission originates, must be enhanced compared to unshocked parts of the cloud. Perhaps, as you suggest for IC443, oxygen charge exchange in this region produces the OH responsible for emission at the most blueshifted velocity.

Drapatz: To compare with the CO surveys one would like a survey of cold dense matter using the S(0) or S(A) rotational transition of H_2 at 28 μm and 17 μm respectively. While pure thermal emission in these lines is undetectably small for clouds of T $\sim$ 10K, shock waves due to cloud-cloud collisions, even for small shock velocities, seem to produce detectable effects ($N(H_2) \gtrsim 10^{21} cm^{-2}$, T $\gtrsim$ 100K) Has such a model already been worked out?

Hollenbach: Not to my knowledge. To produce $N(H_2) \gtrsim 10^{21} cm^{-2}$ at T > 100K would require a shock of speed $V_s \gtrsim$ 3 km s^{-1} penetrating a rather dense region. The exact density would have to be worked out, but I suspect it may be as high as $10^6 cm^{-3}$.

W. Watson: Why is it clear that the IR emission from H_2 is caused by shocks rather than pumping by ultraviolet radiation? Line widths cannot be used since you are now attributing them to multiple components of the gas.

Hollenbach: It is true that far-UV radiation ($\lambda \sim 100$ Å) can excite the 2μ H_2 lines by electronic excitation followed by cascade through the vibrational levels. Two basic questions must be asked about each H_2 source to determine whether this mechanism is important: (i) is the far-UV intensity sufficient and (ii) if more than one line is observed, does the far-UV pump produce the relative strengths of the lines? There is also the problem that, as the far-UV intensity increases and the population of vibrationally-excited H_2 increases, the H_2 can be ionized and destroyed by far-UV photons. Shull has examined this problem and obtains an upper limit to the possible 2μ intensity. All of these agruments show that the H_2 in Orion cannot be pumped by far-UV radiation.

OBSERVATIONS OF SHOCK WAVES IN INTERSTELLAR CLOUDS

Steven Beckwith
Department of Astronomy, Cornell University, Ithaca, N.Y.,
U.S.A.

Abstract: Some techniques for observing shock waves in interstellar clouds are discussed. It is concluded that recent measurements of molecular hydrogen emission provide the best currently available technique for studying shocks. The results of measurements toward the Orion nebula are discussed, and a discussion and summary of the currently known H_2 sources is given.

Shock waves have been recognized as important components of the interstellar medium for many years, a recognition which extends to their presence in molecular clouds (see, e.g., Field et al. 1968). Supersonic gas flows driven by cloud collisions, stellar winds, supernova explosions result in the turbulent heating of these clouds through shock waves. Several theories of sequential star formation rely heavily on shock waves to trigger gravitational collapse (Elmegreen and Lada 1977, Gerola and Seiden 1978, and others), and shocks are considered crucial to the development of spiral structure in galaxies.

Unambiguous observational evidence for shocks in molecular clouds has emerged rather slowly, however, primarily because of ambiguity in the available measurements and limited instrumental sensitivity at infrared wavelengths where most of the radiation is expected.

For some strong shocks such as those proposed to explain Herbig-Haro objects (Schwartz 1975, 1978), easily observable optical lines are excited, but, unfortunately, these lines are not unique to interstellar shock waves. Spitzer and Cochran (1973) found rather high rotational temperatures ~ 1000 K for molecular hydrogen in their ultraviolet absorption line measurements, which Spitzer and Morton (1976) attribute to line of sight gas heated by interstellar shock waves. Their observational technique prohibits mapping, so the extent and power of the shocks has not yet been determined. These observations suffer the additional problem of heavily blended line profiles for the

455

B. H. Andrew (ed.), Interstellar Molecules, 455–463.

different shock components hampering line width analysis. Similar
studies of intermediate velocity gas (50 to 100 km s^{-1}) have been carried out which indicate shocked gas in low density regions (see Shull
1977; Cowie et al. 1979, and references therein).

Several workers have taken a different approach by gathering
indirect evidence for shock waves with observations of abrupt velocity
changes in CO (Lada et al. 1978, Elmegreen and Moran 1979), NH$_3$ (Ho and
Barrett 1978), and neutral hydrogen and OH line profiles (De Noyer
1978, 1979). Most of these observations are ambiguous as regards
shocks, since line of sight velocity discontinuities only suggest the
existence of shock waves. De Noyer's work is the most conclusive of
these studies. Her velocity data demonstrate the existence of shocked
gas which has been confirmed by the detection of molecular hydrogen
emission (Treffers 1979).

The discovery of emission from vibrationally excited molecular
hydrogen by Gautier et al. (1976) was the first detection of lines
which are probably unique to interstellar shocks. Calculations by
Hollenbach and Shull (1977); Kwan (1977); and London, McCray, and Chu
(1977) show these lines can provide the dominant radiative cooling for
interstellar shocks at a variety of densities and shock velocities.
The relevant vibrational states are difficult to excite in great quantities by other means, so the H$_2$ lines are a signature of shock-heated
molecules. Aannestad's (1973) calculations indicate that lines at 6.9
and 9.7μm from rotationally excited molecular hydrogen and a fine
structure line at 63μm from neutral oxygen are equally important for
shocks under different density conditions. The 63μm line has recently
been detected in Orion by Melnick, Gull, and Harwit (1979) and by
Storey, Watson, and Townes (1979), and the rotational lines of H$_2$ are
being actively pursued by at least two groups. Kwan (1977) notes the
probably importance of CO transitions between highly excited rotational states (J $\sim$ 22) to radiative cooling of weak shocks, but instrumental sensitivities are orders of magnitude away from those required
to detect these lines. The near infrared observational techniques
have, however, progressed to a stage where measurements of shock waves
in molecular clouds can be made routinely.

Measurements of molecular hydrogen emission provide the most sensitive currently available technique for directly observing shocked gas
in molecular clouds. In the following two sections we review the
results of the recent observations of this emission and place them in
perspective as they relate to interstellar shocks. The Orion observations are by far the most extensive, so Orion is treated separately in
the first section, and observations of other H$_2$ emission sources are
discussed in the second section. Ironically, the most recent data on
Orion has shed doubt upon the shock wave interpretation where the sup-
portive arguments have previously been the strongest. Nonetheless, the
theoretical considerations still argue strongly that shock heating is
the most plausible means of exciting large quantities of hydrogen

molecules, so, with an eye to the future, we will discuss the H_2 sources in terms of shock excitation.

1. MOLECULAR HYDROGEN EMISSION FROM THE ORION MOLECULAR CLOUD

The original measurements of Gautier et al. (1976) showed level populations for the v = 1; J = 1,2,3, and 4 levels consistent with thermal equilibrium at a temperature between 1000 and 3000 K. More refined measurements which include the v = 2, J = 3 level and corrections for line of sight reddening indicate level populations in thermal equilibrium at a temperature of 1900±300 K. (Beckwith et al. 1978b; Beckwith, Persson, and Neugebauer 1979). Typical H_2 column densities derived from the observations are $\sim 3\times10^{20}$ cm^{-2} assuming complete thermal equilibrium at 2000 K.

Maps by Grasdalen and Joyce (1976) and Beckwith et al. (1978b) show the molecular hydrogen emission comes from a spatially extended region, roughly 1 arc minute (0.15 pc) in diameter. The reddening measurements of Beckwith, Persson, and Neugebauer (1979) and Simon et al. (1979) indicate that the emission region is within the cloud at depths comparable to those of BN and KL. If volume densities inferred from observations of other molecules of $\sim 10^6$ cm^{-3} apply to the H_2 emission region, the observed H_2 column densities of $\sim 3\times10^{20}$ cm^{-2} imply a line of sight extent of 10^{-4} pc for the hot molecular hydrogen. Hollenbach and Shull (1977); Kwan (1977); and London, McCray, and Chu (1977) show the observations result naturally if a shock wave with a velocity between 10 and 25 km s^{-1} moves into the ambient cloud where the density is greater than $\sim 10^7$ cm^{-3}. The larger velocity limit is necessary so that shock does not dissociate all the hydrogen molecules upon passage (Kwan 1977).

Nadeau and Geballe (1979) obtained profiles of the v = 1 → 0 S(1) line which indicate velocities substantially greater than 25 km s^{-1}. Line widths as large as 60 km s^{-1} and blue-shifted wings moving at 90 km s^{-1} relative to the ambient molecular cloud are seen in their spectra. These observations invalidate the interpretation that a single shock wave with a velocity less than 25 km s^{-1} excites the H_2. Indeed, the profiles strongly suggest the molecular hydrogen is excited in a region with a line of sight extent comparable to its projected size undergoing differential expansion, not in a very thin region immediately behind a shock discontinuity. The shock calculations referenced above do not readily account for these observations.

The evidence nonetheless indicates some kind of shock excitation is responsible for the molecular hydrogen emission. The H_2 appears to be thermalized at 2000 K, whereas CO observations show the vast majority of the gas in this region is at less than 100 K (Zuckerman, Kuiper, and Rodriguez-Kuiper (1976; Kwan and Scoville 1976; Phillips et al. 1977) indicating thin regions of hot gas embedded in the molecular cloud.

The extreme velocities seen in the line profiles almost certainly imply the existence of strong shock waves in the cloud in any case. The failure of the calculations to account for all the observations probably results in part from complicated geometrical effects within the gas flows and in part from inaccuracies in the calculations themselves. For example, shocks arising from turbulence within a region of expanding gas might excite H_2. The shock speeds may be less than 25 km^{-1} relative to the gas, but they appear to be larger due to the expansion. A second point has been raised by Hollenbach and McKee (1979). A strong shock wave may dissociate all the hydrogen molecules it encounters which then recombine on grains behind the shock and become thermalized at 2000 K. The 25 km s^{-1} upper limit to the shock velocity does not apply to this situation. Finally, the 25 km s^{-1} limit itself may be incorrect. Dalgarno and Roberge (1979) have shown the dissociation rates used to derive the limit may be too high because of quantum mechanical corrections which have to be taken into account at interstellar cloud densities. A calculation using the revised rates might bring the shock theory into parity with the velocity results.

Perhaps the most interesting feature of the molecular hydrogen observations is that they imply the existence of rather extraordinary energy sources within the Orion molecular cloud. If the hydrogen is heated because energy in a systematic gas flow is converted into thermal energy which is then radiated in molecular hydrogen lines, the total luminosity of all the H_2 line radiation provides a lower limit to the rate of conversion. The observed luminosity of molecular hydrogen emission is of order 1000 $L_\odot$ (Beckwith, Persson, and Neugebauer 1979). The 60 km s^{-1} width of the lines implies a typical flow velocity of 30 km s^{-1}. The observed size of the region is 0.2 pc, which implies this phenomena has proceeded for at least 3,000 years and has thus liberated roughly 10^{48} ergs. If the flow is the result of an explosion as suggested by Kwan and Scoville (1976), then the explosion energy is of supernova proportions. If the flow is the result of strong stellar winds as suggested by a variety of authors, then $1/2 \; \dot{M} v_w^2$ is $\sim 1000 \; L_\odot$, where $\dot{M}$ is the mass loss rate and v_w is a velocity characteristic of the wind. Even if $v = 100$ km s^{-1}, then $\dot{M} \sim 10^{-3} M_\odot$ yr^{-1}. This mass loss rate is uncomfortably large.

Several authors have noted that the spatial coincidence and line of sight proximity of the H_2 emission to the BN and KL infrared sources suggest a casual relationship. The observed mass motions may result from a violent ejection of matter or extraordinary stellar wind which is generated by a star during its premain sequence evolution. This possibly emphasizes the importance of observations which clarify these energy arguments. It is equally important to determine how prevalent this phenomenon is among star formation regions. Attempts to discover another H_2 emission region comparable to Orion have failed to yield conclusive results as discussed in the next section, although W3 and NGC 7538 are promising candidates.

2. ADDITIONAL SOURCES OF MOLECULAR HYDROGEN EMISSION

Thirteen objects in addition to Orion are known to exhibit H_2 emission at this time; they are listed in the table. None of these sources has been observed in sufficient detail to infer shock excitation although it has been argued that other suggested excitation processes appear unlikely (Gautier 1978, Beckwith 1978). If we assume the H_2 is shock heated, then we may estimate radiated energies and mass loss rates by analogy to Orion. By assuming an excitation temperature of 2000 K and using the observed H_2 emission intensities and source sizes in the table, the total H_2 luminosities are computed in the fourth column. Taking 20 km s^{-1} to be a typical velocity and using the source size to obtain minimum ages, the total radiated power and mass loss rates are computed in the fifth and sixth columns. The calculated mass loss rates will of course be substantially smaller if larger wind velocities are assumed. Here, we have taken 20 km s^{-1} to be consistent with the shock model of Kwan (1977).

Object	$I(v=1\to0\ S(1))$[a]	Extent pc	Total L_{H2}[b] ($L_\odot$)	Radiated Energy (ergs)	dM/dt ($M_\odot yr^{-1}$)[e]	Refs.[f]
Orion	100 (4000)[c]	0.2	20 (800)[c]	(5×10^{47})	(2×10^{-2})[c]	5,6,15 17,27,32
W 3 E&W[d]	2	0.9	25	7×10^{46}	8×10^{-4}	14
NGC 7538[d]	3	0.6	6	1×10^{46}	2×10^{-4}	14
NGC 7027	25	0.09	1	3×10^{44}	3×10^{-5}	38
BD +30°3639	7	0.1	0.3	9×10^{43}	$9\ 10^{-6}$	4
Hb 12	7	<0.09	0.2	$<6\times10^{43}$	6×10^{-6}	4
CRL 2688	9	--	--	--	--	4
CRL 618	36	--	--	--	--	4
NGC 6720	2	0.2	0.4	2×10^{44}	1×10^{-5}	4
NGC 2440	2	0.1	0.06	2×10^{43}	2×10^{-6}	2
T Tauri	12	<0.005	0.004	6×10^{40}	1×10^{-7}	3
LkHα 349[4]	3	0.3	1	9×10^{44}	3×10^{-5}	14
IC 443	10	--	--	--	--	37
NGC 1068	6	<900	3×10^{6}	$<8\times10^{54}$	9×10^{1}	36

[a] Normalized to average Orion brightness.
[b] Total L_{H2} = 7.6 $L(v=1\to0\ S(1))$ for an assumed temperature of 2000 K.
[c] Numbers in parentheses have been corrected for extinction.
[d] Preliminary results pending confirmation.
[e] Assumes v_w = 20 km s^{-1}.
[f] Numbers refer to order of reference in bibliography.

The molecular cloud sources W3 and NGC 7538 may be roughly comparable to Orion in apparent H_2 luminosity, although these sources need
confirmation. Because only measurements of the $v = 1 \rightarrow 0$ S(1) line exist,
it is impossible to determine their actual luminosities and flow velocities, so detailed comparison must await more extensive observations.
On the basis of the apparent luminosities, however, it appears the internal turmoil of the Orion source may be a fairly common property of
star formation regions within molecular clouds. The total energies and
mass loss rates estimated for the planetary nebulae compare favorably
with source kinetic energies estimated by other means. For example,
typical kinetic energies of the expanding nebulae are $\sim 10^{44}$ ergs. The
mass loss rate inferred for T Tauri is similar to estimates of the mass
loss rates by the visual spectroscopic measurements of Kuhi (1964). It
is interesting to note, however, that Ulrich (1976) accounts for Kuhi's
measurements with a mass infall model. Molecular hydrogen observations
may help resolve this controversy when the excitation process is
understood.

NGC 1068 is the most surprising source among this sample. Roughly
10^5 clouds each with the apparent H_2 luminosity of Orion are required
to explain the observed emission in NGC 1068. While the Orion phenomena may occur commonly in molecular clouds, the required number is nonetheless large, and indicates that a considerable amount of turbulent
energy is deposited in the molecular medium within this Seyfert galaxy.

We stress that these conclusions are based on the assumption of
collisional excitation of the observed molecular hydrogen. Further
observations are needed to verify this assumption for the emission
sources listed in the table.

3. SUMMARY

A variety of recent observations provide evidence for the existence
of shock waves within molecular clouds. Of the available techniques,
observations of near infrared emission from molecular hydrogen are currently the most promising for direct observations of the shocked gas.
The temperature, extent, and velocity of the gas may be obtained in a
straightforward manner from these observations.

Observations of molecular hydrogen emission from the Orion molecular cloud have pointed out sources of kinetic energy of unexpected proportions. These conclusions are consistent with but entirely independent of similar conclusions inferred from measurements of CO emission
(Zuckerman, Kuiper, and Rodriguez Kuiper 1976; Kwan and Scoville 1976;
Phillips et al. 1977).

ACKNOWLEDGEMENTS

I am grateful to P. Goldreich, D. Hollenbach, E.E. Salpeter, and B. Zuckerman for discussions which contributed to this paper.

REFERENCES

1. Aannestad, P.A.: 1973, Ap. J. Suppl. 25, pp. 223.
2. Beckwith, S.: 1978, Ph.D. Thesis, California Institute of Technology.
3. Beckwith, S., Gatley, I., Matthews, K., and Neugebauer, G.: 1978a, Ap. J. (Letters) 223, L41.
4. Beckwith, S., Persson, S.E., and Gatley, G.: 1978, Ap.J. (Letters) 219, pp. L33.
5. Beckwith, S., Persson, S.E., and Neugebauer, G.: 1979, Ap. J. 227, pp. 436.
6. Beckwith, S., Persson, S.E., Neugebauer, G., and Becklin, E.E.: 1978b, Ap. J. 223, pp. 464.
7. Cowie, L., Laurent, C., Vidal-Madjar, A., and York, D.G.: 1979, Ap. J. (Letters) 119, pp. L81.
8. Dalgarno, A. and Roberge, W.G.: 1979, Ap. J. (Letters) 233, pp. L25.
9. DeNoyer, L.: 1978, Mon. Not. R. astr. Soc. 183, pp. 187.
10. DeNoyer, L.: 1979, Ap. J. (Letters) 228, pp. L41.
11. Elmegreen, B.C. and Lada, C.J.: 1977, Ap. J. 214, pp. 725.
12. Elmegreen, B.C. and Moran, J.M.: 1979, Ap. J. (Letters) 227, pp. L93.
13. Field, G.B., Rather, J.D.G., Aannestad, P.A., and Orszag, S.A.: 1968, Ap. J. 151, pp. 953.
14. Gautier, T.N., III: 1978, Ph.D. Thesis, University of Arizona.
15. Gautier, T.N., III: Fink, V., Treffers, R.R., and Larson, H.P.: 1976, Ap. J. (Letters) 207, pp. L129.
16. Gerola, H. and Seiden, P.E.: 1978, Ap. J. 223, pp. 129.
17. Grasdalen, G.L. and Joyce, R.R.: 1976, Bull. AAS 8, pp. 439.
18. Ho, P.T.P. and Barrett, A.H.: 1978, Ap. J. (Letters) 224, pp. L23.
19. Hollenbach, D.J. and McKee, C.: 1979, Ap. J. Suppl. 41, pp. 555.
20. Hollenbach, D.J. and Shull, J.M.: 1977, Ap. J. 216, pp. 419.
21. Kuhi, L.V.: 1964, Ap. J. 140, pp. 1409.
22. Kwan, J.: 1977, Ap. J. 216, pp. 713.
23. Kwan, J. and Scoville, N.: 1976, Ap. J. (Letters) 210, pp. L39.
24. Lada, C.J., Elmegreen, B.G., Cong, H.-I., and Thaddeus, P.: 1978, Ap. J. (Letters) 226, pp. L39.
25. London, R., McCray, R., and Chu, S.I.: 1977, Ap. J. 217, pp. 442.
26. Melnick, G., Gull, G.E., and Harwit, M.: 1979, Ap. J. (Letters) 227, pp. L29.
27. Nadeau, D. and Geballe, T.R.: 1979, Ap. J. (Letters) 230, pp. L173.
28. Phillips, T.G., Huggins, P.J., Heugebauer, G., and Werner, M.W.: 1977, Ap. J. (Letters) 217, pp. L161.
29. Schwartz, R.D.: 1975, Ap. J. 195, pp. 631.
30. Schwartz, R.D.: 1978, Ap. J. 223, pp. 884.
31. Shull, J.M.: 1977, Ap. J. 216, pp. 414.
32. Simon, M., Righini-Cohen, G., Joyce, R.R., and Simon, T.: 1979, Ap. J. (Letters) 230, pp. L175.

33. Spitzer, L., Jr. and Cochran, W.D.: 1973, Ap. J. (Letters) 186, pp. L23.
34. Spitzer, L., Jr. and Morton, W.A.: 1976, Ap. J. 204, pp. 731.
35. Storey, J.W.V., Watson, D.M., and Townes, C.H.: 1979, Ap. J. 233, pp. 109.
36. Thompson, R.I., Lebofsky, M.J., and Rieke, G.H.: 1978, Ap. J. (Letters) 222, pp. L49.
37. Treffers, R.R.: 1979, Ap. J. (Letters) 233, pp. L17.
38. Treffers, R.R., Fink, V., Larson, H.P., and Gautier, T.N., III: 1976, Ap. J. 209, pp. 793.
39. Ulrich, R.K.: 1976, Ap. J. 210, pp. 377.
40. Zuckerman, B., Kuiper, T.B.H., and Rodriguez Kuiper, E.N.: 1976, Ap. J. (Letters) 209, pp. L137.

DISCUSSION FOLLOWING BECKWITH

Elmegreen: The excited H_2 emission may not originate deep inside the Orion cloud even though the extinction is high: 40 magnitudes of extinction at a density of $10^6 cm^{-3}$ corresponds to a physical depth of only 0.01 pc. In fact, all of the recent star-forming activity in the KL region may be close to the cloud's interface with the Orion Nebula, as if the expansion of this visible nebula directly induced the star formation by compression and gravitational collapse.

Beckwith: Your point is certainly well taken. On the other hand, the total extinction through OMC 1 is estimated from other molecules to be around 200 magnitudes so 40 magnitudes implies a depth ∼1/5 of the total cloud diameter. This distance is much greater than 0.01 pc.

Elmegreen: These shock diagnostics may eventually play an important role as an indirect probe of magnetic field dynamics. If, for example, velocity jumps are seen in the cold molecular emission at locations adjacent to known sources of pressure (bright rims, etc.), and these velocity jumps are less than or equal to the cloud's linewidth, then it is possible that a high temperature shock will not occur (even though the velocity jump may be greater than the gas sound speed) because the compression will propagate into the cloud at speeds less than the Alfvén velocity. This situation may be common near those parts of an HII region where expansion is occurring in a direction perpendicular to the cloud's embedded magnetic field.

Lortet: In your computations of stellar wind, you find a mass loss that is very large indeed ($10^{-3}M_\odot$ yr^{-1}) because you take a very low velocity for the wind ($V_w \simeq 100$ km s^{-1}). Why did you choose so low a velocity?

Beckwith: Most of the energy in the wind is deposited in the H_2, and the wind velocity was simply chosen to equal the highest observed H_2 velocity. Certainly a higher wind velocity will considerably lower the mass loss rate. However, to keep the loss rate low, there has to be a very efficient mechanism for converting the wind energy into the energy radiated by the H_2 molecules. It may be difficult to make such a model consistent with all observations of this region.

Elitzur: Can a velocity of 100 km s^{-1} and H$_2$ rotation-emission both be accommodated without dissociation occurring?

Beckwith: This problem is probably the most difficult we face in explaining the H$_2$ emission as shocked gas. It is possible that the shocks are actually propagating at a velocity <25 km s^{-1} relative to an outflowing wind or expanding envelope which has a very high velocity relative to us. It may be that all the H$_2$ molecules have been dissociated by a fast shock, and we observe H$_2$ which has reformed behind the shock (Hollenbach and McKee, Ap. J. Suppl., October 1979). At present we cannot distinguish between these possibilities, and no other suggestions have yet been made.

Clark: Orion shows evidence of large amounts of angular momentum down to scales of ∿1 arcmin. Your H$_2$ map shows H$_2$ emission decidedly in the *polar* direction. Could a possible explanation involve gravitational infall along the rotational pole, perhaps coupled with a stellar wind, all "meeting", so to speak, at the "centrifugal" barrier? Such an explanation may be consistent with all available data, and would provide a natural anisotropy for the shock.

Beckwith: We have attempted to construct models invoking gravitational collapse, but we find the core masses needed in these models to be implausibly high, of order 10^4 M$_\odot$.

Carruthers: Have searches been made for H$_2$ emission from supernova remnants in highly obscured regions such as Cas A?

Beckwith: Some searches have been made, particularly in IC 443. There have been few sensitive searches toward other supernova remnants, including Cas A.

OBSERVATIONS OF THE V=0 S(2) LINE OF MOLECULAR HYDROGEN AT
12.28μm IN THE ORION MOLECULAR CLOUD

T.R. Geballe
Hale Observatories, Carnegie Institution of Washington

S.C. Beck and J.H. Lacy
Department of Physics, University of California at Berkeley

ABSTRACT

The 12.28μm pure rotational line of molecular hydrogen has been detected in emission from the region of vibration-rotation line emission in Orion. The line shapes, widths, and velocities are similar to those observed in the V=1 → 0 transition at 2.12μm. Constraints imposed by these new results on models of the emitting region are discussed.

INTRODUCTION

Vibration-rotation line emission from molecular hydrogen in the 2μm wavelength region was discovered unexpectedly from an astronomical object about three years ago. Since then these quadrupole transitions have been observed in a growing number of types of objects ranging from molecular clouds and T Tauri stars to planetary nebulae and supernova remnants - objects associated both with the birth and death of stars. The lines are strongest in the Orion molecular cloud where the extinction (Beckwith, Persson, and Neugebauer 1979; Simon et al. 1979) and the spatial distribution of V=1 → 0 S(1) line shapes (Nadeau and Geballe 1979; Nadeau, Neugebauer, and Geballe 1979) indicate that the line arises far behind the optical nebula in a cloud expanding at speeds up to 100 km s^{-1} due to some unknown source or event. From other observations and theoretical work it appears that the acceleration and heating (to 2000 K) of the H_2 results from collisions which are caused by the passage of a shock through the molecular cloud.

At the temperature of 2000 K inferred from the vibration-rotation line measurements, the population of a low rotational level in the V=0 state of H_2 is roughly ten times that for a similar level in the V=1 state. Because the lower rotational levels of the V=0 state will also be populated well below 2000 K it is expected that observations of the associated rotational transitions can provide information about the cooler, post-shock gas. These transitions, however, have A-coefficients typically 100 times less than vibration-rotation transitions. Furthermore,

465

B. H. Andrew (ed.), Interstellar Molecules, 465–468.

every one of them occurs at a wavelength which is difficult for or
inaccessible to ground-based observing and where the terrestrial background is much greater than at the shorter wavelength vibration-rotation
lines. In spite of these difficulties the S(2) line at 12.28μm has been
detected at a number of positions in Orion. The observations were made
in 1979 February and March with a cooled Fabry-Perot and grating
spectrometer on the 2.5m telescope of Las Campanas Observatory. The
results put new constraints on the structure, temperature, and dynamics
of the region of molecular hydrogen emission.

RESULTS AND DISCUSSION

    The S(2) line, observed in a 7" diameter aperture, was seen at
seven positions including all five of the bright peaks of the 2.12μm
V=1 → 0 S(1) line mapped by Beckwith et al. (1978). Typical peak
intensities of the 12.28μm line, which was spectrally resolved, are
$2 - 3 \times 10^{-3}$ erg s^{-1} cm^{-2} sr^{-1}. At four other positions searched
(including that of the Becklin-Neugebauer object) the line was not detected, with 3σ upper limits of 1×10^{-3} erg s^{-1} cm^{-2} sr^{-1}.

    The column densities in the upper level (V=0, J=4) are approximately
3×10^{20} cm^{-2} (after correcting for an estimated 1.5 magnitudes of
extinction). With some variation with position in the cloud, this is
about a factor of five greater than is predicted from the temperature
and column densities observed in the vibrationally excited lines. The
upper levels of the latter transitions lie 6000 K or more above the
ground state, whereas the upper level of the 12.28μm line corresponds
to 1800 K. From this it is apparent that most of the 12.28μm line radiation comes from gas too cool to be vibrationally excited. Qualitatively
that is the expected result for the case of rapid shock-heating of the
gas followed by slower cooling. The V=1 → 0 lines would in that case be
emitted just behind the shock while the 12.28μm line emission would persist further behind (see Fig. 1). The relative intensities of the two
lines depend on both the shock velocity and the abundances of rapidly
radiating molecules such as CO (see Kwan 1977). Such molecules cool the
post-shock gas, thereby decreasing the power emitted by H_2, which is a
very inefficient radiator.

    The large widths (up to 55 km s^{-1}) and the shapes of the 12.28μm
line are very similar to those observed by Nadeau, Neugebauer, and
Geballe (1979) in the 2.12μm line. In all spectra except that at the
position of Beckwith et al.'s (1978) Pk 2 the S(2) line is noticeably
asymmetric with increased emission on the blue side of the line peak.
The velocity of peak intensity at all positions is near + 7 km s^{-1} (LSR),
which is also quite close to that observed for the 2.12μm line. It
should be noted that the rest frequency of the S(2) line used here,
814.452±.005 cm^{-1}, has been calculated from the accurately measured
frequencies of three other pure rotational transitions of H_2. The most
recent laboratory measurement of the S(2) line frequency, made over
20 years ago by Stoicheff (1957), significantly disagrees with this

calculated frequency. It is important that the S(2) line frequency be remeasured in the laboratory.

Detailed comparison of the 2.12μm and 12.28μm line profiles might allow a test of dynamical models of the emitting cloud. In the model of roughly spherically symmetric expansion suggested by Nadeau and Geballe (1979), the observed asymmetry of the 2.12μm S(1) line shape requires extinction of at least 2 magnitudes within the emitting region. At 12.28μm where extinction is less, perhaps by a factor of 2.5 (Becklin et al. 1978), the red side of the S(2) line should extend to larger positive velocities than the S(1) line. Although at a couple of positions this does not seem to be the case, at most positions the signal-to-noise ratio of the S(2) line does not permit a meaningful comparison.

Further details of this work can be found in Beck, Lacy, and Geballe (1979). The work was supported in part by NASA Grant NGL 05-003-272 and NSF Grant AST 78-24453. The Hale Observatories are operated jointly by the Carnegie Institution of Washington and the California Institute of Technology.

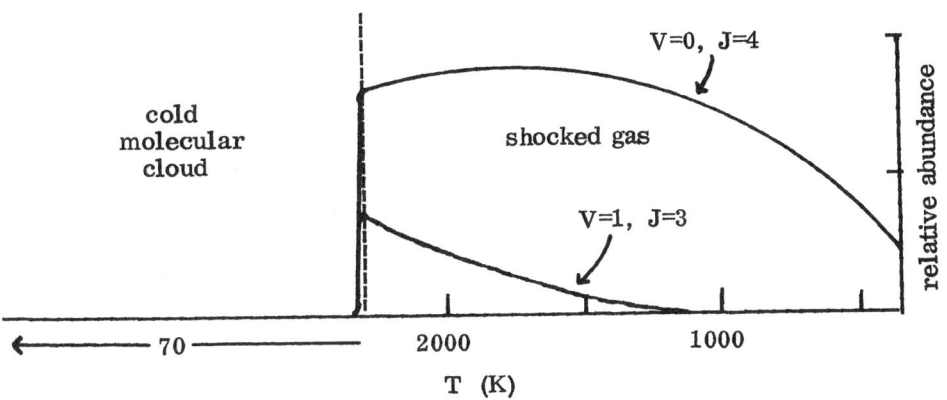

Fig. 1 Effect of a shock on the populations of the upper levels of the V=1 → 0 S(1) and pure rotational S(2) transitions in the Orion molecular cloud.

REFERENCES

Beck, S.C., Lacy, J.H., and Geballe, T.R.: 1979, Astrophys. J. (Letters) 234, pp. L213-L216.
Becklin, E.E., Mathews, K., Neugebauer, G., and Willner, S.P.: 1978, Astrophys. J. 220, pp. 831-833.
Beckwith, S., Persson, S.E., and Neugebauer, G.: 1979, Astrophys. J. 227, pp. 236-240.
Beckwith, S., Persson, S.E., Neugebauer, G., and Becklin, E.E.: 1978, Astrophys. J. 223, pp. 464-470.

Kwan, J.: 1977, Astrophys. J., 216, pp. 713-723.
Nadeau, D., and Geballe, T.R.: 1979, Astrophys. J. 230, pp. L169-L173.
Nadeau, D., Neugebauer, G., and Geballe, T.R.: 1979, paper presented at this symposium.
Simon, M., Righini-Cohen, G., Joyce, R.R., and Simon, T.: 1979, Astrophys J. (Letters) 230, pp. L175-L178.
Stoicheff, B.P.: 1957, Can. J. Phys. 35, 730.

DISCUSSION FOLLOWING GEBALLE

Black: Excess emission in the rotational line is inferred by assuming a galactic center extinction curve. What extinction at 12 microns is derived if it is assumed alternatively that all of the excited H_2 is at the temperature determined for the vibrationally-excited molecules?

Geballe: No extinction at all.

Gilra: Have searches been made for the rotational or vibrational-rotational transitions of HD? The effect of the lower densities of HD will be somewhat offset by higher transition probabilities.

Geballe: There have been none so far as I know. Clearly, there are a large number of molecules which may be detectable in the infrared, given the brightness of the H_2 lines.

Beckwith: The 12.28 µm extinction could be as large as 2 magnitudes a value which is certainly consistent with the curve of Becklin et al. (1978) within the considerable errors. If true, would you not expect the positive velocity edge to be depressed relative to the negative velocity edge, as is the case with the 2 µm lines?

Geballe: Yes, the positive velocity edge would be attenuated, but one might still see a difference in the two profiles if the observations of the 12 µm line had better signal-to-noise ratio. Until such observations are obtained, and until the question of the S(2) rest frequency is answered, an accurate comparison cannot be made.

SPECTRA OF THE 2.12μm QUADRUPOLE LINE OF H₂ IN THE ORION MOLECULAR CLOUD

D. Nadeau, G. Neugebauer   and   T. R. Geballe
Hale Observatories   Hale Observatories
California Institute of Technology   Carnegie Institution of
                                                     Washington

New spectra of the 2.12μm line of H_2 in Orion have been obtained during 1979 February at the Mount Wilson 2.5m and Palomar 5m telescopes. These data provide wider spatial coverage and higher spatial resolution than the observations reported earlier by Nadeau and Geballe (1979). The complete data, displayed here in two maps, are consistent with the earlier interpretation, which is summarized below.

In the south-central part of the emission region the H_2 line is broad and has an extended blue wing. The new data make it clear that in all directions away from there the blue wing contracts and weakens relative to the peak, so that the line becomes narrower and more symmetric. Throughout the entire region the velocity of peak line intensity is +9 ±3 km s^{-1} (LSR), very close to that of the extended and cooler molecular cloud.

The most plausible model of the emitting region is that H_2 has been accelerated and excited by one or more shock waves expanding about a central source. The source, which should have a velocity near +9 km s^{-1} and be located in the direction from which the broad lines are seen, apparently is not identified with any known object. It is unlikely that the H_2 emission is the result of a single shock propagating at velocities less than 24 km s^{-1}, as was predicted by several authors.

From similarities in location, spatial extent, and range of velocities the infrared H_2 lines probably are a manifestation of the "plateau source" of broad radio lines in Orion. The asymmetric H_2 lines, however, contrast with the more symmetric radio lines. The difference may be explained either by an asymmetric expansion of the hottest gas in the "plateau source" or by obscuration of the rear of the expanding cloud by dust mixed with the gas.

REFERENCE

Nadeau, D., and Geballe, T. R.: 1979, Astrophys. J. 230, pp L169-L173.

B. H. Andrew (ed.), Interstellar Molecules, 469–470.
Copyright © 1980 by the IAU.

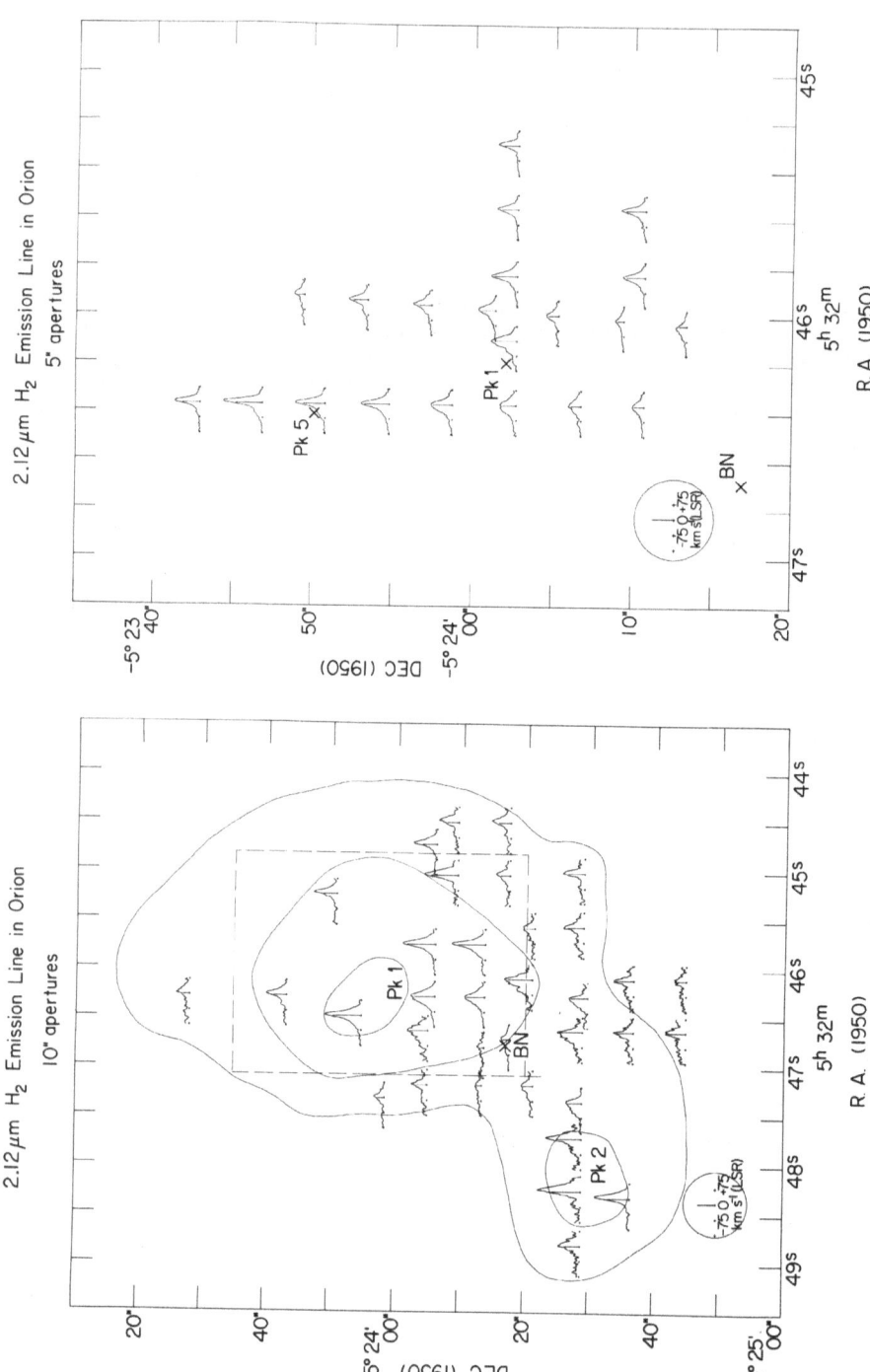

Circles represent aperture sizes. For each spectrum coordinates correspond to position of zero velocity and zero intensity. Dashed rectangle in 10" map is area shown in 5" map.

MOLECULE FORMATION IN THE SEYFERT GALAXY NGC 1068

Wayne J. Carlson and Craig B. Foltz
The Ohio State University, Columbus, Ohio

ABSTRACT

We have constructed models of relatively high-density radiation-bounded filaments near the nucleus of a Seyfert 2 galaxy. The amount of molecular hydrogen predicted by the models for reasonable values of the physical parameters is consistent with observations of the infrared continuum and the quadrupole rotation-vibration lines of H_2 in NGC 1068.

The first detection of extragalactic molecular emission lines was made by Thompson, Lebofsky, and Rieke (1978) with their observation of the $v = 1 \rightarrow 0$, S(1) and S(2) lines of H_2 in NGC 1068, a Seyfert 2 galaxy. The emission, if it is attributed to shocked gas at 2000 K, must come from approximately 6000 $M_\odot$ of shocked molecular hydrogen which is a fairly small fraction of the approximately 10^8 $M_\odot$ of cold material postulated by Jones et al. (1977) to account for the infrared luminosity. As noted in the report of the molecular emission observations, the hydrogen can be expected to lie in cold, relatively dense clouds or filaments.

Detailed models of the line emitting regions of NGC 1068 by Shields and Oke (1975) make use of two regions to explain the spectrum. One has a density of ~ 800 cm^{-3} and a fairly uniform distribution, and the second, more filamentary in structure, has a density of 3×10^5 cm^{-3} (for T=12000 K). If the filaments of the second type of region are radiation bounded, then an even higher-density cold region can be expected to exist beyond the HII-HI transition. This region, in which we expect molecules to exist, is taken to be a thin slab illuminated on one of its faces by a continuum source with either a power law or blackbody distribution as given by Shields and Oke. We assume that no radiation below the Lyman limit is transmitted through the transition region.

Calculations of the photoionization equilibrium for the metals with ionization potentials below 13.6 eV yield an electron density and radiation field as a function of depth into the slab. The molecular

B. H. Andrew (ed.), Interstellar Molecules, 471–472.

equilibrium for hydrogen was calculated using gas-phase reactions. The effect of dust on the calculations was evaluated by computing models which either included or excluded the attenuation of radiation by dust and the formation of molecular hydrogen on grain surfaces (Hollenbach and Salpeter, 1971).

Models were computed at three different densities: 2×10^6, 2×10^7, and 2×10^8 cm^{-3}. The slab is at a distance of 50 pc from the continuum source and a kinetic temperature of 100 K is assumed throughout.

At a sufficient depth into the cloud that the H_2 can be effectively shielded from photodissociation (via predissociation as described by Stecher and Williams, 1967), the hydrogen is all molecular in the cases including dust, and one-third to all molecular even if the grain surface reactions and dust attenuation are not included. The times to reach equilibrium, like the reaction rates, depend directly on the density. These times range from 3 to 300 years for the models including dust, and from 40,000 to 500,000 excluding dust. The shorter times correspond to the higher densities, but the frequency dependence of the radiation does not affect the times significantly.

The absorption of radiation by H_2 was not included in the calculations, so its shielding is due to the continuous absorption of neutral carbon. The formation of carbon monoxide, which eventually dominates the carbon chemistry, would allow into the slab more radiation capable of causing the predissociation of H_2. We considered a more detailed chemistry including CO and found that there is sufficient optical depth in neutral carbon to shield H_2 even in this case.

The column density of H_2 up to the point where the species with the second lowest ionization potential (Mg) became neutral ranges from 2×10^{23} to 1×10^{24} cm^{-2}. For a fraction 1/10 of the sky covered by the filaments, as seen from the source, the total mass of H_2 predicted is 2.5×10^7 $M_\odot$. This is very close to the amount given by Jones et al.; excitation of only a small fraction of it can produce the observed emission.

A more detailed description is to be published in the Astrophysical Journal, vol. 233.

REFERENCES

Hollenbach, D., and Salpeter, E.E.: 1971, Astrophys. J. 163, pp. 165-180
Jones, T.W., Leung, C.M., Gould, R.J., and Stein, W.A.: 1977, Astrophys. J. 212, pp. 52-59.
Shields, G.A., and Oke, J.B.: 1975, Astrophys. J. 197, pp. 5-16.
Stecher, T.P., and Williams, D.A.: 1967, Astrophys. J. (Letters) 149, pp. L29-30.
Thompson, R.I., Lebofsky, M.J., and Rieke, G.H.: 1978, Astrophys. J. (Letters) 222, pp. L49-54.

MOLECULAR CLOUDS NEAR SUPERNOVA REMNANTS

V.I. Slysh
Space Research Institute, Academy of Science
Profsojuznaja 88, 117810 Moscow, USSR

T.L. Wilson, T. Pauls, C. Henkel
Max-Planck-Institut für Radioastronomie
Auf dem Hügel 69, 5300 Bonn 1, FRG

ABSTRACT

A survey of 14 SNR's in the 4.8 GHz absorption line of H_2CO shows that two of them, W28 and W44, possibly interact with molecular clouds. The interaction leads to acceleration of a part of the molecular cloud to a velocity of ~ 5 km s^{-1} without a significant increase in the kinetic temperature or turbulence.

The apparent long-term stability of galactic molecular clouds against gravitational collapse and subsequent star formation has stimulated proposals about possible sources of external pressure, such as shocks, which could upset this equilibrium and lead to collapse. The supernova blast wave was considered as such a shock by Herbst and Assousa (1977). This proposal can be tested by observations. The interaction between a supernova remnant (hereafter SNR) and a molecular cloud may result in a disturbance of the cloud such as changing of its geometry, introducing large velocity gradients, heating etc. Spectral line mapping of the molecular clouds toward SNR's might reveal cases of SNR-molecular cloud interactions and give details of the relevant physical processes. The observations presented here were aimed at (i) a search for clouds interacting with SNR's; (ii) measuring physical parameters of the disturbed molecular gas.

OBSERVATIONS

The observations were made with the Effelsberg 100-m radio telescope in the 4.8 GHz line of H_2CO. The beam width was 3'. The list of the supernova remnants chosen for this survey is given in Table 1. A "+" sign in the third column of the Table 1 indicates that H_2CO absorption was detected, a "+" sign in the fourth column shows which objects were partially mapped in H_2CO line, and the "-" sign in the fifth column shows that no indications of interaction as discussed in the Introduction were

473

B. H. Andrew (ed.), Interstellar Molecules, 473–478.

Table 1.

(1)	(2)	(3)	(4)	(5)
Galactic coordinates	Other name	H$_2$CO absorption	Line map	Physical interaction
6.4 - 0.1	W28	+	-	possible
11.4 - 0.1	-	-	-	
18.8 + 0.3	Kes 67	+	-	
21.8 - 0.6	Kes 69	+	+	-
23.0 - 0.3	W41	+	+	-
24.7 - 0.6	-	-	-	
32.8 - 0.1	Kes 78	+	-	
34.6 - 0.5	W44	+	+	possible
46.8 - 0.3	HC30	-	-	
74.0 - 8.6	Cygnus Loop	-	-	
78.1 + 1.8	DR4	+	+	-
84.2 - 0.8	-	+	+	-
127.3 + 0.7	-	-	-	
189.1 + 2.9	IC443	-	-	

seen on the maps. Negative results in the third column do not necessarily mean that there is no H$_2$CO absorption toward these supernova remnants, since the beam width is much smaller than the SNR size and the observations were limited to several points. Thus we could have missed a molecular cloud occupying only part of a SNR area. Such a case might be IC443 where De Noyer (1979) has found "shocked" OH absorption in the southern part of the remnant. We made no measurements there. Any H$_2$CO absorption is below our detection limit of about 0.05 K.

Two positive detections of interacting molecular clouds in Table 1 are W28 and W44. They were previously proposed as such by Goss et al. (1971) and Pastchenco and Slysh (1974) as a result of OH absorption observations. The present study provides more details on the geometry and kinematics because of our higher angular and velocity resolution, and more complete sampling of the area.

RESULTS

The optical depth map of H$_2$CO in the W28 region is shown in Fig. 1. There are two clouds on the map. One extended cloud at radial velocity 21 km s^{-1} consists of two brighter regions and extends from M20 to the edge of the SNR. It shows little variation of radial velocity and line width, and, on this basis and its geometry, may be a foreground cloud not related to the SNR. The second, more compact, cloud (3' x 15') has a central radial velocity 7 km s^{-1} and is located just on the edge of the SNR. The 6-cm continuum emission contours taken from the survey of Altenhoff et al. (1979) outline the SNR as well as nearby HII regions.

The 7 km s^{-1} cloud is shown in more detail on Fig. 2a, b. Fig. 2(a) is the absorption line equivalent width distribution (km s^{-1}). The dotted line shows the eastern border of the SNR, the crosses show the

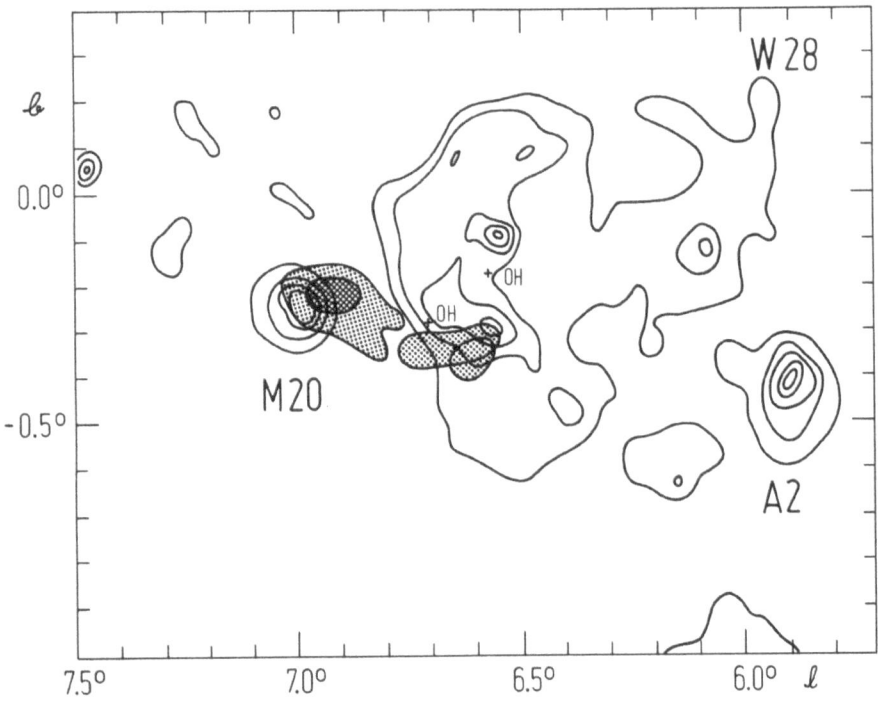

Fig. 1.

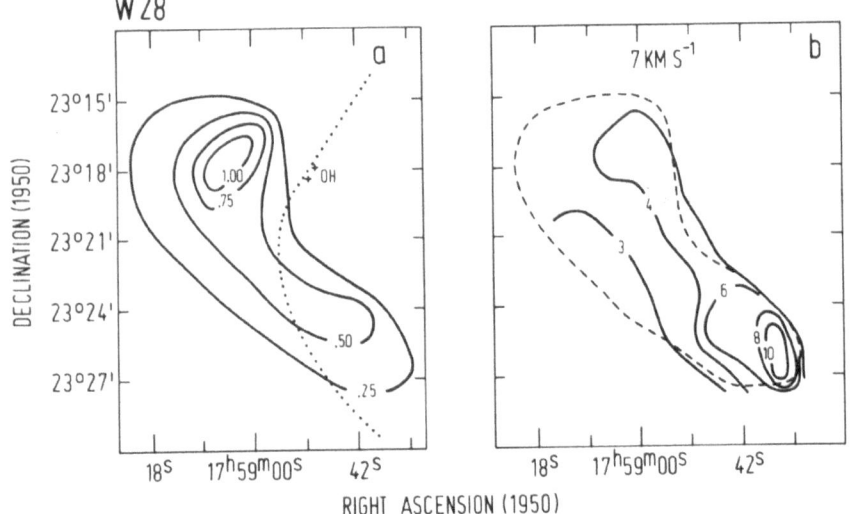

Fig. 2.

positions of 1720 MHz OH masers. Fig. 2b shows line width distribution
(km s⁻¹) across the cloud outlined by the 0.25 km s⁻¹ contour from
Fig. 2a. It is evident that the line width, although otherwise constant
(3.5 km s⁻¹), sharply increases to 10 km s⁻¹ at the lower right where
the cloud is projected on the SNR. The radial velocity in this region
rapidly increases from 4 to 8 km s⁻¹ as one goes from right to left.
The geometry, radial velocity and line width variation are suggestive
of an interaction between the cloud and the SNR.

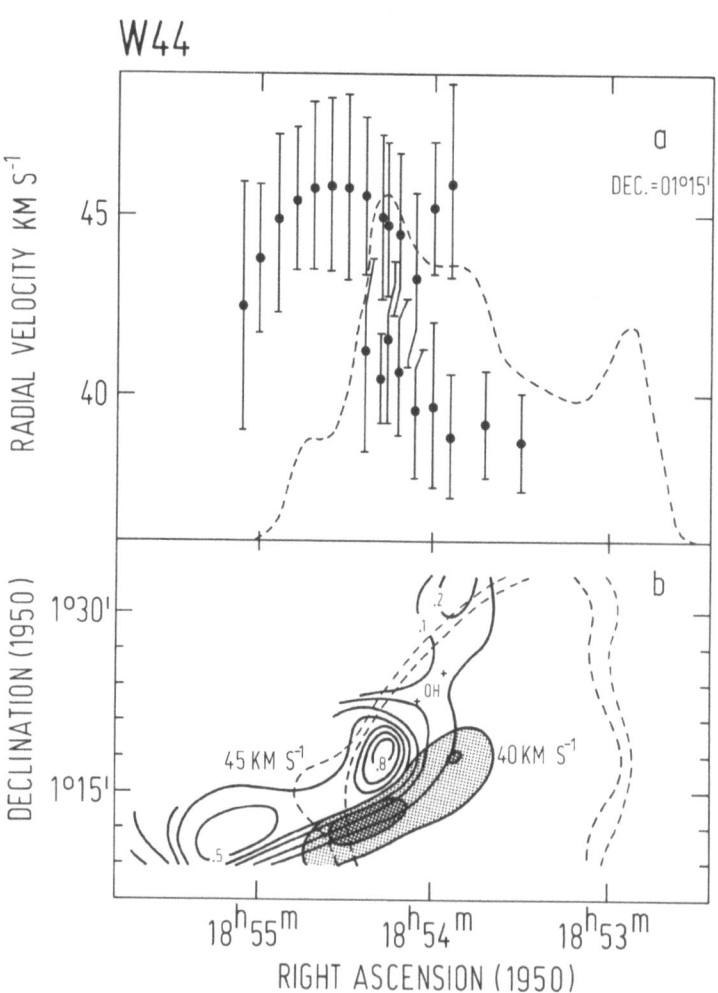

Fig. 3a, b.

The H_2CO optical depth distribution in the W44 region is shown in Fig. 3b where the hatched area shows a velocity feature at about 40 km s^{-1} as compared to a 45 km s^{-1} feature. Dashed lines show contours of 6 cm continuum emission from the SNR W44, crosses are positions of 1720 MHz OH masers. The two velocity features partly overlap just at the eastern edge of the SNR. The velocity variation of the two features across the SNR is shown on Fig. 3a, a right ascension cross-section at declination 01° 15'. Bars indicate the line width. The dashed line is the 6 cm continuum antenna temperature. Going from left to right in the figure, the main 45 km s^{-1} component gradually disappears but its central velocity remains relatively constant. Just at the edge of the SNR a second feature appears with a radial velocity close to that of the first (45 km s^{-1}) feature. Moving still further to the right its radial velocity gradually decreases to 39 km s^{-1} at the center where it also disappears. The behaviour is suggestive of a cosine variation of the radial velocity, which might be expected from spherical expansion with a velocity of about 6 km s^{-1}. Note that the line width of the 40 km s^{-1} feature is the same (or even smaller) than that of 45 km s^{-1} feature.

DISCUSSION

Both W28 and W44 show H_2CO molecular clouds with a variation of radial velocity closely matched to the SNR's geometry, as might be expected if they were disturbed by the SNR's expansion. It appears that there is a parent molecular cloud which presumably existed long before the SN explosion, part of which was accelerated by the SNR expansion to a velocity about 5 km s^{-1}. No extra turbulence was added to this accelerated part of the cloud since the line width has not changed appreciably. Physical parameters of the accelerated part of the cloud may be estimated from the absorption line data as well as from CO observations available for W44 (Wootten, 1977). They are summarized in Table 2.

Table 2

Physical properties of the accelerated part of the molecular clouds

1.	Final acquired velocity	5 km s^{-1}
2.	Velocity dispersion	3 km s^{-1}
3.	Temperature	15 K
4.	Size	2 - 5 pc
5.	Density	2×10^3 cm^{-3}
6.	Mass	$\sim 10^3$ M$_\odot$

Velocity dispersion, temperature and density seem to be the same as in the parent cloud. This is in conflict with current ideas about shock wave heating and compression (Hollenbach, 1979) of the interstellar matter. The 1720 MHz OH masers appear to be intimately related to the interaction between SNR's molecular clouds. Their position, slightly offset from the molecular clouds and within the contours of SNR's, suggests that they are now behind the shock front and may be products of cooling of the shocked material.

REFERENCES

Altenhoff, W.J., Downes, D., Pauls, T., and Schraml, J.: 1979, Astron.
    Astrophys. Suppl. 35, 23.
De Noyer, L.: 1979, Astrophys. J.(Letters) 228, L41.
Goss, W.M., Caswell, J., and Robinson, B.J.: 1971, Astron. & Astrophys.
    14, 481.
Herbst, W., and Assousa, G.E.: 1977, Astrophys. J. 217, 473.
Hollenbach, D.: 1979, this volume.
Pastchenco, M.I., and Slysh, V.I.: 1974, Astron. & Astrophys. 35, 153.
Wootten, H.A.: 1977, Astrophys. J. 216, 440.

DISCUSSION FOLLOWING SLYSH

Radhakrishnan: Is not 5 km s^{-1} a very small velocity?

Slysh: 5 km s^{-1} is correct from the point of view of conservation of momentum, that is to say that a 500 km s^{-1} shock propagating in a medium with 10 cm^{-3} density can accelerate a molecular cloud of 1000 cm^{-3} density to 5 km s^{-1}.

Hollenbach: Although a 5 km s^{-1} shock heats the gas to $\simeq$ 1500 K, would rapid cooling not make most of the post-shock gas cold, as required

Slysh: I am not sure that the time available is sufficient for cooling to 15 K.

Elmegreen: How were the ages of the supernovae determined?

Slysh: The ages come from optical (W28) and radio astronomy data, and are in the range 10,000-50,000 years.

BEHAVIOR AND SIGNIFICANCE OF CIRCUMSTELLAR CLOUDS

B. Zuckerman
University of Maryland and University of Texas

We will discuss molecular envelopes around post main sequence stars. Topics that will be covered include chemical composition, physical properties, mass loss and evolution. In view of the following papers by Drs. Snyder and Goldreich we will have little to say about circumstellar masers here.

Hydrogen is the most abundant element in most stars. We are interested in whether it is in atomic or molecular form in the circumstellar envelopes. The Arecibo radio telescope was used at 21-cm wavelength to search for atomic hydrogen in various evolved stars (Zuckerman et al. 1980). The 21-cm line was not detected in any star: the best limits were obtained for the very cold, infrared sources IRC+10216, CIT6, IRC+10011, and NML Tau. The first two are carbon-stars ($[C]/[O] > 1$) and the latter two are oxygen-rich ($[O]/[C] > 1$). In all cases no more than 10% by mass of the circumstellar hydrogen is in atomic form. The best limit for IRC+10216 is $\lesssim$ 1% H by mass if the model of Kwan and Hill (1977) is adopted or $\lesssim$ 0.2% H by mass if, instead, more recent $J = 2 \rightarrow 1$ CO data (Knapp 1979) are utilized.

That the preponderance of the hydrogen is molecular is not surprising if chemical equilibrium abundances are achieved in the photosphere and then these same abundances are maintained in the circumstellar shell. However, Balmer emission in Miras indicates that at least some of the hydrogen is atomic at least some of the time (possibly due to shocks in the atmosphere). Also, the ambient interstellar radiation field will photodissociate molecular hydrogen far from the star. Thus, the 21-cm limits place constraints on the importance of these and other mechanisms of dissociation. For α Ori the much higher photospheric temperature suggests that much (most) of the hydrogen will be atomic but, unfortunately, strong 21-cm emission from background hydrogen rendered the Arecibo measurement nearly useless in spite of considerable effort to subtract out this background.

For other elements the $[C]/[O]$ ratio is probably the most important in determining the over-all chemical composition in the envelope. From

B. H. Andrew (ed.), Interstellar Molecules, 479–486.
Copyright © 1980 by the IAU.

photospheric spectra one can easily deduce if C/O is greater or less than unity. But quantitative determinations of C/O in M giants are still lacking.

For carbon-stars IRC+10216 is the prototype and 17 molecules have been identified in its envelope either in the radio (CO, CN, CS, HCN, HNC, C_2H, SiS, SiO, HC_3N, HC_5N, HC_7N, C_4H, C_3N, CH_3CN) or in the infrared (C_2H_2, CH_4, NH_3) or in both. Since Dr. McCabe has discussed this star in detail we will add only a few remarks here. Except for a small amount of SiO no oxygen-containing molecules (other than CO of course) have been detected. Indeed, the [SiS]/[SiO] ratio is greater than unity as expected only in a carbon-rich environment (Tsuji 1973). The chemistry is clearly different from the chemistry in the interstellar medium since HCO^+ and N_2H^+ are not seen in IRC+10216. Also [HNC]/[HCN] is $\lesssim 1/100$ as expected for chemical equilibrium at T $\sim$ 1000 K (McCabe et al. 1979) but very different from the interstellar ratio ($\sim$ 1). It is perhaps significant that of 12 polyatomic molecules, 9 are linear.

Wannier and Linke (1978) have measured isotopic ratios for IRC+10216 and find a large $[^{14}N]/[^{15}N]$ ratio indicative of cold CNO processing. Since [C]/[O] > 1 in conjunction with [C]/[H] $\gtrsim$ solar implies carbon production in the 3α process, we have evidence in IRC+10216 for the products of two different nuclear burning processes.

The molecular inventory in envelopes around oxygen-rich stars is sparser: OH, H_2O and SiO masers, thermal mm-λ emission from SiO and CO, and evidence for NH_3 and CO in the infrared. SiS emission is not seen (Palmer and Zuckerman 1978), so [SiS]/[SiO] is less than unity as expected in an oxygen-rich environment. TiO and other simple molecules have yet to be detected at infrared and optical wavelengths.

Calculations of Tsuji (1973) and Vardya (1966) suggest that SiO is the dominant gas phase carrier of silicon in these stars. Results of Morris et al. (1979) and Lambert and Vanden Bout (1978) imply that $\sim$ 99% of the silicon is in the grains in the outer parts of the circumstellar shell but this is unlikely to be the case in the inner shell where the SiO maser is produced. Thus the bulk of the silicon is apparently incorporated into grains between 10^{14} and 10^{16} cm from the central star in agreement with the inner dust boundary determined from 11μm interferometry for a limited sample of stars (Sutton 1979, discussed below).

There seems to be some, although by no means universal, agreement that the grains in oxygen-rich stars are mainly "silicates", (e.g. Mg_2SiO_4), and in carbon-stars mainly graphite and SiC. How large are the grains? Theoretical estimates (e.g. Salpeter 1974) are inconclusive and so are the observations. For example, the presumed graphite and SiC grains in IRC+10216 are estimated to have radii $a \sim$ 1μm from a fit to the far-infrared spectrum (Campbell et al. 1976). But impure grains with enhanced far-IR emissivities could be a lot smaller. In the carbon-rich Egg Nebula (CRL 2688) the very large percentage polarization in the scattered light implies $a < 0.1$μm (Schmidt et al. 1978) whereas if the grains in planetary nebulae (PN's) are graphite then the absence of a $\lambda 2200$ Å feature in their spectra implies $a > 0.04\mu m$ (Mathis 1978). It might be argued that the grains in PN's form in the ionized gas and there-fore may be different from grains that form in the neutral clouds around IRC+10216 and CRL 2688. However, since molecular clumps exist inside PN's (Beckwith et al. 1978) the bulk of the dust emission may be asso-ciated with this neutral gas rather than the ionized gas.

How do circumstellar clouds fare as a source of interstellar grains? Silicate grains are probably produced in proto-stellar nebulae, oxygen-rich red giants and, possibly, planetary nebulae. Graphite is probably produced in carbon-stars and planetary nebulae. We acknowledge the ex-istence of other, more exotic, suggestions for the composition of the grains (e.g. carbyne, organic polymers, and tholins) but do not evaluate them here.

Total mass loss rates, summed over the entire galaxy, are probably comparable for evolved stars (red giants plus PN's) and for proto-stellar nebulae. Estimates are $\sim$ few x $M_\odot$/yr. Therefore, at present, it is difficult to choose between a pre and post main sequence origin for the interstellar dust. It is even conceivable that much of the dust in in-terstellar molecular clouds forms in situ if radiative association rates are very fast for very large molecules (Smith and Adams 1978).

Where does the dust form in the circumstellar shell? Since oxygen-rich giants and supergiants typically have photospheric temperatures be-tween 2000 and 3500 K and silicates condense at $T \lesssim 1200$ K, it is to be expected that most of the dust will form at least a few stellar radii from the center of these stars. For cool carbon-stars the situation is not, a priori, so obvious, since $T_{photosphere} \sim 2000$ K and SiC and graph-ite can condense at $T \sim 1700$ K. Some recent observations bear on this question.

For uniform outflow the gas density drops as r^{-2}. The shape of most radio emission profiles are consistent with $\rho_{gas} \propto r^{-2}$, although in IRC+10216 Wannier et al. (1979) find evidence for an outflow rate de-creasing with time. The dust density distribution around evolved stars was measured first by means of lunar occultations but recently infrared interferometry (e.g. Sutton 1979) has emerged as a powerful technique. Sutton's 11μm visibility curves for a limited sample of stars suggest that if $\rho_{dust} \propto r^{-2}$ then most of the dust condensation begins near the radius at which $T \sim T_{condensation}/2$ (i.e. beyond 10 R_*). Also $\rho_{dust} \propto r^{-2}$ appears to fit the data for IRC+10216 much better than the two shell model suggested by the older lunar occultation data. Alternatively, the visibility curves can be fitted with dust condensation beginning near the radius at which $T \sim T_{condensation}$, provided that $\rho_{dust} \propto r^{-1.5}$.

What might cause an $r^{-1.5}$ dependence in ρ_{dust}? $\dot{M}$ could be declining as a function of time but this seems unlikely to be true in all cases. The gas could be decelerating but this seems unlikely at large r. Per-haps the most likely explanation for an $r^{-1.5}$ dependence is dust formation between 10^{14} and 10^{16} cm from the center of the star. That the dust forms at r $\sim 10^{15}$ cm is consistent with the infrared energy distribution and 10μ silicate depth in the spectra of OH/IR stars (Werner et al. 1979). At larger distances from the star we have the scattered light profiles of McMillan and Tapia (1978) who find ρ_{dust} somewhere in the range $r^{-1.5}$- $r^{-3.0}$ (with r^{-2} preferred) around α Ori. These results apply to r between 4 x 10^{16} and 2 x 10^{17} cm.

Information is available on the azimuthal shapes of some circumstellar clouds. Non spherical shapes may be due to rotation, magnetic fields and non-radial pulsation. Capriotti (1978) has even suggested that the asymmetries in the shapes of PN's may be related to the galaxy since there is a tendancy for the long axis to lie parallel to the plane of the Milky Way. Some less evolved objects (e.g. IRC+10216, VY CMa, the Egg Nebula, OH 0739-14, CRL 618 and M1-92) show very large percentage polarization ($\sim$ 30%) in scattered light with $\lambda \lesssim$ 1μm indicative of asymmetrical dust distributions (the latter four objects are classified as bi-polar nebulae). The grain masses in the scattering nebulae in the Egg Nebula and M1-92 are $\sim 10^{-4}$ $M_\odot$, comparable to those deduced from far IR emission in some PN's (Schmidt et al. 1978). However, polarization is yet to be detected from PN's even though many look similar, superficially at least, to bi-polars.

A λ6000 Å photograph of IRC+10216 shows an elongated 2" x 4" image (Becklin et al. 1969) but no asymmetry is apparent at 11μm on a smaller scale ($\lesssim$ 1", Sutton 1979) or at 1.3 mm on a larger scale ($\sim$ 60", Wannier et al. 1979). (There is a report [McCarthy 1979] of asymmetry at 5μm with, however, a position angle that bears no apparent relation to the 6000 Å image.) The envelope around α Ori appears to be roughly symmetric (Bernat and Lambert 1976; McMillan and Tapia 1978).

A long standing question in the evolution of red-giant stars is what accelerates the gas and causes mass loss. Although radiation pressure on the dust probably is responsible for acceleration of the gas to v_∞, some other mechanism probably initiates the mass loss. As we have seen above, the observational evidence suggests that the bulk of the dust forms outside of 10^{14} cm from the central star. At this distance there is already some evidence (although not completely compelling) for outflow at velocities $\lesssim v_\infty/2$. For early M stars, for example, there are Hα asymmetries (Boesgaard and Hagen 1979); for late M stars there are H_2O and SiO maser velocities. In each case, however, there still remain ambiguities (e.g. the maser velocity spread may be due to turbulence rather than outflow [Moran et al. 1979]) and, indeed, at this symposium Don Hall has suggested a substantially different picture than the one outlined below. Nonetheless, we feel that, at this time, the published literature suggests the following picture. For K and M stars mass loss is initiated by processes other than radiation pressure on dust which only forms at r $\gtrsim 10^{14}$ cm. Between 10^{14} and 10^{16} cm the formation process continues and the gas is accelerated to v_∞ by radiation pressure on the dust. At r $\gtrsim 10^{16}$ cm 1612 MHz OH maser and SiO and CO thermal emission are produced. At much greater distances the molecules are photodissociated by the ambient interstellar radiation field and the circumstellar gas is decelerated in the interstellar medium.

For K and early M stars (which are not pulsating) Mullan (1978) has suggested that large M results from a hydrodynamic expansion of the chromosphere-corona when the sonic point is located in the (high-density) chromosphere. For pulsating late M stars (Miras and SR's) shocks levitate the atmosphere raising the density at the sonic point substantially (Wood 1979). Whether the pulsation is in the fundamental mode or the 1st harmonic is still controversial (Wood 1978; Hill and Willson 1979) and relates to models for the evolution of these stars (Wood and Cahn 1977).

Finally we consider the evolutionary state of stars with circumstellar envelopes. As stars evolve up the red giant branch for the second time it is generally agreed that their core masses, luminosities and mass loss rates steadily increase. For Miras $\sim 10^{-7}$ $M_\odot$/yr is added to the core but, typically, $\gtrsim 10^{-6}$ $M_\odot$/yr is lost via expansion of the circumstellar envelope. So the subsequent state of the red giant star, i.e. planetary nebula or supernova, is determined by its initial mass and the mass loss. Wood (1978) has suggested that V Hya, a carbon-star with CO emission (Zuckerman et al. 1977), may be a pre-supernova. Knapp (1979) suggests that IRC+10216 may be a massive star undergoing extreme ($\sim 10^{-4}$ $M_\odot$/yr) mass loss. On the other hand, R Cr B stars, proposed as precursors of supernovae of type I by Wheeler (1978), do not seem to have molecular envelopes detectable to radio astronomy.

Although we may be looking at a few pre-supernova, most of the red giants with circumstellar envelopes will, no doubt, eventually evolve into planetary nebulae. Which of the many objects now under study by astronomers are actually pre-planetary nebulae (PPN's) is rather controversial. One school of thought is that the PPN's are to be found among the very red giant stars with CO radio emission, most of which are probably carbon-rich (Zuckerman et al. 1978, Zuckerman 1978). Another group prefers peculiar emission line objects such as V 1016 Cyg and HM Sge (Purton 1979). Our view is that if the latter objects are PPN's then they will evolve into only low mass PN's but not the bright ~ 0.2 $M_\odot$ PN's whose pictures appear in elementary astronomy texts. For example, V 1016 Cyg contains only a small amount of ionized gas (Kwok et al. 1978) and apparently little neutral gas since neither CO nor 2μm H_2 emission has been detected.

The matter can be largely resolved if [C]/[O] ratios can be determined for PN's. For the gas the best tool seems to be the UV lines of CIII and CIV. For IC 418 and NGC 7662, C/O ~ 1 is indicated although observational problems still remain (Harrington et al. 1979; Torres-Peimbert et al. 1979). Even if the observational problems are resolved the meaning of C/O ratios near unity may remain ambiguous. If during the PPN phase all of the less abundant of the two elements goes into CO which is later photodissociated by the central star of the PN, and the

bulk of the left-over C or O is incorporated into the dust, then the gas
may always show C/O ∿ 1 largely independent of its origin in C-rich or
O-rich material (Harrington 1979). If so, it will be necessary to determine the composition of the dust as well as the gas. To date, the hand-
ful of PN's (∿ 8) with apparent 10μm silicate or 11μm SiC features are
roughly divided between the two.

This research was partially supported by National Science Foundation
Grants AST 76-17600 and AST 77-28475 to the Universities of Maryland and
Texas, respectively.

REFERENCES

Becklin, E. E., Frogel, J. A., Hyland, A. R., Kristian, J., and Neugebauer
    G.: 1969, Astrophys. J., 158, L133.
Beckwith, S., Persson, S. E. and Gatley, I.: 1978, Astrophys. J., 219,
    L33.
Bernat, A. P. and Lambert, D. L.: 1976, Astrophys. J., 210, 395.
Boesgaard, A. M. and Hagen, W.: 1979, (preprint).
Campbell, M. F., Elias, J. H., Gezari, D. Y., Harvey, P. M., Hoffman,
    W. F., Hudson, H. S., Neugebauer, G., Soifer, B. T., Werner, M. W.,
    and Westbrook, W. E.: 1976, Astrophys. J., 208, 396.
Capriotti, E. R.: 1978, in Y. Terzian (ed.), IAU Symp. #76, Planetary
    Nebulae, Reidel, Dordrecht p. 263.
Harrington, J. P.: 1979, (private communication).
Harrington, J. P.,Lutz, J. H., Seaton, M. J., and Strickland, D. J.:
    1979, (preprint).
Hill, S. J. and Willson, L. A.: 1979, Astrophys. J., 229, 1029.
Knapp, G. R.: 1979, (private communication).
Kwan, J. and Hill, F.: 1977, Astrophys. J., 215, 781.
Kwok, S., Purton, C. R. and Fitzgerald, P. M.: 1978,Astrophys. J.,219,
    L125.
Lambert, D. L. and Vanden Bout, P. A.: 1978, Astrophys. J., 221, 854.
Mathis, J. S.: 1978, in Y. Terzian (ed.) IAU Symp. #76, Planetary Nebulae,
    Reidel, Dordrecht p. 281.
McCabe, E. M., Smith, R. C., and Clegg, R. E. S.: 1979, Nature,
    281, 263.
McCarthy, D. W.: 1979, in J. Davis and W. J. Tango (ed.) IAU Colloquium
    #50, High Angular Resolution Stellar Interferometry, University of
    Sydney, p. 18-1.
McMillan, R. S. and Tapia, S.: 1978, Astrophys. J., 226, L87.
Moran, J. M., Ball, J. A., Predmore, C. R., Lane, A. P., Huguenin, G. R.,
    Reid, M. J. and Hansen, S. S.: 1979, Astrophys. J., 231, L67.
Morris, M., Redman, R., Reid, M. J. and Dickinson, D. F.: 1979, Astrophys.
    J., 229, 257.
Mullan, D. J.: 1978, Astrophys. J., 226, 151.
Palmer, P. and Zuckerman, B.: 1978, (private communication).
Purton, C. R.: 1979, paper presented to Commission 34 at the XVII General
    Assembly of the IAU.

Salpeter, E. E.: 1974, *Astrophys. J.*, <u>193</u>, 579 and 585.
Schmidt, G. D., Angel, J. R. P., Beaver, E. A.: 1978, *Astrophys., J.*, <u>219</u>, 477.
Smith, D. and Adams, N. G.: 1978, *Astrophys. J.*, <u>220</u>, L87.
Sutton, E. C.: 1979, Ph.D. Dissertation, Univ. of California, Berkeley.
Torres-Peimbert, S., Peimbert, M. and Daltabuit, E.: 1979, (preprint).
Tsuji, T.: 1973, *Astron. and Astrophys.*, <u>23</u>, 411.
Vardya, M. S.: 1966, *Mon. Not. Roy. Astron. Soc.*, <u>134</u>, 347.
Wannier, P. G., Leighton, R. B., Knapp, G. R., Redman, R. O., Phillips, T. G., and Huggins, P. J.: 1979, *Astrophys. J.*, <u>230</u>, 149.
Wannier, P. G. and Linke, R. A.: 1978, *Astrophys. J.*, <u>225</u>, 130.
Werner, M. W., Beckwith, S., Gatley, I., Sellgren, K., Berriman, G., and Whiting, D. L.: 1979, (preprint).
Wheeler, J. C.: 1978, *Astrophys. J.*, <u>225</u>, 212.
Wood, P. R.: 1978, in *IAU Colloquium #46, Changing Trends in Variable Star Research*.
Wood, P. R.: 1979, *Astrophys. J.*, <u>227</u>, 220.
Wood, P. R. and Cahn, J. H.: 1977, *Astrophys. J.*, <u>211</u>, 499.
Zuckerman, B.: 1978, in Y. Terzian (ed.), *IAU Symp. #76, Planetary Nebulae*, Reidel, Dordrecht, p. 305.
Zuckerman, B., Palmer, P., Gilra, D. P., Turner, B. E., and Morris, M.: 1978, *Astrophys. J.*, <u>220</u>, L53.
Zuckerman, B., Palmer, P., Morris, M., Turner, B. E., Gilra, D. P., Bowers, P. F., and Gilmore, W.: 1977, *Astrophys. J.*, <u>211</u>, L97.
Zuckerman, B., Silverglate, P., Terzian, Y. and Wolff, M.: 1980, *Astrophys. J.*, (submitted).

DISCUSSION FOLLOWING ZUCKERMAN

Snyder: Professor Zuckerman, do you really believe in circumstellar tholins?

Winnewisser: We do not believe in tholins - let us please have the next question!

Zuckerman: I like Bishun Khare and Carl Sagan.

Elitzur: You have shown numbers for M_H/M_{total}. How did you derive M_{total}?

Zuckerman: For IRC+10011 I used the mass loss rate ($\dot{M}$) suggested by Goldreich and Scoville. For IRC+10216 I took $\dot{M}$ from Kwan and Hill. For CIT6 and NML Tau I assumed $\dot{M} = 10^{-5} M_\odot/yr$.

Greenberg: I could not understand how α Orionis can be spherical. This assertion is inconsistent with some linear polarization measurements of Tinbergen (unpublished) which have been analyzed by him, de Jager and myself, and which seem to provide firm evidence for dust grains distributed non-spherically and even seem to provide a size estimate of $a \simeq 0.05$ μm.

Zuckerman: My remark that α Orionis appears "reasonably" symmetric is based on observations of light scattered by dust (Tapia and McMillan) and on K-line scattering (Bernat and Lambert).

Kwok: Whether dust is the cause or effect of mass loss has been long debated. Clean silicates suffer from an inverse greenhouse effect in the sense that they are relatively transparent in the near IR, but emit more strongly at 10μ. For a 2000 K star, silicate grains can condense at only 0.07 stellar radii above the star (Gilman, unpublished manuscript). The infrared interferometry results are dependent on grain size distribution, grain temperature distribution, etc., and are not obviously incompatible with the above picture.

Zuckerman: I believe that the weight of the observational evidence, combined with the calculations of Jones and Merrill, suggests that the picture I painted is probably correct. However other possibilities may still be viable.

Hall: Why do you believe the dust in the envelope of IRC+10216 is graphite?

Zuckerman: This is the cononical view of the major constituent of carbon-star dust. SiC, for example, appears to be much less abundant. But perhaps the dust is primarily something else that no one has thought of.

Kwok: There are two well observed grain-formation processes: one is condensation in red-giant envelopes and the other is condensation in ionized ejecta from stars, for example, WC stars, novae, and the nuclei of planetary nebulae. The dust usually associated with planetary nebulae is probably formed under a different process than the dust seen in objects like IRC+10216.

Zuckerman: It could be. I acknowledged the possibility in my paper.

Hall: Why do you believe the dust size is ∿1 μm in IRC+10216?

Zuckerman: This conclusion is based on a fit to the far-infrared spectrum obtained by Campbell et al. 1 μm is appropriate for pure graphite or SiC, but smaller sizes would be possible if the grains have impurities with enhanced far-IR emission.

Hall: McCarthy & Low report substantial departures from spherical symmetry in 10216 at 5 and 10 μm. Is this inconsistent with your conclusions?

Zuckerman: My remark that IRC+10216 appears symmetric at 11 μm was based on the work of Sutton et al. If McCarthy and Low did indeed measure an asymmetry, it bears no obvious relationship to that seen in the 6000 Å photograph obtained by the Cal Tech IR group. Therefore, it appears important to confirm McCarthy's measurement.

Clark: Why do you think that the ground vibrational state ("thermal" SiO is situated as far out as the OH?

Zuckerman: I read it in the literature.

Morris: There is no direct observational evidence. The distribution is inferred from the widths, shapes, and intensities of the v=0 SiO lines, all of which are accounted for by models of extended envelopes (Morris & Alcock, Ap. J. 1977).

Silk: Can you set a lower limit on the mean rate of mass ejection of grains into the interstellar medium via mass loss from evolved stars?

Zuckerman: It could probably be done, but I have not yet done it.

OBSERVATIONS OF CIRCUMSTELLAR CLOUDS

P.G. Wannier, R.O. Redman
California Institute of Technology

T.G. Phillips
Bell Laboratories, Murray Hill, New Jersey

R.B. Leighton, G.R. Knapp
California Institute of Technology

P.J. Huggins
New York University

Observations have been made of J=2-1 CO in eleven circumstellar clouds including seven carbon stars and four oxygen-rich stars. Observations in four sources, including IRC+10216 have already been published (Wannier et al. 1979, henceforth Paper I) and the remaining observations are being prepared for publication (Knapp et al. 1980, henceforth Paper II). Several results are discussed below with special emphasis on the implications for two sources, namely IRC+10216 and Mira (o Ceti). The observations of IRC+10216 show CO emission over a diameter of 6 arcmin (~ 0.5pc), a result suggesting a very large mass-loss rate. Mira is unique among the objects studied in displaying a small CO opacity and a high CO excitation temperature. It is suggested that this heating results from the orbital velocity of Mira due to its close binary companion.

1. INTRODUCTION

There is, by now, a considerable list of late-type stars which exhibit measurable CO lines from extended envelopes of ejected material. The observations of thermal millimeter-wave emission lines hold the promise of providing an accurate picture of the entire mass-loss process. In turn, the history of the mass-loss rate $\dot{M}(t)$, the ejection velocity $V_e(t)$ and the composition of the ejecta, should provide additional information about the very uncertain post-main sequence stellar evolution. To date, most of the spatial information about stellar envelopes has been provided by comparing the lines of molecules which have very different excitation requirements. However, such information is indirect and the spatial information so derived is model dependent.

B. H. Andrew (ed.), Interstellar Molecules, 487–493.

The observations discussed below provide high spatial resolution ($\sim$ 25 arcsec) of a line (CO (J=2-1)) which traces out the most extended molecular gas. Spatial resolution of the stellar envelopes provides valuable new information, as evidenced especially by the results from IRC+10216.

2. DISCUSSION

Full details of the observational techniques are given in Paper II, and will not be repeated here. The important point is that the 10m antenna at the Owens Valley Radio Observatory provides a reasonably clean beam with a FWMH beamwidth of 25 arcsec at the observing frequency of 230 GHz. Calibrations of size and intensity were made using Saturn, Jupiter and Mars (Paper I).

In order to appreciate the significance of the maps, it is necessary to understand a few of the simple properties of constant mass outflow. When this is done, we see that the extended emission is far more significant in terms of total mass than is the brighter central source. Complete treatment of such outflow is given, for example, by Kwan and Hill (1977).

The observed outflow velocity far exceeds the escape velocity from the central object so that there can be no significant deceleration of the material. Thus, in the case that $\dot{M}$ and the initial velocity are constant, the space density is expected to fall as $1/r^2$. This gives rise to three effects significant for the observed CO brightness temperature: 1) the tangential column density falls as $1/r$, 2) the collision rate falls as $1/r^2$ and 3) the free expansion gives rise to adiabatic cooling. So long as the CO rotational lines are heavily trapped ($r \lesssim 3 \times 10^{16}$ cm in the case of IRC+10216), the apparent brightness temperature remains quite high. However, beyond a radius of 10^{17} cm the column density falls, the kinetic temperature falls and the collision rate is so low as to yield a very subthermal excitation. For the J=2-1 line, subthermal excitation is especially significant. Thus, the 2-1 CO brightness temperature is expected to fall precipitously with increasing radius.

Another property of circumstellar clouds is that the computational models are likely to be quite accurate. The kinematics, the geometry and the age of the material are reasonably well known. This situation contrasts the case for giant interstellar clouds where clumping, turbulance and a total lack of symmetry can make the application of computational models difficult at best. Our observations are therefore profitably examined in light of the computational model of Kwan and Hill.

2.1 IRC+10216

This well-studied object is an isolated variable carbon star at a distance of 200-300 pc. It displays very intense infra-red emission symptomatic of a thick dust shell. The initial mass of the star is unknown, though it is probably greater than $3M_\odot$. A high mass-loss rate ($\sim 10^{-4.5}\ M_\odot/yr$) has been indicated by millimeter-wave emission lines, and the IR observations are discussed by A.L. Betz and by D.N. Hall and S.T. Ridgeway in these proceedings. The ejected material is strongly affected by evolution in the parent star, as can be seen by its unusual nuclear composition (Wannier and Linke, 1977). However, until the indications in Paper I of a non-constant mass-loss rate, there had been no indications of physical evolution of the central object.

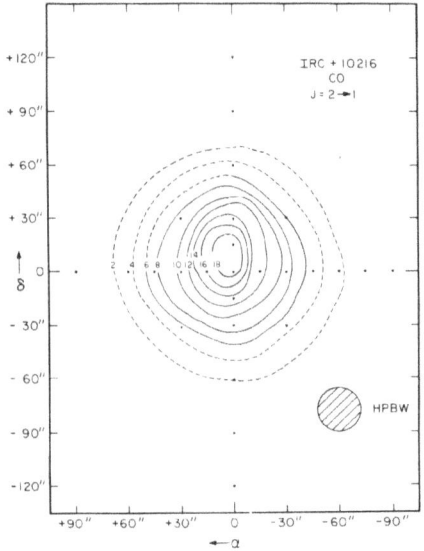

Figure 1. Map of the peak temperature T_A^* for IRC+10216. Map points are indicated by dots and the telescope halfpower beamwidth is shown. A circular symmetry is apparent.

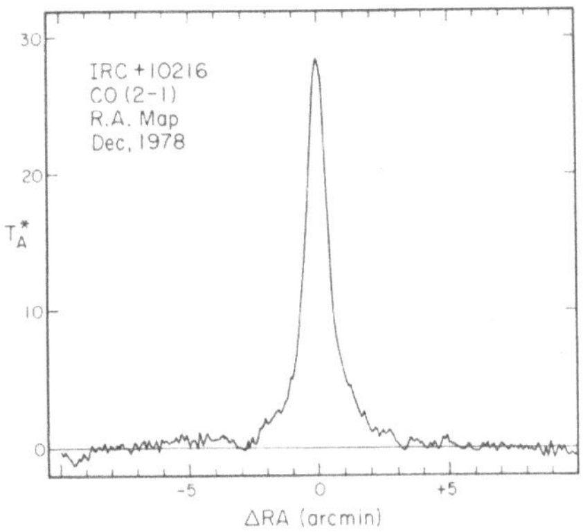

Figure 2. Map of CO (2-1) emission in IRC+10216 at V_{LSR} = -26 km/s.

In Figures 1 and 2, two different types of maps are shown for the 2-1 CO line. The first map, from Paper I, is of peak intensity made from individual 2-1 CO spectra. The second, from Paper II, is a strip map made by scanning the telescope past the source and is of the peak intensity at the central velocity using a total bandwidth of 1 MHz. The first map indicates that the circumstellar shell is spherically symmetric, at least out to a radius of 1.5 arcmin. Emission was detected even at the outermost points of the map. The second map shows that the 2-1 CO emission extends to a radius of at least 3 arcmin. Using an estimated distance of 290 pc (uncertain to at least ± 30%), this yields an apparent source diameter of 1.5×10^{18} cm which corresponds to an age of $\sim 3 \times 10^4$ yrs for the material at the outer observed boundary of the cloud

In light of our discussion about circumstellar clouds, the significance of the extended envelope is apparent. The extended wings, although less bright than the central source, are nontheless unexpectedly intense when compared to the model of Kwan and Hill(1977) by a factor of 3-6, depending on the radius (See Figure 3).

Using an assumed distance of 290 pc, a slightly better fit is obtained from the KH model by increasing the assumed mass-loss rate from 3×10^{-5} $M_\odot$/yr to 10^{-4} $M_\odot$/yr. This however, produces a 2-1 CO brightness which is at once too bright in the center and insufficiently bright in the extended envelope(Kwan, 1979). This extended emission is most naturally explained by assuming an increase in density above the $1/r^2$ relation which results from a constant value of $\dot{M}$ and V_e.

One property of the circumstellar envelope is that beyond a radius of $\sim 3 \times 10^{17}$ cm (the antenna beam diameter), the space density is too low to allow for further evolution of the dust or of the chemical composition of the gas. Thus the fractional CO abundance and the efficiency of

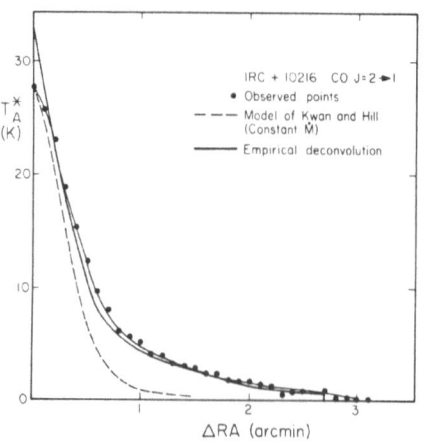

Figure 3. The observations of Fig. 2 are folded and displayed along with the model calculation for $\dot{M} = 2 \times 10^{-5}$ $M_\odot$/yr.

heating by radiation-driven dust must be considered to be constant. If the extended emission results from an unexpectedly large density of material, we must consider secular changes in either of the parameters $\dot{M}$ or V_e, a decrease in the former or an increase in the latter. However, a constant value of V_e is strongly indicated by the data of Kuiper et al. (1976) which show a dramatic high-velocity cut-off in the observed ^{13}CO (1-0) spectrum. We have recently confirmed the steep cut-off with improved (1-0) ^{13}CO observations. Therefore, a positive deviation from the $1/r^2$ density law must result from a larger past value of the mass-loss rate $\dot{M}(t)$.

2.2 Mira

In contrast to IRC+10216, Mira (o Ceti) is an oxygen-rich object exhibiting H_2O, OH and SiO maser emission. It has a much smaller mass-loss rate and is one component of a close binary system. The molecular envelope, whose size is large compared to the binary separation, exhibits a variable broad-velocity CO line (Lo and Bechis, 1977). The distance to Mira is known to be 77pc from measurements of trigonometric parallax.

Our own observations indicate that the CO(2-1) brightness temperature is far in excess of the CO(1-0) temperature measured with a similar telescope. (See Figure 4). Indeed, whereas all other objects in our survey have a CO(2-1)/CO(1-0) brightness ratio of between one and eight, Mira has a ratio in excess of 16 for the non-variable narrow line. Such a large ratio is very difficult to produce. A factor of four enhancement is possible if the source is unresolved, because of the comparable sizes of the OVRO 10.5m antenna used for CO(2-1) and the NRAO 11m used for the CO(1-0). A second factor of four is possible from the ratio of optical depths, so long as both of the lines are optically thin. These two factors could just account for the ratio of 16. Two additional effects, if present, might tend to <u>decrease</u> the CO(2-1) intensity. First, if the hydrogen molecular density were <4000 cm^{-3}, then the CO(2-1) line would become subthermally excited relative to the (1-0) line. Second, if the gas kinetic temperature were $\lesssim$ 20K, the line intensity ratio would be reduced because of the property of T_A^*, the equivalent Rayleigh-Jeans brightness temperature (see the discussion in Paper I). This last effect

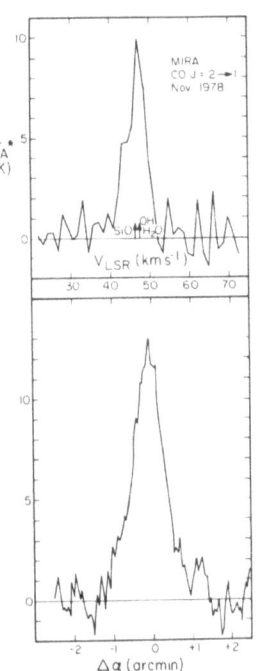

is unlikely in view of the large brightness temperature of the 2-1 CO line. We therefore infer that 1) the source is largely unresolved at CO(2-1), 2) the source is optically thin and 3) the density is $\gtrsim$ 4000 cm^{-3}.

The uniquely large (2-1)/(1-0) CO intensity ratio does not seem to result from the rather unexceptional values for $\dot{M}$ and V_e (2×10^{-6} $\dot{M}$/yr and 5.6 km/s respectively). However, a unique property of Mira is its membership in a close binary pair. The orbital parameters of the binary are not known, but reasonable assumptions for the masses and separation of the two stars ($\sim$ 1M$_\odot$ apiece and $\sim 7 \times 10^{14}$ cm) imply orbital motions of $\sim$3km/s, more than half of the envelope expansion velocity. From the discussion of Paper II, we see that this input of mechanical energy is easily enough to heat the gas to a temperature of 100K at the radius corresponding to our antenna beamwidth.

3.   CONCLUSIONS

We have observed unexpectedly

Figure 4. a) CO(2-1) spectrum of Mira showing velocities of the SiO, OH, H_2O masers. b) RA map at V_{LSR}=+47 km/s

intense CO(2-1) emission from a spatially extended envelope around
IRC+10216. The implied mass-loss rate of $\sim 10^{-4} M_\odot$/yr is larger than
that inferred from molecular line observations of the central source
and may indicate a secular variation of $\dot{M}$ within the past 3×10^4 years.

In Mira we have observed a very large value of ~ 16 for the CO(2-1)/
CO(1-0) line intensity ratio. From this we infer that the CO(2-1) must
be optically thin and that the source is largely unresolved by our 25
arcsec antenna beam. It is suggested that significant heating of the
gas may result from the orbital motion of the central star.

REFERENCES

Knapp, G.R., Phillips, T.G., Huggins, P.J., Leighton, R.B., and
    Wannier, P.G.: 1980, submitted to Ap. J.
Kuiper, T.B.H., Knapp, G.R., Knapp, S.L., and Brown, R.L.: 1976,
    Ap. J. 204, p. 408.
Kwan, J.: 1979, private communication.
Kwan, J., and Hill, F.: 1977, Ap. J. 215, p. 781.
Lo, K.Y., and Bechis, K.P.: 1977, Ap. J. (Letters) 218, p. L27.
Wannier, P.G., and Linke, R.A.: 1978, Ap. J. 225, p. 130.
Wannier, P.G., Leighton, R.B., Knapp, G.R., Redman, R.O., Phillips, T.G.,
    and Huggins, P.J.: 1979, Ap. J. 230, p. 149.

DISCUSSION FOLLOWING WANNIER

Black: The CO molecules in IRC+10216 are exposed to intense infrared radiation. How much does radiative excitation through vibrational
transitions contribute to the formation of the rotational lines?

Wannier: The infrared radiation strongly affects the rotational
excitation of molecules of high dipole moment such as HCN or CS. However
CO is relatively unaffected by IR radiation except in a very small region
$<10^{16}$ cm which does not affect our measurement of even the central
position. For comparison, our diffraction-limited beamwidth corresponds
to $\sim 3 \times 10^{17}$ cm. I refer you to the work of Morris for a discussion of
molecules other than CO.

Willner: If one attributes the excess CO emission to a decreasing
mass-loss rate, by what factor and over what time scale has the rate
decreased? Is the time scale not $r/v = \theta d/v \sim 10^4$ years? There are
very few known objects with mass-loss rates exceeding that of +10216.
Why do we not observe such objects?

Wannier: Your timescale is correct. The lifetime of a mass-loss object is inversely proportional to the mass-loss rate. Since the observed mass-loss rates vary over several orders of magnitude, the implied selection of observed objects is heavily weighted toward those with small rates. The observed list of circumstellar clouds, viewed this way, demonstrates a significant number of objects undergoing catastrophic mass loss.

Betz: In IRC+10216, infrared excitation of molecules such as HCN has been invoked to minimize the required H_2 density at radii <0.5. Is there any direct evidence that the density within this radius is actually as low as the minimum required if infrared pumping is adopted? The mass-loss rate estimated from our NH_3 observations within the central $\sim 1''$ radius also is higher than that used in the Kwan and Hill model.

Wannier: The only evidence against a high central mass-loss rate comes from a comparison of our CO (2-1) map with the model of Kwan and Hill (see section 2.1). A constant mass-loss rate (and a constant fractional abundance of CO) gives a predicted CO intensity which is at once too large in the central regions and too small in the extended envelope.

RADIO DETECTION OF AMMONIA IN IRC+10216

M.B. Bell, Sun Kwok and P.A. Feldman
Herzberg Institute of Astrophysics
National Research Council of Canada, Ottawa.

1. INTRODUCTION

IRC+10216 (CW Leo) is a carbon star surrounded by an expanding circumstellar envelope which is rich in molecules. Recently, Betz et al.[1] reported the detection of rotation-vibration transitions of NH_3 (ν_2 band) in absorption against the infrared continuum of circumstellar dust. We now report the detection of the (1,1) and possibly the (2,2) inversion transitions of (para) NH_3, the first radio detection of ammonia in a star. This information can be used to determine the thermal structure of the envelope of IRC+10216.

2. OBSERVATIONS

The observations were made in May & June 1979 using the 46-meter telescope of the Algonquin Radio Observatory.* The telescope has a beamwidth of 1.4 and an estimated beam efficiency $\eta_B \approx 0.28$ at 23.7 GHz. A cooled parametric amplifier with $T_S \approx 230$K was used with a very wideband (90 MHz) 100-channel filter spectrometer[2]. All observations were made using a new technique designed to provide a stable baseline over the wide observing window. The technique has previously been used successfully at the Algonquin Radio Observatory and is described by Bell[3]. The data were corrected for the variations in antenna gain and atmospheric attenuation with zenith angle, and the averaged spectrum was then smoothed by taking a running mean over two channels.

3. RESULTS

The radio spectrum of IRC+10216 is shown in Figure 1. In addition to the (1,1) and (2,2) inversion lines of NH_3 (para), we also

---

*The Algonquin Radio Observatory is operated by the National Research Council of Canada as a National Radio Astronomy facility.

B. H. Andrew (ed.), Interstellar Molecules, 495–496.

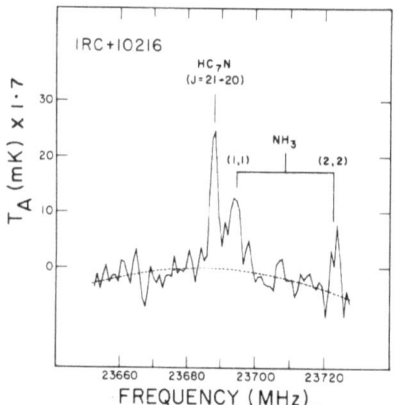

Fig. 1.   Spectrum of IRC+10216

find the J=21→20 rotational transition of HC$_7$N, previously detected in IRC+10216 by Winnewisser and Walmsley[4]. The values of peak antenna temperature, T$_A$*, corrected to outside the atmosphere and for the estimated beam efficiency at 23.7 GHz, are 27±5 and 16±5 mK, respectively, for the (1,1) and (2,2) lines of ammonia.

4.   DISCUSSION

IRC+10216 is believed to be a late-type star undergoing extensive mass loss. If the ammonia molecules are manufactured in the outer atmosphere of the star and subsequently expelled in the form of a stellar wind, the higher rotational states will gradually become de- populated as the molecules move out to where the gas temperature is lower due to adiabatic expansion. Results of our observations are consistent with the (2,2) state being populated only within ~30" of the star, and with a gas temperature distribution of the form T≈350(r/2×10^{15}cm)$^{-0.6}$K, i.e. they are consistent with CO observations (cf ref. 5). Because the (1,1) and (2,2) states have widely different excitation requirements, ammonia can be a very effective probe of the thermal structure of circumstellar envelopes. Assuming a mass-loss rate[5] of 2×10^{-5}M$_\odot$ yr^{-1}, the abundance ratio of NH$_3$ (para) to H$_2$ is found to be 2×10^{-8}, which is similar to the total abundance ratio of ~10^{-7} reported by Betz et al.[1].

REFERENCES

1. Betz, A.L., McLaren, R.A., and Spears, D.L.: 1979, Astrophys. J. (Letters) 229, L97.
2. McLeish, W.C.: 1973, IEEE Trans. Instrument. Measurements IM-22, 279.
3. Bell, M.B.: 1980, Proceedings of Radio Recombination Line Workshop, ed. P. Shaver, D. Reidel Publ. Co, Dordrecht, 259.
4. Winnewisser, G., and Walmsley, C.M.: 1978, Astron. Astrophys. 70, L37.
5. Kwan, J., and Hill, F.: 1977, Astrophys. J. 215, 781.

MOLECULAR ABUNDANCES IN IRC +10216

E.M. McCabe and R. Connon Smith
Astronomy Centre, University of Sussex, Brighton, UK

R.E.S. Clegg
Department of Astronomy, University of Texas
Austin, Texas, U.S.A.

ABSTRACT

The observed molecular column-densities in IRC +10216 can be matched by chemical-equilibrium calculations for $T \sim 1250^\circ K$, $P \sim 100$ dyn cm^{-2} and no graphite grain formation. Condensation of silicon carbide into grains may explain the low observed abundances of SiO and SiS.

INTRODUCTION

The infra-red object IRC +10216 is a barely visible, cool carbon star, surrounded by an expanding, dusty, molecular envelope, and has $T \sim 2230$ K, $L \sim 5 \times 10^4$ $L_\odot$ (Cohen 1979). The temperature of graphite dust is estimated to be 600 K at 10 R_*, falling to 350 K at 60 R_* (Crabtree and Martin 1979). SiC grains are also known to be present from an 11 emission feature (Treffers and Cohen 1974). The molecules CO and HCN are observed out to 4000 R_* and 1200 R_* respectively, while the other 15 molecules identified in IRC +10216 are unresolved. Molecular line widths indicate a terminal velocity of 15 km/s (McCabe et al. 1979, and references therein).

We wish to determine whether the relative abundances of all 17 molecules can be explained by a "freeze-out" model: this supposes that chemical equilibrium holds close to the central star's photosphere, but that as the temperature and density fall in the circumstellar wind molecular reaction rates drop rapidly enough that molecular abundances are "frozen" near their equilibrium abundances at some point near the star. The ratio of the chemical timescale to the expansion timescale is proportional to the distance from the star for bimolecular exchange reactions and a steady wind. This implies a possible freeze-out for $n(total) < 10^{10} cm^{-3}$ (cf. 10^{15} at the photosphere).

The comparison of abundances is tricky because column densities have been derived from infra-red absorption and radio emission lines. Only CO and HCN have been observed in both, and some of the column-

B. H. Andrew (ed.), Interstellar Molecules, 497–502.

densities are model-dependent. We merged the two abundance sets by scaling to the absorption figures to allow for dilution in the larger radio beam (60" compared with 3"). We have adopted abundances relative to CO for comparison with equilibrium calculations. In total there are 12 molecules with measured column-densities and known equilibrium constants.

Table 1. A comparison of molecules observed in IRC +10216 with calculated equilibrium abundances. The adopted relative abundances are with respect to CO = $10^{20} cm^{-3}$ and are taken from the underlined entries. The equilibrium abundances correspond to log P = 2, log T = 3.1.

Molecule	Column density (cm^{-2}) Emission (mm)	Absorption (μm)	Adopted relative abundance	Expansion (kms^{-1}) velocity	$\log\left(\dfrac{\text{observed abundance}}{\text{equilibrium abundance}}\right)$
CO	2×10^{17} 4×10^{17}	5×10^{19} $\underline{10^{20}}$	1	15-16	0 (by definition)
CS	$\underline{1.8\times10^{15}}$ CO/CS $\simeq 10^4$	-	1.8×10^{-5}	13	+0.6
CN	$\underline{10^{15}}$ CN/CO=2.8×10^{-3} CN/CS $\simeq 8$	-	2.8×10^{-3}	14	+6.5
C_3N	$\underline{10^{14} \to 8\times10^{14}}$	-	2.5×10^{-6}	13	+0.3
HCN & HNC	CO/HCN $\simeq 150$ CO/HCN $\simeq 60$ HCN/CN $\simeq 3$	$\underline{\geq 1.5\times10^{18}}$	0.015	-	-0.3
HC_3N	$\underline{1.8\times10^{15}}$	-	1.8×10^{-5}	13	-0.3
HC_5N	4×10^{14}	-	-	-	-
HC_7N	2×10^{14}	-	-	-	-
C_2H	$\underline{3\times10^{14} \to 2\times10^{15}}$	-	10^{-5}	15	+0.2
C_2H_2	-	$\underline{\geq 3\times10^{19}}$	0.3	-	-0.1
SiO	$\underline{4.1\times10^{14}}$	-	4.1×10^{-6}	10	-3.3
SiS	$\underline{1.6\times10^{15}}$	-	1.6×10^{-5}	13	-2.9
CH_3CN	Detection only	-	-	-	-
CH_4	-	$\underline{2.5\times10^{17}}$	2.5×10^{-3}	-	-0.8
C_4H	$\underline{4\times10^{14} \to 3\times10^{15}}$	-	1.2×10^{-5}	-	-0.7
NH_3	-	$\underline{10^{16}-10^{17}}$	$10^{-5.5}$	14	+4.0

CALCULATIONS

The calculations involve 164 molecules containing 25 elements. The 14 IRC +10216 molecules with known equilibrium constants are included. The set of equations for the conservation of each element is solved by Newton-Raphson iteration to give the number density of free neutral atoms for each element (Wyckoff and Clegg 1977). The input parameters are: $1250 \text{ K} < T < 2800 \text{ K}$, $10^{-5} < P < 10^3$ dyn cm^{-2} and $C/O = 1.76$ and 5, typical of conditions expected to exist in carbon star atmospheres. O/H is derived from the CO/H_2 ratio (Kwan and Hill 1977) and other abundances are assumed to be solar. Graphite grains are allowed for in one set of calculations by setting p(C) (the partial pressure of monatomic carbon) equal to p_V (C) (the vapour pressure of monatomic carbon over graphite) for $p(c) > p_V(C)$. An accuracy of around 1 order of magnitude is expected, allowing for 0.3 eV uncertainties in the dissociation energies.

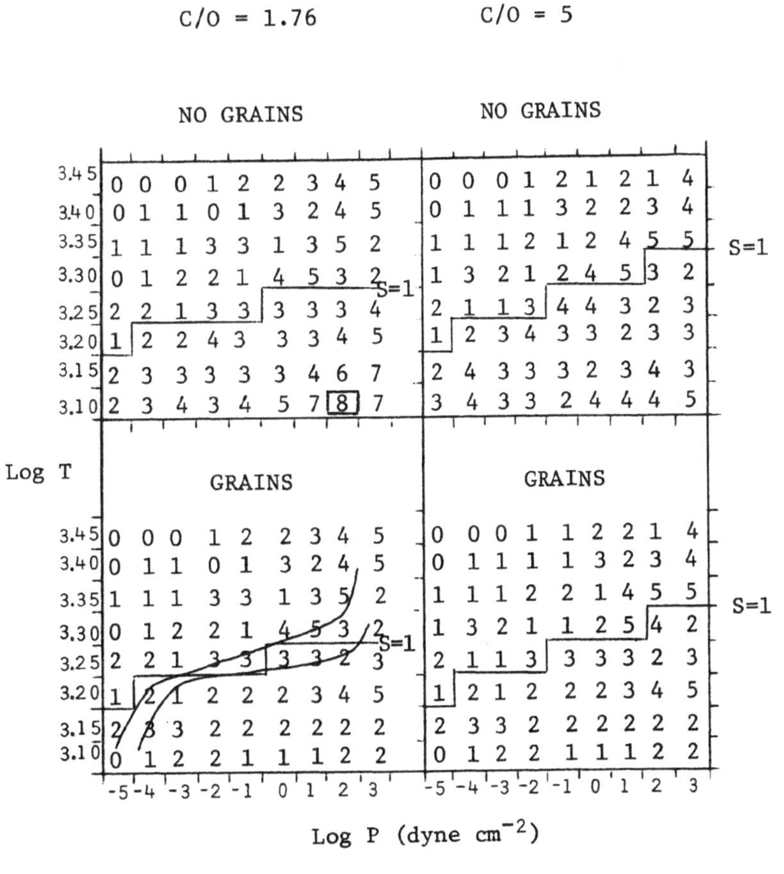

Figure 1. The number of molecules in IRC +10216 agreeing with equilibrium calculations to within an order of magnitude. A total of 12 molecules was considered.

RESULTS

The best agreement occurs for

a) $T \simeq 1250$ K, $P \sim 100$ dyn cm^{-2}
b) $C/O \sim 1.76$
c)   no graphite grains present

Under these circumstances 8 out of 12 molecules agree to within an order of magnitude. Of the other four:

d)   SiS and SiO are below the equilibrium abundance by a factor of about 10^3.
e)   CN and NH_3 are above the equilibrium abundance by factors of about 10^6 and 10^4 respectively.

DISCUSSION OF RESULTS (a) TO (e)

(a)   Such a P-T combination does not fit the Lucy model for cool carbon stars (Lucy 1976) in which the expanding atmosphere is driven primarily by radiation pressure on graphite grains. The P-T relations for that model are shown as the curves in the lower left of Fig. 1. However it is possible to get agreement for 5 molecules at $T \sim 2000$ K, $P \sim 10$ dyn cm^{-2}. Variation of the parameters in the model, in particular the grain composition (see (d)), does not alter this conclusion. A $1/r^2$ density distribution does fit the best-fit conditions if the ideal gas law is obeyed. In that case a model with a heated shock region near the photosphere is more appropriate (Willson 1976).

(b)   The C/O ratios in carbon stars are uncertain. IRC +10216 has a high carbon index in Cohen's 1979 classification. Our results suggest $C/O < 5$, so an intermediate value seems appropriate.

(c) C_2H_2 and HCN agree over a wide range of pressures and temperatures but give poor agreement when graphite grains are present. However the supersaturation ratio $S(C) = p(C)/p_V(C) \sim 30$ when no grains are present, and such a value normally implies the condensation of grains. There already exists evidence that grains do not form until $r = 5R_*$ (Sutton et al. 1979).

(d)   The low abundance of SiO and SiS suggests that SiC grains condense out before graphite. This might happen for a number of reasons:

(i)   For $C/O < 1.05$, $P \sim 10$ dyn cm^{-2} the supersaturation ratio of SiC $[S(SiC) = p(SiC)/p_V(SiC)]$ is greater than $S(C)$.

(ii) $S(SiC)$ rises more sharply than $S(C)$ as the temperature is lowered and may reach some critical value first.

(iii) The stellar radiation tends to inhibit graphite grain formation and promote SiC grain formation (Woolf 1975) - the inverse greenhouse effect.

Efficient SiC grain formation is required, a reasonable proposition in view of estimated graphite and SiC grain masses in the envelope, $3 \times 10^{-7} M_{\odot}$ and $4 \times 10^{-9} M_{\odot}$ respectively (Cohen 1979, Treffers and Cohen 1974). Further, the depletion factor of 10^3 implies initial condensation of SiC at around 1500 K, an acceptable temperature.

(e) The CN/CO ratio is particularly sensitive to temperature. Both CN and NH_3 may well be formed on grain surfaces.

FURTHER POINTS

(1) The abundances of large molecules, e.g. C_2H_4, C_2H_6, C_6H_6, not included in the equations, were found from the atomic abundances. Their inclusion would not affect the calculations.

(2) In the freeze-out model we expect the ratio HCN/HNC to be about 300 from the difference in binding energy between the two molecules. This is consistent with our estimated value from observations by Zuckerman (personal communication).

(3) Crude estimates for the equilibrium constants of HC_5N and HC_7N give a large uncertainty in the values for HC_5N/HC_3N and HC_7N/HC_5N. For a freeze-out at $T \sim 1000$ K the observed values lie within the margins of error (Clegg 1979).

ACKNOWLEDGEMENTS

We thank Leon Lucy, Harry Kroto and David Slavsky for helpful discussions, A. Betz for communicating his results for ammonia before publication and B. Zuckerman for HNC data. RESC acknowledges support from the Robert Welch Foundation of Texas and EMM acknowledges an SRC studentship.

REFERENCES

Clegg, R.: 1979, submitted to Mon. Not. Roy. Astr. Soc.
Cohen, M.: 1979, Mon. Not. Roy. Astr. Soc. 186, pp. 837-852.
Crabtree, D.R. and Martin, P.G.: 1979, Astrophys. J. 227, pp. 900-906.
Kwan, J. and Hill, F.: 1977, Astrophys. J. 215, pp. 781-787.
Lucy, L.B.: 1976, Astrophys. J. 205, pp. 482-491.
McCabe, E.M., Smith, R.E. and Clegg, R.E.S.: 1979, Nature 281, pp. 263-266.
Sutton, E.C., Betz, A. and Storey, J.W.V.: 1979, Astrophys. J. (Letters) 230, pp. L105-L108.
Treffers, R. and Cohen, M.: 1974, Astrophys. J. 188, pp. 545-552.
Willson, L.A.: 1976, Astrophys. J. 205, pp. 172-181.

Woolf, N.J.: 1975, in "The Dusty Universe", Neal Watson Academic
    Publications Inc., New York.
Wyckoff, S. and Clegg, R.: 1977, Mon. Not. Roy. Astr. Soc. 184,
    pp. 127-144.

DISCUSSION FOLLOWING McCABE

Herbst: The JANAF tables are often in error!

McCabe: Updating of equilibrium constants in our calculations has
never altered the relative abundances by significant amounts. We require
an order of magnitude accuracy, typical of the estimated errors in the
observations.

Tatum: In order to compare the results of different equilibrium
calculations, there is a need for us all to use a "standard" set of
equilibrium constants, without necessarily trying to up-date them in the
light of more recent data. For diatomic molecules, for example, one
could calculate "standard" equilibrium constants from the data in the
recent book by Huber and Herzberg.

McCabe: A standard set of equilibrium constants upon which calculations can be based is certainly desirable. For those molecules which
are included in the Tsuji 1964 chemical equilibrium calculations, it has
been possible to compare results over part of the p-T range in the
C/O=5 case. The agreement is found to be adequate for our purposes.
It would, of course, be useful for others to repeat our calculations.

Huebner: Assuming non-equilibrium in your calculations is a very
important advance in stellar atmosphere modelling. Do you have all the
reverse reactions in your program to check if chemical equilibrium is
indeed a good starting condition?

McCabe: Our calculations are for *chemical equilibrium* only. The
individual reactions involved are therefore irrelevant and do not enter
into the program. Crude estimates of reaction rates suggest that to a
first approximation we can consider there to be an abrupt transition
from chemical equilibrium to no reactions at all. Our fit to observed
abundances is based on this "freeze-out" assumption. We are not
attempting to do time-dependent calculations.

INFRARED HETERODYNE SPECTROSCOPY OF CIRCUMSTELLAR MOLECULES

A.L. Betz
Department of Physics, University of California, Berkeley

R.A. McLaren
Department of Astronomy, University of Toronto

ABSTRACT

Ammonia has been detected in the circumstellar envelopes of IRC+10216, VY CMa, VX Sgr, and IRC+10420. A number of absorption lines of $^{14}NH_3$ in the ν_2 vibration-rotation band around 28 THz (950 cm^{-1}) have been observed at a velocity resolution of 0.2 km/s. Typical linewidths are 1 to 4 km/s, and the details of the line profiles provide additional insights on the process of mass loss in these stars.

1. INTRODUCTION

The extension of heterodyne techniques to infrared spectroscopy now permits the vibrational transitions of interstellar and circumstellar molecules to be studied at the high velocity resolution commonly used in microwave line observations. Well-resolved line profiles are especially necessary in characterizing the dynamics of mass loss in circumstellar environments. The interpretation of these profiles is considerably simplified for infrared absorption lines in that only the radial velocity component of circumstellar expansion is seen against a small (<1 arcsec) continuum source. Typical linewidths are 1 to 4 km/s. This is in contrast to microwave emission line observations with ~1 arcmin beamwidths, where line emission over the full range of projected expansion velocities is seen, and wide (~30 km/s) profiles are observed centered about the stellar velocity.

The usefulness of heterodyne techniques in infrared spectroscopy has been demonstrated by the detection of ammonia in the circumstellar envelope of IRC+10216 (Betz et al. 1979). Subsequent observations on this source and several supergiant maser stars show that ammonia is relatively abundant and is an excellent indicator of dynamics throughout the circumstellar cloud. Also, for the maser stars, the velocities of the ammonia absorption lines can be correlated with those of OH-maser emission features determined from VLBI measurements to fix

B. H. Andrew (ed.), Interstellar Molecules, 503–508.
Copyright © 1980 by the IAU.

the location of the central star relative to the extended cloud of maser components.

2. INSTRUMENTATION

Figure 1 shows a simplified schematic of the receiver which is used at the 1.5 m McMath Solar Telescope of Kitt Peak National Observatory. At a wavelength of 10.6µm, the diffraction-limited beam-size is ∿1.5 arcsec. The local oscillator is a CO_2 laser capable of oscillating on any one of a number of discrete vibration-rotation transitions of CO_2 in the 10µm band. Centering of the laser to the peak of the power output curve controls the LO frequency to a fractional accuracy better than 10^{-7}. The mixer is a HgCdTe infrared photodiode with an output current response extending from DC to 1800 MHz. The output of the photomixer is processed by a second mixer and directed into a filterbank of sixty-four 20 MHz filters. Each channel corresponds to a velocity width of ∿0.2 km/s.

A number of close frequency coincidences have been measured between CO_2 laser lines and fundamental transitions of NH_3 in the ν_2 vibration-rotation band. Laboratory measurements have been accomplished with a variety of laser-related techniques, principally laser-Stark spectroscopy (Ueda and Shimoda 1975), laser-microwave double-resonance spectroscopy (Freund and Oka 1976), and laser heterodyne spectroscopy (Hillman et al. 1977). As in microwave astronomy, good rest frequencies must also be known in the infrared before attempting observations. This is especially apparent when the total IF velocity coverage is only ∿14 km/s.

Since the local oscillator frequencies are fixed, stellar sources are fine-tuned into frequency coincidence using the orbital motion of the earth. At a particular time, only a few molecular lines may fall within the 14 km/s "window", and observing times must be selected with this in mind. The orbital motion of the earth shifts the spectra

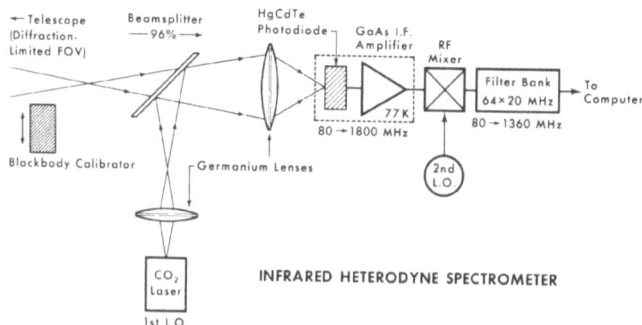

Figure 1

∿1 channel/day, and repeated observations over a few days are generally needed to unambiguously identify the infrared sideband into which an isolated line may fall. Although this technique is not as convenient as having direct frequency control of the local oscillator, it is still practical in that many of the strongest NH_3 lines are accessible for most supergiant stars. It is particularly fortunate that ammonia has such a rich spectrum overlapping the $10\mu m$ CO_2 laser bands. The non-metastable energy levels of NH_3 are especially important in that the collisionally-excited population of these levels is sensitive to H_2 densities of 10^6 to 10^{10} cm^{-3}, which are characteristic of the densities expected in circumstellar clouds.

3. OBSERVATIONS

A) IRC+10216

The identification of NH_3 in IRC+10216 was based on the detection of 3 lines in the ν_2 band: aR(1,1), aQ(2,2) and aQ(6,6) (Betz et al. 1979). These observations were completed in early June 1978, close to the time of maximum infrared brightness. Additional observations were done in May 1979, near minimum infrared brightness, when the continuum intensity was about 2.5 times weaker than at previous maximum. In these latter observations, the aR(0,0) transition illustrated in Figure 2 was also detected. The rotational level of the lower state of this vibrational transition has no 24 GHz inversion transition and hence cannot be detected in microwave observations.

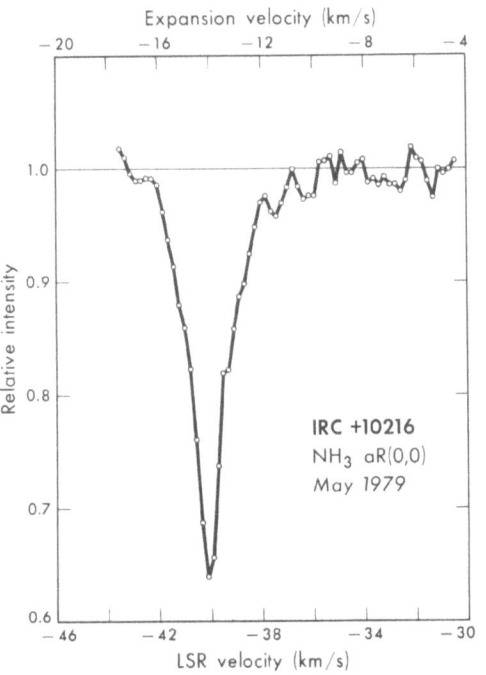

Figure 2: The aR(0,0) line of NH_3 in IRC+10216. The rest frequency of this transition is 28.533534 THz. The expansion velocity for circumstellar gas assumes an intrinsic stellar velocity of -26 km/s.

Both the old and new results indicate that ammonia is relatively abundant, with a column density around 10^{17} cm^{-2} and a fractional abundance of 10^{-7} relative to H_2. Ammonia can be seen both close to the star such that continuum emission from intervening dust partially obscures the lines, and also at large radial distances where the molecular population collects into the lowest rotational levels. The appearances of lines of different rotational excitation are consequently to some extent indicative of gas at different radial distances and in environments of differing temperature and density. Close to the star, within the region of 10μm continuum radiation from dust, a rotational temperature of 400 to 700 K and an H_2 density around 10^9 cm^{-3} are indicated. The central intensities of lines such as $(J,K) = (0,0)$ also reveal that ammonia is detectable in the cooler, <200 K, regions of the circumstellar cloud, well outside the area of 10μm continuum emission. In all the lines, the circumstellar expansion velocity is seen to be remarkably constant. By the time NH_3 is abundant enough to be detectable, the expansion velocity is already close to the terminal velocity of ∿14 km/s. If radiation pressure on grains drives the gas expansion and the hottest grains form about 5 stellar radii (0.2 arcsec) from the central star (Sutton et al. 1979), then the ammonia extends outward from this radius. The relatively large column density of NH_3 and the sharpness of the line profiles suggest that the relative NH_3 abundance is rather more in equilibrium with the dust grain temperature than frozen at the photospheric value. Modeling of the observed line profiles by D. Crabtree at Toronto also indicates that turbulence in the circumstellar gas is small but measureable at about 1 km/s. Future observations should clarify the radial dependence of the NH_3 abundance close to the star and establish more quantitative estimates of the H_2 density throughout the region of NH_3 absorption.

B) OH/IR Supergiants

Ammonia was searched for and detected in 3 OH-maser stars: VY CMa, VX Sgr, and IRC+10420 (McLaren and Betz 1979). In these sources the absorption lines are all on the order of 20 to 30% deep, and the linewidths range from 1 to 4 km/s. In VY CMa the NH_3 is seen at -4.5 km/s (LSR). If the stellar velocity may be taken at +18 km/s from the midpoint of the two OH maser peaks, then the NH_3 is observed at a uniform expansion velocity of about -23 km/s. It is interesting to associate the NH_3 absorption-line velocity with the -4.5 km/s OH maser component seen near the physical center of the masing region in VLBI observations (Reid and Muhleman, 1978; Moran et al. 1977). Presumably, this -4.5 km/s component indicates the gas in direct line-of-sight expansion from the central star. This interpretation is strengthened by similar associations between NH_3 absorption-line velocities in VX Sgr and IRC+10420 and OH maser components seen near the apparant centers of maser groupings in these sources (Moran et al. 1977, Benson et al. 1979). In VX Sgr, NH_3 absorption and OH emission are both seen at -14.3 km/s, while in IRC+10420, NH_3 and OH are found at +45.9 km/s. It thus seems reasonable that the central stars are located directly behind these velocity components. These velocity correlations do not necessarily imply that the NH_3 absorption and OH emission occur at the same radial distance from the star. Gas is already accelerated near the terminal velocity by the time the hottest

NH_3 is seen, and it continues outward at this velocity toward the cooler region where OH emission occurs. However, NH_3 in the low energy rotational states such as (0,0) certainly extends into the OH emission region. Additional observations of NH_3 in a larger number of maser stars will allow more quantitative estimates to be made.

4. FUTURE PROSPECTS

Within the next year or two, expected system improvements in both sensitivity and frequency coverage will greatly expand the applications of heterodyne spectroscopy. Under good seeing conditions, a 3-m class telescope will give a 4-fold increase in sensitivity over the 1.5 m telescope currently in use; and, together with modest receiver improvements, will permit good absorption line spectroscopy at $10\mu m$ on well over 100 sources. With recently improved photomixers and more IF filter channels, the simultaneous velocity coverage can now be extended to ~ 28 km/s. This will not only provide a decent baseline for continuum estimates but also double the efficiency of telescope usage. Laser technology, principally with CO_2, N_2O, and CO gas lasers, is well enough advanced to give thousands of LO frequencies over the 5 to $12\mu m$ region, and thus encourage observations of other important circumstellar molecules such as CO and SiO.

This work is supported in part by NASA Grant NGR 05-003-452 and NSF Grant AST78-24453.

REFERENCES

Benson, J.M., Mutel, R.L., Fix, J.D. and Claussen, M.J., 1979 Astrophys. J., 229, L87-90.
Betz, A.L., McLaren, R.A. and Spears, D.L.,1979, Astrophys. J., 229, L97-100.
Freund, S.M. and Oka, T., 1976, Phys. Rev. A., 13, 2178-2190.
Hillman, J.J., Kostiuk, T., Buhl, D., Faris, J.L., Novoco, J.C. and Mumma, M.J., 1977, Optics Letters, 1, 81-83.
McLaren, R.A. and Betz, A.L. (submitted for publication).
Moran, J.M., Ball, J.A., Yen, J.L., Schwartz, P.R., Johnston, K.J. and Knowles, S.H., 1977, Astrophys. J., 211, 160-169.
Reid, M.J. and Muhleman, D.O., 1978, Astrophys. J., 220, 229-238.
Sutton, E.C., Betz, A.L., Storey, J.W.V. and Spears, D.L., 1979, Astrophys. J., 230, L105-108.
Ueda, Y. and Shimoda, K., in Laser Spectroscopy (II), ed. S. Haroche (Lecture Notes in Physics, Vol. 43) (Springer-Verlag: Berlin, Heidelberg), 1975.

DISCUSSION FOLLOWING BETZ

Snyder: Could you comment on the agreement between your radial velocities and the radial velocities found for radio molecules, for example CO, in IRC+10216.

Betz: We observe NH_3 in absorption at $\sim$ -40 km/s V_{LSR}. The broad radio emission lines of CO and other molecules are centered at -26 km/s and extend from $\sim$ -40 to -12 km/s. Consequently, if we interpret the intrinsic stellar velocity as -26 km/s and the circumstellar expansion velocity as 14 km/s, all the observations are in agreement.

Feldman: Bell, Kwok and I have recently detected both the (1,1) and (2,2) transitions of NH_3 in the radio spectrum of IRC+10216 (this volume). Preliminary results are that the velocity widths are similar to those of other molecules previously found in the circumstellar envelope of IRC+10216, and that the relative abundance of ammonia is $\sim 10^{-7}$ compared to H_2.

McCabe: At what distance in stellar radii from the star are you observing the NH_3 in IRC+10216?

Betz: The ammonia absorption lines are sharp, and indicate that the gas has already been accelerated close to the terminal velocity of $\sim$14 km/s before much of the NH_3 in these lines is formed. Most of the mass-loss acceleration must occur within a few stellar radii, and most of the observable ammonia is outside of this region. NH_3 lines requiring higher excitation, such as the (6,6), are predominantly formed closer to this few-stellar-radii limit than "cold" lines such the (0,0) which extends out into the larger "radio" envelope.

SPECTROSCOPIC STUDIES OF IRC+10216 AND SIMILAR OBJECTS

Stephen T. Ridgway and Donald N. B. Hall
Kitt Peak National Observatory*

ABSTRACT

Spectroscopy of circumstellar molecular species in the 2–14μm range provides evidence for a range of shell optical depths in the +10216 class of stars. In some cases a photospheric spectrum is present. The spectrum of +10216 shows absorption over a range of velocities including two distinct velocities of different excitation temperature. The occurrence of multiple velocities may be common in similar objects.

1. OBSERVATIONS

IRC+10216 is the archetype of a class of cool carbon stars. These very cool, probably Mira type, stars are ejecting substantial quantities of dust and gas. With the techniques of infrared and millimeter spectroscopy and infrared interferometry it is possible to examine the dynamics of the ejection process. These observations will lay the foundations for an interpretation of the ejection mechanisms and the chemical kinetics of the circumstellar and nascent interstellar material.

We have recorded high resolution (2–3 km/sec) infrared spectra of four members of the +10216 class. The objects observed (+10216, +30219, −10236 and +50096) exhibit a range of [2μm–3μm] color temperatures (550, 680, 830 and 975K respectively). As shown below, these objects substantiate the expected correlation between color temperature and optical depth in the circumstellar shell. The spectral range covered includes all terrestrial windows from 2 to 13.5μm. Circumstellar molecular absorption due to CO, HCN, C_2H_2 and CH_4 has been detected in fundamental, overtone and combination bands (e.g. Ridgway et al. 1976, Ridgway et al. 1978, Hall and Ridgway 1978). For the study of temperature–density distributions and dynamics the CO bands are uniquely valuable. Each band provides many spectral lines of various lower state excitations and line strengths.

*Operated by the Association of Universities for Research in Astronomy Inc., under contract with the National Science Foundation.

B. H. Andrew (ed.), Interstellar Molecules, 509–514.

Dynamical information is obtainable from even moderate resolution
spectra. In Figure 1 we show the LSR velocity of line center for each
detected line in just the 2-0 band of CO. (The noise can be estimated
from the scatter between adjacent points.) In each figure the blue edge
of CO microwave thermal emission has been noted (tip of arrow), and also
the microwave line-center if on-scale (origin of arrow). Note the
smooth variation of velocity with excitation (rotational quantum of the
lower state). The observed absorption velocities lie in the range expected for an expanding shell. In the case of +50096, the least obscured of the four objects, the absorption line velocities do not vary
strongly with excitation, but two components appear: one near the velo-

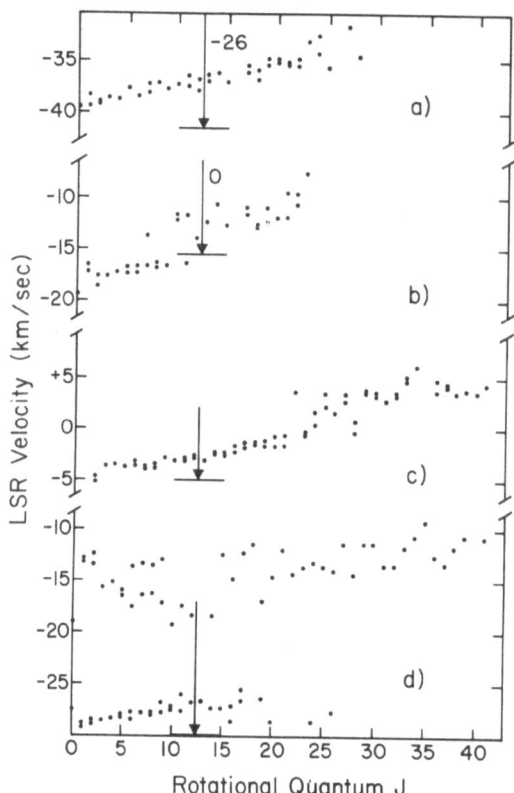

Fig. 1. CO v = 2-0 line positions. a) IRC+10216. b) IRC+30219.
c) IRC-10236. d) IRC+50096. Note two velocity components in part d.
The arrow indicates the velocity of J = 1-0 emission (Wilson et al.
1973, Kuiper et al. 1976, Zuckerman et al. 1977, Zuckerman et al. 1978).
The origin of the arrow indicates 1-0 line center velocity (or labeled
if off-scale), and the tip is at the blue edge. Observation dates are
15 Oct 1978, 17 Oct. 1978, 19 Feb. 1979, and 17 Oct. 1978. To convert
to heliocentric velocity add 6.8, 1.5, 9.8, and 1.6 km/sec.

city of microwave line-center (presumed center of mass velocity) and one at the blue edge (relative expansion velocity).

For a more detailed understanding of the velocities, it is necessary to study the line profiles at higher resolution. In Figure 2 several CO line profiles of +10216 have been collected. At the top is the CO microwave profile measured by Kuiper et al. (1976). For each of the infrared profiles, several CO lines of similar excitation have been

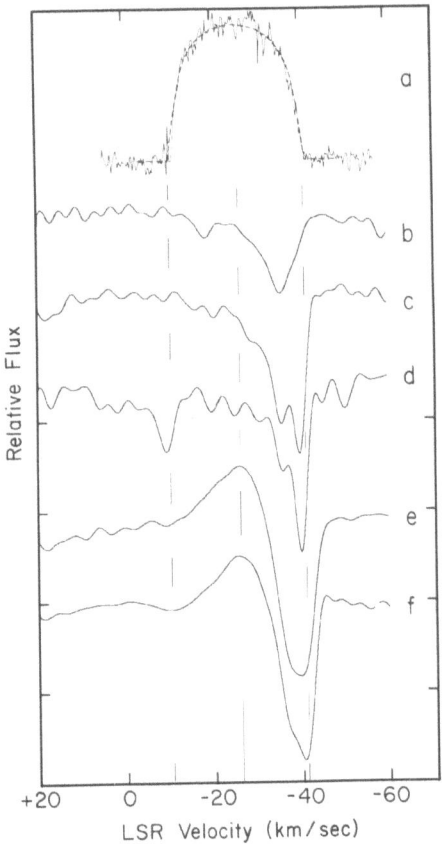

Fig. 2. CO line profiles in +10216. a) J = 1-0 emission from Kuiper et al. (1976). b) v = 2-0 absorption, high excitation profile. c) 2-0 medium excitation profile. d) 2-0 low excitation profile. e) v = 1-0 absorption low excitation profile (from ^{13}CO). f) 1-0 high excitation profile. Zero flux levels for spectra b-e are indicated by tic marks. The fundamental band observation date was 13 Oct. 1978 and overtone observations were 15 Oct. 1978. This was shortly after expected maximum in the light curve of +10216. Terrestrial absorption lines appear in the wings of several of the profiles.

averaged to reduce noise. First consider the medium excitation lines
in the overtone band (Fig. 2c). Line doubling is evident; the splitting
would be more complete in a fully resolved spectrum. The blue component appears in the low excitation overtone line (2d) but not in the
high excitation line (2b). From the distribution of strength with excitation we conclude that the temperature of the gas producing the blue
absorption is $\sim$ 150-250K. From similar considerations, the red component must arise in gas $\sim$ 300-700K. (In addition to observational
error, it is probable that a range of gas temperatures contribute to
each component.) The red (hot) absorption also has a broad red wing
which probably extends to the red side of the center-of-mass velocity.

The fundamental band low and high excitation profiles (Figures 2 e
and f) show the doubling as an asymmetry. In addition, a strong P Cygni
type emission appears in the fundamental. Such an emission is not expected in the overtone, since the fundamental is the preferred route for
emission from excited CO.

2. DISCUSSION

A simple interpretation of the +10216 line shape requires two regimes of absorption with distinct temperatures and velocities. In a
spherically symmetric geometry, we might imagine a warm inner shell expanding at 11 km/sec and a cool outer shell expanding at 16 km/sec. The
redward extension of the overtone absorption to the center-of-mass velocity can be explained by simple geometry. The line of sight to the edge
of the continuum source intercepts radially flowing gas with projected
velocity less than the expansion velocity by an amount dependent on the
radial distance of the gas from the continuum source. For gas near the
continuum source (the hot component) this produces an absorption wing
extending as far to the red as the center of mass velocity. For gas
distant from the continuum source (the cool component) the velocity
spread due to this projection effect is small. These considerations
will not account for absorption to the red side of the center-of-mass
velocity. It may be possible to account for the extended profile if we
hypothesize that a portion of the 2.4μm flux is scattered at least once
so that we see light emitted from the back hemisphere of the continuum
source. The P Cygni emission is probably associated with the hot absorption component, and it appears that the emission approximately
equals the corresponding absorption. This indicates that the scattering
region is mostly within our 2.5 arc-sec aperture, and also that the 0.4
arc-sec continuum source does not substantially occult the scattering
region. The angular diameter of the cool outer shell probably exceeds
the instrument field-of-view; hence, no significant re-emission is expected in the line wing.

It may be difficult to account for a higher expansion velocity in
the outer layers with a steady-state model. If the acceleration is to
be attributed to radiation pressure, it would be necessary for a radiation pressure gradient within the observed column to be contrived so

as to give a discrete boost to the gas velocity. Of course, one might appeal to catastrophism (shell ejection, etc.). But from the similarities between Figures 1a, b, and c, it appears that the combination of gas temperatures and velocities found in +10216 recur in similar objects, weighing against interpretations invoking discrete events. If we abandon spherical symmetry (as we must according to McCarthy et al. 1978) then we may readily sketch diagrams with various segments of disks and shells in the line of sight, each with distinct temperature and velocity. A suitable evaluation of such possibilities requires a comprehensive synthesis of spectral and spatial data.

In +10216 the absence of CO bandheads indicates that less than 6 percent of the flux at 2.3μm is photospheric radiation (direct or scattered). From the known flux, and an assumed stellar temperature > 1000K, it follows that the star is obliterated by a continuum absorption optical depth greater than 5.5. In -10236 and +50096, however, we see evidence for a photospheric spectrum. In both cases the 2μm CO bands extend to line excitation of $J\sim40$ with well-developed band heads. The red components of the lines have velocities near the center-of-mass velocity. In Figure 3, the low excitation line profile in the CO 2-0 band of +50096 substantiates a photospheric origin for this absorption. The hot absorption profile has the broad, flat-bottomed shape of a photospheric line, though diluted by circumstellar continuum emission. A single expanding cool shell produces a blue-shifted line. The photospheric component has an excitation temperature $\sim$1300K, and the circumstellar component $\sim$500K. In -10236 the photospheric spectrum is not resolved from the circumstellar. The resemblance to +50096 suggests a similar situation but with greater circumstellar obscuration and a lower shell expansion velocity (6 vs 13 km/sec). In +50096 and -10236 we estimate absorption optical depths $\sim$ 0.7 and 1.4 respectively.

On the basis of this preliminary study, the +10216 type objects appear to have a range of shell optical depths. The more heavily obscured examples show a double shell velocity structure. Yet cooler, possibly more heavily obscured +10216 objects are known, as well as a

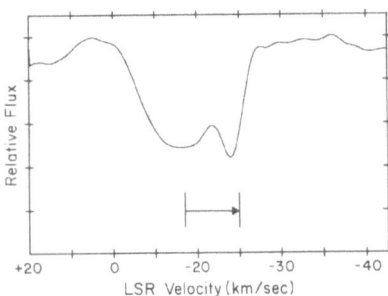

Fig. 3. CO v = 2-0 low excitation line profile in IRC+50096. The origin and tip of arrow indicate center and blue edge of J = 1-0 emission (Zuckerman et al 1977).

range of less obscured stars that span the gap from, for example, the Mira V Cygni to IRC+50096. We plan to extend our survey to these members of the +10216 class. Since each closer look at +10216 reveals yet more complexity, this source also deserves more detailed scrutiny.

REFERENCES

Hall, D. N. B. and Ridgway, S. T. (1978). Nature 273, 281-281.

Kuiper, T. B. H., Knapp, G. R., Knapp, S. L. and Brown, R. L. (1976). Astrophys. J. 204, 408-414.

McCarthy, D. W. (1978). Proc. IAU Colloq. No. 50 (College Park, Maryland).

Ridgway, S. T., Hall, D. N. B., Kleinmann, S. G., Weinberger, D. and Wojslaw, R. S. (1976). Nature 264, 345-346.

Ridgway, S. T., Carbon, D. F. and Hall, D. N. B. (1978). Astrophys. J. 225, 138-147.

Wilson, W. J., Schwartz, P. R. and Epstein, E. E. (1973). Astrophys. J. 183, 871.

Zuckerman, B., Palmer, P., Morris, M., Turner, B. E. and Gilra, D. P. (1977). Astrophys. J. Letters 211, L97-101.

Zuckerman, B., Palmer, P., Gilra, D. P., Turner, B. E. and Morris, M. (1978). Astrophys. J. Letters 220, L53-56.

INFRARED SPECTROSCOPY OF MOLECULES IN CIRCUMSTELLAR MATERIAL

Donald N. B. Hall
Kitt Peak National Observatory*

ABSTRACT

High resolution spectra of red giants and long period variables exhibit lines of infrared CO vibration-rotation bands arising in circumstellar material. In the few such stars so far observed at very high resolution ($\lesssim$ 1 km/s) the circumstellar material appears localized in 3 distinct regimes with temperatures of 800K, 200K and 75K and expansion velocities of 0, 10 and 16 km/s rather than being uniformly distributed.

Many red giants, supergiants and long period variables are known from high resolution observations of atomic resonance lines to be losing mass at a substantial rate and to have enveloped themselves in clouds of circumstellar material. The elucidation of such processes is critical to the understanding of both stellar and galactic evolution, for such stars probably shed enough material to change their evolutionary tracks while at the same time substantially altering the composition of the interstellar medium by ejecting material which has been processed in the star's interior. Many of the circumstellar shells are prolific sources of both thermal and maser lines at millimeter wavelengths while infrared excesses indicate dust coexists with the circumstellar gas. However the mass loss mechanism, factors governing chemical equilibrium in the expanding material and even actual mass loss rates remain unspecified. The subject of mass loss has been reviewed by Goldberg (1979), Reimers (1977), Weymann (1977), Conti (1978) and Cassinelli (1979).

The importance of detecting molecular lines arising in circumstellar material seen in absorption against the central star is evident. One can, in principle, utilize the rotational level distribution both to obtain excitation temperatures and as a depth dependent probe. Attempts to detect molecules such as CN (Weymann, 1962) and

* Operated by the Association of Universities for Research in Astronomy Inc., under contract with the National Science Foundation.

B. H. Andrew (ed.), Interstellar Molecules, 515–524.

TiO (Bernat, 1976; Lambert and Vanden Bout, 1978) at visible wavelengths have yielded negative results. Of all molecules, CO is a
particularly attractive candidate for detection in circumstellar
material because of both its high dissociation potential and the high
cosmic abundances of carbon and oxygen. Although CO does not have any
electronic bands accessible to ground based observers, it does have
vibration rotation bands in convenient regions of the infrared, notably
the fundamental around 4.6μm and the first overtone around 2.3μm. The
strengths and particularly the frequencies of the lines of these bands
have been precisely determined and, in principle, allow accurate determinations of radial velocities and rotational excitation temperatures.
The rotational B value (1.93 cm^{-1}) is such that even at the lowest
circumstellar temperatures a number of levels will be populated, thus
permitting accurate determination of a rotational excitation temperature. In addition, the relatively large isotope shifts of vibration-
rotation lines are favorable to the measurement of carbon and oxygen
isotope ratios in the material being expelled. This is particularly
important in distinguishing circumstellar material from interstellar
clouds which happen to lie in the line of sight.

In the past year or two a number of instrumental techniques have
reached the point where it is possible to observe infrared molecular
transitions arising in circumstellar material at a resolution better
than the 1 km/s necessary to resolve the line profiles. We have used
a 1.4 m Fourier Transform Spectrometer at the coudé focus of the
Mayall 4 m Telescope at Kitt Peak (Hall, et al. 1979) to observe CO
vibration rotation lines arising in circumstellar material around late
type supergiants, long period variables and obscured carbon stars.
Observations in the CO fundamental at 4.6μm are hampered by strong CO
absorption arising from the same transitions in the earth's atmosphere.
However at a spectral resolution of 10 km/s or better, the doppler
shift due to the earth's orbital motion is sufficient to move the lines
of most stars out of the terrestrial absorption at the most favorable
time of the year. Terrestrial CO absorption is not a problem in the
first overtone bands at 2.3μm where the transitions are two orders of
magnitude weaker. The Fourier Transform technique has the decided
advantage of permitting simultaneous observation of an entire molecular
band with very high spectrophotometric accuracy and an extremely well
calibrated frequency scale.

At a spectral resolution of the order of 10 km/s, 4.6μm spectra
of all M supergiants and long period variables we have observed to
date exhibit sharp CO lines with expansion velocities in good agreement with the visible atomic resonance line values (which are generally of the order of 10 km/s). The intensity variation with rotational quantum number implies excitation temperatures of the order of
200K and, as distinct from atomic resonance lines, the radiative rates
are low enough that CO is certainly in collisional equilibrium so this
represents a true gas temperature. The CO results are thus in good
agreement with previous observations; the derived gas temperature of

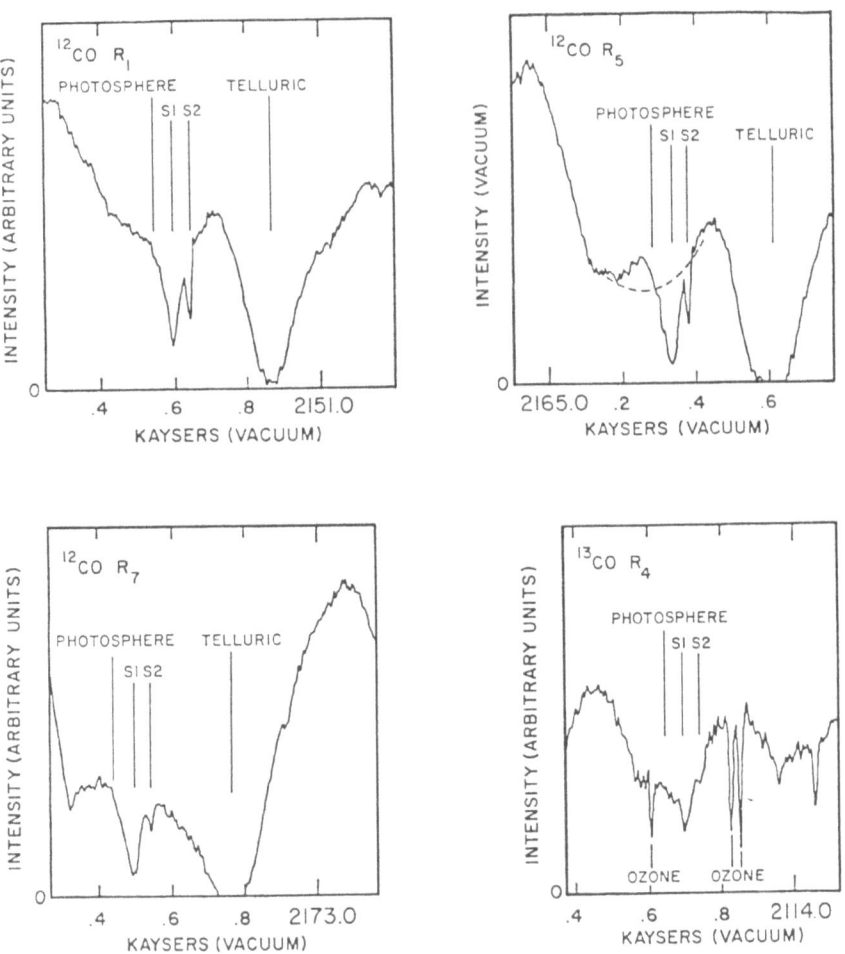

Figure 1:   Selected CO 1-0 lines from a 0.6 km/s resolution spectrum of
α Orionis.   Components due to the stellar photosphere, the two circumstellar shells and telluric absorption are indicated.

200K is reasonable, implying the material is of an average distance of the order of 100 stellar radii.

The situation is, however, considerably more complex in one supergiant, α Orionis, and one long period variable, o Ceti, which we have observed at substantially higher spectral resolution at 4.6μm. At a resolution of 0.6 km/s the circumstellar lines in α Orionis (Bernat et al. 1979) are clearly resolved into two components (Figure 1). The redward component, designated S1, is characterized by a rotational temperature of the order of 200K, an expansion velocity of 11 km/s relative to the center of mass and a column density of the order of 5×10^{17} cm^{-2}; it corresponds to the classical, expanding, circumstellar shell. At this resolution and signal-to-noise ratio a second component, designated S2, is clearly resolved. The extremely narrow (but still fully resolved) S2 lines exhibit an expansion velocity of 18 km/s, a column density of 1.2×10^{16} cm^{-2} and a temperature of 70 ± 10K. These lines might be due to some intervening interstellar cloud except that $^{13}C^{16}O$ features are detected in both components at a strength corresponding to a $^{12}C/^{13}C$ ratio of the order of 6, clearly establishing their stellar origin. Somewhat surprisingly, the curves of growth for each component can be characterized by a single distinct temperature; there is no evidence of material at intervening velocities. Equation of the gas temperatures of the two components to Tsuji's (1979) α Ori dust model temperatures implies that the S1 shell lies at a radius of 150 stellar radii and the S2 shell at the order of 2000 stellar radii. If we assume these shell radii represent the mean distance of the shell, the carbon abundance is solar and all C is tied up in CO, then we infer mass loss rates of 5×10^{-6} and 3.2×10^{-6} M$_\odot$/yr for S1 and S2 respectively. These values for the two shells agree to within model and observational uncertainties and suggest that they are two regimes within a continuous outflow. The initial accelerating force for the gas in the S1 component is likely provided by radiation pressure on dust grains which drag the gas along (Gilman, 1972). Then the abrupt acceleration at large radii from the star might be due to either a change in grain absorption properties or in the gas to dust ratio. It is tantalizing that the observed temperature of the S2 component is in the range where gases such as CO may begin to precipitate as mantles on grains at high enough densities.

A recently obtained, comparably high resolution spectrum of the archetypal long period variable o Ceti exhibits three clearly resolved circumstellar velocity components. The reddest, at a heliocentric velocity of 62.7 km/s (52.9 km/s LSR), probably corresponds to the center of mass velocity. The two other components are then expanding at velocities of 7.5 and 12.2 km/s and preliminary analysis indicates rotational temperatures of 200K and 70K respectively. The velocity of the 200K component is in excellent agreement with those of the OH, H$_2$O and SiO masers in o Ceti (Dickinson, Kollberg and Yngvesson, 1975; Snyder and Buhl, 1975).

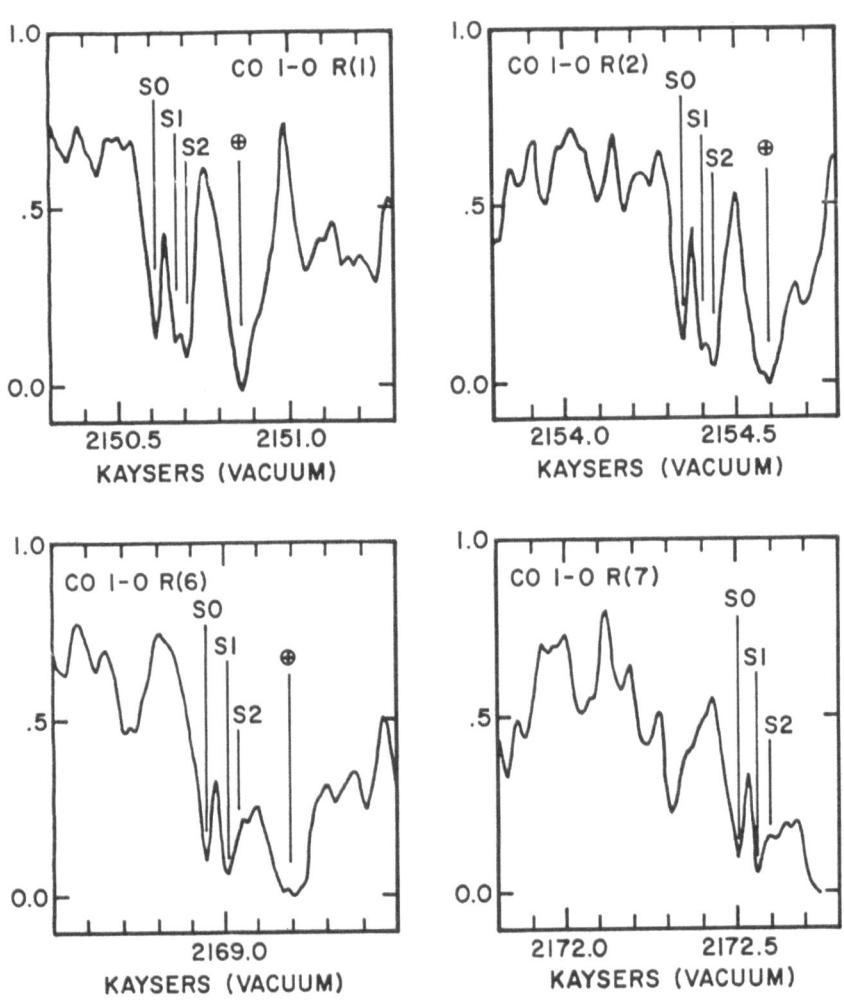

Figure 2: Selected CO 1-0 lines from a 0.6 km/s resolution spectrum of o Ceti. Components due to the three circumstellar shells and telluric absorption are indicated.

Clearly both further observations and more careful modelling of the outflow are required to distinguish between continuous outflow and discrete events. Comparable resolution observations of a number of high mass loss supergiants and long period variables are undoubtedly critical. The occurrence of multiple components with similar velocities and temperatures in many stars would argue strongly for the same physical processes in a continuous outflow. Similarly it is important to see, on a detailed level, whether a continuous outflow model, such as that proposed by Goldreich and Scoville (1976), is compatible with the apparently discrete nature of the components observed in α Orionis and o Ceti.

I would like to turn now to evidence for yet another circumstellar shell revealed by the violence of photospheric motions in the long period variables. From a time series of 1.5 to 2.5μm spectra covering several periods of the long period variable Chi Cygni, Hall, Hinkle and Ridgway (1979) have shown that the stellar photosphere undergoes violent motions which are well correlated at the optical variability. Just before maximum light an outwardly propagating wave, presumably driven by a general stellar oscillation, emerges into the photosphere and over the following month propagates through it. The velocity difference between the gas in front of and behind this wave exceeds the sound speed before the wave emerges from the photosphere. As a result the wave appears as a shock separating regions of differing velocity and temperature. The compressional wave accelerates the stellar surface material to at least 15 km/s, and more probably 20 km/s, outward relative to the center of mass. By phase 0.1 the shock is propagated out into the more tenuous material surrounding the star. The stellar surface layers then appear to decelerate at a constant rate until around phase 0.85 the now infalling material begins to interact with the precursor of the next wave and ceases to accelerate inwards. During a cycle the photospheric CO excitation temperature falls linearly from 3500K to 2400K.

At phases when the photospheric features are shifted away from the center of mass velocity, it is possible to distinguish broad, resolved CO 2-0 lines which are at the center of mass velocity and which exhibit excitation temperatures of approximately 800K. We are unable to detect any significant variations in either the excitation temperature or the radial velocity throughout our entire period of monitoring although the lines are very badly blended with the photospheric features around minimum light. The low excitation temperature of this material implies that it must be circumstellar and simple calculations indicate that it is probably at a distance of 4 to 8 stellar radii from the photosphere. It is presumably some turbulent layer of material which has been ejected from the star and may now be supported by the residual energy of the outwardly propagating shocks. Given the existence this relatively dense material at a temperature highly condusive to the formation of dust grains, and at a distance at which gravitational acceleration is markedly reduced, it seems highly plausible that this material provides the reservoir from which the classical mass loss is accelerated by radiation pressure acting on condensing dust. It is attractive to identify this relatively narrow region of acceleration with the SiO microwave maser. These masers cannot originate near the true stellar photosphere for, if radial, they would mimic the photospheric velocities and, if tangential, their intensities should correlate with the photospheric motions.

One may speculate whether this 800K, circumstellar material is unique to the long period variables, supported by dissipation of the energy associated with the outwardly propagating shock waves, or is instead a common characteristic of all the late type stars exhibiting high mass loss. In the late type supergiants photospheric motions are far too small to separate such circumstellar features from their photospheric counterparts and so it is impossible to obtain the same unambiguous evidence provided in the case of the long period variables. If, however, the variation of line strength with rotational quantum number is plotted for the CO 2-1 lines in the spectra of these stars, the line depth often peaks at a J value of the order of 10, corresponding to a temperature of $\leq$ 1000K. Similar plots for the vibrationally excited bands (3-1 etc.) show no such peaking, instead exhibiting a broad plateau characteristic of photospheric temperatures. It thus appears highly plausible that such 1000K circumstellar shells are characteristic of all late type stars exhibiting high mass loss and probably provide a reservoir from which mass loss is driven by radiation pressure on precipitating dust. The mechanism by which material is injected into these circumstellar shells and supported there remains obscure.

It seems likely that continuing CO observations will greatly improve our knowledge of conditions within circumstellar material around late type stars over the next few years. The availability of a high resolution Fourier Spectrometer on the Kuiper Airborne Observatory offers the possibility extending such measurements several molecules obscured to ground based observers. Particularly important here are the vibration rotation bands of OH and H_2O. Similarly, improvements in 10μm spectroscopic techniques may well open up the possibility of observing circumstellar lines of SiO fundamental vibration rotation bands in the spectra of these stars.

ACKNOWLEDGMENTS

This work was carried out in collaboration with A. Bernat, K. Hinkle and S. Ridgway.

REFERENCES

Bernat, A. P. (1976). Ph.D. thesis, University of Texas.
Bernat, A. P., Hall, D. N., Hinkle, K. H. and Ridgway, S. T. (1979). Astrophys. J. Letters, 233, L135.

Dickinson, D. F., Kollberg, E., and Yngvesson, S. (1975). Astrophys.
    J. 199, 131.
Gilman, R. C. (1972). Astrophys. J. 178, 423.
Goldberg, L. (1979). Quarterly Jnl. Roy. Astron. Soc., in press.
Goldreich, P., and Scoville, N. (1976). Astrophys. J. 205, 384.
Hall, D. N. B., Hinkle, K. H., and Ridgway, S. T. (1979). Proc. IAU
    Colloq. No. 46 (Hamilton, New Zealand).
Lambert, D. L., and Vanden Bout, P. A. (1978). Astrophys. J. 221, 854.
Reimers, D. (1977). Proc. IAU Colloq. No. 42, (Bamberg, West Germany).
Snyder, L. E. and Buhl, D. (1975). Astrophys. J. 197, 329.
Tsuji, T. (1979). Publ. Astron. Soc. Japan 31, 43.
Weymann, R. (1962). Astrophys. J. 136, 844.
Weymann, R. (1977). Proc. IAU Colloq. No. 42 (Bamberg, West Germany).

DISCUSSION FOLLOWING HALL

Dickinson: Observations of thermal and maser SiO in o Ceti (Mira) suggest an expansion velocity of $\sim$5 km s^{-1}, and maser velocities close to the stellar velocity. Can you reconcile these with your model?

Hall: The single sharp maser-feature seen in SiO, CO, OH, and H$_2$O in o Ceti has an LSR velocity of about +46 km/s, in good agreement with our 200 K shell at 45.5 km/s. I choose to associate the 53 km/s component in our spectra with the stellar velocity, as I am not aware of any mm detections of thermal SiO or CO. This assumption implies that the center of mass velocity of the SiO and CO pedestal features is not the stellar velocity of o Ceti.

Snyder: When we observed o Cet last winter we found that part of the SiO maser (v=1, J=1-0) pedestal was missing. Do you see any evidence in the infrared for destruction of SiO maser coherence lengths?

Hall: We have only one spectrum of the 4 μm SiO first overtone bands on o Ceti, and have not observed the SiO fundamental region. These data are inadequate to provide any indication of the SiO maser coherence lengths.

Goldreich: I have some criticisms of your model. First, I think that connecting the dust temperature to the gas temperature in these circumstellar envelopes is almost certainly incorrect. The dust temperature is set by radiative interaction with the star, and there is no reason why the gas temperature should come to equilibrium with the dust temperature at any appreciable distance from the star. In fact, the dust is being driven supersonically through the gas, and the heating effect of the dust is more through the supersonic drag than through the thermal temperature of the dust. Also I think it is very unlikely that the SiO masers are as far from the stars as you propose they are. There are a couple of VLBI experiments which indicate that the SiO masers are essentially in the atmospheres of the stars. Furthermore simple theoretical considerations on keeping a significant population in these excited vibrational levels of SiO indicate that the masers have to operate right at the surfaces of the stars. My final objection is to the idea that there is an appreciable second stage of acceleration at hundreds or even thousands of stellar radii. There would need to be an enormous increase in the opacity per gram of the material in order to have an appreciable acceleration occurring at a hundred or a thousand stellar radii. Perhaps it would have to go up by factors of thousands. So I think that it is much more likely that this evidence for variation or change in velocity either has something to do with the evolutionary state of the star or with a much closer-in region where acceleration is more likely to occur.

Hall: I agree there are real questions about the closeness of the gas and dust temperatures. However, if one assumes that radiative transfer determines the gas temperature, calculations give radii that are within a factor of two or so of those derived from adopting the dust temperature for the gas. So while there is certainly a doubt whether the gas and the dust are in close thermal equilibrium, I do not think we are wrong by an order of magnitude in our estimates of where these shells are. In particular it is very difficult to understand material at 7K being at distances much closer to the star than those we are talking about.

Goldreich: I am not arguing that this material at 7K is very close to the star, I am just arguing that it is fairly evident that there is no good thermal contact between the dust and gas, and that there is no point in using an argument which is incorrect.

Anonymous: We now have a very long time-series of observations of a number of long period variables that have SiO masers associated with them. There is direct evidence now that material is moved many stellar radii during the general oscillation, yet one sees no evidence of these velocity changes in the SiO maser velocities.

Interruption: We see velocity changes all over in the SiO maser. What do you mean by that?

Anonymous: There were no velocity changes that were in any way correlated with the photospheric motions.

Goldreich: There is again no reason to think the principal gain direction of the SiO masers is radial. In fact, if the maser operates in the region where there is a very strong outward radial acceleration of material, the best gain paths will be tangential, so you will see the maser at the limbs of the star, and there will be no velocity change at all, certainly nothing that correlates with the radial velocities that you observe in the photosphere.

Ramsay: Okay, you have a lot to write up there now. Who's going to be next?

Zuckerman: The water masers also indicate sizes smaller than your scale of 10 stellar radii, and material does seem to be moving out at a few kilometers per second or more from both the radio and other observations. Somehow you have gas flowing out and stopping. I do not understand what physical mechanism would cause this deceleration at the shell at 10 stellar radii.

Hall: The problem is that we see no evidence for gas outflow until outside the material at 10 stellar radii.

Zuckerman: Models based on visible and infrared observations also indicate that within this distance of 10 stellar radii there are substantial outflow velocities, not as great as the terminal velocity, but nonetheless substantial.

Hall: I am not aware of any visible observations that show that.

Allamandola: Do you see any absorption in the region of the zero
line of CO between the P and R branches? An absorption there would
indicate that CO has condensed onto the dust grains, and may be the
evidence you are looking for.

Hall: There is no such absorption evident in α Orionis, although
the region is complex because of photospheric CO features. In 5 μm
spectra of early-type stars embedded in molecular clouds, the continuum
is much cleaner, but there is still no indication of such an absorption
feature.

OBSERVATIONAL CHARACTERISTICS OF MASERS ASSOCIATED WITH STARS

Lewis E. Snyder
Department of Astronomy
University of Illinois

OH, H_2O and SiO are the 3 known molecular masers associated with stars. The known transitions, their line profiles, radial velocities, temporal intensity variations and polarization properties are discussed. Current work on determination of circumstellar shell sizes is reviewed. Possible future research topics are outlined.

I. INTRODUCTION

Only 11 years ago, Wilson and Barrett (1968) discovered the first maser emission associated with late-type stars. Their OH observations led the way to other molecular searches of stars and today both early-type emission line stars and late-type stars have been observed. Davis, Seaquist, and Purton (1979) have listed the few early-type objects which have been observed in OH maser emission. Because much more is known about maser emission from circumstellar shells associated with late-type stars, our discussion will be concentrated here. The observations prior to and through 1976 have been summarized by Olnon (1977), Snyder (1977), and Winnberg (1976); hence the more recent developments will be emphasized here. Hydroxyl (OH), water (H_2O) and silicon monoxide (SiO) are the three molecular species found in strong maser emission from circumstellar shells associated with late-type stars. The best values for the maser rest frequencies (Lovas, Snyder and Johnson 1979) are: $OH^2\Pi_{3/2}$ J=3/2, F=1-2, 1612.2310 MHz; F=1-1, 1665.4018 MHz; F=2-2, 1667.3590 MHz: H_2O $J_{K-K+} = 6_{16}-5_{23}$, F=5-4, 22,235.120 MHz (central hyperfine component): SiO v=1, J=3-2, 129,363.262 MHz; J=2-1, 86,243.350 MHz; J=1-0, 43,122.027 MHz; v=2, J=1-0, 42,820.539 MHz; v=3, J=1-0, 42,519.34 MHz. The SiO v=3, J=1-0 transition is the most recently detected (Scalise and Lepine 1978) and hence was not included in previous circumstellar maser reviews. A catalog of more than 300 stellar objects showing maser line radio emission from OH, H_2O and/or SiO has been compiled by Engels (1979). About 200 of these objects have been identified with optical or infrared stars (mostly M-supergiants, Mira or semiregular variables), and they may be described by Turner's (1970) OH classification scheme as extended by Wilson (1973).

B. H. Andrew (ed.), Interstellar Molecules, 525–533.

Type II OH maser stars have stronger 1612 MHz emission than main line emission; 1720 MHz OH emission is never observed. These stars tend to have excess infrared radiation at 10 μm, and there may be no detectable H_2O maser emission. There usually are two OH emission groups separated by 20 km s^{-1} or more. Type I stars have stronger main lines than 1612 MHz emission; again, 1720 MHz emission is not observed. There is less excess radiation at 10 μm and, usually, detectable H_2O maser emission. The two OH emission groups are typically separated by about 10 km s^{-1}. A third group includes the supergiant OH masers such as VY CMa, VX Sgr and NML Cyg. Representative spectra from Type II masers stars are shown in Figure 1.

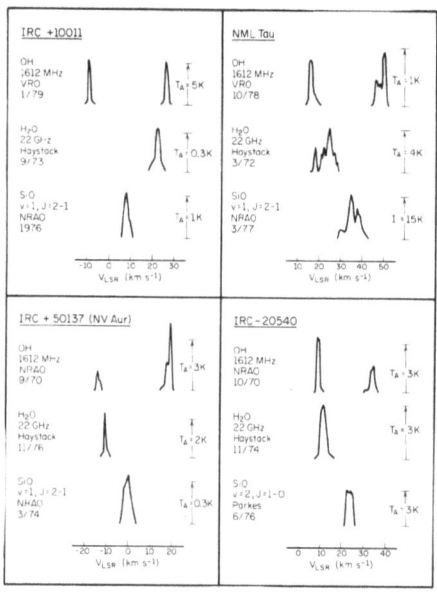

Figure 1. Representative OH, H_2O and SiO maser emission spectra from Type II maser stars. Ordinates: antenna temperature T_A. Abscissae: radial velocity with respect to the LSR.

2. CURRENT OBSERVATIONAL PROGRAMS

2.1 Line profile and radial velocity studies

The work of Reid (1976), Reid and Dickinson (1976) and Dickinson et al. (1978) has shown that the two OH emission groups are displaced symmetrically about the stellar radial velocity. This evidence supports an expanding shell model for late-type maser stars and has helped define the goals of many current observational programs. Earlier, Dickinson, Kollberg, and Yngvesson (1975) had confirmed the correlation between the velocity separation of the characteristic double-peaked OH spectrum and stellar period. This correlation can be used to establish the approximate periods and types for OH stars. The situation for H_2O and SiO maser line profile studies is not as well defined. Knowles and Batchelor (1976) have identified symmetrical H_2O radial velocity structure in VY CMa and Orion as well as in several H II region H_2O maser sources. Both these profiles and symmetrical SiO maser profiles in VY CMa and possibly NML Cyg (Buhl et al. 1974; Snyder and Buhl 1975) have been studied by Van Blerkom and Auer (1976) and Van Blerkom (1978) as possible examples of rotating disk geometries where the disk fragments into a system of concentric rings. These studies suggest that perhaps the existence of emerging planetary systems could be inferred from the maser line profiles they produce, with the precaution that a symmetrical maser spectrum does not uniquely define the ring geometry. In other studies, Snyder et al. (1978) have found broad, weak SiO maser emission pedestals underlying the principal SiO maser peaks in the v=1, J=1-0 transition. This emission pedestal appears to be an intrinsic part of the spectral signature of SiO maser emission from late-type stars: the width of the pedestal is usually somewhat less than the expansion velocity and often its center velocity may be used to determine stellar radial velocities. Dinger, Dickinson and Snyder (1978) have also observed the SiO pedestal in the v=1, J=2-1 transition where it appears to be an assembly of several individual features. Schwartz, Waak and Bologna (1979) have detected broad emission pedestals in both v=1 and v=2, J=1-0 SiO maser transitions but none has been found in the v=3 J=1-0 transitions detected to date. Elitzur (1979) interprets the weak SiO maser emission pedestal as emission from the circumstellar shell, while the more easily observed, strong, sharp main emission spikes would come from convective cells located much closer, possibly in the upper atmosphere of the star itself. Fortunately, extensive radial velocity measurements of high dispersion optical spectra now exist for many of the important maser emission stars (e.g., Wallerstein 1975, 1977; Hagen 1978). These optical data are being utilized in conjunction with the radio maser velocity data to build consistent physical pictures of the excitation processes in the atmospheres of these stars.

2.2 Intensity variations and intensity correlation studies

The optical, infrared and OH maser emission from late-type stars usually varies in intensity in a regular or semi-regular manner. Harvey et al. (1974) found that the intensity of the 1612 MHz satellite line of OH followed the infrared luminosity changes closely and agreed qualitatively with available optical data. A correlation has been established between the intensity variations of the OH main lines at 1665 and 1667 MHz and the optical emission (Fillit, Proust and Lepine 1977; Jewell et al. 1979). These studies support the theory that the OH level populations are inverted by a radiative pump which is intimately related to the optical output of the central exciting star. The temporal intensity variations of H_2O and SiO have not behaved as nicely. Schwartz, Harvey and Barrett (1974) studied the time variability of both H_2O and 2.2 μm emission from 8 late M stars. For the 3 strongest

H_2O emitters (R Aql, U Her and W Hya) they found a correlation between optical and infrared variations and no phase difference greater than about 30 days between H_2O and infrared variations. Cox and Parker (1979) included these 3 stars in their study of H_2O emission from 9 stars during the period 1974 September-1977 May. In several the H_2O intensities were very different from those found by earlier observers and they concluded that stellar H_2O masers are often not stable for more than a few cycles of the stellar luminosity. The SiO maser time variation studies reported to date with regular sample monitoring over part or all of a cycle include only o Cet, R Cas and R Leo. Spencer and Schwartz (1975) found phase dependent SiO time variations in o Cet and R Cas, similar to OH and H_2O variations, which suggest a partially saturated maser. Hjalmarson and Olofsson (1979) found good correlation between SiO and the 2.7 μm infrared intensity as well as a distinct phase lag with respect to the visual light curve for R Leo and o Cet. On the other hand, Troland et al. (1979) observed 7 regular variable stars 3 years after Kaifu, Buhl and Snyder (1975) and found substantial changes in the line profiles and integrated fluxes. Most of the changes were thought to be real (and not due to the linear polarization effects discussed in 2.4), but they could not be related to the optical phases of the stars. Hence time variation studies of the SiO maser often are complicated by polarization and perhaps by the pedestal-spike line shape discussed in 2.1. Time variations have also been observed in the Orion SiO maser by Snyder et al. (1978), Schwartz, Waak and Bologna (1979), and Troland et al. (1979).

     Maser luminosities have been used successfully in several statistical correlation studies. Nguyen-Q-Rieu et al. (1979) point out that about 75% of the OH Mira masers detected within $\sim$ 1 kpc of the sun are Type I sources. They found that OH intrinsic luminosity varies from star to star and that Type II OH Mira masers generally are more distant and have higher OH luminosities than Type I OH Mira masers. This generally supports the results of systematic surveys which suggest that many optically unidentified Type II OH/IR maser sources are situated at large distances in the central part of the galaxy (e.g. Johansson et al. 1977; Bowers 1978; Baud et al. 1979). Cahn and Wyatt (1978) have developed a period-luminosity-spectral type scheme which had led to a successful correlation between SiO maser luminosities and stellar luminosities in long-period variables (Cahn 1977; Cahn and Elitzur 1979). This correlation suggests that SiO masers are radiatively pumped by direct stellar radiation, saturated, and occur at roughly the same distance from the star.

2.3 Determination of circumstellar shell sizes

     OH VLBI observations of late-type stars, both M supergiants and long-period variables, have been made by Masheder, Booth and Davies (1974), Moran et al. (1977), Reid et al. (1977), Reid and Muhleman (1978), Reid et al. (1979), and Mutel et al. (1979). Typical OH maser shell radii for supergiants are $\sim 10^{17}$ cm for 1612 MHz emission and apparently somewhat less for the main-line emission. Typical Mira variable stars have 1612 MHz radii of $\gtrsim 3 \times 10^{15}$ cm. H_2O VLBI observations

have been made by Rosen et al. (1978) and by Spencer et al. (1979).
Typical late-type stars and the supergiant VY CMa have an H_2O masing
region with radius $\sim 10^{15}$ cm. Figure 2 (from Moran et al. 1979b) shows
the distribution of 1612 MHz OH and H_2O masers near VX Sgr.

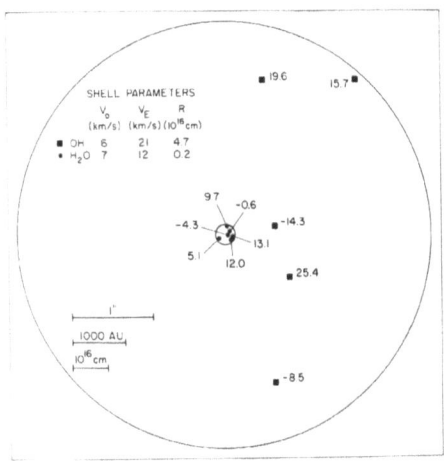

Figure 2.   The distribution of 1612 MHz OH and H_2O masers near VX Sgr
(from Moran et al. 1979b).

Recently Moran et al. (1979a) performed the first successful VLBI measurements of SiO stars. They found the radius of the SiO masering
region to be $\sim 10^{14}$ cm for the Mira variable R Cas and $\sim 10^{15}$ for VY
CMa. The VLBI technique measures the angular diameter of the maser
emission region. Recently another technique has been used to attempt
to measure the OH radial diameter. If the OH at all points in the
shell emits and varies in phase relative to the star, then emission
from different parts of the shell will be out of phase relative to
Earth because of the difference in distance the emission travels. Thus
the blueshifted line, from the front of the star, should lead the red-
shifted line, from the back of the star, by a phase corresponding to
the light travel time across the shell. Through careful monitoring,
it should be possible to determine shell diameters. Shultz, Sherwood,
and Winnberg (1978) investigated a number of maser sources but, owing
to their small number of data, only an indication of shell sizes could
be determined. Jewell et al. (1979) have monitored 8 OH maser stars
over a time base of several years, and their data suggest OH shell
dimensions on the order of a few $\times 10^{16}$ cm. However, both sets of observations demonstrate that it is imperative to obtain a high density
of data points, perhaps 20 or more over a cycle, if reliable measurements with small standard errors are to be forthcoming.

2.4 Polarization studies

Maser polarization measurements are necessary for correctly cali-brated line intensities and for placing constraints on proposed pumping models. To date, there have been no sensitive, large scale surveys of the polarization of stellar masers associated with late-type stars. On the basis of somewhat limited observational sampling, it has been found that some OH stellar masers have linear and/or circular polarization; stellar H_2O maser emission shows little or no polarization and SiO masers have linear but no observable circular polarization. Reid et al. (1979) have measured the 1612 MHz OH polarization of U Ori and IRC+ 10420. Their results suggest that the magnetic fields of the central exciting stars in stellar OH masers are of the order of 10-100 gauss. Troland et al. (1979) have determined the Stokes parameters for 8 SiO masers using the v=1, J=2-1 transition. No circular polarization was found and the typical linear polarization was 15 to 30% except that R Cas has one feature with $\sim$ 100% linear polarization and Stokes param-eters which changed over $\sim$ 1 day. At this meeting, Clark et al. (1979) have reported linearly polarized SiO v=1, J=2-1 emission spread over 12 km s^{-1} from R Leo. The position angle of linear polarization changes uniformly over the underlying emission pedestal but undergoes an abrupt change at the strong principal emission peak; hence this promises to be an important technique for separating the spectrum of the SiO pedestal from that of the stronger main emission spike.

3. FUTURE RESEARCH

Several interesting directions are suggested for future research. New molecular masers are hard to find. Morris (1979) interprets the CO emission from CIT-6 as a possible weak maser. Buxton et al. (1977) searched 132 sources without finding methanol maser emission. Orion A remains the only known methanol maser but it may not be intrinsically unique because the other sources may be too far or too weak to be de-tected. The Orion region has other interesting properties. Of all the known SiO maser sources, only the Orion A maser is associated with a molecular cloud region. None of the other OH/H_2O maser IR sources associated with molecular clouds have been found to be SiO maser sources. Bieging et al. (1979) have interferometrically determined that the Orion SiO maser is coincident with the infrared point source IRC2, in agreement with the suggestion of Genzel et al. (1979). It re-mains to be determined whether IRC2 really is a unique object or just another late-type star. Other interesting objects have been revealed by OH time monitoring. One such object, U Ori, has been found to be-have like a damped oscillator in its 1612 MHz OH emission (Jewell et al. 1979). Undoubtedly other important results will be found from the SiO maser monitoring programs now underway at places such as the Naval Re-search Laboratory, the Five College Radio Astronomy Observatory and the Onsala Space Observatory. The results from programs like these might be used to search for multiple systems associated with late M giants, as suggested by Zuckerman (1979). Finally, we need to know more observational detail about mass loss in maser stars. For example, it has been suggested (e.g., Bowers and Kerr, 1977) that OH maser stars have high mass loss rates while non-OH maser stars have low mass loss rates. It is clear from discussions presented at this meeting that this is a rich area for further investigation. This work was supported in part by NSF Grant AST 79-07830.

REFERENCES

Baud, B., Habing, H. J., Matthews, H. E., and Winnberg, A.: 1979, Astron. Astrophys. Suppl 36, pp. 193-211.
Bieging, J., Plambeck, R., Thornton, D., Welch, W., and Wright, M.: 1979, this volume.
Bowers, P. F.: 1978, Astron. Astrophys. 64, pp. 307-318.
Bowers, P.F., and Kerr, F.J.: 1977, Astron. Astrophys. 57, pp. 115-123.
Buhl, D., Snyder, L. E., Lovas, F. J., and Johnson, D. R.: 1974, Astrophys. J. 192, pp. L97-L100.
Buxton, R. B., Barrett, A. H., Ho, P. T. P., and Schneps, M. H.: 1977, Astron. J. 82, pp. 985-988.
Cahn, J. H.: 1977, Astrophys. J. 212, pp. L135-L137.
Cahn, J. H., and Elitzur, M.: 1979, Astrophys. J. 231, pp. 124-127.
Cahn, J. H., and Wyatt, S. P.: 1978, Astrophys. J. 221, pp. 163-174.
Clark, F.O., Johnson, D.R., Troland, T.H., and Heiles, C.E.: 1979, this volume.
Cox, G., and Parker, E.A.: 1979, M.N.R.A.S. 186, pp. 197-215.
Davis, L.E., Seaquist, E.R., and Purton, C.R.: 1979, Astrophys. J. 230, pp. 434-441.
Dickinson, D. F., Kollberg, E., and Yngvesson, S.: 1975, Astrophys. J. 199, pp. 131-134.
Dickinson, D. F., Reid, M. J., Morris, M., and Redman, R.: 1978, Astrophys. J. 220, pp. L113-L116.
Dinger, A. St. C., Dickinson, D. F., and Snyder, L. E.: 1978, Bull. A.A.S. 10, p. 392.
Elitzur, M.: 1979, private communication.
Engels, D.: 1979, Astron. Astrophys. Suppl. 36, pp. 337-345.
Fillit, R., Proust, D., and Lepine, J. R.: 1977, Astron. Astrophys. 58, pp. 281-286.
Genzel, R., Moran, J. M., Lane, A. P., Predmore, C. R., Ho, P. T. P., Hansen, S.S., and Reid, M.J.: 1979, Astrophys. J. 231, pp. L73-L76.
Hagan, W.: 1978, Astrophys. J. 222, pp. L37-L40.
Harvey, P. M., Bechis, K. B., Wilson, W. J., and Ball, J. A.: 1974, Astrophys. J. Suppl. 27, pp. 331-357.
Hjalmarson, Å., and Olofsson, H.: 1979, preprint.
Jewell, P. R., Elitzur, M., Webber, J. C., and Snyder, L. E.: 1979, Astrophys. J. Suppl. 41, pp. 191-207.
Johansson, L. E. B., Andersson, C., Goss, W. M., and Winnberg, A.: 1977, Astron. Astrophys. 54, pp. 323-334.
Kaifu, N., Buhl, D. and Snyder, L.E.: 1975, Astrophys. J. 195, pp. 359-366.
Knowles, S. H., and Batchelor, R. A.: 1976, M.N.R.A.S. 174, pp. 69P-73P.
Lovas, F.J., Snyder, L.E. and Johnson, D.R.: 1979, Astrophys. J. Suppl. 41, pp. 451-480.

Masheder, M. R. W., Booth, R. S., and Davies, R. D.: 1974, M.N.R.A.S. 166, pp. 561-583.

Moran, J.M., Ball, J.A., Predmore, C.R., Lane, A.P., Huguenin, G. R., Reid, M.J., and Hansen, S.S.: 1979a, Astrophys. J. 231, pp. L67-L71.

Moran, J. M., Ball, J. A., Yen, J. L., Schwartz, P. R., Johnston, K. J., and Knowles, S. H.: 1977, Astrophys. J. 211, pp. 160-169.

Moran,J.M.,Lichten,S.,Reid,M.,Huguenin,R., and Predmore,R.:1979b,in prep.

Morris, M.: 1979, preprint.

Mutel, R. L., Fix, J. D., Benson, J. M., and Webber, J. C.: 1979, Astrophys. J. 228, pp. 771-779.

Nguyen-Q-Rieu, Laury-Micoulaut, C., Winnberg, A., and Schultz, G. V.: 1979, Astron. Astrophys. 75, pp. 351-364.

Olnon, F. M.: 1977, Ph.D. thesis, Leiden University.

Reid, M. J.: 1976, Astrophys. J. 207, pp. 784-798.

Reid, M. J., and Dickinson, D. F.: 1976, Astrophys. J. 209, pp. 505-508.

Reid,M.J.,Moran,J.M.,Leach,R.W.,Ball,J.A.,Johnston,K.J.,Spencer,J.H., and Swenson,G.W.: 1979, Astrophys. J. 227, pp. L89-L92.

Reid, M.J., and Muhleman, D.O.: 1978, Astrophys. J. 220, pp. 229-238.

Reid, M. J., Muhleman, D. O., Moran, J. M., Johnston, K. J., and Schwartz, P. R.: 1977, Astrophys. J. 214, pp. 60-77.

Rosen,B.R.,Moran,J.M.,Reid,M.J.,Walker,R.C.,Burke,B.F.,Johnston,K.J., and Spencer,J.H.: 1978, Astrophys. J. 222, pp. 132-139.

Scalise, E., Jr., and Lepine, J. R. D.: 1978, Astron. Astrophys. 65, pp. L7-L8.

Schultz, G. V., Sherwood, W. A., and Winnberg, A.: 1978, Astron. Astrophys. 63, pp. L5-L7.

Schwartz, P. R., Harvey, P. M., and Barrett, A. H.: 1974, Astrophys. J. 187, pp. 491-496.

Schwartz, P. R., Waak, J. A. and Bologna, J. M.: 1979, preprint.

Snyder, L. E.: 1977, in "Topics in Interstellar Matter" (ed. H. van Woerden, Dordrecht-Holland: D. Reidel), pp. 97-104.

Snyder, L. E., and Buhl, D.: 1975, Astrophys. J. 197, pp. 329-340.

Snyder, L. E., Dickinson, D. F., Brown, L. W., and Buhl, D.: 1978, Astrophys. J. 224, pp. 512-519.

Spencer, J. H., Johnston, K. J., Moran, J. M., Reid, M. J., and Walker, R. C.: 1979, Astrophys. J. 230, pp. 449-455.

Spencer, J.H., and Schwartz, P.R.: 1975, Astrophys. J. 199,pp.L111-L113.

Troland, T.H., Heiles, C., Johnson, D. R., and Clark, F. O.: 1979, Astrophys. J. 232, pp. 143-157.

Turner, B. E.: 1970, J. Roy. Astron. Soc. Canada 64, pp. 221-304.

Van Blerkom, D.: 1978, Astrophys. J. 223, pp. 835-839.

Van Blerkom, D., and Auer, L.: 1976, Astrophys. J. 204, pp. 775-780.

Wallerstein, G.: 1975, Astrophys. J. Suppl. 29, pp. 375-396.

Wallerstein, G.: 1977, Astrophys. J. 211, pp. 170-177.

Wilson, W. J.: 1973, in "Molecules in the Galactic Environment" (ed. M. A. Gordon Jr. and L.E. Snyder, New York: John Wiley & Sons), pp. 165-171.

Wilson, W. J., and Barrett, A. H.: 1978, Science 161, pp. 778-779.

Winnberg, A.: 1976, paper presented at the I.A.U. at Grenoble.

Zuckerman, B.: 1979, Astrophys. J. 230, pp. 442-448.

DISCUSSION FOLLOWING SNYDER

Zuckerman: The SiO masers could be used to study the frequency of binary systems containing late M giants if accurate maser positions could be obtained with conventional or VLB interferometers.

Slysh: Mira is a binary star. What influence might the white dwarf companion Mira B have on the maser emission?

Snyder: We have found some unusual temporal behavior in the intensities and velocities of the SiO maser emission from Mira (o Cet). The period of Mira B appears to be too long to explain these changes. However we are still examining the possibility.

Winnewisser: You mentioned that it would be important to have predictions for new masers. Are you prepared to make a more qualitative statement about what you think is the most important area for predictions?

Snyder: Yes.

Booth: Careful comparison of the spectra of SiO, H_2O and OH in Orion show several features which overlap in velocity and which fall into distinct velocity ranges. Furthermore, our interferometric measurements at Jodrell Bank show that Orion is the only HII region where OH and H_2O emissions are coincident. Do you think we are seeing a late-type star in the Orion complex?

Snyder: To my knowledge, all of the other known SiO masers can be associated with some sort of late-type star. Also, none of the other large molecular clouds associated with HII regions have shown SiO maser emission. These points suggest that if the SiO maser in Orion is not associated with a late-type star, then it is a unique object.

Gold: Let us be quite clear and specific, in this session, whether we are discussing the physical size of a masering region, or what we measure with radio telescopes and interferometers here on Earth, the curvature of a phase-front and its irregularities. The two quantities may be totally different, or have little relation to each other for sources of coherent radiation. The angular size of a maser region as observed by interferometer refers in reality only to the irregularities within the region, which cause a slight scatter in direction of the amplified wavefronts. It has no direct relation to the physical size.

Snyder: I assumed that the listeners are familiar with these points about coherence. Those who are not may wish to read Professor Gold's discussion on p. 747 of Interstellar Ionized Hydrogen, ed. Y. Terzian (W.A. Benjamin, Inc., New York 1968).

VLBI OBSERVATIONS OF THE V=1 AND V=2 SiO MASERS IN W HYDRA AND
VX SAGITTARIUS

A. P. Lane, P. T. P. Ho, C. R. Predmore
University of Massachusetts
J. M. Moran and R. Genzel
Harvard-Smithsonian Center for Astrophysics
S. S. Hansen and M. J. Reid
National Radio Astronomy Observatory

Our previous observations[1] established the small angular size and
high brightness temperature of emission from the v=1, J=1-0 transition
of SiO from the circumstellar envelopes of the supergiant VX Sgr and
the Mira variable R Cas. We performed a second VLBI experiment on the
SiO masers in several late type stars on 31 Oct.-2 Nov. 1978 to compare
the physical characteristics of the SiO masers in the v=1 and v=2 states.
With an energy separation of 1258 cm^{-1} (an equivalent temperature of
1753 K) between the two vibrational states, differences in excitation
and pumping of the maser states may lead to different maser properties.

The interferometer elements were the 13.7 m antenna of the Five
College Radio Astronomy Observatory and the 36.6 m antenna of the Haystack Observatory. The 75 km baseline provides a fringe spacing of
0".02-0".03. Instrumental setup and observational procedures are discussed in detail by Genzel et al.[2] The data were recorded on the Mark
II VLBI system and processed at the National Radio Astronomy Observatory.

We present in Fig. 1 the total-power and cross-power spectra of the
v=1 and v=2 lines observed toward VX Sgr and W Hya. The cross-power
spectra were calculated from the data by coherently averaging for 45
minutes using a strong spectral feature as a phase reference. The total
bandwidth was 2 MHz (14 km s^{-1}) and the velocity resolution (with
Hanning weighting) was 0.30 km s^{-1}.

We find a remarkable agreement between the v=1 and v=2 masers toward VX Sgr in their cross power spectra. The fringe visibilities of
the strongest features are about 0.4 in both transitions. Assuming a
Gaussian source model for individual features, we estimate an angular
size of 0".010 for the maser spots, corresponding to a linear apparent
size of 2 x 10^{14} cm at a distance of 1500 pc. The implication is that
both the v=1 and v=2 masers are at the same distance from the star with
similar excitation. Velocity structure across the spectrum is complex
with blending of many narrow components of width a few times 0.1 km s^{-1}.

In the case of W Hya, differences between the two vibrational
states are evident in the velocities of partially resolved features.

535

B. H. Andrew (ed.), Interstellar Molecules, 535–536.

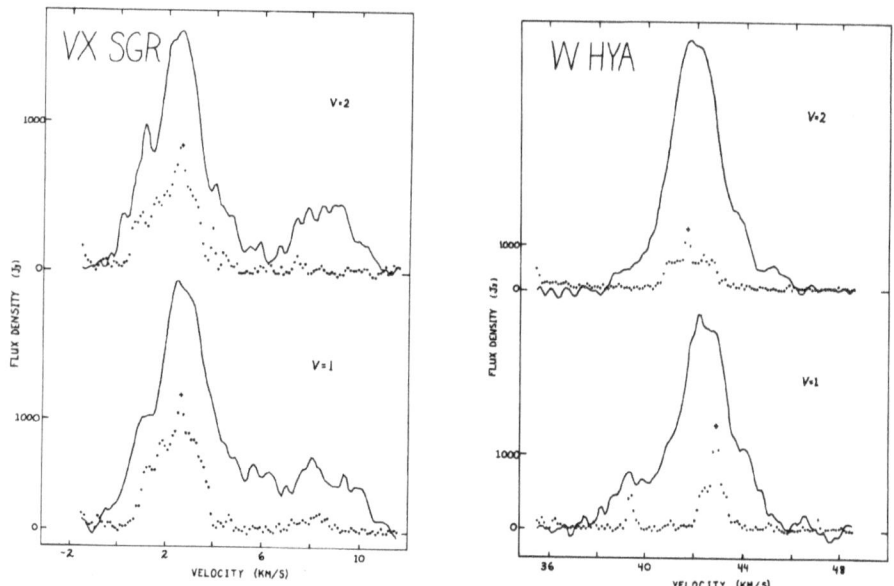

Figure 1: Total-power (solid line) and cross-power (dotted line) spectra for VX Sgr and W Hya. Velocity is relative to the local standard of rest. The upper spectrum for each source is the v=2, J=1-0 transition (rest frequency 42.820510 GHz); lower spectra are v=1, J=1-0 (rest frequency 43.122027 GHz). The v=1 and v=2 spectra for each source were taken on consecutive days at similar hour angles. Phase referencing overestimates the amplitude of the reference feature (plotted as +).

Peak visibilities, however, are comparable, ~ 0.2, implying an angular size of 0".016 and a linear size of 4 X 10^{13} cm at a distance of 154 pc. One may conclude in this case that the masering clouds are probably distinct objects in the two vibrational lines. This may be related to the fact that the W Hya masers appear to be smaller by a factor of five than the VX Sgr masers, which in an expanding (or contracting) shell model implies a smaller distance from the central exciting source.

The Five College Radio Astronomy Observatory is operated with support from the National Science Foundation under grant AST 76-24610. The Haystack Observatory is operated by the Massachusetts Institute of Technology with support from the National Science Foundation.

REFERENCES

1. Moran, J.M., Ball,J.A., Predmore,C.R., Lane,A.P., Huguenin,G.R., Reid,M.J., Hansen,S.S.: 1979, Ap. J. (Letters) 231, pp. L67-L72.
2. Genzel,R., Moran,J.M., Lane,A.P., Predmore,C.R., Ho,P.T.P., Hansen, S.S., Reid,M.J.: 1979, Ap. J. (Letters) 231, pp. L73-L76.

VIBRATIONALLY EXCITED SILICON MONOXIDE MASERS

D. Buhl[1], F.O. Clark[2], G. Chin[1], D. Glenar[1], T. Kostiuk[1],
M.J. Mumma[1], and F.J. Lovas[3]
[1]NASA-GSFC IR and Radio Astronomy Branch, Greenbelt, MD 20771
[2]University of Kentucky, Lexington, KY 40506
[3]National Bureau of Standards, Washington, D.C. 20234

ABSTRACT

   Published data on a select group of SiO maser sources have been
analyzed for velocity variations as a function of phase. No apparent
correlation was found to a level of about 2 km/s. This places constraints
on the location of the maser molecules. Such a correlation should be
present at some level. The implications for future high resolution
infrared measurements are discussed.

   We have carefully analyzed published data on the stellar maser
sources R Leo, chi Cyg, W Hya, o Ceti and R Cas to learn more about the
environment of SiO masers. The intent is to find the probable properties
of infrared spectral lines of SiO that are to be observed in the near
future with an infrared heterodyne spectrometer, and to determine what
new information the infrared measurements may yield.

   Most published SiO maser spectra have been carefully analyzed. The
velocity of the emission peak and the velocity extent (width) of emission
have been examined as a function of stellar phase; there were no apparent
correlations to within 2 km/s. This result is somewhat surprising since
Hinkle (1978) has measured velocity pulsations of 16 and 27 km/s of the
2 micron OH and CO emission from the star R Leo. There are two tentative
conclusions to be drawn:

1. The agreement of the velocity of the narrow SiO maser feature with
   that of the ground state SiO centroid and with center of the OH
   peaks, plus the apparent lack of correlation noted above, suggests
   that the narrow SiO features are tangentially amplified, and should
   therefore mark the stellar velocity to within a few km/s.

2. Further, the apparent lack of an observable systematic effect in
   the line wings (maximum and minimum extent) is consistent with the
   apparent lack of a blue shift of the wings (or "broad feature")
   with respect to the peak. This implies that the maser feature

537

B. H. Andrew (ed.), Interstellar Molecules, 537–538.
Copyright © 1980 by the IAU.

originates far enough from the star that geometrical blockage of emission from the far side is negligible. It is interesting in this regard that each successive vibrational state ought to be excited closer to the star. The observations of Scalise and Lepine (1978) and Snyder et al. (1980) of simultaneous v=1,2, and 3 maser SiO lines show the v=3 to be substantially blue shifted from v=1 and 2. Could this be an example of geometrical blockage?

Consideration of the above plus the dynamical scale height would appear to imply that the SiO emission originates at radial distances greater than about 1.8 stellar radii in typical type I objects. In principle limits can also be placed on the maximum distance from the stars at which maser emission can occur. Under average conditions of uniform mass loss near the star, the gas density should decrease as R^{-2}. When the dust grains condense, they are accelerated and carry the gas along. Conservation of mass implies that the gas density will decrease even more sharply in this accelerated region, i.e. as R^{-3}. In addition, the fractional abundance of gaseous SiO is expected to drop dramatically as SiO condenses onto dust grains. Thus there should be far less gaseous SiO beyond the grain condensation radius, so that the maser would be expected to occur within this region.

Therefore these two lines of inquiry indicate that the SiO maser originates somewhere roughly beyond the dynamical scale height but not significantly further than the grain condensation radius. Quantitatively this should be about 2-3 stellar radii for type I sources.

The lack of observable velocity pulsations is very convenient for the infrared heterodyne measurements, which have inherently high resolutions. However, such correlations of velocity with phase are expected at some level. It would appear that the rotational level populations for each vibrational state should be directly determinable from the infrared measurements, so that the range of possible pumping mechanisms would be tightly constrained. It also appears probable that the infrared measurements will supply a wealth of data on the stellar environment.

REFERENCES

Hinkle, K.H.: 1978, Astrophys. J. 220, 210.
Scalise, E., and Lepine, J.R.D.: 1978, Astron. & Astrophys. 65, L7.
Snyder, L.E., Buhl, D., Dickson, D., and Dickinson, A.: 1980, in
    preparation.

TIME VARIATION OF SiO MASER EMISSIONS

Nobuharu Ukita
Department of Astronomy, University of Tokyo, and
Norio Kaifu
Tokyo Astronomical Observatory

We report a series of observations of the v=1, J=2-1 transition of SiO (86243.28 MHz) toward four Mira variables (o Cet, R Leo, W Hya, and R Cas) and Orion A over the period September 1976 - February 1978. Observations were made with the 6-m mm-wave telescope of the Tokyo Astronomical Observatory. Most of the spectra were taken with an acousto-optical spectrometer with an effective resolution of about 57 kHz.

Among these sources, only o Cet tended to show in-phase variation during our observing period. We could find no clear periodicity in the variations of either fluxes or profiles of the other three. In particular we observed irregular variations in W Hya, the spectrum of which consisted of a narrow emission component and a broad wing component (Figure 1). The narrow components showed a rapid decrease by a factor 3 from the beginning to the end of June 1977 (at phase 0.8). A simple radiative pumping model alone cannot account satisfactorily for the variation of the narrow components. The wing component could be seen throughout the observed period; its integrated flux seemed to be small near minimum (Figure 2). It varied independently of the narrow component, and its degree of variation was rather small. These two components may be formed in different regions of the envelope. The broad line shape of the wing component suggests that it may be formed in some region where the maser gain is similar in all directions, such as a region where the dust and gas are rapidly accelerated outward, or the flow is turbulent.

The double emission feature of the Orion SiO maser has steep outer edges and gradually tapering inner edges which resemble the shapes of OH maser lines emitted in the circumstellar envelopes of OH/IR stars. We could notice in the spectra a slight bridge between the double emission features (Figure 3), which seems to show us that these two velocity components originate in the envelope of single star, and not from two independent late-type stars as suggested by Snyder et al. (1978). The relative intensity of the -6.5 km/s component and +16.2 km/s component changed from 0.4 on September 1976 to 1.25 on July 1978. The relative intensity of these two components since their first discovery has been plotted in Figure 4.

539

B. H. Andrew (ed.), Interstellar Molecules, 539–540.
Copyright © 1980 by the IAU.

The change of the relative intensity in the J=2-1 transition within the
observed period is smaller than that in the J=1-0 transition(Moran et al.
1977, Balister et al. 1977, Snyder et al. 1978, and Lepine et al. 1978).

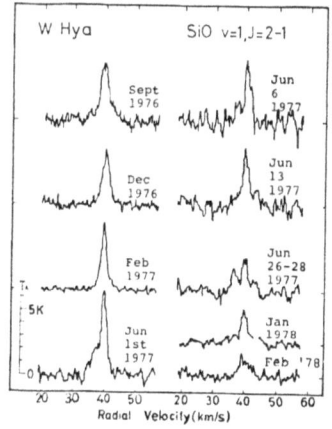

Fig. 1   Spectra of W Hya

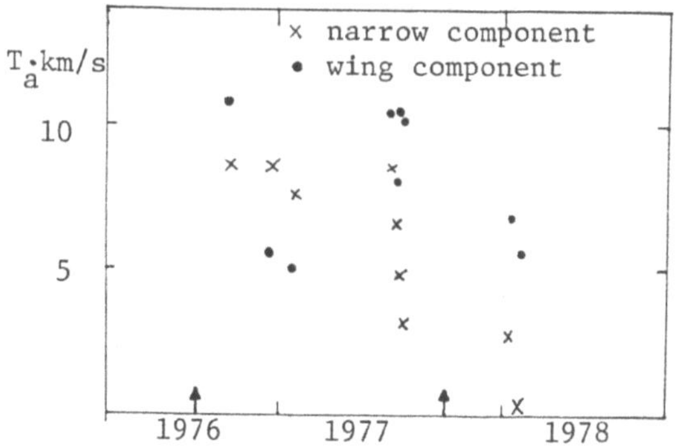

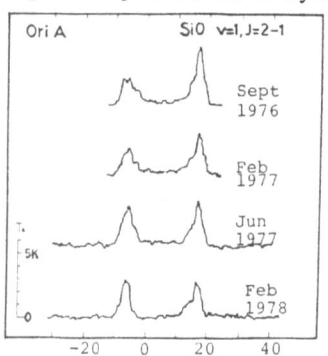

Fig. 3   Spectra of Ori A

Fig. 2   Variation of integrated fluxes of W Hya

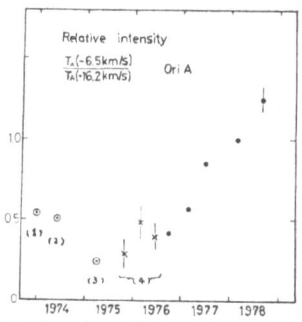

Fig. 4   Relative intensity of double emission
feature of Orion A; (1) Snyder et al.(1974),
(2) Ulich and Haas (1976),(3) Kaifu(1975),
(4) observation with 1 MHz filter bank at TAO

REFERENCES

Balister, M., Batchelor, R.A., Haynes, R.F., Knowles, S.H., McCulloch,
    M.G., Robinson, B.J., Wellington, K.J., and Yabslay, D.E.: 1977,
    Monthly Notices Roy. Astron. Soc. 180, 415.
Kaifu, N.: 1975, private communication.
Lepine, J.R.D., Le Squeren, A.M., and Scalise Jr., E.: 1978, Astrophys.
    J. 225, 880.
Moran, J.M., Johnston, K.J., Spencer, J.H., and Schwartz, P.R.: 1977,
    Astrophys. J. 217, 434.
Snyder, L.E., and Buhl, D.: 1974, Astrophys. J. (Letters) 189, L31.
Snyder, L.E., Dickinson, D.F., Brown, L.W., and Buhl, D.: 1978,
    Astrophys. J. 224, 519.
Ulich, B.L., and Haas, R.W.: 1976, Astrophys. J. Suppl. Ser. 30, 247.

TIME VARIABILITY OF THE ORION A, R LEO AND o CETI SiO (v=1, J=2-1) MASERS

Å. Hjalmarson and H. Olofsson
Onsala Space Observatory, S-430 34 Onsala, Sweden

We here report on observations of 86 GHz SiO (v=1, J=2-1) maser emission from the Mira variables R Leo and o Ceti and from Orion A, made between December 1977 and June 1979 with the new Onsala 20 m millimeter wave telescope equipped with a room temperature mixer. The SiO fluxes from R Leo and o Ceti appear to be correlated with their near infrared intensities, and to have a distinct phase lag with respect to the visual light curve. An irregular behaviour of the R Leo maser, at a time when the star approached an unusually bright maximum, is very probably a manifestation of the extreme sensitivity of the maser process to any disturbances. Definite intensity variations on a time scale of a day have been observed. The total integrated flux of the Orion A SiO maser is relatively stable but there are considerable relative intensity variations between the two main components, and weak emission appears to be present in the entire interval between the two strong features. Pumping considerations indicate that a very efficient, probably radiative pump is needed.

In figure 1 we present integrated SiO (v=1, J=2-1) and peak 2.7μm flux densities and visual magnitudes as functions of time for (a) R Leo and (b) o Ceti; (c) Integrated SiO (v=1, J=2-1) flux densities, their ratio and the peak flux ratio for the low- and high-velocity features in the Orion A maser. The errors in the integrated SiO fluxes are conservatively estimated to be smaller than ± 20%.

B. H. Andrew (ed.), Interstellar Molecules, 541–542.
Copyright © 1980 by the IAU.

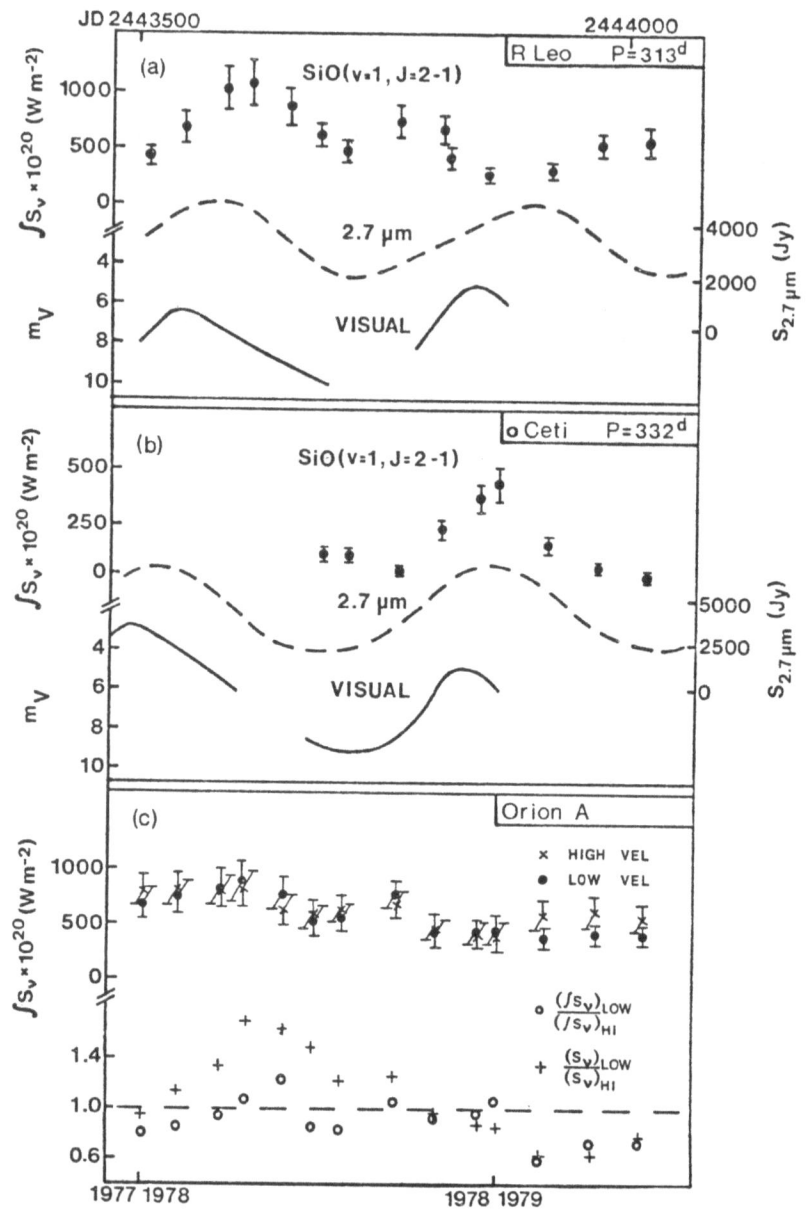

Fig. 1

POLARIZED EMISSION IN THE BROAD SiO FEATURE FROM R LEO

F.O. Clark, University of Kentucky
D.R. Johnson, National Bureau of Standards
T.H. Troland and C.E. Heiles, University of
California at Berkeley

ABSTRACT

Linearly polarized SiO emission spread over 12 km/s has been detected from the star R Leo. The position angle of polarized emission
varies systematically with respect to the spectral line center. Interpreted in terms of radiative transfer theory, this change in position
angle may be due to magnetorotation, which allows the determination of
the magnetic field ($9 \times 10^{-3}/\cos \theta$ Gauss), and the SiO systemic velocity
(-1 ± 2 km/s).

In 1978 Snyder et al. reported a weak maser emission pedestal
associated with SiO emission from the first vibrational state. They
argued that this emission, which is quite spread out in frequency, is
non-thermal because they could not detect ground vibrational state
emission in the same objects at appropriate levels for thermal
emission. Recent observations are reported of linearly polarized
emission spread over 12 km/s from the SiO maser source R Leo, as large
a velocity spread as has been reported for this source.

Data were taken for the SiO maser source R Leo using the $v = 1$,
$J = 2-1$ transition during December 1978, April 1979, and May 1979,
consecutive phases of 0.1, 0.4, and 0.7. Equipment and technique
are described by Troland et al. (1979). During this period, the
peak intensity decreased less than a factor of 2, which is consistent
with saturated maser theory. In the May 1979 data linearly polarized
radiation is detectable over a large range of velocity, approximately
12 km/s. Figure 1 displays Stokes parameters I, $(Q^2 + U^2)^{1/2}$, and
position angle of the linearly polarized radiation. Resolution is
100 kHz or about 0.3 km/s. The abscissa is channel number centered
on -2 km/s, and velocity increases to the right.

An immediate conclusion is the confirmation that the broad
radiation is maser emission, because of the presence of polarization
(first suggested by Snyder et al., 1978).

543

B. H. Andrew (ed.), Interstellar Molecules, 543–544.
Copyright © 1980 by the IAU.

Goldreich et al. (1973) predicted a rotation of the plane of polarization which should be most apparent for unsaturated off-resonant transfer, appropriate for the broad emission component:

$$\Delta\Phi \propto B\cos\theta \,/\, (\nu-\nu_0)^2$$

where Φ is position angle, B is magnetic field, ν is frequency, and θ is the angle between the magnetic field and the line of sight. This result appears to be a reasonable explanation for the position angle data reported, where the position angle of the narrow feature is not considered. This equation allows a calculation of the magnetic field in the region where the broad emission originates:

$$B \sim 9 \times 10^{-3}/\cos\theta \text{ Gauss.}$$

There is an even more exciting implication of the Goldreich et al. theory. If this interpretation is correct, the centroid of the position angle change denotes the systemic velocity, the stellar velocity of R Leo. The theory predicts a rapid change near resonance. However the centroid may be reasonably estimated from the wings as -1 km/s with an estimated error of 2 km/s. This is an independent line of evidence in support of the stellar velocity as the centroid of the broad SiO emission (Reid and Dickinson 1976).

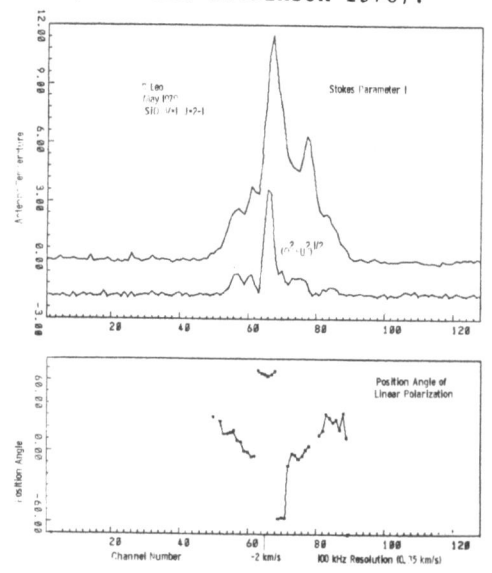

REFERENCES

Goldreich, P., Keeley, D.A., and Kwan, J.Y.: 1973, Ap. J., 179, 111.
Reid, M.J. and Dickinson, D.F.: 1976, Ap. J., 209, 505.
Snyder, L.E., Dickinson, D.F., Brown, L.W. and Buhl, D.: 1978, Ap. J., 224, 512.
Troland, T.H., Heiles, C.E., Johnson, D.R. and Clark, F.O.: 1979, Ap. J. 232, 143.

SiO EMISSION FROM THE ORION NEBULA

B. Baud, J.H. Bieging, R.L. Plambeck, D.D. Thornton,
    W.J. Welch, and M.C.H. Wright
Radio Astronomy Lab, University of California, Berkeley

ABSTRACT

SiO emission in both a strong maser transition in the first excited vibrational state and in a weaker transition in the ground vibrational state both arise from the same small region in the Kleinmann-Low Nebula in Orion. Within the errors, the source position coincides with that of the infrared source IRc2.

INTRODUCTION

Radiation from the SiO molecule emanates from the Orion molecular cloud both in strong maser emission in the first excited vibrational state and, more weakly, from the ground vibrational state. SiO maser emission is generally observed only in red giant stars, and Orion is unique among molecular clouds in exhibiting this radiation. Because of this particular circumstance, it is of special interest to locate the source (or sources) of the radiation with respect to other objects in the nebula. We have used the Hat Creek Interferometer to locate the emission from both the excited and ground states in Orion.

OBSERVATIONS OF MASER EMISSION

The angular extent of the maser source is very small (Genzel et al., 1979), and we have measured its position in the v=1, J=2-1 transition at 86.2 GHz. The technique is that described by Forster et al. (1978), in which the source phase near transit is compared with that of a calibrator. East-West spacings of 150 and 300 meters were used for the right ascension determination and a North-South spacing of 180 meters was used for the declination determination. Figure 1 shows the source spectrum at the time of observation and the series of phase measurements on the source and calibrators near transit for the 150-meter East-West spacing. At this spacing, a 2π phase change corresponds to a 4 arcsecond

545

B. H. Andrew (ed.), Interstellar Molecules, 545–548.
Copyright © 1980 by the IAU.

546

B. BAUD ET AL.

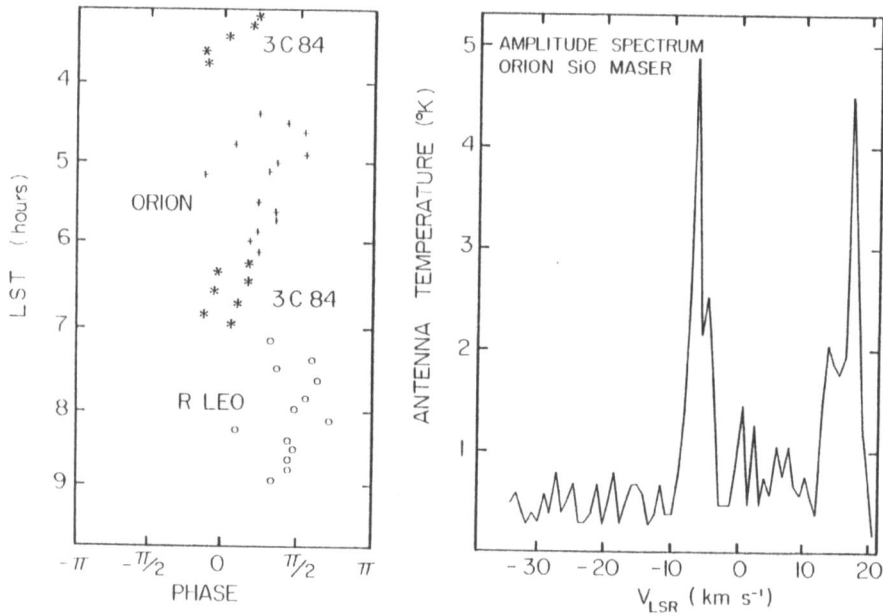

Figure 1.   The Orion SiO maser at 86.243 GHz on 12 Dec 1978.

position change.   The inferred source position is:

R.A.(1950) = $5^h32^m47^s.0 \pm 0^s.03$ Dec.(1950) = $-5°24'23'' \pm 1''$.

The two main features, at -6 and +16 km/sec, are coincident within 0".2.

OBSERVATIONS OF GROUND STATE EMISSION

In the ground vibrational state, the SiO emission is much weaker and somewhat broader in spectrum than the maser. Its spectrum is more similar to the "plateau" sources seen in CO and HCN. Hence, it was included in an aperture synthesis program intended to yield maps of a 2 arcminute field of view centered on the Kleinmann-Low Nebula with an angular resolution of 10 arcseconds. The SiO was observed in the v=0, J=2-1 transition at 86.8 GHz. A preliminary inspection of the data shows that the visibility amplitude for the SiO is essentially constant out to the longest spacing,and the phase is constant and equal to that of the maser. This indicates that the emission arises in a single source which is no larger than 10" in diameter and is centered on the position of the maser. The measurements of Dickinson et al. (1976) with the 11-meter telescope are consistent with a small source at this position.

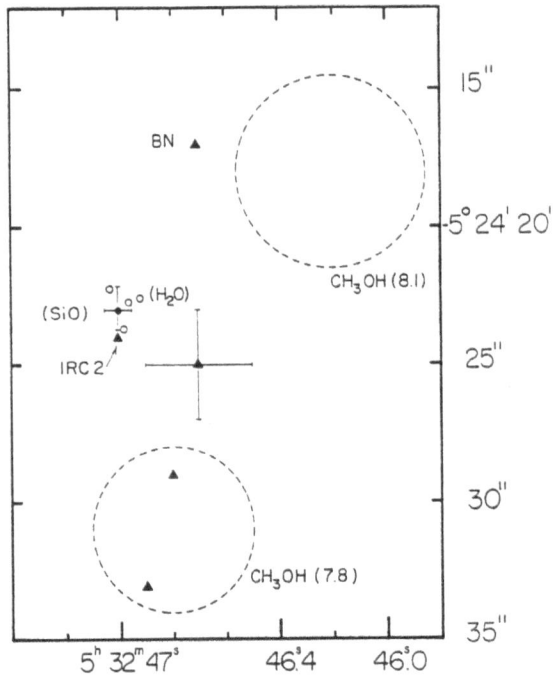

Figure 2.   Position of the SiO emission in Orion.

DISCUSSION

        Figure 2 shows the observed SiO position along with a number of
other objects in the neighborhood of the Kleinmann-Low Nebula.   The
open circles are water vapor masers (Genzel et al., 1979; Forster et al.,
1979) whose positional uncertainties are comparable to that of the SiO.
The triangles are infrared sources (Rieke et al., 1973).   Their relative
positions are somewhat more accurate than their absolute placement,
which is uncertain by about 2" (Rieke, 1979).   The dotted circles represent methanol maser positions and approximate sizes (Matsakis, 1978).
The velocities of the SiO and water vapor masers are the same, and
their positions are coincident to within the errors.   They also coincide
with IRc2, one of the infrared objects.   However, the positional uncertainties of the infrared objects are sufficiently large that coincidence
with the more westerly source IRc3 is also possible, although less
likely.

        In all probability the emission in both the v=0 and v=1 levels
arises from deep in the atmosphere of IRc2.   The color temperature of
this source is 335 K and its blackbody size is about 0".1 (50 a.u. at
the distance of Orion).   As compared with the red giant stars, the
observed 8 micron and 10 micron fluxes are a little too weak for adequate
radiative pumping of the SiO maser.   However, the foreground dust absorption is very large, and there should be adequate flux deeper in the

atmosphere of IRc2. The deep silicate absorption at 10 microns implies a foreground column density of about 3×10^{-4} g cm^2, for example. In any case, the observed 10 micron flux is adequate to pump the methanol masers, and it would be useful to determine whether there is a correlation between the time variations of the SiO and methanol fluxes. The nature of IRc2 remains obscure. Its close proximity to the other IR objects makes it unlikely that it is an evolved star like the other SiO sources. Observations with higher spatial resolution, particularly at longer infrared wavelengths, may be essential to explain its nature. This work was supported by NSF grant AST 75-13511.

REFERENCES

Dickinson, D.F., Gottlieb, C.A., Gottlieb, E.W., and Litvak, M.M.: 1976, Ap. J., 206, pp. 79-84.
Forster, J.R., Welch, W.J., Wright, M.C.H., and Baudry, A.: 1978, Ap. J., 221, pp. 137-144.
Genzel, R., Moran, J.M., Lane, A.P., Predmore, C.R., Ho, P.T.P., Hansen, S.S., and Reid, M.J.: 1979, Ap. J. (Letters), 231, pp. L73-L76.
Matsakis, D.N.: 1978, Ph.D. thesis, University of California, Berkeley.
Rieke, G.H., Low, F.J., and Kleinmann, D.E.: 1973, Ap. J., 186, pp. L7-L12.
Rieke, G.H.: 1979, private communication.

DISCUSSION FOLLOWING WELCH

Elmegreen: Can you think of any way to prove that the SiO maser in Orion is not a background late-type star?
Welch: No. The best present argument against its being a background star is the small angular separation between the maser source and the other four infrared sources in the infrared nebula. The projected separations between all these objects are only a few thousand A.U. The coincidence with a late type star seems unlikely.
Zuckerman: One way to rule out a background star would be to show that plateau emission from molecules such as SO is also associated with the source star (presumably IRC 2).
Welch: Yes, if IRC 2 turns out to be the "plateau" source, it will be distinctly different from the late-type maser stars.
Hjalmarson: I think we are all very impressed by your work. What is your position accuracy in the case of HCN?
Welch: The positional accuracy is about 1".
Winnewisser: Do you have any evidence for clumping in HCN in Orion? Over which area did you map HCN and what are the typical values for the brightness temperature?
Welch: Yes, visibility amplitudes at various baselines that we have used show that there is structure in HCN at scales between 1" and 10". At this point we do not know whether there is one clump or several. The field of view is about 2 arc minutes centered on IRC 2. HCN at scale between 1" and 10" would have to have brightness temperatures in excess of 100K.

THE OH CIRCUMSTELLAR MASER IN LATE-TYPE STARS

Nguyen-Q-Rieu, V. Bujarrabal, J. Guibert, A. Omont
Observatoire de Meudon, France

A recent investigation based upon the OH luminosity distribution (Nguyen-Q-Rieu et al. 1979) has shown that Type I OH sources associated with Mira variables (OH Miras) are weak OH sources and are therefore only detected within $\sim$1 kpc from the Sun. Type II OH-Miras, which are more intense and rarer than Type I OH-Miras, are probably more distant objects. The group of unidentified Type II OH-IR sources probably consists of Type II OH-Miras of high OH luminosity. The IR colour index, which is usually higher for Type II sources, suggests that they have a colder and denser dust shell.

High maser gains ($\tau \sim -20$) can be obtained for the 1612 MHz line (Type II sources) from a model of radiative pumping by far-IR radiation from cold dust (grain temperature $T_g \sim 120$ K) (Elitzur et al. 1976). The pumping of the main lines (Type I sources) is difficult to achieve without invoking some asymmetrical effects which invert the lambda-doublets (Bujarrabal et al. 1980a). The most powerful inversion process is based on the IR line overlap (see also Lucas 1979). We have investigated such an effect in the case of a large-scale velocity field in an expanding circumstellar shell of radius R and thickness ΔR, using the Sobolev approximation to treat the radiative transfer (Bujarrabal et al. 1980b). The velocity field is $V(r) = V_M (r/R)^{\epsilon}$. The IR radiation field due to dust can be expressed as

$$I_\lambda = (2h\nu^3 \, W/c^2) \ (80/\lambda\mu)^P \ (e^{h\nu/kTg}-1)^{-1} .$$

W represents the averaged optical depth of dust grains at 80 μ.

The results of calculations are given in the Figure as a series of curves, displaying the OH optical depth as a function of $n_{OH}r/V$; n_{OH} is the OH volume density. The following conditions have been assumed:

$$\epsilon = 0.25; \ W = 5.10^{-4}; \ p = 2; \ T_g = 350 \text{ K}; \ V_M = 5 \text{ km s}^{-1}; \ \Delta R = 0.5 \text{ R}$$

The kinetic temperature T_K and the hydrogen density n_{H_2} are taken equal to 100 K and 10^5 cm^{-3}, respectively. In these conditions, the

B. H. Andrew (ed.), Interstellar Molecules, 549–550.

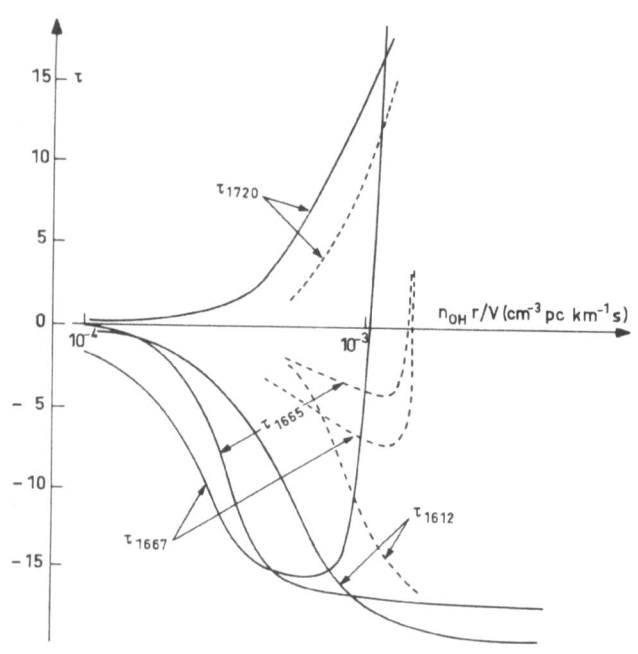

influence of various collisional laws is of little importance, since the pumping is essentially radiative (see also Omont 1979). We have assumed equal downward transitions between magnetic sub-levels. The Figure shows that when IR overlap is taken into account (full line curves), the optical depths of the main lines and the 1612 MHz line are increased by a factor of 2 to 3 with respect to the case of no overlap (dashed line curves). The efficiency of the pumping by line overlap can be estimated by calculating the 35μ flux required to achieve a main line optical depth $\tau \sim -15$. The IR flux F is equal to $I_{35\mu}\Omega$; Ω corresponds to the angular size of the OH shell. For a linear diameter $\sim 10^{16}$ cm and a heliocentric distance ~ 1 kpc, $F_{35\mu}$ is ~ 7.5 Jy, which is much smaller than the IR flux expected to be observed in Mira variables (Nguyen-Q-Rieu et al. 1979). In the absence of overlap, the 35μ flux would be ~ 10 times higher. Therefore, the line overlap pumping can be very efficient in the envelope of Mira variables.

REFERENCES

Bujarrabal, V., Destombes, J.L., Guibert, J., Marliere, C., Nguyen-Q-Rieu, Omont, A.: 1980, Astron. Astrophys. (in press).
Bujarrabal, V., Guibert, J., Nguyen-Q-Rieu, Omont, A.: 1980, Astron. Astrophys. (submitted).
Elitzur, M., Goldreich, P., Scoville, N.: 1976, Astrophys. J. 205, 384.
Lucas, R.: 1979, paper published in these Symposium Proceedings.
Nguyen-Q-Rieu, Laury-Micoulaut, C., Winnberg, A., Schultz, G.V.: 1979, Astron. Astrophys. 75, 351.
Omont, A.: 1979, paper published in these Symposium Proceedings.

INTERPRETATION OF CIRCUMSTELLAR MASERS

Peter Goldreich
California Institute of Technology

1. INTRODUCTION

    In a short review paper one is often forced to survey the subject
in a broad way and to ignore many, perhaps even important, details.
Such is the case here. My remarks are intended to indicate how I view
the current status of the field. In general, I shall avoid qualifying
my statements, even when I am clearly guilty of oversimplification.
If someone finds my paper provocative, so much the better.

    My comments are strictly directed toward the interpretation of
masers which operate in the circumstellar envelopes about oxygen-rich
Mira variables. Most of them also apply to the masers associated with
oxygen-rich supergiants.

1.1 Types of masers

    Maser action in circumstellar envelopes has been observed in
certain microwave transitions from the SiO, H_2O and OH molecules.
There is a distinct possibility that additional types of masers await
discovery.

    The SiO, H_2O and OH masers form a sequence along which the excita-
tion required to adequately populate the maser levels decreases. The
same sequence also orders the masers in terms of increasing distance
from the central star, increasing mass of the circumstellar envelope
in which they operate most effectively, and decreasing variability.

1.2 Models of masers

    Interstellar masers are both basic astronomical phenomena and
probes of the physical conditions in regions where they operate. The
latter aspect alone makes their study a valuable one. However, much
effort has been and is continuing to be devoted to developing detailed
models for maser pump cycles. I doubt whether this game is really
worth playing in most cases.

551

B. H. Andrew (ed.), Interstellar Molecules, 551–558.

To accurately calculate a maser pump cycle, we require knowledge of the number densities of maser molecules, hydrogen molecules and dust grains; the kinetic temperature and velocity at each point in the maser cloud; the external radiation field incident upon the cloud; the surface area and wavelength dependent emissivity of the dust grains; the cross sections for all transitions of the maser molecules induced by collisions with hydrogen molecules. Given this information, we could solve the coupled equations of radiative transfer and statistical equilibrium to obtain at every point in the cloud the population in all of the levels of the maser molecule and the radiation field in each of the lines which couple these levels. To solve these equations would be an enormous undertaking, but in principal it would enable us to represent nature closely enough to reveal how the maser is pumped.

Of course, the program we have outlined cannot, in general, be carried out. We lack too much of the input information. However, masers associated with circumstellar envelopes of envolved stars are far more amenable to theoretical interpretation than those associated with regions of star formation. We know much more about the environment where the former operate than about that where the latter do. In particular, the radiation field, velocity, temperature and molecule and dust grain number densities in these envelopes may all be reasonably estimated. Lack of detailed knowledge of collision cross sections and limited computing power are among the more serious obstacles to determining the pump schemes.

2. CIRCUMSTELLAR ENVELOPES

2.1 Dynamics

Infrared and microwave observations imply that the maser stars are losing 10^{-6}-10^{-4} $M_\odot$ yr^{-1}. It is unclear whether the initial acceleration of material away from the star is primarily due to a periodic shock wave driven by the stellar pulsation (1,2) or to radiation pressure acting on dust grains (3,4).

VLBI observations of masers suggest that substantial acceleration of the circumstellar material continues until at least ten stellar radii. This far from the star the acceleration can only be due to radiation pressure and requires that the dust grains continue to grow out to at least $10R_*$ (5).

2.2 Thermodynamics

A crude model of the thermal structure of the circumstellar envelope (6) shows that the major heat input is due to collisions between gas molecules and dust grains which are driven by radiation pressure at supersonic speed through the gas. Adiabatic expansion and the emission of line radiation by abundant molecules, especially H_2O, are the principal sources of cooling. For a mass loss rate

$\dot{M} \approx 3 \times 10^{-5}$ $M_\odot$ yr^{-1} and $T \simeq 2 \times 10^3$ K, at $r = R_*$, the model yields $T \approx 10^3$ K at $r = 10R_*$ and $T \simeq 3 \times 10^2$ K at $r = 10^2 R_*$.

The stellar luminosity variation forces a variation in the speed at which the dust grains stream through the gas. This in turn gives rise to a small variation in the gas temperature in the outer part of the circumstellar envelope. The temperature variation lags behind the luminosity variation by between $\pi/2$ and π radians.

2.3 Chemistry

In the atmospheres of cool, oxygen-rich Miras, most hydrogen exists as H_2. Silicon is predominately found in SiO. The oxygen which is not tied up in CO is bound in H_2O. Given this atmospheric composition, the source of the SiO and H_2O maser molecules is obvious. The origin of the OH maser molecules is less apparent.

At temperatures above 500 K, OH molecules are destroyed on the expansion timescale by the exothermic reaction OH + H_2 → H_2O + H. However, an adequate supply of OH molecules may be produced in the cool outer part of the circumstellar envelope by the dissociation of H_2O molecules either by interstellar ultraviolet photons or by high speed dust grains (6).

3. MASER AMPLIFICATION

3.1 Basic questions

Short of determining a detailed model for the maser we might hope to obtain answers to the following questions

i) Where in the circumstellar envelope does the maser operate?

ii) What is the directionality of the maser gain and maser radiation?

iii) What is the principal input to the maser? Is it spontaneous emission, the stellar continuum or the cosmic background radiation?

iv) Is the maser unsaturated or saturated?

v) Is the maser radiatively, collisionally or chemically pumped?

3.2 Answers to basic questions

Only partial answers to the basic questions are available. They are given below.

i) VLBI observations of the spread in angular position of the different types of masers imply that the SiO and H_2O masers are located within a few radii of the central star (7,8). The SiO masers probably operate somewhat closer to the star than the H_2O masers do, as is indicated by observations of masers associated with supergiants. The OH masers are clearly very distant from the central star, $r \geqslant 10^2$ R_* (9). The available VLBI data shows that the main-line, 1665 and 1667 MHz OH masers are a bit closer to the star than the satellite line, 1612 MHz maser.

There is a hint from monitoring programs of the 1612 MHz OH maser variations that the high-velocity flux shows a phase-lag with respect to the low-velocity flux (10,11). If real, the time-lags imply that the masers are located between 10^{16} and 10^{17} cm from the central stars.

The maser locations determined from VLBI observations are in accord with the conclusions of theoretical models of maser pump cycles.

ii) The maser gain in direction $\hat{l}$ is proportional to the path length along which velocity coherence is maintained. In other words, the gain is inversely proportional to dv/dl. For a spherically symmetric flow

$$\frac{dv}{dl} = (1 - \mu^2) \frac{v}{r} + \mu^2 \frac{dv}{dr}$$

where $\mu = \hat{r} \cdot \hat{l}$. Thus the maximum gain path is radial if $dv/dr < v/r$ and tangential if $dv/dr > v/r$. Clearly, we expect dv/dr to be smaller than v/r except perhaps very close to the star where the major acceleration takes place.

The directionality of the maser radiation depends upon the directionality of both the gain and the radiation input to the maser.

iii) The relative importance of spontaneous emission, the cosmic background radiation and the stellar continuum as input sources for radial amplification may be assessed by comparing $|T_{ex}|$, T_{BB} and $T_* \Omega_* / \Omega_M$ (12). Here T_{ex} is the negative excitation temperature of the maser transition, $T_{BB} \simeq 2.7$ K, and $T_* \simeq 2000$ K is the stellar brightness temperature at the maser frequency. The solid angles are

$$\Omega_* \equiv \pi \left(\frac{R_*}{r}\right)^2 \quad \text{and} \quad \Omega_M \equiv \frac{\pi}{|\tau|} \frac{d\ln v}{d\ln r}$$

where τ is the negative optical depth in the radial direction.

Theoretical considerations indicate that spontaneous emission and the stellar continuum are the most important inputs to the SiO and H_2O masers. The principal input to OH masers is usually sponta- neous emission, but both the background radiation and the stellar continuum may be significant as well.

The OH masers often have two velocity peaks which come from the front and back of the circumstellar envelope (2, 12, 13, 14). There is no tendency for the blue-shifted peak to be stronger than the red-shifted one. This proves that the stellar continuum is not the dominant source of input to these masers.

VLBI observers should look for the amplified image of the central star in their maps. This would be especially worthwhile for OH masers which usually have angular sizes large compared to that of the stellar disk.

iv)   Theoretical calculations suggest that circumstellar masers are generally saturated. Observations of the variation of maser luminosity with stellar luminosity for OH and H_2O masers (15,16) support this conclusion.

v)   There is little doubt that the OH masers are radiatively pumped. They vary in phase with the stellar luminosity unlike the local gas temperature. The situation for SiO and H_2O masers is less clear. In particular, theoretical considerations suggest that collisions as well as radiation are important in the SiO pump cycle.

4.   PUMP MODELS

4.1   1612 MHz OH masers

Detailed models of these masers show that the maser radiation is narrowly beamed in the radial direction, both inward and outward (2, 12). The masers are pumped by far infrared radiation emitted by hot dust grains. The rotationally excited OH molecules radiatively decay down to the ground rotational state. If the lines which connect the $^2\Pi_{1/2}$ J = 1/2 state to the $^2\Pi_{3/2}$ J = 3/2 ground state are optically thick, the 1612 MHz transition is inverted (12). Numerical calculations predict that about 4 FIR photons are absorbed in each pump line for every 1612 MHz maser photon which is emitted. Recent observations have confirmed this prediction (17).

The 1612 MHz OH masers are a particularly good case for theory. The line strength of the 1612 MHz transition is much smaller than that of the main-lines at 1665 and 1667 MHz. Nevertheless, the 1612 MHz masers are the strongest ones in stars with dense circumstellar envelopes. Furthermore, these masers operate very far from the central star where the density is so low that collisionally induced transitions are negligible.

4.2   Main-line OH masers

Radiative pump models for these masers rely on the parity dependent hyperfine splitting which breaks the symmetry between the upper

and lower halves of the lambda doublets (18-22). The most sophisticated
models take into account the overlaps between different components of
hyperfine-split rotational lines. In circumstellar envelopes, line
overlap due to the expansion velocity may be modeled in considerable
detail (22). A recent study indicates that pumping by FIR radiation
together with the effects of line overlap probably accounts for the
essential features of the main-line pump (22).

4.3 H_2O masers

The H_2O molecule has a far more complicated rotational spectrum
than the OH molecule. Also, to date, it has been observed in a single
maser line. These factors make the H_2O masers less attractive to
model than the OH masers. Nevertheless, some models have been put
forth (23,24). As far as I can see, it is not yet clear whether the
basic excitation is due to collisions or to radiation.

4.4 SiO masers

The SiO masers involve rotational transitions in excited
vibrational states. There are several different transitions which
show maser action. Although the maser molecules are highly excited,
the level structure of the SiO molecule is so simple that detailed
modeling of the pump mechanism may be rewarding. An early model
showed that radiative decays from excited vibrational levels would
tend to invert the lowest rotational transitions in these levels, if
the decay lines were optically thick (25). More recent studies (26,
27) indicate that both collisional and radiative transitions are of
importance in the pump cycle. A definitive model of SiO masers has
yet to appear.

ACKNOWLEDGMENTS

I am indebted to C. Alcock for much valuable advice and to
M. Litvak for his pioneering contributions to the field.

This work was supported by NSF Grant AST 78-21453.

BIBLIOGRAPHY

1.   Maehara, H.: 1971, Publ. Astron. Soc. Japan 23, p. 503.
2.   Slutz, S.: 1976, Ap. J. 210, p. 750.
3.   Salpeter, E.E.: 1974, Ap. J. 193, p. 585.
4.   Kwok, S.: 1975, Ap. J. 198, p. 583.
5.   Zuckerman, B.: 1979, Invited talk on Circumstellar Envelopes at
     IAU Symposium No. 87 Published in this volume.
6.   Goldreich, P. and Scoville, N.Z.: 1976, Ap. J. 205, p. 144.
7.   Moran, J.M., Ball, J.A. Predmore, C.R., Lane, A.P., Huguenin, G.,
     Reid, M.J. and Hansen, S.S.: 1979, Ap.J. Letters 231, p. L67.

8. Spencer, J.H., Johnston, K.J., Moran, J.M., Reid, M.J. and
    Walker, R.C.: 1979, Ap. J. 230, p. 449.
9. Reid, M.J., Muhleman, D.O., Moran, J.M., Johnston, K.J. and
    Schwartz, P.R.: 1977, Ap. J. 214, p. 60.
10. Schultz, G.V., Sherwood, W.A. and Winnberg, A.: 1978, Astron.
    Astrophys. 63, p. L5.
11. Jewell, P.R., Elitzur, M., Webber, J.C. and Synder, L.E.: 1979,
    Ap. J. Suppl. 41, p. 191.
12. Elitzur, M., Goldreich, P. and Scoville, N.Z.: 1976, Ap. J.
    205, p. 384.
13. Reid, M.J.: 1976, Ap. J. 207, p. 784.
14. Reid, M.J. and Dickinson, D.F.: 1976, Ap. J. 209, p. 505.
15. Harvey, P.M., Bechis, K.T., Wilson, W.J. and Ball, J.A.: 1974,
    Ap. J. Suppl. 27, p. 331.
16. Schwartz, P.R., Harvey, P.M. and Barrett, A.H.: 1974, Ap. J.
    187, p. 491.
17. Werner, M.W., Beckwith, S., Gatley, I., Sellgren, K., Berriman,
    G. and Whiting, D.L.: 1979, preprint.
18. Litvak, M.M.: 1969, Ap. J. 156, p. 471.
19. Pelling, M.: 1977, M.N.R.A.S. 178, p. 441.
20. Lucas, R.: 1979, Astron. Astrophys. In press.
21. Elitzur, M.: 1978, Astron. Astrophys. 62, p. 305.
22. Bujarrabal, V., Guibert, J., Nguyen-Q-Rieu and Omont, A.: 1979,
    Astron. Astrophys., in press.
23. Litvak, M.M.: 1969, Science 165, p. 855.
24. Deguchi, S.: 1977, Publ. Astron. Soc. Japan 29, p. 669.
25. Kwan, J. and Scoville, N.Z.: 1974, Ap.J. Letters 194, p. L97.
26. Deguchi, S. and Iguchi, T.: Publ. Astron. Soc. Japan 28, p. 307.
27. Alcock, C.: 1977, Ph.D. Thesis, California Insitute of Technology.

DISCUSSION FOLLOWING GOLDREICH

Dickinson: Your placing of the H_2O and SiO masers near the star is
consistent with the emission velocities, which are near the stellar
velocity, and with the familiar double-peaked OH, which is out in the
shell at higher velocities. However, we often see in H_2O, and occasionally in SiO as well, "shell" components similar to the OH. Is there
anything in your theory which would explain why there are two velocity
domains - near the star and in the shell - with no intervening emission?

Goldreich: "Shell" components probably signify radial amplification,
which can be important for the H_2O and SiO masers as well as for the OH
masers. However, the H_2O and SiO masers may also operate so near the
star that the best gain paths are tangential rather than radial because
of the radial acceleration of the outflowing gas.

Elitzur: It is not difficult to show that a radiative pump cannot
be the explanation of the SiO masers. From very general properties of
cross-sections for vibration-rotation excitations, it follows that
collisions can provide an adequate explanation. I propose that the SiO
maser emission has two components; the weak maser pedestal and the broad

$v = 0$ thermal emission are sampling the shell, whereas the strong maser spikes arise from the upper atmosphere, where the temperature is high enough. The velocity structure, the time variation of the spikes, and the general properties of polarization, can all be well explained if the emission arises from large convective cells of the type proposed by Schwarzchild in 1975, for other reasons.

Gold: If there is enough gain, then surely a line can be made to drift gradually in frequency by a medium with a radial velocity gradient. For each step, a small but finite velocity shift can be tolerated, and the emerging wave will have the additional power supplied at the frequency appropriate to the new velocity. Together with such gradual frequency shifts, there may be slight changes in direction of the wave. These effects will be particularly important in cases of saturated masers that may have a great deal more gain available than can be used in view of their total power limitation.

Goldreich: The effects you refer to are not significant in the masers associated with IR stars. Although such effects can exist in principal, the radiation field intensities in these masers are far too low for these nonlinear phenomena to be important.

Burdjuzha: Why do you not also discuss the pumping mechanism of OH masers through the $J=5/2$ level of the $^2\pi_{3/2}$ band? The optical depth is high, is it not?

Goldreich: We do include pumping through the $J=5/2$, $^2\pi_{3/2}$ level in our numberical calculations. The optical depths of the transitions between states in this level and the ground level are indeed high.

PUMPING MECHANISMS OF OH MASERS

A. Omont[1,2], J. Guibert[1], S. Guilloteau[1,2], V. Bujarrabal[1],
Nguyen-Q-Rieu[1]

[1]Observatoire de Meudon, France
[2]Ecole Normale Supérieure, Paris

ABSTRACT

We analyze in detail the pumping of 18 cm OH masers by the overlap
of far infra-red lines in HII/OH regions and in circumstellar shells.
We present some results of a model of pumping of HII/OH masers by
thermal overlap. Pumping by overlap appears to be quite general but
very complicated. It can account for the different observed main-line
masers. However, other types of pumping, mainly by collisions with H
or by near infra-red radiation, are possible if certain conditions are
met for the physical and astrophysical parameters.

1. INTRODUCTION

Although the pumping of type II OH masers appears to be well understood (Litvak 1969, Elitzur et al. 1976, Elitzur 1976), at least for the
1612 MHz line, the pumping mechanism of the OH main lines is still
unclear. It is generally agreed that UV or chemical pumping is unlikely
because the needed rates are unrealistically large (see e.g. Elitzur and
de Jong 1978). Much progress has been made recently in the treatment of
most of the other possible pumping mechanisms: collisional pumping
(Shapiro and Kaplan 1979, Dixon and Field 1979a,b,c, Flower 1979,
Elitzur 1979); and pumping by far infra-red radiation (Elitzur 1978,
Lucas 1979a,b, Bujarrabel et al. 1979a,b, Nguyen-Q-Rieu et al. 1979),
or by near infra-red radiation (Cimerman and Scoville 1979). Litvak
1974 and Cook 1977 give reviews of previous work. The purpose of this
paper is to try to draw some conclusions about the most likely pumping
mechanisms of OH masers in circumstellar envelopes and in HII/OH regions,
with special emphasis on the details of the pumping by far infra-red
line overlap and on its efficiency compared to collisional pumping by
atomic hydrogen.

2. FAR INFRA-RED LINE OVERLAP

The hyperfine structures of the rotational far infra-red lines of

559

B. H. Andrew (ed.), Interstellar Molecules, 559–564.

OH (35-120 μm) range from a
few tenths of km/s to a few
km/s (see figure 1 of Lucas
1979b), and are thus of the
same order of magnitude as
the large scale and the
thermal velocities in the
masering regions. As a result the intensity in one
infra-red line strongly depends on the absorption or
emission in other lines
either at the same location
or in other parts of the
region. As the hyperfine
structures of the IR lines
connected to the two halves
of the Λ-doublet are quite
different, the IR rates from
and to the main-line masering levels can be completely
different. The potentially
high efficiency of this
process has been pointed out
a long time ago (Litvak
1969, Pelling 1977). However,
because of the very large
number of possible overlaps,
the situation is very intricate and requires a detailed
treatment of the different
arrangements of the velocity
fields in the masering regions.

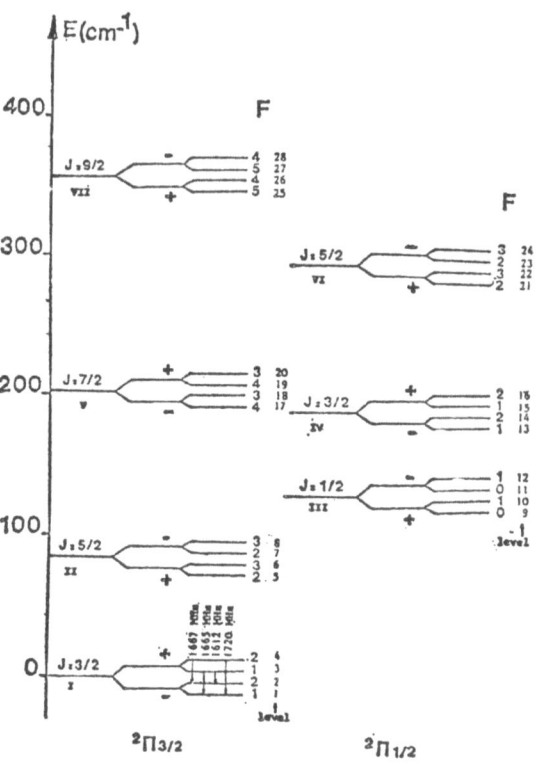

Figure. Levels of OH

2.1. Circumstellar Shells (Bujarrabal et al. 1979b).

In an expanding shell a line formed at one point on the shell is
seen red-shifted by an observer located at another point. The results
of a model for the pumping of circumstellar masers (Bujarrabal et al.
1979b) are presented in a communication at this Symposium by
Nguyen-Q-Rieu et al. It can be seen in their figure that the inversion
of the main lines by far-IR overlap appears to be quite general under
the conditions of circumstellar main-line masers, and has a very good
efficiency. We will measure the efficiency of a pumping mechanism by
the value $\Delta n'/n = 2(n_2-n_1)'/(n_2+n_1)$ of the inversion achieved by
artifically setting to zero the intensity of the maser lines. In this
case $\Delta n'/n$ can reach 7×10^{-2}. A detailed analysis (Bujarrabal et al.
1979b) shows that a large number of overlaps are important for the
pumping of each line: the 1667 MHz line is inverted mainly by the overlaps 5-2, 6-2 and 14-4, 13-3, while the pair 22-2, 21-1 has an important
anti-inverting effect. On the other hand, the inversion of the 1665 MHz
line is mainly due to the overlap 23-3, 24-4.

2.2. HII/OH Regions (Lucas 1979, Guilloteau et al. 1979)

As discussed by Lucas at this Symposium, very strong inversions ($\Delta n'/n \sim 1$) are easily achieved by pumping the OH in a small cloud ($N_{OH} \sim 10^{15} cm^{-2}$) by radiation from a larger cloud with a high excitation temperature (Lucas 1979a,b). At larger column densities ($N_{OH} \sim 10^{16}$–$10^{18} cm^{-2}$) at which this external pumping is less efficient, an alternative pumping mechanism is provided without any external cloud by the thermal overlaps of the IR lines within the masering cloud (Guilloteau et al. 1979). The overlap of the thermal profiles of the IR lines strongly perturbs the transfer of the IR radiation even when the gaussian widths of the lines correspond to temperatures as low as 100–200 K. These thermal overlaps have two main pumping effects: i) they increase the general trapping of the photons of the two overlapping lines, and hence they reduce the radiative repopulation of the two lowest levels, and ii) they mix the populations of the levels of the two lines; in particular they tend to equalize the populations of the two upper levels, and hence they overpopulate the lower level of the line which had the larger initial population in its upper level. Besides, the thermal overlaps in the masering cloud also obviously affect the pumping of the molecules by the external continuum IR radiation.

These intricate effects result in relatively efficient pumping. For instance, $\Delta n'/n$ is of the order 10^{-2} for typical conditions (spherical model, $n_{H_2} = 10^6$ cm^{-3}, T = 200 K) with $N_{OH} \sim 10^{16} cm^{-2}$ for the 1667 MHz line and $N_{OH} \gtrsim 10^{17} cm^{-2}$ for the 1665 MHz line. The corresponding brightness temperature of the 1665 MHz line ($T_B > 10^{10} K$) is comparable to the observed T_B. It is smaller in the 1667 MHz line. The power observed in this line can be accounted for either by a cylindrical maser, or by very large column densities ($N_{OH} \gtrsim 10^{18} cm^{-2}$), or by a relatively intense IR continuum ($T_R \sim 100$ K, dilution factor $W \sim 0.1$) which is not unlikely in HII/OH regions. Of course, the analysis of the details of these complicated maser mechanisms is not simple. However, it appears that the following effects are the most important: i) selective enhancement of the trapping of 120 μm photons at low density leading to the inversion of the 1667 MHz line, and ii) population mixing in the overlap 15-1, 16-2 (53 μm) inverting the 1665 MHz line.

In summary, because of the very rich level structure of the OH molecule, pumping by overlap appears to be both very general and very complicated. It is very sensitive, first to the velocity fields, but also to the other parameters of the masering regions: OH column density, total density, kinetic temperature, IR continuum, etc. Accordingly, detailed models of observed masers are very difficult to build, although we feel that this type of pumping could account for the different observed masers of type I.

3. COMPARISON WITH OTHER PUMPING MECHANISMS

3.1. Collisional Pumping

We can disregard collisional transitions between the components of the Λ-doublet which are caused by collisions with electron streams (Johnston 1967) or with ion streams (Elitzur 1979): the velocity required for electrons is very unlikely, and a collision rate asymmetry is unlikely for ions (Bouloy and Omont 1979). The possibility of pumping through rotational excitation by H_2 and e^- is less easily discounted: the asymmetries are probably small and possibly in the wrong sense, and the rotation rates are probably smaller than the Λ-doublet rates. Precise rates are urgently needed to clarify this question (see Bertojo et al. 1976, Shapiro and Kaplan 1979, Dixon and Field 1979a,b,c, Flower 1979).

It seems that atomic hydrogen can provide the most powerful collisional pumping. The asymmetry in the excitation rates to the first rotational level seems well established as 10% in the direction favouring 18 cm OH maser pumping (Shapiro and Kaplan 1979, Dixon and Field 1979a,b,c). However, there is a strong disagreement between these authors about the values of the rotation rates and hence about their ratio to the Λ-doublet rates. Shapiro and Kaplan's values result in a rather poor pumping efficiency ($\Delta n'/n < 3 \ 10^{-3}$) which should disqualify this type of pumping. However, as discussed by Dixon and Field 1979a,b, the rotation rates are probably much larger. A pumping efficiency comparable to the one by overlap would require very strict conditions on the hydrogen abundance: $n_H/n_{H2} > 20\%$, and collisional rates comparable to radiative excitation rates but not large enough to cause rotational thermalization (Elitzur 1979).

3.2. Pumping by Near Infra-red

Initially proposed by Litvak 1969, this mechanism has been recently revived for circumstellar masers by Cimerman and Scoville 1979, who point out that there are coincidences between IR transitions of H_2O and OH at 2.8 μm. The relative efficiencies of pumping by far-IR overlap and by 2.8 μm will depend on the amount of absorption by dust between the star and OH.

REFERENCES

Bertojo, M., Cheung, A.C., and Townes, C.H.: 1976, Astrophys. J. 208, 914.
Bouloy, D., and Omont, A.: 1979, Astron. Astrophys. Supp. 38, 101.
Bujarrabal, V., Destombes, J.L., Guibert, J., Marlière, C., Nguyen-Q-Rieu, and Omont, A.: 1979a, Astron. Astrophys., in press.
Bujarrabal, V., Guibert, J., Nguyen-Q-Rieu, and Omont, A.: 1979b, Astron. Astrophys., in press.
Cimerman, A.H., and Scoville, N.: 1979, preprint.
Cook, A.: 1977, 'Celestial Masers', Cambridge University Press.

Dixon, R.N., Field, D.: 1979a, Proc. R. Soc. A. 368, 99.
Dixon, R.N., and Field, D.: 1979b, M.N.R.A.S. 189, 583.
Dixon, R.N., and Field, D.: 1979c, Proc. IAU Symposium No. 87 (this
    volume).
Elitzur, M.: 1976, Astrophys. J. 203, 124.
Elitzur, M.: 1978, Astron. Astrophys. 62, 305.
Elitzur, M.: 1979, Astron. Astrophys. 73, 322.
Elitzur, M., and de Jong, T.: 1978, Astron. Astrophys. 67, 323.
Elitzur, M., Goldreich, P., and Scoville, N.: 1976, Astrophys. J.
    205, 384.
Flower, D.: 1979, preprint.
Guilloteau, S., Lucas, R., and Omont, A.: 1979, in preparation.
Johnston, I.D.: 1967, Astrophys. J. 150, 33.
Litvak, M.M.: 1969, Astrophys. J. 156, 471.
Litvak, M.M.: 1974, Ann. Rev. Astron. and Astrophys. 12, 97.
Lucas, R.: 1979a, Astron. Astrophys., in press.
Lucas, R.: 1979b, Proc. IAU Symposium No. 87 (this volume).
Nguyen-Q-Rieu, Bujarrabal, V., Guibert, J., Guilloteau, S., and
    Omont, A.: 1979, Proc. IAU Symposium No. 87 (this volume).
Pelling, M.: 1977, M.N.R.A.S. 178, 441.
Shapiro, M., and Kaplan, H.: 1979, J. Chem. Phys. 71, 2182.

DISCUSSION FOLLOWING OMONT

Field: The results of Shapiro and Kaplan on H + OH rotational
excitation were obtained from calculations using a potential surface
which did not include medium to long range anisotropy. They obtained
such low cross-sections because these anisotropies are all-important in
determining absolute rates of energy transfer. Furthermore in modelling
H + OH it is important to note the gas kinetic rate of the atom exchange
reaction H' + OH → OH' + OH.

Omont: Although Shapiro and Kaplan do not give the details of the
potential they use, I probably agree with your comment. It should be
noted that Dixon and Field do not take into account the long range part
of the potential (R^{-4} dipole-quadrupole, R^{-6} dispersion, etc. (see
Flower 1979)).

Field: With respect to high rates of transfer directly *across* the
Λ-doublet, there are good experimental and theoretical grounds for
believing that these should not be extremely large - indeed no larger
than rotational excitation rates.

Omont: I agree that the question of the relative magnitude of
Λ-doublet and rotational rates remains open, and should be checked
experimentally and theoretically.

Elitzur: A problem arises from the line-overlap model in trying
to explain all masers with the same underlying mechanism. The properties
of mainline masers in IR/OH stars and HII/OH regions are very different.
In one the 1667 line is stronger and in the other the 1665. Also, polarization is usually complete in HII/OH regions, but not in IR/OH stars.
I think we should be guided by the observations with regard to the pumps.

In W3(OH) the maser spots appear in two distinct clusters, almost
diametrically opposed. When we consider also the strong polarization,
which is an inherent feature, it seems evident that we should always try
to make the magnetic field a basic ingredient of the inversion mechanism

   Omont: It is obvious that IR/OH masers and HII/OH masers are very
different, and their pumping mechanisms are not necessarily the same.
The details of the operation of pumping by line overlap are quite
different in the two situations. It seems quite possible for this
pumping mechanism to reproduce in general terms the ratios of the two
main lines in the different cases. I do not see any realistic mechanism
where the magnetic field is a basic ingredient of the inversion mechanism
itself. There are other ways to account for the polarization (e.g. Cook
1977).

OBSERVATIONS OF MASERS IN REGIONS OF STAR FORMATION

D. Downes, Max-Planck-Institut für Radioastronomie, Bonn and
R. Genzel, Center for Astrophysics, Cambridge, Mass.

1. INTRODUCTION

This review covers recent progress in observations of masers in regions of star formation, especially in the years since the reviews by Burke (1975) and Lequeux (1977). Four regions of special interest are described in more detail: Orion, W51, W49 and W3.

1.1 Type of Source Observed. There are four types of strong masers observed to date in regions of star formation: OH, H_2O, SiO and methanol (CH_3OH). Methanol masers near 25 GHz have been seen only in Orion, and Orion is also the only example of a region of star formation containing an SiO maser source. OH masers are widespread throughout the Galaxy, and are associated on a scale of ~ 1 pc with compact radio continuum sources, IR objects and H_2O masers (see e.g. Evans et al. 1979). The main new OH surveys are by Turner (1979) and Caswell et al. (1979).

1.2 Discovery of New H_2O Sources. During the past three years, many new H_2O masers have been found in regions of star formation. The most extensive surveys are by Batchelor et al. (1979), Kaufmann et al. (1977) and Scalise and Braz (1979) for the southern hemisphere, and by Genzel and Downes (1977a, 1979) for the northern hemisphere. Additional H_2O masers have recently been found by Blitz and Lada (1979), Cesarsky et al. (1978), and by Rodriguez et al. (1978, 1979). These papers list nearly all of the H_2O masers known to date in regions of star formation. Figure 1 shows the distribution in galactic longitude of 168 H_2O masers listed in these papers. The histogram includes only H_2O sources which seem to lie in regions of star formation, and not the weaker class of H_2O masers coinciding with late-type stars. Although the diagram has not been corrected for observational selection, it shows major concentrations of H_2O masers at $\ell \sim 32^0$ and $\ell \sim 327^0$, that is, in directions tangent to spiral arms. There is no significant concentration of H_2O masers in the galactic center.

B. H. Andrew (ed.), Interstellar Molecules, 565–577.

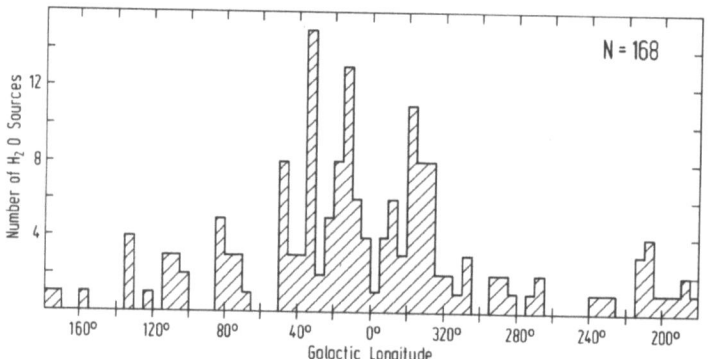

Figure 1:
Distribution of
168 H_2O masers
in galactic
longitude. H_2O
masers coinciding
with late-type
stars have not
been included.

RECENT STUDIES OF PARTICULAR REGIONS

2. ORION-KL

Figure 2 shows the current information on H_2O masers in the Orion-KL region. The small black dots indicate the positions derived from VLBI measurements (Genzel et al. 1978 and in prep.) of the strong, "low-velocity" emission between V_{LSR} -10 to +30 km s^{-1}. The crosses show the positions of weak, high-velocity H_2O lines between -102 and +80 km s^{-1} measured with the Effelsberg 100-m telescope (Genzel and Downes 1977b and further, unpublished measurements). The VLBI relative positions can be tied to absolute coordinates by means of the accurate H_2O position measured for the 10.8 km s^{-1} line with the Berkeley interferometer (Forster et al. 1978).

This map is an improvement over previously published versions in that VLBI positions have now been obtained for the high velocity features near 67 to 70 km s^{-1} and 46 km s^{-1} in an experiment involving telescopes at Simeis, Crimea; Onsala, Sweden; Effelsberg, Haystack and NRAO.

The low velocity H_2O lines are clustered in nine or more "centers of activity" of extent 1 to 2 10^{16} cm shown in Fig. 2. The newest maps indicate that at least three of the H_2O centers of activity coincide with infrared sources (Table 1). Up to now the situation has been confused by resolution effects in VLBI measurements.

2.1 The Infrared Source IRc4 (Rieke et al. 1973) has the richest H_2O spectrum in number of lines and in the strongest integrated emission. It dominates H_2O VLBI maps on baselines ≥600 km, such as Onsala-Bonn and Haystack-NRAO. We have designated it "Source A" in previous papers.

2.2 The Infrared Source IRS2 (Wynn-Williams and Becklin 1974) has not yet received much attention in the literature. The new H_2O VLBI maps show that it has low velocity emission near 6-7 km s^{-1} and high velocity emission at 68.6, and 70.5 km s^{-1}. It is also located near the peak of

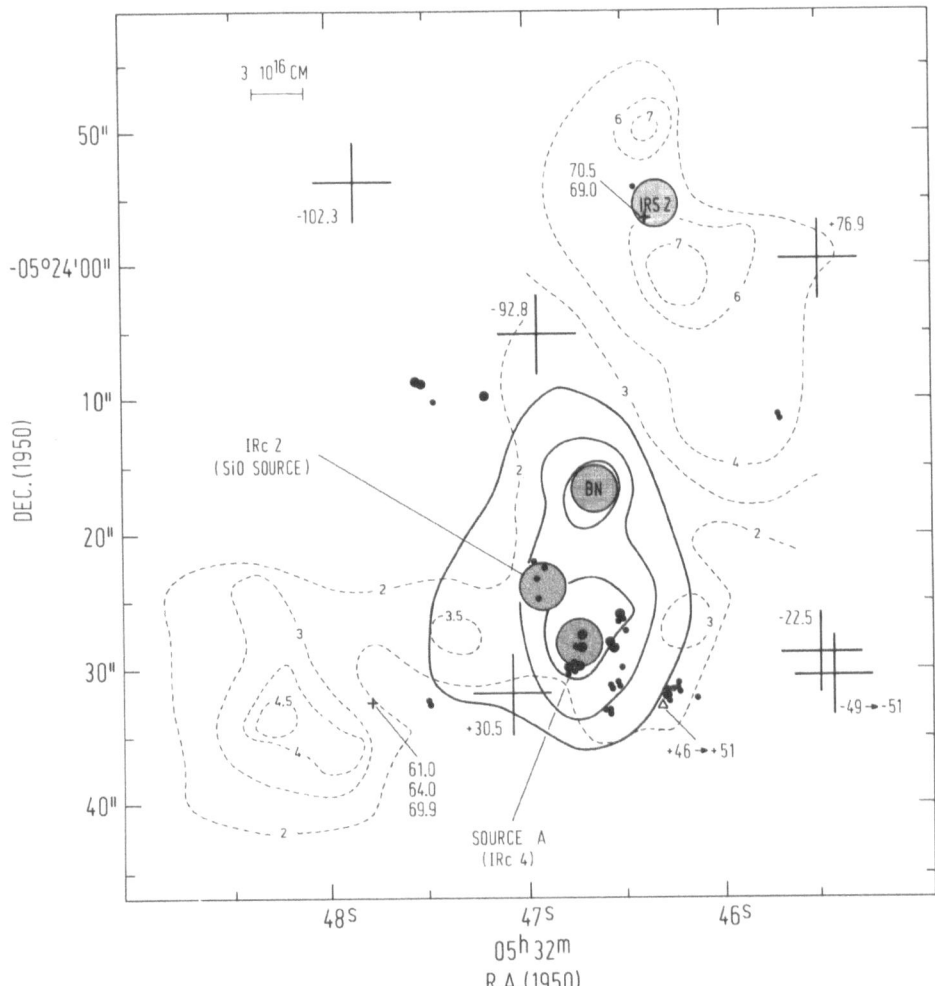

Figure 2: Positions of H_2O masers in the Kleinmann-Low nebula in Orion. The VLBI positions of the low-velocity H_2O features are shown as black dots (Genzel et al. 1978), superimposed on the contours of 20 μm infrared radiation (Wynn-Williams and Becklin 1974; Rieke et al. 1973). Shaded circles represent compact infrared objects. Crosses show the positions of weak, high-velocity H_2O masers measured with the 100-m telescope (Genzel and Downes 1977b and unpublished). High velocity H_2O features over 46 to 70.5 km s⁻¹ have also been measured by VLBI. The dashed contours are those of the v = 1→0 S(1) transition of molecular hydrogen (Beckwith et al. 1978).

Table 1. Apparent Coincidences of Infrared and Maser Sources in Orion

Infrared Source	IR Ref.	Position (1950)	Maser Source	Maser Ref.
IRc2 =IRS3	RLK73 WWB74	$05^h32^m47.0^s$ $-05^o24'24''$	H_2O "shell source" (75 km VLBI) SiO maser (v=0,1,2) possibly thermal SiO as well. OH maser at 17-23 km s^{-1}	H79 G79a NBM79
IRc4	RLK73	$05^h32^m46.8^s$ $-05^o24'29''$	H_2O "Source A" ($\sim10^3$ km VLBI) OH maser 2-11 km s^{-1}	G78 NBM79
IRS2	WWB74	$05^h32^m46.3^s$ $-05^o23'55''$	H_2O low velocity (6-7 km s^{-1}) H_2O high velocity (68-70 km s^{-1})	

Refs.: RLK = Rieke et al.; WWB = Wynn-Williams and Becklin;
       NBM = Norris et al.; G = Genzel et al.; H = Hansen

the molecular hydrogen emission at 2 μm, the contours of which are shown in Fig. 2, after Beckwith et al. 1978. The H_2 emission has the broadest lines near IRS2, IRc2 and IRc4 (Nadeau and Geballe 1979), suggesting that the high velocity maser emission, the broad plateau emission seen in molecular lines, and the H_2 emission at 2 μm are all related to mass-loss from the stars in the KL cluster.

2.3 The Infrared Source IRc2 (Rieke et al. 1973 = IRS3, Wynn-Williams and Becklin 1974) appears to be a different sort of object. In inter-ferometer observations, the H_2O emission near -6 and +19 km s^{-1} is strong on the Haystack-U. Mass. 75 km baseline (Hansen 1979), is weak on the Haystack-NRL and Onsala-Bonn 600 km baselines (Genzel et al. 1978), and is completely absent on the Crimea-Bonn baseline. This in-dicates that the H_2O source is larger than most of the other H_2O masers in Orion-KL. The apparent angular diameter of the H_2O features is 0.03" to 0.13", about ten times larger than all the other H_2O maser features in Orion.

2.4 The SiO Maser Source in Orion = IRc2

The velocities of the H_2O features from IRc2 are the same as those of the SiO maser emission and on the basis of a simultaneous VLBI ex-periment on the 22 GHz H_2O and 43 GHz SiO lines Genzel et al. (1979a) have identified the SiO maser with IRc2. This result has now been con-firmed by the Berkeley interferometer, operating on the 86 GHz SiO lines (Bieging et al., this Symposium). The VLBI observations of Genzel et al. show that the v=1,J=1-0 SiO lines at 43 GHz are partially resolved, have apparent angular diameters of 0.02", (1.5 10^{14}cm), and that the separate

velocity components are coincident to within 0.05". Additional SiO
observations with the Effelsberg 100-m telescope at 43 GHz have been
made by Genzel (Center for Astrophysics), Schwartz and Spencer (NRL)
and Pankonin, Baars and Downes (MPI). These observations of the J=1-0
rotational lines in the v=0,1 and 2 vibrational levels, show that the
v=0 emission is unresolved (θ<10") and comes from the same position as
the v=1 and v=2 maser lines. The v=0 emission has a sharp spike at
-6 km s^{-1}, indicating that there is also maser emission in the ground
vibrational state. In addition, there is weak (T_a* $\sim$3 K) emission over
a range of $\sim$37 km s^{-1} in the v=0,1 and 2 levels, which may be either
thermal emission or weak maser emission (Genzel et al. 1980; see also
Zuckerman 1979).

SiO emission was also searched for with the 100-m telescope in 27
other regions of star formation, all with negative results to $\sim$5 Jy.
This is an order of magnitude better than previous limits, and an SiO
maser with the same luminosity as the Orion source would have been detected in the sources searched, up to a distance of $\sim$8 kpc.

Since SiO emission is otherwise seen only from Mira variables or
M supergiants, the presence of the SiO maser source in Orion has been
puzzling for some time now. The velocity spread of the Orion SiO maser
lies midway between that of the M giants and the M supergiants (see
discussions by Snyder and Buhl 1974, Snyder et al. 1975, 1978), but the
derived mass loss rate appears to be about two orders of magnitude higher.

The source IRc2 thus appears to be either an evolved giant or supergiant right at the core of the Orion molecular cloud or else is a unique
object in the Galaxy.

2.5 OH and Methanol Masers in Orion

The main new OH maps are by Hansen et al. (1977) and by Norris et al.
(1979). The latter map shows two concentrations of OH masers, which
coincide with IRc2 and IRc4.

The positions of the 1.2 cm methanol masers in Orion have been
measured by Matsakis (1979) with maser radiometers on the Berkeley
interferometer. The CH$_3$OH masers seem to be much weaker than and not
coincident with the H$_2$O, OH or SiO masers in Orion. Matsakis finds 10
concentrations, 6 of which were unresolved (in R.A.). The others had
sizes of 6" - 8". The line brightness temperatures or lower limits were
1000 to 4000 K. The velocities are in the range 7 - 10 km s^{-1}, that is,
in the normal range for the Orion molecular cloud. They may therefore
be denser clumps in the general cloud material, excited by the infrared
emission from objects in the vicinity. They do not appear to be in the
circumstellar envelopes themselves, as do the H$_2$O, OH and SiO masers.

3. THE W51 REGION

The H₂O masers in W51 have been the object of several recent VLBI
investigations (Walker et al. 1978, Mader et al. 1978, Genzel et al.
1978, 1979b, Downes et al. 1979).

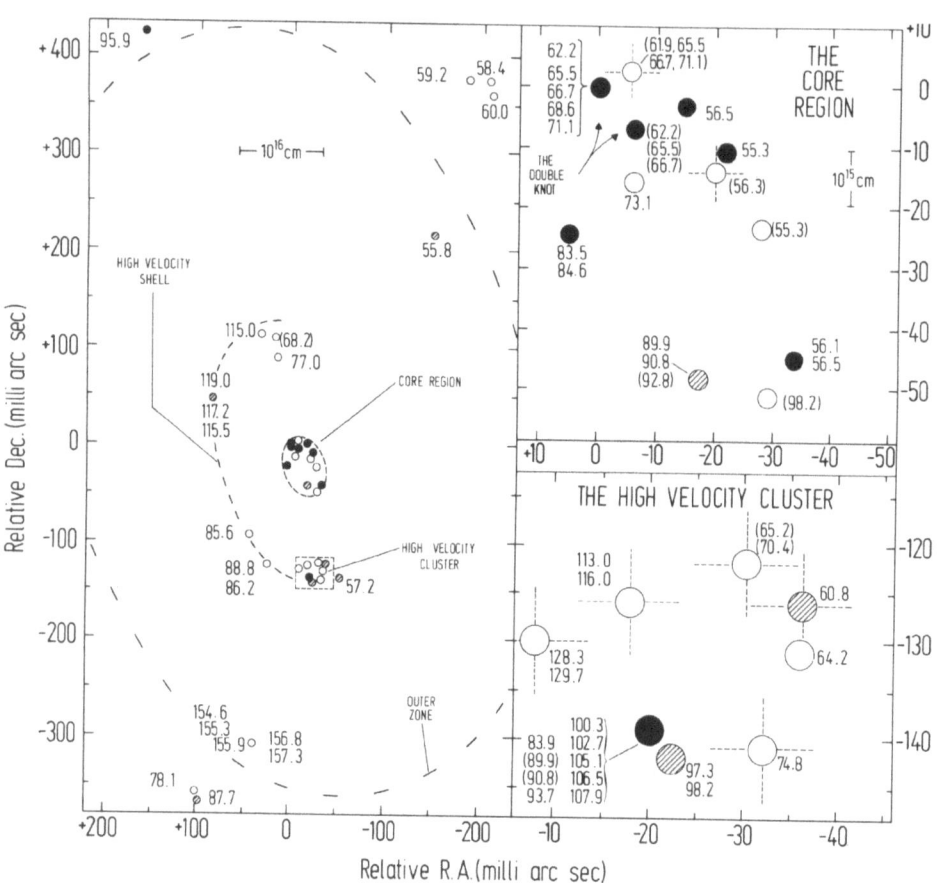

Figure 3. (from Genzel et al. 1979b). Positions of H₂O masers in W51
Main in December 1977, relative to the feature at 62.2 km s⁻¹. Labels
are LSR velocities in km s⁻¹. The coordinate zero is 19ʰ21ᵐ26.20ˢ±0.02ˢ.
14°24'43.6"±0.4" (1950)(Forster et al. 1978). Relative positional uncertainties (2σ) are indicated by black circles for strong (>100 Jy) lines,
shaded circles for intermediate lines (>50 Jy) and open circles for weak
lines. The diagrams at right show enlargements of the "Core Region" and
the "High Velocity Cluster". Most of the low-velocity emission comes from
the maser spots marked as the "Double Knot",which may be the location of
the exciting star.

3.1 W51 Main, the strongest source, is near (2.2"±0.8") but not coincident with a super-compact HII region (Scott 1978, Forster et al. 1978).
The main new results are (i) a proper H_2O aperture synthesis of the
intense, low velocity "double knot" at the core of the source (Walker
et al. 1978),which results provide spectra across the source at intervals of 1 milli arc sec, and (ii) a series of VLBI maps of the structure
of the entire source made by Genzel et al. (1979b). A map from the latter
study is shown in Fig. 3. The strong low-velocity emission in the range
50 to 76 km s^{-1} appears to come from a core region of size $\sim$5 10^{15} cm,
while redshifted high velocity features are in a "shell" or "arc" with a
diameter of 3 10^{16} cm surrounding the core. A few of the high velocity
features (e.g. near $V_{LSR}\sim$155 km s^{-1}) come from an "outer zone" of diameter 9 10^{15} cm. Between November 1977 and March 1978 the source was
observed by VLBI at monthly intervals, and was observed to undergo a
nearly simultaneous flare in numerous high velocity features at different
points on the map. These circumstances suggests an impulse of pumping
energy from the star at the center of the low velocity emission. Independent of the pumping mechanism, the rise time of $\sim$5 10^6 sec of the
simultaneous flare and the diameter of typical maser cloudlets of 10^{14}cm
imply a lower limit of 200 km s^{-1} for the propagation velocity of the
pumping agent through the maser. As with the variations in W3(OH)
(Haschick et al. 1977) this rules out the excitation of H_2O masers by
ionization or shock fronts with velocities of 10 km s^{-1}. Genzel et al.
(1979b) suggest that W51 Main is a massive, young O star surrounded by
a disk of 5 10^{15}to 3 10^{16} cm which emits the low velocity H_2O emission.
A stellar wind escapes from the disk (possibly in the manner described
by Elmegreen and Morris 1979) and plows into a dense molecular cloud
behind W51 Main. The redshifted high-velocity H_2O masers with velocities
of 100 km s^{-1} appear in dense clumps at the wind/cloud interface.

3.2 W51 North is located 1.5' northwest of W51 Main and is a good example
of the clustering of H_2O sources into "centers of activity". It has six
such "centers" spread out in a 2" x 6" region (2 x 6 10^{17} cm), and a total
velocity spread in the H_2O spectrum of 170 km s^{-1}. The dominant center of
activity seems to have a "core"/"shell" structure very similar to that
of W51 Main, although in contrast to that source the high velocity
emission is mainly blueshifted (diameters in W51N are: core = 1 10^{15} cm,
shell = 7 10^{15} cm, "outer zone" = 2 10^{16} cm; Downes et al. 1979).

4. THE W49 REGION

The main new results have been (i) an accurate position for the H_2O
source (Dieter et al. 1979), (ii) studies of the H_2O variability (Little
et al. 1977 and White 1979), and (iii) VLBI maps (Walker et al. 1977,
Genzel et al. 1978, Mader et al. 1978). Figure 4 shows the spectrum of
W49 N and the distribution of H_2O maser sources in a region $\sim$3 10^{17} cm
across. The clustering of H_2O sources in W49 N seems to be typical for
H_2O maser regions. Presumably the clusters represent Trapezium-type groups
of newly-formed O stars. In W49 N the H_2O cluster is well separated from
the nearest compact HII regions and the main OH source. Figure 4 (lower)

shows schematically the low velocity "centers of activity as large circle (strongest ones shaded). Several of the "centers" are surrounded by high velocity features (small dots; Walker et al. 1977) which presumably indicate that these stars are losing mass in the same manner as the stars in the K-L nebula, although with much greater luminosity and at about three times greater velocities. A new, extensive study of the high velocity H_2O in W49 N is in progress by Walker et al.

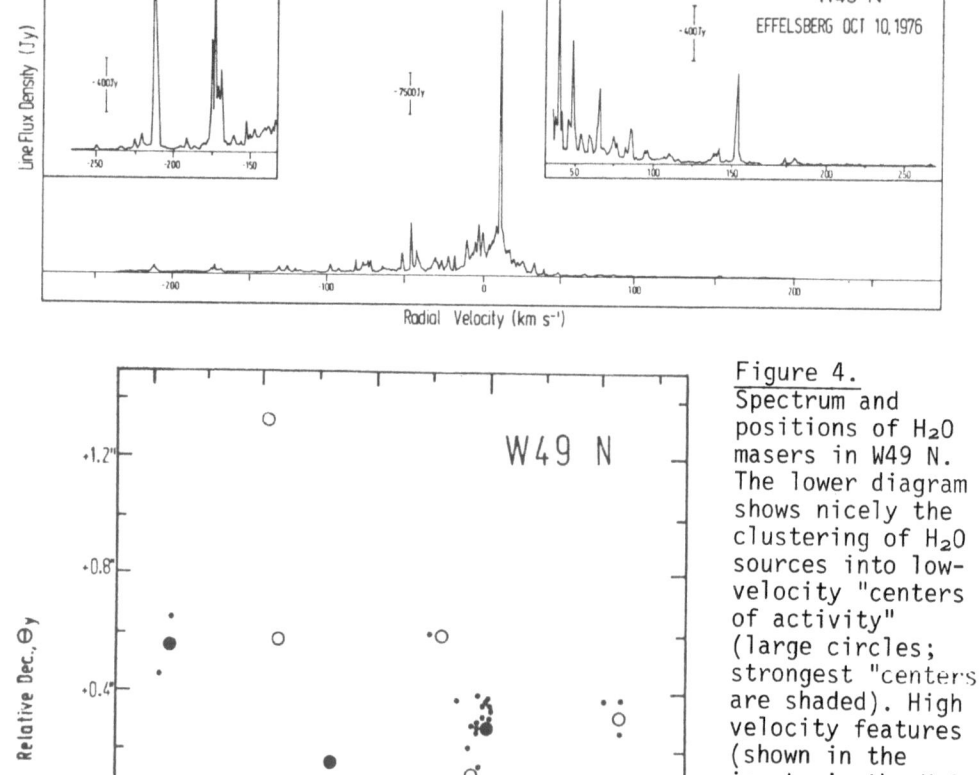

Figure 4.
Spectrum and positions of H_2O masers in W49 N. The lower diagram shows nicely the clustering of H_2O sources into low-velocity "centers of activity" (large circles; strongest "centers" are shaded). High velocity features (shown in the insets in the H_2O spectrum at top) are found in swarms (small black dots) around some of the low velocity "centers" For H_2O velocities represented by the individual dots, see Walker et al. 1977, Genzel et al. 1978. The coordinate zero is $19^h07^m49.77^s\pm0.05^s$, $+09°01'17.1"\pm0.5"$ (Dieter et al. 1979).

5. THE W3 REGION

The region around the W3 radio continuum sources contains at least 5 H_2O maser sources; an overview of the maser positions is given in Downes and Genzel 1979.

5.1 W3(OH). The main new result has been the accurate position of the H_2O (Forster et al. 1977) which is about 7" east of the compact HII region. New VLBI maps of the H_2O source (Walker et al. 1978, Genzel et al. 1978) show that 90% of the H_2O intensity comes from a "center of activity" to the east of a "line" of weak H_2O spots, each with a single velocity. An extensive series of VLBI maps of this line of sources has been obtained by Giuffrida (1977) and collaborators.

The -50.3 km s^{-1} feature in the H_2O "center of activity" flared up in May-June 1977. The 4-day risetime was followed by a 25-day decay (Haschick et al. 1977). The pattern has been analyzed by Burke et al. (1980) whose calculations favor a collisionally pumped, cool (100-200 K) saturated maser with a large injection of energy (10^{39}-10^{41}erg) into a cloud of dimension 10^{14}-10^{15} cm in one or two days time at most.

The main new result in OH has been an accurate position obtained from the excited-state OH lines at 6035 MHz, (Moran et al. 1978): it shows that the OH masers are exactly coincident with the compact HII region. Since the W3(OH) HII region is optically thick at 6 GHz, the masers must be on the front side of the source. Moran et al. find that the maser spots occur in spatial pairs of right and left-circularly polarized features, and thus provide probably the best evidence to date for Zeeman splitting in OH. Moran et al. derive magnetic field strengths of 2 to 9 milligauss.

A remarkable aperture synthesis has been done with VLBI data in the OH ground state lines by Reid et al. (in preparation) who find numerous Zeeman OH pairs scattered over the face of the source. When the apparent velocities of the Zeeman pairs are averaged together, one obtains a velocity close to -45 km s^{-1} with a slow variation of $\sim\pm3$ km s^{-1} across the face of the HII region. Since this is also the velocity of NH_3 absorption and 23-GHz OH absorption toward the continuum source, the OH masers may occur in a relatively quiescent, dense cloud somewhat external to the HII region, and not in a turbulent shock close to the ionization front.

5.2 W3-IRS5 appears from H_2O VLBI maps to consist of two or three sources and may be a stellar group like the cluster in Orion-KL. Cal Tech infrared observations (see Wynn-Williams 1976) had indicated that the source was at least double, separated by 1", but recent speckle inter-ferometry at 4.8 μm by Chelli et al. (1979) has resolved only a single source, roughly in the north-south direction, with a diameter of 0.28"±0.08". It would be of interest to pursue the investigation of source structure in the infrared to determine its relation to the H_2O masers.

6. OBSERVATIONAL PROSPECTS FOR THE FUTURE

A common trend of the maser, infrared and radio continuum observations has been to show the occurence of these sources in clusters
somewhat similar to the Trapezium in Orion (cf Beichman et al. 1979,
Evans et al. 1979, Habing and Israel 1979). For most of the sources
the positional accuracy must still be improved to see whether the
sources in various wavelength regions actually coincide on scales <0.1 pc.

Relative positions accurate to 0.5 milli arc sec can now be achieved
for strong H_2O sources as a matter of course by the fringe rate method
described by Moran (1976). We have attempted improvements by phase
mapping and have obtained repeatable measurements to 0.1 milli arc sec
for a very few high velocity features in W51 Main. The results are in
general disappointing, however, and not in proportion to the effort
required. Nevertheless, with the standard fringe-rate technique, it
should be possible within the next year to obtain measurements of the
relative proper motion of H_2O masers.

ACKNOWLEDGEMENTS

The VLBI and SiO work reported here has mainly resulted from
collaboration with the following colleagues: J. M. Moran, A. Haschick
(Center for Astrophysics); K. J. Johnston, J. H. Spencer, P. Schwartz
(NRL); L. I. Matveyenko, L. R. Kogan, V. I. Kostenko (Space Research
Institute, Moscow); B. Rönnang, O. E. H. Rydbeck (Onsala Space
Observatory); I.G. Moiseev (Crimean Astrophysical Observatory);
B. F. Burke, T. Giuffrida (M.I.T.); R. C. Walker (Cal Tech); M. Reid
(NRAO) and V. Pankonin, J. Baars (MPIfR).

REFERENCES

Batchelor, R.A., Caswell, J.L., Goss, W.M., Haynes, R.F., Knowles, S.H.,
    and Wellington, K.J.: 1979, Austral. J. Phys., in press.
Beckwith, S., Persson, S.E., Neugebauer, G., and Becklin, E.E.: 1978,
    Astrophys. J. 223, 464.
Beichman, C.A., Becklin, E.E., and Wynn-Williams, C.G.: 1979, Astrophys.
    J. (Letters) 232, L47.
Blitz, L., and Lada, C.J.: 1979, Astrophys. J. 227, 152.
Burke, B.F.: 1975, in HII Regions and Related Topics, ed. T.L. Wilson
    and D. Downes, Springer, Berlin, p. 188.
Burke, B.F., Giuffrida, T.S., and Haschick, A.D.: 1980, Astrophys. J.,
    in press.
Caswell, J.L., Haynes, R.F., and Goss, W.M.: 1979, Austral. J. Phys.,
    in press.
Cesarsky, C.J., Cesarsky, D.A., Churchwell, E., Lequeux, J.: 1978,
    Astron. Astrophys., in press.
Chelli, A., Lena, P., and Sibille, F.: 1979, Nature 278, 143.
Dieter, N.H., Welch, W.J., and Wright, M.C.H.: 1979, Astrophys. J.
    230, 768.

Downes, D., and Genzel, R.: 1979, Proceedings of the Gregynon Symposium on Giant Molecular Clouds ed. P. Solomon and M. Edmunds, Pergamon Press, in press.

Downes, D., Genzel, R., Moran, J.M., Johnston, K.J., Matveyenko, L.I., Kogan, L.R., Kostenko, V.I., and Rönnäng, B.: 1979, Astron. Astrophys. 79, 233.

Elmegreen, B., and Morris, M.: 1979, Astrophys. J. 229, 593.

Evans, N.J., Beckwith, S., Brown, R.L., and Gilmore, W.: 1978, Astrophys. J. 227, 450.

Forster, J.R., Welch, W.J., and Wright, M.C.H.: 1977, Astrophys. J. (Letters) 215, L121.

Forster, J.R., Welch, W.J., Wright, M.C.H., and Baudry, A.: 1978, Astrophys. J. 221, 137.

Genzel, R., and Downes, D.: 1977a, Astron. Astrophys. Suppl. 30, 145.

Genzel, R., and Downes, D.: 1977b, Astron. Astrophys. 61, 117.

Genzel, R., and Downes, D.: 1979, Astron. Astrophys. 72, 234.

Genzel, R., Downes, D., Moran, J.M., Johnston, K.J., Spencer, J.H., Walker, R.C., Haschick, A., Matveyenko, L.I., Kogan, L.R., Kostenko, V.I., Rönnäng, B., Rydbeck, O.E.H., and Moiseev, I.G.: 1978, Astron. Astrophys. 66, 13.

Genzel, R., Moran, J.M., Lane, A.P., Predmore, C.R., Ho, P.T.P., Hansen, S.S., and Reid, M.J.: 1979a, Astrophys. J. (Letters) 231, L73.

Genzel, R., Downes, D., Moran, J.M., Johnston, K.J., Spencer, J.H., Matveyenko, L.I., Kogan, L.R., Kostenko, V.I., Rönnäng, B., Haschick, A.D., Reid, M.J., Walker, R.C., Giuffrida, T.S., Burke, B.F., and Moiseev, I.G.: 1979b, Astron. Astrophys. 78, 239.

Giuffrida, T.S.: 1977, Ph.D. Thesis, Mass. Inst. of Tech.

Habing, H.J., and Israel, F.P.: 1979, Ann. Rev. Astron. Astrophys., in press.

Hansen, S.S.: 1979, Ph.D. Thesis, Univ. of Mass.

Hansen, S.S., Moran, J.M., Reid, M.J., Johnston, K.J., Spencer, J.H., and Walker, R.C.: 1977, Astrophys. J. (Letters) 218, L65.

Haschick, A.D., Burke, B.F., and Spencer, J.H.: 1977, Science 198, 1153.

Kaufmann, P., Zisk, S., Scalise, E., Schaal, R.E., and Gammon, R.H.: 1977, Astron. J. 82, 577.

Lequeux, J.: 1977, in Star Formation, IAU Symposium 75, ed T. de Jong and A. Maeder, Reidel, Dordrecht, p. 69.

Little, L.T., White, G.J., and Riley, P.W.: 1977, Mon. Not. Roy. Astron. Soc. 180, 639.

Mader, G.L., Johnston, K.J., and Moran, J.M.: 1978, Astrophys. J. 224, 115.

Matsakis, D.N.: 1979, Ph.D. Thesis, University of California, Berkeley.

Moran, J.M.: 1976, in Methods of Experimental Physics, Vol. 12c, Ed. M.L. Meeks, New York, Academic Press, p. 228.

Moran, J.M., Reid, M.J., Lada, C.J., Yen, J.L., Johnston, K.J., and Spencer, J.H.: 1978, Astrophys. J. (Letters) 224, L67.

Nadeau, D., and Geballe, T.R.: 1979, Astrophys. J. (Letters) 230, L169.

Norris, R.P., Booth, R.S., and McLaughlin, W.: 1979, in prep.

Rieke, G.H., Low, F.J., and Kleinmann, D.E.: 1973, Astrophys. J. (Letters) 186, L7.

Rodriguez, L.F., Moran, J.M., Dickinson, D.F., and Gyulbudaghian, A.L.: 1978, Astrophys. J. 226, 115.

Rodriguez, L.F., Moran, J.M., Ho, P.T.P., and Gottlieb, E.W.: 1979,
    Astrophys. J., in press.
Scalise, E., and Braz, M.A.: 1979, Astron. Astrophys., in press.
Scott, P.F.: 1978, Mon. Not. Roy. Astron. Soc., 183, 435.
Snyder, L.E., and Buhl, D.: 1974, Astrophys. J. (Letters) 189, L31.
Snyder, L.E., Hollis, J.M., Ulich, B.L., Lovas, F.J., Johnson, D.R.,
    and Buhl, D.: 1975, Astrophys. J. (Letters) 198, L81.
Snyder, L.E., Dickinson, D.F., Brown, L.W., and Buhl, D.: 1978, Astrophys
    J. 224, 512.
Turner, B.E.: 1979, Astron. Astrophys. Suppl. 37, 1.
Walker, R.C., Johnston, K.J., Burke, B.F., and Spencer, J.H.: 1977,
    Astrophys. J. (Letters) 211, L135.
Walker, R.C., Burke, B.F., Haschick, A.D., Crane, P.C., Moran, J.M.,
    Johnston, K.J., Lo, K.Y., Yen, J.L., Broten, N.W., Legg, T.H.,
    Greisen, E.W., and Hansen, S.S.: 1978, Astrophys. J. 226, 95.
White, G.J.: 1979, Mon. Not. Roy. Astron. Soc. 186, 377.
Wynn-Williams, C.G.: 1976, Observatory 96, 6.
Wynn-Williams, C.G., and Becklin, E.E.: 1974, Pub. Astron. Soc. Pacific
    86, 5.
Zuckerman, B.: 1979, Astrophys. J. 230, 442.

DISCUSSION FOLLOWING DOWNES

Elitzur: It was suggested once by the group at Leiden that the
SiO maser in Orion could come from infall of material. It seems to me
that the observations you presented support this idea. The standing
shock resulting from the infall could be the location of all the maser
features of the different molecules which coincide in position and
velocity.

Downes: The possibility certainly deserves further thought.
However the VLBI distribution of H_2O maser "spots" from this source
indicates radii of 10^{15}-10^{16} cm from the star. These radii, together
with the velocity half-range of 11 to 17 km s^{-1}, imply a very large
central mass if the velocities come from infall. Remember also that
the upper limit of the infrared luminosity of the star is $\sim$2 x $10^4 L_\odot$,
which suggests a limit to the mass of the star of 10-20 $M_\odot$.

Bieging: What lower limit can you place on the brightness temperature of the SiO V=0, J=1-0 line in Orion, which you suggest may be a
maser?

Downes: For the spike at -5 km s^{-1} in the V=0 source, the lower
limit to the brightness temperature is 350±50K. For the broad feature,
it is 150±30K.

Elmegreen: Various models of maser geometries, such as shells or
protoplanetary-type disks, cannot be distinguished on the basis of
velocity-position maps alone. Proper motions and/or accelerations are
needed as well. What are the prospects for eventually detecting
accelerations of individual components?

Downes: We expect to be able to establish proper motions in about
a year. Accelerations are more difficult to observe because an apparent
change in velocity may simply be due to a change in the relative inten-

sities of two nearby, blended lines. It might be possible to attack this problem by searching for velocity changes among the features which show proper motions.

Booth: Comparison of the spectra of W3(OH) at 1665 MHz taken over the past ten years shows that a feature, now at -43.9 km s^{-1}, has moved in velocity by 0.2 km s^{-1}. It has decreased steadily in intensity during the same period, and its position (relative to the feature at -45.1 km s^{-1}) appears to have changed by 0.05 arc seconds. This may be an example of acceleration of material in the source, although we can interpret it as a beaming effect.

ABSOLUTE POSITIONS OF OH MASERS ASSOCIATED WITH HII REGIONS

R. S. Booth and R. P. Norris
University of Manchester,
Nuffield Radio Astronomy Laboratories, Jodrell Bank,
Macclesfield, Cheshire, SK11 9DL

We have measured the absolute positions of five OH maser sources associated with HII regions, using the Jodrell Bank phase compensated radio linked interferometer (Norris, Booth & Davis, 1980); these are reproduced in Table 1. In addition, we have mapped the relative positions of the features in the OH spectra at 1665 MHz for all of the sources and at 1667 MHz for 3 of them (W3OH, NGC 7538 and W49). We are now able to compare the distribution of the OH masers in a given source with its radio continuum map and to investigate the proximity of masers in the 2 main line transitions and the H_2O masers.

The highest accuracy was achieved for the source W3OH shown in Figure 1. On the 15 GHz continuum map of Harris & Scott (1976) we have plotted the positions of the 1665 and 1667 MHz masers as well as the 6035 MHz features published by Moran et al. (1978). Several points emerge:

1. In all cases the OH masers are offset from the main continuum centre although they are usually located within the continuum envelope; W3OH (fig.1) has several features outside this envelope. Several authors, e.g. Elitzur & de Jong (1978), have suggested that the OH masers lie in a thin shell of dense dust and gas lying between shock and ionization fronts. The present observations are not entirely consistent with this idea and suggest that inhomogeneities due to magnetic

TABLE 1   OH SOURCE POSITIONS (1665 MHz)

OH Source (Velocity)	R.A. (1950)	Dec (1950)
W3OH (-45.1)	$02^h 23^m 16^s.317 + .011$	$61°38'57".78 + .13$
NGC 7538 (-59.9)	23 11 36.60 $\mp$ .02	61 11 49.67 $\mp$ .15
Orion A (7.1)	05 32 46.78 $\mp$ .02	-5 24 25.9 $\mp$ 1.7
ON2 (2.2)	20 19 51.93 $\mp$ .02	37 16 59.9 $\mp$ .3
W49(1) (20.9)	19 07 49.60 $\mp$ .03	09 01 16.1 $\mp$ 3.0
W49(2) (16.0)	19 07 58.04 $\mp$ .02	09 01 05.4 $\mp$ 2.9

B. H. Andrew (ed.), Interstellar Molecules, 579–580.

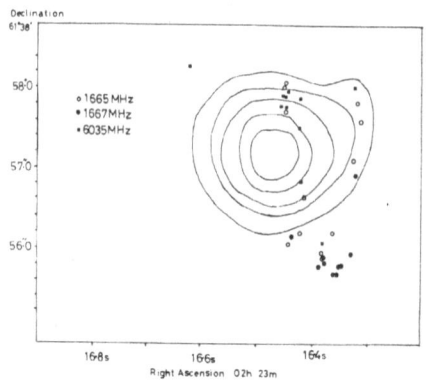

Fig. 1   The OH maser features plotted on the 15 GHz continuum Cambridge
map of W3OH.   The 6035 components are from Moran et al. (1978).   All
main line components are plotted relative to the 1665 MHz feature at
-45.1 km s^{-1}.

fields and turbulence are important.

2.   The 1665 and 1667 MHz masers are generally separate but show a small
amount of overlap.   Where this overlap occurs there is no obvious
general relationship between component velocities.   The same can be
said of the relationship with the 6035 MHz masers in W3OH (Fig.1).

3.   In general the OH and H$_2$O masers are displaced by a small but significant amount.   However, in Orion there is strong evidence for a close
association in position and velocity between a clump of OH features and
an H$_2$O centre of activity (Norris, 1978).

4.   The most likely pump mechanism for the OH masers is that due to
collisions (e.g. Elitzur & de Jong 1978).   Radiative mechanisms must be
ruled out since the UV and IR photon rates at the distance of the masers
are too low by at least an order of magnitude.   Finally, chemical
schemes are unlikely because of the general non-coincidence of OH and
H$_2$O.

REFERENCES

Elitzur, M. and de Jong, T.: 1978, Astron.Astrophys. 67, pp323-332.
Harris, S. and Scott, P.F.: 1976, Monthly Notices Roy.Astron.Soc.
    175, pp371-379.
Moran, J.M., Reid, M.J., Ladar, C.J., Yen, J.L., Johnston, K.L. and
    Spencer, J.H.: 1978, Astrophys.J. 224, ppL64-L71.
Norris, R.P.: 1978, Ph.D. Thesis, University of Manchester.
Norris, R.P., Booth, R.S. and Davis, R.J.: 1980, Monthly Notices
    Roy.Astron.Soc. 190, pp. 163-167.

THE PUMPING OF INTERSTELLAR OH MAIN LINE MASERS: AN EFFICIENT MECHANISM

Robert Lucas
Groupe de Radioastronomie, Ecole Normale Supérieure,and
Département de Radioastronomie, Observatoire de Meudon

Hyperfine components of the far-infrared lines of OH are separated by velocity intervals ranging from about .6 to 8.0 km s^{-1} (fig. 1). As early as 1969 Litvak had suggested that overlapping of these components due to either thermal or systematic motions could be the source of OH main line inversion. We investigate here the effects of overlapping due to systematic motions.

Our study is restricted to the rather simplified scheme of a quiescent maser cloud pumped by a hot neighbouring cloud moving with a velocity v (-10 to +10 km s^{-1}) with respect to the maser cloud. Any line overlap due to thermal or microturbulent random motions inside the maser cloud is neglected. The spectrum emitted by the pumping cloud is calculated assuming LTE (temperature T_p), Doppler thermal broadening, and a column density of molecules N_p. The 28 lowest levels of OH are included in the computations. Dust emission and collisions with charged and neutral particles are treated in the same way as by Guibert et al. (1978). The radiative transfer in the maser cloud is treated in the following way: for non-maser transitions we use the standard escape probability method (Goldreich and Kwan 1974); for the maser transitions we use a generalization of the spherical maser solution of Goldreich and Keeley (1972) to compute the maser brightness temperature as a function of the unsaturated maser parameters (gain and source function).

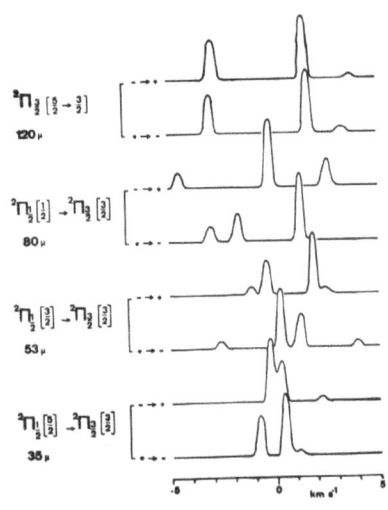

Figure 1: the hyperfine structure of the far-infrared transitions linking the OH ground state to low-lying rotational states

Results are shown (fig. 2) for a pumping cloud of $N_p=3\times10^{16}$ cm^{-2} molecules

581

B. H. Andrew (ed.), Interstellar Molecules, 581–582.

at temperature T_p=300 K filling 10% of the sky as seen from the maser cloud. The column density of the maser cloud is 10^{15} cm^{-2} its hydrogen number density 10^6 cm^{-3}, and its temperature 100 K. The mechanism is very efficient since unsaturated inversion ratios (percentages of lower masing state molecules pumped to the upper masing state in the absence of saturation) are as high as 60% for the main lines and 90% for the satellite lines. Inversion is mainly obtained from excitation of the $2\,\Pi\,1/2(J=3/2)$ rotational state, as is expected at the temperature of the pumping cloud (300 K).

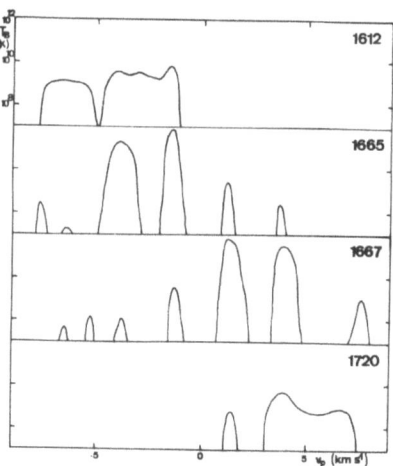

The velocities needed to obtain an efficient inversion are not too large (-8 to +8 km s^{-1}). Velocities of this order of magnitude are certainly present since the velocity intervals between the maser features are generally a few km s^{-1}. Expansion movements would enhance 1665 MHz emission with respect to 1667 MHz emission, as is observed.

Figure 2: predicted brightness temperature of a spherical maser, as a function of pumping velocity v_p

A high column density (3 10^{16}cm^{-2}) and temperature (300 K) are needed for the pumping cloud. Winnberg et al. (1978) found absorption in W3(OH) in the $2\,\Pi\,3/2(J=9/2)$ lines of OH with a rather large optical depth (.06) and inferred a column density of 10^{16} to 10^{17} cm^{-2}, assuming a rotational temperature of 100 K. Thus evidence may be presented of an excited medium of high OH abundance.

We think that our model may lead to a relatively simple account of OH maser phenomena close to HII regions. A similar model is proposed to explain the main line emission of OH/IR stars (Nguyen-Q.-Rieu 1979).

REFERENCES

Goldreich, P., Keeley, D.: 1972, Astrophys. J. 174, p. 517.
Goldreich, P., Kwan, J.D.: 1974, Astrophys. J. 189, p. 441.
Guibert, J., Elitzur, M., Nguyen-Q.-Rieu: 1978, Astron. Astrophys. 66, p. 395.
Litvak, M.M.: 1969, Astrophys. J. 156, p. 471.
Nguyen-Q.-Rieu: 1979, paper presented to this symposium.
Winnberg, A., Walmsley, C.M., Churchwell, E.: 1978, Astron. Astrophys. 66, p. 431.

Λ-DOUBLET POPULATION INVERSION IN COLLISIONS OF OH, OD, CH, CD AND NH$^+$

R.N. Dixon and D. Field
School of Chemistry, University of Bristol, Bristol BS8 1TS U.K.

The results of a new approach to the problem of the collisional step in the pumping cycle for OH and CH masers are reported. Rotationally in-elastic collisions of OH and CH with both open and closed shell collision partners are considered using an expression derived from scattering theory. It is shown how Λ-doublet population inversion may arise in OH and CH. H-atoms and H$_2$ may show opposite behaviour with CH, one partner cooling where collisions with the other lead to inversion. Implications for maser action are discussed and reveal excellent qualitative agreement with observation. Λ-doublet population inversion in OD, CD and NH$^+$ is also considered.

INTRODUCTION

This contribution is concerned with the collisional step in the mechanism by which population inversion may be induced within the Λ-doublets of OH and CH. Previous treatments have related collisional behaviour to the electronic structure of the isolated OH and CH radicals[1,2]. Different cross-sections for rotational excitation have been assigned to pairs of Λ-doublets on the grounds that the spatial orientation of the unpaired electron differs within the doublets, leading to different interaction potentials and thus excitation probabilities.[3] Whilst this model embodies an important element of the physics of the interaction, it contains only a partial description of the inelastic collisions of interest. We adopt the well-established scattering theoretic approach to rotationally inelastic collisions, and the present work is thus fundamentally different from that of previous authors. A further new element that we wish to introduce is that exchange forces between two open shell collision partners impart a strong anisotropy to the inter-action potential. Calculated potentials[3] show this clearly. In conse-quence collisions with H atoms have larger cross-sections than those with H$_2$ or He.

B. H. Andrew (ed.), Interstellar Molecules, 583–587.

THE COLLISIONAL MODEL

We have extended standard scattering theory to include the case of rotational excitation of $^2\Pi$-state molecules by open shell atoms.[4a,1] The non-axially symmetric effective Hamiltonian, involving the electronic angle about the molecular axis, is essentially similar to that of Green and Zare[5] but also now incorporates spin correlation between the collision partners.

The interaction potential is expressed as an expansion of a product of radial functions $V_{\lambda\mu}(R)$ and spherical harmonics $Y_{\lambda\mu}(\Omega_0)$ in the angles between the molecular axes and the collision partner. There are selection rules for collisional matrix elements:

$$J''(+) \rightarrow J'(+) \text{ and } J''(-) \rightarrow J'(-) : \lambda = \text{even only}$$
$$J''(+) \rightarrow J'(-) \text{ and } J''(-) \rightarrow J'(+) : \lambda = \text{odd only}.$$

For Π-states the projection μ down the molecular axis of the pole of interaction λ is limited to the values 0 and ± 2. To develop an expression for excitation cross sections we have used the restricted distorted wave Born approximation. Experimental results on the analogous system $H + NH_2$ lend strong support to the validity of the method here.

We refer to any non-equality of the cross-sections for the excitation of pairs of transitions between Λ-doublets as 'parity discrimination'. Differences in excitation probabilities arise largely from terms involving $\lambda = 2$ whereas terms for $\lambda = 0$ or 1 do not contribute at this level of approximation. Experimental results[4c] suggest that the expansion of the potential can be meaningfully truncated at $\lambda = 2$. Parity discrimination is then determined by the interference between the radial integrals of $V_{20}(R)$ and $V_{22}(R)$. The absolute degree of discrimination is quite sensitive to the relative magnitudes of $<V_{20}>$ and $<V_{22}>$ but, more significantly, their relative sign determines the sense of the discrimination. By the 'sense' is conveyed which of the transitions $(+)\rightarrow(+)$ or $(-)\rightarrow(-)$ has the larger cross-section for a given pair of J levels. From the available potential surfaces, together with an empirical model of directed valence very much in the spirit of the original Gwinn and Townes model, the relative signs of V_{20} and V_{22} could be deduced. The sense of the parity discrimination is the same for OH or OD with H, H_2 and He. However for CH, CD and NH^+ H-atoms show an opposite sense of discrimination to H_2 or He.

DISCUSSION

Simple collisional radiative cycles are considered with a view to making some qualitative predictions of population inversion in various levels of OH, OD, CH, CD and NH^+. Table 1 shows the favoured parities for the various transitions originating from the lowest rotational levels of OH and CH for collisions involving H atoms. Energy levels are shown in fig. 1. The radiative step in the cycle is known to have only a very small parity discrimination and thus the $\lambda = 2$ contribution to the collisional step is all-important. There is no term in the interaction Hamiltonian which is dependent on nuclear spin: accordingly the eventual choice of hyperfine component that mases will be determined by the properties of the cloud and the radiation field but not of the collision.

OH: Collisions of H, H_2 and He will preferentially populate the level of (-) symmetry of $F_1(2\frac{1}{2})$. Spontaneous emission results in inversion in the $F_1(1\frac{1}{2})$ ground state. The inversion predicted in $F_1(2\frac{1}{2})$ and $(3\frac{1}{2})$ may also lead to maser action in the levels. Populations of $F_2(\frac{1}{2})$ are cooled according to the present calculations. Inverted populations of $F_2(\frac{1}{2})$ may be formed by absorption of a resonant photon at 126 cm^{-1} transferring the inverted population of $F_1(1\frac{1}{2})$ to $F_2(\frac{1}{2})$. Our calcul- ations would therefore suggest that inversion in ground and excited states of OH would result from collisions in diffuse, intermediate and dense clouds in accord with observation.

CH: Let us first consider excitation by H atoms. The low temperature excitation of $F_1(1\frac{1}{2})$ results in the inversion of the ground state. At higher temperatures CH maser action may be quenched by excitation of $F_2(1\frac{1}{2})$. Collisions with H_2, He will tend to cool the ground state of CH at low temperature.

Our results predict collisional population inversion of the CH ground state in low temperature clouds in which hydrogen is at least partly atomic in form. Emission should be seen from the ground state only (see below). CH emission should not be seen from dense clouds. Further, CH emission should not be seen in diffuse or intermediate clouds accompanied by OH emission, on the grounds of excitation temper- ature. All these features are supported by observation.[6] Our results would suggest that it would be worth searching for CH emission from the Λ-doublet transition in $F_2(1\frac{1}{2})$ of CH.

OD, CD, NH^+: OD population inversion is predicted in precisely the same circumstances as OH. However OD is closer to Hund's case (a) and for this reason will show a weaker parity discrimination in collisions. The pattern for CD and NH^+ inversion is identical to that of CH. CD emission is unlikely to be seen since isotopic fractionation enhance- ment should not be operative in diffuse or intermediate clouds.

Table 1: Parity Discrimination in collisions of OH and CH with H-atoms

OH	CH	Favoured Parity
$F_1(1\frac{1}{2}) \rightarrow F_1(2\frac{1}{2})$	$F_2(\frac{1}{2}) \rightarrow F_2(1\frac{1}{2})$	$- \rightarrow -$
$F_1(3\frac{1}{2})$	$F_2(2\frac{1}{2})$	$+ \rightarrow +$
$F_2(\frac{1}{2})$	$F_1(1\frac{1}{2})$	$+ \rightarrow +$
$F_2(1\frac{1}{2})$	$F_1(2\frac{1}{2})$	$- \rightarrow -$

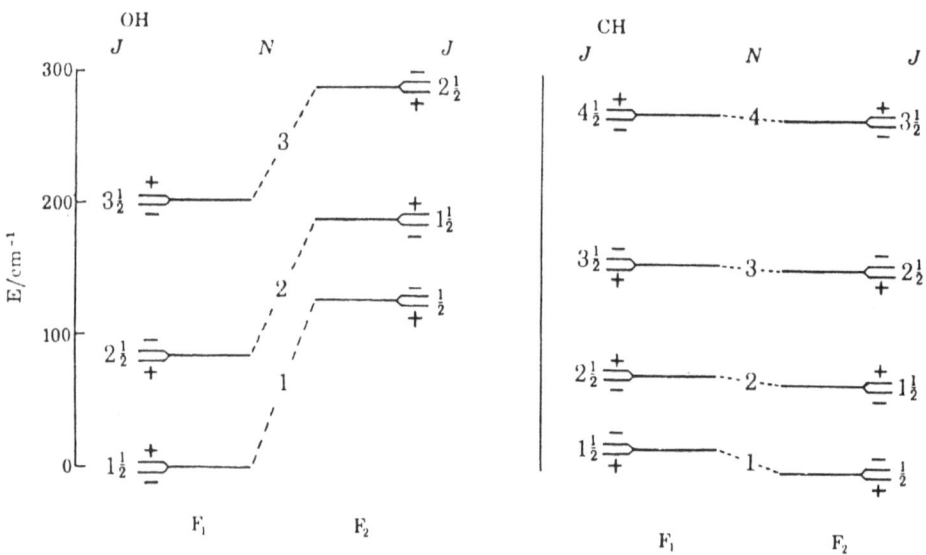

Fig. 1:   Rotational Energy Levels of OH and CH

REFERENCES

1.   Townes, C.H.: 1968, Quantum Electronics Conference, Miami, Florida
2.   Gwinn, W.D., Turner, B.E., Goss, W.M., and Blackman, G.L.: 1973,
        Astrophys. J. 179, 789.
3.   Bertojo, M., Cheung, A.C., and Townes, C.H.: 1976, Astrophys. J.
        208, 914.
4.   a) Dixon, R.N., and Field, D.: 1979, Proc. Roy. Soc. A 366, 225.
     b) Dixon, R.N., and Field, D.: 1979, Proc. Roy. Soc. A 368, 99.
     c) Dixon, R.N., and Field, D.: 1979, Proc. Roy. Soc. A 366, 247.
5.   Green, S., and Zare, R.N.: 1975, Chem. Phys. 7, 62.
6.   Turner, B.E., and Zuckerman, B.: 1973, Bull. A.A.S. 5, 240.

DISCUSSION FOLLOWING FIELD

Robinson: We observe the $^2\pi_{1/2}$ CH 9 cm triplet in *emission* and weakly masering in low density regions. As density increases towards the centres of clouds, the transitions are cooled, and the lines progressively change to *absorption*. This behaviour agrees qualitatively with your suggestion of collisional pumping of CH.

Field: My interpretation of the results presented in Zuckerman and Turner (1975, Ap. J. 197, 123) was precisely the same, and it is very nice to hear it confirmed in your work too.

Elitzur: If collisions with H_2 are so efficient in producing inversion, we have a problem with the observations. If your cross-sections are right the $^2\pi_{3/2}$ (J=5/2) state is sufficiently excited to lead to main-line inversion for temperatures larger than about 30 K. Main-line inversions are certainly not as widespread as that.

Field: I certainly appreciate this problem, and I hope that further experimental and theoretical investigation both in astrophysics and in the laboratory will serve to resolve it.

Hjalmarson: At which densities do you think we will be unable to observe CH maser inversion? At Onsala we see an indication of CH maser emission in the dust cloud L1500, by observations towards and around 3C123. The indication also seems to be confirmed by Effelsberg CH data taken in the same direction.

Field: Without accurate cross-sections and cloud parameters, it is difficult to say. A rough estimate might be made as follows: if the cross-section for excitation of CH by H is 5 to 10 times that for excitation by H_2, perhaps when $[H]/[H_2]$ drops below .2 to .1, the CH $F_2(\frac{1}{2})$ Λ-doublet will begin to cool. This may be the case around a number density of $10^3 \mathrm{cm}^{-3}$.

Goldreich: Non-LTE level populations occur under a wide variety of conditions if radiative rates are greater than or comparable to collisional rates. It is probably not worth making detailed models of pump cycles unless the maser has very special observational characteristics. It is only in this case that one might hope to obtain conclusive and unambiguous results. While it is admirable that people are working in this subject, their energies might be better spent in other ways.

Field: In the first place, the work I have presented is not a detailed pumping model but the most simple one that could be thought of. In the second place, there is no doubt whatsoever that collision cross-sections are data of the greatest value in astrophysics, and are eagerly awaited by many workers in the field. Before deciding on the critical parameters in a model for any particular maser, it is essential to have basic molecular data such as the collision cross-sections.

Irvine: Sume and I at Onsala searched for the $^2\pi_{3/2}$, J=5/2, Λ-doublet transition of CH at about 4.8 GHz. Unless the rest frequency is considerably outside the expected range, we can rule out any *strong* maser in several sources, but not a weak maser action like that seen in the ground state Λ-doublet (Astron. Astrophysics, published).

Field: That is certainly an interesting observation and seems to agree with our predictions.

COLLISIONAL INVERSION OF THE POPULATIONS OF Λ-DOUBLETS IN CH AND OH:
A CRITICAL STUDY

D.R. Flower
Physics Department, University of Durham, DH1 3LE, England

ABSTRACT

    Astronomical studies of the CH and OH masers have been based on
the hypotheses of both radiative and collisional pumping. The mechanism
of collisional inversion of the ground state Λ-doublets by neutral
perturbers was first described by Bertojo et al. (1976) who made a semi-
quantitative analysis of the collision process, based on calculations
of potential energy curves. More recent work (Kaplan and Shapiro, 1979;
Dixon and Field, 1979; Flower, 1980) has, however, been critical of
some of the assumptions made by Bertojo et al. Consideration of the
symmetry and the form of the interaction of CH and OH with the
spherically symmetric perturbers para-H_2 and He shows that current
knowledge of the interaction potential is insufficient for quantitative
results to be obtained.

    Studies of the interstellar CH and OH masers, based upon the
assumption of collisional inversion by neutral perturbers, generally
derive from the work of Bertojo et al. (1976). These authors calculated
the relevant potential energy curves for a perturber approaching
perpendicular to the intramolecular axis, and deduced the sense and the
magnitude of the inversion produced by H, H_2 and He. However, the
analysis, particularly of the collision process, was based on a number
of assumptions, and certain of these have been criticised in the recent
literature (Kaplan and Shapiro, 1979; Dixon and Field, 1979; Flower,
1980). For example, the implicit assumption that the whole of the
interaction energy can give rise to rotational excitation is invalid;
only the anisotropic part of the potential can contribute to rotationally
inelastic scattering, and the magnitude of the corresponding terms must
be determined before the efficiency of the collisional pumping
mechanism may be assessed.

    This fact may most simply be illustrated by considering the
interaction between an atom and a rigid diatomic molecule. In this case,
the potential may be expanded in terms of Legendre polynomials, $P_\lambda(\cos\theta)$,

B. H. Andrew (ed.), Interstellar Molecules, 589–590.

which are functions of the angle θ between the inter- and intramolecular axes:

$$V(\rho,\theta) = \sum_{\lambda} v_{\lambda}(\rho) \, P_{\lambda}(\cos\theta),$$

where ρ is the distance of the atom from the centre of mass of the molecule. Retaining terms in the expansion up to $\lambda = 2$, which are likely to be the most important, we obtain

$$V(\rho,\theta) = v_0(\rho) + v_1(\rho) \cos\theta + v_2(\rho) \frac{3\cos^2\theta-1}{2}$$

(In the case of CH and OH, there are additional second order terms, arising from the interaction of the atom with the unpaired electron; see Flower, 1979).

Bertojo et al. made calculations of $V(\rho,\theta)$ for $\theta = \pi/2$. In this case,

$$V(\rho,\pi/2) = v_0(\rho) - v_2(\rho)/2.$$

In order to determine the contributions of v_1 and v_2 to the anisotropy of the potential, the interaction energy must also be known for collinear approaches of the perturber:

$$V(\rho,0) = v_0(\rho) + v_1(\rho) + v_2(\rho),$$

$$V(\rho,\pi) = v_0(\rho) - v_1(\rho) + v_2(\rho).$$

Potential energy calculations are planned for the additional collision geometries which will enable quantitative conclusions to be drawn regarding the efficiency of H_2 as a collisional pumping agent. In the meantime, only qualitative statements may be made regarding the relative efficiency of the different perturbers. In the case of para - H_2, the interaction with the radical is likely to become more rapidly repulsive with ρ for collinear than for perpendicular approaches, owing to the greater extent of the electron charge distribution along the internuclear axis of the radical. Under these circumstances, it may be shown that the degree of selective excitation of a given Λ-doublet component will be relatively small. Similar remarks apply to He. However, H is intrinsically different, in that one of the singlet state potential energy curves is attractive at intermediate range. For these reasons, H, but not para - H_2 or He, may be expected to induce significant departures of the level populations of the radical from their values in LTE.

REFERENCES

Bertojo, M., Cheung, A.C., and Townes, C.H.: 1976, Astrophys. J. 208, pp. 914-922.
Dixon, R.N., and Field, D.: 1979, Proc. Roy. Soc. A. 368, pp. 99-123.
Flower, D.R.: 1980, Astron. Astrophys., in press.
Kaplan, H., and Shapiro, M.: 1979, Astrophys. J. 229, pp.L91-L96.

PUMPING OF STRONG H_2O COSMIC MASERS

V.S. Strelnitsky
Astronomical Council, USSR Academy of Sciences

ABSTRACT

All the existing models of H_2O masers fail to explain such a strong source as W49 N. Observed and theoretical quantities are related by: $n_{H_2O} W_p \ \ell^3 \gtrsim 10^{46}$ S, where S is the maser flux density (in Janskys), n_{H_2O} is the H_2O number density (cm^{-3}), W_p is pump rate (s^{-1}), and ℓ is the length of amplification region on the line of sight (cm). Models involving vibrational activation (or deactivation) of H_2O by H_2 (Goldreich and Kwan, 1974; Norman and Silk, 1979), with the usual cross-section $\sigma^\nu \lesssim 10^{-19} cm^2$, require $\ell > 10^{16}$ cm for the strongest H_2O features $(\sim 10^4$ Jy), which is unacceptable in view of the VLBI results. Besides, because σ^ν is so small, it is questionable if vibrational pumping could control rotational level populations at all. Depending on the energy source and sink there are four possible schemes of rotational pumping: CR, RC, RR, and CC (C - collisional, R - radiative). The first was modelled by de Jong (1973) and by Shmeld et al. (1976). Though difficulties with the sink (Goldreich and Kwan, 1974; 1979) are avoidable in the model by Shmeld et al. (Strelnitsky, 1979), $\ell \gtrsim 10^{15} - 10^{16}$ cm is still required for the strongest features. Therefore other possibilities of rotational pumping are being investigated. One CC-model is presented below.

CC-pumping is possible if two kinds of particles with different temperatures are present. These can be electrons and H_2 molecules which at $T \sim 1000$ K compete in rotational excitation of H_2O when $10^{-6} \lesssim n_e/n_H \lesssim 10^{-4}$ (cf. Itikawa, 1972). In principle, both $T_H > T_e$ and $T_H < T_e$ may suffice for the pump, but numerical investigation (Bolgova, 1979) favors $T_H > T_e$ for $H_2O \ 6_{16} - 5_{23}$ inversion. CC-pumping may be realized as follows. A strong stellar wind (e.g. $\dot{M} \sim 10^{-5} M_\odot/yr$, $v_w \sim 400$ km/s) from a pre-MS star is stopped by clumps either generated by the interaction of the wind with the circumstellar nebula (Norman and Silk, 1979) or present in the nebula before the wind is switched on. At $\sim 10^{15}$ cm from the star the clumps are compressed to $n_H \sim 10^{11} cm^{-3}$ by the ram pressure of the wind and heated by MHD-turbulence (Norman and Silk, 1979)

591

B. H. Andrew (ed.), Interstellar Molecules, 591–592.
Copyright © 1980 by the IAU.

or by shock waves alternating with expansion waves. In both cases the average energy input reaches $\sim 10\,\rho\,v^3/d$, v being velocity of the gas at greatest scale d. With v $\sim$ 10 km/s at d $\sim 10^{14}$ cm (e.g. shock velocity at indicated density) the input is $\sim 10^{-8}$ ergs/cm$^3\cdot$s and gas with $n_{H_2O} \sim 10^{-3}\,n_H$, $n_e \sim 10^{-6}\,n_H$ is maintained at 1000 - 1500 K, its cooling being due primarily to H_2O (and/or CO, CH_4) vibrational photons generated by electron impact and leaked from the clump or lost on the cold dust. In shocks and in Alfvenic turbulence heavy particles are heated first and electrons will remain 5 - 10% cooler than H_2 owing to greater energy losses. Then W_p is $\sim 3.10^{-2}$ s^{-1} and $\ell \sim 3.10^{14}$ cm is sufficient to explain the strongest H_2O features. The fast (5-10 MeV) protons accelerated by the plasma turbulence in the compressed stellar wind or less energetic particles generated in the turbulent clump itself should ensure the required ionization losses ($\gtrsim 10^{-5}$ of the total energy input). The temperature of the gas is controlled by the (variable) stellar radiation, so the observed correlated variations of maser features (Gammon, 1976; White, 1979) could be explained by synchronous variations of the maser sink efficiency. These can also be explained by changes in n_e/n_H due to a variable flux of ionizing particles from the star's flare-ups.

Maser clumps may have protoplanetary origins. The accreted envelope of a very massive ($\sim 0.01\ M_\odot$) protoplanet (Perri and Cameron, 1973) at $\sim 10^{15}$ cm from a star of solar mass would have a radius of $\sim 10^{14}$ cm and would be torn off almost completely by the assumed stellar wind. This process gives birth to the described strong CC-pumped masers and subsequently to high velocity clouds (Strelnitsky and Sunjaev, 1972; Norman and Silk, 1979) of weaker maser emission pumped by rotational CC, CR or RC. However, the stability of the envelopes around massive protoplanets is open to question (Perri and Cameron, 1974).

REFERENCES

Bolgova, G.T.: 1979, submitted to Nauchn. Inf. Astron. Council.
Goldreich, P., and Kwan, J.: 1974, Astrophys. J. 191, pp. 93-100.
Goldreich, P., and Kwan, J.: 1979, Astrophys. J. 227, pp. 150-151.
Gammon, R.H.: 1976, Astron. Astrophys. 50, pp. 297-313.
Itikawa, I.: 1972, J. Phys. Soc. Japan 32, pp. 217-226.
de Jong, T.: 1973, Astron. Astrophys. 26, pp. 297-313.
Norman, C., and Silk, J.: 1979, Astrophys. J. 228, pp. 197-205.
Perri, F., and Cameron, A.G.W.: 1974, Icarus 22, pp. 416-425.
Shmeld, I.K., Strelnitsky, V.S., and Muzylev, V.V.: 1976, Astron. Zh.
    53, pp. 728-741.
Strelnitsky, V.S.: 1979, submitted to Astron. Zh.
Strelnitsky, V.S., and Sunjaev, R.A.: 1972, Astron. Zh. 49, p. 704.
White, C.J.: 1979, Mon. Not. Roy. Astron. Soc. 186, pp. 377-381.

TIME VARIATIONS OF INTERSTELLAR WATER MASERS IN HII REGIONS

G.J.White[1] and G.H.Macdonald[2]
[1]Physics Dept., Queen Mary College, University of London, U.K.
[2]Electronics Laboratory, University of Kent at Canterbury, U.K.

1. Introduction

Previous studies of time variations in water vapour sources associated with HII regions by Sullivan (1971, 1973) and Gammon (1976) have revealed variability on time-scales of 10^6-10^7s. Sullivan monitored nine sources over a period of 14 months and Gammon observed W49 (H_2O) for 13 months. In an attempt to gather more complete information on the variability of interstellar water masers we have monitored 18 sources for the 32-month period 1974 September-1977 May. The sources selected are listed in Table 1. They were the strongest known at the beginning of the programme above declination $-20°$.

Table 1. The 18 H_2O maser sources monitored in the period 1974 September - 1977 May.

SOURCE	RIGHT ASCENSION (1950)			DECLINATION (1950)			DISTANCE (kpc)	MAXIMUM FLUX (Jy) 1969-70	MAXIMUM FLUX (Jy) 1974-77	VELOCITY RANGE OBSERVED (km s^{-1})	VELOCITY WIDTH (km s^{-1})	NO. OF COMPONENTS 1974-77
	h	m	n	o	'	"						
W3	02	21	53	61	52	22	3	3000	1560	-48 to -25	11	> 7
W30H	02	23	17	61	38	57	2.5	10000	8000	-56 to -46	15	blended
NGC 1333	03	25	59	31	06	05	0.5		160	-40 to 0	3	·2
ORION A	05	32	47	-05	24	08	0.5	15000	35000	-10 to +30	37	>10
S255	06	09	58	18	00	02	1 - 3		780	-10 to +40	7	2
S269	06	11	47	13	50	24	2		200	0 to +40	1	1
W31	18	07	34	-19	5	3	6		1420	-10 to +10	6	4
W33	18	10	54	-18	03	12	4.5		235	+48 to +68	1	1
M17	18	17	29	-16	13	42	2.5	3200	150	-17 to +27	1	1
W49N	19	07	50	9	01	15	14	80000	40000	-220 to +160	380	many
W51M	19	21	26	14	24	41	8	4000	12000	0 to +160	160	many
ON1	20	08	10	31	22	39	6	500	515	-10 to +30	9	4
ON2	20	19	51	37	16	24	5.5		400	-21 to +23	20	> 5
CRL 2591	20	27	36	40	01	11	>2		580	-40 to 0	17	5
W75S	20	37	15	42	12	00	3	5000	655	-20 to +20	9	4
S140	22	17	42	63	03	45	-		530	-20 to +20	26	·3
S152	22	55	38	58	33	02	3.2		55	-61 to -42	1	1
NGC 7538	23	11	36	61	10	20	3.5		540	-65 to -45	10	3

B. H. Andrew (ed.), Interstellar Molecules, 593–598.

2. Long-term variability of individual sources

The present data can be compared with that obtained by Sullivan (1971, 1973) during 1969-70 for the sources W3, W3(OH), Orion A, W49N, W51M, ON1 and W75S. The integrated flux from the averaged spectra over these two periods is given in Table 2 together with the ratio of integrated flux between the two periods for each source.

Table 2. A comparison of the average integrated flux of W3, W3(OH), Orion A, W49N, W51M, ON1 and W75S during 1975-76 with Sullivan's data of 1969-70.

Source	Velocity range (km/s)	Integrated flux 1969-70 (Jy km/s)	Integrated flux 1975-76 (Jy km/s)	Ratio of integrated flux (1975-76) : (1969-70)
W3	$-45 \rightarrow -30$	2.5×10^3	1.2×10^3	0.48
W3(OH)	$-9 \rightarrow -45$	1.2×10^4	1.2×10^4	1.0
Orion A	$-9 \rightarrow 20$	6.1×10^4	8.1×10^4	1.33
W49N	$-15 \rightarrow 30$	2.3×10^5	1.3×10^5	0.57
W51M	$52 \rightarrow 72$	1.5×10^4	3.5×10^4	2.33
ON1	$2 \rightarrow 22$	1.1×10^3	1.1×10^3	1.0
W75S	$-6.5 \rightarrow 6.5$	1.3×10^3	4.7×10^2	0.36

It is apparent from Table 2 that significant long-term variations are found in W3, W49N, W51M and W75S and that while some sources have decreased (W3, W49N, W75S) others have increased (Orion A, W51M) in integrated flux.

Long-term variations imply that the source of excitation is common to many maser components. Little et al. (1977) have discussed the implications of these variations for models of W51M and W49N. They conclude that the flux of radiation from the common exciting object must vary on a time-scale of ~10 or <100 years, depending on whether the masers are saturated or unsaturated. Recent VLBI observations (Genzel et al. 1978) have shown that at least 10 separate centres of activity exist in W49N. If each of these centres indicates the site of a young stellar object responsible for pumping the nearby maser compo-nents, it is surprising that systematic long-term variability should be detected. However, it appears from the present data that only the low-velocity components have varied significantly. Inspection of the VLBI map of Genzel et al. (1978) shows that most of the integrated low-velocity flux in W49N is emitted by four of the 10 centres of activity. It is therefore plausible that the observed long-term variations are due to changes in one or more of these centres of activity.

This conclusion is strongly supported by the recent detection of correlated short-term variability in the maser components associated with one of these centres (White, 1979). This striking result is illustrated in Figure 1 which shows the time variations during the

period 1975 January-1976 October for the six velocity components found
to be spatially associated in a VLBI map made from observations in 1976
March (Walker et al., 1977). All six components showed significant
increases at this time with one, at -45 km s^{-1} velocity, becoming the
second most intense component in the W49 spectrum by 1976 October.

VLBI maps of W51M (Genzel et al., 1978; Walker et al., 1977) show
that in this source most of the emission arises from a single centre of
activity. Changes in the pump rate of the central object would readily
account for the long-term variations observed.

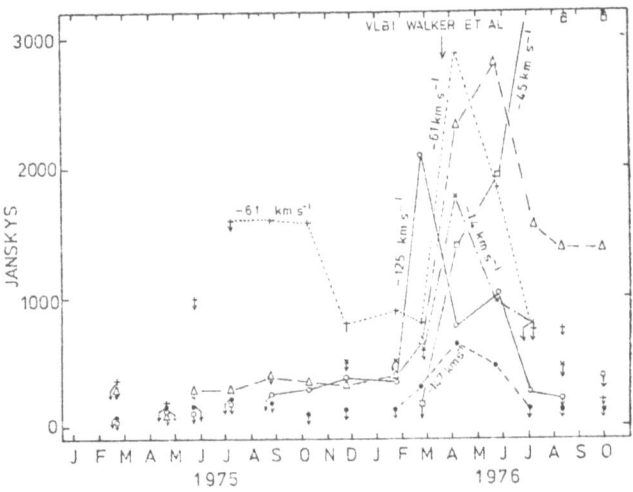

Figure 1. Intensity variations in those velocity components spatially
associated with one centre of activity in W49.

3.   Short-term variability in the whole sample

A statistical analysis of the short-term variability in the whole
sample was made by computing the percentage change in certain selected
velocity components between consecutive observations, which were 46 ± 4
days apart. Strict selection criteria were applied in choosing velocity
components for this analysis to avoid the effects of blending, particularly in low-velocity components. Histograms showing the percentage
variations over the interval 46 ± 4 days for the velocity components
selected are given in Figure 2.

It can be seen that large flux variations (>100%) over the ~46-day
interval are found in sources other than W49N although the largest
variations are found in this source which showed changes >100% in ~25%
of samples.

The largest short-term flux variation observed was in the -19 km s^{-1}
component in NGC 1333 which decreased by a factor of 2.5 over 3 days.

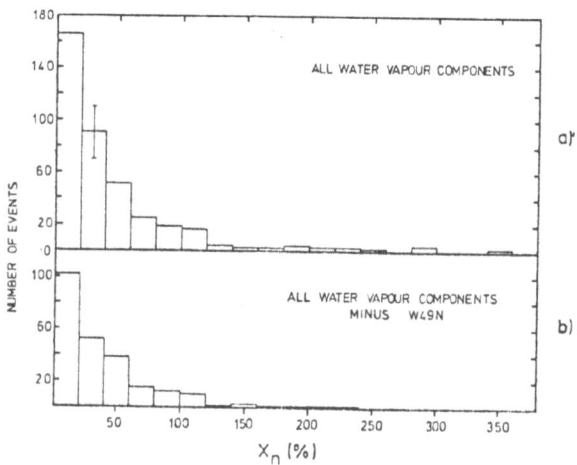

Figure 2. a) Histogram showing the short-term variability in all water
maser components in the present sample.
b) As a) but without data on W49N.

4.   Related variability in features of similar velocity

Genzel and Downes (1977) have noted related variability in pairs
of low-velocity components in several sources (W51M, W75N, W75S(3),
10.5 + 0.0) which from their double-peaked spectra they believe to be
shell sources. They interpret this behaviour in terms of a saturated
travelling-wave maser with several modes and suggest the term "mode
switching" for the phenomenon.

We have observed this effect in two sources. The −84 and −82 km s^{-1}
high-velocity features in W49N peaked within six weeks of each other and
the −4.6 and −5.7 km s^{-1} components in W75S interchanged intensity
between 1975 November and 1976 February.

These results may be interpreted by the mechanism proposed by
Genzel and Downes (1977) in terms of the kinematics of circumstellar
shells. Alternatively, if the pump radiation is beamed and the direction of this beam varies, possibly due to movement of an obscuring dust
cloud between the pump source and the masing regions, then different
clouds of water vapour may be successively excited.

5.   Interpretation

The time-dependent equations of radiative transfer for a masing
region longer than it is wide have been solved recently by Salem and
Middleton (1978). They find that there is no inherent feature in these
non-linear equations which leads to variations in the output of the
maser, provided that physical conditions such as the number density and
pump rate remain constant. It follows that the observed time variations

in astrophysical masers must be due to changes in pump rate or source geometry or possibly a combination of both.

Changes in source geometry can give rise to an effective change in pump rate where the pump radiation is beamed, for example by changes in opacity in an obscuring dust cloud moving between the pump source and maser volumes. Alternatively, relative movement of water vapour clouds can give rise to favourable velocity alignments such that the effective length of the maser region is enhanced for an interval of time.

Of the various pumping schemes proposed for water maser sources, none is found to satisfy completely the constraints imposed by the present observational data. The shock front excited, radiative pump model of Litvak (1969) cannot supply sufficient pump power for maser sources with photon emission rates $>10^{45}s^{-1}$ (Goldreich and Kwan, 1974). The collisional pump model of de Jong (1973) can provide more pump power but predicts dimensions of the maser regions which are several orders of magnitude larger than those observed. The hot-dust-cool-gas model of Goldreich and Kwan (1974) predicts levels of infrared radiation from the hot dust significantly greater than observed. Oka (1973) has proposed a mechanism of selective predissociation by absorption of ultraviolet radiation, which is particularly attractive for water masers in the vicinity of HII regions, but an important feature of this model is the presence of OH more than 10 times more abundant than H_2O. There is some doubt that the OH can exist in the strong ultraviolet radiation field necessary to selectively dissociate the H_2O.

It must be concluded that at present no completely satisfactory model exists which combines defined geometrical and physical properties with the dynamics of maser regions in the vicinity of an early-type stellar object.

REFERENCES

Gammon, R.H. 1976. Astron. Astrophys., 50, pp. 71-77.
Genzel, R. and Downes, D., 1977. Astron. Astrophys. Suppl., 30,
     pp. 145-168.
Genzel, R.,et al., 1978. Astron. Astrophys., 66, pp. 13-29.
Goldreich, P. and Kwan, J., 1974. Astrophys. J., 191, pp. 93-100.
de Jong, T., 1973. Astron. Astrophys., 26, pp. 297-313.
Little, L.T., White, G.J. and Riley, P.W., 1977. Mon.Not.R.astr.Soc.,
     180, pp. 639-656.
Litvak, M.M., 1969. Science, 165, pp. 855-861.
Oka, T., 1973. Molecules in the Galactic Environment, pp. 258-266,
     Wiley Interscience, New York.
Salem, M. and Middleton, M.S., 1978. Mon.Not.R.astr.Soc., 183,
     pp. 491-500.
Sullivan, W.T., 1971. Astrophys. J., 166, pp. 321-332.
Sullivan, W.T., 1973. Astrophys. J. Suppl., 25, pp. 393-432.
Walker, R.G., et al., 1977. Astrophys. J., 211, L135-L138.
White, G.J., 1979. Mon.Not.R.astr.Soc., 186, pp. 377-381.

DISCUSSION FOLLOWING MACDONALD

Gold: You spoke of the conflict with respect to the size of the maser. I presume, again, that you had no knowledge of the physical size at all, but that you merely observed some phase-front scatter.

Macdonald: Although I agree one cannot interpret the angular size of maser "hot spots" in terms of physical dimensions, I believe we can reliably derive a physical size from the overall angular extent of the maser spots on a VLBI map.

Scalise: Did you study the polarization of these H_2O maser sources? How did it vary in the long term?

Macdonald: We have made some observations of these H_2O masers in linear polarization, but not over a long period. This work has not been published since we were not confident of our calibration of polarization angle.

Dickinson: You mentioned one "center of activity", as defined by VLBI observations, where the six velocity components all varied in the same way. It sounds as though you have delineated the pump source for these six lines. Have you other examples of this kind?

Macdonald: After applying selection criteria to avoid the effects of blending of velocity components, we had sufficient data on only two of the four "centres of activity" responsible for the majority of the low-velocity emission in W49. Of these two centres, only one showed the remarkable correlated variability I have described.

Schwartz: Is there any evidence for variations of any of the IR sources associated with the HII-region masers you have studied, and if so, on what time scales?

Macdonald: I do not know of any systematic monitoring of IR sources associated with H_2O masers in HII regions that would be comparable to the work done on H_2O masers associated with late-type stars. Such a study would be worthwhile.

FURTHER OBSERVATIONS OF THE H_2O EMISSION FROM NGC 4945

J.R.D. Lépine and P. Marques dos Santos
Centro de Radioastronomia do Observatório Nacional
Rua Pará 277 - 01243 - São Paulo, Brasil

We wish to report briefly here the results of new observations of the H_2O maser emission from the galaxy NGC 4945 made in 1979 July with the Itapetinga radio telescope. The H_2O emission that we detected in 1978 September, shown in fig. 1, represented an intrinsic intensity about ten times that of W 49, the strongest source in our Galaxy, and raised the question of its nature (Marques dos Santos and Lepine, 1979). The main purpose of the new observations was to obtain a more accurate position for the source and to look for variability.

NGC 4945 is a barred spiral galaxy seen almost edge-on, with the major axis at about 45° position angle and dimensions about 17' x 3'. We obtained in 1979 July a series of spectra, displacing the 4' beam by steps of 2' along the major axis. The spectra are presented in fig. 2. The spectrum corresponding to zero displacement in fig. 2 should be compared with the spectrum of the first detection in 1978 September (fig. 1), since both were obtained in the direction of the nucleus. In the new spectrum we have not removed the continuum, which is attributed to the radio source located at the nucleus (Mathewson and Rome, 1963). Although the noise level is greater in the present observations, the narrow peak at 674 km s^{-1} has obviously disappeared, while the features at 700 km s^{-1} and 714 km s^{-1} seem to be still present at about the same intensity. This result cannot be considered as evidence that the narrow feature originated from a different source, since similar strong variability of one peak relative to others has also been observed in galactic sources, e.g. W 49 (Gammon, 1976).

The spectrum obtained at 2' S-W of the nucleus (-2') also shows some indication of emission features, while nothing can be seen at 4' S-W and 2' N-E of the nucleus. The positive average flux that we obtained at -2' is likely to be due to emission features, and not to be instrumental, since with our beam-switching technique only marginal zero-level offsets are expected. There is, then, a suggestion from the data that the source may be displaced from the nucleus (0') towards the south-west (-2'). If so, it may be coincident with a bright patch, probably a spiral arm, that is easily seen on the plates in this

599

B. H. Andrew (ed.), Interstellar Molecules, 599–601.

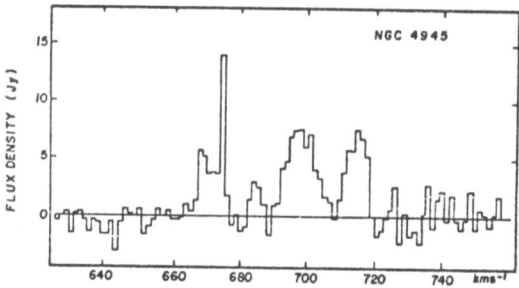

Fig. 1 - 6_{16}-5_{23} H_2O spectrum of NGC 4945 obtained in 1978 September, already published in ref. 1. Velocity is with respect to L.S.R.

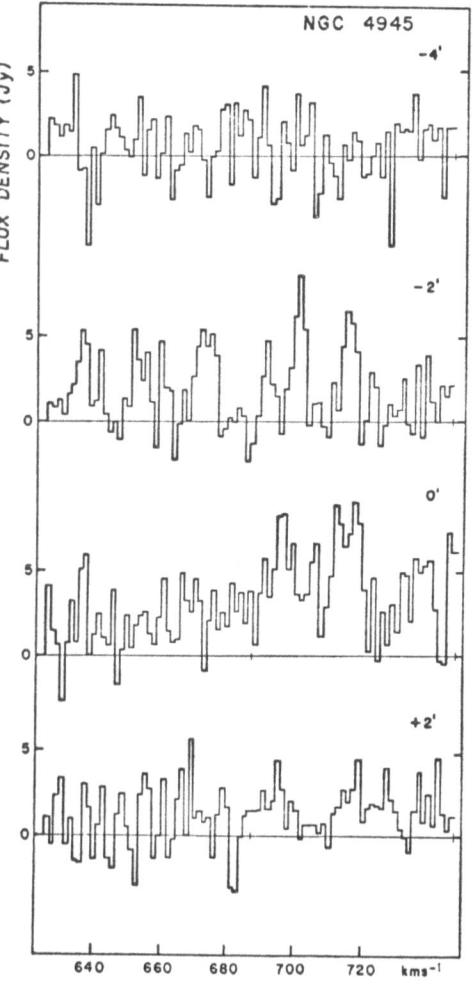

Fig. 2 - 6_{16}-5_{23} H_2O spectra of NGC 4945 obtained in 1979 July. Displacements with respect to the nucleus, along the major axis, are indicated; negative values correspond to the south-west. Velocities are with respect to L.S.R.

region, and is also prominent on the isophote maps by Sérsic (1968). It seems possible that the very strong H_2O emission features that we have detected in NGC 4945 are associated with giant HII regions which are typical of Sc galaxies. It should be noted, however, that it is somewhat surprising to find a maximum of emission around 700 km s^{-1} to the south-west of the nucleus. Such velocities do not fit the rotational motion that can be inferred from HI observations (Whiteoak and Gardner, 1977).

Further observations are planned, and a more detailed account of our investigations, including negative results for other galaxies, will be published later in Astronomy and Astrophysics.

REFERENCES

Marques dos Santos, P., and Lépine, J.R.D.: 1979, Nature 278, 34.
Mathewson, D.S., and Rome, J.M.: 1963, Observatory 83, 21.
Gammon, R.H.: 1976, Astron. and Astrophys. 50, 71.
Sérsic, L.: 1968, Catálogo de Galaxias Australes, Univ. de Cordoba.
Whiteoak, J.B., and Gardner, F.F.: 1977, Aust. J. Phys. 30, 187.

DISCUSSION FOLLOWING LÉPINE

Gillespie: Have you observed any sources in the Magellanic Clouds?
Lépine: Scalise has searched for H_2O emission in the Magellanic Clouds. I think he can give you more information.
Scalise: We searched for water masers in more than 40 positions in the Large Magellanic Cloud, covering the velocity range from 200 to 330 km s^{-1}. Special attention was given to 30Dor (Tarantula Nebula) and Henize's N159, particularly this latter source where OH, H_2CO and CO have been detected. No emission above 0.2 K antenna temperature has been detected so far. Several HII regions in the Small Magellanic Cloud were searched, again with negative results. We intend to continue searching for water masers in the Magellanic Clouds.
Gillespie: It does not surprise me that you see nothing in the Small Magellanic Cloud since I have failed to detect CO there, but it does surprise me that you found nothing in N159 where other molecules have been found.
Dickinson: Could you tell us the distance to NGC 4945?
Lépine: NGC 4945 is a member of the Centaurus group, estimated to be at 4 Mpc.

THE INTERPRETATION OF HIGH VELOCITY H_2O MASERS

V.V. Burdyuzha
Space Research Institute
Academy of Sciences of the USSR
Moscow, USSR

ABSTRACT

Kinematic and physical models of high velocity H_2O masers are briefly discussed.

INTRODUCTION

In some sources of maser emission associated with compact HII regions H_2O maser features are spread in velocity up to ±200 km s^{-1} from the central features. If these high velocity features (HVF) are attributed to the Doppler effect, the corresponding sources cannot be gravitationally bound to a central body as the central mass required, $\sim10^5 M_\odot$, would be too big. This mass would in some way or other manifest itself, which is not the case. Therefore the HVF are an interesting phenomenon in the physics of galactic masers. The latest VLBI observations (Genzel et al. 1978, 1979) have discovered new peculiarities of the spatial distributions of the maser features, namely their grouping around centers of activity. The low velocity (±15 km s^{-1}) most intense maser features form the centers of activity. High velocity H_2O maser features cluster around the centers of activity at both positive and negative velocities, occupying a large spatial area. The HVF vary in time (Morris 1976, Gammon 1976, Little et al. 1977, White 1979, Genzel et al. 1979). It is now obvious (Genzel and Downes 1977a) that our first naive models of maser sources related to HII regions should be considered as an evolutionary sequence. Maser sources may exist in the expanding shell of a recently born star of high luminosity. It is evident that the maser sources are cloudlets formed due to fragmentation of the medium in which they are immersed. The medium can fragment as the result of thermal instability caused by the passage of a shock wave (Burdyuzha and Ruzmaikina 1974, 1975) or due to Rayleigh-Taylor instability (Norman and Silk 1979). H_2O maser behaviour increases the inhomogeneity of the medium because the maser emission is a quite powerful cooling source (Burdyuzha and Ruzmaikina 1977, Norman and Silk 1979).

B. H. Andrew (ed.), Interstellar Molecules, 603–609.

In addition to the assumption that the low velocity and high velocity features are clouds radiating as masers (Strelnitskii and Sunyaev 1972, Norman and Silk 1979, Elmegreen and Morris 1979) another interpretation was proposed involving a various and sometimes amazing distribution of radial velocities. This interpretation is associated with the physics of line splitting (Stark effect-Slysh 1973; induced by Compton scattering - Galeev and Sunyaev 1972, Montes 1977; Raman scattering - Radhakrishman et al. 1975, Boyd 1977; and scattering from plasma noise - Fernandez and Reinisch 1978, Burdyuzha 1978, Burdyuzha et al. 1979).

All models of "high velocity H_2O masers" (about ten) can be divided into two types: kinematic, and physical.

KINEMATIC MODELS

The basis for the interpretation of all kinematic models is the Doppler effect. The high velocity clouds radiating as masers may be accelerated either by a stellar wind or by radiation pressure. Strelnitskii and Sunyaev (1972) were the first to draw attention to this fact.

As has been shown by many authors, a young massive star of high luminosity radiates an intense stellar wind: the outflow rate is $10^{-5}-10^{-6}M_\odot$/year. The stellar wind can accelerate maser clouds up to rather high velocities $\sim$200-300 km s^{-1} and carry them from a distance of $R_0 \sim 10^{14}$cm from the central star to a distance of $R \sim 10^{17}$cm and farther. $R_0 \sim 10^{14}$cm appears to be the minimum distance at which the maser clouds could originate due to the fragmentation of the medium, and to be the minimum distance at which there are appropriate conditions for the H_2O maser to operate ($n \sim 10^7-10^{11}$cm^{-3} and $T \sim 3 \cdot 10^2-10^3$K). The criterion for this mechanism to be efficient is

$$\dot{M}V\sigma/2\pi R_0 mv^2 \sim 1 \qquad (1)$$

where $\dot{M}$ is the outflow rate (g s^{-1}), V is the stellar wind velocity (cm s^{-1}), σ is the cross-sectional area of the maser cloud (cm^2), m is the maser cloud mass (g), and v is the maser cloud velocity (cm s^{-1}).

In the case of acceleration by radiation pressure, the criterion may be written as

$$L_*\sigma/2\pi R_0 cmv^2 \sim 1 \qquad (2)$$

where L_* is the central star luminosity (erg s^{-1}), and c is the speed of light (cm s^{-1}).

Quite recently, more specific and elaborate models have been suggested. In the paper by Norman and Silk (1979) a new mechanism was proposed in which the magnetic field plays an essential part in the acceleration of the cloud. When the intense stellar wind interacts with gas in the magnetic field of the shell, a Rayleigh-Taylor instability is developed which leads both to fragmentation of the medium and to the acceleration of the clouds thereby formed. In these processes the magnetic field is the energy source. Later these high velocity clouds ("bullets") are ejected into the environment, thus providing an explanation of the phenomenon of Herbig-Haro objects. A rather large magnetic field ($B \geqslant 10^{-2}$ G) is required to make this mechanism efficient. With increasing distance from the central star the "bullets" slowly lose their velocity. However this prediction is contradicted by the observations in Orion (Genzel and Downes 1977b), making the interpretation of the high velocity H_2O masers as "bullets" seem rather doubtful.

Shell and disk kinematic models of cloud acceleration are similar in many ways. In these models the clouds are also accelerated by either the stellar wind or the radiation pressure. In the shell model, the spherical shell is expanding together with its low velocity features (LVF). In the disk model, the disk with the immersed LVF is also expanding and rotates differentially. The authors of the models of this type are R. Genzel and D. Downes.

The most developed model among the kinematic models is that of a Keplerian disk (Elmegreen and Morris 1979). In this model, the LVF are maser regions moving in orbit together with the disk around a star of mass $M \sim 30$ $M_\odot$. The HVF are clouds blown by the stellar wind out of the inner parts of the Keplerian disk, which is a stable formation as far as its expansion or compression is concerned. In this model the high velocity clouds should be either smaller in size or warmer than the low velocity clouds.

Summing up this brief survey of the kinematic models, I should like to note that in Doppler interpretations of the H_2O HVF a few aspects still remain obscure:
1. Why do the HVF appear only in the H_2O molecule emission? (Of course there always remains the possibility that moving H_2O features can radiate as masers).
2. With several years or even months of attentive "patrolling" of the high velocity H_2O features of a source, changes in the distance between them as well as the velocity of their separation should be determinable. It is not clear why these motions have not yet been detected.
3. It is not quite clear how the high velocity H_2O maser operates when there are rather high velocity gradients in the cloud.

PHYSICAL MODELS

Physical models of high velocity H_2O masers are based either on there being a splitting and a shift of features in frequency caused by the influence of electric or magnetic fields, or on there emerging satellite lines as a result of interaction between the maser emission and a plasma. For H_2O maser emission, 1 MHz separation in frequency corresponds to 13.5 km s^{-1} in velocity.

The hyperfine structure of the $6_{16} \to 5_{23}$ line of water vapor has been investigated by Kukolich (1969). The interaction between rotational angular momentum and nuclear spin splits the 22 GHz transition into six lines with a total spread in velocity of ~ 5.9 km s^{-1}. Note that of the six hyperfine components, three correspond to a transition with the same value of the projection of the total angular momentum F, and have an intensity two orders smaller than that of the basic components, so it is more correct to speak about the hyperfine interaction resulting in a split in velocity of order 1 km s^{-1}.

The splitting in frequency caused by the Zeeman effect is also very small in the magnetic fields (~ 10mG) that are characteristic of maser regions (Moran et al. 1978).

According to Slysh's calculations (1973) the electric component of the electromagnetic field of maser radiation may be set equal to

$$E_0 = (4\pi W^t)^{1/2} = (8\pi k\nu^2 \Delta\nu_{dop}\ T_b \Omega/c^3)^{1/2} \sim 10^{-2} \text{esu} \tag{3}$$

where W^t is the radiation energy density (if we assume an isotropic maser at $T_b \sim 10^{15}$K). Satellite lines due to the Stark effect are observed at $\pm\Delta\nu_{st} = \mu E_0/\hbar$, and their splitting may exceed the Doppler width $\Delta\nu_{dop} \sim 50$ kHz $\mu = 0.125 \times 10^{-18}$ esu is the electric dipole moment of the H_2O molecule. Since it has become apparent that the H_2O masers are anisotropic ($\Omega/4\pi \sim 10^{-2}$) a splitting of a few km s^{-1} is difficult to explain by the Stark effect (much larger brightness temperatures are required, but are not observed).

Since $T_b >> T_e$, and the density of plasma through which the maser radiation passes is close to that required for the generation of Compton solitons, Compton scattering may be essential for the creation of the high velocity maser features and for their variability (Galeev and Sunyaev 1972, Montes 1977). It should be noted that Compton solitons are formed only at positive velocities (red features) while the maser features are observed both at positive and negative velocities relative to the central structures.

Radhakrishnan et al. (1975) noted that satellite radiation that is essentially HVF may appear as a result of stimulated Raman scattering of maser radiation. Raman scattering is the inelastic scattering of radiation from molecules. A photon can gain or lose an energy ΔE in this scattering process, where ΔE is the energy separation of two of the molecular levels. Boyd (1977) investigated this idea. He found that the best candidate for the scattering agent is the ammonia molecule. He showed that there are some difficulties in the Raman interpretation of high velocity H_2O features. The main difficulty seems to lie in the fact that there is very little chance of the features occurring at negative velocity (anti-Stokes scattering). Besides, a very high density of ammonia ($n_{NH_3} \sim 10^{11}$cm^{-3}) is required, and the population of the hyperfine levels must be in equilibrium. The models of HVF with both Compton solitons and stimulated Raman scattering have one more important limitation connected to the fact that the HVF produced in such a way must be generated in the same spatial region as the low velocity maser features, a requirement that again contradicts the observations (Genzel et al. 1978).

Burdyuzha (1978), Fernandez and Reinisch (1978), and Burdyuzha et al. (1979) suggest that HVF are the Langmuir satellites of maser radiation at frequencies $\omega_0 \pm \omega_e$ (ω_0 is the cyclic frequency of the maser radiation, ω_e is the plasma frequency). Langmuir satellites arise from the interaction between maser emission and turbulent plasma as a result of merging and decay processes. In the paper by Fernandez and Reinisch (1978), the possibility of satellite generation by maser emission itself is considered within the limits of a one-dimensional model without blueshifted satellites. We have considered the case where the plasma is excited by external turbulence i.e. a shock wave. We resort to the shock wave model because the plasma turbulence level W^ℓ which is excited by maser emission is very small ($W^\ell \sim 6 \times 10^{-25}$erg cm^{-3} for $W^t \sim 10^{-11}$erg cm^{-3}). A plasma turbulence level $W^\ell \sim 10^{-3} n_e k T_e$ is produced by the shock waves which occur in the vicinity of a newborn O star (Burdyuzha and Ruzmaikina 1974, Cochran and Ostriker 1977). We assume that the symmetric LVF and HVF were formed while maser emission passed through the turbulent medium. Various electron densities from $n_e \sim 10^4$ to $n_e \sim 10^6$cm^{-3} are necessary. The relation between the plasma density and the separation of features in velocity takes the form:

$$n_e \sim 0.7 \times 10^2 \times (\Delta V \text{ km s}^{-1})^2 \text{cm}^{-3} \qquad (4)$$

The approximate formula correlating the turbulent plasma length with its density n_e and temperature T_e is

$$\Delta Z \sim 2.5 \times 10^{24} \times A^{-1} B \ n_e^{-3/2} T_e^{-1/2} \text{ cm} \qquad (5)$$

where A is the degree of excitation of the cosmic plasma, and B is the ratio of the intensity of the H2O satellites to the intensity of the radiation at the basic frequency. The observed ratio of intensities is $\sim 10^{-2}$ (Goss et al. 1976). The Langmuir satellites propagate in a direction which differs from the direction of propagation of the basic signal. The approximate formula describing the angle between the direction of propagation of scattered and incident radiation is

$$\theta/2 \sim 0.005 n_e^{1/2} T_e^{-1/2} \qquad (6)$$

An H$_2$O maser with the dimensions $(10^{13} \times 10^{13} \times 10^{15})$ cm appears as a small elongated cloudlet in the gas-dust shell of a young O star. The H$_2$O maser radiation propagating in a region with $n_e \sim 10^4$cm^{-3} and $T_e \sim 10^3$K produces symmetric features with $\Delta V \sim 12$ km s^{-1}. With $n_e \sim 10^4$cm^{-3} and $T_e \sim 10^3$K the angle of scattering is still rather small, and the interferometric picture shows the low velocity maser features in regions with an area of $10^{15} \times 10^{16}$cm^2 (the centers of activity). As the maser radiation propagates through a region with higher density, $n_e \sim 10^6$cm^{-3} (the region behind the dust front of the shock wave) the angle of scattering and the plasma frequency ω_e increase. This creates HVF at frequencies $\pm\omega_e$, i.e. velocities $\Delta V=\pm40$ to 200 km s^{-1}. These HVF are spread over a much larger area ($10^{16}\times10^{17}$ cm^2 and more) around the LVF. The observed asymmetry in the distribution of LVF (Goss et al. 1976, Morris 1976) may be due to the inhomogeneity of the electron density and plasma velocity behind the shock wave front at 10^{16}-10^{17} cm in the spatial plane. An observer sees the integrated effect caused by inhomogeneities of density or velocity. Of all the predictions made on the basis of this model the most interesting is that superhigh velocity features with $\Delta V \sim \pm 1200$ km s^{-1} must be present in the H$_2$O radiation from IR stars.

Summing up this brief survey, I should like to say that for a better understanding of the phenomenon of "high velocity H$_2$O masers" we need more information on the movement of both high velocity and low velocity features.

REFERENCES

Boyd, R.W.: 1977, Publ. Astron. Soc. Pacific 89, pp. 141-146.
Burdyuzha, V.V., and Ruzmaikina, T.V.: 1974, Astron. Zh. 51, pp. 346-353.
Burdyuzha, V.V., Ruzmaikina, T.V.: 1974, Astron. Zh. 51, pp. 346-353.
Burdyuzha, V.V., Ruzmaikina, T.V.: 1975, Astron. Astrophys. 40,
    pp. 233-236.
Burdyuzha, V.V., Ruzmaikina, T.V.: 1977, Proceedings of IAU
    pp. 53-56. Eds. E. Basinska-Grzesik, M. Mayor.
Burdyuzha, V.V.: 1978, Pis'ma Astron. Zh. 4, pp. 555-558.
Burdyuzha, V.V., Charugin, V.M., Tomozov, V.M.: 1979, Astron. Astrophys.
    (in press).
Cochran, W.D., Ostriker, J.P.: 1977, Astrophys. J. 211, pp. 392-399.
Elmegreen, B.G., Morris, M.: 1979, Astrophys. J. 229, pp. 593-603.
Fernandez, J.C., Reinisch, G.: 1978, Astron. Astrophys. 67, pp. 163-174.
Galeev, A.A., and Sunyaev, R.A.: 1972, Zh. Eksper. Teor. Fis. 63,
    pp. 1266-1282.
Gammon, R.H.: 1976, Astron. Astrophys. 50, pp. 71-77.
Genzel, R., Downes, D., Moran, J.M., Johnston, K.J., Spencer, J.H.,
    Walker, R.C., Haschick, A., Matveyenko, L.I., Kogan, L.R.,
    Kostenko, V.I., Ronnang, B., Rydbeck, O.E.H., Moiseev, I.G.: 1978,
    Astron. Astrophys. 66, pp. 13-29.
Genzel, R., Downes, D., Moran, J.M., Johnston, K.J., Spencer, J.H.,
    Matveyenko, L.I., Kogan, L.R., Kostenko, V.I., Ronnang, B.,
    Haschick, A.D., Reid, M.J., Walker, R.C., Giuffrida, T.S., Burke,
    B.F., Moiseev, I.G.: 1979, Astron. Astrophys. 78, pp. 239-247.
Genzel, R., Downes, D.: 1977a, Astron. Astrophys. Suppl. 30, pp. 145-168.
Genzel, R., Downes, D.: 1977b, Astron. Astrophys. 61, pp. 117-126.
Goss, W.M., Knowles, S.H., Balister, M., Batchelor, R.A.: 1976, Mon. Not.
    R. astr. Soc. 174, pp. 541-549.
Kukolich, S.G.: 1969, J. Chem. Phys. 50, pp. 3751-3756.
Little, L.T., White, G.I., Riley, P.W.: 1977, Mon. Not. R. astr. Soc.
    180, pp. 639-656.
Montes, C.: 1977, Astrophys. J. 216, pp. 329-337.

Morris, M.: 1976, Astrophys. J. 210, pp. 100-107.
Moran, J.M., Reid, M.J., Lada, C.J., Yen, J.L., Johnston, K.J.,
    Spencer, J.H.: 1978, Astrophys. J. (Letters) 224, pp. L67-L71.
Norman, C., Silk, J.: 1979, Astrophys. J. 228, pp. 197-205.
Radhakrishnan, V., Goss, W.M., Bhandari, R.: 1975, Pramana 5, pp. 49-57.
Slysh, V.I.: 1973, Astrophys. Letters 14, pp. 213-216.
Strelnitskii, V.C., Sunyaev, R.A.: 1972, Astron. Zh. 49, pp. 704-711.
White, G.J.: 1979, Mon. Not. R. astr. Soc. 186, pp. 377-381.

DISCUSSION FOLLOWING BURDJUZHA

Field: If I understand you correctly, you postulate a high concentration of electrons and H_2O molecules. It is known that vibrational excitation of H_2O by electrons proceeds at a high rate. Would it not be corroboration of your theory if vibrational transitions in H_2O could be observed? Furthermore, could it possibly be that some of the rotational lines seen in H_2O maser sources (e.g. those with apparently very high Doppler velocities) arise from stimulated emission in vibrationally excited H_2O (1,0,0; 0,1,0 or 0,0,1)?

Burdjuzha: The scattering of H_2O maser radiation occurs in the external plasma, where there may be no H_2O molecules. Thus to speak about excitation by electrons may not be necessary.

Lepine: I made a rough estimate of the frequencies of the 6_{16}-5_{23} rotational transition in excited vibrational states of H_2O. These frequencies fall off by several GHz from the ground vibrational transition at 22 GHz, and thus cannot explain high velocity features. However a search for these lines could be interesting, if accurate vibration-rotation interaction constants were available.

Burdjuzha: This idea is interesting, but it requires accurate calculation. Possibly the shift will be only to one side. A shift of a few GHz is very big.

de Jong: The central region in your model has an electron density of $\sim 10^4 cm^{-3}$ and a radius of $\sim 10^{16} cm$. Such a region should be a strong source of free-free radio emission, which in spite of several attempts has never been detected. How do you account for this?

Burdjuzha: If the low velocity features are interpreted as satellite lines of the maser emission then really rather high emission measures ($\sim 3.10^5 cm^{-6} pc$) must be present. In general it is more realistic to say that the low velocity features are identified with the cloudlets and that only the high velocity features are satellite lines. However, according to this model there must be weak free-free radiation from maser sources.

FAR-ULTRAVIOLET OBJECTIVE SPECTROGRAPHIC SURVEYS FOR MAPPING OF INTERSTELLAR H$_2$, H, AND CO

George R. Carruthers
Naval Research Laboratory, Washington, DC 20375

A far-ultraviolet, wide-field camera/spectrograph investigation has been proposed for Spacelab missions. By use of an objective grating, this experiment could survey large areas of the sky, obtaining spectra of early-type stars as faint as 12th visual magnitude in the 950-2000 Å wavelength range, with 1 to 2 Å spectral resolution.

A major objective of this proposed objective-spectrographic survey would be mapping of the distributions of interstellar atomic and molecular hydrogen over large areas and in many different regions of the sky. An objective-grating survey can achieve this with much greater observational efficiency than possible with a conventional telescope/spectrometer (which can observe only a single star at a time).

To determine the optimum spectral resolution of such a survey, we have computer-degraded typical <u>Copernicus</u> U2 spectra (0.2 Å resolution) to equivalent resolutions of 1 to 32 Å. We find that degradation to 1 or 2 Å resolution still permits acceptably accurate determination of H and H$_2$ column densities in the directions of moderately reddened stars, while greatly improving the limiting sensitivity for a given instrumental aperture.

Such a survey would also be useful for mapping interstellar CO in the directions of highly reddened stars if a higher-dispersion objective grating were used. The survey could also map the distribution of inter-stellar dust with greater sensitivity than possible from the ground if sensitivity were provided in the middle ultraviolet (including the interstellar extinction "bump" near 2200 Å). The distribution of dust could then be directly correlated with the atomic and molecular column densities.

The astrophysical significance of such a survey is that it would provide a statistical basis for comparing with theoretical predictions the observed relationships between H$_2$/H ratio, local gas and dust density, local ultraviolet radiation intensity, etc., that would be much

B. H. Andrew (ed.), Interstellar Molecules, 611–612.
Copyright © 1980 by the IAU.

better than the one derived from the present, limited Copernicus surveys, which include somewhat more than 100 stars. The proposed survey could reach several thousand stars, including many highly reddened and/or distant galactic stars, plus the brighter early-type stars in the Magellanic Clouds.

I thank the Princeton University Copernicus team for use of the U2 spectra and Dr. Chet Opal for computing and plotting the resolution-degraded spectra.

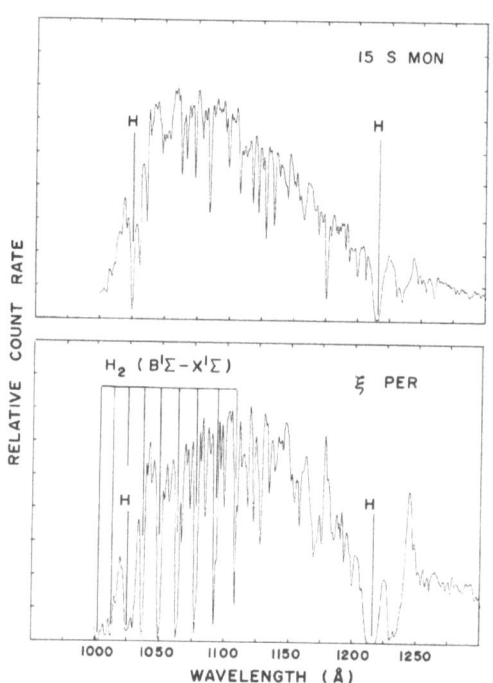

Comparison of Copernicus U2 spectra, computer-degraded to 2 Å spectral resolution (equivalent to that of the proposed objective-grating survey), of two stars: (Top) 15 Mon, a nearly unreddened O7 star (E(B-V)=.07) and a reddened O7 star, ξ Persei (E(B-V)=0.32). Note the strong interstellar H_2 absorptions in the latter spectrum.

FUTURE POSSIBILITIES FOR ULTRAVIOLET OBSERVATIONS OF INTERSTELLAR MOLECULES

George R. Carruthers
Naval Research Laboratory, Washington, D.C. 20375

Ultraviolet observations of interstellar molecules are currently being obtained primarily with two astronomical satellites, *Copernicus* (OAO-3) and *International Ultraviolet Explorer* (IUE). The former covers the wavelength range down to 900 Å with 0.05 or 0.2 Å resolution. IUE can observe fainter or more highly reddened stars than *Copernicus*, with 0.1 Å resolution, but its wavelength range does not include the resonance absorptions of the important interstellar molecules H_2, HD, and N_2.

The only currently approved future investigation which will make a major contribution to ultraviolet observations of interstellar molecules is the High Resolution Spectrograph (HRS) on the Space Telescope, currently planned for launch in late 1983. The HRS will have spectral resolution modes of 2×10^4 (equivalent to *Copernicus*) and 1×10^5 (five times better than *Copernicus*). The wavelength range accessible to the HRS is 1100-3200 Å. It will be able to observe much more distant and/or more highly reddened stars than previously accessible (about V=15 at medium resolution, and V=12 at high resolution, unreddened AO V stars at 2400 Å). The HRS will be able to observe interstellar CO, OH, and at least the (0,0) and (1,0) Lyman bands of H_2. Other molecules which may be observable include H_2O, C_2, SiO, CS, and CH_2. The high-resolution mode will allow study of the absorption line profiles and velocity shifts, and of the rotational distributions in the bands of the heavier molecules. The high sensitivity in the medium-resolution mode will allow observations of some stars with color excesses as large as E(B-V)=2, vs a maximum of about 0.4 for *Copernicus*.

Improvements in the wavelength range 900-1100 Å, which includes most of the H_2, HD, and N_2 transitions, will require a spectrograph and telescope which have optical reflective coatings optimized for this wavelength range, and the use of windowless detectors. Two possibilities are (a) a high-resolution Rowland spectrograph with the *Starlab* telescope for Spacelab, which has been studied by NASA; and (b) an objective-echelle-grating, high-resolution spectrograph proposed for Spacelab by Princeton University. The former instrument would have

B. H. Andrew (ed.), Interstellar Molecules, 613-614.
Copyright © 1980 by the IAU.

$R \sim 4 \times 10^4$ and, with the 1-meter aperture of *Starlab*, and integration times of 30 minutes, could observe (with comparable S/N) stars at least 4 magnitudes fainter than *Copernicus*. The latter instrument could obtain at resolution 1.3×10^5 spectra over the 925-1315 Å range of stars as faint or fainter than observable with *Copernicus*. This improvement in resolution would allow studies of line profiles, and permit detection of weaker lines than possible with *Copernicus*. An alternative, low-resolution (R=1000) mode using only the concave cross-dispersing grating Wadsworth mode, could reach stars as faint as V=12. This would allow, for example, measurements of interstellar H_2 toward the brighter early-type stars in the Magellanic Clouds.

Finally, the possibility of wide-field, objective-grating surveys of interstellar atomic and molecular hydrogen at low (R=500 to 1000) resolution with a far-ultraviolet Schmidt camera in Spacelab missions is discussed elsewhere in this volume.

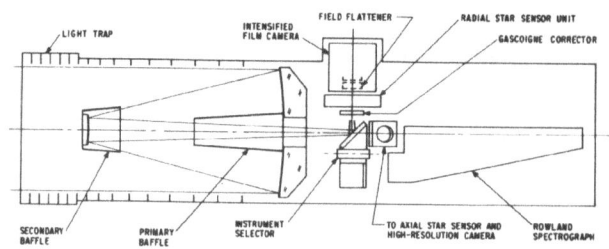

Fig. 1.   The *Starlab* 1-meter UV/Optical Facility Telescope for Spacelab.

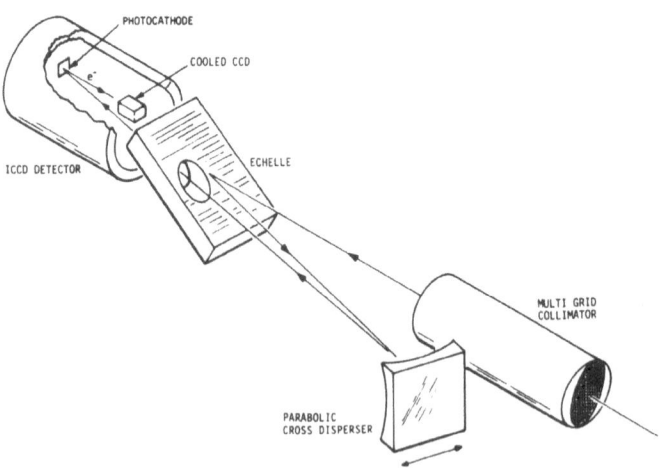

Fig. 2.   The Princeton IMAPS (Interstellar Medium Absorption Profile Spectrograph) proposed for Spacelab.

MASER AMPLIFIERS

E. Kollberg
Chalmers Institute of Technology
S-402 20 Göteborg 5, Sweden

When the lowest possible noise is required the maser amplifier remains the unquestionable amplifier choice for moderate bandwidth microwave and long wavelength millimeterwave receivers. The reason for the outstanding low noise properties of the maser is partly that it is cooled to a few degrees K ($\simeq$ 4 K) but also that the amplification process is the most fundamental one, i.e. amplification takes place when quanta (photons) are added directly to the signal field by stimulated emission of radiation from energetically excited particles.

As practically useful devices masers are today used between 1 GHz and 40 GHz at various places around the globe. In Table 1 data are given for a few typical masers in use in radio astronomy.

Table 1. Typical Maser Data

Tunable freq. range GHz	Inst. bandw. MHz	Receiver Noise temp. K	Active material	Operated by
3.2 - 3.5	17	< 10	rutile	Onsala
18 - 26	240	< 15	ruby	NRAO
29 - 35	60	$\simeq$ 35	rutile	Onsala

The noise temperature of a maser measured at the input port of the interaction circuit is typically only a few degrees K. It is therefore very important to design the waveguide section between the horn-antenna and the amplifier for minimum loss i.e. minimum noise contribution.

Masers operating up to about 50 GHz should be possible with existing techniques and materials. In the longer time perspective there are a number of possible and very challenging questions in maser development.

615

B. H. Andrew (ed.), Interstellar Molecules, 615–617.

- Can rutile masers be made to work properly for frequencies between 50 - 150 GHz? An 80 - 90 GHz rutile maser is near completion at Univ. of Massachusetts.

- Are there other paramagnetic materials that can be used in masers for frequencies above 50 GHz, maybe up to a couple of hundred GHz? There are a great number of materials that are not available as large synthetic crystals today, which, however, may be more efficient than ruby or rutile, not only for millimeter waves but also for microwaves.

- Are there other types of materials where phenomena other than para-magnetism can be utilized for maser amplification of microwaves and millimeter waves?

It should be emphasized that the amount of research done on maser materials and also on the design of maser interaction circuits is not very large, so there is certainly room for surprises in the future.

Parametric amplifiers cooled to 4 K are the only type of amplifier that can compete with the maser where low noise is concerned. Such an amplifier has been proposed for 22 GHz with a bandwidth of about 2 GHz and with a noise temperature which is about 30 K at the room temperature input flange. Cooled FET amplifiers with an input noise temperature below 40 K can be built for a few GHz. However, the noise properties for frequencies above 10 GHz become an order of magnitude worse than for masers. Parametric amplifiers also become less competitive for increasing frequencies.

For frequencies above 50 GHz the predominant receiver type today is the Schottky barrier diode mixer. In Table 2 are listed data for existing cooled mixer receivers for about 100 GHz and for room-temperature mixer receivers for 230 GHz. The figures within parentheses indicate what might be obtainable within one or two years using improved Schottky-barrier diodes recently developed at Bell Laboratories and the University of Virginia.

Two different types of superconducting mixers are worked on, the Josephson mixer and the quasiparticle mixer. The latter type particularly is receiving a lot of attention and there is hope that this mixer repre-sents a breakthrough for low noise receivers for short millimeter and sub-millimeter waves. Experiments carried out at Bell Laboratories and University of California indicate noise properties around 100 GHz superior to those of Schottky mixers, and in Table 2 are given probable single sideband data that may be obtained within a year or so. It should be pointed out that the superconducting mixer is operated with a very low pump power, which is a great advantage at very high frequencies. The bandwidths are limited by the if-amplifier used, which is a parametric amplifier for the Schottky mixer and a FET amplifier for the super-conducting mixers. To obtain the system noise temperature one has to add the atmospheric and antenna noise, which add 50 - 100 K around 100 GHz.

Table 2.

Mixer type	Freq. GHz	Bandwidth MHz	Rec. Noise K SSB
Cooled Schottky	$\sim$ 100	$\sim$ 500	$\sim$ 300 ($\sim$ 200)
Room Temp. Schottky	$\sim$ 240	$\sim$ 500	$\sim$ 2000
Josephson	$\sim$ 100	($\gtrsim$ 200)	($\sim$ 100 - 150)
Quasi part.	$\sim$ 100	($\gtrsim$ 200)	($\sim$ 150 - 200)

DISCUSSION FOLLOWING KOLLBERG

Morimoto: What limits the bandwidth of the various types of front ends?

Kollberg: The bandwidth of the maser is limited by the natural linewidth of the paramagnetic transition used. It is possible to increase the maser bandwidth by broadening the paramagnetic resonance line using a tuning magnetic field which is inhomogeneous (staggertuned) over the maser crystal, as is done with the maser in the 18-26 GHz band listed in Table 1. The bandwidth of the various mixers is mainly limited by the if-amplifiers used. For the Schottky diode mixers one usually uses a parametric amplifier with about 500 MHz bandwidth. The superconducting mixers are easily saturated by the pump power of a parametric amplifier and therefore for these mixers one uses FET amplifiers with about 100 MHz bandwidth (which might be improved in the future).

Thaddeus: I would like to ask Dr. Phillips at what level the quasiparticle mixer is expected to saturate as compared to the Josephson-junction mixer.

Phillips: It should be noted that the InSb mixer operates to 500 GHz with noise temperatures of 200-500 K, but with a bandwidth of only $\sim$2 MHz. Josephson-junction mixers require a prefilter of about 2-5 GHz to prevent saturation. They saturate at a power about 10^5 lower than the quasiparticle photon-assisted tunneling diodes.

WIDEBAND SPECTROMETERS FOR MILLIMETRE WAVELENGTHS

B.J. Robinson
Division of Radiophysics, CSIRO, Sydney, Australia

FREQUENCY RESOLUTION AND COVERAGE NEEDED

Millimetre-wave spectral lines show a wide range of linewidths, from 0.1 km s^{-1} in dark clouds to 30 km s^{-1} in some massive molecular clouds; at 100 GHz the corresponding Doppler widths range from 33 kHz to 10 MHz. SiO masers at 86 GHz have two lines typically 1 MHz wide spaced by 6 MHz. For most applications a resolution of 100 or 250 kHz is suitable.

Frequency coverage is also important. 115 GHz CO observations near the galactic centre call for a coverage of 300 km s^{-1} ≡ 115 MHz. A wide frequency coverage has also proved valuable for line searches, allowing simultaneous observations of known lines near the search frequency. Wide coverage is needed for extragalactic observations, ideally 500 km s^{-1}.

How do the filter bank, the digital correlator and the acousto-optical spectrograph satisfy the frequency resolution and coverage implied in these examples, or corresponding values scaled for other frequencies?

FILTER BANKS

Many observatories have long experience with filter banks, which have excellent performance in terms of sensitivity stability and flexibility. The NRAO design (Mauzy, 1974) is widely used; at Kitt Peak the options available are:

Frequency Resolution	Frequency Coverage
1 MHz	256 MHz
500 kHz	128 MHz
250 kHz	64 MHz
100 kHz	25.6 MHz
30 kHz	3 MHz

B. H. Andrew (ed.), Interstellar Molecules, 619–623.

The 250 kHz filters are widely used, offering 0.75 km s^{-1} resolution and 192 km s^{-1} coverage at 100 GHz. The Kitt Peak filter bank is normally used in the total-power mode with an off-source reference taken at intervals of a few minutes. The noise has been found to decrease as (time)$^{-\frac{1}{2}}$ for up to 24 h.

DIGITAL CORRELATORS

With modern ICs a 1000-channel digital correlator is not difficult to design. For one-bit sampling there is a $\pi/2$ degradation in sensitivity to 64% of filter bank sensitivity. Noise decreases as (time)$^{-\frac{1}{2}}$ for many hours. The resolution can be changed very simply, but the coverage is limited to half the sampling frequency. The special IC chips built for the VLA can operate at a sampling frequency of 160 MHz; for 1024-channels this would allow (at 100 GHz) a resolution of 0.5 km s^{-1} and a coverage of 240 km s^{-1}. A 1024-channel correlator (Mark IV) using the VLA chips is currently being tested at NRAO (Shalloway, private communication).

The asynchronous correlator proposed by Ables is being developed in Sydney (Frater, private communication). This would greatly simplify the construction of a high-speed correlator since it is not necessary to distribute the clock pulses. The clock signal for each module is derived from the preceding module, so each operates at its own local clock phase.

ACOUSTO-OPTICAL SPECTROGRAPH

The acousto-optical spectrograph (AOS) was invented by Lambert (1962). It has taken many years to be developed into an operational spectrometer for millimetre wavelengths, mainly because of problems of vibration and thermal instability. There are operational AOS in Australia (Milne and Cole, 1979) and Japan (Kaifu et al., 1977). The performance of these AOS is listed in Table I.

TABLE I. AOS PERFORMANCE

Parameter	AOS model			
	CSIRO 1	CSIRO 2	Tokyo 1	Tokyo 2
Ultrasonic material	Quartz	LiNbO$_3$	TeO$_2$	TeO$_2$
Centre frequency (MHz)	135	405	65	360
Overall bandwidth (MHz)	90	270	41	200
Number of channels	512	256	1728	1728
Frequency resolution (kHz)	240	1050*	38	250
I.F. drive power (mW)	200	500	10	500
Frequency drift (kHz h^{-1})	7	–	1	2
Linearity	<1%	–	<2%	<3%
Readout noise (dB below saturation)	-24	–	–	–

*Set by number of channels; expected maximum resolution about 500 kHz (interaction length 1.8 μs). This AOS has so far only been used for pulsar and scintillation observations.

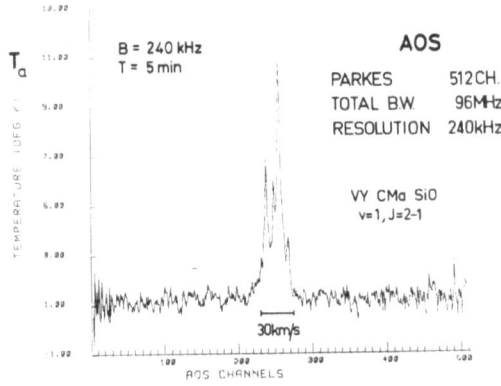

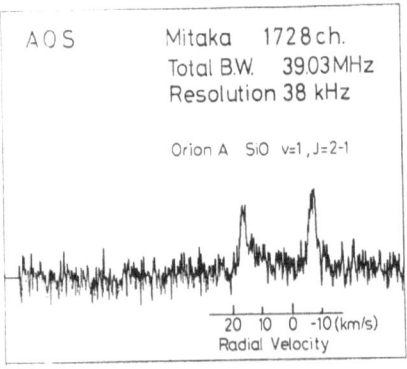

Fig. 1 - 86 GHz SiO maser emission
from VY CMa recorded with 512-
channel AOS on 17-m telescope.

Fig. 2 - 86 GHz SiO maser emission
from Orion with 1728-channel AOS
on 6-m telescope.

Spectra of 86 GHz SiO masers obtained with CSIRO 1 and Tokyo 1 are
shown in Figures 1 and 2. In Figure 1 the ultrasonic modulator response
is down 9 and 6 dB at the left and right ends of the spectrum and readout
noise is becoming significant. With beam switching at a 2 s rate the
sensitivity of CSIRO 1 over the central 80 MHz band (3 dB modulator
response) has been measured as 1.5 times that of the Parkes digital
correlator under the same operating conditions (an ideal AOS would be a
factor $(0.64)^{-1} = 1.56$ better than an ideal one-bit correlator). The
noise of CSIRO 1 decreases as $(time)^{-\frac{1}{2}}$ for an interval of at least 3 h.

For long switching times (≥ 10 s) a fraction of the profiles recorded with CSIRO 1 have sloping or curved baselines. The curvature
appears when T_{sky} is varying rapidly. Tests are under way to establish
whether the link between baseline shape and T_{sky} variations results
from a relatively slow sequential readout of the photodiodes.

Maintenance on an AOS is extremely simple because of the small
number of components (laser, beam-spreading optics, ultrasonic modulator,
Fourier transform lens, photodiode array and serial readout circuitry).
CSIRO 1 has required no maintenance in the last year of operation. The
reliability and ruggedness of the photodiode array are extremely high.

A LiNbO$_3$ ultrasonic modulator with a bandwidth of 1 GHz has been
made by ITEK (Hecht, 1977). The same paper describes a GaP modulator
with a bandwidth of 630 MHz with much higher efficiency. The number N
of resolution elements is limited by the transit time of the ultrasonic
wave along the aperture; for the 1 GHz modulator N $\doteq$ 1100, while for
the 630 MHz modulator N $\doteq$ 400.

SPECTROMETER COSTS

Figures obtained from various groups lead to approximate component costs per channel and man-years for construction as follows:

Type	Cost per channel	Man years
NRAO-type filter bank	$50 to $80	2
Digital correlator	$40 to $50	7*
Acousto-optical spectrograph	$15 to $30	2

*Includes design effort.

These cost estimates do not include a minicomputer or interfacing hardware. Because the AOS has serial readout, interfacing to a computer is particularly simple.

REFERENCES

Hecht, D.L.: 1977, IEEE Ultrasonics Symposium Proceedings.
Kaifu, N., Ukita, N., Chikada, Y., and Miyaji, T.: 1977, Publ. Astron. Soc. Japan 29, p. 429.
Lambert, L.E.: 1962, I.R.E. Int. Conv. Rec. 10, Pt. 6, p. 69.
Mauzy, B.: 1974, NRAO Electronics Division Report No. 146.
Milne, D.K., and Cole, T.W.: 1979, Proc. Inst. Radio Electron. Eng. Aust. 40, p. 43.

DISCUSSION FOLLOWING ROBINSON

Gillespie: What is the dynamic range of the acousto-optical spectrometers?

Robinson: For the AOS used on the 4 metre telescope at Epping the dynamic range from readout noise to photodiode saturation is 24 db. This gives a 5 dB margin below the $\sqrt{B\tau}$ noise with B=240 kHz and τ=30 ms. The margin could be improved by reducing the interval τ between readouts of the photodiodes; a compensating increase in I.F. drive power would be required. The linearity of the device is very good but when saturation occurs there is a very sharp cut-off. A variety of attenuators must be used to keep the signal within range if a number of services of greatly different intensity are being observed.

Linke: Is it possible to build an AOS with adjustable resolutions?

Robinson: The resolution is the reciprocal of the transit time of the acoustic wave along the aperture. We have changed the resolution by changing the Bragg cell (e.g. Cole and Ables, Astron. Astrophys. 34, 149 (1974)). At Tokyo Astronomical Observatory they have built *two* AOS giving a 6:1 ratio of resolution (see Kaifu et al. 1977). The AOS thus lacks the flexibility in resolution of the digital correlator.

Morimoto: In our case we use the AOS at maximum resolution and maximum bandwidth.

Robinson: I understand that an AOS is also in use at Kisarazu Technical College for a CO survey. This AOS has 230 MHz bandwidth, but the resolution is only 2 MHz, while 250 kHz should be attainable with more readout channels.

Bieging: A number of firms are developing integrated optical circuits to incorporate the components of an acousto-optic spectrograph onto a monolithic silicon substrate, including detectors. If this work is successful, it will almost certainly reduce dramatically the cost per channel, compared to acousto-optic spectrographs constructed with discrete components.

Robinson: This is a most significant development. One fringe benefit of the compact monolithic construction should be a reduced sensitivity to vibration, which has been a problem with the "optical bench" type of AOS.

R. Wilson: I would add one device to your list. P.S. Henry has built a spectrum expander which in effect allows one to have a filter bank with simply adjustable resolution. Our expander lets us use 128 channels of our 0.25 MHz per channel filter bank at resolutions of 12.5 kHz to 100 kHz per cahnnel. The device is relatively inexpensive to build.

Robinson: Henry's device (Rev. Sc. Instrum. 50, 185 (1979)) uses a recirculating loop memory to achieve higher resolution when only a coarse filter bank is available. We await a real-time device to expand the bandwidth instead!

T. Wilson: The NRAO autocorrelator achieves its 80 MHz bandwidth by offsetting 2 samplers, which are driven at 80 MHz, by 12.5 μs. This causes a reduction in the number of channels, from 1024 to 512. At bandwidths of 40 MHz and less, 1024 channels are available.

Robinson: You are quite correct. My figures are in error. On the subject of digital correlators, I should add that in Sydney a scheme was devised by A. Bos to build a wideband digital correlator using a high-speed sampler with the samples multiplexed down parallel chains of low-speed logic. However, the sensitivity is degraded. Dr. Morimoto informs me that in Tokyo a sophisticated hard-wired FFT system (see Yen, Astron. Astrophys. Suppl. 15, 483 (1974)) has been designed to process the output of a 160 MHz sampler with 10 MHz logic without loss of sensitivity.

ACOUSTO-OPTIC RADIOSPRECTROMETERS FOR MM-WAVE SPECTROSCOPY

Yoshihiro Chikada[1], Nobuharu Ukita[1], Junji Inatani[1],
Norio Kaifu[2], and Shinji Kodaira[3]
[1]Department of Astronomy, University of Tokyo
[2]Tokyo Astronomical Observatory
[3]Kisarazu Technical College

In mm-wave spectroscopy, the acousto-optic spectrometer (AOS hereafter) compares favourably with conventional spectrometers such as filter-
banks or autocorrelators. Although its frequency resolution is limited
to 20 kHz and its instantaneous bandwidth is limited to several hundred
MHz, it has 10^3 resolvable points. Because of the simplicity and stability of the AOS, it is not difficult to construct and maintain a system
of 10^4 resolvable points by the parallel operation of spectrometers. At
Tokyo Astronomical Observatory (TAO) a new AOS has been constructed which
is equipped with two different types of TeO_2 AO deflectors and provides
high resolution (38 kHz) or wide band (220 MHz) spectra. It has 1728
frequency channels (figure 1). Another AOS has been built at Kisarazu
Technical College (KTC) for a CO survey at 115 GHz. It has 256 frequency
channels and a bandwidth of 230 MHz (figure 2).

The principle of the operation is simple. It was first described
by Lambert (1962). The intermediate frequency signal from the receiver
is fed to a piezo-electric transducer ($LiNbO_3$) bonded on AO material
(TeO_2) and is converted to an ultrasonic wave. The wave travels through
the material, maintaining the spectral information of the radio frequency
signal. A He-Ne laser beam is diffracted by the ultrasonic wave and is
focused on an image-sensor to produce a spectral image. The detected
light intensity as a function of diffraction angle is proportional to
the radio-wave power as a function of frequency. The performance of the
spectrometer is determined by the AO light deflector and the image-sensing
circuitry. The piezo-electric transducer and the AO material limit the
band width and the frequency resolution (Uchida and Niizeki 1973; Kaifu
et al. 1977). The TeO_2 deflector has the high efficiency of the AO interaction and very slow sound velocity. This makes it possible to build a
compact and therefore stable spectrometer with 10^3 resolvable points.
The image-sensing circuitry itself contains undesirable noise sources.
The noise arises mainly from the photo-detector array, which has a
signal-to-noise voltage ratio (R) of 300:1 to 1000:1, measured after
square-law detection. In order not to degrade the S/N ratio of the
receiver, the following inequality must be satisfied (Chikada 1978):

B. H. Andrew (ed.), Interstellar Molecules, 625–626.

$$D^2 \times \Delta f \times \tau \ll R^2,$$

where Δf and $1/\tau$ are the frequency resolution and the frame rate of the image-sensor scanning, respectively, and D stands for the dynamic range over the total bandwidth. The pass-band is not flat over the full frequency span and, if a deviation of 3 dB is allowed, D^2 becomes greater than 4. The performances of the spectrometers are summarized in the table below.

		TAO		KTC
Bandwidth	(MHz)	39	220	230
Resolution	(kHz)	38	275	2000
Freq. channels	(bins)	1728	1728	256
Freq. spacing	(kHz)	23	127	1000
S/N degradation*	(dB)		1	1

*Worst case with weak signal.

The Large Radio Telescope Project is now at the stage of constructing antennas. The acousto-optic spectrometers are to form a part of a receiver back-end system with an instantaneous bandwidth of 2 GHz and 10^4 frequency channels.

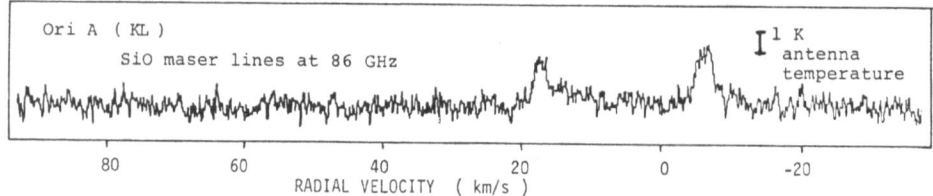

Figure 1.   SiO maser lines observed with the 1728 channel AOS of TAO.

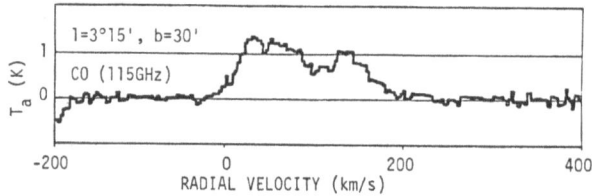

Figure 2.   CO emission observed with the 256 channel AOS of KTC.

REFERENCES

Chicada, Y.: 1978, private communication (in Japanese).
Kaifu, N., Ukita, N., Chikada, Y., and Miyaji, T.: 1977, Publ. Astron. Soc. Japan 29, pp. 429-435.
Lambert, L.B.: 1962, Inst. Radio Engrs. Intl. Conv. Rec. Pt. 6,10,p. 69.
Uchida, N., and Niizeki, N.: 1973, Proc. Inst. Elec. Electron. Engrs. 61, p. 1073.

FUTURE SPECTRAL LINE RESEARCH WITH THE VLA

K.J. Johnston
E.O. Hulbert Center for Space Research, Naval Research
Laboratory, Washington, D.C. 20375

The Very Large Array (VLA) is presently being constructed on the Plains of San Augustine near Socorro, New Mexico by the National Radio Astronomy Observatory. The purpose for which this instrument is being constructed is to produce "radio images" of resolution comparable to that of large optical telescopes. There have already been several "test" observations, some successful, using the partially completed instrument to study the molecular species of OH, H_2O, and NH_3 with wide bandwidths (200-1500 kHz).

Let me briefly describe the VLA for those unfamiliar with it. When completed the VLA will consist of twenty-seven 25-meter antennas arranged in an equiangular wye configuration. The collecting area of these telescopes is equivalent to a 130 meter telescope. One arm of the wye is almost due north (azimuth 354°59'42"). Each arm of the wye will have nine antennas arranged in a power law configuration to give good coverage in the U-V plane. The antennas can be moved along the wye to give four configurations, denoted by A, B, C, and D, of a maximum length from the center of the wye of approximately 21, 6.4, 1.9, and 0.6 kilometers respectively. This allows a range in resolution of 35:1. The operating frequencies of the array are in the L, C, U, and K microwave bands. For example, the resolution at L band (1600 MHz) will vary from 1".1 (A configuration) to 37" (D configuration) while at K band (22235 MHz) the corresponding resolutions will be 0".08 and 2".7. Putting this into galactic perspective, operation at L and K bands in the A array will give spatial resolutions varying from 10^{15} to 10^{14} cm respectively for stellar objects such as OH and H_2O masing late-type stars located at 100 pc. For the large molecular cloud complexes, located $\sim$ 5 Kpc, the more sensitive D or C configuration will yield resolutions of 3-9 x 10^{17} cm for studies of H_2CO (4830 MHz) and of 0.6-2 x 10^{17} cm for studies of NH_3 ($\sim$ 23 GHz).

There are four IF channels which in the normal operating mode carry the opposite senses of circular polarization and their cross products. The signals from the antennas are transmitted via a waveguide transmission systems to the control building where the signals are processed.

B. H. Andrew (ed.), Interstellar Molecules, 627–629.

The maximum IF bandwidth is 50 MHz. The bandwidth may be reduced and by use of a recirculating correlator the number of channels may be increased. At 50 MHz IF bandwidth 16 channels are available. The number of channels doubles each time the bandwidth is halved, reaching a maximum of 512 at 1.5625 MHz. The minimum IF bandwidth will be 97.7 kHz. A present limitation of 256 channels is set by the capacity of the asynchronous computer. Two hundred and fifty-six is the total number of channels available for all four IF channels. If one wishes to observe polarization using all four IF channels, the resolution will be 64 channels at present. This corresponds to a maximum frequency resolution ranging from 0.3-0.02 km s^{-1} at L and K bands, respectively, using the narrow 97.7 kHz IF bandwidth.

Although the operating frequencies were chosen primarily to include protected radio bands, they contain many spectral lines. In the L-band waveguide range (1340-1730 MHz) atomic hydrogen, together with the molecules of formamide, formic acid and OH, is present. In the C band waveguide range (4500-5000 MHz) OH and formamide are present along with formaldehyde. At U band (14400-15000) there is again formaldehyde, while in the K band range (22000-24000 MHz) there is water and ammonia. In all four frequency bands the recombination lines of hydrogen and helium can be observed, ranging from the H169α to H65α hydrogen lines.

The capabilities of the VLA should lead to improved studies of masing as well as thermal lines. Studies of masing regions of OH and H_2O will lead to improved positions exceeding 0".1 in accuracy for these objects. For the ground state OH ($2\pi_{3/2}$ J = 3/2) and H_2O masers the positional accuracy will probably be limited only by the structure of the individual masers themselves, and will perhaps allow the measurement of the parallax and proper motion of late-type stars along with motions caused by stellar companions. Many of these masers such as the H_2O masers associated with late-type stars are large in apparent size (10^{14} cm). VLA measurements will allow the study of all the maser radiation. In the past, VLBI observations, which are sensitive to components of size 0".1-0".001, have been unable to detect large features.

Study of atomic hydrogen will lead to knowledge of galactic structure on the 10^{15}-10^{17} cm scales. These studies will be complimented by studies of high velocity clouds and OH absorption.

Molecular species such as formaldehyde and ammonia will yield information on temperatures of molecular clouds without worry about filling factors. With regard to ammonia in the galactic center, the question of clumping or size structure on the arc second scale can be answered.

The kinematics of clouds may be studied via the recombination lines or molecular transitions. The hydrogen/helium abundance may be studied in detail in regions such as the galactic center at a resolution of 10^{16} cm. The recombination lines of carbon may also be studied. The positional accuracy will exceed one arc second in all cases.

In addition to galactic studies, the sensitivity of the VLA will make it the premier instrument for studying extragalactic radiation. Combining all the antennae in a single dish mode will be equivalent to a 130 meter instrument, making it the most sensitive antenna at frequencies as high as K band.

In summary the VLA will allow significant advances in the study of molecular clouds through its increased sensitivity and resolution allowing direct comparison of transitions found in the radio range of the spectrum with those in the infrared, optical and UV.

DISCUSSION FOLLOWING JOHNSTON

Snyder: Has the VLA been used successfully for any molecular observations? If so, can you comment on the positional accuracy of the molecular observations?

Johnston: The VLA has been used successfully to observe OH masers at 1612 MHz. These sources are very intense, making them easy to detect when observed with a bandwidth much larger than the inherent linewidth. After September 1979 narrower bandwidths will be available, leading to an operational spectral line system by the end of 1980. The current accuracy of VLA positions should be calculable from observations of OH masers in May 1979 by Reid, Moran and myself. The rms error from least squares fits to position from the observed phases was about 0."1, but we have not yet evaluated the systematic errors in the observations. The positional accuracy of the VLA when it is in full operation should easily be better than 0."1 for maser sources and probably should approach 0."01.

T. Wilson: What are the plans for the VLA spectral line system?

Johnston: In my talk I limited myself to describing the basic operating system, making no reference to the processing of the data after the observations are completed. By the end of 1980 there should be a limited capacity at the VLA for making maps of spectral line sources. Plans are now being formulated for handling the large volume of data resulting from spectral line observations. Editing of data and the displaying and cleaning of maps of specific lines at specific radial velocities will be possible.

NEW EXPERIMENTAL POSSIBILITIES AND FUTURE PROSPECTS FOR 1-5μm INFRARED SPECTROSCOPY OF INTERSTELLAR MOLECULES

Donald N. B. Hall
Kitt Peak National Observatory*

During the past week we have heard a considerable number of papers dealing with spectroscopic observations in the 1-5μm region of the infrared. I predict that as instruments and detectors continue to improve, such observations will play a major role in the study of interstellar molecules in both molecular clouds and circumstellar clouds around evolved stars. Fourier Transform spectrometers and Fabry Perot interferometers have already yielded spectra of such sources with spectral resolution and radial velocity precision fully comparable to millimeter wave observations. The critical need is to improve the limiting magnitude of such observations by 3-5 stellar magnitudes so that one can study large numbers of sources rather than the few brightest in each class.

Rapid advances in the technology of infrared self-scanned array detectors hold promise of such improvements. In particular, 32 x 32 pixel InSb Charge Injection Devices are now available commercially and, even with present dark current and read noise, will provide a substantial improvement in the 3-5μm range. There is an excellent chance that further developments will extend this down to 1.5μm or even 1.0μm. I expect that these detectors will be used with either two or three etalon Fabry Perot interferometers (for studies of extended emission line sources) or an echelle spectrograph in series with some form of interferometer (for studies of point sources and the interstellar material seen in absorption against them). In either case the entire instrument would be cooled to cryogenic temperatures and used with a low background, infrared optimized telescope.

If this can all be realized then the improvement in observational capability is enormous. At 4.6μm one would be able to observe pre-main sequence objects such as the Becklin Neugebauer source and evolved sources such as α Orionis or IRC+10°216 anywhere in the galaxy.

*Operated by the Association of Universities for Research in Astronomy Inc., under contract with the National Science Foundation.

B. H. Andrew (ed.), Interstellar Molecules, 631–632.

Similarly one would observe a number of sources embedded in nearby molecular clouds such as OMC 1 and thus probe the kinematics and physical conditions in the intervening cloud material. Such observations are crucial to our understanding of both pre and post main sequence evolution and also of molecular clouds and circumstellar shells.

DISCUSSION FOLLOWING HALL

Bok: Areal devices (like the Charge Coupled Device) provide great opportunities for 1 to 3 μm observations of globules with diameters of 3' to 5' and of sections of larger molecular clouds. Using the CCD at 1.2 μ at Cerro Tololo Inter-American Observatory, we recorded in 20 minutes with the 36-inch reflector several stars that could not be seen on a 200-minute hypersensitized IV N photograph taken at the prime focus of the 4-meter reflector. With the proper areal receivers operating at 1 to 3 μm and covering areas of 3' to 5' in the sky, one should be able to obtain a complete census of the distribution of dust (density gradient) inside a globule, and one should moreover detect any imbedded objects or concentrations. Such observations can readily penetrate globules with central values of A_v up to 30 and 40 magnitudes! The near-infrared observers should be able to obtain without special effort pictures of distributions comparable in quality to photographs of the stars in globular clusters. Such researches can be done from earth (or high-flying airplanes) well before the Space Telescope goes into full operation. Data of this sort should prove invaluable to radio and far-infrared astronomers engaged upon the study of molecular clouds.

Chaffee: What read-out noise do you think will be achieved with the IR CID's that you mentioned?

Hall: Elements in a prototype linear InSb CID have been operated under laboratory conditions with 50 electron read-noise; reduction to a figure of 20 electrons seems possible. In the two dimensional arrays reduction by a further factor ∿5 may be feasible using multiple non-destructive reads, as is now done with Si CID's.

Gillespie: It may be of interest to the radio astronomers to know that the array receivers are viable in the 200 to 300 GHz region using an InSb array of the type proposed by Phillips and myself recently.

A 10 MICRON HETERODYNE RECEIVER FOR ULTRA HIGH RESOLUTION ASTRONOMICAL SPECTROSCOPY

D. Buhl, G. Chin, J. Faris, T. Kostiuk and M.J. Mumma
Infrared and Radio Astronomy Branch
NASA/Goddard Space Flight Center
Greenbelt, MD 20771
D. Zipoy
Department of Astronomy, University of Maryland

Infrared heterodyne spectroscopy is extremely useful in determining by precise measurements of atomic and molecular infrared line spectra the molecular abundances, velocity structure, and excitation conditions of the interstellar medium, and the temperature- and pressure-profiles of planetary atmospheres (cf. Kostiuk et al. 1977, Betz et al. 1976, 1977, Abbas et al. 1978).

Heterodyne detection is achieved by combining infrared source radiation with a coherent laser local oscillator output at a non-linear detector - e.g. HgCdTe photo-mixer. The difference frequency or inter-mediate-frequency (IF) recovers the spectral characteristics of the source radiation and may be measured by radio frequency techniques to an arbitrarily high resolving power.

Our present CO_2 laser heterodyne spectrometer (Fig. 1) includes several improvements over our first generation system described by Mumma et al. 1978 and Kostiuk et al. 1977: the present system uses reflective optics to eliminate re-focusing at different wavelengths, and the local oscillator is a line-center-stabilized isotopic CO_2 laser built at GSFC. Easy and rapid selection of over 50 transitions per isotope of CO_2 is made possible by a tunable diffraction grating. A GSFC built visible and infrared star-tracker can be used for automatic guiding on astrono-nomical sources. The IF (0-1.6 GHz) from the HgCdTe photomixer is analysed by a 128-channel filter bank. A tunable bank of 64 5-MHz filters and a fixed set of 64 25-MHz RF filters provide resolving powers of $\sim 10^6$-10^7 and velocity resolution of 50-250 m/sec. The output of the 128 filters is synchronously detected, integrated, multiplexed and stored as signal (S) and reference (R) in a buffer memory for the desired inte-gration period. The result is presented as (S-R)/R, and stored on floppy disks for subsequent data reduction. A CRT graphics terminal provides on-line display with the printer and hard-copy unit providing more permanent records (Fig. 2).

The observations were taken on the main telescope of the McMath Solar Observatory at Kitt Peak National Observatory in June 1979. Line

B. H. Andrew (ed.), Interstellar Molecules, 633–636.

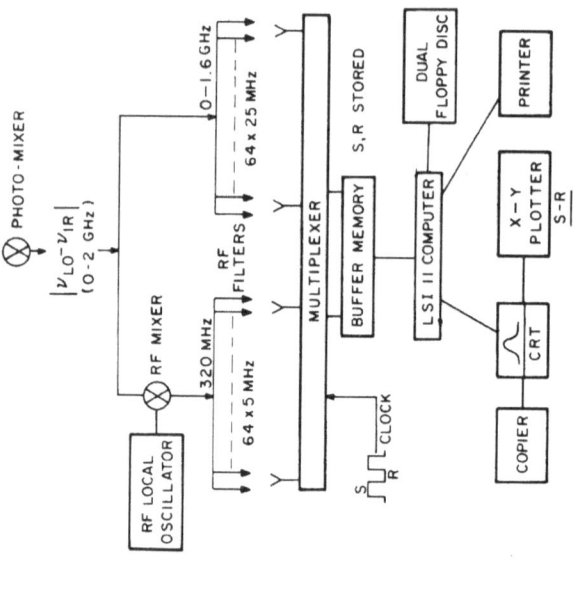

Fig. 2. The receiver IF(0-2 GHz) is analysed by 128 filters in sets of 64 25-MHz-wide and a tunable set of 64 5-MHz-wide banks. The RF signals are digitized and stored in signal (S) and reference (R) memories in a digital multiplexor. The output of the multiplexor is read out to a LSI-11 microcomputer system where (S-R)/R is calculated, stored on floppy disks, and displayed.

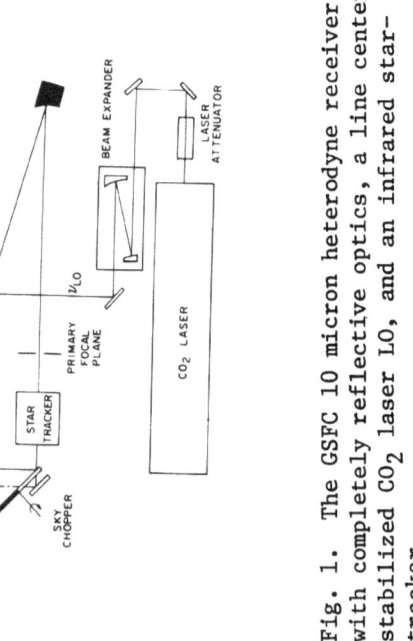

Fig. 1. The GSFC 10 micron heterodyne receiver with completely reflective optics, a line center stabilized CO_2 laser LO, and an infrared star-tracker.

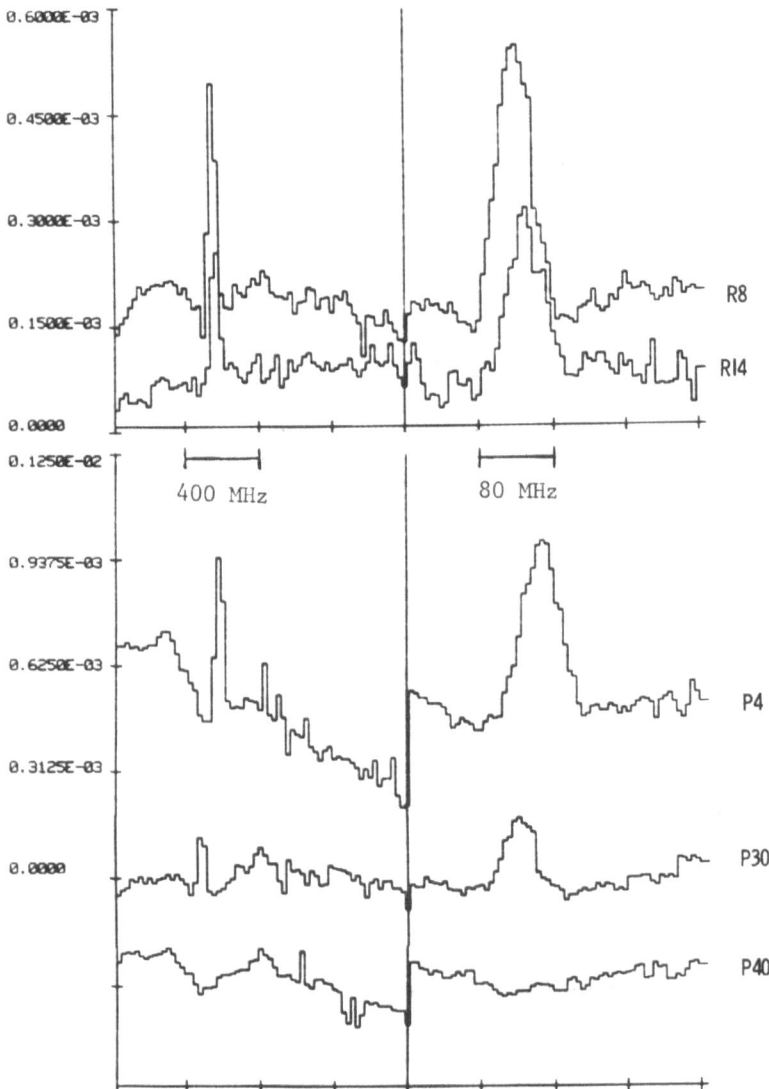

Fig. 3. Observations of CO_2 on Venus. These preliminary
results with the GSFC 10 micron heterodyne system show
that the central core CO_2 emission varies with J values.
The lower Venusian atmosphere can be sampled in scans with
an absence of core emission while scans with core emission
may be used to fit the temperature and pressure of the
Venusian mesosphere and upper atmosphere.

profiles from the R8, R14, and P4 through P44 transitions of CO_2 were observed from the Venusian atmosphere. Fig. 3 shows several of the observed spectra. These scans are uncorrected for atmospheric attenuation or Doppler shifts of individual unstacked data. Solar scans at each observed Venusian line were also made as a measure of the amount of telluric attenuation. We see that higher J transitions in the P branch in Venus have weaker or negligible emission cores and may be used to deconvolve the pressure- and temperature-profiles of the lower atmosphere. Lower J transitions in the P branch have sizeable emission cores and are probes of the mesosphere and upper atmosphere (Betz et al. 1976). The R8 and R14 transitions show evidence of a non-thermal spike above a thermal plateau. This thermal pedestal would imply the existence of a mesospheric temperature inversion in the Venusian atmosphere.

The NH_3 transitions in near coincidence with the 626 and 636 laser were searched for in α Sco, R Cas, and α Her. However there was a sizeable continuum (S/N ∿ 10) and no spectral features were observed above the noise. Expected modifications will increase the sensitivity ten-fold at 1.5 GHz.

These Kitt Peak observations with the GSFC infrared heterodyne receiver show the wide spectral coverage, wide mixer and electronics bandwidth and high sensitivity of our system. The manifold of P- and R- branch transitions on Venus will allow us to make theoretical models of the lower, mesospheric, and upper levels of the atmosphere after the data have been corrected for telluric attenuation and Doppler shifts. Improvements underway at GSFC include the development of a (0 - 2.5 GHz) photo-mixer with an increased bandwidth response, a cooled matched low-noise preamplifier and back-end filter, the development of a tunable diode laser local oscillator heterodyne front-end, and the extension of operating wavelengths to 30 μm.

REFERENCES

Abbas, M.M., Kostiuk, T., Mumma, M.J., Buhl, D., Kunde, V., and Brown,
    L.W.: 1978, Geophysical Research Letters 5, 317.
Betz, A.L., Johnson, M.A., McLaren, R.A., and Sutton, E.C.: 1976,
    Astrophys. J. (Letters) 208, L141.
Betz, A.L., McLaren, R.A., Sutton, E.C., and Johnson, M.A.: 1977, Icarus
    30, 650.
Kostiuk, T., Mumma, M.J., Hillman, J.J., Buhl, D., and Brown, L.W.:
    1977, Infrared Physics 17, 431.
Mumma, M.J., Kostiuk, T., and Buhl, D.: 1978, Opt. Eng. 17, 50.

NEW EXPERIMENTAL POSSIBILITIES AND THE FUTURE AT FAR IR WAVELENGTHS

C.H. Townes
Department of Physics, University of California, Berkeley

ABSTRACT

The observation of molecular phenomena in interstellar clouds by spectroscopy in the longer wavelength region of the infrared is discussed. This is a region where sensitivity is almost universally affected by background continua, appropriate techniques may be either coherent or incoherent, and work above most of the atmosphere is required for much of the spectral range. So far, only a very limited amount of molecular work has been done in the far-infrared, but one can expect fairly rapid growth and important results.

The infrared region from a few microns to about one millimeter is full of molecular frequencies and, from the point of view of astronomy, is relatively unexploited. While the microwave region contains a large fraction of the lowest rotational transitions of molecules, the submillimeter range includes the low rotational states of simple hydrides and many higher rotational states of heavier molecules. The medium infrared is, of course, characteristic of the rotation-vibration spectra, which extend between about 2 and 50 microns. Rather hot objects, such as stellar atmospheres or shocked interstellar gases, are needed to excite spectra in the shorter wavelength infrared range. However, the molecules themselves need not necessarily be so hot, since absorption spectra can be observed against a hot continuum. In addition, warm interstellar gas of temperatures in the range of 50 to 100K can excite the far-infrared, in particular rotational transitions of light molecules in the region 50 to 500 microns. So far, a modest number of molecular transitions have been studied in the 10 micron atmospheric window and at shorter wavelengths, and at longer wavelengths a very few lines have been studied in the submillimeter range from platforms above most of the atmosphere. Clearly, there remains to be obtained from the infrared region a great deal of information about interstellar molecules which is of a different nature and complimentary to what we already know from microwave work.

There are, of course, technical difficulties which have heretofore impeded progress. One is the lack of atmospheric transparency except

637

B. H. Andrew (ed.), Interstellar Molecules, 637–644.

for windows at certain well-known wavelength regions. This may be
obviated by airplane, balloon, and satellite work. Another has been
the lack of sensitive detectors. Recently the required detector developments have been getting under way and there is already some success. A third and major difficulty is the competing background radiation from all that is around us. At wavelengths longer than about 10
microns, apparatus such as spectrometers, lenses, and telescopes at
room temperature emit quanta copiously, and so also does the atmosphere.
Detection of astronomically interesting radiation must at times be
achieved in the presence of a background which is as much as 10^6 times
greater than the radiation being detected. While microwave astronomy
has some problems with background radiation, infrared astronomy faces
much more severe ones.

Since my colleague Don Hall has discussed the shorter IR region,
I shall concentrate attention on the longer wavelength range, where
background radiation from objects at normal temperature is omni-present
and dictates the experimental techniques which can be successful. In
this region, the background radiation is proportional to bandwidth,
and hence it is important to use spectral resolution which is as fine
as the line widths to be detected in order to obtain the best discrimination between lines and background. This is true not only because of
interfering background from our apparatus on the Earth, but also in
many cases from interfering continuum of the astronomical objects
themselves.

Ideally, if the background is completely constant, noise is
usually proportional to the square root of the number of quanta received and in this case a Fourier transform spectrometer would be
exactly as sensitive as a very narrow bandpass spectrometer which is
swept in frequency over the same range of the spectrum which might be
observed by Fourier transform techniques. In actuality, the noise is
not of this character. There are systematic fluctuations in background, for example, time or spatial variations in the temperature of
the sky or of the telescope being used, which produce excess noise and
which dominate if the bandwidth is large. Hence in high backgrounds,
Fourier transform spectroscopy, which is so generally valuable in the
shorter infrared region, cannot usefully be employed with its full
potential bandwidth. A narrow band system may itself use Fourier
transform spectroscopy with a narrow band filter, or it may be simply
a very narrow band monochromator which is swept in frequency just
enough to cover a line or a few lines which are to be observed. Thus
gratings and Fabry-Perot systems are frequently favored.

Background radiation in the short infrared region gives fluctuations which are proportional to the square root of the total number of
photons, $\sqrt{n}$, or to the square root of the power incident on the detector. In the microwave region, where Bose-Einstein condensation of
photons is prominent, the noise fluctuations are directly porportional
to incident power, being given, for example, by $kT\sqrt{\Delta\nu}$. The far-infrared
is in the transition region between these, and an exact calculation of

quantum noise fluctuations needs to come from a more general expression. Since this transition behavior is not commonly treated, it is discussed in Appendix I. Usually, near the transition region an estimate from either a $\sqrt{n}$ or $kT\sqrt{\Delta\nu}$ type of approximation is not far wrong.

Bolometers were the detectors which opened up infrared astronomy. However, they are primarily useful for broadband detection and will probably not be very widely used in the future for molecular spectroscopy. Fortunately quantum detectors, photoconductors, and photodiodes have now been developed with good quantum efficiency for selected spectral regions varying from the shortest infrared wavelengths to those as long as about 150 microns. Under favorable conditions, these provide little excess noise above the fundamental quantum noise of the background photons which strike them.

At the longer wavelengths, heterodyne detection is a favorable technique, particularly for high spectral resolution. Of course, for wavelengths longer than one millimeter it is almost universally used. For wavelengths slightly shorter than one millimeter a variety of non-linear elements are used as mixers, including Schottky diodes and other devices common at longer wavelengths. Phillips' interesting hot electron heterodyne system is quite successful in this wavelength range, though its bandwidth is narrower than is desired. Superconducting devices, including superconducting amplifiers, are being developed and already are useful in limited ways. In the shorter wavelength range, from about 100 microns to the shorter infrared, photoconductors and photodiodes allow the possibility of good quality heterodyne detection. A moderate amount of heterodyne molecular spectorscopy has been done in the 10 micron region, as reported here by Betz, and this general style of spectroscopy should be useful for the study of stellar atmospheres at wavelengths as short as four to five microns. In the 100 micron region, there is as yet no very good heterodyne detection system. However, there are promising methods which are being explored, and perhaps good heterodyne detection in this region is not very far off. If it can be achieved, it can be expected to replace direct detection by photoconductors and photodiodes, even though these are relatively successful.

Principles required for sensitive spectroscopy in the longer wavelength infrared may be illustrated by the schematic in Figure 1 of a 100 micron spectrometer built at Berkeley, primarily by Storey and Watson, and used in NASA's Kuiper Observatory. The instrument involves a fixed Fabry-Perot interferometer in tandem with a tunable or variable Fabry-Perot. Plates of these Fabry-Perots are made of fine metal mesh appropriate to a wavelength of about 100 microns. The entire interior of the spectrometer is cooled, so that radiation reaching the detector is essentially limited to that seen by the instrument through the telescope within the narrow band representing the resolution of the instrument. While the variable or tunable Fabry-Perot illustrated in the figure is not at low temperature,

Figure 1: Schematic of the
Berkeley cooled-optics,
narrow band spectrometer
for the 100 micron region,
built by J.Storey and
D.Watson

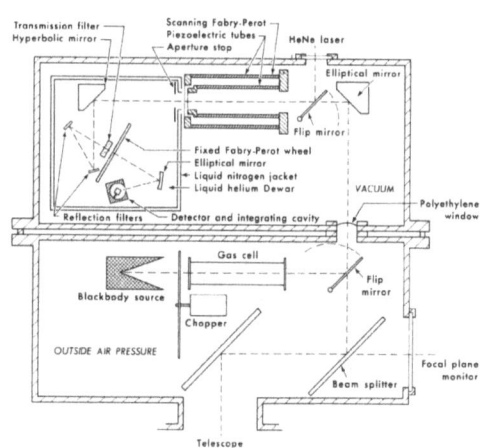

it is a good reflector at all wavelengths except those which are
transmitted, and hence appears to be cool so far as radiation seen by
the detector is concerned. This spectrometer has a limiting resolving
power of about 5000, which corresponds approximately to the broader
molecular features in interstellar clouds. It would be desirable to
have still somewhat higher resolution. On the other hand, the spectrometer is very effective and I believe adequately sensitive to
detect a number of molecular lines in warm clouds.

For the present, no molecular spectra have been detected from
interstellar clouds between the near submillimeter range and the 10
micron window. In atmospheric windows at 10 micron wavelengths and
shorter, photoconductive and photodiode detectors have successfully
detected a variety of molecules in circumstellar materials, as well as
CO and hydrogen in a few special interstellar regions. The Fourier
transform system of Hall and Ridgeway has, for example, been quite
successful at the shorter IR wavelengths. Ten micron heterodyne spectroscopy has successfully detected a number of molecular lines in circumstellar material, although not yet directly in interstellar clouds.
Heterodyne detection working in the submillimeter range has been used
successfully from the ground to detect a number of molecular lines,
for example the second and third transitions of CO. A limited but
valuable program of submillimeter heterodyne observations has been
carried out in NASA's high flying C141 airplane by Phillips and his
associates.

While work so far has been rather limited, it seems inevitable
that the infrared, particularly the far-infrared which has not yet
been properly exploited, will provide a great deal of valuable information about interstellar clouds. The high rotational transitions of
simple linear molecules such as CO should reveal much about the

Figure 2: NASA's Kuiper
Astronomical Observatory.
A 36" telescope looks
through the rectangular
hole in the upper left
side of the fuselage of a
C141 airplane.

Figure 3: Interior of the
Kuiper Astronomical Observatory, showing a star
field being tracked and
presentation of observatory
parameters.

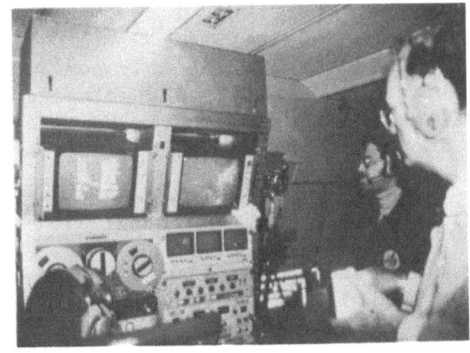

This work was supported in part by NASA Grants NGL 05-003-272 and NGR 05-003-511

cooling of clouds and the excitation of molecules. The lower rotational transitions of hydrides, including HD, should be detectable in the far-infrared. At the moment, with microwaves we usually measure so few molecular parameters that we can generally successfully fit observations to broad and simple models of clouds. Undoubtedly, as additional transitions in the far-infrared are measured the situation will in many cases appear more complicated and more difficult to fit with simple models, but at least we should be closer to an accurate view of cloud behavior. The mid and far infrared have already produced a good deal of valuable information from observations of the fine structure of ions, and as the far-infrared begins to detect molecular transitions as well, it should be especially valuable in understanding the warmer regions in clouds and the interfaces between ionized and molecular regions.

Where will far-infrared astronomy be done? Certainly at dry sites. These will include the best dry sites presently known for observations from the ground, and also observations above the atmosphere. Already some spectroscopy has been done from balloons, incidental to the study of the short wavelength end of the isotropic radiation. More has been done from NASA's Gerard Kuiper Observatory, a C141 outfitted with a 36" telescope which looks directly out of the side of the airplane. The C141 Observatory with an opening for the 36" telescope is illustrated in Figure 2 and its interior in Figure 3. The latter figure shows the shirt-sleeve environment in which crew and astronomers work, and an oscilloscope display of the star field used both for finding and for some forms of tracking. In the longer run, infrared spectral observations should be made from satellites, particulary for the study of H_2O, which still shows intense absorption lines even at high altitude. So far, plans for satellite observations of the far-infrared involve only continuum measurements, but in time the additional sensitivity and pointing precision and the freedom from absorption lines which can be obtained from satellites should play an important role in molecular observations.

Appendix I

The theoretical sensitivity of a system using a photodiode detector and limited only by photon noise is given by an NEP (signal power equal to noise with 1 Hz post-detection bandwidth) as follows:

$$\text{NEP} = \frac{1}{1-\varepsilon}\sqrt{\frac{2P_B h\nu}{\eta}\left(\frac{P_B \lambda^2 \eta}{2h\nu A\Omega\Delta\nu} + 1\right)}\bigg/\sqrt{\text{Hz}} \qquad (1)$$

where $h\nu$ is the quantal energy, λ the wavelength, P_B the continuum power falling on the detector, η the detector's quantum efficiency, $A\Omega\Delta\nu$ respectively the area, solid angle, and bandwidth accepted by the detector, and ε the fractional transmission loss in the optical path from source to detector. The quantity $\dfrac{2A\Omega\Delta\nu}{\lambda^2}$ is simply the number m of propagation modes (spatial, polarization, and temporal), accepted by the detector and the 1 Hz bandwidth. Hence $n = \dfrac{P_B \lambda^2 \eta}{2h\nu A\Omega\Delta\nu}$ is the number of photoelectrons produced per mode.

If $n \gg 1$, (1) becomes $NEP_{n \gg 1} = \dfrac{\sqrt{2\ P_B}}{\sqrt{m}(1-\varepsilon)}$.

For the case of a diffraction limited telescope and a single polarization, $\dfrac{2A\Omega}{\lambda^2}$ is unity and one has $NEP_{n \gg 1} = \dfrac{\sqrt{2}kT_B\sqrt{\Delta\nu}}{1-\varepsilon}$. This corresponds to the familiar noise power of $kT_B \sqrt{\Delta\nu}$ for a one second average, where T_B is the effective temperature of the incident radiation. It is the appropriate form in the microwave region where Bose–Einstein condensation is important, and the NEP is independent of quantum efficiency as long as η is large enough so that $n \gg 1$.

If $n \ll 1$, (1) becomes

$$NEP_{n \ll 1} = \frac{1}{1-\varepsilon}\sqrt{\frac{2P_B h\nu}{\eta}} = \frac{h\nu}{1-\varepsilon}\sqrt{\frac{2P_B}{\eta h\nu}} , \text{ which is a familiar form for}$$

short wavelengths where photons are individually countable, since $\dfrac{P_B}{h\nu}$ is the total number of photons/sec. In this case, the quantum efficiency η is quite important.

Another form of expression (1) can be useful in the infrared. If the background radiation power P_B comes from a warm, partially transmitting optical path of temperature T, one has

$$NEP = \frac{2h\nu}{1-\varepsilon}\sqrt{\frac{A\Omega\Delta\nu\varepsilon/\eta}{\lambda^2\ e^{\frac{h\nu}{kT}}-1}}\ \sqrt{\frac{e^{\frac{h\nu}{kT}}-1+\varepsilon\ \eta}{e^{\frac{h\nu}{kT}}-1}}\ \bigg/ \sqrt{Hz} \qquad (2)$$

where ε is the fractional loss in the warm optical path, or $\varepsilon = 1 - e^{-\tau}$ where τ is the optical depth. This form assumes there are no other losses in the system. It applies, as does (1), for the general case independent of the occurrence of Bose–Einstein condensation.

If a photoconductive detector is used rather than a photodiode, there are noise fluctuations due to electron recombination as well as generation, and all of the above expressions must be multiplied by $\sqrt{2}$. If beam chopping is used, the effective NEP is multiplied further by a factor of 2.

DISCUSSION FOLLOWING TOWNES

Snyder: Can you comment on the apparent absence of hydrides?

Townes: Of course, the hydrides of C, O, and N have been detected in the microwave region. Other hydrides, such as metal hydrides, have not yet been found, but I expect that, as soon as good detection of far-infrared molecular lines is possible, hydrides will be among the molecules found. We hope on our next flight to concentrate on and detect molecular lines in the 100μ region.

Goldsmith: What is the resolution of the 100μ spectrometer system?

Townes: We have used it with frequency resolutions between about 0.15 cm^{-1} and 0.05 cm^{-1}. The limiting resolution without further modification is about 0.02 cm^{-1}.

Thaddeus: Do plans exist to search for the fundamental rotational line of HD at 112 microns? It appears to be our best hope of directly observing molecular hydrogen in fairly cool molecular clouds.

Townes: Yes. We have in fact already searched for the 112μm line of HD, but with mediocre sensitivity which provided an upper limit of modest significance. We are hoping to improve the sensitivity of our equipment in a winter flight, and estimates indicate that the line should be detected, at least in the warmer clouds. There is little hope of finding this line, I believe, in the very cool clouds.

Churchwell: Have you been able to detect the OIII 88μm or 52μm lines in the galactic center? If so, is the intensity much lower than in local HII regions?

Townes: We detected the 52μm OIII line from the galactic center. It is far weaker than in most HII regions. This result is consistent with our previously reported low intensity for ArIII and SIV radiation, and implies that there is a cool ionizing source at the galactic center. The 88μm OIII line was not detected; we have only a low upper limit.

Wynn-Williams: How big an advantage would a cooled space-borne telescope give you for molecular infrared spectroscopy?

Townes: A cooled space-borne spectroscope could essentially eliminate background continuum radiation, and, in principle, gain some orders of magnitude in sensitivity. The gain would not be so great, however, for lines from warm clouds where there would be background noise due to the continuum radiation from the clouds themselves. In such cases the increase in sensitivity would be limited to perhaps one order of magnitude. The amount of gain would depend also, of course, on the resolution.

Drapatz: Experiment proposals from German scientists concerning infrared research on Spacelab were combined to form the "German Infra-Red Laboratory" (GIRL), a superfluid helium-cooled telescope (40 cm diameter) with five focal-plane instruments. The instrument, which will also be used for observations of interstellar molecules, is a Michelson interferometer ($\lambda \gtrsim 20$ m, resolution ~ 0.05 cm^{-1}, field of view 30 - 300 arcmin, limiting sensitivity $\sim 10^{-19}$W cm^{-2} line^{-1} in 10^3 sec integration time). The project was approved in 1978, and the first flight is planned for 1984. More details can be found in the proceedings of the SPIE, Electro-Optical Technical Symposium & Workshop, Huntsville, 197'

DETECTION OF INTERSTELLAR BS IN THE CIRRUS DARK CLOUD OF THE NUMBBUM ASSOCIATION
I. AN INTUITIVE MODEL AND ITS SUBSEQUENT OBSERVATION

J.J. Charfman
Bora-Bora Radio Observatory
Tahiti, South Pacific

ABSTRACT

Previous suspicions of large quantities of BS in large astronomical associations have been confirmed by observation.

Interstellar boron sulfide, BS, has been predicted as a fragment of the detailed ion-molecule chain outlined by Dack and Balgarno (1980). Its significance in the grand cosmochemical scheme of Charfman, Houdini, Aardvaark, Raoul, and Finzi (1974, hereinafter referred to as CHARF) was pointed out by Sollenoid and Haltpeter (1978) in predicting the correlation between radio map contours and chemical reaction routes. Figure 1 illustrates this correlation.

Searches for the gaseous phase were first carried out in 1963 by Zen Buckerman while seated in the lotus position at the 2-m Charfman Memorable Telescope on Maiden-knoll in Tahiti. We have subsequently repeated these observations with increased sensitivity. Our search was restricted to the Cirrus dark cloud centered on the BO star HD 1, located in the Numbbum Association and referred to as OMCTMCGMC-280Z, and noted for its intense wind. Figure 2 illustrates a typical line of BS. There is a composite photograph of the BS source taken (B. and D. Greenerelm, 1981) with broad-band alphabetical filters at the front of this book.

Alternate theories of BS formation on solid surfaces by Mayonnaise Hamburger (1977) have been considered, but only gas-produced BS can account for the amount reported here. However the yellow stuff reported by Hamburger (this volume) may well be mustard.

BS emission was generally observed in mornings and evenings, but was finally detected on Friday afternoon at a 1.8σ level. Its lifetime was estimated to be approximately 5 days, but the level of emission was generally variable. The detection of this molecule has considerable impact on our understanding of Giant Gas Clouds which will be discussed in a later paper, but we want to note the most important of these at

B. H. Andrew (ed.), Interstellar Molecules, 645–648.
Copyright © 1980 by the IAU.

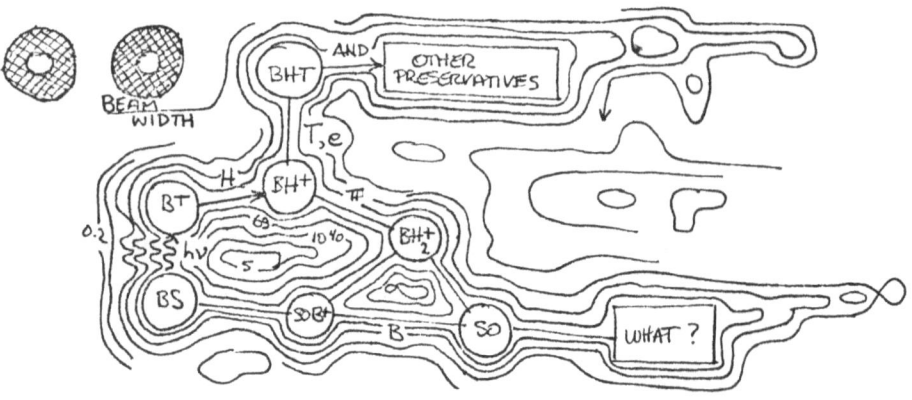

Figure 1

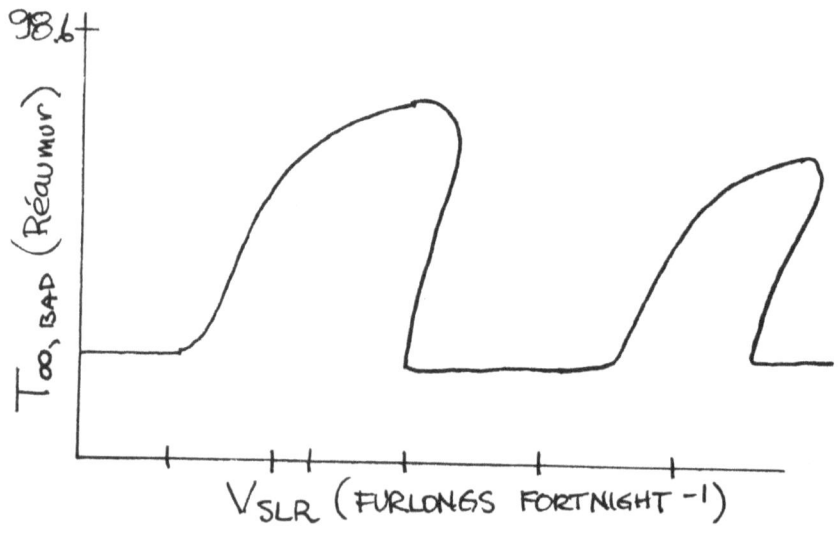

Figure 2

this time: as predicted by B. Bok (1979) "the only way you play the piano is at all ends-simultaneously."

This research was supported in part by the Office of Navel Research and GNASSAU, and was completed despite the advice and assistance of J.H. Black, P.C. Frish, B.L. Lutz, and P.L. Smith et al.

REFERENCES

Bok, B.: 1979, post-prandial remarks.
Buckerman, Z.: 1963, Readers' Digest, 13794, 32.
Charfman, J.J., Houdini, G.G., Aardvaark, B., Raoul, F., and Finzi, A.N. 1974, Ap. J. part V (submitted).
Dack, J.H., and Balgarno, A.: 1980, Nat. Enquirer, 382, 1.
Greenerelm, B., and Greenerelm, D.: 1981, Arizona Highways 87, 29.
Hamburger, M.: 1977, Pub. McDonald's Obs. 17, 24.
Sollenoid, D., and Haltpeter, E.E.: 1978, Biennial Review of the Follies Berg. 3, 28.

DISCUSSION FOLLOWING CHARFMAN

Moustachio: You seem not to have considered the effects of magnetic fields on the formation and subsequent evolution of clouds of boron sulfide, or as the British would say, boron sulphide. As you may be aware I have long advocated magnetism, particularly the personal variety; one of my numerous speeches on the subject is quoted at length in the Proceedings of last year's Eggnog Symposium on the Search for Insulin in Interstellar Space (ed. by Oil and Wickerarmchair). Would you care to comment?

Charfman: Your point is well taken. What was it?

Theoretiksen: You mentioned that the boron sulfide profile showed evidence of self-absorption. Could you tell us whether the absorbing material lies in front of or behind the source, and if so, what it is likely to consist of?

Charfman: Yes.

Ripoff: I wish to point out that I and my student who did all the work but whose name I forget have searched extensively for the ring analog of boron sulfide, achieving upper limits of $\lesssim$ 5 mK. Allowing for beam dilution and the effects of the partition function, the equivalent upper limit to the molecular density is $10^{18} cm^{-3}$, which indicates that Avagadro's Number is not applicable to the interstellar medium, at least for boron sulphide.

Charfman: I agree.

Effusiveus: I can't overemphasize how terribly important your discovery is for interstellar chemistry. It opens up all kinds of new avenues of research. Quite obviously there is a whole plethora of molecules out there all bumping and grinding in the dark and getting excited. We were having a go with our four-footer on Broadway just yesterday and we got all sorts of new lines, especially from cab drivers.

In fact, sorry about that, I'll pick it up in a minute, what was the bloody microphone doing there anyway, in fact we find it's a very good observing site, we've seen a lot of fine structure, particularly on hot summer days. No, I don't want to write this down on one of your beastly pieces of paper, and I hope you're not recording this either.

Charfman: Thank you. I think.

Sock: Have you looked for boron sulfide in any Southern Hemisphere globules?

Charfman: No.

Sock: You dumb bonehead.

Solemnambulous: I will be giving a talk tomorrow afternoon which involves giant clouds of BS. Those of you who have arranged to play tennis can read a report of it in yesterday's New York Times.

Buckerman: As some of you may know, Pat Panhandler and I recently went on a trip to the Himalayas in search of the Yeti. We didn't actually find the Yeti but maybe he found us because something big got into our sleeping bags one night and anyway it was all very interesting and we think you'll enjoy it and we're going to be showing slides of our trip for our friends and anybody else who might be interested downstairs in the Kandahar room after dinner tonight probably about 8 o'clock I think if we can find a projector. It probably won't go on past midnight because I have to start preparing my two-hour invited address I'm giving first thing tomorrow morning.

Snide: Have you been sniffing something?

Charfman: Would you guys shut up? Who the hell's talk is this anyhow?

Maxplanckenstein: Gentlemen, this session is already three hours and eleven minutes behind time. I think we'd better move on to the next speaker. Would somebody wake up the next speaker please? That's him over there slumped against the radiator. Meanwhile Dr. Android from the Local Organising Committee has some brief announcements.

Android: Somebody lost his scoreboard after yesterday's golf tournament. Unfortunately it seems to have been vandalised, but we've managed to piece it together and the owner can have it back by correctly identifying his score and answering a skill-testing question. Also we would like a few volunteers to form a search party to look for Drs. Panhandler and Shermantank who are overdue from yesterday's fun run around the lake. Finally, Dr. Moresaki wants to challenge whoever it was he played at table tennis in the bar last night to another game this afternoon, when he thinks he would be less likely to keep falling off the table.

Maxplanckenstein: Thank you, Dr. Android. And now, the next speaker, Dr. Hamburger, whose talk is entitled "Gas Phase Theory Goes against the Grain."

INDEX OF ASTRONOMICAL OBJECTS

Objects designated by a prefix and the name of a constellation are listed according to the constellation e.g. HL Tau is listed under T. English characters are given precedence over Greek characters. Page numbers refer to the title page of the article in which the object is mentioned.

MOLECULES INDEX

Molecules are grouped according to the number of constituent atoms.
Within each group they are listed alphabetically (CH appears ahead of
CN). Subscripts are used to differentiate between alphabetically identical molecules. The smallest subscript is ranked first, and precedence
is given in order from the left (CH_2 appears ahead of C_2H). Molecular
ions are listed immediately following the equivalent molecule (CH_2
appears ahead of CH_2^+) and are given precedence over isotopes (CH_2^+
appears ahead of $^{13}CH_2$). Isotopes are listed following the parent molecule and are given precedence in the order in which the molecule is
written ($^{13}CH_2$ appears ahead of CDH). Thus the molecules CH_2, CH_2^+,
CDH, CD_2, C_2H, C_2H^+, $C^{13}CH$, C_2D would appear in that order.

Molecules containing 6 atoms or less are written in their commonly-
used form e.g. CH_3OH is not rearranged to be CH_4O as is the practice in
some indices. However the editor, to preserve his sanity and disguise
his ignorance, has wantonly destroyed the chemical identity of molecules
having more than 6 atoms, except in the case of the easily recognisable
cyanopolyyne series which was frequently discussed at the Symposium.

Page numbers refer to the title page of the article in which the
listed molecule is mentioned.

670

MOLECULES INDEX

4 atoms

5 atoms

CH_4	41, 77, 283, 289, 291, 307, 311, 317, 323, 325, 337, 341, 355, 365, 367, 373, 387, 479, 497, 509, 591
CD_4	307
C_2H_3	337
$C_2H_3^+$	67, 299, 307, 311, 323
$C_2D_2H^+$	307
C_3H_2	337
C_4H	47, 479, 497
CH_2CN	59, 337
CH_2CO	317, 337
CH_3I	59
CH_2N_2	337
CH_2NH	311, 337
CH_2OH	337
C_4N	45
C_4O	45
HC_3N	1, 25, 47, 59, 71, 77, 81, 91, 299, 307, 317, 325, 331, 337, 397, 427, 479, 497
HC_3N^+	307
$H^{13}CC_2N$	397
$HC^{13}CCN$	59
$HC_2^{13}CN$	59
H_3CO	239, 337
H_3CO^+	323, 439
H_2DCO^+	439
$HCOOH$	337, 355, 365, 627
H_3CS	337
NH_4^+	291, 307, 311, 323, 331, 341
NH_2CN	365

6 atoms

CH_5^+	291, 311, 317, 323, 337
C_2H_4	291, 311, 337, 367, 497
$C_2H_4^+$	323
C_4H_2	325, 337
CH_2C_2H	337
CH_3CN	307, 325, 365, 479
CH_3CO	337
CH_3CO^+	317, 337
CH_2NH_2	311
$CH_2NH_2^+$	311, 337
CH_3OH	1, 117, 317, 323, 337, 365, 373, 524, 525, 565
C_5N	45
H_2C_3N	59
$H_2C_3N^+$	299, 307

SUBJECT INDEX

Page numbers are those of the title page of the article in which the subject is mentioned.

675

SUBJECT INDEX

702